Mathematical Formulae

Ken Kubota

2017

Author's website:
www.kenkubota.de

Publisher's website:
www.owlofminerva.net

Website for this work:

Kubota – Mathematical Formulae
http://doi.org/10.4444/100.3

Owl of Minerva Press
First edition
Berlin 2017
doi: 10.4444/100.3
ISBN 978-3-943334-07-4

Contents

Chapter 1

Introduction

This work contains the formula part of the presentation of the mathematical logic $\mathcal{R}_0$, a further development of Peter B. Andrews' logic $\mathcal{Q}_0$. The syntactic features provided by $\mathcal{R}_0$ are type variables (*polymorphic type theory*), the binding of type variables with the abstraction operator and single variable binder λ (*type abstraction*), and (some of) the means necessary for dependent types (*dependent type theory*). Mathematical entities may have different types, following the standard interpretation of types as (non-empty) sets, where entities may be the member of more than one set. The resulting formal language allows one to naturally and precisely express mathematical concepts without any circumlocutions. It follows Andrews' concept of *expressiveness* (I also use the term *reducibility*), which aims at the ideal and natural language of formal logic and mathematics.

In addition to the language of $\mathcal{Q}_0$, $\mathcal{R}_0$ has a type of types τ containing all types, and a universal type ω containing all mathematical entities. This means that types are not a separate syntactic category anymore, but terms of type τ. Vice versa, all types (including τ and ω) are also mathematical entities of type ω, which allows expressions like $\tau = \tau$. In philosophy, it is well known that self-reference alone is not sufficient for obtaining a paradox, but the classical philosophical antinomy always requires *both* properties: self-reference *and* negativity (for example, Russell's paradox: the set of all sets that are *not* members of *themselves*; the philosophical notion "negativity" includes not only the logical negation, but also any further specification, for example, as used in the Burali-Forti paradox, in the sense of Spinoza's principle "omnis determinatio est negatio"). Furthermore, $\mathcal{R}_0$ provides a simple and intuitive method of type introduction, and for type abstraction an internal type referencing mechanism necessary for the proper treatment of dependencies by use of the means of the language only (i.e., syntactically). Whenever a theorem of the form $p_{o\alpha}e_\alpha$ is inferred, meaning that a mathematical entity (or object, element) e of any type α has property p (which in set theory is expressed by $e \in p$), first, the now provably non-empty set p is acknowledged as a type by attaching type τ to it (p_τ), and second, the new type p is attached to the entity (e_p) unless e is a variable, since variables do not denote a concrete mathematical entity, but are intended for latter substitution and therefore can only have a single type. The first part (p_τ) even holds for the case that e is a variable (e.g., $p_{o\alpha}v_\alpha$), due to the fact that in higher-order logic, free variables are implicitly universally quantified ([Andrews, 2002, p. 221 (5215 = $\forall$I) and p. 222 (5220 = Gen)]), and the fact that, since α is a type, it must be non-empty. An internal type referencing mechanism is implemented which allows for properly resolving the dependencies *within* the type (subscript) level only (e.g., in the definition of the universal quantifier with type abstraction $\forall := [\lambda t_\tau.[\lambda p_{ot}.(=_{o(ot)(ot)}[\lambda x_t.T_o]p_{ot})_o]_{(o(ot))}] = [\lambda t.[\lambda p.(=[\lambda x.T]\,p)]]_{o(o\backslash 3)\tau}$, with depth/nesting index $\backslash 3$ in $o(o\backslash 3)\tau$ referring to τ, hence $\forall\,\omega$ is of type $o(o\backslash 3)[\omega\,/\,\backslash 3] = o(o\omega)$).

A prerequisite for the understanding of $\mathcal{R}_0$ is the thorough study of the syntactic part [Andrews, 2002, pp. 201–237] of chapter 5 of Andrews' 2002 textbook. This part, in which $\mathcal{Q}_0$ is presented [Andrews,

2002, pp. 210–215] and elementary logic developed, is clearly the masterpiece of mathematical scientific literature as of today. An introductory text on $\mathcal{R}_0$ will be published under the title *On the Theory of Mathematical Forms*, and the software implementation will be made available online (license restrictions apply). For more information, including an online version of this book and possible future updates, please visit the following website on $\mathcal{R}_0$: http://doi.org/10.4444/100.10.

Since $\mathcal{R}_0$ is, like $\mathcal{Q}_0$, a Hilbert-style system, it has theorems and metatheorems. Files with the extension "r0" contain (definitions and) proofs of regular theorems, and files with the extension "r0t" contain proof templates for metatheorems. For the demonstration of metatheorems, sometimes the assumption of additional axioms is required, which is allowed in files with the extensions "r0a", and in those with the extension "r0e" that test error detection (type violations, etc.).

As $\mathcal{R}_0$ does not provide any automation (every single step has to be specified manually), it was not possible for the author to formalize within a reasonable amount of time the proof required for the use of the recursion operator R [Andrews, 2002, pp. 282–284 (6400)], or the further proofs [Andrews, 2002, pp. 262 ff. (6103 ff.)] that Andrews' definition of natural numbers [Andrews, 2002, p. 260] satisfies the Peano axioms, ending with Theorem 6102 as the last one formalized within the $\mathcal{R}_0$ implementation. Of course, the latter proofs may not be necessary as with type abstraction, a definition of natural numbers (using the Peano axioms directly) that is more general and canonical than Andrews' original definition is possible. Such a canonical definition exploiting the expressiveness of type abstraction is the definition of groups *Grp* further below, using the three group axioms only [p. 362].

For full *dependent type theory* with structurally dependent types, substitution at the type (subscript) level might be desirable. For example, given the dimension, the function fuv returns the first unit vector $\vec{e}_1$, yielding a result where its dimension, and therefore its type, depend on the argument:

$$\vec{e}_1^{\,1} = \begin{bmatrix}1\end{bmatrix} = \mathrm{fuv}(1) = (\mathrm{fuv}\ 1)_{\mathbb{R}^1},\ \vec{e}_1^{\,2} = \begin{bmatrix}1\\0\end{bmatrix} = \mathrm{fuv}(2) = (\mathrm{fuv}\ 2)_{\mathbb{R}^2},\ \vec{e}_1^{\,3} = \begin{bmatrix}1\\0\\0\end{bmatrix} = \mathrm{fuv}(3) = (\mathrm{fuv}\ 3)_{\mathbb{R}^3},\ \text{etc.},$$

with fuv(n) of type $\mathbb{R}^n$. As shown further below [pp. 388 ff.], the recursion operator R [Andrews, 2002, pp. 281 f., 284] can be used to define a function which obtains the type of a vector of a given dimension. With such a function vectype(n), the type of fuv(n) could be specified in a way preserving the dependency, e.g., $\mathrm{fuv}(n) := [\lambda n_{\mathbb{N}}.\,\mathrm{fuvbase}]_{(\mathrm{vectype}(\backslash 3))\mathbb{N}}$, such that the type of $\mathrm{fuv}(2) = [\lambda n_{\mathbb{N}}.\,\mathrm{fuvbase}]_{(\mathrm{vectype}(\backslash 3))\mathbb{N}}\ 2_{\mathbb{N}}$ reduces to vectype(2), and since $\mathrm{vectype}(2) = \mathbb{R}^2$, one would like to infer from $\mathrm{fuv}(2)_{\mathrm{vectype}(2)}$ to $\mathrm{fuv}(2)_{\mathbb{R}^2}$. An implementation of a substitution rule directly at the type (subscript) level appeared too experimental to the author and was removed, since further case studies (here using the recursion operator R) should be undertaken.

The turnstile ($\vdash$) was eliminated and replaced by the logical implication ($\supset$) [e.g., cf. pp. 303 ff. (K8025 = Deduction Theorem) and Andrews, 2002, pp. 228 f. (5240 = Deduction Theorem)]. Proper rules are denoted by §s for substitution [Andrews, 2002, p. 213 (Rule R)], §\ for lambda-conversion, or, more exactly, β-reduction [Andrews, 2002, pp. 213 f. (Axiom Schemata 4_1 - $4_5 \triangleq 5207$) and pp. 218 f. (5207)], §= for reflexivity [Andrews, 2002, p. 215 (5200)], and prime variants like §s' for the case of hypotheses [Andrews, 2002, p. 214 (Rule R′)]. Improper rules are denoted by §r for the renaming of bound variables [Andrews, 2002, pp. 217 f. (5206) and p. 219 (α-conversion)], and §! for the introduction of new axioms (allowed only with a special flag provided on the invocation of the $\mathcal{R}_0$ implementation). Theorems are referenced by %0 (the last theorem inferred), %1 (the second to last inferred), etc., well-formed formulae by 0, 1, 2, etc., and subformulae by /1, /2, etc.

A first draft of $\mathcal{R}_0$ [Kubota, 2015] has been circulated privately since May 2015. The author would like to thank Peter for his groundbreaking work, in particular the logic $\mathcal{Q}_0$, and for his kind support.

Chapter 2

Mathematical Formulae of $\mathcal{R}_0$

2.1 Results

2.1.1 Results for File A5200t.r0.txt

```
##
##   Proof A5200t:   T   (special case of A = A)
##
##
##   Source: [Andrews 2002 (ISBN 1-4020-0763-9), p. 215]
##
##   Copyright (c) 2017 Owl of Minerva Press GmbH. All rights reserved.
##   Written by Ken Kubota (<mail@kenkubota.de>).
##
##   This file is part of the publication of the mathematical logic R0.
##   For more information, visit: <http://doi.org/10.4444/100.10>
##

<< definitions1.r0.txt


##
##   Proof
##

§= =
#                ===       :=  T

:=  A5200t  %0
# wff    12  :       ===_o      :=  A5200t   T


##
##   Q.E.D.
##
```

%0
$===$:= $A5200t$ T
$=_{o\omega\omega}=_{\omega}=_{\omega}$:= $A5200t$ T

2.1.2 Results for File A5201b.r0a.txt

##
Proof Template A5201b (Swap): A = B → B = A
for any A, B of any type T
##
Source: [Andrews 2002 (ISBN 1-4020-0763-9), p. 215]
##

Written by Ken Kubota (<mail@kenkubota.de>).
##
This file is part of the publication of the mathematical logic $\mathcal{R}_0$.
For more information, visit: <http://doi.org/10.4444/100.10>
##

##
Assumptions and Resulting Syntactical Variables
##

the assumption as last theorem on stack (%0)
§! $=_{ooo}a_o b_o$
$= a\,b$

##
Include Proof Template
##

<<< A5201b.r0t.txt
Include begin (A5201b.r0t.txt) [oldfile=(A5201b.r0a.txt)]
##
Proof Template A5201b (Swap): A = B → B = A
for any A, B of any type T
##
Source: [Andrews 2002 (ISBN 1-4020-0763-9), p. 215]
##

Written by Ken Kubota (<mail@kenkubota.de>).
##
This file is part of the publication of the mathematical logic $\mathcal{R}_0$.
For more information, visit: <http://doi.org/10.4444/100.10>
##

##
Proof Template
##

use polymorphic identity relation with type of right side of given equation ({**%0})
§= $_o$ /5
$= a\, a$

now replace left hand side of new equation
§s %0 5 %1
$= b\, a$
Include end (A5201b.r0t.txt) [newfile=(A5201b.r0a.txt)]
>>>

##
Q.E.D.
##

%0
$= b\, a$
$=_{ooo} b_o a_o$

2.1.3 Results for File A5201bH.r0a.txt

##
Proof Template A5201bH (SwapH): H ⊃ (A = B) → H ⊃ (B = A)
for any A, B of any type T
##
Source: [Andrews 2002 (ISBN 1-4020-0763-9), p. 215]
##

Written by Ken Kubota (<mail@kenkubota.de>).
##
This file is part of the publication of the mathematical logic $\mathcal{R}_0$.
For more information, visit: <http://doi.org/10.4444/100.10>
##

<< basics.r0.txt

##
Assumptions and Resulting Syntactical Variables
##

the assumption as last theorem on stack (%0)
§! $\supset_{ooo} h_o (=_{ooo} a_o b_o)$
$\supset h\, (= a\, b)$

```
##
##   Include Proof Template
##

## <<< A5201bH.r0t.txt
## Include begin (A5201bH.r0t.txt) [oldfile=(A5201bH.r0a.txt)]
##
##   Proof Template A5201bH (SwapH):   H ⊃ (A = B)   →   H ⊃ (B = A)
##        for any A, B of any type T
##
##   Source: [Andrews 2002 (ISBN 1-4020-0763-9), p. 215]
##
##   Copyright (c) 2017 Owl of Minerva Press GmbH. All rights reserved.
##   Written by Ken Kubota (<mail@kenkubota.de>).
##
##   This file is part of the publication of the mathematical logic R0.
##   For more information, visit: <http://doi.org/10.4444/100.10>
##


##
##   Exception: Forward Reference
##
##   Because of the different rules of inference, unlike in Q0,
##   this theorem (with hypothesis) cannot be inferred from
##   previous theorems only, but depends on new theorems.
##
##   Dependencies (selection):
##
##             K8003 << K8000a, A5219b, A5221
##                      K8000a << A5222, A5229a, A5229c
##                                        A5221 << A5220
##                                                   A5229c << A5227 << A5226 << A5225
##


##
##   Proof Template
##

:=  $TMPswapH  %0
# wff     212  :          ⊃ h (= a b)_o       :=   $TMPswapH

## use polymorphic identity relation with type of right side of given equation ({**%0})
§='   o /5
#                   = a a
```

use Proof Template K8003 (Intro): A $\rightarrow$ H $\supset$ A
:= $\$A8003$ %0
wff 213 : $= a\, a_o$:= $\$A8003$
:= $\$H8003$ %1/5
wff 208 : h_o := $\$H8003$
<< K8003.r0t.txt
:= $\$A8003$
:= $\$H8003$
%0
$\supset h\, (= a\, a)$
$\supset_{ooo} h_o (=_{ooo} a_o a_o)$

%$\$TMPswapH$
$\supset h\, (= a\, b)$:= $\$TMPswapH$
$\supset_{ooo} h_o (=_{ooo} a_o b_o)$:= $\$TMPswapH$
:= $\$TMPswapH$

now replace left hand side of new equation
§s' %1 5 %0
$\supset h\, (= b\, a)$
Include end (A5201bH.r0t.txt) [newfile=(A5201bH.r0a.txt)]
>>>

##
Q.E.D.
##

%0
$\supset h\, (= b\, a)$
$\supset_{ooo} h_o (=_{ooo} b_o a_o)$

2.1.4 Results for File A5205.r0.txt

##
Proof Template A5205: f = [\y.fy]
for any y of type b and f of type ab
##
Source: [Andrews 2002 (ISBN 1-4020-0763-9), p. 217]
##

Written by Ken Kubota (<mail@kenkubota.de>).
##
This file is part of the publication of the mathematical logic $\mathcal{R}_0$.
For more information, visit: <http://doi.org/10.4444/100.10>
##

##

Define Syntactical Variables
##

the variable of type b{^} to be used
:= $Y5205 y_b
wff 12 : y_{b_τ} := $Y5205

##
Include Proof Template
##

<<< A5205.r0t.txt
Include begin (A5205.r0t.txt) [oldfile=(A5205.r0.txt)]
##
Proof Template A5205: f = [\y.fy]
for any y of type b and f of type ab
##
Source: [Andrews 2002 (ISBN 1-4020-0763-9), p. 217]
##
Copyright (c) 2017 Owl of Minerva Press GmbH. All rights reserved.
Written by Ken Kubota (<mail@kenkubota.de>).
##
This file is part of the publication of the mathematical logic $\mathcal{R}_0$.
For more information, visit: <http://doi.org/10.4444/100.10>
##

<< axioms.r0.txt

##
Exception: Forward Reference
##
The original proof 5205 uses some of the Axiom Schemata 4_1 - 4_5
(indirectly via 5203 and 5204), which are not available in $\mathcal{R}_0$,
but replaced by Rule 2 (Lambda Conversion) [5207].
Therefore the use of the Rule of Substitution (A5221) is required here.
##
For historical purposes, and since the proof did not change otherwise,
the proof number 5205 was not altered.
##

##
Proof Template
##

.1

%$A3$
\# $=(=f\,g)\,(\forall\, b\,[\lambda x.(=(f\,x)\,(g\,x))])$:= $A3$
\# $=_{ooo}(=_{o(ab)(ab)}f_{ab}g_{ab})(\forall_{o(o\backslash 3)\tau}b_{\tau}[\lambda x_{b}.(=_{oaa}(f_{ab}x_{b})(g_{ab}x_{b}))_{o}])$:= $A3$

\## use Proof Template A5221 (Sub): B → B [x/A]
:= $\$B5221$ %0
\# wff 125 : $=(=f\,g)\,(\forall\, b\,[\lambda x.(=(f\,x)\,(g\,x))])_{o}$:= $\$B5221$ $A3$
:= $\$T5221$ ab
\# wff 107 : ab_{τ} := $\$T5221$
:= $\$X5221$ $g_{\$T5221}$
\# wff 112 : $g_{\$T5221}$:= $\$X5221$
:= $\$A5221$ $[\lambda\$Y5205_{b}.(f_{\$T5221}\$Y5205_{b})_{a}]$
\# wff 132 : $[\lambda\$Y5205.(f\,\$Y5205)]_{\$T5221}$:= $\$A5221$
<< A5221.r0t.txt
:= $\$B5221$
:= $\$T5221$
:= $\$X5221$
:= $\$A5221$
%0
\# $=(=f\,[\lambda\$Y5205.(f\,\$Y5205)])\,(\forall\, b\,[\lambda x.(=(f\,x)\,([\lambda\$Y5205.(f\,\$Y5205)]\,x))])$
\# $=_{ooo}(=_{o(ab)(ab)}f_{ab}[\lambda\$Y5205_{b}.(f_{ab}\$Y5205_{b})_{a}])\ldots$
$\ldots(\forall_{o(o\backslash 3)\tau}b_{\tau}[\lambda x_{b}.(=_{oaa}(f_{ab}x_{b})([\lambda\$Y5205_{b}.(f_{ab}\$Y5205_{b})_{a}]x_{b}))_{o}])$

\## .2

§\ $[\lambda\$Y5205_{b}.(f_{ab}\$Y5205_{b})_{a}]x_{b}$
\# $=([\lambda\$Y5205.(f\,\$Y5205)]\,x)\,(f\,x)$

\## .3

§s %1 31 %0
\# $=(=f\,[\lambda\$Y5205.(f\,\$Y5205)])\,(\forall\, b\,[\lambda x.(=(f\,x)\,(f\,x))])$
:= $\$TMP5205$ %0
\# wff 775 : $=(=f\,[\lambda\$Y5205.(f\,\$Y5205)])\,(\forall\, b\,[\lambda x.(=(f\,x)\,(f\,x))])_{o}$:= $\$TMP5205$

\## .4

%$A3$
\# $=(=f\,g)\,(\forall\, b\,[\lambda x.(=(f\,x)\,(g\,x))])$:= $A3$
\# $=_{ooo}(=_{o(ab)(ab)}f_{ab}g_{ab})(\forall_{o(o\backslash 3)\tau}b_{\tau}[\lambda x_{b}.(=_{oaa}(f_{ab}x_{b})(g_{ab}x_{b}))_{o}])$:= $A3$

\## use Proof Template A5221 (Sub): B → B [x/A]
:= $\$B5221$ %0
\# wff 125 : $=(=f\,g)\,(\forall\, b\,[\lambda x.(=(f\,x)\,(g\,x))])_{o,\ldots}$:= $\$B5221$ $A3$
:= $\$T5221$ ab
\# wff 107 : ab_{τ} := $\$T5221$
:= $\$X5221$ $g_{\$T5221}$
\# wff 112 : $g_{\$T5221}$:= $\$X5221$

:= $\$A5221\ f_{\$T5221}$
wff 110 : $f_{\$T5221}$:= $\$A5221$
<< A5221.r0t.txt
:= $\$B5221$
:= $\$T5221$
:= $\$X5221$
:= $\$A5221$
%0
$=(=f\,f)\,(\forall\,b\,[\lambda x.(=(f\,x)\,(f\,x))])$
$=_{ooo}(=_{o(ab)(ab)}f_{ab}f_{ab})(\forall_{o(o\backslash 3)\tau}b_\tau[\lambda x_b.(=_{oaa}(f_{ab}x_b)(f_{ab}x_b))_o])$

.5

%$\$TMP5205$
$=(=f\,[\lambda\$Y5205.(f\,\$Y5205)])\,(\forall\,b\,[\lambda x.(=(f\,x)\,(f\,x))])$:= $\$TMP5205$
$=_{ooo}(=_{o(ab)(ab)}f_{ab}[\lambda\$Y5205_b.(f_{ab}\$Y5205_b)_a])\ldots$
$\ldots(\forall_{o(o\backslash 3)\tau}b_\tau[\lambda x_b.(=_{oaa}(f_{ab}x_b)(f_{ab}x_b))_o])$:= $\$TMP5205$
:= $\$TMP5205$

use Proof Template A5201b (Swap): A = B → B = A
<< A5201b.r0t.txt
%0
$=(\forall\,b\,[\lambda x.(=(f\,x)\,(f\,x))])\,(=f\,[\lambda\$Y5205.(f\,\$Y5205)])$
$=_{ooo}(\forall_{o(o\backslash 3)\tau}b_\tau[\lambda x_b.(=_{oaa}(f_{ab}x_b)(f_{ab}x_b))_o])(=_{o(ab)(ab)}f_{ab}[\lambda\$Y5205_b.(f_{ab}\$Y5205_b)_a])$

§s %4 3 %0
$=(=f\,f)\,(=f\,[\lambda\$Y5205.(f\,\$Y5205)])$

§= $_{ab}$ f_{ab}
$=f\,f$
§s %0 1 %1
$=f\,[\lambda\$Y5205.(f\,\$Y5205)]$
Include end (A5205.r0t.txt) [newfile=(A5205.r0.txt)]
>>>
:= $A5205$ %0
wff 761 : $=f\,[\lambda\$Y5205.(f\,\$Y5205)]_{o,\ldots}$:= $A5205$

##
Undefine Syntactical Variables
##

:= $\$Y5205$

##
Q.E.D.
##

%0
$= f\,[\lambda y.(f\,y)]$:= $A5205$
$=_{o(ab)(ab)} f_{ab}[\lambda y_b.(f_{ab}y_b)_a]$:= $A5205$

2.1.5 Results for File A5209.r0a.txt

##
Proof Template A5209 (incl. A5204): B = C → (B = C) [x/A]
(Substitution of a Free Variable on Both Sides of an Equation)
##
Source: [Andrews 2002 (ISBN 1-4020-0763-9), p. 220 (217)]
##

Written by Ken Kubota (<mail@kenkubota.de>).
##
This file is part of the publication of the mathematical logic $\mathcal{R}_0$.
For more information, visit: <http://doi.org/10.4444/100.10>
##

##
Define Syntactical Variables
##

type of both sides of the equation (of b and c)
:= $\$M5209\ m_\tau$
wff 11 : m_τ := $\$M5209$

type of the variable and the substitution term
:= $\$T5209\ t_\tau$
wff 4 : t_τ := $\$T5209$

the variable to be replaced
:= $\$X5209\ x_{\$T5209}$
wff 12 : $x_{\$T5209}$:= $\$X5209$

substitution term
:= $\$A5209\ a_{\$T5209}$
wff 13 : $a_{\$T5209}$:= $\$A5209$

assumption (equation b=c)
:= $\$E5209\ =_{o\,\$M5209\,\$M5209}(bs_{\$M5209\,\$T5209}\$X5209_{\$T5209})(cs_{\$M5209\,\$T5209}\$X5209_{\$T5209})$
wff 22 : $= (bs\,\$X5209)\,(cs\,\$X5209)_o$:= $\$E5209$

##
Assumptions and Resulting Syntactical Variables
##

§! $\$E5209$
$\qquad = (bs\ \$X5209)\ (cs\ \$X5209) \qquad := \quad \$E5209$

##
Include Proof Template
##

<<< A5209.r0t.txt
Include begin (A5209.r0t.txt) [oldfile=(A5209.r0a.txt)]
##
Proof Template A5209 (incl. A5204): B = C → (B = C) [x/A]
(Substitution of a Free Variable on Both Sides of an Equation)
##
Source: [Andrews 2002 (ISBN 1-4020-0763-9), p. 220 (217)]
##
Copyright (c) 2017 Owl of Minerva Press GmbH. All rights reserved.
Written by Ken Kubota (<mail@kenkubota.de>).
##
This file is part of the publication of the mathematical logic $\mathcal{R}_0$.
For more information, visit: <http://doi.org/10.4444/100.10>
##

##
Proof Template
##

extract b and c
:= $\$B5209 \ =_{o\,\$M5209\,\$M5209}(bs_{\$M5209\,\$T5209}\$X5209_{\$T5209})(cs_{\$M5209\,\$T5209}\$X5209_{\$T5209})/5$
wff 18 : $bs\ \$X5209_{\$M5209} \quad := \quad \$B5209$
:= $\$C5209 \ =_{o\,\$M5209\,\$M5209}\$B5209_{\$M5209}(cs_{\$M5209\,\$T5209}\$X5209_{\$T5209})/3$
wff 21 : $cs\ \$X5209_{\$M5209,\ldots} \quad := \quad \$C5209$

.1

§= $_{\$M5209}\ [\lambda\$X5209_{\$T5209}.\$B5209_{\$M5209}]\$A5209_{\$T5209}$
$\qquad = ([\lambda\$X5209.\$B5209]\ \$A5209)\ ([\lambda\$X5209.\$B5209]\ \$A5209)$

.2

%$\$E5209$
$\qquad = \$B5209\ \$C5209 \quad := \quad \$E5209$
$\qquad =_{o\,\$M5209\,\$M5209}\$B5209_{\$M5209}\$C5209_{\$M5209} \quad := \quad \$E5209$
§s %1 13 %0
$\qquad = ([\lambda\$X5209.\$B5209]\ \$A5209)\ ([\lambda\$X5209.\$C5209]\ \$A5209)$

.3

§\ /5
$= ([\lambda\$X5209.\$B5209]\ \$A5209)\ (bs\ \$A5209)$

.4

§\ %1/3
$= ([\lambda\$X5209.\$C5209]\ \$A5209)\ (cs\ \$A5209)$

.5

§s %2 5 %1
$= (bs\ \$A5209)\ ([\lambda\$X5209.\$C5209]\ \$A5209)$
§s %0 3 %1
$= (bs\ \$A5209)\ (cs\ \$A5209)$

undefine local variables
:= $\$B5209$
:= $\$C5209$
Include end (A5209.r0t.txt) [newfile=(A5209.r0a.txt)]
>>>

##
Undefine Syntactical Variables
##

:= $\$M5209$
:= $\$E5209$
:= $\$T5209$
:= $\$X5209$
:= $\$A5209$

##
Q.E.D.
##

%0
$= (bs\ a)\ (cs\ a)$
$=_{omm} (bs_{mt} a_t)(cs_{mt} a_t)$

2.1.6 Results for File A5210.r0.txt

##
Proof Template A5210: T = (B = B)
for any B of any type
##
Source: [Andrews 2002 (ISBN 1-4020-0763-9), p. 220]
##

##
This file is part of the publication of the mathematical logic $\mathcal{R}_0$.
For more information, visit: <http://doi.org/10.4444/100.10>
##

##
Define Syntactical Variables
##

type of the wff
:= $\$T5210$ t_τ
wff 4 : t_τ := $\$T5210$

the wff
:= $\$B5210$ $b_{\$T5210}$
wff 11 : $b_{\$T5210}$:= $\$B5210$

##
Include Proof Template
##

<<< A5210.r0t.txt
Include begin (A5210.r0t.txt) [oldfile=(A5210.r0.txt)]
##
Proof Template A5210: T = (B = B)
for any B of any type
##
Source: [Andrews 2002 (ISBN 1-4020-0763-9), p. 220]
##

##
This file is part of the publication of the mathematical logic $\mathcal{R}_0$.
For more information, visit: <http://doi.org/10.4444/100.10>
##

##
Proof Template
##

.1

use Proof Template: Axiom 3 Substitutions
:= $\$AA3$ t_τ

wff 4 : t_τ := $\$AA3$ $\$T5210$
:= $\$BA3$ t_τ
wff 4 : t_τ := $\$AA3$ $\$BA3$ $\$T5210$
:= $\$FA3$ $[\lambda y_{\$AA3}.y_{\$AA3}]$
wff 13 : $[\lambda y.y]_{\$AA3\,\$AA3}$:= $\$FA3$
:= $\$GA3$ $[\lambda y_{\$AA3}.y_{\$AA3}]$
wff 13 : $[\lambda y.y]_{\$AA3\,\$AA3}$:= $\$FA3$ $\$GA3$
<< axiom3_substitutions.r0t.txt
:= $\$AA3$
:= $\$BA3$
:= $\$FA3$
:= $\$GA3$
%0
$=(=[\lambda y.y]\,[\lambda y.y])\,(\forall\,\$T5210\,[\lambda x.(=([\lambda y.y]\,x)\,([\lambda y.y]\,x))])$
$=_{ooo}(=_{o(\$T5210\,\$T5210)(\$T5210\,\$T5210)}[\lambda y_{\$T5210}.y_{\$T5210}][\lambda y_{\$T5210}.y_{\$T5210}])\ldots$
$\ldots(\forall_{o(o\backslash 3)\tau}\$T5210_\tau[\lambda x_{\$T5210}.(=_{o\,\$T5210\,\$T5210}([\lambda y_{\$T5210}.y_{\$T5210}]x_{\$T5210})([\lambda y_{\$T5210}.y_{\$T5210}]x_{\$T5210}))_o])$

.2

§= $_{\$T5210\,\$T5210}$ $[\lambda y_{\$T5210}.y_{\$T5210}]$
$=[\lambda y.y]\,[\lambda y.y]$
§s %0 1 %1
$\forall\,\$T5210\,[\lambda x.(=([\lambda y.y]\,x)\,([\lambda y.y]\,x))]$
§\ $[\lambda y_{\$T5210}.y_{\$T5210}]x_{\$T5210}$
$=([\lambda y.y]\,x)\,x$
§s %1 29 %0
$\forall\,\$T5210\,[\lambda x.(=x\,([\lambda y.y]\,x))]$
§s %0 15 %1
$\forall\,\$T5210\,[\lambda x.(=x\,x)]$
§\ $\forall_{o(o\backslash 3)\tau}\$T5210_\tau$
$=(\forall\,\$T5210)\,[\lambda p.(=[\lambda x.T]\,p)]$
§s %1 2 %0
$[\lambda p.(=[\lambda x.T]\,p)]\,[\lambda x.(=x\,x)]$
§\ $[\lambda p_{o\,\$T5210}.(=_{o(o\,\$T5210)(o\,\$T5210)}[\lambda x_{\$T5210}.T_o]p_{o\,\$T5210})_o][\lambda x_{\$T5210}.(=_{o\,\$T5210\,\$T5210}x_{\$T5210}x_{\$T5210})_o]$
$=([\lambda p.(=[\lambda x.T]\,p)]\,[\lambda x.(=x\,x)])\,(=[\lambda x.T]\,[\lambda x.(=x\,x)])$
§s %1 1 %0
$=[\lambda x.T]\,[\lambda x.(=x\,x)]$

.3

:= $\$LxT5210$ $[\lambda x_{\$T5210}.T_o]$
wff 29 : $[\lambda x.T]_{o\,\$T5210}$:= $\$LxT5210$
§= $_o$ $\$LxT5210_{o\,\$T5210}\$B5210_{\$T5210}$
$=(\$LxT5210\,\$B5210)\,(\$LxT5210\,\$B5210)$
§s %0 6 %1
$=(\$LxT5210\,\$B5210)\,([\lambda x.(=x\,x)]\,\$B5210)$

.4

```
§\ /5
#                  = ($LxT5210 $B5210) T
§\ %1/3
#                  = ([λx.(= x x)] $B5210) (= $B5210 $B5210)
§s %2 5 %1
#                  = T ([λx.(= x x)] $B5210)
§s %0 3 %1
#                  = T (= $B5210 $B5210)

## undefine local variables
:=  $LxT5210
## Include end (A5210.r0t.txt) [newfile=(A5210.r0.txt)]
>>>


##
##  Undefine Syntactical Variables
##

:=  $T5210
:=  $B5210


##
##  Q.E.D.
##

%0
#                  = T (= b b)
#                  =_{ooo}T_o(=_{ott}b_t b_t)
```

2.1.7 Results for File A5211.r0.txt

```
##
##  Proof A5211:  (T ∧ T) = T
##
##
##  Source: [Andrews 2002 (ISBN 1-4020-0763-9), p. 220]
##
##  Copyright (c) 2017 Owl of Minerva Press GmbH. All rights reserved.
##  Written by Ken Kubota (<mail@kenkubota.de>).
##
##  This file is part of the publication of the mathematical logic R0.
##  For more information, visit: <http://doi.org/10.4444/100.10>
##

<< axioms.r0.txt
```

##
Proof
##

.1

%$A1$
$=(\wedge (g\,T)\,(g\,F))\,(\forall\, o\,[\lambda x.(g\,x)])$:= $A1$
$=_{ooo}(\wedge_{ooo}(g_{oo}T_o)(g_{oo}F_o))(\forall_{o(o\backslash 3)\tau}o_\tau[\lambda x_o.(g_{oo}x_o)_o])$:= $A1$

use Proof Template A5209 (incl. A5204): B = C → (B = C) [x/A]
:= $\$M5209$ o
wff 2 : o_τ := $\$M5209$
:= $\$E5209$ %0
wff 90 : $=(\wedge (g\,T)\,(g\,F))\,(\forall\, o\,[\lambda x.(g\,x)])_o$:= $\$E5209$ $A1$
:= $\$T5209$ oo
wff 13 : oo_τ := $\$T5209$
:= $\$X5209$ $g_{\$T5209}$
wff 80 : $g_{\$T5209}$:= $\$X5209$
:= $\$A5209$ $[\lambda y_o.T_o]$
wff 130 : $[\lambda y.T]_{\$T5209}$:= $\$A5209$
<< A5209.r0t.txt
:= $\$M5209$
:= $\$E5209$
:= $\$T5209$
:= $\$X5209$
:= $\$A5209$
%0
$=(\wedge\,([\lambda y.T]\,T)\,([\lambda y.T]\,F))\,(\forall\, o\,[\lambda x.([\lambda y.T]\,x)])$
$=_{ooo}(\wedge_{ooo}([\lambda y_o.T_o]T_o)([\lambda y_o.T_o]F_o))(\forall_{o(o\backslash 3)\tau}o_\tau[\lambda x_o.([\lambda y_o.T_o]x_o)_o])$

.2

§\ $[\lambda y_o.T_o]T_o$
$=([\lambda y.T]\,T)\,T$
§s %1 21 %0
$=(\wedge\, T\,([\lambda y.T]\,F))\,(\forall\, o\,[\lambda x.([\lambda y.T]\,x)])$
§\ $[\lambda y_o.T_o]F_o$
$=([\lambda y.T]\,F)\,T$
§s %1 11 %0
$=(\wedge\, T\,T)\,(\forall\, o\,[\lambda x.([\lambda y.T]\,x)])$
§\ $[\lambda y_o.T_o]x_o$
$=([\lambda y.T]\,x)\,T$
§s %1 15 %0
$=(\wedge\, T\,T)\,(\forall\, o\,[\lambda x.T])$
:= $\$ATMP5211$ %0
wff 166 : $=(\wedge\, T\,T)\,(\forall\, o\,[\lambda x.T])_o$:= $\$ATMP5211$

.3

§= $\forall_{o(o\backslash 3)\tau} o_\tau [\lambda x_o . T_o]$
\# $= (\forall\, o\, [\lambda x.T])\, (\forall\, o\, [\lambda x.T])$
§\ $\forall_{o(o\backslash 3)\tau} o_\tau$
\# $= (\forall\, o)\, [\lambda p.(= [\lambda x.T]\, p)]$
§s %1 10 %0
\# $= ([\lambda p.(= [\lambda x.T]\, p)]\, [\lambda x.T])\, (\forall\, o\, [\lambda x.T])$
§\ $_o$ $[\lambda p_{oo}.(=_{o(oo)(oo)} [\lambda x_o.T_o] p_{oo})_o][\lambda x_o.T_o]$
\# $= ([\lambda p.(= [\lambda x.T]\, p)]\, [\lambda x.T])\, (= [\lambda x.T]\, [\lambda x.T])$
§s %1 5 %0
\# $= (= [\lambda x.T]\, [\lambda x.T])\, (\forall\, o\, [\lambda x.T])$
:= $\$BTMP5211$ %0
\# wff 181 : $= (= [\lambda x.T]\, [\lambda x.T])\, (\forall\, o\, [\lambda x.T])_o$:= $\$BTMP5211$

\#\# use Proof Template A5210: T = (B = B)
:= $\$T5210$ oo
\# wff 13 : oo_τ := $\$T5210$
:= $\$B5210$ $[\lambda x_o.T_o]$
\# wff 17 : $[\lambda x.T]_{\$T5210}$:= $\$B5210$
<< A5210.r0t.txt
:= $\$T5210$
:= $\$B5210$
%0
\# $= T\, (= [\lambda x.T]\, [\lambda x.T])$
\# $=_{ooo} T_o (=_{o(oo)(oo)} [\lambda x_o.T_o][\lambda x_o.T_o])$

%$\$BTMP5211$
\# $= (= [\lambda x.T]\, [\lambda x.T])\, (\forall\, o\, [\lambda x.T])$:= $\$BTMP5211$
\# $=_{o\omega\omega} (=_{o(oo)(oo)} [\lambda x_o.T_o][\lambda x_o.T_o])(\forall_{o(o\backslash 3)\tau} o_\tau [\lambda x_o.T_o])$:= $\$BTMP5211$
§s %1 3 %0
\# $= T\, (\forall\, o\, [\lambda x.T])$

\#\# .4

%$\$ATMP5211$
\# $= (\wedge\, T\, T)\, (\forall\, o\, [\lambda x.T])$:= $\$ATMP5211$
\# $=_{ooo} (\wedge_{ooo} T_o T_o)(\forall_{o(o\backslash 3)\tau} o_\tau [\lambda x_o.T_o])$:= $\$ATMP5211$
§= T
\# $= T\, T$
§s %0 5 %2
\# $= (\forall\, o\, [\lambda x.T])\, T$
§s %2 3 %0
\# $= (\wedge\, T\, T)\, T$

:= $A5211$ %0
\# wff 318 : $= (\wedge\, T\, T)\, T_o$:= $A5211$

\#\# undefine local variables
:= $\$ATMP5211$

:= $\$BTMP5211$

```
##
## Q.E.D.
##
```

%0
\# $=(\wedge T\,T)\,T$:= $A5211$
\# $=_{ooo}(\wedge_{ooo}T_oT_o)T_o$:= $A5211$

2.1.8 Results for File A5212.r0.txt

```
##
## Proof A5212: T ∧ T
##
##
## Source: [Andrews 2002 (ISBN 1-4020-0763-9), p. 220]
##
```

```
## Written by Ken Kubota (<mail@kenkubota.de>).
##
## This file is part of the publication of the mathematical logic R0.
## For more information, visit: <http://doi.org/10.4444/100.10>
##
```

<< A5200t.r0.txt
<< A5211.r0.txt

```
##
## Proof
##
```

%$A5211$
\# $=(\wedge T\,T)\,T$:= $A5211$
\# $=_{ooo}(\wedge_{ooo}T_oT_o)T_o$:= $A5211$
\## use Proof Template A5201b (Swap): A = B → B = A
<< A5201b.r0t.txt
%0
\# $=T\,(\wedge T\,T)$
\# $=_{ooo}T_o(\wedge_{ooo}T_oT_o)$
%T
\# $===$:= $A5200t\ \ T$
\# $=_{o\omega\omega}=_\omega=_\omega$:= $A5200t\ \ T$
§s %0 1 %1
\# $\wedge T\,T$

:= $A5212$ %0

wff 160 : $\wedge T\, T_{o,\ldots}$:= $A5212$

##
Q.E.D.
##

%0
$\wedge T\, T$:= $A5212$
$\wedge_{ooo} T_o T_o$:= $A5212$

2.1.9 Results for File A5213.r0a.txt

##
Proof Template A5213: A = B and C = D → (A = B) ∧ (C = D)
for any A, B of type T and any C, D of type U
##
Source: [Andrews 2002 (ISBN 1-4020-0763-9), pp. 220 f.]
##

Written by Ken Kubota (<mail@kenkubota.de>).
##
This file is part of the publication of the mathematical logic $\mathcal{R}_0$.
For more information, visit: <http://doi.org/10.4444/100.10>
##

##
Define Syntactical Variables
##

type of A, B
:= $T5213 t_τ
wff 4 : t_τ := $T5213

A = B
:= $AB5213 $=_{o\omega\omega} a_\omega b_\omega$
wff 14 : $= a\, b_o$:= $AB5213

type of C, D
:= $U5213 u_τ
wff 15 : u_τ := $U5213

C = D
:= $CD5213 $=_{o\omega\omega} c_\omega d_\omega$
wff 19 : $= c\, d_o$:= $CD5213

##

Assumptions and Resulting Syntactical Variables
##

§! $\$AB5213$
$= a\,b$:= $\$AB5213$
§! $\$CD5213$
$= c\,d$:= $\$CD5213$

##
Include Proof Template
##

<<< A5213.r0t.txt
Include begin (A5213.r0t.txt) [oldfile=(A5213.r0a.txt)]
##
Proof Template A5213: A = B and C = D $\rightarrow$ (A = B) $\wedge$ (C = D)
for any A, B of type T and any C, D of type U
##
Source: [Andrews 2002 (ISBN 1-4020-0763-9), pp. 220 f.]
##

Written by Ken Kubota (<mail@kenkubota.de>).
##
This file is part of the publication of the mathematical logic $\mathcal{R}_0$.
For more information, visit: <http://doi.org/10.4444/100.10>
##

<< A5212.r0.txt

##
Proof Template
##

.1

%$\$AB5213$
$= a\,b$:= $\$AB5213$
$=_{o\omega\omega} a_\omega b_\omega$:= $\$AB5213$

.2

use Proof Template A5210: T = (B = B)
:= $\$T5210$ t_τ
wff 4 : t_τ := $\$T5210$ $\$T5213$
:= $\$B5210$ $=_{o\omega\omega} a_\omega b_\omega / 5$
wff 11 : $a_{\$T5213}$:= $\$B5210$

<< A5210.r0t.txt
:= $\$T5210$
:= $\$B5210$
%0
$=T\,(=a\,a)$
$=_{ooo}T_o(=_{o\$T5213\,\$T5213}a_{\$T5213}a_{\$T5213})$

%$\$AB5213$
$=a\,b$:= $\$AB5213$
$=_{o\omega\omega}a_\omega b_\omega$:= $\$AB5213$
§s %1 7 %0
$=T\,(=a\,b)$
:= $\$TMP5213$ %0
wff 452 : $=T\,(=a\,b)_o$:= $\$TMP5213$

.3

%$\$CD5213$
$=c\,d$:= $\$CD5213$
$=_{o\omega\omega}c_\omega d_\omega$:= $\$CD5213$

.4

use Proof Template A5210: T = (B = B)
:= $\$T5210$ u_τ
wff 15 : u_τ := $\$T5210$ $\$U5213$
:= $\$B5210$ $=_{o\omega\omega}c_\omega d_\omega/5$
wff 16 : $c_{\$U5213}$:= $\$B5210$
<< A5210.r0t.txt
:= $\$T5210$
:= $\$B5210$
%0
$=T\,(=c\,c)$
$=_{ooo}T_o(=_{o\$U5213\,\$U5213}c_{\$U5213}c_{\$U5213})$

%$\$CD5213$
$=c\,d$:= $\$CD5213$
$=_{o\omega\omega}c_\omega d_\omega$:= $\$CD5213$
§s %1 7 %0
$=T\,(=c\,d)$

.5

%$A5212$
$\wedge T\,T$:= $A5212$
$\wedge_{ooo}T_oT_o$:= $A5212$
%$\$TMP5213$
$=T\,(=a\,b)$:= $\$TMP5213$
$=_{ooo}T_o(=_{o\$T5213\,\$T5213}a_{\$T5213}b_{\$T5213})$:= $\$TMP5213$

§s %1 5 %0
$\wedge (= a\, b)\, T$
§s %0 3 %3
$\wedge (= a\, b)\, (= c\, d)$

undefine local variables
:= $\$TMP5213$
Include end (A5213.r0t.txt) [newfile=(A5213.r0a.txt)]
>>>

##
Undefine Syntactical Variables
##

:= $\$T5213$
:= $\$AB5213$
:= $\$U5213$
:= $\$CD5213$

##
Q.E.D.
##

%0
$\wedge (= a\, b)\, (= c\, d)$
$\wedge_{ooo}(=_{ott} a_t b_t)(=_{ouu} c_u d_u)$

2.1.10 Results for File A5214.r0.txt

##
Proof A5214: (T ∧ F) = F
##
##
Source: [Andrews 2002 (ISBN 1-4020-0763-9), p. 221]
##
Copyright (c) 2017 Owl of Minerva Press GmbH. All rights reserved.
Written by Ken Kubota (<mail@kenkubota.de>).
##
This file is part of the publication of the mathematical logic $\mathcal{R}_0$.
For more information, visit: <http://doi.org/10.4444/100.10>
##

<< axioms.r0.txt

##
Proof

\##

\## .1

%$A1$
\# $= (\wedge\,(g\,T)\,(g\,F))\,(\forall\,o\,[\lambda x.(g\,x)])$:= $A1$
\# $=_{ooo}(\wedge_{ooo}(g_{oo}T_o)(g_{oo}F_o))(\forall_{o(o\backslash 3)\tau}o_\tau[\lambda x_o.(g_{oo}x_o)_o])$:= $A1$

\## use Proof Template A5209 (incl. A5204): B = C → (B = C) [x/A]
:= $\$M5209\ o$
\# wff 2 : o_τ := $\$M5209$
:= $\$E5209$ %0
\# wff 90 : $= (\wedge\,(g\,T)\,(g\,F))\,(\forall\,o\,[\lambda x.(g\,x)])_o$:= $\$E5209$ $A1$
:= $\$T5209\ oo$
\# wff 13 : oo_τ := $\$T5209$
:= $\$X5209\ g_{\$T5209}$
\# wff 80 : $g_{\$T5209}$:= $\$X5209$
:= $\$A5209\ [\lambda x_o.x_o]$
\# wff 19 : $[\lambda x.x]_{\$T5209}$:= $\$A5209$
<< A5209.r0t.txt
:= $\$M5209$
:= $\$E5209$
:= $\$T5209$
:= $\$X5209$
:= $\$A5209$
%0
\# $= (\wedge\,([\lambda x.x]\,T)\,([\lambda x.x]\,F))\,(\forall\,o\,[\lambda x.([\lambda x.x]\,x)])$
\# $=_{ooo}(\wedge_{ooo}([\lambda x_o.x_o]T_o)([\lambda x_o.x_o]F_o))(\forall_{o(o\backslash 3)\tau}o_\tau[\lambda x_o.([\lambda x_o.x_o]x_o)_o])$

\## .2

§\ $[\lambda x_o.x_o]T_o$
\# $= ([\lambda x.x]\,T)\,T$
§s %1 21 %0
\# $= (\wedge\,T\,([\lambda x.x]\,F))\,(\forall\,o\,[\lambda x.([\lambda x.x]\,x)])$
§\ $[\lambda x_o.x_o]F_o$
\# $= ([\lambda x.x]\,F)\,F$
§s %1 11 %0
\# $= (\wedge\,T\,F)\,(\forall\,o\,[\lambda x.([\lambda x.x]\,x)])$
§\ $[\lambda x_o.x_o]x_o$
\# $= ([\lambda x.x]\,x)\,x$
§s %1 15 %0
\# $= (\wedge\,T\,F)\,(\forall\,o\,[\lambda x.x])$

§= $\forall_{o(o\backslash 3)\tau}o_\tau[\lambda x_o.x_o]$
\# $= (\forall\,o\,[\lambda x.x])\,(\forall\,o\,[\lambda x.x])$
§\ $\forall_{o(o\backslash 3)\tau}o_\tau$
\# $= (\forall\,o)\,[\lambda p.(= [\lambda x.T]\,p)]$
§s %1 6 %0

\# $= (\forall\, o\, [\lambda x.x])\, ([\lambda p.(= [\lambda x.T]\, p)]\, [\lambda x.x])$
§\ $[\lambda p_{oo}.(=_{o(oo)(oo)}[\lambda x_o.T_o]p_{oo})_o][\lambda x_o.x_o]$
\# $= ([\lambda p.(= [\lambda x.T]\, p)]\, [\lambda x.x])\, F$
§s %1 3 %0
\# $= (\forall\, o\, [\lambda x.x])\, F$
§s %5 3 %0
\# $= (\wedge\, T\, F)\, F$

:= $A5214$ %0
\# wff 178 : $= (\wedge\, T\, F)\, F_o$:= $A5214$

\##
\## Q.E.D.
\##

%0
\# $= (\wedge\, T\, F)\, F$:= $A5214$
\# $=_{ooo}(\wedge_{ooo}T_oF_o)F_o$:= $A5214$

2.1.11 Results for File A5215.r0a.txt

\##
\## Proof Template A5215 (∀ I): ∀ x: B → B [x/a]
\## (Universal Instantiation)
\##
\## Source: [Andrews 2002 (ISBN 1-4020-0763-9), p. 221]
\##

\## Written by Ken Kubota (<mail@kenkubota.de>).
\##
\## This file is part of the publication of the mathematical logic $\mathcal{R}_0$.
\## For more information, visit: <http://doi.org/10.4444/100.10>
\##

\##
\## Define Syntactical Variables
\##

<< definitions1.r0.txt

\## type of the variable and the substitution term
:= $\$T5215$ t_τ
\# wff 4 : t_τ := $\$T5215$

\## the variable to be replaced
:= $\$X5215$ $x_{\$T5215}$
\# wff 24 : $x_{\$T5215}$:= $\$X5215$

substitution term
:= $A5215 $a_{\$T5215}$
wff 80 : $a_{\$T5215}$:= $A5215

hypothesis: ∀ x of type t: B (in this example, B is defined as x=x)
:= $H5215 $\forall_{o(o\backslash 3)\tau}\$T5215_{\tau}[\lambda\$X5215_{\$T5215}.(=_{o\omega\omega}\$X5215_{\omega}\$X5215_{\omega})_o]$
wff 85 : $\forall\,\$T5215\,[\lambda\$X5215.(=\$X5215\,\$X5215)]_o$:= $H5215

##
Assumptions and Resulting Syntactical Variables
##

§! $H5215
$\forall\,\$T5215\,[\lambda\$X5215.(=\$X5215\,\$X5215)]$:= $H5215

##
Include Proof Template
##

<<< A5215.r0t.txt
Include begin (A5215.r0t.txt) [oldfile=(A5215.r0a.txt)]
##
Proof Template A5215 (∀ I): ∀ x: B → B [x/a]
(Universal Instantiation)
##
Source: [Andrews 2002 (ISBN 1-4020-0763-9), p. 221]
##
Copyright (c) 2017 Owl of Minerva Press GmbH. All rights reserved.
Written by Ken Kubota (<mail@kenkubota.de>).
##
This file is part of the publication of the mathematical logic $\mathcal{R}_0$.
For more information, visit: <http://doi.org/10.4444/100.10>
##

<< A5200t.r0.txt

##
Proof Template
##

.1

%$H5215
$\forall\,\$T5215\,[\lambda\$X5215.(=\$X5215\,\$X5215)]$:= $H5215
$\forall_{o(o\backslash 3)\tau}\$T5215_{\tau}[\lambda\$X5215_{\$T5215}.(=_{o\omega\omega}\$X5215_{\omega}\$X5215_{\omega})_o]$:= $H5215

§\ $\forall_{o(o\backslash 3)\tau} \$T5215_\tau$
$= (\forall\, \$T5215)\, [\lambda p.(= [\lambda \$X5215.T]\, p)]$
§s %1 2 %0
$[\lambda p.(= [\lambda \$X5215.T]\, p)]\, [\lambda \$X5215.(= \$X5215\, \$X5215)]$
§\ $[\lambda p_{o\,\$T5215}.(=_{o(o\,\$T5215)(o\,\$T5215)} [\lambda \$X5215_{\$T5215}.T_o] p_{o\,\$T5215})_o] \ldots$
$\ldots [\lambda \$X5215_{\$T5215}.(=_{o\omega\omega} \$X5215_\omega \$X5215_\omega)_o]$
$= ([\lambda p.(= [\lambda \$X5215.T]\, p)]\, [\lambda \$X5215.(= \$X5215\, \$X5215)]) \ldots$
$\ldots (= [\lambda \$X5215.T]\, [\lambda \$X5215.(= \$X5215\, \$X5215)])$
§s %1 1 %0
$= [\lambda \$X5215.T]\, [\lambda \$X5215.(= \$X5215\, \$X5215)]$

.2

§= $[\lambda \$X5215_{\$T5215}.T_o] \$A5215_{\$T5215}$
$= ([\lambda \$X5215.T]\, \$A5215)\, ([\lambda \$X5215.T]\, \$A5215)$
§s %0 6 %1
$= ([\lambda \$X5215.T]\, \$A5215)\, ([\lambda \$X5215.(= \$X5215\, \$X5215)]\, \$A5215)$

.3

§\ $[\lambda \$X5215_{\$T5215}.T_o] \$A5215_{\$T5215}$
$= ([\lambda \$X5215.T]\, \$A5215)\, T$
§s %1 5 %0
$= T\, ([\lambda \$X5215.(= \$X5215\, \$X5215)]\, \$A5215)$
§\ $[\lambda \$X5215_{\$T5215}.(=_{o\omega\omega} \$X5215_\omega \$X5215_\omega)_o] \$A5215_{\$T5215}$
$= ([\lambda \$X5215.(= \$X5215\, \$X5215)]\, \$A5215)\, (= \$A5215\, \$A5215)$
§s %1 3 %0
$= T\, (= \$A5215\, \$A5215)$

.4

%T
$= = =$:= $A5200t$ T
$=_{o\omega\omega} =_\omega =_\omega$:= $A5200t$ T
§s %0 1 %1
$= \$A5215\, \$A5215$
Include end (A5215.r0t.txt) [newfile=(A5215.r0a.txt)]
>>>

##
Undefine Syntactical Variables
##

:= $\$T5215$
:= $\$X5215$
:= $\$A5215$
:= $\$H5215$

##
Q.E.D.
##

%0
$= a\, a$
$=_{o\omega\omega} a_\omega a_\omega$

2.1.12 Results for File A5215H.r0a.txt

##
Proof Template A5215H (∀ I): H ⊃ ∀ x: B → H ⊃ B [x/a]
(Universal Instantiation)
##
Source: [Andrews 2002 (ISBN 1-4020-0763-9), p. 221]
##

Written by Ken Kubota (<mail@kenkubota.de>).
##
This file is part of the publication of the mathematical logic $\mathcal{R}_0$.
For more information, visit: <http://doi.org/10.4444/100.10>
##

##
Define Syntactical Variables
##

<< definitions1.r0.txt

type of the variable and the substitution term
:= $\$T5215H\ \ t_\tau$
wff 4 : t_τ := $\$T5215H$

the variable to be replaced
:= $\$X5215H\ \ x_{\$T5215H}$
wff 24 : $x_{\$T5215H}$:= $\$X5215H$

substitution term
:= $\$A5215H\ \ a_{\$T5215H}$
wff 80 : $a_{\$T5215H}$:= $\$A5215H$

hypothesis: H ⊃ ∀ x of type t: B (in this example, B is defined as x=x)
:= $\$H5215H\ \ \supset_{ooo} h_o (\forall_{o(o\backslash 3)\tau} \$T5215H_\tau [\lambda \$X5215H_{\$T5215H}.(=_{o\omega\omega} \$X5215H_\omega \$X5215H_\omega)_o])$
wff 88 : $\supset h\, (\forall\, \$T5215H\, [\lambda \$X5215H.(= \$X5215H\, \$X5215H)])_o$:= $\$H5215H$

##

```
##  Assumptions and Resulting Syntactical Variables
##

§! $H5215H
#                    ⊃ h (∀ $T5215H [λ$X5215H.(= $X5215H $X5215H)])       :=  $H5215H


##
##  Include Proof Template
##

## <<< A5215H.r0t.txt
## Include begin (A5215H.r0t.txt) [oldfile=(A5215H.r0a.txt)]
##
##  Proof Template A5215H (∀ I):  H ⊃ ∀ x: B  →  H ⊃ B [x/a]
##       (Universal Instantiation)
##
##  Source: [Andrews 2002 (ISBN 1-4020-0763-9), p. 221]
##
##  Copyright (c) 2017 Owl of Minerva Press GmbH. All rights reserved.
##  Written by Ken Kubota (<mail@kenkubota.de>).
##
##  This file is part of the publication of the mathematical logic R0.
##  For more information, visit: <http://doi.org/10.4444/100.10>
##

<< A5200t.r0.txt


##
##  Exception: Forward Reference
##
##  Because of the different rules of inference, unlike in Q0,
##  this theorem (with hypothesis) cannot be inferred from
##  previous theorems only, but depends on new theorems.
##
##  Dependencies (selection):
##
##  K8004 << K8003
##             K8003 << K8000a, A5219b, A5221
##                        K8000a << A5222, A5229a, A5229c
##                                          A5221 << A5220
##                                                    A5229c << A5227 << A5226 << A5225
##


##
##  Proof Template
##
```

.1

%$\$H5215H$
$\supset h\,(\forall\,\$T5215H\,[\lambda\$X5215H.(=\$X5215H\,\$X5215H)])$:= $\$H5215H$
$\supset_{ooo}h_o(\forall_{o(o\backslash 3)\tau}\$T5215H_\tau[\lambda\$X5215H_{\$T5215H}.(=_{o\omega\omega}\$X5215H_\omega\$X5215H_\omega)_o])$
:= $\$H5215H$
§\ $\forall_{o(o\backslash 3)\tau}\$T5215H_\tau$
$=(\forall\,\$T5215H)\,[\lambda p.(=[\lambda\$X5215H.T]\,p)]$
§s %1 6 %0
$\supset h\,([\lambda p.(=[\lambda\$X5215H.T]\,p)]\,[\lambda\$X5215H.(=\$X5215H\,\$X5215H)])$
§\ $[\lambda p_{o\,\$T5215H}.(=_{o(o\,\$T5215H)(o\,\$T5215H)}[\lambda\$X5215H_{\$T5215H}.T_o]p_{o\,\$T5215H})_o]\ldots$
$\ldots[\lambda\$X5215H_{\$T5215H}.(=_{o\omega\omega}\$X5215H_\omega\$X5215H_\omega)_o]$
$=([\lambda p.(=[\lambda\$X5215H.T]\,p)]\,[\lambda\$X5215H.(=\$X5215H\,\$X5215H)])\ldots$
$\ldots(=[\lambda\$X5215H.T]\,[\lambda\$X5215H.(=\$X5215H\,\$X5215H)])$
§s %1 3 %0
$\supset h\,(=[\lambda\$X5215H.T]\,[\lambda\$X5215H.(=\$X5215H\,\$X5215H)])$
:= $\$ATMP5215H$ %0
wff 96 : $\supset h\,(=[\lambda\$X5215H.T]\,[\lambda\$X5215H.(=\$X5215H\,\$X5215H)])_o$:=
$\$ATMP5215H$

.2

§= $[\lambda\$X5215H_{\$T5215H}.T_o]\$A5215H_{\$T5215H}$
$=([\lambda\$X5215H.T]\,\$A5215H)\,([\lambda\$X5215H.T]\,\$A5215H)$

use Proof Template K8004 (Trans): (H ⊕ A), B → H ⊃ B
:= $\$HA8004$ $\supset_{ooo}h_o(\forall_{o(o\backslash 3)\tau}\$T5215H_\tau[\lambda\$X5215H_{\$T5215H}.(=_{o\omega\omega}\$X5215H_\omega\$X5215H_\omega)_o])$
wff 88 : $\supset h\,(\forall\,\$T5215H\,[\lambda\$X5215H.(=\$X5215H\,\$X5215H)])_o$:= $\$H5215H$
$\$HA8004$
:= $\$B8004$ %0
wff 99 : $=([\lambda\$X5215H.T]\,\$A5215H)\,([\lambda\$X5215H.T]\,\$A5215H)_o$:= $\$B8004$
<< K8004.r0t.txt
:= $\$HA8004$
:= $\$B8004$
%0
$\supset h\,(=([\lambda\$X5215H.T]\,\$A5215H)\,([\lambda\$X5215H.T]\,\$A5215H))$
$\supset_{ooo}h_o\ldots$
$\ldots(=_{o\omega\omega}([\lambda\$X5215H_{\$T5215H}.T_o]\$A5215H_{\$T5215H})([\lambda\$X5215H_{\$T5215H}.T_o]\$A5215H_{\$T5215H}))$

:= $\$BTMP5215H$ %0
wff 1430 : $\supset h\,(=([\lambda\$X5215H.T]\,\$A5215H)\,([\lambda\$X5215H.T]\,\$A5215H))_{o,\ldots}$:=
$\$BTMP5215H$

shorthand to avoid overlong line
:= $\$A2TMP5215H$ $[\lambda\$X5215H_{\$T5215H}.(=_{o\omega\omega}\$X5215H_\omega\$X5215H_\omega)_o]$
wff 86 : $[\lambda\$X5215H.(=\$X5215H\,\$X5215H)]_{o\,\$T5215H}$:= $\$A2TMP5215H$

%$\$ATMP5215H$

$\supset h\,(=[\lambda\$X5215H.T]\,\$A2TMP5215H)$:= $\$ATMP5215H$
$\supset_{ooo}h_o\ldots$
$\ldots(=_{o(o\$T5215H)(o\$T5215H)}[\lambda\$X5215H_{\$T5215H}.T_o]\$A2TMP5215H_{o\$T5215H})$:= $\$ATMP5215H$
:= $\$ATMP5215H$
%$\$BTMP5215H$
$\supset h\,(=([\lambda\$X5215H.T]\,\$A5215H)\,([\lambda\$X5215H.T]\,\$A5215H))$:=
$\$BTMP5215H$
$\supset_{ooo}h_o\ldots$
$\ldots(=_{o\omega\omega}([\lambda\$X5215H_{\$T5215H}.T_o]\$A5215H_{\$T5215H})([\lambda\$X5215H_{\$T5215H}.T_o]\$A5215H_{\$T5215H}))$
:= $\$BTMP5215H$
:= $\$BTMP5215H$
§s' %0 6 %1
$\supset h\,(=([\lambda\$X5215H.T]\,\$A5215H)\,(\$A2TMP5215H\,\$A5215H))$

undefine shorthand
:= $\$A2TMP5215H$

.3

§\ $[\lambda\$X5215H_{\$T5215H}.T_o]\$A5215H_{\$T5215H}$
$=([\lambda\$X5215H.T]\,\$A5215H)\,T$
§s %1 13 %0
$\supset h\,(=T\,([\lambda\$X5215H.(=\$X5215H\,\$X5215H)]\,\$A5215H))$
§\ $[\lambda\$X5215H_{\$T5215H}.(=_{o\omega\omega}\$X5215H_\omega\$X5215H_\omega)_o]\$A5215H_{\$T5215H}$
$=([\lambda\$X5215H.(=\$X5215H\,\$X5215H)]\,\$A5215H)\,(=\$A5215H\,\$A5215H)$
§s %1 7 %0
$\supset h\,(=T\,(=\$A5215H\,\$A5215H))$
:= $\$ATMP5215H$ %0
wff 1537 : $\supset h\,(=T\,(=\$A5215H\,\$A5215H))_o$:= $\$ATMP5215H$

.4

use Proof Template K8004 (Trans): (H ⊕ A), B → H ⊃ B
:= $\$HA8004$ $\supset_{ooo}h_o(\forall_{o(o\backslash 3)\tau}\$T5215H_\tau[\lambda\$X5215H_{\$T5215H}.(=_{o\omega\omega}\$X5215H_\omega\$X5215H_\omega)_o])$
wff 88 : $\supset h\,(\forall\,\$T5215H\,[\lambda\$X5215H.(=\$X5215H\,\$X5215H)])_o$:= $\$H5215H$
$\$HA8004$
:= $\$B8004$ $=_{o\omega\omega}=_\omega=_\omega$
wff 12 : $===_{o,\ldots}$:= $\$B8004$ $A5200t$ T
<< K8004.r0t.txt
:= $\$HA8004$
:= $\$B8004$
%0
$\supset h\,T$
$\supset_{ooo}h_oT_o$

:= $\$BTMP5215H$ %0
wff 1538 : $\supset h\,T_{o,\ldots}$:= $\$BTMP5215H$
%$\$ATMP5215H$
$\supset h\,(=T\,(=\$A5215H\,\$A5215H))$:= $\$ATMP5215H$

\# $\supset_{ooo} h_o (=_{o\omega\omega} T_\omega (=_{o\omega\omega} \$A5215H_\omega \$A5215H_\omega))$:= $\$ATMP5215H$
:= $\$ATMP5215H$
%$\$BTMP5215H$
\# $\supset h T$:= $\$BTMP5215H$
\# $\supset_{ooo} h_o T_o$:= $\$BTMP5215H$
:= $\$BTMP5215H$
§s' %0 1 %1
\# $\supset h (= \$A5215H\, \$A5215H)$
Include end (A5215H.r0t.txt) [newfile=(A5215H.r0a.txt)]
>>>

##
Undefine Syntactical Variables
##

:= $\$T5215H$
:= $\$X5215H$
:= $\$A5215H$
:= $\$H5215H$

##
Q.E.D.
##

%0
\# $\supset h (= a\, a)$
\# $\supset_{ooo} h_o (=_{o\omega\omega} a_\omega a_\omega)$

2.1.13 Results for File A5216.r0.txt

##
Proof Template A5216: (T ∧ A) = A
##
##
Source: [Andrews 2002 (ISBN 1-4020-0763-9), p. 221]
##

Written by Ken Kubota (<mail@kenkubota.de>).
##
This file is part of the publication of the mathematical logic $\mathcal{R}_0$.
For more information, visit: <http://doi.org/10.4444/100.10>
##

##
Define Syntactical Variables
##

the proposition
:= $A5216 a_o
wff 11 : a_o := $A5216

##
Include Proof Template
##

<<< A5216.r0t.txt
Include begin (A5216.r0t.txt) [oldfile=(A5216.r0.txt)]
##
Proof Template A5216: (T ∧ A) = A
##
##
Source: [Andrews 2002 (ISBN 1-4020-0763-9), p. 221]
##

Written by Ken Kubota (<mail@kenkubota.de>).
##
This file is part of the publication of the mathematical logic $\mathcal{R}_0$.
For more information, visit: <http://doi.org/10.4444/100.10>
##

<< A5211.r0.txt
<< A5214.r0.txt

##
Proof Template
##

.1

%$A1$
$= (\wedge\, (g\, T)\, (g\, F))\, (\forall\, o\, [\lambda x.(g\, x)])$:= $A1$
$=_{ooo}(\wedge_{ooo}(g_{oo}T_o)(g_{oo}F_o))(\forall_{o(o\backslash 3)\tau}o_\tau[\lambda x_o.(g_{oo}x_o)_o])$:= $A1$

use Proof Template A5209 (incl. A5204): B = C → (B = C) [x/A]
:= $M5209 o
wff 2 : o_τ := $M5209
:= $E5209 %0
wff 90 : $= (\wedge\, (g\, T)\, (g\, F))\, (\forall\, o\, [\lambda x.(g\, x)])_o$:= $E5209 $A1$
:= $T5209 oo
wff 14 : oo_τ := $T5209
:= $X5209 $g_{\$T5209}$
wff 80 : $g_{\$T5209}$:= $X5209
:= $A5209 $[\lambda x_o.(=_{\$T5209\, o}(\wedge_{\$T5209\, o}T_o x_o)x_o)_o]$

wff 363 : $[\lambda x.(=(\wedge\, T\, x)\, x)]_{\$T5209}$:= $\$A5209$
<< A5209.r0t.txt
:= $\$M5209$
:= $\$E5209$
:= $\$T5209$
:= $\$X5209$
:= $\$A5209$
%0
$=(\wedge\,([\lambda x.(=(\wedge\, T\, x)\, x)]\, T)\,([\lambda x.(=(\wedge\, T\, x)\, x)]\, F))\,(\forall\, o\,[\lambda x.([\lambda x.(=(\wedge\, T\, x)\, x)]\, x)])$
$=_{ooo}(\wedge_{ooo}([\lambda x_o.(=_{ooo}(\wedge_{ooo}T_o x_o)x_o)_o]T_o)([\lambda x_o.(=_{ooo}(\wedge_{ooo}T_o x_o)x_o)_o]F_o))\ldots$
$\ldots(\forall_{o(o\backslash 3)\tau}o_\tau[\lambda x_o.([\lambda x_o.(=_{ooo}(\wedge_{ooo}T_o x_o)x_o)_o]x_o)_o])$

.2

§\ $[\lambda x_o.(=_{ooo}(\wedge_{ooo}T_o x_o)x_o)_o]T_o$
$=([\lambda x.(=(\wedge\, T\, x)\, x)]\, T)\; A5211$
§s %1 21 %0
$=(\wedge\; A5211\,([\lambda x.(=(\wedge\, T\, x)\, x)]\, F))\,(\forall\, o\,[\lambda x.([\lambda x.(=(\wedge\, T\, x)\, x)]\, x)])$
§\ $[\lambda x_o.(=_{ooo}(\wedge_{ooo}T_o x_o)x_o)_o]F_o$
$=([\lambda x.(=(\wedge\, T\, x)\, x)]\, F)\; A5214$
§s %1 11 %0
$=(\wedge\; A5211\; A5214)\,(\forall\, o\,[\lambda x.([\lambda x.(=(\wedge\, T\, x)\, x)]\, x)])$
§\ $[\lambda x_o.(=_{ooo}(\wedge_{ooo}T_o x_o)x_o)_o]x_o$
$=([\lambda x.(=(\wedge\, T\, x)\, x)]\, x)\,(=(\wedge\, T\, x)\, x)$
§s %1 15 %0
$=(\wedge\; A5211\; A5214)\,(\forall\, o\,[\lambda x.(=(\wedge\, T\, x)\, x)])$
:= $\$TMP5216$ %0
wff 397 : $=(\wedge\; A5211\; A5214)\,(\forall\, o\,[\lambda x.(=(\wedge\, T\, x)\, x)])_o$:= $\$TMP5216$

.3

use Proof Template A5213: A = B and C = D → (A = B) ∧ (C = D)
:= $\$T5213\;\; o$
wff 2 : o_τ := $\$T5213$
:= $\$AB5213\;\; =_{ooo}(\wedge_{ooo}T_oT_o)T_o$
wff 318 : $=(\wedge\, T\, T)\, T_{o,\ldots}$:= $\$AB5213$ $A5211$
:= $\$U5213\;\; o$
wff 2 : o_τ := $\$T5213$ $\$U5213$
:= $\$CD5213\;\; =_{ooo}(\wedge_{ooo}T_oF_o)F_o$
wff 359 : $=(\wedge\, T\, F)\, F_{o,\ldots}$:= $\$CD5213$ $A5214$
<< A5213.r0t.txt
:= $\$T5213$
:= $\$AB5213$
:= $\$U5213$
:= $\$CD5213$
%0
$\wedge\; A5211\; A5214$
$\wedge_{ooo}A5211_oA5214_o$

.4

%$TMP5216
$= (\wedge\ A5211\ A5214)\ (\forall\ o\ [\lambda x.(= (\wedge\ T\ x)\ x)])$:= $TMP5216
$=_{ooo}(\wedge_{ooo}A5211_o A5214_o)(\forall_{o(o\backslash 3)\tau} o_\tau [\lambda x_o.(=_{ooo}(\wedge_{ooo}T_o x_o)x_o)_o])$:= $TMP5216
§s %1 1 %0
$\forall\ o\ [\lambda x.(= (\wedge\ T\ x)\ x)]$

.5

use Proof Template A5215 (∀ I): ∀ x: B → B [x/a]
:= $T5215 o
wff 2 : o_τ := $T5215
:= $X5215 x_o
wff 17 : x_o := $X5215
:= $A5215 a_o
wff 11 : a_o := $A5215 $A5216
:= $H5215 %0
wff 396 : $\forall\ o\ [\lambda \$X5215.(= (\wedge\ T\ \$X5215)\ \$X5215)]_{o,\ldots}$:= $H5215
<< A5215.r0t.txt
:= $T5215
:= $X5215
:= $A5215
:= $H5215
%0
$= (\wedge\ T\ \$A5216)\ \$A5216$
$=_{ooo}(\wedge_{ooo}T_o \$A5216_o)\$A5216_o$

undefine local variables
:= $TMP5216
Include end (A5216.r0t.txt) [newfile=(A5216.r0.txt)]
>>>

##
Undefine Syntactical Variables
##

:= $A5216

##
Q.E.D.
##

%0
$= (\wedge\ T\ a)\ a$
$=_{ooo}(\wedge_{ooo}T_o a_o)a_o$

2.1.14 Results for File A5217.r0.txt

##
Proof A5217: (T = F) = F
##
##
Source: [Andrews 2002 (ISBN 1-4020-0763-9), pp. 221 f.]
##

Written by Ken Kubota (<mail@kenkubota.de>).
##
This file is part of the publication of the mathematical logic $\mathcal{R}_0$.
For more information, visit: <http://doi.org/10.4444/100.10>
##

<< axioms.r0.txt

##
Proof
##

.1

%$A1$
$= (\wedge\,(g\,T)\,(g\,F))\,(\forall\,o\,[\lambda x.(g\,x)])$:= $A1$
$=_{ooo}(\wedge_{ooo}(g_{oo}T_o)(g_{oo}F_o))(\forall_{o(o\backslash 3)_\tau}o_\tau[\lambda x_o.(g_{oo}x_o)_o])$:= $A1$

use Proof Template A5209 (incl. A5204): B = C → (B = C) [x/A]
:= $\$M5209$ o
wff 2 : o_τ := $\$M5209$
:= $\$E5209$ %0
wff 90 : $= (\wedge\,(g\,T)\,(g\,F))\,(\forall\,o\,[\lambda x.(g\,x)])_o$:= $\$E5209$ $A1$
:= $\$T5209$ oo
wff 13 : oo_τ := $\$T5209$
:= $\$X5209$ $g_{\$T5209}$
wff 80 : $g_{\$T5209}$:= $\$X5209$
:= $\$A5209$ $[\lambda x_o.(=_{\$T5209\,o}T_o x_o)_o]$
wff 132 : $[\lambda x.(=T\,x)]_{\$T5209}$:= $\$A5209$
<< A5209.r0t.txt
:= $\$M5209$
:= $\$E5209$
:= $\$T5209$
:= $\$X5209$
:= $\$A5209$
%0
$= (\wedge\,([\lambda x.(=T\,x)]\,T)\,([\lambda x.(=T\,x)]\,F))\,(\forall\,o\,[\lambda x.([\lambda x.(=T\,x)]\,x)])$
$=_{ooo}(\wedge_{ooo}([\lambda x_o.(=_{ooo}T_o x_o)_o]T_o)([\lambda x_o.(=_{ooo}T_o x_o)_o]F_o))\ldots$
$\ldots(\forall_{o(o\backslash 3)_\tau}o_\tau[\lambda x_o.([\lambda x_o.(=_{ooo}T_o x_o)_o]x_o)_o])$

.2

§\ $[\lambda x_o.(=_{ooo}T_o x_o)_o]T_o$
\# $= ([\lambda x.(= T\, x)]\, T)\, (= T\, T)$
§s %1 21 %0
\# $= (\wedge\, (= T\, T)\, ([\lambda x.(= T\, x)]\, F))\, (\forall\, o\, [\lambda x.([\lambda x.(= T\, x)]\, x)])$
§\ $[\lambda x_o.(=_{ooo}T_o x_o)_o]F_o$
\# $= ([\lambda x.(= T\, x)]\, F)\, (= T\, F)$
§s %1 11 %0
\# $= (\wedge\, (= T\, T)\, (= T\, F))\, (\forall\, o\, [\lambda x.([\lambda x.(= T\, x)]\, x)])$
§\ $[\lambda x_o.(=_{ooo}T_o x_o)_o]x_o$
\# $= ([\lambda x.(= T\, x)]\, x)\, (= T\, x)$
§s %1 15 %0
\# $= (\wedge\, (= T\, T)\, (= T\, F))\, (\forall\, o\, [\lambda x.(= T\, x)])$
:= $\$ATMP5217$ %0
\# wff 170 : $= (\wedge\, (= T\, T)\, (= T\, F))\, (\forall\, o\, [\lambda x.(= T\, x)])_o$:= $\$ATMP5217$

.3

use Proof Template A5210: T = (B = B)
:= $\$T5210$ o
\# wff 2 : o_τ := $\$T5210$
:= $\$B5210$ $=_{o\omega\omega}=_\omega=_\omega$
\# wff 12 : $===_o$:= $\$B5210$ T
<< A5210.r0t.txt
:= $\$T5210$
:= $\$B5210$
%0
\# $= T\, (= T\, T)$
\# $=_{ooo}T_o(=_{ooo}T_o T_o)$

%$\$ATMP5217$
\# $= (\wedge\, (= T\, T)\, (= T\, F))\, (\forall\, o\, [\lambda x.(= T\, x)])$:= $\$ATMP5217$
\# $=_{ooo}(\wedge_{ooo}(=_{ooo}T_o T_o)(=_{ooo}T_o F_o))\ldots$
$\ldots(\forall_{o(o\backslash 3)\tau}o_\tau[\lambda x_o.(=_{ooo}T_o x_o)_o])$:= $\$ATMP5217$
§= T
\# $= T\, T$
§s %0 5 %2
\# $= (= T\, T)\, T$
§s %2 21 %0
\# $= (\wedge\, T\, (= T\, F))\, (\forall\, o\, [\lambda x.(= T\, x)])$

:= $\$BTMP5217$ %0
\# wff 294 : $= (\wedge\, T\, (= T\, F))\, (\forall\, o\, [\lambda x.(= T\, x)])_o$:= $\$BTMP5217$

.4

use Proof Template A5216: (T ∧ A) = A

:= $\$A5216\ =_{ooo}T_oF_o$
wff 162 : $=T\,F_{o,\ldots}$:= $\$A5216$
<< A5216.r0t.txt
:= $\$A5216$
%0
$=(\wedge\,T\,(=T\,F))\,(=T\,F)$
$=_{ooo}(\wedge_{ooo}T_o(=_{ooo}T_oF_o))(=_{ooo}T_oF_o)$

%$\$BTMP5217$
$=(\wedge\,T\,(=T\,F))\,(\forall\,o\,[\lambda x.(=T\,x)])$:= $\$BTMP5217$
$=_{ooo}(\wedge_{ooo}T_o(=_{ooo}T_oF_o))(\forall_{o(o\backslash 3)\tau}o_\tau[\lambda x_o.(=_{ooo}T_ox_o)_o])$:= $\$BTMP5217$
§s %0 5 %1
$=(=T\,F)\,(\forall\,o\,[\lambda x.(=T\,x)])$

:= $\$CTMP5217$ %0
wff 593 : $=(=T\,F)\,(\forall\,o\,[\lambda x.(=T\,x)])_o$:= $\$CTMP5217$

.5

use Proof Template: Axiom 3 Substitutions
:= $\$AA3\ o$
wff 2 : o_τ := $\$AA3$
:= $\$BA3\ o$
wff 2 : o_τ := $\$AA3\ \ \$BA3$
:= $\$FA3\ [\lambda x_o.T_o]$
wff 17 : $[\lambda x.T]_{oo}$:= $\$FA3$
:= $\$GA3\ [\lambda x_o.x_o]$
wff 19 : $[\lambda x.x]_{oo}$:= $\$GA3$
<< axiom3_substitutions.r0t.txt
:= $\$AA3$
:= $\$BA3$
:= $\$FA3$
:= $\$GA3$
%0
$=F\,(\forall\,o\,[\lambda x.(=([\lambda x.T]\,x)\,([\lambda x.x]\,x))])$
$=_{ooo}F_o(\forall_{o(o\backslash 3)\tau}o_\tau[\lambda x_o.(=_{ooo}([\lambda x_o.T_o]x_o)([\lambda x_o.x_o]x_o))_o])$

.6

§\ $[\lambda x_o.T_o]x_o$
$=([\lambda x.T]\,x)\,T$
§s %1 61 %0
$=F\,(\forall\,o\,[\lambda x.(=T\,([\lambda x.x]\,x))])$
§\ $[\lambda x_o.x_o]x_o$
$=([\lambda x.x]\,x)\,x$
§s %1 31 %0
$=F\,(\forall\,o\,[\lambda x.(=T\,x)])$

.7

use Proof Template A5201b (Swap): A = B → B = A
<< A5201b.r0t.txt
%0
$=(\forall o\,[\lambda x.(=T\,x)])\,F$
$=_{ooo}(\forall_{o(o\backslash 3)\tau}o_{\tau}[\lambda x_o.(=_{ooo}T_o x_o)_o])F_o$
%$\$CTMP5217$
$=(=T\,F)\,(\forall o\,[\lambda x.(=T\,x)])$:= $\$CTMP5217$
$=_{ooo}(=_{ooo}T_oF_o)(\forall_{o(o\backslash 3)\tau}o_{\tau}[\lambda x_o.(=_{ooo}T_o x_o)_o])$:= $\$CTMP5217$
§s %0 3 %1
$=(=T\,F)\,F$

:= $A5217$ %0
wff 638 : $=(=T\,F)\,F_o$:= $A5217$

undefine local variables
:= $\$ATMP5217$
:= $\$BTMP5217$
:= $\$CTMP5217$

##
Q.E.D.
##

%0
$=(=T\,F)\,F$:= $A5217$
$=_{ooo}(=_{ooo}T_oF_o)F_o$:= $A5217$

2.1.15 Results for File A5218.r0.txt

##
Proof Template A5218: (T = A) = A
##
##
Source: [Andrews 2002 (ISBN 1-4020-0763-9), p. 222]
##

Written by Ken Kubota (<mail@kenkubota.de>).
##
This file is part of the publication of the mathematical logic $\mathcal{R}_0$.
For more information, visit: <http://doi.org/10.4444/100.10>
##

##
Define Syntactical Variables
##

the bool wff
:= $\$A5218\ \ a_o$
wff 11 : a_o := $\$A5218$

##
Include Proof Template
##

<<< A5218.r0t.txt
Include begin (A5218.r0t.txt) [oldfile=(A5218.r0.txt)]
##
Proof Template A5218: (T = A) = A
##
##
Source: [Andrews 2002 (ISBN 1-4020-0763-9), p. 222]
##

Written by Ken Kubota (<mail@kenkubota.de>).
##
This file is part of the publication of the mathematical logic $\mathcal{R}_0$.
For more information, visit: <http://doi.org/10.4444/100.10>
##

<< A5217.r0.txt

##
Proof Template
##

.1

%$A1$
$=(\wedge\,(g\,T)\,(g\,F))\,(\forall\,o\,[\lambda x.(g\,x)])$:= $A1$
$=_{ooo}(\wedge_{ooo}(g_{oo}T_o)(g_{oo}F_o))(\forall_{o(o\backslash 3)_\tau}o_\tau[\lambda x_o.(g_{oo}x_o)_o])$:= $A1$

use Proof Template A5209 (incl. A5204): B = C → (B = C) [x/A]
:= $\$M5209\ \ o$
wff 2 : o_τ := $\$M5209$
:= $\$E5209$ %0
wff 90 : $=(\wedge\,(g\,T)\,(g\,F))\,(\forall\,o\,[\lambda x.(g\,x)])_o$:= $\$E5209\ \ A1$
:= $\$T5209\ \ oo$
wff 14 : oo_τ := $\$T5209$
:= $\$X5209\ \ g_{\$T5209}$
wff 80 : $g_{\$T5209}$:= $\$X5209$
:= $\$A5209\ \ [\lambda x_o.(=_{\$T5209\,o}(=_{\$T5209\,o}T_o x_o)x_o)_o]$
wff 641 : $[\lambda x.(=(=T\,x)\,x)]_{\$T5209}$:= $\$A5209$
<< A5209.r0t.txt

:= $\$E5209$
:= $\$T5209$
:= $\$X5209$
:= $\$A5209$
%0
$=(\wedge\,([\lambda x.(=(=T\,x)\,x)]\,T)\,([\lambda x.(=(=T\,x)\,x)]\,F))\,(\forall\,o\,[\lambda x.([\lambda x.(=(=T\,x)\,x)]\,x)])$
$=_{ooo}(\wedge_{ooo}([\lambda x_o.(=_{ooo}(=_{ooo}T_o x_o)x_o)_o]T_o)([\lambda x_o.(=_{ooo}(=_{ooo}T_o x_o)x_o)_o]F_o))\ldots$
$\ldots(\forall_{o(o\backslash 3)\tau}o_\tau[\lambda x_o.([\lambda x_o.(=_{ooo}(=_{ooo}T_o x_o)x_o)_o]x_o)_o])$

§\ $[\lambda x_o.(=_{ooo}(=_{ooo}T_o x_o)x_o)_o]T_o$
$=([\lambda x.(=(=T\,x)\,x)]\,T)\,(=(=T\,T)\,T)$
§s %1 21 %0
$=(\wedge\,(=(=T\,T)\,T)\,([\lambda x.(=(=T\,x)\,x)]\,F))\,(\forall\,o\,[\lambda x.([\lambda x.(=(=T\,x)\,x)]\,x)])$
§\ $[\lambda x_o.(=_{ooo}(=_{ooo}T_o x_o)x_o)_o]F_o$
$=([\lambda x.(=(=T\,x)\,x)]\,F)\,A5217$
§s %1 11 %0
$=(\wedge\,(=(=T\,T)\,T)\,A5217)\,(\forall\,o\,[\lambda x.([\lambda x.(=(=T\,x)\,x)]\,x)])$
§\ $[\lambda x_o.(=_{ooo}(=_{ooo}T_o x_o)x_o)_o]x_o$
$=([\lambda x.(=(=T\,x)\,x)]\,x)\,(=(=T\,x)\,x)$
§s %1 15 %0
$=(\wedge\,(=(=T\,T)\,T)\,A5217)\,(\forall\,o\,[\lambda x.(=(=T\,x)\,x)])$
:= $\$TMP5218$ %0
wff 677 : $=(\wedge\,(=(=T\,T)\,T)\,A5217)\,(\forall\,o\,[\lambda x.(=(=T\,x)\,x)])_o$:= $\$TMP5218$

.2

use Proof Template A5210: T = (B = B)
:= $\$T5210\ o$
wff 2 : o_τ := $\$M5209$ $\$T5210$
:= $\$B5210\ =_{o\omega\omega}=_\omega=_\omega$
wff 13 : $===_{o,\ldots}$:= $\$B5210$ $A5200t$ T
<< A5210.r0t.txt
:= $\$T5210$
:= $\$B5210$
%0
$=T\,(=T\,T)$
$=_{ooo}T_o(=_{ooo}T_oT_o)$

.3

use Proof Template A5201b (Swap): A = B → B = A
<< A5201b.r0t.txt
%0
$=(=T\,T)\,T$
$=_{ooo}(=_{ooo}T_oT_o)T_o$

use Proof Template A5213: A = B and C = D → (A = B) ∧ (C = D)
:= $\$T5213\ o$
wff 2 : o_τ := $\$T5213$

:= $\$AB5213$ %0
wff 663 : $=(=T\,T)\,T_{o,\dots}$:= $\$AB5213$
:= $\$U5213$ o
wff 2 : o_τ := $\$T5213$ $\$U5213$
:= $\$CD5213$ $=_{ooo}(=_{ooo}T_oF_o)F_o$
wff 638 : $=(=T\,F)\,F_{o,\dots}$:= $\$CD5213$ $A5217$
<< A5213.r0t.txt
:= $\$T5213$
:= $\$AB5213$
:= $\$U5213$
:= $\$CD5213$
%0
$\wedge\,(=(=T\,T)\,T)\,A5217$
$\wedge_{ooo}(=_{ooo}(=_{ooo}T_oT_o)T_o)A5217_o$

.4

%$\$TMP5218$
$=(\wedge\,(=(=T\,T)\,T)\,A5217)\,(\forall\,o\,[\lambda x.(=(=T\,x)\,x)])$:= $\$TMP5218$
$=_{ooo}(\wedge_{ooo}(=_{ooo}(=_{ooo}T_oT_o)T_o)A5217_o)(\forall_{o(o\backslash 3)\tau}o_\tau[\lambda x_o.(=_{ooo}(=_{ooo}T_ox_o)x_o)_o])$:=
$\$TMP5218$
§s %1 1 %0
$\forall\,o\,[\lambda x.(=(=T\,x)\,x)]$

.5

use Proof Template A5215 (∀ I): ∀ x: B → B [x/a]
:= $\$T5215$ o
wff 2 : o_τ := $\$T5215$
:= $\$X5215$ x_o
wff 17 : x_o := $\$X5215$
:= $\$A5215$ a_o
wff 11 : a_o := $\$A5215$ $\$A5218$
:= $\$H5215$ %0
wff 676 : $\forall\,o\,[\lambda\$X5215.(=(=T\,\$X5215)\,\$X5215)]_{o,\dots}$:= $\$H5215$
<< A5215.r0t.txt
:= $\$T5215$
:= $\$X5215$
:= $\$A5215$
:= $\$H5215$
%0
$=(=T\,\$A5218)\,\$A5218$
$=_{ooo}(=_{ooo}T_o\$A5218_o)\$A5218_o$

undefine local variables
:= $\$TMP5218$
Include end (A5218.r0t.txt) [newfile=(A5218.r0.txt)]
>>>

##
Undefine Syntactical Variables
##

:= $\$A5218$

##
Q.E.D.
##

%0
$= (= T\ a)\ a$
$=_{ooo}(=_{ooo}T_o a_o)a_o$

2.1.16 Results for File A5219a.r0a.txt

##
Proof Template A5219a (Rule T): A → T = A
##
##
Source: [Andrews 2002 (ISBN 1-4020-0763-9), p. 222]
##

Written by Ken Kubota (<mail@kenkubota.de>).
##
This file is part of the publication of the mathematical logic $\mathcal{R}_0$.
For more information, visit: <http://doi.org/10.4444/100.10>
##

##
Define Syntactical Variables
##

<< basics.r0.txt

the assumption
:= $\$A5219a\ a_o$
wff 54 : a_o := $\$A5219a$

##
Assumptions and Resulting Syntactical Variables
##

§! $\$A5219a$
a := $\$A5219a$

```
##
##   Include Proof Template
##

## <<< A5219a.r0t.txt
## Include begin (A5219a.r0t.txt) [oldfile=(A5219a.r0a.txt)]
##
##   Proof Template A5219a (Rule T):   A   →   T = A
##
##
##   Source: [Andrews 2002 (ISBN 1-4020-0763-9), p. 222]
##
##   Copyright (c) 2017 Owl of Minerva Press GmbH. All rights reserved.
##   Written by Ken Kubota (<mail@kenkubota.de>).
##
##   This file is part of the publication of the mathematical logic R0.
##   For more information, visit: <http://doi.org/10.4444/100.10>
##




##   Empty lines are needed for comparison between A5219a-A5219d and A5219aH-A5219dH.


##
##   Proof Template
##

## use Proof Template A5218:   (T = A) = A
:=  $A5218 a_o
# wff     54  :      a_o      :=  $A5218  $A5219a
<< A5218.r0t.txt
:=  $A5218
%0
#                    = (= T $A5219a) $A5219a
#                    =_ooo(=_oooT_o$A5219a_o)$A5219a_o

## use Proof Template A5201b (Swap):   A = B   →   B = A
<< A5201b.r0t.txt
%0
#                    = $A5219a (= T $A5219a)
#                    =_ooo$A5219a_o(=_oooT_o$A5219a_o)
```

%$A5219a
\# a := $A5219a
\# a_o := $A5219a

§s %0 1 %1
\# $=T$ $A5219a
\#\# Include end (A5219a.r0t.txt) [newfile=(A5219a.r0a.txt)]
>>>

\#\#
\#\# Undefine Syntactical Variables
\#\#

:= $A5219a

\#\#
\#\# Q.E.D.
\#\#

%0
\# $=T\ a$
\# $=_{ooo}T_oa_o$

2.1.17 Results for File A5219aH.r0a.txt

\#\#
\#\# Proof Template A5219aH (Rule T): H ⊃ A → H ⊃ (T = A)
\#\#
\#\#
\#\# Source: [Andrews 2002 (ISBN 1-4020-0763-9), p. 222]
\#\#

\#\# Written by Ken Kubota (<mail@kenkubota.de>).
\#\#

This file is part of the publication of the mathematical logic $\mathcal{R}_0$.
For more information, visit: <http://doi.org/10.4444/100.10>
##

##
Define Syntactical Variables
##

<< basics.r0.txt

the assumption
:= $A5219aH $\supset_{ooo} h_o a_o$
wff 210 : $\supset h\, a_o$:= $A5219aH

##
Assumptions and Resulting Syntactical Variables
##

§! $A5219aH
$\supset h\, a$:= $A5219aH

##
Include Proof Template
##

<<< A5219aH.r0t.txt
Include begin (A5219aH.r0t.txt) [oldfile=(A5219aH.r0a.txt)]
##
Proof Template A5219aH (Rule T): H ⊃ A → H ⊃ (T = A)
##
##
Source: [Andrews 2002 (ISBN 1-4020-0763-9), p. 222]
##

Written by Ken Kubota (<mail@kenkubota.de>).
##
This file is part of the publication of the mathematical logic $\mathcal{R}_0$.
For more information, visit: <http://doi.org/10.4444/100.10>
##

##
Exception: Forward Reference
##
(See comment in Proof Template A5215H.)
##

Empty lines are needed for comparison between A5219a-A5219d and A5219aH-A5219dH.

##
Proof Template
##

use Proof Template A5218: (T = A) = A
:= $\$A5218\ \supset_{ooo}h_o a_o/3$
wff 54 : a_o := $\$A5218$
<< A5218.r0t.txt
:= $\$A5218$
%0
$=(=T\,a)\,a$
$=_{ooo}(=_{ooo}T_o a_o)a_o$

use Proof Template A5201b (Swap): A = B → B = A
<< A5201b.r0t.txt
%0
$=a\,(=T\,a)$
$=_{ooo}a_o(=_{ooo}T_o a_o)$

use Proof Template K8004 (Trans): (H ⊕ A), B → H ⊃ B
:= $\$HA8004\ \supset_{ooo}h_o a_o$
wff 210 : $\supset h\,a_o$:= $\$A5219aH$ $\$HA8004$
:= $\$B8004$ %0
wff 802 : $=a\,(=T\,a)_o$:= $\$B8004$
<< K8004.r0t.txt
:= $\$HA8004$
:= $\$B8004$
%0
$\supset h\,(=a\,(=T\,a))$
$\supset_{ooo}h_o(=_{ooo}a_o(=_{ooo}T_o a_o))$

%$\$A5219aH$
$\supset h\,a$:= $\$A5219aH$
$\supset_{ooo}h_o a_o$:= $\$A5219aH$

§s' %0 1 %1
$\supset h\,(=T\,a)$
Include end (A5219aH.r0t.txt) [newfile=(A5219aH.r0a.txt)]
>>>

```
##
##  Undefine Syntactical Variables
##

:=  $A5219aH


##
##  Q.E.D.
##

%0
#                ⊃ h (= T a)
#                ⊃ooo h_o (=ooo T_o a_o)
```

2.1.18 Results for File A5219b.r0a.txt

```
##
##  Proof Template A5219b (Rule T):  A  →  A = T
##
##
##  Source: [Andrews 2002 (ISBN 1-4020-0763-9), p. 222]
##
##  Copyright (c) 2017 Owl of Minerva Press GmbH. All rights reserved.
##  Written by Ken Kubota (<mail@kenkubota.de>).
##
##  This file is part of the publication of the mathematical logic R0.
##  For more information, visit: <http://doi.org/10.4444/100.10>
##


##
##  Define Syntactical Variables
##

<< basics.r0.txt

## the assumption
:=  $A5219b a_o
# wff      54  :        a_o       :=  $A5219b


##
##  Assumptions and Resulting Syntactical Variables
##

§! $A5219b
#                 a       :=  $A5219b
```

##
Include Proof Template
##

<<< A5219b.r0t.txt
Include begin (A5219b.r0t.txt) [oldfile=(A5219b.r0a.txt)]
##
Proof Template A5219b (Rule T): A $\to$ A = T
##
##
Source: [Andrews 2002 (ISBN 1-4020-0763-9), p. 222]
##

Written by Ken Kubota (<mail@kenkubota.de>).
##
This file is part of the publication of the mathematical logic $\mathcal{R}_0$.
For more information, visit: <http://doi.org/10.4444/100.10>
##

Empty lines are needed for comparison between A5219a-A5219d and A5219aH-A5219dH.

##
Proof Template
##

use Proof Template A5218: (T = A) = A
:= $\$A5218\ a_o$
\# wff 54 : a_o := $\$A5218\ \ \$A5219b$
<< A5218.r0t.txt
:= $\$A5218$
%0
\# $= (= T\ \$A5219b)\ \$A5219b$
\# $=_{ooo}(=_{ooo}T_o\$A5219b_o)\$A5219b_o$

use Proof Template A5201b (Swap): A = B $\to$ B = A
<< A5201b.r0t.txt
%0
\# $= \$A5219b\ (= T\ \$A5219b)$
\# $=_{ooo}\$A5219b_o(=_{ooo}T_o\$A5219b_o)$

%$\$A5219b$
a := $\$A5219b$
a_o := $\$A5219b$

§s %0 1 %1
$= T\,\$A5219b$

use Proof Template A5201b (Swap): A = B → B = A
<< A5201b.r0t.txt
%0
$= \$A5219b\,T$
$=_{ooo}\$A5219b_o T_o$
Include end (A5219b.r0t.txt) [newfile=(A5219b.r0a.txt)]
>>>

##
Undefine Syntactical Variables
##

:= $\$A5219b$

##
Q.E.D.
##

%0
$= a\,T$
$=_{ooo}a_o T_o$

2.1.19 Results for File A5219bH.r0a.txt

##
Proof Template A5219bH (Rule T): H ⊃ A → H ⊃ (A = T)
##

```
##
##   Source: [Andrews 2002 (ISBN 1-4020-0763-9), p. 222]
##
##   Copyright (c) 2017 Owl of Minerva Press GmbH. All rights reserved.
##   Written by Ken Kubota (<mail@kenkubota.de>).
##
##   This file is part of the publication of the mathematical logic $\mathcal{R}_0$.
##   For more information, visit: <http://doi.org/10.4444/100.10>
##


##
##   Define Syntactical Variables
##

<< basics.r0.txt

## the assumption
:=  $A5219bH  $\supset_{ooo} h_o a_o$
# wff      210  :          $\supset h\, a_o$        :=  $A5219bH


##
##   Assumptions and Resulting Syntactical Variables
##

§! $A5219bH
#                     $\supset h\, a$        :=  $A5219bH


##
##   Include Proof Template
##

## <<< A5219bH.r0t.txt
## Include begin (A5219bH.r0t.txt) [oldfile=(A5219bH.r0a.txt)]
##
##   Proof Template A5219bH (Rule T):   H ⊃ A   →   H ⊃ (A = T)
##
##
##   Source: [Andrews 2002 (ISBN 1-4020-0763-9), p. 222]
##
##   Copyright (c) 2017 Owl of Minerva Press GmbH. All rights reserved.
##   Written by Ken Kubota (<mail@kenkubota.de>).
##
##   This file is part of the publication of the mathematical logic $\mathcal{R}_0$.
##   For more information, visit: <http://doi.org/10.4444/100.10>
##
```

##
Exception: Forward Reference
##
(See comment in Proof Template A5215H.)
##

Empty lines are needed for comparison between A5219a-A5219d and A5219aH-A5219dH.

##
Proof Template
##

use Proof Template A5218: (T = A) = A
:= $\$A5218\ \supset_{ooo}h_o a_o/3$
wff 54 : a_o := $\$A5218$
<< A5218.r0t.txt
:= $\$A5218$
%0
$=(=T\,a)\,a$
$=_{ooo}(=_{ooo}T_o a_o)a_o$

use Proof Template A5201b (Swap): A = B → B = A
<< A5201b.r0t.txt
%0
$=a\,(=T\,a)$
$=_{ooo}a_o(=_{ooo}T_o a_o)$

use Proof Template K8004 (Trans): (H ⊕ A), B → H ⊃ B
:= $\$HA8004\ \supset_{ooo}h_o a_o$
wff 210 : $\supset h\,a_o$:= $\$A5219bH$ $\$HA8004$
:= $\$B8004$ %0
wff 802 : $=a\,(=T\,a)_o$:= $\$B8004$
<< K8004.r0t.txt
:= $\$HA8004$
:= $\$B8004$
%0
$\supset h\,(=a\,(=T\,a))$
$\supset_{ooo}h_o(=_{ooo}a_o(=_{ooo}T_o a_o))$

%$\$A5219bH$
$\supset h\,a$:= $\$A5219bH$
$\supset_{ooo}h_o a_o$:= $\$A5219bH$

§s' %0 1 %1
$\supset h\,(=T\,a)$

use Proof Template A5201bH (SwapH): H ⊃ (A = B) → H ⊃ (B = A)
<< A5201bH.r0t.txt
%0
$\supset h\,(=a\,T)$
$\supset_{ooo}h_o(=_{ooo}a_oT_o)$
Include end (A5219bH.r0t.txt) [newfile=(A5219bH.r0a.txt)]
>>>

##
Undefine Syntactical Variables
##

:= $\$A5219bH$

##
Q.E.D.
##

%0
$\supset h\,(=a\,T)$
$\supset_{ooo}h_o(=_{ooo}a_oT_o)$

2.1.20 Results for File A5219c.r0a.txt

##
Proof Template A5219c (Rule T): T = A → A
##
##
Source: [Andrews 2002 (ISBN 1-4020-0763-9), p. 222]
##
Copyright (c) 2017 Owl of Minerva Press GmbH. All rights reserved.
Written by Ken Kubota (<mail@kenkubota.de>).
##
This file is part of the publication of the mathematical logic $\mathcal{R}_0$.
For more information, visit: <http://doi.org/10.4444/100.10>
##

##
Define Syntactical Variables
##

<< basics.r0.txt

the assumption
:= $\$A5219c =_{ooo}T_oa_o$
wff 209 : $=T\,a_o$:= $\$A5219c$

##
Assumptions and Resulting Syntactical Variables
##

§! $\$A5219c$
$=T\,a$:= $\$A5219c$

##
Include Proof Template
##

<<< A5219c.r0t.txt
Include begin (A5219c.r0t.txt) [oldfile=(A5219c.r0a.txt)]
##
Proof Template A5219c (Rule T): T = A → A
##
##
Source: [Andrews 2002 (ISBN 1-4020-0763-9), p. 222]
##

Written by Ken Kubota (<mail@kenkubota.de>).
##
This file is part of the publication of the mathematical logic $\mathcal{R}_0$.
For more information, visit: <http://doi.org/10.4444/100.10>
##

Empty lines are needed for comparison between A5219a-A5219d and A5219aH-A5219dH.

##
Proof Template
##

use Proof Template A5218: (T = A) = A
:= $\$A5218 =_{ooo}T_oa_o/3$

wff 54 : a_o := $\$A5218$
<< A5218.r0t.txt
:= $\$A5218$
%0
$= \$A5219c\, a$
$=_{ooo}\$A5219c_o a_o$

%$\$A5219c$
$= T\, a$:= $\$A5219c$
$=_{ooo} T_o a_o$:= $\$A5219c$

§s %0 1 %1
a
Include end (A5219c.r0t.txt) [newfile=(A5219c.r0a.txt)]
>>>

##
Undefine Syntactical Variables
##

:= $\$A5219c$

##
Q.E.D.
##

%0
a
a_o

2.1.21 Results for File A5219cH.r0a.txt

##
Proof Template A5219cH (Rule T): H ⊃ (T = A) → H ⊃ A
##
##
Source: [Andrews 2002 (ISBN 1-4020-0763-9), p. 222]
##

Written by Ken Kubota (<mail@kenkubota.de>).
##
This file is part of the publication of the mathematical logic $\mathcal{R}_0$.
For more information, visit: <http://doi.org/10.4444/100.10>
##

##
Define Syntactical Variables
##

<< basics.r0.txt

the assumption
:= $\$A5219cH\ \supset_{ooo}h_o(=_{ooo}T_oa_o)$
wff 212 : $\supset h\,(=T\,a)_o$:= $\$A5219cH$

##
Assumptions and Resulting Syntactical Variables
##

§! $\$A5219cH$
$\supset h\,(=T\,a)$:= $\$A5219cH$

##
Include Proof Template
##

<<< A5219cH.r0t.txt
Include begin (A5219cH.r0t.txt) [oldfile=(A5219cH.r0a.txt)]
##
Proof Template A5219cH (Rule T): H ⊃ (T = A) → H ⊃ A
##
##
Source: [Andrews 2002 (ISBN 1-4020-0763-9), p. 222]
##

Written by Ken Kubota (<mail@kenkubota.de>).

##
This file is part of the publication of the mathematical logic $\mathcal{R}_0$.
For more information, visit: <http://doi.org/10.4444/100.10>
##

##
Exception: Forward Reference
##
(See comment in Proof Template A5215H.)
##

Empty lines are needed for comparison between A5219a-A5219d and A5219aH-A5219dH.

##
Proof Template
##

use Proof Template A5218: (T = A) = A
:= $\$A5218$ $\supset_{ooo} h_o(=_{ooo} T_o a_o)/7$
wff 54 : a_o := $\$A5218$
<< A5218.r0t.txt
:= $\$A5218$
%0
$=(=T\,a)\,a$
$=_{ooo}(=_{ooo} T_o a_o) a_o$

use Proof Template K8004 (Trans): (H ⊕ A), B → H ⊃ B
:= $\$HA8004$ $\supset_{ooo} h_o(=_{ooo} T_o a_o)$
wff 212 : $\supset h\,(=T\,a)_o$:= $\$A5219cH$ $\$HA8004$
:= $\$B8004$ %0
wff 797 : $=(=T\,a)\,a_{o,\ldots}$:= $\$B8004$
<< K8004.r0t.txt
:= $\$HA8004$
:= $\$B8004$
%0
$\supset h\,(=(=T\,a)\,a)$
$\supset_{ooo} h_o(=_{ooo}(=_{ooo} T_o a_o) a_o)$

%$\$A5219cH$
$\supset h\,(=T\,a)$:= $\$A5219cH$
$\supset_{ooo} h_o(=_{ooo} T_o a_o)$:= $\$A5219cH$

$\S s'$ %0 1 %1
\# $\supset h\, a$
\## Include end (A5219cH.r0t.txt) [newfile=(A5219cH.r0a.txt)]
>>>

\##
\## Undefine Syntactical Variables
\##

:= $\$A5219cH$

\##
\## Q.E.D.
\##

%0
\# $\supset h\, a$
\# $\supset_{ooo} h_o a_o$

2.1.22 Results for File A5219d.r0a.txt

\##
\## Proof Template A5219d (Rule T): A = T → A
\##
\##
\## Source: [Andrews 2002 (ISBN 1-4020-0763-9), p. 222]
\##
\## Copyright (c) 2017 Owl of Minerva Press GmbH. All rights reserved.
\## Written by Ken Kubota (<mail@kenkubota.de>).
\##
\## This file is part of the publication of the mathematical logic $\mathcal{R}_0$.
\## For more information, visit: <http://doi.org/10.4444/100.10>
\##

\##
\## Define Syntactical Variables
\##

<< basics.r0.txt

\## the assumption
:= $\$A5219d\ =_{ooo} a_o T_o$

```
# wff     209  :      =aT_o      :=  $A5219d


##
##  Assumptions and Resulting Syntactical Variables
##

§! $A5219d
#                  =aT     :=  $A5219d


##
##  Include Proof Template
##

## <<< A5219d.r0t.txt
## Include begin (A5219d.r0t.txt) [oldfile=(A5219d.r0a.txt)]
##
##  Proof Template A5219d (Rule T):  A = T  →  A
##
##
##  Source: [Andrews 2002 (ISBN 1-4020-0763-9), p. 222]
##
##  Copyright (c) 2017 Owl of Minerva Press GmbH. All rights reserved.
##  Written by Ken Kubota (<mail@kenkubota.de>).
##
##  This file is part of the publication of the mathematical logic R0.
##  For more information, visit: <http://doi.org/10.4444/100.10>
##








##  Empty lines are needed for comparison between A5219a-A5219d and A5219aH-A5219dH.


##
##  Proof Template
##

## use Proof Template A5218:  (T = A) = A
:=  $A5218 =_ooo a_o T_o/5
# wff    54  :      a_o     :=  $A5218
<< A5218.r0t.txt
:=  $A5218
```

%0
$= (= T\, a)\, a$
$=_{ooo}(=_{ooo}T_o a_o)a_o$

:= $\$TMP5219d$ %0
wff 796 : $= (= T\, a)\, a_{o,\dots}$:= $\$TMP5219d$
%$\$A5219d$
$= a\, T$:= $\$A5219d$
$=_{ooo}a_o T_o$:= $\$A5219d$
use Proof Template A5201b (Swap): A = B $\rightarrow$ B = A
<< A5201b.r0t.txt
%0
$= T\, a$
$=_{ooo}T_o a_o$
%$\$TMP5219d$
$= (= T\, a)\, a$:= $\$TMP5219d$
$=_{ooo}(=_{ooo}T_o a_o)a_o$:= $\$TMP5219d$
:= $\$TMP5219d$

§s %1 1 %0
a
Include end (A5219d.r0t.txt) [newfile=(A5219d.r0a.txt)]
>>>

##
Undefine Syntactical Variables
##

:= $\$A5219d$

##
Q.E.D.
##

%0
a

\# a_o

2.1.23 Results for File A5219dH.r0a.txt

\##
\## Proof Template A5219dH (Rule T): H $\supset$ (A = T) $\rightarrow$ H $\supset$ A
\##
\##
\## Source: [Andrews 2002 (ISBN 1-4020-0763-9), p. 222]
\##

\## Written by Ken Kubota (<mail@kenkubota.de>).
\##
\## This file is part of the publication of the mathematical logic $\mathcal{R}_0$.
\## For more information, visit: <http://doi.org/10.4444/100.10>
\##

\##
\## Define Syntactical Variables
\##

<< basics.r0.txt

\## the assumption
:= $\$A5219dH\ \supset_{ooo} h_o (=_{ooo} a_o T_o)$
\# wff 212 : $\supset h\,(= a\,T)_o$:= $\$A5219dH$

\##
\## Assumptions and Resulting Syntactical Variables
\##

§! $\$A5219dH$
\# $\supset h\,(= a\,T)$:= $\$A5219dH$

\##
\## Include Proof Template
\##

\## <<< A5219dH.r0t.txt
\## Include begin (A5219dH.r0t.txt) [oldfile=(A5219dH.r0a.txt)]
\##
\## Proof Template A5219dH (Rule T): H $\supset$ (A = T) $\rightarrow$ H $\supset$ A
\##
\##
\## Source: [Andrews 2002 (ISBN 1-4020-0763-9), p. 222]
\##

##
This file is part of the publication of the mathematical logic $\mathcal{R}_0$.
For more information, visit: <http://doi.org/10.4444/100.10>
##

##
Exception: Forward Reference
##
(See comment in Proof Template A5215H.)
##

Empty lines are needed for comparison between A5219a-A5219d and A5219aH-A5219dH.

##
Proof Template
##

use Proof Template A5218: (T = A) = A
:= $\$A5218 \ \supset_{ooo}h_o(=_{ooo}a_oT_o)/13$
wff 54 : a_o := $\$A5218$
<< A5218.r0t.txt
:= $\$A5218$
%0
$=(=T\,a)\,a$
$=_{ooo}(=_{ooo}T_oa_o)a_o$

use Proof Template K8004 (Trans): (H ⊕ A), B → H ⊃ B
:= $\$HA8004 \ \supset_{ooo}h_o(=_{ooo}a_oT_o)$
wff 212 : $\supset h\,(=a\,T)_o$:= $\$A5219dH$ $\$HA8004$
:= $\$B8004$ %0
wff 799 : $=(=T\,a)\,a_{o,\ldots}$:= $\$B8004$
<< K8004.r0t.txt
:= $\$HA8004$
:= $\$B8004$
%0
$\supset h\,(=(=T\,a)\,a)$
$\supset_{ooo}h_o(=_{ooo}(=_{ooo}T_oa_o)a_o)$

:= $\$TMP5219dH$ %0
wff 1429 : $\supset h\,(=(=T\,a)\,a)_{o,\ldots}$:= $\$TMP5219dH$
%$\$A5219dH$

$\supset h\,(= a\,T)$:= $\$A5219dH$
$\supset_{ooo}h_o(=_{ooo}a_oT_o)$:= $\$A5219dH$
use Proof Template A5201bH (SwapH): H ⊃ (A = B) → H ⊃ (B = A)
<< A5201bH.r0t.txt
%0
$\supset h\,(= T\,a)$
$\supset_{ooo}h_o(=_{ooo}T_oa_o)$
%$\$TMP5219dH$
$\supset h\,(= (= T\,a)\,a)$:= $\$TMP5219dH$
$\supset_{ooo}h_o(=_{ooo}(=_{ooo}T_oa_o)a_o)$:= $\$TMP5219dH$
:= $\$TMP5219dH$

§s' %1 1 %0
$\supset h\,a$
Include end (A5219dH.r0t.txt) [newfile=(A5219dH.r0a.txt)]
>>>

##
Undefine Syntactical Variables
##

:= $\$A5219dH$

##
Q.E.D.
##

%0
$\supset h\,a$
$\supset_{ooo}h_oa_o$

2.1.24 Results for File A5220.r0a.txt

##
Proof Template A5220 (Gen): A → ∀ x: A
for any x of any type (Rule of Universal Generalization)
##
Source: [Andrews 2002 (ISBN 1-4020-0763-9), p. 222]
##

Written by Ken Kubota (<mail@kenkubota.de>).
##
This file is part of the publication of the mathematical logic $\mathcal{R}_0$.
For more information, visit: <http://doi.org/10.4444/100.10>
##

##
Define Syntactical Variables
##

type of variable
:= $\$T5220$ t_τ
wff 4 : t_τ := $\$T5220$

the variable
:= $\$X5220$ $x_{\$T5220}$
wff 11 : $x_{\$T5220}$:= $\$X5220$

the proposition
:= $\$A5220$ a_o
wff 12 : a_o := $\$A5220$

##
Assumptions and Resulting Syntactical Variables
##

§! $\$A5220$
a := $\$A5220$

##
Include Proof Template
##

<<< A5220.r0t.txt
Include begin (A5220.r0t.txt) [oldfile=(A5220.r0a.txt)]
##
Proof Template A5220 (Gen): A → ∀ x: A
for any x of any type (Rule of Universal Generalization)
##
Source: [Andrews 2002 (ISBN 1-4020-0763-9), p. 222]
##

Written by Ken Kubota (<mail@kenkubota.de>).
##
This file is part of the publication of the mathematical logic $\mathcal{R}_0$.
For more information, visit: <http://doi.org/10.4444/100.10>
##

##
Proof Template
##

.1

%$\$A5220$
a := $\$A5220$
a_o := $\$A5220$

.2

use Proof Template A5219a (Rule T): A → T = A
:= $\$A5219a$ %0
wff 12 : a_o := $\$A5219a$ $\$A5220$
<< A5219a.r0t.txt
:= $\$A5219a$
%0
$= T\,\$A5220$
$=_{ooo}T_o\$A5220_o$

.3

§= $_{o\,\$T5220}$ $[\lambda \$X5220_{\$T5220}.T_o]$
$= [\lambda\$X5220.T]\,[\lambda\$X5220.T]$
§r /5 $\$X5220$
$= [\lambda\$X5220.T]\,[\lambda\$X5220.T]$
§s %1 5 %0
$= [\lambda\$X5220.T]\,[\lambda\$X5220.T]$

.4

§s %0 7 %3
$= [\lambda\$X5220.T]\,[\lambda\$X5220.\$A5220]$
§= $\forall_{o(o\backslash 3)\tau}\$T5220_\tau[\lambda\$X5220_{\$T5220}.\$A5220_o]$
$= (\forall\,\$T5220\,[\lambda\$X5220.\$A5220])\,(\forall\,\$T5220\,[\lambda\$X5220.\$A5220])$
§\ $\forall_{o(o\backslash 3)\tau}\$T5220_\tau$
$= (\forall\,\$T5220)\,[\lambda p.(=[\lambda\$X5220.T]\,p)]$
§s %1 10 %0
$= ([\lambda p.(=[\lambda\$X5220.T]\,p)]\,[\lambda\$X5220.\$A5220])\,(\forall\,\$T5220\,[\lambda\$X5220.\$A5220])$
§\ $[\lambda p_{o\,\$T5220}.(=_{o(o\,\$T5220)(o\,\$T5220)}[\lambda\$X5220_{\$T5220}.T_o]p_{o\,\$T5220})_o][\lambda\$X5220_{\$T5220}.\$A5220_o]$
$= ([\lambda p.(=[\lambda\$X5220.T]\,p)]\,[\lambda\$X5220.\$A5220])\,(=[\lambda\$X5220.T]\,[\lambda\$X5220.\$A5220])$
§s %1 5 %0
$= (=[\lambda\$X5220.T]\,[\lambda\$X5220.\$A5220])\,(\forall\,\$T5220\,[\lambda\$X5220.\$A5220])$
§s %5 1 %0
$\forall\,\$T5220\,[\lambda\$X5220.\$A5220]$
Include end (A5220.r0t.txt) [newfile=(A5220.r0a.txt)]
>>>

##
Undefine Syntactical Variables
##

:= $\$T5220$
:= $\$X5220$
:= $\$A5220$

##
Q.E.D.
##

%0
$\forall t\, [\lambda x.a]$
$\forall_{o(o\backslash 3)\tau} t_\tau [\lambda x_t.a_o]$

2.1.25 Results for File A5220H.r0a.txt

##
Proof Template A5220H (Gen): (H ⊃ A) → (H ⊃ ∀ x: A)
for any x of any type (Rule of Universal Generalization), provided x is not free in H
##
Source: [Andrews 2002 (ISBN 1-4020-0763-9), p. 222]
##
Copyright (c) 2017 Owl of Minerva Press GmbH. All rights reserved.
Written by Ken Kubota (<mail@kenkubota.de>).
##
This file is part of the publication of the mathematical logic $\mathcal{R}_0$.
For more information, visit: <http://doi.org/10.4444/100.10>
##

##
Define Syntactical Variables
##

<< basics.r0.txt

type of variable
:= $\$T5220H\ t_\tau$
wff 4 : t_τ := $\$T5220H$

the variable
:= $\$X5220H\ x_{\$T5220H}$
wff 24 : $x_{\$T5220H}$:= $\$X5220H$

the proposition
:= $\$A5220H\ \supset_{ooo} h_o a_o$
wff 210 : $\supset h\, a_o$:= $\$A5220H$

##
Assumptions and Resulting Syntactical Variables
##

§! $\$A5220H$
$\supset h\,a$:= $\$A5220H$

##
Include Proof Template
##

<<< A5220H.r0t.txt
Include begin (A5220H.r0t.txt) [oldfile=(A5220H.r0a.txt)]
##
Proof Template A5220H (Gen): (H ⊃ A) → (H ⊃ ∀ x: A)
for any x of any type (Rule of Universal Generalization), provided x is not free in H
##
Source: [Andrews 2002 (ISBN 1-4020-0763-9), p. 222]
##

Written by Ken Kubota (<mail@kenkubota.de>).
##
This file is part of the publication of the mathematical logic $\mathcal{R}_0$.
For more information, visit: <http://doi.org/10.4444/100.10>
##

##
Exception: Forward Reference
##
(See comment in Proof Template A5215H.)
##

##
Proof Template
##

.1

%$\$A5220H$
$\supset h\,a$:= $\$A5220H$
$\supset_{ooo} h_o a_o$:= $\$A5220H$

.2

use Proof Template A5219aH (Rule T): H ⊃ A → H ⊃ (T = A)
:= $\$A5219aH$ %0

wff 210 : $\supset h\, a_o$:= $\$A5219aH$ $\$A5220H$
<< A5219aH.r0t.txt
:= $\$A5219aH$
%0
$\supset h\,(= T\, a)$
$\supset_{ooo} h_o (=_{ooo} T_o a_o)$

:= $\$HTMP5220H$ %0
wff 1526 : $\supset h\,(= T\, a)_o$:= $\$HTMP5220H$

.3

§= $_{o\$T5220H}$ $[\lambda \$X5220H_{\$T5220H}.T_o]$
$= [\lambda \$X5220H.T]\,[\lambda \$X5220H.T]$
§r /5 $\$X5220H$
$= [\lambda \$X5220H.T]\,[\lambda \$X5220H.T]$
§s %1 5 %0
$= [\lambda \$X5220H.T]\,[\lambda \$X5220H.T]$
:= $\$TTMP5220H$ %0
wff 1527 : $= [\lambda \$X5220H.T]\,[\lambda \$X5220H.T]_o$:= $\$TTMP5220H$

use Proof Template K8004 (Trans): (H ⊕ A), B → H ⊃ B
:= $\$HA8004$ $\supset_{ooo} h_o (=_{ooo} T_o a_o)$
wff 1526 : $\supset h\,(= T\, a)_o$:= $\$HA8004$ $\$HTMP5220H$
:= $\$B8004$ $=_{o(o\$T5220H)(o\$T5220H)} [\lambda \$X5220H_{\$T5220H}.T_o][\lambda \$X5220H_{\$T5220H}.T_o]$
wff 1527 : $= [\lambda \$X5220H.T]\,[\lambda \$X5220H.T]_o$:= $\$B8004$ $\$TTMP5220H$
:= $\$TTMP5220H$
<< K8004.r0t.txt
:= $\$HA8004$
:= $\$B8004$
%0
$\supset h\,(= [\lambda \$X5220H.T]\,[\lambda \$X5220H.T])$
$\supset_{ooo} h_o (=_{o(o\$T5220H)(o\$T5220H)} [\lambda \$X5220H_{\$T5220H}.T_o][\lambda \$X5220H_{\$T5220H}.T_o])$

.4

%$\$HTMP5220H$
$\supset h\,(= T\, a)$:= $\$HTMP5220H$
$\supset_{ooo} h_o (=_{ooo} T_o a_o)$:= $\$HTMP5220H$
:= $\$HTMP5220H$
§s' %1 7 %0
$\supset h\,(= [\lambda \$X5220H.T]\,[\lambda \$X5220H.a])$
:= $\$HTMP5220H$ %0
wff 1566 : $\supset h\,(= [\lambda \$X5220H.T]\,[\lambda \$X5220H.a])_o$:= $\$HTMP5220H$
§= $\forall_{o(o\backslash 3)\tau} \$T5220H_\tau [\lambda \$X5220H_{\$T5220H}.a_o]$
$= (\forall\, \$T5220H\, [\lambda \$X5220H.a])\, (\forall\, \$T5220H\, [\lambda \$X5220H.a])$

use Proof Template K8004 (Trans): (H ⊕ A), B → H ⊃ B
:= $\$HA8004$ $\supset_{ooo} h_o (=_{o(o\$T5220H)(o\$T5220H)} [\lambda \$X5220H_{\$T5220H}.T_o][\lambda \$X5220H_{\$T5220H}.a_o])$

wff 1566 : $\supset h\,(=[\lambda \$X5220H.T]\,[\lambda \$X5220H.a])_o$:= $\$HA8004$ $\$HTMP5220H$
:= $\$B8004$ %0
wff 1569 : $=(\forall\,\$T5220H\,[\lambda \$X5220H.a])\,(\forall\,\$T5220H\,[\lambda \$X5220H.a])_o$:= $\$B8004$
<< K8004.r0t.txt
:= $\$HA8004$
:= $\$B8004$
%0
$\supset h\,(=(\forall\,\$T5220H\,[\lambda \$X5220H.a])\,(\forall\,\$T5220H\,[\lambda \$X5220H.a]))$
$\supset_{ooo} h_o \ldots$
$\ldots(=_{o\omega\omega}(\forall_{o(o\backslash 3)\tau}\$T5220H_\tau[\lambda \$X5220H_{\$T5220H}.a_o])(\forall_{o(o\backslash 3)\tau}\$T5220H_\tau[\lambda \$X5220H_{\$T5220H}.a_o]))$

§\ $\forall_{o(o\backslash 3)\tau}\$T5220H_\tau$
$=(\forall\,\$T5220H)\,[\lambda p.(=[\lambda \$X5220H.T]\,p)]$
§s %1 26 %0
$\supset h\,(=([\lambda p.(=[\lambda \$X5220H.T]\,p)]\,[\lambda \$X5220H.a])\,(\forall\,\$T5220H\,[\lambda \$X5220H.a]))$
§\ $[\lambda p_{o\,\$T5220H}.(=_{o(o\,\$T5220H)(o\,\$T5220H)}[\lambda \$X5220H_{\$T5220H}.T_o]p_{o\,\$T5220H})_o][\lambda \$X5220H_{\$T5220H}.a_o]$
$=([\lambda p.(=[\lambda \$X5220H.T]\,p)]\,[\lambda \$X5220H.a])\,(=[\lambda \$X5220H.T]\,[\lambda \$X5220H.a])$
§s %1 13 %0
$\supset h\,(=(=[\lambda \$X5220H.T]\,[\lambda \$X5220H.a])\,(\forall\,\$T5220H\,[\lambda \$X5220H.a]))$
%$\$HTMP5220H$
$\supset h\,(=[\lambda \$X5220H.T]\,[\lambda \$X5220H.a])$:= $\$HTMP5220H$
$\supset_{ooo} h_o(=_{o(o\,\$T5220H)(o\,\$T5220H)}[\lambda \$X5220H_{\$T5220H}.T_o][\lambda \$X5220H_{\$T5220H}.a_o])$
:= $\$HTMP5220H$
:= $\$HTMP5220H$
§s' %0 1 %1
$\supset h\,(\forall\,\$T5220H\,[\lambda \$X5220H.a])$
Include end (A5220H.r0t.txt) [newfile=(A5220H.r0a.txt)]
>>>

##
Undefine Syntactical Variables
##

:= $\$T5220H$
:= $\$X5220H$
:= $\$A5220H$

##
Q.E.D.
##

%0
$\supset h\,(\forall\,t\,[\lambda x.a])$
$\supset_{ooo} h_o(\forall_{o(o\backslash 3)\tau}t_\tau[\lambda x_t.a_o])$

2.1.26 Results for File A5221.r0a.txt

```
##
## Proof Template A5221 (Sub): B → B [x/A]
##       (Rule of Substitution)
##
## Source: [Andrews 2002 (ISBN 1-4020-0763-9), pp. 222 f.]
##
## Copyright (c) 2017 Owl of Minerva Press GmbH. All rights reserved.
## Written by Ken Kubota (<mail@kenkubota.de>).
##
## This file is part of the publication of the mathematical logic R0.
## For more information, visit: <http://doi.org/10.4444/100.10>
##


##
## Define Syntactical Variables
##

<< basics.r0.txt

## assumption
:= $B5221 g_oo x_o
# wff    167 :       g x_o      := $B5221

## type of the variable and the substitution term
:= $T5221 o
# wff    2 :      o_τ      := $T5221

## the variable to be replaced
:= $X5221 x_o
# wff    16 :      x_o      := $X5221

## substitution term
:= $A5221 =_o(oo)(oo) [λ$X5221_o.T_o][λ$X5221_o.$X5221_o]
# wff    20 :      = [λ$X5221.T] [λ$X5221.$X5221]_o      := $A5221 F


##
## Assumptions and Resulting Syntactical Variables
##

§! $B5221
#               g $X5221      := $B5221


##
## Include Proof Template
```

```
##

## <<< A5221.r0t.txt
## Include begin (A5221.r0t.txt) [oldfile=(A5221.r0a.txt)]
##
##  Proof Template A5221 (Sub):  B  →  B [x/A]
##        (Rule of Substitution)
##
##  Source: [Andrews 2002 (ISBN 1-4020-0763-9), pp. 222 f.]
##
##  Copyright (c) 2017 Owl of Minerva Press GmbH. All rights reserved.
##  Written by Ken Kubota (<mail@kenkubota.de>).
##
##  This file is part of the publication of the mathematical logic R_0.
##  For more information, visit: <http://doi.org/10.4444/100.10>
##


##
##  Proof Template
##

## .1

%$B5221
#              g $X5221       :=  $B5221
#              g_{oo}$X5221_o      :=  $B5221

## .2

## use Proof Template A5220 (Gen):  A  →  ∀ x: A
:=  $T5220  o
# wff    2  :      o_τ      :=  $T5220  $T5221
:=  $X5220  x_o
# wff    16  :      x_o      :=  $X5220  $X5221
:=  $A5220  %0
# wff    167  :      g $X5221_o      :=  $A5220  $B5221
<< A5220.r0t.txt
:=  $T5220
:=  $X5220
:=  $A5220
%0
#              ∀ o [λ$X5221.$B5221]
#              ∀_{o(o\3)τ} o_τ [λ$X5221_o.$B5221_o]

## .3

## use Proof Template A5215 (∀ I):  ∀ x: B  →  B [x/a]
:=  $T5215  o
```

wff 2 : o_τ := $T5215 $T5221
:= $X5215 x_o
wff 16 : x_o := $X5215 $X5221
:= $A5215 $=_{o(oo)(oo)} [\lambda \$X5221_o.T_o][\lambda \$X5221_o.\$X5221_o]$
wff 20 : $= [\lambda \$X5221.T]\, [\lambda \$X5221.\$X5221]_{o,\ldots}$:= $A5215 $A5221 F
:= $H5215 %0
wff 169 : $\forall o\, [\lambda \$X5221.\$B5221]_{o,\ldots}$:= $H5215
<< A5215.r0t.txt
:= $T5215
:= $X5215
:= $A5215
:= $H5215
%0
$g\,F$
$g_{oo}F_o$
Include end (A5221.r0t.txt) [newfile=(A5221.r0a.txt)]
>>>

##
Undefine Syntactical Variables
##

:= $B5221
:= $T5221
:= $X5221
:= $A5221

##
Q.E.D.
##

%0
$g\,F$
$g_{oo}F_o$

2.1.27 Results for File A5221H.r0a.txt

##
Proof Template A5221H (Sub): H ⊃ B → H ⊃ B [x/A]
(Rule of Substitution)
##
Source: [Andrews 2002 (ISBN 1-4020-0763-9), pp. 222 f.]
##

Written by Ken Kubota (<mail@kenkubota.de>).
##
This file is part of the publication of the mathematical logic $\mathcal{R}_0$.

For more information, visit: <http://doi.org/10.4444/100.10>
##

##
Define Syntactical Variables
##

<< basics.r0.txt

assumption
:= $\$B5221H\ \supset_{ooo} h_o (g_{oo} x_o)$
\# wff 210 : $\supset h\,(g\,x)_o$:= $\$B5221H$

type of the variable and the substitution term
:= $\$T5221H\ o$
\# wff 2 : o_τ := $\$T5221H$

the variable to be replaced
:= $\$X5221H\ x_o$
\# wff 16 : x_o := $\$X5221H$

substitution term
:= $\$A5221H\ =_{o(oo)(oo)} [\lambda \$X5221H_o.T_o][\lambda \$X5221H_o.\$X5221H_o]$
\# wff 20 : $= [\lambda \$X5221H.T]\,[\lambda \$X5221H.\$X5221H]_o$:= $\$A5221H$ F

##
Assumptions and Resulting Syntactical Variables
##

§! $\$B5221H$
\# $\supset h\,(g\,\$X5221H)$:= $\$B5221H$

##
Include Proof Template
##

<<< A5221H.r0t.txt
Include begin (A5221H.r0t.txt) [oldfile=(A5221H.r0a.txt)]
##
Proof Template A5221H (Sub): H ⊃ B → H ⊃ B [x/A]
(Rule of Substitution)
##
Source: [Andrews 2002 (ISBN 1-4020-0763-9), pp. 222 f.]
##

Written by Ken Kubota (<mail@kenkubota.de>).

##
This file is part of the publication of the mathematical logic $\mathcal{R}_0$.
For more information, visit: <http://doi.org/10.4444/100.10>
##

##
Proof Template
##

.1

%$\$B5221H$
$\supset h\,(g\,\$X5221H)$:= $\$B5221H$
$\supset_{ooo}h_o(g_{oo}\$X5221H_o)$:= $\$B5221H$

.2

use Proof Template A5220H (Gen): (H ⊃ A) → (H ⊃ ∀ x: A)
:= $\$T5220H$ o
wff 2 : o_τ := $\$T5220H$ $\$T5221H$
:= $\$X5220H$ x_o
wff 16 : x_o := $\$X5220H$ $\$X5221H$
:= $\$A5220H$ %0
wff 210 : $\supset h\,(g\,\$X5221H)_o$:= $\$A5220H$ $\$B5221H$
<< A5220H.r0t.txt
:= $\$T5220H$
:= $\$X5220H$
:= $\$A5220H$
%0
$\supset h\,(\forall\, o\,[\lambda\$X5221H.(g\,\$X5221H)])$
$\supset_{ooo}h_o(\forall_{o(o\backslash 3)\tau}o_\tau[\lambda\$X5221H_o.(g_{oo}\$X5221H_o)_o])$

.3

use Proof Template A5215H (∀ I): H ⊃ ∀ x: B → H ⊃ B [x/a]
:= $\$T5215H$ o
wff 2 : o_τ := $\$T5215H$ $\$T5221H$
:= $\$X5215H$ x_o
wff 16 : x_o := $\$X5215H$ $\$X5221H$
:= $\$A5215H$ $=_{o(oo)(oo)}[\lambda\$X5221H_o.T_o][\lambda\$X5221H_o.\$X5221H_o]$
wff 20 : $=[\lambda\$X5221H.T]\,[\lambda\$X5221H.\$X5221H]_{o,\ldots}$:= $\$A5215H$ $\$A5221H$ F
:= $\$H5215H$ %0
wff 1606 : $\supset h\,(\forall\, o\,[\lambda\$X5221H.(g\,\$X5221H)])_o$:= $\$H5215H$
<< A5215H.r0t.txt
:= $\$T5215H$
:= $\$X5215H$
:= $\$A5215H$
:= $\$H5215H$

%0
\# $\supset h\,(g\,F)$
\# $\supset_{ooo} h_o (g_{oo} F_o)$
\## Include end (A5221H.r0t.txt) [newfile=(A5221H.r0a.txt)]
>>>

\##
\## Undefine Syntactical Variables
\##

:= $\$B5221H$
:= $\$T5221H$
:= $\$X5221H$
:= $\$A5221H$

\##
\## Q.E.D.
\##

%0
\# $\supset h\,(g\,F)$
\# $\supset_{ooo} h_o (g_{oo} F_o)$

2.1.28 Results for File A5222.r0a.txt

\##
\## Proof Template A5222 (Rule of Cases): [\x.A]T, [\x.A]F → A
\## for any x of type bool
\##
\## Source: [Andrews 2002 (ISBN 1-4020-0763-9), p. 223]
\##

\## Written by Ken Kubota (<mail@kenkubota.de>).
\##
\## This file is part of the publication of the mathematical logic $\mathcal{R}_0$.
\## For more information, visit: <http://doi.org/10.4444/100.10>
\##

<< basics.r0.txt

\##
\## Define Syntactical Variables
\##

\## the lambda abstraction
:= $\$L5222$ $[\lambda x_o.a_o]$

\# wff 208 : $[\lambda x.a]_{oo}$:= $\$L5222$

\## the variable to be used in place of the one abstracted
:= $\$X5222\ x_o$
\# wff 16 : x_o := $\$X5222$

\## assumption 1
:= $\$T5222\ \$L5222_{oo}T_o$
\# wff 209 : $\$L5222\,T_o$:= $\$T5222$

\## assumption 2
:= $\$F5222\ \$L5222_{oo}F_o$
\# wff 210 : $\$L5222\,F_o$:= $\$F5222$

\##
\## Assumptions and Resulting Syntactical Variables
\##

§! $\$T5222$
\# $\$L5222\,T$:= $\$T5222$
§! $\$F5222$
\# $\$L5222\,F$:= $\$F5222$

\##
\## Include Proof Template
\##

\## <<< A5222.r0t.txt
\## Include begin (A5222.r0t.txt) [oldfile=(A5222.r0a.txt)]
\##
\## Proof Template A5222 (Rule of Cases): [\x.A]T, [\x.A]F → A
\## for any x of type bool
\##
\## Source: [Andrews 2002 (ISBN 1-4020-0763-9), p. 223]
\##

\## Written by Ken Kubota (<mail@kenkubota.de>).
\##
\## This file is part of the publication of the mathematical logic $\mathcal{R}_0$.
\## For more information, visit: <http://doi.org/10.4444/100.10>
\##

<< A5212.r0.txt

\##
\## Proof Template

```
##

## .1

%$T5222
#               $L5222 T        :=  $T5222
#               $L5222_ooT_o     :=  $T5222

## use Proof Template A5219a (Rule T):  A  →  T = A
:=  $A5219a %0
# wff    209  :      $L5222 T_o       :=  $A5219a  $T5222
<< A5219a.r0t.txt
:=  $A5219a
%0
#               = T $T5222
#               =_oooT_o$T5222_o

:=  $ATMP5222 %0
# wff    795  :      = T $T5222_o,...      :=  $ATMP5222

## .2

%$F5222
#               $L5222 F        :=  $F5222
#               $L5222_ooF_o     :=  $F5222

## use Proof Template A5219a (Rule T):  A  →  T = A
:=  $A5219a %0
# wff    210  :      $L5222 F_o       :=  $A5219a  $F5222
<< A5219a.r0t.txt
:=  $A5219a
%0
#               = T $F5222
#               =_oooT_o$F5222_o

## .3

%A5212
#               ∧ T T       :=  A5212
#               ∧_oooT_oT_o     :=  A5212

## .4

§s %0 3 %1
#               ∧ T $F5222
%$ATMP5222
#               = T $T5222       :=  $ATMP5222
#               =_oooT_o$T5222_o      :=  $ATMP5222
§s %1 5 %0
```

\# $\wedge \$T5222 \$F5222$

:= $\$BTMP5222$ %0
\# wff 821 : $\wedge \$T5222 \$F5222_o$:= $\$BTMP5222$

\## .5

$\%A1$
\# $= (\wedge (g\,T)\,(g\,F))\,(\forall\, o\,[\lambda \$X5222.(g\,\$X5222)])$:= $A1$
\# $=_{ooo}(\wedge_{ooo}(g_{oo}T_o)(g_{oo}F_o))(\forall_{o(o\backslash 3)\tau}o_\tau[\lambda \$X5222_o.(g_{oo}\$X5222_o)_o])$:= $A1$

\## use Proof Template A5221 (Sub): B → B [x/A]
:= $\$B5221$ %0
\# wff 170 : $= (\wedge (g\,T)\,(g\,F))\,(\forall\, o\,[\lambda \$X5222.(g\,\$X5222)])_o$:= $\$B5221$ $A1$
:= $\$T5221$ oo
\# wff 13 : oo_τ := $\$T5221$
:= $\$X5221$ $g_{\$T5221}$
\# wff 160 : $g_{\$T5221}$:= $\$X5221$
:= $\$A5221$ $[\lambda \$X5222_o.a_o]$
\# wff 208 : $[\lambda \$X5222.a]_{\$T5221}$:= $\$A5221$ $\$L5222$
<< A5221.r0t.txt
:= $\$B5221$
:= $\$T5221$
:= $\$X5221$
:= $\$A5221$
%0
\# $= \$BTMP5222\,(\forall\, o\,[\lambda \$X5222.(\$L5222\,\$X5222)])$
\# $=_{ooo}\$BTMP5222_o(\forall_{o(o\backslash 3)\tau}o_\tau[\lambda \$X5222_o.(\$L5222_{oo}\$X5222_o)_o])$

§\ $\$L5222_{oo}\$X5222_o$
\# $= (\$L5222\,\$X5222)\,a$
§s %1 15 %0
\# $= \$BTMP5222\,(\forall\, o\,\$L5222)$

\## .6

%$\$BTMP5222$
\# $\wedge \$T5222 \$F5222$:= $\$BTMP5222$
\# $\wedge_{ooo}\$T5222_o\$F5222_o$:= $\$BTMP5222$
§s %0 1 %1
\# $\forall\, o\,\$L5222$

\## .7

\## use Proof Template A5215 (∀ I): ∀ x: B → B [x/a]
:= $\$T5215$ o
\# wff 2 : o_τ := $\$T5215$
:= $\$X5215$ x_o
\# wff 16 : x_o := $\$X5215$ $\$X5222$

:= $\$A5215$ x_o
wff 16 : x_o := $\$A5215$ $\$X5215$ $\$X5222$
:= $\$H5215$ %0
wff 872 : $\forall o \$L5222_{o,\ldots}$:= $\$H5215$
<< A5215.r0t.txt
:= $\$T5215$
:= $\$X5215$
:= $\$A5215$
:= $\$H5215$
%0
a
a_o

undefine local variables
:= $\$ATMP5222$
:= $\$BTMP5222$
Include end (A5222.r0t.txt) [newfile=(A5222.r0a.txt)]
>>>

##
Undefine Syntactical Variables
##

:= $\$L5222$
:= $\$X5222$
:= $\$T5222$
:= $\$F5222$

##
Q.E.D.
##

%0
a
a_o

2.1.29 Results for File A5223.r0.txt

##
Proof A5223: (T ⊃ y) = y
with y of type o
##
Source: [Andrews 2002 (ISBN 1-4020-0763-9), pp. 223 f.]
##

Written by Ken Kubota (<mail@kenkubota.de>).
##

This file is part of the publication of the mathematical logic $\mathcal{R}_0$.
For more information, visit: <http://doi.org/10.4444/100.10>
##

<< basics.r0.txt

##
Proof
##

.1

§= $_o \supset_{ooo} T_o y_o$
$=(\supset T\, y)\,(\supset T\, y)$
§\ $\supset_{ooo} T_o$
$=(\supset T)\,[\lambda y.(=T\,(\wedge T\, y))]$
§s %1 6 %0
$=(\supset T\, y)\,([\lambda y.(=T\,(\wedge T\, y))]\, y)$
§\ $[\lambda y_o.(=_{ooo} T_o(\wedge_{ooo} T_o y_o))_o] y_o$
$=([\lambda y.(=T\,(\wedge T\, y))]\, y)\,(=T\,(\wedge T\, y))$
§s %1 3 %0
$=(\supset T\, y)\,(=T\,(\wedge T\, y))$
:= $ATMP5223 %0
wff 223 : $=(\supset T\, y)\,(=T\,(\wedge T\, y))_o$:= $ATMP5223

.2

use Proof Template A5218: (T = A) = A
:= $A5218 $\wedge_{ooo} T_o y_o$
wff 215 : $\wedge T\, y_o$:= $A5218
<< A5218.r0t.txt
:= $A5218
%0
$=(=T\,(\wedge T\, y))\,(\wedge T\, y)$
$=_{ooo}(=_{ooo} T_o(\wedge_{ooo} T_o y_o))(\wedge_{ooo} T_o y_o)$

%$ATMP5223
$=(\supset T\, y)\,(=T\,(\wedge T\, y))$:= $ATMP5223
$=_{ooo}(\supset_{ooo} T_o y_o)(=_{ooo} T_o(\wedge_{ooo} T_o y_o))$:= $ATMP5223
§s %0 3 %1
$=(\supset T\, y)\,(\wedge T\, y)$
:= $BTMP5223 %0
wff 810 : $=(\supset T\, y)\,(\wedge T\, y)_o$:= $BTMP5223

.3

use Proof Template A5216: (T ∧ A) = A
:= $A5216 y_o

wff 34 : y_o := $\$A5216$
<< A5216.r0t.txt
:= $\$A5216$
%0
$=(\wedge T\, y)\, y$
$=_{ooo}(\wedge_{ooo} T_o y_o) y_o$

%$\$BTMP5223$
$=(\supset T\, y)\,(\wedge T\, y)$:= $\$BTMP5223$
$=_{ooo}(\supset_{ooo} T_o y_o)(\wedge_{ooo} T_o y_o)$:= $\$BTMP5223$
§s %0 3 %1
$=(\supset T\, y)\, y$

:= $A5223$ %0
wff 823 : $=(\supset T\, y)\, y_o$:= $A5223$

undefine local variables
:= $\$ATMP5223$
:= $\$BTMP5223$

##
Q.E.D.
##

%0
$=(\supset T\, y)\, y$:= $A5223$
$=_{ooo}(\supset_{ooo} T_o y_o) y_o$:= $A5223$

2.1.30 Results for File A5224.r0a.txt

##
Proof A5224 (MP): A, (A ⊃ B) → B
(Modus Ponens)
##
Source: [Andrews 2002 (ISBN 1-4020-0763-9), p. 224]
##

Written by Ken Kubota (<mail@kenkubota.de>).
##
This file is part of the publication of the mathematical logic $\mathcal{R}_0$.
For more information, visit: <http://doi.org/10.4444/100.10>
##

##
Define Syntactical Variables
##

```
<< basics.r0.txt

## the proposition A
:=  $A5224 a_o
# wff     54  :      a_o      :=  $A5224

## the proposition A ⊃ B
:=  $AB5224 ⊃_ooo$A5224_ob_o
# wff     209  :      ⊃ $A5224 b_o      :=  $AB5224


##
##  Assumptions and Resulting Syntactical Variables
##

§! $A5224
#               a       :=  $A5224
§! $AB5224
#               ⊃ $A5224 b      :=  $AB5224


##
##  Include Proof Template
##

## <<< A5224.r0t.txt
## Include begin (A5224.r0t.txt) [oldfile=(A5224.r0a.txt)]
##
##  Proof A5224 (MP):  A, (A ⊃ B)  →  B
##        (Modus Ponens)
##
##  Source: [Andrews 2002 (ISBN 1-4020-0763-9), p. 224]
##
##  Copyright (c) 2017 Owl of Minerva Press GmbH. All rights reserved.
##  Written by Ken Kubota (<mail@kenkubota.de>).
##
##  This file is part of the publication of the mathematical logic R0.
##  For more information, visit: <http://doi.org/10.4444/100.10>
##

<< A5223.r0.txt


##
##  Proof Template
##

## .1
```

%$\$AB5224$
$\supset \$A5224\, b$:= $\$AB5224$
$\supset_{ooo}\$A5224_o b_o$:= $\$AB5224$

.2

use Proof Template A5219b (Rule T): A $\rightarrow$ A = T
:= $\$A5219b$ a_o
wff 54 : a_o := $\$A5219b$ $\$A5224$
<< A5219b.r0t.txt
:= $\$A5219b$
%0
$= \$A5224\, T$
$=_{ooo}\$A5224_o T_o$

.3

%$\$AB5224$
$\supset \$A5224\, b$:= $\$AB5224$
$\supset_{ooo}\$A5224_o b_o$:= $\$AB5224$
§s %0 5 %1
$\supset T\, b$

:= $\$TMP5224$ %0
wff 843 : $\supset T\, b_o$:= $\$TMP5224$

.4

use Proof Template A5221 (Sub): B $\rightarrow$ B [x/A]
:= $\$B5221$ $=_{ooo}(\supset_{ooo} T_o y_o) y_o$
wff 825 : $= (\supset T\, y)\, y_o$:= $\$B5221$ $A5223$
:= $\$T5221$ o
wff 2 : o_τ := $\$T5221$
:= $\$X5221$ y_o
wff 34 : y_o := $\$X5221$
:= $\$A5221$ %0/3
wff 58 : b_o := $\$A5221$
<< A5221.r0t.txt
:= $\$B5221$
:= $\$T5221$
:= $\$X5221$
:= $\$A5221$
%0
$= \$TMP5224\, b$
$=_{ooo}\$TMP5224_o b_o$

%$\$TMP5224$
$\supset T\, b$:= $\$TMP5224$
$\supset_{ooo} T_o b_o$:= $\$TMP5224$

```
:=  $TMP5224
§s %0 1 %1
#                b
## Include end (A5224.r0t.txt) [newfile=(A5224.r0a.txt)]
>>>


##
##  Undefine Syntactical Variables
##

:=  $AB5224
:=  $A5224


##
##  Q.E.D.
##

%0
#                    b
#                    b_o
```

2.1.31 Results for File A5224H.r0a.txt

```
##
##  Proof A5224H (MP):  H ⊃ A, H ⊃ (A ⊃ B)  →  H ⊃ B
##        (Modus Ponens)
##
##  Source: [Andrews 2002 (ISBN 1-4020-0763-9), p. 224]
##
##  Copyright (c) 2017 Owl of Minerva Press GmbH. All rights reserved.
##  Written by Ken Kubota (<mail@kenkubota.de>).
##
##  This file is part of the publication of the mathematical logic R0.
##  For more information, visit: <http://doi.org/10.4444/100.10>
##


##
##  Define Syntactical Variables
##

<< basics.r0.txt

## the proposition H ⊃ A
:=  $A5224H  ⊃_ooo h_o a_o
# wff    210  :      ⊃ h a_o      :=  $A5224H
```

the proposition H ⊃ (A ⊃ B)
:= $\$AB5224H$ $\supset_{ooo}h_o(\supset_{ooo}a_ob_o)$
wff 213 : $\supset h\,(\supset a\,b)_o$:= $\$AB5224H$

##
Assumptions and Resulting Syntactical Variables
##

§! $\$A5224H$
$\supset h\,a$:= $\$A5224H$
§! $\$AB5224H$
$\supset h\,(\supset a\,b)$:= $\$AB5224H$

##
Include Proof Template
##

<<< A5224H.r0t.txt
Include begin (A5224H.r0t.txt) [oldfile=(A5224H.r0a.txt)]
##
Proof A5224H (MP): H ⊃ A, H ⊃ (A ⊃ B) → H ⊃ B
(Modus Ponens)
##
Source: [Andrews 2002 (ISBN 1-4020-0763-9), p. 224]
##
Copyright (c) 2017 Owl of Minerva Press GmbH. All rights reserved.
Written by Ken Kubota (<mail@kenkubota.de>).
##
This file is part of the publication of the mathematical logic $\mathcal{R}_0$.
For more information, visit: <http://doi.org/10.4444/100.10>
##

<< A5223.r0.txt

##
Proof Template
##

.1

%$\$AB5224H$
$\supset h\,(\supset a\,b)$:= $\$AB5224H$
$\supset_{ooo}h_o(\supset_{ooo}a_ob_o)$:= $\$AB5224H$

.2

use Proof Template A5219bH (Rule T): H ⊃ A → H ⊃ (A = T)
:= $\$A5219bH\ \supset_{ooo}h_oa_o$
wff 210 : $\supset h\, a_o$:= $\$A5219bH$ $\$A5224H$
<< A5219bH.r0t.txt
:= $\$A5219bH$
%0
$\supset h\,(=a\,T)$
$\supset_{ooo}h_o(=_{ooo}a_oT_o)$

.3

%$\$AB5224H$
$\supset h\,(\supset a\,b)$:= $\$AB5224H$
$\supset_{ooo}h_o(\supset_{ooo}a_ob_o)$:= $\$AB5224H$
§s' %0 5 %1
$\supset h\,(\supset T\,b)$

:= $\$TMP5224H$ %0
wff 1601 : $\supset h\,(\supset T\,b)_o$:= $\$TMP5224H$

.4

use Proof Template A5221 (Sub): B → B [x/A]
:= $\$B5221\ =_{ooo}(\supset_{ooo}T_oy_o)y_o$
wff 829 : $=(\supset T\,y)\,y_o$:= $\$B5221$ $A5223$
:= $\$T5221\ o$
wff 2 : o_τ := $\$T5221$
:= $\$X5221\ y_o$
wff 34 : y_o := $\$X5221$
:= $\$A5221$ %0/7
wff 58 : b_o := $\$A5221$
<< A5221.r0t.txt
:= $\$B5221$
:= $\$T5221$
:= $\$X5221$
:= $\$A5221$
%0
$=(\supset T\,b)\,b$
$=_{ooo}(\supset_{ooo}T_ob_o)b_o$

use Proof Template K8004 (Trans): (H ⊕ A), B → H ⊃ B
:= $\$HA8004\ \supset_{ooo}h_o(\supset_{ooo}T_ob_o)$
wff 1601 : $\supset h\,(\supset T\,b)_o$:= $\$HA8004$ $\$TMP5224H$
:= $\$B8004$ %0
wff 1643 : $=(\supset T\,b)\,b_{o,\ldots}$:= $\$B8004$
<< K8004.r0t.txt
:= $\$HA8004$
:= $\$B8004$
%0

$\supset h\,(=(\supset T\,b)\,b)$
$\supset_{ooo}h_o(=_{ooo}(\supset_{ooo}T_ob_o)b_o)$

%$TMP5224H
$\supset h\,(\supset T\,b)$:= $TMP5224H
$\supset_{ooo}h_o(\supset_{ooo}T_ob_o)$:= $TMP5224H
:= $TMP5224H
§s′ %0 1 %1
$\supset h\,b$
Include end (A5224H.r0t.txt) [newfile=(A5224H.r0a.txt)]
>>>

##
Undefine Syntactical Variables
##

:= $AB5224H
:= $A5224H

##
Q.E.D.
##

%0
$\supset h\,b$
$\supset_{ooo}h_ob_o$

2.1.32 Results for File A5225.r0.txt

##
Proof A5225: ∀ x: f ⊃ f x
for any x of any type a and any f of any type oa
##
Source: [Andrews 2002 (ISBN 1-4020-0763-9), p. 224]
##

Written by Ken Kubota (<mail@kenkubota.de>).
##
This file is part of the publication of the mathematical logic $\mathcal{R}_0$.
For more information, visit: <http://doi.org/10.4444/100.10>
##

<< axioms.r0.txt

##
Proof

##

.1

use Proof Template: Axiom 2 Substitutions
:= $\$AA2\ oa$
wff 92 : oa_τ := $\$AA2$
:= $\$HA2\ [\lambda f_{\$AA2}.(f_{\$AA2} x_a)_o]$
wff 132 : $[\lambda f.(f\,x)]_{o\$AA2}$:= $\$HA2$
:= $\$XA2\ [\lambda x_a.T_o]$
wff 134 : $[\lambda x.T]_{\$AA2}$:= $\$XA2$
:= $\$YA2\ f_{\$AA2}$
wff 130 : $f_{\$AA2}$:= $\$YA2$
<< axiom2_substitutions.r0t.txt
:= $\$AA2$
:= $\$HA2$
:= $\$XA2$
:= $\$YA2$
%0
$\supset (= [\lambda x.T]\, f)\, (= ([\lambda f.(f\,x)]\, [\lambda x.T])\, ([\lambda f.(f\,x)]\, f))$
$\supset_{ooo}(=_{o(oa)(oa)}[\lambda x_a.T_o]f_{oa})(=_{ooo}([\lambda f_{oa}.(f_{oa}x_a)_o][\lambda x_a.T_o])([\lambda f_{oa}.(f_{oa}x_a)_o]f_{oa}))$

§= $\forall_{o(o\backslash 3)\tau} a_\tau f_{oa}$
$= (\forall\, a\, f)\, (\forall\, a\, f)$
§\ $\forall_{o(o\backslash 3)\tau} a_\tau$
$= (\forall\, a)\, [\lambda p.(= [\lambda x.T]\, p)]$
§s %1 6 %0
$= (\forall\, a\, f)\, ([\lambda p.(= [\lambda x.T]\, p)]\, f)$
§\ $[\lambda p_{oa}.(=_{o(oa)(oa)}[\lambda x_a.T_o]p_{oa})_o]f_{oa}$
$= ([\lambda p.(= [\lambda x.T]\, p)]\, f)\, (= [\lambda x.T]\, f)$
§s %1 3 %0
$= (\forall\, a\, f)\, (= [\lambda x.T]\, f)$
§= $\forall_{o(o\backslash 3)\tau} a_\tau f_{oa}$
$= (\forall\, a\, f)\, (\forall\, a\, f)$
§s %0 5 %1
$= (= [\lambda x.T]\, f)\, (\forall\, a\, f)$
§s %7 5 %0
$\supset (\forall\, a\, f)\, (= ([\lambda f.(f\,x)]\, [\lambda x.T])\, ([\lambda f.(f\,x)]\, f))$

.2

§\ $[\lambda f_{oa}.(f_{oa}x_a)_o][\lambda x_a.T_o]$
$= ([\lambda f.(f\,x)]\, [\lambda x.T])\, ([\lambda x.T]\, x)$
§s %1 13 %0
$\supset (\forall\, a\, f)\, (= ([\lambda x.T]\, x)\, ([\lambda f.(f\,x)]\, f))$
§\ $[\lambda x_a.T_o]x_a$
$= ([\lambda x.T]\, x)\, T$
§s %1 13 %0
$\supset (\forall\, a\, f)\, (= T\, ([\lambda f.(f\,x)]\, f))$

§\ $[\lambda f_{oa}.(f_{oa}x_a)_o]f_{oa}$
$= ([\lambda f.(f\,x)]\,f)\,(f\,x)$
§s %1 7 %0
$\supset (\forall\, a\, f)\,(=T\,(f\,x))$
:= $\$TMP5225$ %0
wff 973 : $\supset (\forall\, a\, f)\,(=T\,(f\,x))_o$:= $\$TMP5225$

.3

use Proof Template A5218: (T = A) = A
:= $\$A5218$ $f_{oa}x_a$
wff 131 : $f\,x_{o,\ldots}$:= $\$A5218$
<< A5218.r0t.txt
:= $\$A5218$
%0
$= (=T\,(f\,x))\,(f\,x)$
$=_{ooo}(=_{ooo}T_o(f_{oa}x_a))(f_{oa}x_a)$

%$\$TMP5225$
$\supset (\forall\, a\, f)\,(=T\,(f\,x))$:= $\$TMP5225$
$\supset_{ooo}(\forall_{o(o\backslash 3)\tau}a_\tau f_{oa})(=_{ooo}T_o(f_{oa}x_a))$:= $\$TMP5225$
§s %0 3 %1
$\supset (\forall\, a\, f)\,(f\,x)$

:= $A5225$ %0
wff 986 : $\supset (\forall\, a\, f)\,(f\,x)_o$:= $A5225$

undefine local variables
:= $\$TMP5225$

##
Q.E.D.
##

%0
$\supset (\forall\, a\, f)\,(f\,x)$:= $A5225$
$\supset_{ooo}(\forall_{o(o\backslash 3)\tau}a_\tau f_{oa})(f_{oa}x_a)$:= $A5225$

2.1.33 Results for File A5226.r0a.txt

##
Proof Template A5226: ∀ x: B ⊃ B [x/a]
for any x of any type a and any A, B of type oa
##
Source: [Andrews 2002 (ISBN 1-4020-0763-9), p. 224]
##

Written by Ken Kubota (<mail@kenkubota.de>).

```
##
##  This file is part of the publication of the mathematical logic R0.
##  For more information, visit: <http://doi.org/10.4444/100.10>
##


##
##  Define Syntactical Variables
##

## type of the variable
:=  $T5226  t_τ
# wff     4  :        t_τ       :=  $T5226

## the variable to be replaced
:=  $X5226  x_$T5226
# wff     11  :       x_$T5226        :=  $X5226

## substitution term
:=  $A5226  a_$T5226
# wff     12  :       a_$T5226        :=  $A5226

## the proposition (in this example, B is defined as x=x)
:=  $B5226  =_oωω$X5226_ω$X5226_ω
# wff     14  :       =$X5226 $X5226_o       :=  $B5226


##
##  Assumptions and Resulting Syntactical Variables
##

§! $B5226
#                   =$X5226 $X5226       :=  $B5226


##
##  Include Proof Template
##

## <<< A5226.r0t.txt
## Include begin (A5226.r0t.txt) [oldfile=(A5226.r0a.txt)]
##
##  Proof Template A5226:  ∀ x: B  ⊃  B [x/a]
##        for any x of any type a and any A, B of type oa
##
##  Source: [Andrews 2002 (ISBN 1-4020-0763-9), p. 224]
##
##  Copyright (c) 2017 Owl of Minerva Press GmbH. All rights reserved.
##  Written by Ken Kubota (<mail@kenkubota.de>).
```

##
This file is part of the publication of the mathematical logic $\mathcal{R}_0$.
For more information, visit: <http://doi.org/10.4444/100.10>
##

<< A5225.r0.txt

##
Proof Template
##

%$A5225$
$\supset (\forall\, a\, f)\,(f\, x)$:= $A5225$
$\supset_{ooo}(\forall_{o(o\backslash 3)\tau} a_\tau f_{oa})(f_{oa} x_a)$:= $A5225$

.1a Replace type a in A5225

use Proof Template A5221 (Sub): B $\rightarrow$ B [x/A]
:= $\$B5221$ %0
wff 989 : $\supset (\forall\, a\, f)\,(f\, x)_o$:= $\$B5221$ $A5225$
:= $\$T5221$ τ
wff 0 : τ_τ := $\$T5221$
:= $\$X5221$ a_τ
wff 94 : a_τ := $\$X5221$
:= $\$A5221$ t_τ
wff 4 : t_τ := $\$A5221$ $\$T5226$
<< A5221.r0t.txt
:= $\$B5221$
:= $\$T5221$
:= $\$X5221$
:= $\$A5221$
%0
$\supset (\forall\, \$T5226\, f)\,(f\, \$X5226)$
$\supset_{ooo}(\forall_{o(o\backslash 3)\tau} \$T5226_\tau f_{o\,\$T5226})(f_{o\,\$T5226} \$X5226_{\$T5226})$

.1b Replace variable x in A5225

use Proof Template A5221 (Sub): B $\rightarrow$ B [x/A]
:= $\$B5221$ %0
wff 1030 : $\supset (\forall\, \$T5226\, f)\,(f\, \$X5226)_{o,\dots}$:= $\$B5221$
:= $\$T5221$ t_τ
wff 4 : t_τ := $\$T5221$ $\$T5226$
:= $\$X5221$ $x_{\$T5226}$
wff 11 : $x_{\$T5226}$:= $\$X5221$ $\$X5226$
:= $\$A5221$ $a_{\$T5226}$
wff 12 : $a_{\$T5226}$:= $\$A5221$ $\$A5226$
<< A5221.r0t.txt
:= $\$B5221$

:= $\$T5221$
:= $\$X5221$
:= $\$A5221$
%0
$\supset (\forall \$T5226\, f)\, (f\, \$A5226)$
$\supset_{ooo}(\forall_{o(o\backslash 3)\tau}\$T5226_\tau f_{o\$T5226})(f_{o\$T5226}\$A5226_{\$T5226})$

.1c Replace variable f in A5225

use Proof Template A5221 (Sub): B $\rightarrow$ B [x/A]
:= $\$B5221$ %0
wff 1074 : $\supset (\forall \$T5226\, f)\, (f\, \$A5226)_{o,\ldots}$:= $\$B5221$
:= $\$T5221$ $o\,\$T5226$
wff 5 : $o\,\$T5226_\tau$:= $\$T5221$
:= $\$X5221$ $f_{\$T5221}$
wff 1026 : $f_{\$T5221}$:= $\$X5221$
:= $\$A5221$ $[\lambda \$X5226_{\$T5226}.\$B5226_o]$
wff 1077 : $[\lambda \$X5226.\$B5226]_{\$T5221}$:= $\$A5221$
<< A5221.r0t.txt
:= $\$B5221$
:= $\$T5221$
:= $\$X5221$
:= $\$A5221$
%0
$\supset (\forall \$T5226\, [\lambda \$X5226.\$B5226])\, ([\lambda \$X5226.\$B5226]\, \$A5226)$
$\supset_{ooo}(\forall_{o(o\backslash 3)\tau}\$T5226_\tau[\lambda \$X5226_{\$T5226}.\$B5226_o])\ldots$
$\ldots([\lambda \$X5226_{\$T5226}.\$B5226_o]\$A5226_{\$T5226})$

.2

§\ $[\lambda \$X5226_{\$T5226}.\$B5226_o]\$A5226_{\$T5226}$
$= ([\lambda \$X5226.\$B5226]\, \$A5226)\, (= \$A5226\, \$A5226)$
§s %1 3 %0
$\supset (\forall \$T5226\, [\lambda \$X5226.\$B5226])\, (= \$A5226\, \$A5226)$
Include end (A5226.r0t.txt) [newfile=(A5226.r0a.txt)]
>>>

##
Undefine Syntactical Variables
##

:= $\$T5226$
:= $\$X5226$
:= $\$A5226$
:= $\$B5226$

##

```
##  Q.E.D.
##

%0
#                    ⊃ (∀ t [λx.(= x x)]) (= a a)
#                    ⊃ooo(∀o(o\3)τ tτ [λxt.(=oωω xω xω)o])(=oωω aω aω)
```

2.1.34 Results for File A5227.r0.txt

```
##
##  Proof A5227:   F ⊃ x
##        with x of type o
##
##  Source: [Andrews 2002 (ISBN 1-4020-0763-9), p. 224]
##
##  Copyright (c) 2017 Owl of Minerva Press GmbH. All rights reserved.
##  Written by Ken Kubota (<mail@kenkubota.de>).
##
##  This file is part of the publication of the mathematical logic R0.
##  For more information, visit: <http://doi.org/10.4444/100.10>
##



##
## Proof
##

## use Proof Template A5226:   ∀ x: B   ⊃   B [x/a]
:=  $T5226  o
# wff    2  :        oτ      :=  $T5226
:=  $X5226  xo
# wff    11  :        xo      :=  $X5226
:=  $A5226  xo
# wff    11  :        xo      :=  $A5226  $X5226
:=  $B5226  xo
# wff    11  :        xo      :=  $A5226  $B5226  $X5226
<< A5226.r0t.txt
:=  $T5226
:=  $X5226
:=  $A5226
:=  $B5226
%0
#                    ⊃ (∀ o [λx.x]) x
#                    ⊃ooo(∀o(o\3)τ oτ [λxo.xo])xo

§\  ∀o(o\3)τ oτ
#                    = (∀ o) [λp.(= [λx.T] p)]
§s  %1  10  %0
#                    ⊃ ([λp.(= [λx.T] p)] [λx.x]) x
```

§\ $[\lambda p_{oo}.(=_{o(oo)(oo)}[\lambda x_o.T_o]p_{oo})_o][\lambda x_o.x_o]$
\# $= ([\lambda p.(= [\lambda x.T]\, p)]\, [\lambda x.x])\, F$
§s %1 5 %0
\# $\supset F\, x$

:= $A5227$ %0
\# wff 1095 : $\supset F\, x_o$:= $A5227$

\##
\## Q.E.D.
\##

%0
\# $\supset F\, x$:= $A5227$
\# $\supset_{ooo} F_o x_o$:= $A5227$

2.1.35 Results for File A5228.r0.txt

\##
\## Proof A5228: (T ⊃ T) = T; (T ⊃ F) = F; (F ⊃ T) = T; (F ⊃ F) = T
\##
\##
\## Source: [Andrews 2002 (ISBN 1-4020-0763-9), p. 224]
\##

\## Written by Ken Kubota (<mail@kenkubota.de>).
\##
\## This file is part of the publication of the mathematical logic $\mathcal{R}_0$.
\## For more information, visit: <http://doi.org/10.4444/100.10>
\##

<< A5223.r0.txt
<< A5227.r0.txt

\##
\## Proof
\##

\## .a: (T ⊃ T) = T

\## use Proof Template A5221 (Sub): B → B [x/A]
:= $B5221 $=_{ooo}(\supset_{ooo}T_o y_o)y_o$
\# wff 823 : $= (\supset T\, y)\, y_o$:= $B5221 $A5223$
:= $T5221 o
\# wff 2 : o_τ := $T5221
:= $X5221 y_o
\# wff 34 : y_o := $X5221

:= $\$A5221\ =_{o\omega\omega}=_{\omega}=_{\omega}$
wff 12 : $===_{o,\ldots}$:= $\$A5221\quad A5200t\quad T$
<< A5221.r0t.txt
:= $\$B5221$
:= $\$T5221$
:= $\$X5221$
:= $\$A5221$
%0
$=(\supset T\,T)\,T$
$=_{ooo}(\supset_{ooo}T_oT_o)T_o$

:= $A5228a$ %0
wff 1253 : $=(\supset T\,T)\,T_{o,\ldots}$:= $A5228a$

.b: (T ⊃ F) = F

use Proof Template A5221 (Sub): B → B [x/A]
:= $\$B5221\ =_{ooo}(\supset_{ooo}T_oy_o)y_o$
wff 823 : $=(\supset T\,y)\,y_{o,\ldots}$:= $\$B5221\quad A5223$
:= $\$T5221\ o$
wff 2 : o_τ := $\$T5221$
:= $\$X5221\ y_o$
wff 34 : y_o := $\$X5221$
:= $\$A5221\ =_{o(oo)(oo)}[\lambda x_o.T_o][\lambda x_o.x_o]$
wff 20 : $=[\lambda x.T]\,[\lambda x.x]_{o,\ldots}$:= $\$A5221\quad F$
<< A5221.r0t.txt
:= $\$B5221$
:= $\$T5221$
:= $\$X5221$
:= $\$A5221$
%0
$=(\supset T\,F)\,F$
$=_{ooo}(\supset_{ooo}T_oF_o)F_o$

:= $A5228b$ %0
wff 1266 : $=(\supset T\,F)\,F_{o,\ldots}$:= $A5228b$

.c: (F ⊃ T) = T

use Proof Template A5221 (Sub): B → B [x/A]
:= $\$B5221\ \supset_{ooo}F_ox_o$
wff 1213 : $\supset F\,x_o$:= $\$B5221\quad A5227$
:= $\$T5221\ o$
wff 2 : o_τ := $\$T5221$
:= $\$X5221\ x_o$
wff 16 : x_o := $\$X5221$
:= $\$A5221\ =_{o\omega\omega}=_{\omega}=_{\omega}$
wff 12 : $===_{o,\ldots}$:= $\$A5221\quad A5200t\quad T$
<< A5221.r0t.txt

:= $\$B5221$
:= $\$T5221$
:= $\$X5221$
:= $\$A5221$
%0
$\supset F\, T$
$\supset_{ooo} F_o T_o$

use Proof Template A5219b (Rule T): A → A = T
:= $\$A5219b$ %0
wff 1300 : $\supset F\, T_{o,\dots}$:= $\$A5219b$
<< A5219b.r0t.txt
:= $\$A5219b$
%0
$=(\supset F\, T)\, T$
$=_{ooo}(\supset_{ooo} F_o T_o) T_o$

:= $A5228c$ %0
wff 1319 : $=(\supset F\, T)\, T_o$:= $A5228c$

.d: (F ⊃ F) = T

use Proof Template A5221 (Sub): B → B [x/A]
:= $\$B5221\ \supset_{ooo} F_o x_o$
wff 1213 : $\supset F\, x_{o,\dots}$:= $\$B5221$ $A5227$
:= $\$T5221$ o
wff 2 : o_τ := $\$T5221$
:= $\$X5221$ x_o
wff 16 : x_o := $\$X5221$
:= $\$A5221\ =_{o(oo)(oo)}[\lambda \$X5221_o.T_o][\lambda \$X5221_o.\$X5221_o]$
wff 20 : $=[\lambda \$X5221.T]\,[\lambda \$X5221.\$X5221]_{o,\dots}$:= $\$A5221$ F
<< A5221.r0t.txt
:= $\$B5221$
:= $\$T5221$
:= $\$X5221$
:= $\$A5221$
%0
$\supset F\, F$
$\supset_{ooo} F_o F_o$

use Proof Template A5219b (Rule T): A → A = T
:= $\$A5219b$ %0
wff 1324 : $\supset F\, F_{o,\dots}$:= $\$A5219b$
<< A5219b.r0t.txt
:= $\$A5219b$
%0
$=(\supset F\, F)\, T$
$=_{ooo}(\supset_{ooo} F_o F_o) T_o$

:= $A5228d$ %0
wff 1343 : $=(\supset F\,F)\,T_o$:= $A5228d$

##
Q.E.D.
##

%A5228a
%$A5228a$
$=(\supset T\,T)\,T$:= $A5228a$
$=_{ooo}(\supset_{ooo}T_oT_o)T_o$:= $A5228a$

%A5228b
%$A5228b$
$=(\supset T\,F)\,F$:= $A5228b$
$=_{ooo}(\supset_{ooo}T_oF_o)F_o$:= $A5228b$

%A5228c
%$A5228c$
$=(\supset F\,T)\,T$:= $A5228c$
$=_{ooo}(\supset_{ooo}F_oT_o)T_o$:= $A5228c$

%A5228d
%$A5228d$
$=(\supset F\,F)\,T$:= $A5228d$
$=_{ooo}(\supset_{ooo}F_oF_o)T_o$:= $A5228d$

2.1.36 Results for File A5229.r0.txt

##
Proof A5229: (T ∧ T) = T; (T ∧ F) = F; (F ∧ T) = F; (F ∧ F) = F
##
##
Source: [Andrews 2002 (ISBN 1-4020-0763-9), p. 225]
##

Written by Ken Kubota (<mail@kenkubota.de>).
##
This file is part of the publication of the mathematical logic $\mathcal{R}_0$.
For more information, visit: <http://doi.org/10.4444/100.10>
##

<< A5227.r0.txt

##
Proof
##

.a: (T ∧ T) = T

use Proof Template A5216: (T ∧ A) = A
:= $\$A5216 \; =_{o\omega\omega}=_{\omega}=_{\omega}$
wff 13 : $===_{o,\ldots}$:= $\$A5216 \quad A5200t \quad T$
<< A5216.r0t.txt
:= $\$A5216$
:= $A5229a$ %0
wff 474 : $=A5212\,T_{o,\ldots}$:= $A5211 \quad A5229a$

.b: (T ∧ F) = F

use Proof Template A5216: (T ∧ A) = A
:= $\$A5216 \; =_{o(oo)(oo)}[\lambda x_o.T_o][\lambda x_o.x_o]$
wff 20 : $=[\lambda x.T]\,[\lambda x.x]_{o,\ldots}$:= $\$A5216 \quad F$
<< A5216.r0t.txt
:= $\$A5216$

:= $A5229b$ %0
wff 515 : $=(\wedge\,T\,F)\,F_{o,\ldots}$:= $A5214 \quad A5229b$

.c: (F ∧ T) = F

%$A5227$
$\supset F\,x$:= $A5227$
$\supset_{ooo}F_o x_o$:= $A5227$

use Proof Template A5221 (Sub): B → B [x/A]
:= $\$B5221$ %0
wff 1095 : $\supset F\,x_o$:= $\$B5221 \quad A5227$
:= $\$T5221 \; o$
wff 2 : o_τ := $\$T5221$
:= $\$X5221 \; x_o$
wff 11 : x_o := $\$X5221$
:= $\$A5221 \; =_{o\omega\omega}=_{\omega}=_{\omega}$
wff 13 : $===_{o,\ldots}$:= $\$A5221 \quad A5200t \quad T$
<< A5221.r0t.txt
:= $\$B5221$
:= $\$T5221$
:= $\$X5221$
:= $\$A5221$
%0
$\supset F\,T$
$\supset_{ooo}F_o T_o$

§\ $\supset_{ooo}F_o$
$=(\supset F)\,[\lambda y.(=F\,(\wedge\,F\,y))]$
§s %1 2 %0

$[\lambda y.(= F\,(\wedge F\,y))]\,T$
§\ $[\lambda y_o.(=_{ooo}F_o(\wedge_{ooo}F_o y_o))_o]T_o$
$= ([\lambda y.(= F\,(\wedge F\,y))]\,T)\,(= F\,(\wedge F\,T))$
§s %1 1 %0
$= F\,(\wedge F\,T)$
§= $_o$ F
$= F\,F$
§s %0 5 %1
$= (\wedge F\,T)\,F$

:= $A5229c$ %0
wff 1153 : $= (\wedge F\,T)\,F_o$:= $A5229c$

.d: (F ∧ F) = F

%$A5227$
$\supset F\,x$:= $A5227$
$\supset_{ooo}F_o x_o$:= $A5227$

use Proof Template A5221 (Sub): B → B [x/A]
:= $\$B5221$ %0
wff 1095 : $\supset F\,x_{o,\ldots}$:= $\$B5221$ $A5227$
:= $\$T5221$ o
wff 2 : o_T := $\$T5221$
:= $\$X5221$ x_o
wff 11 : x_o := $\$X5221$
:= $\$A5221$ $=_{o(oo)(oo)}[\lambda \$X5221_o.T_o][\lambda \$X5221_o.\$X5221_o]$
wff 20 : $= [\lambda \$X5221.T]\,[\lambda \$X5221.\$X5221]_{o,\ldots}$:= $\$A5221$ F
<< A5221.r0t.txt
:= $\$B5221$
:= $\$T5221$
:= $\$X5221$
:= $\$A5221$
%0
$\supset F\,F$
$\supset_{ooo}F_o F_o$

§\ $\supset_{ooo}F_o$
$= (\supset F)\,[\lambda y.(= F\,(\wedge F\,y))]$
§s %1 2 %0
$[\lambda y.(= F\,(\wedge F\,y))]\,F$
§\ $[\lambda y_o.(=_{ooo}F_o(\wedge_{ooo}F_o y_o))_o]F_o$
$= ([\lambda y.(= F\,(\wedge F\,y))]\,F)\,(= F\,(\wedge F\,F))$
§s %1 1 %0
$= F\,(\wedge F\,F)$
§= $_o$ F
$= F\,F$
§s %0 5 %1
$= (\wedge F\,F)\,F$

:= $A5229d$ %0
\# wff 1167 : $=(\wedge F\,F)\,F_o$:= $A5229d$

\##
\## Q.E.D.
\##

\## %A5229a
%$A5211$
\# $=A5212\,T$:= $A5211$ $A5229a$
\# $=_{ooo}A5212_oT_o$:= $A5211$ $A5229a$

\## %A5229b
%$A5214$
\# $=(\wedge T\,F)\,F$:= $A5214$ $A5229b$
\# $=_{ooo}(\wedge_{ooo}T_oF_o)F_o$:= $A5214$ $A5229b$

\## %A5229c
%$A5229c$
\# $=(\wedge F\,T)\,F$:= $A5229c$
\# $=_{ooo}(\wedge_{ooo}F_oT_o)F_o$:= $A5229c$

\## %A5229d
%$A5229d$
\# $=(\wedge F\,F)\,F$:= $A5229d$
\# $=_{ooo}(\wedge_{ooo}F_oF_o)F_o$:= $A5229d$

2.1.37 Results for File A5230.r0.txt

\##
\## Proof A5230: (T = T) = T; (T = F) = F; (F = T) = F; (F = F) = T
\##
\##
\## Source: [Andrews 2002 (ISBN 1-4020-0763-9), p. 225]
\##

\## Written by Ken Kubota (<mail@kenkubota.de>).
\##
\## This file is part of the publication of the mathematical logic $\mathcal{R}_0$.
\## For more information, visit: <http://doi.org/10.4444/100.10>
\##

<< basics.r0.txt
<< A5229.r0.txt

\##

Proof
##

.a: (T = T) = T

use Proof Template A5218: (T = A) = A
:= $\$A5218\ =_{o\omega\omega}=_{\omega}=_{\omega}$
wff 12 : $===_{o,\ldots}$:= $\$A5218\ \ A5200t\ \ T$
<< A5218.r0t.txt
:= $\$A5218$
%0
$=(=T\,T)\,T$
$=_{ooo}(=_{ooo}T_oT_o)T_o$

:= $A5230a$ %0
wff 746 : $=(=T\,T)\,T_{o,\ldots}$:= $A5230a$

.b: (T = F) = F

use Proof Template A5218: (T = A) = A
:= $\$A5218\ =_{o(oo)(oo)}[\lambda x_o.T_o][\lambda x_o.x_o]$
wff 20 : $=[\lambda x.T]\,[\lambda x.x]_{o,\ldots}$:= $\$A5218\ \ F$
<< A5218.r0t.txt
:= $\$A5218$
%0
$=(=T\,F)\,F$:= $A5217$
$=_{ooo}(=_{ooo}T_oF_o)F_o$:= $A5217$

:= $A5230b$ %0
wff 721 : $=(=T\,F)\,F_{o,\ldots}$:= $A5217\ \ A5230b$

.c: (F = T) = F

.1

use Proof Template: Axiom 2 Substitutions
:= $\$AA2\ o$
wff 2 : o_τ := $\$AA2$
:= $\$HA2\ [\lambda x_o.(=_{ooo}x_oF_o)_o]$
wff 1253 : $[\lambda x.(=x\,F)]_{oo}$:= $\$HA2$
:= $\$XA2\ =_{o(oo)(oo)}[\lambda x_o.T_o][\lambda x_o.x_o]$
wff 20 : $=[\lambda x.T]\,[\lambda x.x]_{o,\ldots}$:= $\$XA2\ \ F$
:= $\$YA2\ =_{o\omega\omega}=_{\omega}=_{\omega}$
wff 12 : $===_{o,\ldots}$:= $\$YA2\ \ A5200t\ \ T$
<< axiom2_substitutions.r0t.txt
:= $\$AA2$
:= $\$HA2$
:= $\$XA2$
:= $\$YA2$

%0
$\supset (= F\, T)\, (= ([\lambda x.(= x\, F)]\, F)\, ([\lambda x.(= x\, F)]\, T))$
$\supset_{ooo}(=_{ooo}F_oT_o)(=_{ooo}([\lambda x_o.(=_{ooo}x_oF_o)_o]F_o)([\lambda x_o.(=_{ooo}x_oF_o)_o]T_o))$

.2

§\ $[\lambda x_o.(=_{ooo}x_oF_o)_o]F_o$
$= ([\lambda x.(= x\, F)]\, F)\, (= F\, F)$
§s %1 13 %0
$\supset (= F\, T)\, (= (= F\, F)\, ([\lambda x.(= x\, F)]\, T))$
§\ $[\lambda x_o.(=_{ooo}x_oF_o)_o]T_o$
$= ([\lambda x.(= x\, F)]\, T)\, (= T\, F)$
§s %1 7 %0
$\supset (= F\, T)\, (= (= F\, F)\, (= T\, F))$

:= $\$ATMP5230$ %0
wff 1405 : $\supset (= F\, T)\, (= (= F\, F)\, (= T\, F))_o$:= $\$ATMP5230$

.3a

use Proof Template A5210: T = (B = B)
:= $\$T5210$ o
wff 2 : o_τ := $\$T5210$
:= $\$B5210$ $=_{o(oo)(oo)}[\lambda x_o.T_o][\lambda x_o.x_o]$
wff 20 : $= [\lambda x.T]\, [\lambda x.x]_{o,\ldots}$:= $\$B5210$ F
<< A5210.r0t.txt
:= $\$T5210$
:= $\$B5210$
%0
$= T\, (= F\, F)$
$=_{ooo}T_o(=_{ooo}F_oF_o)$

use Proof Template A5201b (Swap): A = B → B = A
<< A5201b.r0t.txt
%0
$= (= F\, F)\, T$
$=_{ooo}(=_{ooo}F_oF_o)T_o$

:= $\$BTMP5230$ %0
wff 1414 : $= (= F\, F)\, T_o$:= $\$BTMP5230$

.3b

use Proof Template A5218: (T = A) = A
:= $\$A5218$ $=_{o(oo)(oo)}[\lambda x_o.T_o][\lambda x_o.x_o]$
wff 20 : $= [\lambda x.T]\, [\lambda x.x]_{o,\ldots}$:= $\$A5218$ F
<< A5218.r0t.txt
:= $\$A5218$
%0

$=(=T\,F)\,F$:= $A5217$ $A5230b$
$=_{ooo}(=_{ooo}T_oF_o)F_o$:= $A5217$ $A5230b$

:= $\$CTMP5230$ %0
wff 721 : $=(=T\,F)\,F_{o,\ldots}$:= $\$CTMP5230$ $A5217$ $A5230b$

.3c

%$\$ATMP5230$
$\supset(=F\,T)\,(=(=F\,F)\,(=T\,F))$:= $\$ATMP5230$
$\supset_{ooo}(=_{ooo}F_oT_o)(=_{ooo}(=_{ooo}F_oF_o)(=_{ooo}T_oF_o))$:= $\$ATMP5230$
%$\$BTMP5230$
$=(=F\,F)\,T$:= $\$BTMP5230$
$=_{ooo}(=_{ooo}F_oF_o)T_o$:= $\$BTMP5230$
§s %1 13 %0
$\supset(=F\,T)\,(=T\,(=T\,F))$
%$A5217$
$=(=T\,F)\,F$:= $\$CTMP5230$ $A5217$ $A5230b$
$=_{ooo}(=_{ooo}T_oF_o)F_o$:= $\$CTMP5230$ $A5217$ $A5230b$
§s %1 7 %0
$\supset(=F\,T)\,(=T\,F)$
§s %0 3 %1
$\supset(=F\,T)\,F$

.4

§\ $\supset_{ooo}(=_{ooo}F_oT_o)$
$=(\supset(=F\,T))\,[\lambda y.(=(=F\,T)\,(\wedge(=F\,T)\,y))]$
§s %1 2 %0
$[\lambda y.(=(=F\,T)\,(\wedge(=F\,T)\,y))]\,F$
§\ $[\lambda y_o.(=_{ooo}(=_{ooo}F_oT_o)(\wedge_{ooo}(=_{ooo}F_oT_o)y_o))_o]F_o$
$=([\lambda y.(=(=F\,T)\,(\wedge(=F\,T)\,y))]\,F)\,(=(=F\,T)\,(\wedge(=F\,T)\,F))$
§s %1 1 %0
$=(=F\,T)\,(\wedge(=F\,T)\,F)$

:= $\$DTMP5230$ %0
wff 1429 : $=(=F\,T)\,(\wedge(=F\,T)\,F)_{o,\ldots}$:= $\$DTMP5230$

.5

use Proof Template A5222 (Rule of Cases): [\x.A]T, [\x.A]F → A
:= $\$L5222$ $[\lambda x_o.(=_{ooo}(\wedge_{ooo}x_oF_o)F_o)_o]$
wff 1434 : $[\lambda x.(=(\wedge\,x\,F)\,F)]_{oo}$:= $\$L5222$
:= $\$X5222$ x_o
wff 16 : x_o := $\$X5222$
:= $\$T5222$ $\$L5222_{oo}T_o$
wff 1435 : $\$L5222\,T_o$:= $\$T5222$
:= $\$F5222$ $\$L5222_{oo}F_o$
wff 1436 : $\$L5222\,F_o$:= $\$F5222$

```
## Case T
§\  $T5222
#                  = $T5222 A5214
## use Proof Template A5201b (Swap):   A = B   →   B = A
<< A5201b.r0t.txt
%0
#                  = A5214 $T5222
#                  =_oωω A5214_ω $T5222_ω
%A5214
#                  = (∧ T F) F        :=    A5214    A5229b
#                  =_ooo (∧_ooo T_o F_o) F_o        :=    A5214    A5229b
§s  %0  1  %1
#                  $L5222 T        :=    $T5222

## Case F
§\  $F5222
#                  = $F5222 A5229d
## use Proof Template A5201b (Swap):   A = B   →   B = A
<< A5201b.r0t.txt
%0
#                  = A5229d $F5222
#                  =_oωω A5229d_ω $F5222_ω
%A5229d
#                  = (∧ F F) F        :=    A5229d
#                  =_ooo (∧_ooo F_o F_o) F_o        :=    A5229d
§s  %0  1  %1
#                  $L5222 F        :=    $F5222

<< A5222.r0t.txt
:=   $L5222
:=   $X5222
:=   $T5222
:=   $F5222
%0
#                  = (∧ x F) F
#                  =_ooo (∧_ooo x_o F_o) F_o

## .6

## use Proof Template A5221 (Sub):   B   →   B [x/A]
:=   $B5221  %0
# wff      1433   :         = (∧ x F) F_o,...        :=    $B5221
:=   $T5221  o
# wff      2   :        o_τ        :=   $T5221
:=   $X5221  x_o
# wff      16   :         x_o        :=   $X5221
:=   $A5221  =_ooo F_o T_o
# wff      1390   :         = F T_o        :=    $A5221
```

<< A5221.r0t.txt
:= $\$B5221$
:= $\$T5221$
:= $\$X5221$
:= $\$A5221$
%0
$=(\wedge\,(=F\,T)\,F)\,F$
$=_{ooo}(\wedge_{ooo}(=_{ooo}F_oT_o)F_o)F_o$

.7

%$\$DTMP5230$
$=(=F\,T)\,(\wedge\,(=F\,T)\,F)$:= $\$DTMP5230$
$=_{ooo}(=_{ooo}F_oT_o)(\wedge_{ooo}(=_{ooo}F_oT_o)F_o)$:= $\$DTMP5230$
use Proof Template A5201b (Swap): A = B → B = A
<< A5201b.r0t.txt
%0
$=(\wedge\,(=F\,T)\,F)\,(=F\,T)$
$=_{ooo}(\wedge_{ooo}(=_{ooo}F_oT_o)F_o)(=_{ooo}F_oT_o)$
§s %4 5 %0
$=(=F\,T)\,F$

:= $A5230c$ %0
wff 1576 : $=(=F\,T)\,F_o$:= $A5230c$

.d: (F = F) = T

use Proof Template A5210: T = (B = B)
:= $\$T5210$ o
wff 2 : o_τ := $\$T5210$
:= $\$B5210$ $=_{o(oo)(oo)}[\lambda x_o.T_o][\lambda x_o.x_o]$
wff 20 : $=[\lambda x.T]\,[\lambda x.x]_{o,\ldots}$:= $\$B5210$ F
<< A5210.r0t.txt
:= $\$T5210$
:= $\$B5210$
%0
$=T\,(=F\,F)$
$=_{ooo}T_o(=_{ooo}F_oF_o)$

use Proof Template A5201b (Swap): A = B → B = A
<< A5201b.r0t.txt
%0
$=(=F\,F)\,T$:= $\$BTMP5230$
$=_{ooo}(=_{ooo}F_oF_o)T_o$:= $\$BTMP5230$

:= $A5230d$ %0
wff 1414 : $=(=F\,F)\,T_o$:= $\$BTMP5230$ $A5230d$

undefine local variables

:= $\$ATMP5230$
:= $\$BTMP5230$
:= $\$CTMP5230$
:= $\$DTMP5230$

##
Q.E.D.
##

%A5230a
%$A5230a$
$=(=T\,T)\,T$:= $A5230a$
$=_{ooo}(=_{ooo}T_oT_o)T_o$:= $A5230a$

%A5230b
%$A5217$
$=(=T\,F)\,F$:= $A5217$ $A5230b$
$=_{ooo}(=_{ooo}T_oF_o)F_o$:= $A5217$ $A5230b$

%A5230c
%$A5230c$
$=(=F\,T)\,F$:= $A5230c$
$=_{ooo}(=_{ooo}F_oT_o)F_o$:= $A5230c$

%A5230d
%$A5230d$
$=(=F\,F)\,T$:= $A5230d$
$=_{ooo}(=_{ooo}F_oF_o)T_o$:= $A5230d$

2.1.38 Results for File A5231.r0.txt

##
Proof A5231: $\sim$ T = F; $\sim$ F = T
##
##
Source: [Andrews 2002 (ISBN 1-4020-0763-9), p. 225]
##

Written by Ken Kubota (<mail@kenkubota.de>).
##
This file is part of the publication of the mathematical logic $\mathcal{R}_0$.
For more information, visit: <http://doi.org/10.4444/100.10>
##

<< A5230.r0.txt

##

Proof
##

.a: $\sim$ T = F

§\ $_o$ $\sim_{oo}T_o$
$= (\sim T)\,(= F\,T)$
%$A5230c$
$= (= F\,T)\,F$:= $A5230c$
$=_{ooo}(=_{ooo}F_oT_o)F_o$:= $A5230c$
§s %1 3 %0
$= (\sim T)\,F$

:= $A5231a$ %0
wff 1580 : $= (\sim T)\,F_o$:= $A5231a$

.b: $\sim$ F = T

§\ $_o$ $\sim_{oo}F_o$
$= (\sim F)\,(= F\,F)$
%$A5230d$
$= (= F\,F)\,T$:= $A5230d$
$=_{ooo}(=_{ooo}F_oF_o)T_o$:= $A5230d$
§s %1 3 %0
$= (\sim F)\,T$

:= $A5231b$ %0
wff 1584 : $= (\sim F)\,T_o$:= $A5231b$

##
Q.E.D.
##

%A5231a
%$A5231a$
$= (\sim T)\,F$:= $A5231a$
$=_{ooo}(\sim_{oo}T_o)F_o$:= $A5231a$

%A5231b
%$A5231b$
$= (\sim F)\,T$:= $A5231b$
$=_{ooo}(\sim_{oo}F_o)T_o$:= $A5231b$

2.1.39 Results for File A5232.r0.txt

##
Proof A5232: T ∨ T = T; T ∨ F = T; F ∨ T = T; F ∨ F = F
##
##
Source: [Andrews 2002 (ISBN 1-4020-0763-9), p. 225]
##

Written by Ken Kubota (<mail@kenkubota.de>).
##
This file is part of the publication of the mathematical logic $\mathcal{R}_0$.
For more information, visit: <http://doi.org/10.4444/100.10>
##

<< A5231.r0.txt

##
Proof
##

.a: T ∨ T = T

§= $_o \vee_{ooo} T_o T_o$
$= (\vee\, T\, T)\, (\vee\, T\, T)$

§\ $\vee_{ooo} T_o$
$= (\vee\, T)\, [\lambda b.(\sim (\wedge\, (\sim T)\, (\sim b)))]$
§s %1 6 %0
$= (\vee\, T\, T)\, ([\lambda b.(\sim (\wedge\, (\sim T)\, (\sim b)))]\, T)$
§\ $[\lambda b_o.(\sim_{oo}(\wedge_{ooo}(\sim_{oo}T_o)(\sim_{oo}b_o)))_o]T_o$
$= ([\lambda b.(\sim (\wedge\, (\sim T)\, (\sim b)))]\, T)\, (\sim (\wedge\, (\sim T)\, (\sim T)))$
§s %1 3 %0
$= (\vee\, T\, T)\, (\sim (\wedge\, (\sim T)\, (\sim T)))$

%*A5231a*
$= (\sim T)\, F$:= *A5231a*
$=_{ooo}(\sim_{oo}T_o)F_o$:= *A5231a*
§s %1 29 %0
$= (\vee\, T\, T)\, (\sim (\wedge\, F\, (\sim T)))$

%*A5231a*
$= (\sim T)\, F$:= *A5231a*
$=_{ooo}(\sim_{oo}T_o)F_o$:= *A5231a*
§s %1 15 %0
$= (\vee\, T\, T)\, (\sim (\wedge\, F\, F))$

%*A5229d*

\# $=(\wedge F\,F)\,F$:= $A5229d$
\# $=_{ooo}(\wedge_{ooo}F_oF_o)F_o$:= $A5229d$
§s %1 7 %0
\# $=(\vee T\,T)\,(\sim F)$

%$A5231b$
\# $=(\sim F)\,T$:= $A5231b$
\# $=_{ooo}(\sim_{oo}F_o)T_o$:= $A5231b$
§s %1 3 %0
\# $=(\vee T\,T)\,T$

:= $A5232a$ %0
\# wff 1608 : $=(\vee T\,T)\,T_o$:= $A5232a$

\## .b: T ∨ F = T

§= $_o$ $\vee_{ooo}T_oF_o$
\# $=(\vee T\,F)\,(\vee T\,F)$

§\ $\vee_{ooo}T_o$
\# $=(\vee T)\,[\lambda b.(\sim(\wedge(\sim T)\,(\sim b)))]$
§s %1 6 %0
\# $=(\vee T\,F)\,([\lambda b.(\sim(\wedge(\sim T)\,(\sim b)))]\,F)$
§\ $[\lambda b_o.(\sim_{oo}(\wedge_{ooo}(\sim_{oo}T_o)(\sim_{oo}b_o)))_o]F_o$
\# $=([\lambda b.(\sim(\wedge(\sim T)\,(\sim b)))]\,F)\,(\sim(\wedge(\sim T)\,(\sim F)))$
§s %1 3 %0
\# $=(\vee T\,F)\,(\sim(\wedge(\sim T)\,(\sim F)))$

%$A5231a$
\# $=(\sim T)\,F$:= $A5231a$
\# $=_{ooo}(\sim_{oo}T_o)F_o$:= $A5231a$
§s %1 29 %0
\# $=(\vee T\,F)\,(\sim(\wedge F\,(\sim F)))$

%$A5231b$
\# $=(\sim F)\,T$:= $A5231b$
\# $=_{ooo}(\sim_{oo}F_o)T_o$:= $A5231b$
§s %1 15 %0
\# $=(\vee T\,F)\,(\sim(\wedge F\,T))$

%$A5229c$
\# $=(\wedge F\,T)\,F$:= $A5229c$
\# $=_{ooo}(\wedge_{ooo}F_oT_o)F_o$:= $A5229c$
§s %1 7 %0
\# $=(\vee T\,F)\,(\sim F)$

%$A5231b$
\# $=(\sim F)\,T$:= $A5231b$
\# $=_{ooo}(\sim_{oo}F_o)T_o$:= $A5231b$

§s %1 3 %0
$= (\vee T\, F)\, T$

:= $A5232b$ %0
wff 1625 : $= (\vee T\, F)\, T_o$:= $A5232b$

.c: F ∨ T = T

§= $_o$ $\vee_{ooo} F_o T_o$
$= (\vee F\, T)\, (\vee F\, T)$

§\ $\vee_{ooo} F_o$
$= (\vee F)\, [\lambda b.(\sim (\wedge (\sim F)\, (\sim b)))]$
§s %1 6 %0
$= (\vee F\, T)\, ([\lambda b.(\sim (\wedge (\sim F)\, (\sim b)))]\, T)$
§\ $[\lambda b_o.(\sim_{oo}(\wedge_{ooo}(\sim_{oo}F_o)(\sim_{oo}b_o)))_o]T_o$
$= ([\lambda b.(\sim (\wedge (\sim F)\, (\sim b)))]\, T)\, (\sim (\wedge (\sim F)\, (\sim T)))$
§s %1 3 %0
$= (\vee F\, T)\, (\sim (\wedge (\sim F)\, (\sim T)))$

%$A5231b$
$= (\sim F)\, T$:= $A5231b$
$=_{ooo}(\sim_{oo}F_o)T_o$:= $A5231b$
§s %1 29 %0
$= (\vee F\, T)\, (\sim (\wedge T\, (\sim T)))$

%$A5231a$
$= (\sim T)\, F$:= $A5231a$
$=_{ooo}(\sim_{oo}T_o)F_o$:= $A5231a$
§s %1 15 %0
$= (\vee F\, T)\, (\sim (\wedge T\, F))$

%$A5214$
$= (\wedge T\, F)\, F$:= $A5214$ $A5229b$
$=_{ooo}(\wedge_{ooo}T_oF_o)F_o$:= $A5214$ $A5229b$
§s %1 7 %0
$= (\vee F\, T)\, (\sim F)$

%$A5231b$
$= (\sim F)\, T$:= $A5231b$
$=_{ooo}(\sim_{oo}F_o)T_o$:= $A5231b$
§s %1 3 %0
$= (\vee F\, T)\, T$

:= $A5232c$ %0
wff 1649 : $= (\vee F\, T)\, T_o$:= $A5232c$

.d: F ∨ F = F

§= $_o$ $\vee_{ooo} F_o F_o$
$= (\vee F F) (\vee F F)$

§\ $\vee_{ooo} F_o$
$= (\vee F) [\lambda b.(\sim (\wedge (\sim F) (\sim b)))]$
§s %1 6 %0
$= (\vee F F) ([\lambda b.(\sim (\wedge (\sim F) (\sim b)))] F)$
§\ $[\lambda b_o.(\sim_{oo}(\wedge_{ooo}(\sim_{oo}F_o)(\sim_{oo}b_o)))_o]F_o$
$= ([\lambda b.(\sim (\wedge (\sim F) (\sim b)))] F) (\sim (\wedge (\sim F) (\sim F)))$
§s %1 3 %0
$= (\vee F F) (\sim (\wedge (\sim F) (\sim F)))$

%$A5231b$
$= (\sim F) T$:= $A5231b$
$=_{ooo}(\sim_{oo}F_o)T_o$:= $A5231b$
§s %1 29 %0
$= (\vee F F) (\sim (\wedge T (\sim F)))$

%$A5231b$
$= (\sim F) T$:= $A5231b$
$=_{ooo}(\sim_{oo}F_o)T_o$:= $A5231b$
§s %1 15 %0
$= (\vee F F) (\sim A5212)$

%$A5211$
$= A5212\,T$:= $A5211$ $A5229a$
$=_{ooo}A5212_o T_o$:= $A5211$ $A5229a$
§s %1 7 %0
$= (\vee F F) (\sim T)$

%$A5231a$
$= (\sim T) F$:= $A5231a$
$=_{ooo}(\sim_{oo}T_o)F_o$:= $A5231a$
§s %1 3 %0
$= (\vee F F) F$

:= $A5232d$ %0
wff 1666 : $= (\vee F F) F_o$:= $A5232d$

##
Q.E.D.
##

%A5232a
%$A5232a$
$= (\vee T T) T$:= $A5232a$
$=_{ooo}(\vee_{ooo}T_o T_o)T_o$:= $A5232a$

%A5232b
$\%A5232b$
\# $=(\vee T\,F)\,T$:= $A5232b$
\# $=_{ooo}(\vee_{ooo}T_oF_o)T_o$:= $A5232b$

%A5232c
$\%A5232c$
\# $=(\vee F\,T)\,T$:= $A5232c$
\# $=_{ooo}(\vee_{ooo}F_oT_o)T_o$:= $A5232c$

%A5232d
$\%A5232d$
\# $=(\vee F\,F)\,F$:= $A5232d$
\# $=_{ooo}(\vee_{ooo}F_oF_o)F_o$:= $A5232d$

2.1.40 Results for File A5245.r0a.txt

```
##
##  Proof Template A5245 (Rule C):  H ⊃ ∃ x: B, (H ∧ (B [x/y])) ⊃ A  →  H ⊃ A
##      for any x, y of any type, provided y is not free in H, ∃ x: B or A
##
##  Source: [Andrews 2002 (ISBN 1-4020-0763-9), p. 230 (5245)]
##
##  Copyright (c) 2017 Owl of Minerva Press GmbH. All rights reserved.
##  Written by Ken Kubota (<mail@kenkubota.de>).
##
##  This file is part of the publication of the mathematical logic R0.
##  For more information, visit: <http://doi.org/10.4444/100.10>
##

<< basics.r0.txt


##
##  Define Syntactical Variables
##

## type of variable
```

:= $\$T5245\ t_\tau$
\# wff 4 : t_τ := $\$T5245$

name of variable in assumption 1
:= $\$X5245\ x_{\$T5245}$
\# wff 24 : $x_{\$T5245}$:= $\$X5245$

name of variable in assumption 2
:= $\$Y5245\ y_{\$T5245}$
\# wff 105 : $y_{\$T5245}$:= $\$Y5245$

assumption 1: H ⊃ ∃ x: B
:= $\$B5245\ \supset_{ooo} h_o(\exists_{o(o\backslash 3)\tau}\$T5245_\tau[\lambda\$X5245_{\$T5245}.(b_{o\,\$T5245}\$X5245_{\$T5245})_o])$
wff 214 : $\supset h\,(\exists\,\$T5245\,[\lambda\$X5245.(b\,\$X5245)])_o$:= $\$B5245$

assumption 2: (H ∧ (B [x/y])) ⊃ A
:= $\$A5245\ \supset_{ooo}(\wedge_{ooo}h_o(b_{o\,\$T5245}\$Y5245_{\$T5245}))a_o$
wff 219 : $\supset(\wedge\,h\,(b\,\$Y5245))\,a_o$:= $\$A5245$

##
Assumptions and Resulting Syntactical Variables
##

§! $\$B5245$
$\supset h\,(\exists\,\$T5245\,[\lambda\$X5245.(b\,\$X5245)])$:= $\$B5245$
§! $\$A5245$
$\supset(\wedge\,h\,(b\,\$Y5245))\,a$:= $\$A5245$

##
Include Proof Template
##

<<< A5245.r0t.txt
Include begin (A5245.r0t.txt) [oldfile=(A5245.r0a.txt)]
##
Proof Template A5245 (Rule C): H ⊃ ∃ x: B, (H ∧ (B [x/y])) ⊃ A → H ⊃ A
for any x, y of any type, provided y is not free in H, ∃ x: B or A
##
Source: [Andrews 2002 (ISBN 1-4020-0763-9), p. 230 (5245)]
##

Written by Ken Kubota (<mail@kenkubota.de>).
##
This file is part of the publication of the mathematical logic $\mathcal{R}_0$.
For more information, visit: <http://doi.org/10.4444/100.10>
##

##
Proof Template
##

.1

%$\$A5245$
$\supset(\wedge\,h\,(b\,\$Y5245))\,a$:= $\$A5245$
$\supset_{ooo}(\wedge_{ooo}h_o(b_{o\,\$T5245}\$Y5245_{\$T5245}))a_o$:= $\$A5245$

.2

use Proof Template K8030 (∃ Rule): (H ∧ B) ⊃ A → (H ∧ ∃ x: B) ⊃ A
:= $\$T8030\ t_\tau$
wff 4 : t_τ := $\$T5245$ $\$T8030$
:= $\$X8030\ y_{\$T5245}$
wff 105 : $y_{\$T5245}$:= $\$X8030$ $\$Y5245$
:= $\$A8030$ %0
wff 219 : $\supset (\wedge\, h\, (b\, \$Y5245))\, a_o$:= $\$A5245$ $\$A8030$
<< K8030.r0t.txt
:= $\$T8030$
:= $\$X8030$
:= $\$A8030$

%0
$\supset (\wedge\, h\, (\exists\, \$T5245\, [\lambda \$Y5245.(b\, \$Y5245)]))\, a$
$\supset_{ooo}(\wedge_{ooo} h_o(\exists_{o(o\backslash 3)\tau} \$T5245_\tau [\lambda \$Y5245_{\$T5245}.(b_{o\,\$T5245} \$Y5245_{\$T5245})_o]))a_o$

.3

use Proof Template K8025 (Deduction Theorem): (H ∧ I) ⊃ A → H ⊃ (I ⊃ A)
<< K8025.r0t.txt
%0
$\supset h\, (\supset (\exists\, \$T5245\, [\lambda \$Y5245.(b\, \$Y5245)])\, a)$
$\supset_{ooo} h_o(\supset_{ooo}(\exists_{o(o\backslash 3)\tau} \$T5245_\tau [\lambda \$Y5245_{\$T5245}.(b_{o\,\$T5245} \$Y5245_{\$T5245})_o])a_o)$

.4

§r /27 $\$X5245$
$= [\lambda \$Y5245.(b\, \$Y5245)]\, [\lambda \$X5245.(b\, \$X5245)]$
§s %1 27 %0
$\supset h\, (\supset (\exists\, \$T5245\, [\lambda \$X5245.(b\, \$X5245)])\, a)$

.5

%$\$B5245$
$\supset h\, (\exists\, \$T5245\, [\lambda \$X5245.(b\, \$X5245)])$:= $\$B5245$
$\supset_{ooo} h_o(\exists_{o(o\backslash 3)\tau} \$T5245_\tau [\lambda \$X5245_{\$T5245}.(b_{o\,\$T5245} \$X5245_{\$T5245})_o])$:= $\$B5245$

.6

use Proof Template A5224H (MP): H ⊃ A, H ⊃ (A ⊃ B) → H ⊃ B
:= $\$A5224H$ %0
wff 214 : $\supset h\, (\exists\, \$T5245\, [\lambda \$X5245.(b\, \$X5245)])_o$:= $\$A5224H$ $\$B5245$
:= $\$AB5224H$ %1
wff 5135 : $\supset h\, (\supset (\exists\, \$T5245\, [\lambda \$X5245.(b\, \$X5245)])\, a)_o$:= $\$AB5224H$
<< A5224H.r0t.txt
:= $\$AB5224H$
:= $\$A5224H$

%0
$\supset h\ a$
$\supset_{ooo} h_o a_o$
Include end (A5245.r0t.txt) [newfile=(A5245.r0a.txt)]
>>>

##
Undefine Syntactical Variables
##

:= $\$T5245$
:= $\$X5245$
:= $\$Y5245$
:= $\$B5245$
:= $\$A5245$

##
Q.E.D.
##

%0
$\supset h\ a$
$\supset_{ooo} h_o a_o$

2.1.41 Results for File A5304.r0.txt

##
Proof A5304: $\exists_1$ y: P y = $\exists$ y: P = (= y)
##
##
Source: [Andrews 2002 (ISBN 1-4020-0763-9), p. 233]
##
Copyright (c) 2017 Owl of Minerva Press GmbH. All rights reserved.
Written by Ken Kubota (<mail@kenkubota.de>).
##
This file is part of the publication of the mathematical logic $\mathcal{R}_0$.
For more information, visit: <http://doi.org/10.4444/100.10>
##

<< basics.r0.txt
<< A5205.r0.txt

##
Proof
##

.1

§= $_o$ $\exists_{1o(o\backslash 3)\tau} t_\tau [\lambda y_t.(p_{ot} y_t)_o]$
\# $= (\exists_1\, t\, [\lambda y.(p\, y)])\; (\exists_1\, t\, [\lambda y.(p\, y)])$
§\ $\exists_{1o(o\backslash 3)\tau} t_\tau$
\# $= (\exists_1\, t)\; [\lambda p.(\exists\, t\, [\lambda y.(= p\, (= y))])]$
§s %1 6 %0
\# $= (\exists_1\, t\, [\lambda y.(p\, y)])\; ([\lambda p.(\exists\, t\, [\lambda y.(= p\, (= y))])]\; [\lambda y.(p\, y)])$
§\ $[\lambda p_{ot}.(\exists_{o(o\backslash 3)\tau} t_\tau [\lambda y_t.(=_{o(ot)(ot)} p_{ot} (=_{ott} y_t))_o])_o][\lambda y_t.(p_{ot} y_t)_o]$
\# $= ([\lambda p.(\exists\, t\, [\lambda y.(= p\, (= y))])]\; [\lambda y.(p\, y)])\; (\exists\, t\, [\lambda y.(= [\lambda y.(p\, y)]\; (= y))])$
§s %1 3 %0
\# $= (\exists_1\, t\, [\lambda y.(p\, y)])\; (\exists\, t\, [\lambda y.(= [\lambda y.(p\, y)]\; (= y))])$
:= $\$TMP5304$ %0
\# wff 887 : $= (\exists_1\, t\, [\lambda y.(p\, y)])\; (\exists\, t\, [\lambda y.(= [\lambda y.(p\, y)]\; (= y))])_o$:= $\$TMP5304$

.2

use Proof Template: A5205 Substitutions
:= $\$AA5205$ o
\# wff 2 : o_τ := $\$AA5205$
:= $\$BA5205$ t_τ
\# wff 4 : t_τ := $\$BA5205$
:= $\$FA5205$ $p_{o\$BA5205}$
\# wff 21 : $p_{o\$BA5205}$:= $\$FA5205$
<< a5205_substitutions.r0t.txt
:= $\$AA5205$
:= $\$BA5205$
:= $\$FA5205$
%0
\# $= p\, [\lambda y.(p\, y)]$
\# $=_{o(ot)(ot)} p_{ot} [\lambda y_t.(p_{ot} y_t)_o]$

use Proof Template A5201b (Swap): A = B → B = A
<< A5201b.r0t.txt
%0
\# $= [\lambda y.(p\, y)]\, p$
\# $=_{o(ot)(ot)} [\lambda y_t.(p_{ot} y_t)_o] p_{ot}$

%$\$TMP5304$
\# $= (\exists_1\, t\, [\lambda y.(p\, y)])\; (\exists\, t\, [\lambda y.(= [\lambda y.(p\, y)]\; (= y))])$:= $\$TMP5304$
\# $=_{ooo} (\exists_{1o(o\backslash 3)\tau} t_\tau [\lambda y_t.(p_{ot} y_t)_o]) (\exists_{o(o\backslash 3)\tau} t_\tau [\lambda y_t.(=_{o(ot)(ot)} [\lambda y_t.(p_{ot} y_t)_o] (=_{ott} y_t))_o])$
:= $\$TMP5304$
:= $\$TMP5304$
§s %0 61 %1
\# $= (\exists_1\, t\, [\lambda y.(p\, y)])\; (\exists\, t\, [\lambda y.(= p\, (= y))])$

:= $A5304$ %0
\# wff 1049 : $= (\exists_1\, t\, [\lambda y.(p\, y)])\; (\exists\, t\, [\lambda y.(= p\, (= y))])_o$:= $A5304$

```
##
## Q.E.D.
##
```

%0

$= (\exists_1 t\, [\lambda y.(p\, y)])\, (\exists t\, [\lambda y.(= p\, (= y))])$:= $A5304$

$=_{ooo}(\exists_{1o(o\backslash 3)\tau} t_\tau [\lambda y_t.(p_{ot} y_t)_o])(\exists_{o(o\backslash 3)\tau} t_\tau [\lambda y_t.(=_{o(ot)(ot)} p_{ot} (=_{ott} y_t))_o])$:= $A5304$

2.1.42 Results for File A5305.r0.txt

```
##
## Proof A5305: ∃1 y: P y = ∃ y: ∀ z: P z = (y = z)
##
##
## Source: [Andrews 2002 (ISBN 1-4020-0763-9), p. 233]
##
## Copyright (c) 2017 Owl of Minerva Press GmbH. All rights reserved.
## Written by Ken Kubota (<mail@kenkubota.de>).
##
## This file is part of the publication of the mathematical logic R0.
## For more information, visit: <http://doi.org/10.4444/100.10>
##
```

```
<< basics.r0.txt
<< A5304.r0.txt
```

```
##
## Proof
##
```

```
## .1
```

use Proof Template: Axiom 3 Substitutions

:= $\$AA3$ o

wff 2 : o_τ := $\$AA3$

:= $\$BA3$ t_τ

wff 4 : t_τ := $\$BA3$

:= $\$FA3$ $p_{o\$BA3}$

wff 21 : $p_{o\$BA3}$:= $\$FA3$

:= $\$GA3$ $=_{o\$BA3\$BA3} y_{\$BA3}$

wff 107 : $= y_{o\$BA3}$:= $\$GA3$

<< axiom3_substitutions.r0t.txt

:= $\$AA3$

:= $\$BA3$

:= $\$FA3$

:= $\$GA3$

%0

```
#                  = (= p (= y)) (∀ t [λx.(= (p x) (= y x))])
```
$\#\quad =_{ooo}(=_{o(ot)(ot)}p_{ot}(=_{ott}y_t))(\forall_{o(o\backslash 3)\tau}t_\tau[\lambda x_t.(=_{ooo}(p_{ot}x_t)(=_{ott}y_tx_t))_o])$

\## .2

%A5304

$\#\quad = (\exists_1\, t\,[\lambda y.(p\, y)])\,(\exists\, t\,[\lambda y.(= p\,(= y))])\quad := \quad A5304$

$\#\quad =_{ooo}(\exists_{1o(o\backslash 3)\tau}t_\tau[\lambda y_t.(p_{ot}y_t)_o])(\exists_{o(o\backslash 3)\tau}t_\tau[\lambda y_t.(=_{o(ot)(ot)}p_{ot}(=_{ott}y_t))_o])\quad := \quad A5304$

§s %0 15 %1

$\#\quad = (\exists_1\, t\,[\lambda y.(p\, y)])\,(\exists\, t\,[\lambda y.(\forall\, t\,[\lambda x.(= (p\, x)\,(= y\, x))])])$

§r /31 z_t

$\#\quad = [\lambda x.(= (p\, x)\,(= y\, x))]\,[\lambda z.(= (p\, z)\,(= y\, z))]$

§s %1 31 %0

$\#\quad = (\exists_1\, t\,[\lambda y.(p\, y)])\,(\exists\, t\,[\lambda y.(\forall\, t\,[\lambda z.(= (p\, z)\,(= y\, z))])])$

:= $A5305$ %0

\# wff 1120 : $= (\exists_1\, t\,[\lambda y.(p\, y)])\,(\exists\, t\,[\lambda y.(\forall\, t\,[\lambda z.(= (p\, z)\,(= y\, z))])])_o \quad := \quad A5305$

\##
\## Q.E.D.
\##

%0

$\#\quad = (\exists_1\, t\,[\lambda y.(p\, y)])\,(\exists\, t\,[\lambda y.(\forall\, t\,[\lambda z.(= (p\, z)\,(= y\, z))])])\quad := \quad A5305$

$\#\quad =_{ooo}(\exists_{1o(o\backslash 3)\tau}t_\tau[\lambda y_t.(p_{ot}y_t)_o])\ldots$

$\ldots(\exists_{o(o\backslash 3)\tau}t_\tau[\lambda y_t.(\forall_{o(o\backslash 3)\tau}t_\tau[\lambda z_t.(=_{ooo}(p_{ot}z_t)(=_{ott}y_tz_t))_o])_o])\quad := \quad A5305$

2.1.43 Results for File A5310.r0.txt

\##
\## Proof A5310: (∀ z: P z = (y = z)) ⊃ (ι P = y)
\##
\##
\## Source: [Andrews 2002 (ISBN 1-4020-0763-9), p. 235]
\##

\## Written by Ken Kubota (<mail@kenkubota.de>).
\##
\## This file is part of the publication of the mathematical logic $\mathcal{R}_0$.
\## For more information, visit: <http://doi.org/10.4444/100.10>
\##

<< basics.r0.txt
<< K8005.r0.txt

\##
\## Proof

##

.1

%$K8005$
$\supset x\, x$:= $K8005$
$\supset_{ooo} x_o x_o$:= $K8005$

use Proof Template A5221 (Sub): B → B [x/A]
:= $\$B5221$ %0
wff 1357 : $\supset x\, x_{o,\ldots}$:= $\$B5221$ $K8005$
:= $\$T5221$ o
wff 2 : o_τ := $\$T5221$
:= $\$X5221$ x_o
wff 16 : x_o := $\$X5221$
:= $\$A5221$ $\forall_{o(o\backslash 3)\tau} t_\tau [\lambda z_t.(=_{ooo}(p_{ot} z_t)(=_{ott} y_t z_t))_o]$
wff 1376 : $\forall t\, [\lambda z.(=(p\, z)\, (=y\, z))]_o$:= $\$A5221$
<< A5221.r0t.txt
:= $\$B5221$
:= $\$T5221$
:= $\$X5221$
:= $\$A5221$

:= $\$TMP5310$ %0
wff 1413 : $\supset (\forall t\, [\lambda z.(=(p\, z)\, (=y\, z))])\, (\forall t\, [\lambda z.(=(p\, z)\, (=y\, z))])_{o,\ldots}$:= $\$TMP5310$

.2

use Proof Template: Axiom 3 Substitutions
:= $\$AA3$ o
wff 2 : o_τ := $\$AA3$
:= $\$BA3$ t_τ
wff 4 : t_τ := $\$BA3$
:= $\$FA3$ $p_{o\$BA3}$
wff 21 : $p_{o\$BA3}$:= $\$FA3$
:= $\$GA3$ $=_{o\$BA3\$BA3} y_{\$BA3}$
wff 107 : $=y_{o\$BA3}$:= $\$GA3$
<< axiom3_substitutions.r0t.txt
:= $\$AA3$
:= $\$BA3$
:= $\$FA3$
:= $\$GA3$
%0
$=(=p\,(=y))\,(\forall t\, [\lambda x.(=(p\, x)\, (=y\, x))])$
$=_{ooo}(=_{o(ot)(ot)} p_{ot}(=_{ott} y_t))(\forall_{o(o\backslash 3)\tau} t_\tau [\lambda x_t.(=_{ooo}(p_{ot} x_t)(=_{ott} y_t x_t))_o])$

§r /7 z_t
$=[\lambda x.(=(p\, x)\, (=y\, x))]\, [\lambda z.(=(p\, z)\, (=y\, z))]$
§s %1 7 %0

\# $= (= p\,(= y))\,(\forall t\,[\lambda z.(= (p\,z)\,(= y\,z))])$

\#\# use Proof Template A5201b (Swap): A = B → B = A
<< A5201b.r0t.txt
%0
\# $= (\forall t\,[\lambda z.(= (p\,z)\,(= y\,z))])\,(= p\,(= y))$
\# $=_{ooo}(\forall_{o(o\backslash 3)\tau} t_\tau [\lambda z_t.(=_{ooo}(p_{ot} z_t)(=_{ott} y_t z_t))_o])(=_{o(ot)(ot)} p_{ot}(=_{ott} y_t))$

%$TMP5310
\# $\supset (\forall t\,[\lambda z.(= (p\,z)\,(= y\,z))])\,(\forall t\,[\lambda z.(= (p\,z)\,(= y\,z))])$:= $TMP5310
\# $\supset_{ooo}(\forall_{o(o\backslash 3)\tau} t_\tau [\lambda z_t.(=_{ooo}(p_{ot} z_t)(=_{ott} y_t z_t))_o])\ldots$
$\ldots(\forall_{o(o\backslash 3)\tau} t_\tau [\lambda z_t.(=_{ooo}(p_{ot} z_t)(=_{ott} y_t z_t))_o])$:= $TMP5310
:= $TMP5310
§s %0 3 %1
\# $\supset (\forall t\,[\lambda z.(= (p\,z)\,(= y\,z))])\,(= p\,(= y))$
:= $TMP5310 %0
\# wff 1481 : $\supset (\forall t\,[\lambda z.(= (p\,z)\,(= y\,z))])\,(= p\,(= y))_o$:= $TMP5310

\#\# .3

§= t_τ $\iota_{t(ot)} p_{ot}$
\# $= (\iota\,p)\,(\iota\,p)$

\#\# use Proof Template K8004 (Trans): (H ⊕ A), B → H ⊃ B
:= $HA8004 %1
\# wff 1481 : $\supset (\forall t\,[\lambda z.(= (p\,z)\,(= y\,z))])\,(= p\,(= y))_o$:= $HA8004 $TMP5310
:= $B8004 %0
\# wff 1484 : $= (\iota\,p)\,(\iota\,p)_o$:= $B8004
<< K8004.r0t.txt
:= $HA8004
:= $B8004
%0
\# $\supset (\forall t\,[\lambda z.(= (p\,z)\,(= y\,z))])\,(= (\iota\,p)\,(\iota\,p))$
\# $\supset_{ooo}(\forall_{o(o\backslash 3)\tau} t_\tau [\lambda z_t.(=_{ooo}(p_{ot} z_t)(=_{ott} y_t z_t))_o])(=_{ott}(\iota_{t(ot)} p_{ot})(\iota_{t(ot)} p_{ot}))$

%$TMP5310
\# $\supset (\forall t\,[\lambda z.(= (p\,z)\,(= y\,z))])\,(= p\,(= y))$:= $TMP5310
\# $\supset_{ooo}(\forall_{o(o\backslash 3)\tau} t_\tau [\lambda z_t.(=_{ooo}(p_{ot} z_t)(=_{ott} y_t z_t))_o])(=_{o(ot)(ot)} p_{ot}(=_{ott} y_t))$:=
$TMP5310
:= $TMP5310
§s' %1 7 %0
\# $\supset (\forall t\,[\lambda z.(= (p\,z)\,(= y\,z))])\,(= (\iota\,p)\,(\iota\,(= y)))$
%$A5$
\# $= (\iota\,(= y))\,y$:= $A5$
\# $=_{ott}(\iota_{t(ot)}(=_{ott} y_t))y_t$:= $A5$
§s %1 7 %0
\# $\supset (\forall t\,[\lambda z.(= (p\,z)\,(= y\,z))])\,(= (\iota\,p)\,y)$

##
Q.E.D.
##

%0
$\supset (\forall t\, [\lambda z.(=(p\, z)\, (=y\, z))])\, (=(\iota\, p)\, y)$
$\supset_{ooo}(\forall_{o(o\backslash 3)\tau} t_\tau [\lambda z_t.(=_{ooo}(p_{ot} z_t)(=_{ott} y_t z_t))_o])(=_{ott}(\iota_{t(ot)} p_{ot}) y_t)$

2.1.44 Results for File A5311.r0.txt

##
Proof A5311: ($\exists_1$ y: P y) ⊃ (P (ι P))
##
##
Source: [Andrews 2002 (ISBN 1-4020-0763-9), p. 235]
##

Written by Ken Kubota (<mail@kenkubota.de>).
##
This file is part of the publication of the mathematical logic $\mathcal{R}_0$.
For more information, visit: <http://doi.org/10.4444/100.10>
##

<< basics.r0.txt
<< A5304.r0.txt
<< K8000.r0.txt
<< K8005.r0.txt

##
Proof
##

.1

%$K8005$
$\supset x\, x$:= $K8005$
$\supset_{ooo} x_o x_o$:= $K8005$

use Proof Template A5221 (Sub): B → B [x/A]
:= $\$B5221$ %0
wff 1732 : $\supset x\, x_{o,\ldots}$:= $\$B5221$ $K8005$
:= $\$T5221$ o
wff 2 : o_τ := $\$T5221$
:= $\$X5221$ x_o
wff 16 : x_o := $\$X5221$
:= $\$A5221$ $=_{o(ot)(ot)} p_{ot}(=_{ott} y_t)$
wff 108 : $=p\,(=y)_o$:= $\$A5221$

<< A5221.r0t.txt
:= $\$B5221$
:= $\$T5221$
:= $\$X5221$
:= $\$A5221$

:= $\$TMP5311$ %0
wff 1782 : $\supset (=p\,(=y))\,(=p\,(=y))_{o,\ldots}$:= $\$TMP5311$
:= $\$LTMP5311$ %0
wff 1782 : $\supset (=p\,(=y))\,(=p\,(=y))_{o,\ldots}$:= $\$LTMP5311$ $\$TMP5311$

.2

§= $_o$ $p_{ot}y_t$
$=(p\,y)\,(p\,y)$

use Proof Template K8004 (Trans): (H ⊕ A), B → H ⊃ B
:= $\$HA8004$ %1
wff 1782 : $\supset (=p\,(=y))\,(=p\,(=y))_{o,\ldots}$:= $\$HA8004$ $\$LTMP5311$ $\$TMP5311$
:= $\$B8004$ %0
wff 1786 : $=(p\,y)\,(p\,y)_o$:= $\$B8004$
<< K8004.r0t.txt
:= $\$HA8004$
:= $\$B8004$
%0
$\supset (=p\,(=y))\,(=(p\,y)\,(p\,y))$
$\supset_{ooo}(=_{o(ot)(ot)}p_{ot}(=_{ott}y_t))(=_{ooo}(p_{ot}y_t)(p_{ot}y_t))$

%$\$TMP5311$
$\supset (=p\,(=y))\,(=p\,(=y))$:= $\$LTMP5311$ $\$TMP5311$
$\supset_{ooo}(=_{o(ot)(ot)}p_{ot}(=_{ott}y_t))(=_{o(ot)(ot)}p_{ot}(=_{ott}y_t))$:= $\$LTMP5311$ $\$TMP5311$
:= $\$TMP5311$
§s' %1 6 %0
$\supset (=p\,(=y))\,(=(p\,y)\,(=y\,y))$

.3

use Proof Template A5201bH (SwapH): H ⊃ (A = B) → H ⊃ (B = A)
<< A5201bH.r0t.txt
%0
$\supset (=p\,(=y))\,(=(=y\,y)\,(p\,y))$
$\supset_{ooo}(=_{o(ot)(ot)}p_{ot}(=_{ott}y_t))(=_{ooo}(=_{ott}y_ty_t)(p_{ot}y_t))$

:= $\$TMP5311$ %0
wff 1884 : $\supset (=p\,(=y))\,(=(=y\,y)\,(p\,y))_o$:= $\$TMP5311$
§= t_τ y_t
$=y\,y$

use Proof Template K8004 (Trans): (H ⊕ A), B → H ⊃ B

:= $\$HA8004$ %1
wff 1884 : $\supset (=p\,(=y))\,(=(=y\,y)\,(p\,y))_o$:= $\$HA8004$ $\$TMP5311$
:= $\$B8004$ %0
wff 1879 : $=y\,y_o$:= $\$B8004$
<< K8004.r0t.txt
:= $\$HA8004$
:= $\$B8004$
%0
$\supset (=p\,(=y))\,(=y\,y)$
$\supset_{ooo}(=_{o(ot)(ot)}p_{ot}(=_{ott}y_t))(=_{ott}y_ty_t)$

%$\$TMP5311$
$\supset (=p\,(=y))\,(=(=y\,y)\,(p\,y))$:= $\$TMP5311$
$\supset_{ooo}(=_{o(ot)(ot)}p_{ot}(=_{ott}y_t))(=_{ooo}(=_{ott}y_ty_t)(p_{ot}y_t))$:= $\$TMP5311$
:= $\$TMP5311$
§s' %1 1 %0
$\supset (=p\,(=y))\,(p\,y)$
:= $\$TMP5311$ %0
wff 1918 : $\supset (=p\,(=y))\,(p\,y)_o$:= $\$TMP5311$

.4

%$A5$
$=(\iota\,(=y))\,y$:= $A5$
$=_{ott}(\iota_{t(ot)}(=_{ott}y_t))y_t$:= $A5$

use Proof Template A5201b (Swap): A = B $\rightarrow$ B = A
<< A5201b.r0t.txt
%0
$=y\,(\iota\,(=y))$
$=_{ott}y_t(\iota_{t(ot)}(=_{ott}y_t))$

%$\$TMP5311$
$\supset (=p\,(=y))\,(p\,y)$:= $\$TMP5311$
$\supset_{ooo}(=_{o(ot)(ot)}p_{ot}(=_{ott}y_t))(p_{ot}y_t)$:= $\$TMP5311$
:= $\$TMP5311$
§s %0 7 %1
$\supset (=p\,(=y))\,(p\,(\iota\,(=y)))$
:= $\$TMP5311$ %0
wff 1922 : $\supset (=p\,(=y))\,(p\,(\iota\,(=y)))_o$:= $\$TMP5311$

.5

%$\$LTMP5311$
$\supset (=p\,(=y))\,(=p\,(=y))$:= $\$LTMP5311$
$\supset_{ooo}(=_{o(ot)(ot)}p_{ot}(=_{ott}y_t))(=_{o(ot)(ot)}p_{ot}(=_{ott}y_t))$:= $\$LTMP5311$
:= $\$LTMP5311$

use Proof Template A5201bH (SwapH): H $\supset$ (A = B) $\rightarrow$ H $\supset$ (B = A)

<< A5201bH.r0t.txt
%0
$\supset (= p\,(= y))\,(= (= y)\,p)$
$\supset_{ooo}(=_{o(ot)(ot)}p_{ot}(=_{ott}y_t))(=_{o(ot)(ot)}(=_{ott}y_t)p_{ot})$

%$TMP5311
$\supset (= p\,(= y))\,(p\,(\iota\,(= y)))$:= $TMP5311
$\supset_{ooo}(=_{o(ot)(ot)}p_{ot}(=_{ott}y_t))(p_{ot}(\iota_{t(ot)}(=_{ott}y_t)))$:= $TMP5311
:= $TMP5311
§s′ %0 7 %1
$\supset (= p\,(= y))\,(p\,(\iota\,p))$
:= $TMP5311 %0
wff 1962 : $\supset (= p\,(= y))\,(p\,(\iota\,p))_o$:= $TMP5311

.6

%$K8000b$
$= (\wedge\,T\,x)\,x$:= $K8000b$
$=_{ooo}(\wedge_{ooo}T_o x_o)x_o$:= $K8000b$

use Proof Template A5221 (Sub): B → B [x/A]
:= $B5221 %0
wff 594 : $= (\wedge\,T\,x)\,x_{o,\ldots}$:= $B5221 $K8000b$
:= $T5221 o
wff 2 : o_τ := $T5221
:= $X5221 x_o
wff 16 : x_o := $X5221
:= $A5221 %1/5
wff 108 : $= p\,(= y)_{o,\ldots}$:= $A5221
<< A5221.r0t.txt
:= $B5221
:= $T5221
:= $X5221
:= $A5221
%0
$= (\wedge\,T\,(= p\,(= y)))\,(= p\,(= y))$
$=_{ooo}(\wedge_{ooo}T_o(=_{o(ot)(ot)}p_{ot}(=_{ott}y_t)))(=_{o(ot)(ot)}p_{ot}(=_{ott}y_t))$

use Proof Template A5201b (Swap): A = B → B = A
<< A5201b.r0t.txt
%0
$= (= p\,(= y))\,(\wedge\,T\,(= p\,(= y)))$
$=_{ooo}(=_{o(ot)(ot)}p_{ot}(=_{ott}y_t))(\wedge_{ooo}T_o(=_{o(ot)(ot)}p_{ot}(=_{ott}y_t)))$

%$TMP5311
$\supset (= p\,(= y))\,(p\,(\iota\,p))$:= $TMP5311
$\supset_{ooo}(=_{o(ot)(ot)}p_{ot}(=_{ott}y_t))(p_{ot}(\iota_{t(ot)}p_{ot}))$:= $TMP5311
:= $TMP5311
§s %0 5 %1

\# $\supset(\wedge T\,(=p\,(=y)))\,(p\,(\iota\,p))$

\## use Proof Template K8030 (∃ Rule): (H ∧ B) ⊃ A → (H ∧ ∃ x: B) ⊃ A
:= $\$T8030\ t_\tau$
\# wff 4 : t_τ := $\$T8030$
:= $\$X8030\ y_{\$T8030}$
\# wff 105 : $y_{\$T8030}$:= $\$X8030$
:= $\$A8030$ %0
\# wff 1996 : $\supset(\wedge T\,(=p\,(=\$X8030)))\,(p\,(\iota\,p))_o$:= $\$A8030$
<< K8030.r0t.txt
:= $\$T8030$
:= $\$X8030$
:= $\$A8030$

:= $\$TMP5311$ %0
\# wff 5448 : $\supset(\wedge T\,(\exists\, t\,[\lambda y.(=p\,(=y))]))\,(p\,(\iota\,p))_{o,\ldots}$:= $\$TMP5311$

%$K8000b$
\# $=(\wedge T\,x)\,x$:= $K8000b$
\# $=_{ooo}(\wedge_{ooo}T_o x_o)x_o$:= $K8000b$

\## use Proof Template A5221 (Sub): B → B [x/A]
:= $\$B5221$ %0
\# wff 594 : $=(\wedge T\,x)\,x_{o,\ldots}$:= $\$B5221$ $K8000b$
:= $\$T5221\ o$
\# wff 2 : o_τ := $\$T5221$
:= $\$X5221\ x_o$
\# wff 16 : x_o := $\$X5221$
:= $\$A5221$ %1/11
\# wff 110 : $\exists\, t\,[\lambda y.(=p\,(=y))]_{o,\ldots}$:= $\$A5221$
<< A5221.r0t.txt
:= $\$B5221$
:= $\$T5221$
:= $\$X5221$
:= $\$A5221$
%0
\# $=(\wedge T\,(\exists\, t\,[\lambda y.(=p\,(=y))]))\,(\exists\, t\,[\lambda y.(=p\,(=y))])$
\# $=_{ooo}(\wedge_{ooo}T_o(\exists_{o(o\backslash 3)\tau}t_\tau[\lambda y_t.(=_{o(ot)(ot)}p_{ot}(=_{ott}y_t))_o]))\ldots$
$\ldots(\exists_{o(o\backslash 3)\tau}t_\tau[\lambda y_t.(=_{o(ot)(ot)}p_{ot}(=_{ott}y_t))_o])$

%$\$TMP5311$
\# $\supset(\wedge T\,(\exists\, t\,[\lambda y.(=p\,(=y))]))\,(p\,(\iota\,p))$:= $\$TMP5311$
\# $\supset_{ooo}(\wedge_{ooo}T_o(\exists_{o(o\backslash 3)\tau}t_\tau[\lambda y_t.(=_{o(ot)(ot)}p_{ot}(=_{ott}y_t))_o]))(p_{ot}(\iota_{t(ot)}p_{ot}))$:=
$\$TMP5311$
:= $\$TMP5311$
§s %0 5 %1
\# $\supset(\exists\, t\,[\lambda y.(=p\,(=y))])\,(p\,(\iota\,p))$
:= $\$LTMP5311$ %0

wff 5395 : $\supset(\exists\, t\, [\lambda y.(=p\,(=y))])\,(p\,(\iota\, p))_{o,\ldots}$:= $\$LTMP5311$

.7

%$A5304$
$=(\exists_1\, t\, [\lambda y.(p\, y)])\,(\exists\, t\, [\lambda y.(=p\,(=y))])$:= $A5304$
$=_{ooo}(\exists_{1o(o\backslash 3)\tau}t_\tau[\lambda y_t.(p_{ot}y_t)_o])(\exists_{o(o\backslash 3)\tau}t_\tau[\lambda y_t.(=_{o(ot)(ot)}p_{ot}(=_{ott}y_t))_o])$:= $A5304$

use Proof Template A5201b (Swap): A = B → B = A
<< A5201b.r0t.txt
%0
$=(\exists\, t\, [\lambda y.(=p\,(=y))])\,(\exists_1\, t\, [\lambda y.(p\, y)])$
$=_{ooo}(\exists_{o(o\backslash 3)\tau}t_\tau[\lambda y_t.(=_{o(ot)(ot)}p_{ot}(=_{ott}y_t))_o])(\exists_{1o(o\backslash 3)\tau}t_\tau[\lambda y_t.(p_{ot}y_t)_o])$

%$\$LTMP5311$
$\supset(\exists\, t\, [\lambda y.(=p\,(=y))])\,(p\,(\iota\, p))$:= $\$LTMP5311$
$\supset_{ooo}(\exists_{o(o\backslash 3)\tau}t_\tau[\lambda y_t.(=_{o(ot)(ot)}p_{ot}(=_{ott}y_t))_o])(p_{ot}(\iota_{t(ot)}p_{ot}))$:= $\$LTMP5311$
:= $\$LTMP5311$
§s %0 5 %1
$\supset(\exists_1\, t\, [\lambda y.(p\, y)])\,(p\,(\iota\, p))$

:= $A5311$ %0
wff 5467 : $\supset(\exists_1\, t\, [\lambda y.(p\, y)])\,(p\,(\iota\, p))_o$:= $A5311$

##
Q.E.D.
##

%0
$\supset(\exists_1\, t\, [\lambda y.(p\, y)])\,(p\,(\iota\, p))$:= $A5311$
$\supset_{ooo}(\exists_{1o(o\backslash 3)\tau}t_\tau[\lambda y_t.(p_{ot}y_t)_o])(p_{ot}(\iota_{t(ot)}p_{ot}))$:= $A5311$

2.1.45 Results for File A5312.r0.txt

##
Proof A5312: $\exists_1$ y: P y ⊃ ∀ z: P z = (ι P = z)
##
##
Source: [Andrews 2002 (ISBN 1-4020-0763-9), p. 235]
##

Written by Ken Kubota (<mail@kenkubota.de>).
##
This file is part of the publication of the mathematical logic $\mathcal{R}_0$.
For more information, visit: <http://doi.org/10.4444/100.10>
##

<< basics.r0.txt

<< K8005.r0.txt
<< A5304.r0.txt

##
Proof
##

.1

:= $\$HYP5312\ =_{o(ot)(ot)}p_{ot}(=_{ott}y_t)$
wff 108 : $=p\,(=y)_o$:= $\$HYP5312$

$\%K8005$
$\supset x\,x$:= $K8005$
$\supset_{ooo}x_o x_o$:= $K8005$

use Proof Template A5221 (Sub): B → B [x/A]
:= $\$B5221$ %0
wff 1357 : $\supset x\,x_{o,\ldots}$:= $\$B5221$ $K8005$
:= $\$T5221\ o$
wff 2 : o_τ := $\$T5221$
:= $\$X5221\ x_o$
wff 16 : x_o := $\$X5221$
:= $\$A5221\ =_{o(ot)(ot)}p_{ot}(=_{ott}y_t)$
wff 108 : $=p\,(=y)_o$:= $\$A5221$ $\$HYP5312$
<< A5221.r0t.txt
:= $\$B5221$
:= $\$T5221$
:= $\$X5221$
:= $\$A5221$
%0
$\supset\$HYP5312\,\$HYP5312$
$\supset_{ooo}\$HYP5312_o\$HYP5312_o$

:= $\$ATMP5312$ %0
wff 1651 : $\supset\$HYP5312\,\$HYP5312_{o,\ldots}$:= $\$ATMP5312$

.2

$\%A5$
$=(\iota\,(=y))\,y$:= $A5$
$=_{ott}(\iota_{t(ot)}(=_{ott}y_t))y_t$:= $A5$

use Proof Template K8003 (Intro): A → H ⊃ A
:= $\$A8003$ %0
wff 207 : $=(\iota\,(=y))\,y_o$:= $\$A8003$ $A5$
:= $\$H8003\ \supset_{ooo}\$HYP5312_o\$HYP5312_o/5$
wff 108 : $=p\,(=y)_o$:= $\$H8003$ $\$HYP5312$

<< K8003.r0t.txt
:= $\$A8003$
:= $\$H8003$
%0
$\supset \$HYP5312\, A5$
$\supset_{ooo} \$HYP5312_o A5_o$

:= $\$LTMP5312$ %0
wff 1785 : $\supset \$HYP5312\, A5_{o,\ldots}$:= $\$LTMP5312$

%$\$ATMP5312$
$\supset \$HYP5312\, \$HYP5312$:= $\$ATMP5312$
$\supset_{ooo} \$HYP5312_o \$HYP5312_o$:= $\$ATMP5312$

use Proof Template A5201bH (SwapH): H ⊃ (A = B) → H ⊃ (B = A)
<< A5201bH.r0t.txt
%0
$\supset \$HYP5312\, (= (= y)\, p)$
$\supset_{ooo} \$HYP5312_o (=_{o(ot)(ot)} (=_{ott} y_t) p_{ot})$

%$\$LTMP5312$
$\supset \$HYP5312\, A5$:= $\$LTMP5312$
$\supset_{ooo} \$HYP5312_o A5_o$:= $\$LTMP5312$
:= $\$LTMP5312$
§s' %0 11 %1
$\supset \$HYP5312\, (= (\iota\, p)\, y)$

:= $\$BTMP5312$ %0
wff 1917 : $\supset \$HYP5312\, (= (\iota\, p)\, y)_o$:= $\$BTMP5312$

.3

use Proof Template A5201bH (SwapH): H ⊃ (A = B) → H ⊃ (B = A)
<< A5201bH.r0t.txt
%0
$\supset \$HYP5312\, (= y\, (\iota\, p))$
$\supset_{ooo} \$HYP5312_o (=_{ott} y_t (\iota_{t(ot)} p_{ot}))$

%$\$ATMP5312$
$\supset \$HYP5312\, \$HYP5312$:= $\$ATMP5312$
$\supset_{ooo} \$HYP5312_o \$HYP5312_o$:= $\$ATMP5312$
§s' %0 7 %1
$\supset \$HYP5312\, (= p\, (= (\iota\, p)))$

:= $\$CTMP5312$ %0
wff 1956 : $\supset \$HYP5312\, (= p\, (= (\iota\, p)))_o$:= $\$CTMP5312$

.4

use Proof Template: Axiom 3 Substitutions
:= $\$AA3\ o$
wff 2 : o_τ := $\$AA3$
:= $\$BA3\ t_\tau$
wff 4 : t_τ := $\$BA3$
:= $\$FA3\ \supset_{ooo}\$HYP5312_o(=_{o(o\$BA3)(o\$BA3)}p_{o\$BA3}(=_{o\$BA3\$BA3}(\iota_{\$BA3(o\$BA3)}p_{o\$BA3})))/13$
wff 21 : $p_{o\$BA3}$:= $\$FA3$
:= $\$GA3\ \supset_{ooo}\$HYP5312_o(=_{o(o\$BA3)(o\$BA3)}\$FA3_{o\$BA3}(=_{o\$BA3\$BA3}(\iota_{\$BA3(o\$BA3)}\$FA3_{o\$BA3})))/7$
wff 1915 : $=(\iota\,\$FA3)_{o\$BA3}$:= $\$GA3$
<< axiom3_substitutions.r0t.txt
:= $\$AA3$
:= $\$BA3$
:= $\$FA3$
:= $\$GA3$
%0
$=(=p\,(=(\iota\,p)))\,(\forall\,t\,[\lambda x.(=(p\,x)\,(=(\iota\,p)\,x))])$
$=_{ooo}(=_{o(ot)(ot)}p_{ot}(=_{ott}(\iota_{t(ot)}p_{ot})))(\forall_{o(o\backslash 3)\tau}t_\tau[\lambda x_t.(=_{ooo}(p_{ot}x_t)(=_{ott}(\iota_{t(ot)}p_{ot})x_t))_o])$

use Proof Template K8003 (Intro): A → H ⊃ A
:= $\$A8003$ %0
wff 2013 : $=(=p\,(=(\iota\,p)))\,(\forall\,t\,[\lambda x.(=(p\,x)\,(=(\iota\,p)\,x))])_o$:= $\$A8003$
:= $\$H8003\ \supset_{ooo}\$HYP5312_o\$HYP5312_o/5$
wff 108 : $=p\,(=y)_{o,\dots}$:= $\$H8003$ $\$HYP5312$
<< K8003.r0t.txt
:= $\$A8003$
:= $\$H8003$
%0
$\supset\$HYP5312\,(=(=p\,(=(\iota\,p)))\,(\forall\,t\,[\lambda x.(=(p\,x)\,(=(\iota\,p)\,x))]))$
$\supset_{ooo}\$HYP5312_o\dots$
$\dots(=_{ooo}(=_{o(ot)(ot)}p_{ot}(=_{ott}(\iota_{t(ot)}p_{ot})))(\forall_{o(o\backslash 3)\tau}t_\tau[\lambda x_t.(=_{ooo}(p_{ot}x_t)(=_{ott}(\iota_{t(ot)}p_{ot})x_t))_o]))$

%$\$CTMP5312$
$\supset\$HYP5312\,(=p\,(=(\iota\,p)))$:= $\$CTMP5312$
$\supset_{ooo}\$HYP5312_o(=_{o(ot)(ot)}p_{ot}(=_{ott}(\iota_{t(ot)}p_{ot})))$:= $\$CTMP5312$
§s' %0 1 %1
$\supset\$HYP5312\,(\forall\,t\,[\lambda x.(=(p\,x)\,(=(\iota\,p)\,x))])$

§r /7 z_t
$=[\lambda x.(=(p\,x)\,(=(\iota\,p)\,x))]\,[\lambda z.(=(p\,z)\,(=(\iota\,p)\,z))]$
§s %1 7 %0
$\supset\$HYP5312\,(\forall\,t\,[\lambda z.(=(p\,z)\,(=(\iota\,p)\,z))])$

:= $\$DTMP5312$ %0
wff 2057 : $\supset\$HYP5312\,(\forall\,t\,[\lambda z.(=(p\,z)\,(=(\iota\,p)\,z))])_o$:= $\$DTMP5312$

.5

use Proof Template A5216: (T ∧ A) = A
:= $\$A5216\ =_{o(ot)(ot)}p_{ot}(=_{ott}y_t)$

wff 108 : $=p\,(=y)_{o,\dots}$:= $\$A5216$ $\$HYP5312$
<< A5216.r0t.txt
:= $\$A5216$
%0
$=(\wedge\,T\,\$HYP5312)\,\$HYP5312$
$=_{ooo}(\wedge_{ooo}T_o\$HYP5312_o)\$HYP5312_o$

§= $\wedge_{ooo}T_o\$HYP5312_o$
$=(\wedge\,T\,\$HYP5312)\,(\wedge\,T\,\$HYP5312)$
§s %0 5 %1
$=\$HYP5312\,(\wedge\,T\,\$HYP5312)$

%$\$DTMP5312$
$\supset \$HYP5312\,(\forall\,t\,[\lambda z.(=(p\,z)\,(=(\iota\,p)\,z))])$:= $\$DTMP5312$
$\supset_{ooo}\$HYP5312_o(\forall_{o(o\backslash 3)\tau}t_\tau[\lambda z_t.(=_{ooo}(p_{ot}z_t)(=_{ott}(\iota_{t(ot)}p_{ot})z_t))_o])$:=
$\$DTMP5312$
§s %0 5 %1
$\supset(\wedge\,T\,\$HYP5312)\,(\forall\,t\,[\lambda z.(=(p\,z)\,(=(\iota\,p)\,z))])$

use Proof Template K8030 (∃ Rule): (H ∧ B) ⊃ A → (H ∧ ∃ x: B) ⊃ A
:= $\$T8030$ t_τ
wff 4 : t_τ := $\$T8030$
:= $\$X8030$ $y_{\$T8030}$
wff 105 : $y_{\$T8030}$:= $\$X8030$
:= $\$A8030$ %0
wff 2072 : $\supset(\wedge\,T\,\$HYP5312)\,(\forall\,\$T8030\,[\lambda z.(=(p\,z)\,(=(\iota\,p)\,z))])_o$:= $\$A8030$
<< K8030.r0t.txt
:= $\$T8030$
:= $\$X8030$
:= $\$A8030$

%0
$\supset(\wedge\,T\,(\exists\,t\,[\lambda y.\$HYP5312]))\,(\forall\,t\,[\lambda z.(=(p\,z)\,(=(\iota\,p)\,z))])$
$\supset_{ooo}(\wedge_{ooo}T_o(\exists_{o(o\backslash 3)\tau}t_\tau[\lambda y_t.\$HYP5312_o]))\dots$
$\dots(\forall_{o(o\backslash 3)\tau}t_\tau[\lambda z_t.(=_{ooo}(p_{ot}z_t)(=_{ott}(\iota_{t(ot)}p_{ot})z_t))_o])$

:= $\$LTMP5312$ %0
wff 5524 : $\supset(\wedge\,T\,(\exists\,t\,[\lambda y.\$HYP5312]))\,(\forall\,t\,[\lambda z.(=(p\,z)\,(=(\iota\,p)\,z))])_{o,\dots}$:=
$\$LTMP5312$

use Proof Template A5216: (T ∧ A) = A
:= $\$A5216$ %0/11
wff 110 : $\exists\,t\,[\lambda y.\$HYP5312]_{o,\dots}$:= $\$A5216$
<< A5216.r0t.txt
:= $\$A5216$
%0
$=(\wedge\,T\,(\exists\,t\,[\lambda y.\$HYP5312]))\,(\exists\,t\,[\lambda y.\$HYP5312])$
$=_{ooo}(\wedge_{ooo}T_o(\exists_{o(o\backslash 3)\tau}t_\tau[\lambda y_t.\$HYP5312_o]))(\exists_{o(o\backslash 3)\tau}t_\tau[\lambda y_t.\$HYP5312_o])$

%$\$LTMP5312$
\# $\supset (\wedge\, T\, (\exists\, t\, [\lambda y.\$HYP5312])) \, (\forall\, t\, [\lambda z.(= (p\, z)\, (= (\iota\, p)\, z))])$:= $\$LTMP5312$
\# $\supset_{ooo}(\wedge_{ooo}T_o(\exists_{o(o\backslash 3)\tau}t_\tau[\lambda y_t.\$HYP5312_o]))\ldots$
$\ldots(\forall_{o(o\backslash 3)\tau}t_\tau[\lambda z_t.(=_{ooo}(p_{ot}z_t)(=_{ott}(\iota_{t(ot)}p_{ot})z_t))_o])$:= $\$LTMP5312$
:= $\$LTMP5312$
§s %0 5 %1
\# $\supset (\exists\, t\, [\lambda y.\$HYP5312]) \, (\forall\, t\, [\lambda z.(= (p\, z)\, (= (\iota\, p)\, z))])$

\## .6

%$A5304$
\# $= (\exists_1\, t\, [\lambda y.(p\, y)]) \, (\exists\, t\, [\lambda y.\$HYP5312])$:= $A5304$
\# $=_{ooo}(\exists_{1o(o\backslash 3)\tau}t_\tau[\lambda y_t.(p_{ot}y_t)_o])(\exists_{o(o\backslash 3)\tau}t_\tau[\lambda y_t.\$HYP5312_o])$:= $A5304$

§= $\exists_{1o(o\backslash 3)\tau}t_\tau[\lambda y_t.(p_{ot}y_t)_o]$
\# $= (\exists_1\, t\, [\lambda y.(p\, y)]) \, (\exists_1\, t\, [\lambda y.(p\, y)])$
§s %0 5 %1
\# $= (\exists\, t\, [\lambda y.\$HYP5312]) \, (\exists_1\, t\, [\lambda y.(p\, y)])$

§s %3 5 %0
\# $\supset (\exists_1\, t\, [\lambda y.(p\, y)]) \, (\forall\, t\, [\lambda z.(= (p\, z)\, (= (\iota\, p)\, z))])$

:= $A5312$ %0
\# wff 5545 : $\supset (\exists_1\, t\, [\lambda y.(p\, y)]) \, (\forall\, t\, [\lambda z.(= (p\, z)\, (= (\iota\, p)\, z))])_o$:= $A5312$

\## undefine local variables
:= $\$HYP5312$
:= $\$ATMP5312$
:= $\$BTMP5312$
:= $\$CTMP5312$
:= $\$DTMP5312$

\##
\## Q.E.D.
\##

%0
\# $\supset (\exists_1\, t\, [\lambda y.(p\, y)]) \, (\forall\, t\, [\lambda z.(= (p\, z)\, (= (\iota\, p)\, z))])$:= $A5312$
\# $\supset_{ooo}(\exists_{1o(o\backslash 3)\tau}t_\tau[\lambda y_t.(p_{ot}y_t)_o])(\forall_{o(o\backslash 3)\tau}t_\tau[\lambda z_t.(=_{ooo}(p_{ot}z_t)(=_{ott}(\iota_{t(ot)}p_{ot})z_t))_o])$
:= $A5312$

2.1.46 Results for File A5313.r0.txt

\##
\## Proof A5313: (C_t_x_y_T = x) ∧ (C_t_x_y_F = y)
\##
\##
\## Source: [Andrews 2002 (ISBN 1-4020-0763-9), pp. 235 f.]

This file is part of the publication of the mathematical logic $\mathcal{R}_0$.
For more information, visit: <http://doi.org/10.4444/100.10>
##

##
"C[t]xyp can be read 'if p then x, else y'." [Andrews 2002, p. 235]
##

<< basics.r0.txt
<< A5205.r0.txt
<< A5231.r0.txt
<< K8000.r0.txt
<< K8001.r0.txt
<< K8010.r0.txt

:= $COND$ ...
... $[\lambda t_\tau.[\lambda x_t.[\lambda y_t.[\lambda p_o.(\iota_{t(ot)}[\lambda q_t.(\vee_{ooo}(\wedge_{ooo}p_o(=_{ott}x_t q_t))(\wedge_{ooo}(\sim_{oo}p_o)(=_{ott}y_t q_t)))_o])_t]_{(to)}]_{(tot)}]_{(tott)}]$
wff 2080 : $[\lambda t.[\lambda x.[\lambda y.[\lambda p.(\iota\,[\lambda q.(\vee\,(\wedge\,p\,(=x\,q))\,(\wedge\,(\sim p)\,(=y\,q)))])]]]]_{\backslash 4o\backslash 3\backslash 2\tau}$:=
$COND$

##
Proof
##

.1

§= t_τ $COND_{\backslash 4o\backslash 3\backslash 2\tau} t_\tau x_t y_t T_o$
$=(COND\,t\,x\,y\,T)\,(COND\,t\,x\,y\,T)$
§\ $COND_{\backslash 4o\backslash 3\backslash 2\tau} t_\tau$
$=(COND\,t)\,[\lambda x.[\lambda y.[\lambda p.(\iota\,[\lambda q.(\vee\,(\wedge\,p\,(=x\,q))\,(\wedge\,(\sim p)\,(=y\,q)))])]]]$
§s %1 24 %0
$=(COND\,t\,x\,y\,T)\,([\lambda x.[\lambda y.[\lambda p.(\iota\,[\lambda q.(\vee\,(\wedge\,p\,(=x\,q))\,(\wedge\,(\sim p)\,(=y\,q)))])]]]\,x\,y\,T)$
§\ $[\lambda x_t.[\lambda y_t.[\lambda p_o.(\iota_{t(ot)}[\lambda q_t.(\vee_{ooo}(\wedge_{ooo}p_o(=_{ott}x_t q_t))(\wedge_{ooo}(\sim_{oo}p_o)(=_{ott}y_t q_t)))_o])_t]_{(to)}]_{(tot)}]x_t$
$=([\lambda x.[\lambda y.[\lambda p.(\iota\,[\lambda q.(\vee\,(\wedge\,p\,(=x\,q))\,(\wedge\,(\sim p)\,(=y\,q)))])]]]\,x)$ ...
... $[\lambda y.[\lambda p.(\iota\,[\lambda q.(\vee\,(\wedge\,p\,(=x\,q))\,(\wedge\,(\sim p)\,(=y\,q)))])]]$
§s %1 12 %0
$=(COND\,t\,x\,y\,T)\,([\lambda y.[\lambda p.(\iota\,[\lambda q.(\vee\,(\wedge\,p\,(=x\,q))\,(\wedge\,(\sim p)\,(=y\,q)))])]]\,y\,T)$
§\ $[\lambda y_t.[\lambda p_o.(\iota_{t(ot)}[\lambda q_t.(\vee_{ooo}(\wedge_{ooo}p_o(=_{ott}x_t q_t))(\wedge_{ooo}(\sim_{oo}p_o)(=_{ott}y_t q_t)))_o])_t]_{(to)}]y_t$
$=([\lambda y.[\lambda p.(\iota\,[\lambda q.(\vee\,(\wedge\,p\,(=x\,q))\,(\wedge\,(\sim p)\,(=y\,q)))])]]\,y)$ ...
... $[\lambda p.(\iota\,[\lambda q.(\vee\,(\wedge\,p\,(=x\,q))\,(\wedge\,(\sim p)\,(=y\,q)))])]$
§s %1 6 %0

$=(COND\,t\,x\,y\,T)\,([\lambda p.(\iota\,[\lambda q.(\vee\,(\wedge\,p\,(=x\,q))\,(\wedge\,(\sim p)\,(=y\,q)))])]\,T)$
§\ $[\lambda p_o.(\iota_{t(ot)}[\lambda q_t.(\vee_{ooo}(\wedge_{ooo}p_o(=_{ott}x_tq_t))(\wedge_{ooo}(\sim_{oo}p_o)(=_{ott}y_tq_t)))_o])_t]T_o$
$=([\lambda p.(\iota\,[\lambda q.(\vee\,(\wedge\,p\,(=x\,q))\,(\wedge\,(\sim p)\,(=y\,q)))])]\,T)\ \ldots$
$\ldots(\iota\,[\lambda q.(\vee\,(\wedge\,T\,(=x\,q))\,(\wedge\,(\sim T)\,(=y\,q)))])$
§s %1 3 %0
$=(COND\,t\,x\,y\,T)\,(\iota\,[\lambda q.(\vee\,(\wedge\,T\,(=x\,q))\,(\wedge\,(\sim T)\,(=y\,q)))])$
:= $\$LTMP5313$ %0
wff 2115 : $=(COND\,t\,x\,y\,T)\,(\iota\,[\lambda q.(\vee\,(\wedge\,T\,(=x\,q))\,(\wedge\,(\sim T)\,(=y\,q)))])_o$:= $\$LTMP5313$

.2

§= $_o$ /15
$=(\vee\,(\wedge\,T\,(=x\,q))\,(\wedge\,(\sim T)\,(=y\,q)))\,(\vee\,(\wedge\,T\,(=x\,q))\,(\wedge\,(\sim T)\,(=y\,q)))$

%$A5231a$
$=(\sim T)\,F$:= $A5231a$
$=_{ooo}(\sim_{oo}T_o)F_o$:= $A5231a$
§s %1 29 %0
$=(\vee\,(\wedge\,T\,(=x\,q))\,(\wedge\,(\sim T)\,(=y\,q)))\,(\vee\,(\wedge\,T\,(=x\,q))\,(\wedge\,F\,(=y\,q)))$
:= $\$TMP5313$ %0
wff 2120 : $=(\vee\,(\wedge\,T\,(=x\,q))\,(\wedge\,(\sim T)\,(=y\,q)))\,(\vee\,(\wedge\,T\,(=x\,q))\,(\wedge\,F\,(=y\,q)))_o$:= $\$TMP5313$

%$K8001b$
$=(\wedge\,F\,x)\,F$:= $K8001b$
$=_{ooo}(\wedge_{ooo}F_ox_o)F_o$:= $K8001b$

use Proof Template A5221 (Sub): B → B [x/A]
:= $\$B5221$ %0
wff 1805 : $=(\wedge\,F\,x)\,F_{o,\ldots}$:= $\$B5221$ $K8001b$
:= $\$T5221$ o
wff 2 : o_τ := $\$T5221$
:= $\$X5221$ x_o
wff 16 : x_o := $\$X5221$
:= $\$A5221$ %1/15
wff 2069 : $=y\,q_o$:= $\$A5221$
<< A5221.r0t.txt
:= $\$B5221$
:= $\$T5221$
:= $\$X5221$
:= $\$A5221$
%0
$=(\wedge\,F\,(=y\,q))\,F$
$=_{ooo}(\wedge_{ooo}F_o(=_{ott}y_tq_t))F_o$

%$\$TMP5313$
$=(\vee\,(\wedge\,T\,(=x\,q))\,(\wedge\,(\sim T)\,(=y\,q)))\,(\vee\,(\wedge\,T\,(=x\,q))\,(\wedge\,F\,(=y\,q)))$:= $\$TMP5313$

\# $=_{ooo}(\vee_{ooo}(\wedge_{ooo}T_o(=_{ott}x_tq_t))(\wedge_{ooo}(\sim_{oo}T_o)(=_{ott}y_tq_t)))\ldots$
$\ldots(\vee_{ooo}(\wedge_{ooo}T_o(=_{ott}x_tq_t))(\wedge_{ooo}F_o(=_{ott}y_tq_t)))$:= $\$TMP5313$
:= $\$TMP5313$
§s %0 7 %1
\# $=(\vee\,(\wedge\,T\,(=x\,q))\,(\wedge\,(\sim T)\,(=y\,q)))\,(\vee\,(\wedge\,T\,(=x\,q))\,F)$
:= $\$TMP5313$ %0
\# wff 2155 : $=(\vee\,(\wedge\,T\,(=x\,q))\,(\wedge\,(\sim T)\,(=y\,q)))\,(\vee\,(\wedge\,T\,(=x\,q))\,F)_o$:= $\$TMP5313$

%$K8000b$
\# $=(\wedge\,T\,x)\,x$:= $K8000b$
\# $=_{ooo}(\wedge_{ooo}T_ox_o)x_o$:= $K8000b$

\#\# use Proof Template A5221 (Sub): B → B [x/A]
:= $\$B5221$ %0
\# wff 594 : $=(\wedge\,T\,x)\,x_{o,\ldots}$:= $\$B5221$ $K8000b$
:= $\$T5221$ o
\# wff 2 : o_τ := $\$T5221$
:= $\$X5221$ x_o
\# wff 16 : x_o := $\$X5221$
:= $\$A5221$ %1/43
\# wff 2064 : $=x\,q_o$:= $\$A5221$
<< A5221.r0t.txt
:= $\$B5221$
:= $\$T5221$
:= $\$X5221$
:= $\$A5221$
%0
\# $=(\wedge\,T\,(=x\,q))\,(=x\,q)$
\# $=_{ooo}(\wedge_{ooo}T_o(=_{ott}x_tq_t))(=_{ott}x_tq_t)$

%$\$TMP5313$
\# $=(\vee\,(\wedge\,T\,(=x\,q))\,(\wedge\,(\sim T)\,(=y\,q)))\,(\vee\,(\wedge\,T\,(=x\,q))\,F)$:= $\$TMP5313$
\# $=_{ooo}(\vee_{ooo}(\wedge_{ooo}T_o(=_{ott}x_tq_t))(\wedge_{ooo}(\sim_{oo}T_o)(=_{ott}y_tq_t)))(\vee_{ooo}(\wedge_{ooo}T_o(=_{ott}x_tq_t))F_o)$
:= $\$TMP5313$
:= $\$TMP5313$
§s %0 13 %1
\# $=(\vee\,(\wedge\,T\,(=x\,q))\,(\wedge\,(\sim T)\,(=y\,q)))\,(\vee\,(=x\,q)\,F)$
:= $\$TMP5313$ %0
\# wff 2191 : $=(\vee\,(\wedge\,T\,(=x\,q))\,(\wedge\,(\sim T)\,(=y\,q)))\,(\vee\,(=x\,q)\,F)_o$:= $\$TMP5313$

%$K8010a$
\# $=(\vee\,x\,F)\,x$:= $K8010a$
\# $=_{ooo}(\vee_{ooo}x_oF_o)x_o$:= $K8010a$

\#\# use Proof Template A5221 (Sub): B → B [x/A]
:= $\$B5221$ %0
\# wff 2028 : $=(\vee\,x\,F)\,x_o$:= $\$B5221$ $K8010a$
:= $\$T5221$ o
\# wff 2 : o_τ := $\$T5221$

:= $\$X5221\ x_o$
wff 16 : x_o := $\$X5221$
:= $\$A5221$ %1/13
wff 2064 : $= x\, q_{o,\dots}$:= $\$A5221$
<< A5221.r0t.txt
:= $\$B5221$
:= $\$T5221$
:= $\$X5221$
:= $\$A5221$
%0
$= (\vee\, (= x\, q)\, F)\, (= x\, q)$
$=_{ooo}(\vee_{ooo}(=_{ott}x_t q_t)F_o)(=_{ott}x_t q_t)$

%$\$TMP5313$
$= (\vee\, (\wedge\, T\, (= x\, q))\, (\wedge\, (\sim T)\, (= y\, q)))\, (\vee\, (= x\, q)\, F)$:= $\$TMP5313$
$=_{ooo}(\vee_{ooo}(\wedge_{ooo}T_o(=_{ott}x_t q_t))(\wedge_{ooo}(\sim_{oo}T_o)(=_{ott}y_t q_t)))(\vee_{ooo}(=_{ott}x_t q_t)F_o)$:= $\$TMP5313$
:= $\$TMP5313$
§s %0 3 %1
$= (\vee\, (\wedge\, T\, (= x\, q))\, (\wedge\, (\sim T)\, (= y\, q)))\, (= x\, q)$

.3

%$\$LTMP5313$
$= (COND\, t\, x\, y\, T)\, (\iota\, [\lambda q.(\vee\, (\wedge\, T\, (= x\, q))\, (\wedge\, (\sim T)\, (= y\, q)))])$:= $\$LTMP5313$
$=_{ott}(COND_{\backslash 4o\backslash 3\backslash 2\tau}t_\tau x_t y_t T_o)\dots$
$\dots(\iota_{t(ot)}[\lambda q_t.(\vee_{ooo}(\wedge_{ooo}T_o(=_{ott}x_t q_t))(\wedge_{ooo}(\sim_{oo}T_o)(=_{ott}y_t q_t)))_o])$:= $\$LTMP5313$
:= $\$LTMP5313$
§s %0 15 %1
$= (COND\, t\, x\, y\, T)\, (\iota\, [\lambda q.(= x\, q)])$
:= $\$TMP5313$ %0
wff 2230 : $= (COND\, t\, x\, y\, T)\, (\iota\, [\lambda q.(= x\, q)])_o$:= $\$TMP5313$

.4

use Proof Template: A5205 Substitutions
:= $\$AA5205\ o$
wff 2 : o_τ := $\$AA5205$
:= $\$BA5205\ t_\tau$
wff 4 : t_τ := $\$BA5205$
:= $\$FA5205$ $=_{o\$BA5205\$BA5205}x_{\$BA5205}$
wff 115 : $= x_{o\$BA5205}$:= $\$FA5205$
<< a5205_substitutions.r0t.txt
:= $\$AA5205$
:= $\$BA5205$
:= $\$FA5205$
%0
$= (= x)\, [\lambda y.(= x\, y)]$
$=_{o(ot)(ot)}(=_{ott}x_t)[\lambda y_t.(=_{ott}x_t y_t)_o]$

§r /3 q_t
$= [\lambda y.(= x\, y)]\, [\lambda q.(= x\, q)]$
§s %1 3 %0
$= (= x)\, [\lambda q.(= x\, q)]$

use Proof Template A5201b (Swap): A = B $\rightarrow$ B = A
<< A5201b.r0t.txt
%0
$= [\lambda q.(= x\, q)]\, (= x)$
$=_{o(ot)(ot)}[\lambda q_t.(=_{ott}x_t q_t)_o](=_{ott}x_t)$

%$\$TMP5313$
$= (COND\, t\, x\, y\, T)\, (\iota\, [\lambda q.(= x\, q)])$:= $\$TMP5313$
$=_{ott}(COND_{\backslash 4o\backslash 3\backslash 2\tau}t_\tau x_t y_t T_o)(\iota_{t(ot)}[\lambda q_t.(=_{ott}x_t q_t)_o])$:= $\$TMP5313$
:= $\$TMP5313$
§s %0 7 %1
$= (COND\, t\, x\, y\, T)\, (\iota\, (= x))$
:= $\$TMP5313$ %0
wff 2376 : $= (COND\, t\, x\, y\, T)\, (\iota\, (= x))_o$:= $\$TMP5313$

.5

%$A5$
$= (\iota\, (= y))\, y$:= $A5$
$=_{ott}(\iota_{t(ot)}(=_{ott}y_t))y_t$:= $A5$

use Proof Template A5221 (Sub): B $\rightarrow$ B [x/A]
:= $\$B5221$ %0
wff 207 : $= (\iota\, (= y))\, y_o$:= $\$B5221$ $A5$
:= $\$T5221$ t_τ
wff 4 : t_τ := $\$T5221$
:= $\$X5221$ $y_{\$T5221}$
wff 105 : $y_{\$T5221}$:= $\$X5221$
:= $\$A5221$ $x_{\$T5221}$
wff 24 : $x_{\$T5221}$:= $\$A5221$
<< A5221.r0t.txt
:= $\$B5221$
:= $\$T5221$
:= $\$X5221$
:= $\$A5221$
%0
$= (\iota\, (= x))\, x$
$=_{ott}(\iota_{t(ot)}(=_{ott}x_t))x_t$

%$\$TMP5313$
$= (COND\, t\, x\, y\, T)\, (\iota\, (= x))$:= $\$TMP5313$
$=_{ott}(COND_{\backslash 4o\backslash 3\backslash 2\tau}t_\tau x_t y_t T_o)(\iota_{t(ot)}(=_{ott}x_t))$:= $\$TMP5313$
:= $\$TMP5313$

§s %0 3 %1
$=(COND\,t\,x\,y\,T)\,x$
:= \$$LTMP5313$ %0
wff 2424 : $=(COND\,t\,x\,y\,T)\,x_o$:= \$$LTMP5313$

.6

§= $_{t_\tau}$ $COND_{\backslash 4o\backslash 3\backslash 2\tau}t_\tau x_t y_t F_o$
$=(COND\,t\,x\,y\,F)\,(COND\,t\,x\,y\,F)$
§\ $COND_{\backslash 4o\backslash 3\backslash 2\tau}t_\tau$
$=(COND\,t)\,[\lambda x.[\lambda y.[\lambda p.(\iota\,[\lambda q.(\vee\,(\wedge\,p\,(=x\,q))\,(\wedge\,(\sim p)\,(=y\,q)))])]]]$
§s %1 24 %0
$=(COND\,t\,x\,y\,F)\,([\lambda x.[\lambda y.[\lambda p.(\iota\,[\lambda q.(\vee\,(\wedge\,p\,(=x\,q))\,(\wedge\,(\sim p)\,(=y\,q)))])]]]\,x\,y\,F)$
§\ $[\lambda x_t.[\lambda y_t.[\lambda p_o.(\iota_{t(ot)}[\lambda q_t.(\vee_{ooo}(\wedge_{ooo}p_o(=_{ott}x_tq_t))(\wedge_{ooo}(\sim_{oo}p_o)(=_{ott}y_tq_t)))_o])_t]_{(to)}]_{(tot)}]x_t$
$=([\lambda x.[\lambda y.[\lambda p.(\iota\,[\lambda q.(\vee\,(\wedge\,p\,(=x\,q))\,(\wedge\,(\sim p)\,(=y\,q)))])]]]\,x)$ ...
... $[\lambda y.[\lambda p.(\iota\,[\lambda q.(\vee\,(\wedge\,p\,(=x\,q))\,(\wedge\,(\sim p)\,(=y\,q)))])]]$
§s %1 12 %0
$=(COND\,t\,x\,y\,F)\,([\lambda y.[\lambda p.(\iota\,[\lambda q.(\vee\,(\wedge\,p\,(=x\,q))\,(\wedge\,(\sim p)\,(=y\,q)))])]]\,y\,F)$
§\ $[\lambda y_t.[\lambda p_o.(\iota_{t(ot)}[\lambda q_t.(\vee_{ooo}(\wedge_{ooo}p_o(=_{ott}x_tq_t))(\wedge_{ooo}(\sim_{oo}p_o)(=_{ott}y_tq_t)))_o])_t]_{(to)}]y_t$
$=([\lambda y.[\lambda p.(\iota\,[\lambda q.(\vee\,(\wedge\,p\,(=x\,q))\,(\wedge\,(\sim p)\,(=y\,q)))])]]\,y)$ ...
... $[\lambda p.(\iota\,[\lambda q.(\vee\,(\wedge\,p\,(=x\,q))\,(\wedge\,(\sim p)\,(=y\,q)))])]$
§s %1 6 %0
$=(COND\,t\,x\,y\,F)\,([\lambda p.(\iota\,[\lambda q.(\vee\,(\wedge\,p\,(=x\,q))\,(\wedge\,(\sim p)\,(=y\,q)))])]\,F)$
§\ $[\lambda p_o.(\iota_{t(ot)}[\lambda q_t.(\vee_{ooo}(\wedge_{ooo}p_o(=_{ott}x_tq_t))(\wedge_{ooo}(\sim_{oo}p_o)(=_{ott}y_tq_t)))_o])_t]F_o$
$=([\lambda p.(\iota\,[\lambda q.(\vee\,(\wedge\,p\,(=x\,q))\,(\wedge\,(\sim p)\,(=y\,q)))])]\,F)$ ...
... $(\iota\,[\lambda q.(\vee\,(\wedge\,F\,(=x\,q))\,(\wedge\,(\sim F)\,(=y\,q)))])$
§s %1 3 %0
$=(COND\,t\,x\,y\,F)\,(\iota\,[\lambda q.(\vee\,(\wedge\,F\,(=x\,q))\,(\wedge\,(\sim F)\,(=y\,q)))])$

%$A5231b$
$=(\sim F)\,T$:= $A5231b$
$=_{ooo}(\sim_{oo}F_o)T_o$:= $A5231b$
§s %1 125 %0
$=(COND\,t\,x\,y\,F)\,(\iota\,[\lambda q.(\vee\,(\wedge\,F\,(=x\,q))\,(\wedge\,T\,(=y\,q)))])$
:= \$$TMP5313$ %0
wff 2447 : $=(COND\,t\,x\,y\,F)\,(\iota\,[\lambda q.(\vee\,(\wedge\,F\,(=x\,q))\,(\wedge\,T\,(=y\,q)))])_o$:= \$$TMP5313$

%$K8001b$
$=(\wedge\,F\,x)\,F$:= $K8001b$
$=_{ooo}(\wedge_{ooo}F_ox_o)F_o$:= $K8001b$

use Proof Template A5221 (Sub): B → B [x/A]
:= \$$B5221$ %0
wff 1805 : $=(\wedge\,F\,x)\,F_{o,\ldots}$:= \$$B5221$ $K8001b$
:= \$$T5221$ o
wff 2 : o_τ := \$$T5221$
:= \$$X5221$ x_o
wff 16 : x_o := \$$X5221$
:= \$$A5221$ %1/123

\# wff 2064 : $=x\,q_{o,\ldots}$:= $\$A5221$
<< A5221.r0t.txt
:= $\$B5221$
:= $\$T5221$
:= $\$X5221$
:= $\$A5221$
%0
\# $=(\wedge\,F\,(=x\,q))\,F$
\# $=_{ooo}(\wedge_{ooo}F_o(=_{ott}x_tq_t))F_o$

%$\$TMP5313$
\# $=(COND\,t\,x\,y\,F)\,(\iota\,[\lambda q.(\vee\,(\wedge\,F\,(=x\,q))\,(\wedge\,T\,(=y\,q)))])$:= $\$TMP5313$
\# $=_{ott}(COND_{\backslash 4o\backslash 3\backslash 2\tau}t_\tau x_ty_tF_o)\ldots$
$\ldots(\iota_{t(ot)}[\lambda q_t.(\vee_{ooo}(\wedge_{ooo}F_o(=_{ott}x_tq_t))(\wedge_{ooo}T_o(=_{ott}y_tq_t)))_o])$:= $\$TMP5313$
:= $\$TMP5313$
§s %0 61 %1
\# $=(COND\,t\,x\,y\,F)\,(\iota\,[\lambda q.(\vee\,F\,(\wedge\,T\,(=y\,q)))])$
:= $\$TMP5313$ %0
\# wff 2459 : $=(COND\,t\,x\,y\,F)\,(\iota\,[\lambda q.(\vee\,F\,(\wedge\,T\,(=y\,q)))])_o$:= $\$TMP5313$

%$K8000b$
\# $=(\wedge\,T\,x)\,x$:= $K8000b$
\# $=_{ooo}(\wedge_{ooo}T_ox_o)x_o$:= $K8000b$

\#\# use Proof Template A5221 (Sub): B → B [x/A]
:= $\$B5221$ %0
\# wff 594 : $=(\wedge\,T\,x)\,x_{o,\ldots}$:= $\$B5221$ $K8000b$
:= $\$T5221$ o
\# wff 2 : o_τ := $\$T5221$
:= $\$X5221$ x_o
\# wff 16 : x_o := $\$X5221$
:= $\$A5221$ %1/63
\# wff 2069 : $=y\,q_o$:= $\$A5221$
<< A5221.r0t.txt
:= $\$B5221$
:= $\$T5221$
:= $\$X5221$
:= $\$A5221$
%0
\# $=(\wedge\,T\,(=y\,q))\,(=y\,q)$
\# $=_{ooo}(\wedge_{ooo}T_o(=_{ott}y_tq_t))(=_{ott}y_tq_t)$

%$\$TMP5313$
\# $=(COND\,t\,x\,y\,F)\,(\iota\,[\lambda q.(\vee\,F\,(\wedge\,T\,(=y\,q)))])$:= $\$TMP5313$
\# $=_{ott}(COND_{\backslash 4o\backslash 3\backslash 2\tau}t_\tau x_ty_tF_o)(\iota_{t(ot)}[\lambda q_t.(\vee_{ooo}F_o(\wedge_{ooo}T_o(=_{ott}y_tq_t)))_o])$:=
$\$TMP5313$
:= $\$TMP5313$
§s %0 31 %1
\# $=(COND\,t\,x\,y\,F)\,(\iota\,[\lambda q.(\vee\,F\,(=y\,q))])$

:= $\$TMP5313$ %0
wff 2471 : $=(COND\,t\,x\,y\,F)\,(\iota\,[\lambda q.(\vee\,F\,(=y\,q))])_o$:= $\$TMP5313$

%$K8010b$
$=(\vee\,F\,x)\,x$:= $K8010b$
$=_{ooo}(\vee_{ooo}F_o x_o)x_o$:= $K8010b$

use Proof Template A5221 (Sub): B → B [x/A]
:= $\$B5221$ %0
wff 2058 : $=(\vee\,F\,x)\,x_o$:= $\$B5221$ $K8010b$
:= $\$T5221$ o
wff 2 : o_τ := $\$T5221$
:= $\$X5221$ x_o
wff 16 : x_o := $\$X5221$
:= $\$A5221$ %1/31
wff 2069 : $=y\,q_{o,\ldots}$:= $\$A5221$
<< A5221.r0t.txt
:= $\$B5221$
:= $\$T5221$
:= $\$X5221$
:= $\$A5221$
%0
$=(\vee\,F\,(=y\,q))\,(=y\,q)$
$=_{ooo}(\vee_{ooo}F_o(=_{ott}y_t q_t))(=_{ott}y_t q_t)$

%$\$TMP5313$
$=(COND\,t\,x\,y\,F)\,(\iota\,[\lambda q.(\vee\,F\,(=y\,q))])$:= $\$TMP5313$
$=_{ott}(COND_{\backslash 4o\backslash 3\backslash 2\tau}t_\tau x_t y_t F_o)(\iota_{t(ot)}[\lambda q_t.(\vee_{ooo}F_o(=_{ott}y_t q_t))_o])$:= $\$TMP5313$
:= $\$TMP5313$
§s %0 15 %1
$=(COND\,t\,x\,y\,F)\,(\iota\,[\lambda q.(=y\,q)])$
:= $\$TMP5313$ %0
wff 2509 : $=(COND\,t\,x\,y\,F)\,(\iota\,[\lambda q.(=y\,q)])_o$:= $\$TMP5313$

.7

use Proof Template: A5205 Substitutions
:= $\$AA5205$ o
wff 2 : o_τ := $\$AA5205$
:= $\$BA5205$ t_τ
wff 4 : t_τ := $\$BA5205$
:= $\$FA5205$ $=_{o\$BA5205\$BA5205}z_{\$BA5205}$
wff 2510 : $=z_{o\$BA5205}$:= $\$FA5205$
<< a5205_substitutions.r0t.txt
:= $\$AA5205$
:= $\$BA5205$
:= $\$FA5205$
%0
$=(=z)\,[\lambda y.(=z\,y)]$

\# $=_{o(ot)(ot)}(=_{ott}z_t)[\lambda y_t.(=_{ott}z_ty_t)_o]$

§r /3 q_t
\# $=[\lambda y.(=z\,y)]\,[\lambda q.(=z\,q)]$
§s %1 3 %0
\# $=(=z)\,[\lambda q.(=z\,q)]$

\#\# use Proof Template A5221 (Sub): B → B [x/A]
:= $\$B5221$ %0
\# wff 2529 : $=(=z)\,[\lambda q.(=z\,q)]_o$:= $\$B5221$
:= $\$T5221$ t_τ
\# wff 4 : t_τ := $\$T5221$
:= $\$X5221$ $z_{\$T5221}$
\# wff 83 : $z_{\$T5221}$:= $\$X5221$
:= $\$A5221$ $y_{\$T5221}$
\# wff 105 : $y_{\$T5221}$:= $\$A5221$
<< A5221.r0t.txt
:= $\$B5221$
:= $\$T5221$
:= $\$X5221$
:= $\$A5221$
%0
\# $=(=y)\,[\lambda q.(=y\,q)]$
\# $=_{o(ot)(ot)}(=_{ott}y_t)[\lambda q_t.(=_{ott}y_tq_t)_o]$

\#\# use Proof Template A5201b (Swap): A = B → B = A
<< A5201b.r0t.txt
%0
\# $=[\lambda q.(=y\,q)]\,(=y)$
\# $=_{o(ot)(ot)}[\lambda q_t.(=_{ott}y_tq_t)_o](=_{ott}y_t)$

%$\$TMP5313$
\# $=(COND\,t\,x\,y\,F)\,(\iota\,[\lambda q.(=y\,q)])$:= $\$TMP5313$
\# $=_{ott}(COND_{\backslash 4o\backslash 3\backslash 2\tau}t_\tau x_ty_tF_o)(\iota_{t(ot)}[\lambda q_t.(=_{ott}y_tq_t)_o])$:= $\$TMP5313$
:= $\$TMP5313$
§s %0 7 %1
\# $=(COND\,t\,x\,y\,F)\,(\iota\,(=y))$

%$A5$
\# $=(\iota\,(=y))\,y$:= $A5$
\# $=_{ott}(\iota_{t(ot)}(=_{ott}y_t))y_t$:= $A5$
§s %1 3 %0
\# $=(COND\,t\,x\,y\,F)\,y$

\#\# .8

\#\# use Proof Template K8020: A, B → A ∧ B
:= $\$A8020$ $=_{ott}(COND_{\backslash 4o\backslash 3\backslash 2\tau}t_\tau x_ty_tT_o)x_t$
\# wff 2424 : $=(COND\,t\,x\,y\,T)\,x_o$:= $\$A8020$ $\$LTMP5313$

:= $\$LTMP5313$
:= $\$B8020$ %0
wff 2579 : $=(COND\,t\,x\,y\,F)\,y_o$:= $\$B8020$
<< K8020.r0t.txt
:= $\$A8020$
:= $\$B8020$

:= $A5313$ %0
wff 2614 : $\wedge\,(=(COND\,t\,x\,y\,T)\,x)\,(=(COND\,t\,x\,y\,F)\,y)_o$:= $A5313$

##
Q.E.D.
##

%0
$\wedge\,(=(COND\,t\,x\,y\,T)\,x)\,(=(COND\,t\,x\,y\,F)\,y)$:= $A5313$
$\wedge_{ooo}(=_{ott}(COND_{\backslash 4o\backslash 3\backslash 2\tau}t_\tau x_t y_t T_o)x_t)(=_{ott}(COND_{\backslash 4o\backslash 3\backslash 2\tau}t_\tau x_t y_t F_o)y_t)$:=
$A5313$

2.1.47 Results for File A53X08.r0a.txt

##
Proof A53X08: AC [t/b] ⊃ ∀ x: ∃ y: p_x_y = ∃ f: ∀ x: p_x_(f_x)
##
##
Source: [cf. Andrews 2002 (ISBN 1-4020-0763-9), p. 237 (X5308)]
##

Written by Ken Kubota (<mail@kenkubota.de>).
##
This file is part of the publication of the mathematical logic $\mathcal{R}_0$.
For more information, visit: <http://doi.org/10.4444/100.10>
##

##
Axioms
##

<< axiom_of_choice.r0a.txt

##
Proof
##

.1

%AC

\# $\exists\,(t(ot))\,[\lambda j.(\forall\,(ot)\,[\lambda p.(\supset(\exists\,t\,[\lambda x.(p\,x)])\,(p\,(j\,p)))])]$:= AC

\# $\exists_{o(o\backslash 3)\tau}(t(ot))_\tau\ldots$

$\ldots[\lambda j_{t(ot)}.(\forall_{o(o\backslash 3)\tau}(ot)_\tau[\lambda p_{ot}.(\supset_{ooo}(\exists_{o(o\backslash 3)\tau}t_\tau[\lambda x_t.(p_{ot}x_t)_o])(p_{ot}(j_{t(ot)}p_{ot})))_o])_o]$:= AC

\#\# .2

\#\# use Proof Template A5221 (Sub): B → B [x/A]

:= $\$B5221$ $\exists_{o(o\backslash 3)\tau}(t(ot))_\tau[\lambda j_{t(ot)}.(\forall_{o(o\backslash 3)\tau}(ot)_\tau[\lambda p_{ot}.(\supset_{ooo}(\exists_{o(o\backslash 3)\tau}t_\tau[\lambda x_t.(p_{ot}x_t)_o])(p_{ot}(j_{t(ot)}p_{ot})))_o])_o]$

\# wff 96 : $\exists\,(t(ot))\,[\lambda j.(\forall\,(ot)\,[\lambda p.(\supset(\exists\,t\,[\lambda x.(p\,x)])\,(p\,(j\,p)))])]_o$:= $\$B5221$ AC

:= $\$T5221$ τ

\# wff 0 : τ_τ := $\$T5221$

:= $\$X5221$ t_τ

\# wff 4 : t_τ := $\$X5221$

:= $\$A5221$ u_τ

\# wff 97 : u_τ := $\$A5221$

<< A5221.r0t.txt

:= $\$B5221$

:= $\$T5221$

:= $\$X5221$

:= $\$A5221$

%0

\# $\exists\,(u(ou))\,[\lambda j.(\forall\,(ou)\,[\lambda p.(\supset(\exists\,u\,[\lambda x.(p\,x)])\,(p\,(j\,p)))])]$

\# $\exists_{o(o\backslash 3)\tau}(u(ou))_\tau\ldots$

$\ldots[\lambda j_{u(ou)}.(\forall_{o(o\backslash 3)\tau}(ou)_\tau[\lambda p_{ou}.(\supset_{ooo}(\exists_{o(o\backslash 3)\tau}u_\tau[\lambda x_u.(p_{ou}x_u)_o])(p_{ou}(j_{u(ou)}p_{ou})))_o])_o]$

:= $\$AC53X08$ %0

\# wff 802 : $\exists\,(u(ou))\,[\lambda j.(\forall\,(ou)\,[\lambda p.(\supset(\exists\,u\,[\lambda x.(p\,x)])\,(p\,(j\,p)))])]_{o,\ldots}$:= $\$AC53X08$

\#\# left-hand side of the equation (equivalence)

:= $\$A53X08$ $\forall_{o(o\backslash 3)\tau}t_\tau[\lambda x_t.(\exists_{o(o\backslash 3)\tau}u_\tau[\lambda y_u.(p_{out}x_ty_u)_o])_o]$

\# wff 814 : $\forall\,t\,[\lambda x.(\exists\,u\,[\lambda y.(p\,x\,y)])]_o$:= $\$A53X08$

\#\# right-hand side of the equation (equivalence)

:= $\$B53X08$ $\exists_{o(o\backslash 3)\tau}(ut)_\tau[\lambda f_{ut}.(\forall_{o(o\backslash 3)\tau}t_\tau[\lambda x_t.(p_{out}x_t(f_{ut}x_t))_o])_o]$

\# wff 825 : $\exists\,(ut)\,[\lambda f.(\forall\,t\,[\lambda x.(p\,x\,(f\,x))])]_o$:= $\$B53X08$

\#\# .3

<< A53X08a.r0a.txt

:= $\$ATMP53X08$ %0

\# wff 7655 : $\supset\$AC53X08\,(\supset\$A53X08\,\$B53X08)_{o,\ldots}$:= $\$ATMP53X08$

\#\# .4

<< A53X08b.r0a.txt

:= $\$BTMP53X08$ %0

\# wff 9980 : $\supset\$AC53X08\,(\supset\$B53X08\,\$A53X08)_{o,\ldots}$:= $\$BTMP53X08$

.5

%$\$ATMP53X08$
$\supset \$AC53X08\,(\supset \$A53X08\,\$B53X08)$:= $\$ATMP53X08$
$\supset_{ooo}\$AC53X08_o(\supset_{ooo}\$A53X08_o\$B53X08_o)$:= $\$ATMP53X08$
:= $\$ATMP53X08$
%$\$BTMP53X08$
$\supset \$AC53X08\,(\supset \$B53X08\,\$A53X08)$:= $\$BTMP53X08$
$\supset_{ooo}\$AC53X08_o(\supset_{ooo}\$B53X08_o\$A53X08_o)$:= $\$BTMP53X08$
:= $\$BTMP53X08$

use Proof Template K8013H: H ⊃ (A ⊃ B), H ⊃ (B ⊃ A) → H ⊃ (A = B)
:= $\$H8013H$ $\exists_{o(o\backslash 3)\tau}(u(ou))_\tau \ldots$
$\ldots[\lambda j_{u(ou)}.(\forall_{o(o\backslash 3)\tau}(ou)_\tau[\lambda p_{ou}.(\supset_{ooo}(\exists_{o(o\backslash 3)\tau}u_\tau[\lambda x_u.(p_{ou}x_u)_o])(p_{ou}(j_{u(ou)}p_{ou})))_o])_o]$
wff 802 : $\exists\,(u(ou))\,[\lambda j.(\forall\,(ou)\,[\lambda p.(\supset(\exists\,u\,[\lambda x.(p\,x)])\,(p\,(j\,p)))])]_{o,\ldots}$:= $\$AC53X08$ $\$H8013H$
:= $\$A8013H$ $\forall_{o(o\backslash 3)\tau}t_\tau[\lambda x_t.(\exists_{o(o\backslash 3)\tau}u_\tau[\lambda y_u.(p_{out}x_ty_u)_o])_o]$
wff 814 : $\forall\,t\,[\lambda x.(\exists\,u\,[\lambda y.(p\,x\,y)])]_{o,\ldots}$:= $\$A53X08$ $\$A8013H$
:= $\$B8013H$ $\exists_{o(o\backslash 3)\tau}(ut)_\tau[\lambda f_{ut}.(\forall_{o(o\backslash 3)\tau}t_\tau[\lambda x_t.(p_{out}x_t(f_{ut}x_t))_o])_o]$
wff 825 : $\exists\,(ut)\,[\lambda f.(\forall\,t\,[\lambda x.(p\,x\,(f\,x))])]_{o,\ldots}$:= $\$B53X08$ $\$B8013H$
<< K8013H.r0t.txt
:= $\$H8013H$
:= $\$A8013H$
:= $\$B8013H$
%0
$\supset \$AC53X08\,(= \$A53X08\,\$B53X08)$
$\supset_{ooo}\$AC53X08_o(=_{ooo}\$A53X08_o\$B53X08_o)$

##
Q.E.D.
##

%0
$\supset \$AC53X08\,(= \$A53X08\,\$B53X08)$
$\supset_{ooo}\$AC53X08_o(=_{ooo}\$A53X08_o\$B53X08_o)$

##
Undefine Syntactical Variables
##

:= $\$AC53X08$
:= $\$A53X08$
:= $\$B53X08$

2.1.48 Results for File A53X08a.r0a.txt

##
Proof A53X08a (Part A ⊃ B): AC [t/b] ⊃ ∀ x: ∃ y: p_x_y = ∃ f: ∀ x: p_x_(f_x)
##
##
Source: [cf. Andrews 2002 (ISBN 1-4020-0763-9), p. 237 (X5308)]
##

Written by Ken Kubota (<mail@kenkubota.de>).
##
This file is part of the publication of the mathematical logic $\mathcal{R}_0$.
For more information, visit: <http://doi.org/10.4444/100.10>
##

<< K8005.r0.txt

##
Axioms
##

<< axiom_of_choice.r0a.txt

##
Define Syntactical Variables
##

left-hand side of the equation (equivalence)
:= $\$A53X08A$ $\forall_{o(o\backslash 3)\tau} t_\tau [\lambda x_t.(\exists_{o(o\backslash 3)\tau} u_\tau [\lambda y_u.(p_{out} x_t y_u)_o])_o]$
wff 1399 : $\forall t\, [\lambda x.(\exists u\, [\lambda y.(p\, x\, y)])]_o$:= $\$A53X08A$

right-hand side of the equation (equivalence)
:= $\$B53X08A$ $\exists_{o(o\backslash 3)\tau} (ut)_\tau [\lambda f_{ut}.(\forall_{o(o\backslash 3)\tau} t_\tau [\lambda x_t.(p_{out} x_t (f_{ut} x_t))_o])_o]$
wff 1410 : $\exists\, (ut)\, [\lambda f.(\forall t\, [\lambda x.(p\, x\, (f\, x))])]_o$:= $\$B53X08A$

##
Proof
##

.1

use Proof Template A5221 (Sub): B → B [x/A]
:= $\$B5221$ $\exists_{o(o\backslash 3)\tau} (t(ot))_\tau [\lambda j_{t(ot)}.(\forall_{o(o\backslash 3)\tau} (ot)_\tau [\lambda p_{ot}.(\supset_{ooo} (\exists_{o(o\backslash 3)\tau} t_\tau [\lambda x_t.(p_{ot} x_t)_o])(p_{ot}(j_{t(ot)} p_{ot})))_o])_o]$
wff 1386 : $\exists\, (t(ot))\, [\lambda j.(\forall\, (ot)\, [\lambda p.(\supset (\exists t\, [\lambda x.(p\, x)])\, (p\, (j\, p)))])]_o$:= $\$B5221$ AC
:= $\$T5221$ τ
wff 0 : τ_τ := $\$T5221$

:= $\$X5221\ t_\tau$
wff 4 : t_τ := $\$X5221$
:= $\$A5221\ u_\tau$
wff 1387 : u_τ := $\$A5221$
<< A5221.r0t.txt
:= $\$B5221$
:= $\$T5221$
:= $\$X5221$
:= $\$A5221$
:= $\$AC53X08A$ %0
wff 1474 : $\exists\,(u(ou))\,[\lambda j.(\forall\,(ou)\,[\lambda p.(\supset(\exists\,u\,[\lambda x.(p\,x)])\,(p\,(j\,p)))])]_{o,\ldots}$:= $\$AC53X08A$

part of the Axiom of Choice (after the existential quantifier)
:= $\$C53X08A$ %0/7
wff 1472 : $\forall\,(ou)\,[\lambda p.(\supset(\exists\,u\,[\lambda x.(p\,x)])\,(p\,(j\,p)))]_o$:= $\$C53X08A$

hypotheses
:= $\$HYP1\ \wedge_{ooo}(\wedge_{ooo}\$AC53X08A_o\$A53X08A_o)\$C53X08A_o$
wff 1480 : $\wedge\,(\wedge\,\$AC53X08A\,\$A53X08A)\,\$C53X08A_o$:= $\$HYP1$
:= $\$HYP2\ \wedge_{ooo}(\wedge_{ooo}\$AC53X08A_o\$B53X08A_o)\$C53X08A_o$
wff 1483 : $\wedge\,(\wedge\,\$AC53X08A\,\$B53X08A)\,\$C53X08A_o$:= $\$HYP2$

.2

%$K8005$
$\supset x\,x$:= $K8005$
$\supset_{ooo}x_o x_o$:= $K8005$

use Proof Template A5221 (Sub): B → B [x/A]
:= $\$B5221$ %0
wff 1357 : $\supset x\,x_{o,\ldots}$:= $\$B5221$ $K8005$
:= $\$T5221\ o$
wff 2 : o_τ := $\$T5221$
:= $\$X5221\ x_o$
wff 16 : x_o := $\$X5221$
:= $\$A5221\ \wedge_{ooo}(\wedge_{ooo}\$AC53X08A_o\$A53X08A_o)\$C53X08A_o$
wff 1480 : $\wedge\,(\wedge\,\$AC53X08A\,\$A53X08A)\,\$C53X08A_o$:= $\$A5221$ $\$HYP1$
<< A5221.r0t.txt
:= $\$B5221$
:= $\$T5221$
:= $\$X5221$
:= $\$A5221$
%0
$\supset\$HYP1\,\$HYP1$
$\supset_{ooo}\$HYP1_o\$HYP1_o$

use Proof Template K8019H: H ⊃ (A ∧ B) → H ⊃ A, H ⊃ B
:= $\$H8019H$ %0
wff 1520 : $\supset\$HYP1\,\$HYP1_{o,\ldots}$:= $\$H8019H$

```
<< K8019H.r0t.txt
:=  $H8019H
:=  $ATMP53X08A ⊃_{ooo}$HYP1_o$C53X08A_o
# wff     1886  :          ⊃ $HYP1 $C53X08A_o        :=  $ATMP53X08A  $B8019H
%$A8019H
#                       ⊃ $HYP1 (∧ $AC53X08A $A53X08A)        :=  $A8019H
#                       ⊃_{ooo}$HYP1_o(∧_{ooo}$AC53X08A_o$A53X08A_o)        :=  $A8019H
:=  $A8019H
:=  $B8019H

## .3

## use Proof Template K8019H:  H ⊃ (A ∧ B)  →  H ⊃ A, H ⊃ B
:=  $H8019H  %0
# wff     1814  :          ⊃ $HYP1 (∧ $AC53X08A $A53X08A)_o        :=  $H8019H
<< K8019H.r0t.txt
:=  $H8019H
%$B8019H
#                       ⊃ $HYP1 $A53X08A        :=  $B8019H
#                       ⊃_{ooo}$HYP1_o$A53X08A_o        :=  $B8019H
:=  $A8019H
:=  $B8019H

## use Proof Template A5215H (∀ I):  H ⊃ ∀ x: B  →  H ⊃ B [x/a]
:=  $T5215H  t_τ
# wff     4  :        t_τ        :=  $T5215H
:=  $X5215H  x_{$T5215H}
# wff     24  :        x_{$T5215H}        :=  $X5215H
:=  $A5215H  x_{$T5215H}
# wff     24  :        x_{$T5215H}        :=  $A5215H  $X5215H
:=  $H5215H  %0
# wff     1951  :          ⊃ $HYP1 $A53X08A_o        :=  $H5215H
<< A5215H.r0t.txt
:=  $T5215H
:=  $X5215H
:=  $A5215H
:=  $H5215H
%0
#                       ⊃ $HYP1 (∃ u [λy.(p x y)])
#                       ⊃_{ooo}$HYP1_o(∃_{o(o\3)τ}u_τ[λy_u.(p_{out}x_t y_u)_o])

:=  $BTMP53X08A  %0
# wff     2010  :          ⊃ $HYP1 (∃ u [λy.(p x y)])_o        :=  $BTMP53X08A

%$ATMP53X08A
#                       ⊃ $HYP1 $C53X08A        :=  $ATMP53X08A
#                       ⊃_{ooo}$HYP1_o$C53X08A_o        :=  $ATMP53X08A
:=  $ATMP53X08A
```

.4

use Proof Template A5215H (∀ I): H ⊃ ∀ x: B → H ⊃ B [x/a]
:= $\$T5215H\ ou$
\# wff 1389 : ou_τ := $\$T5215H$
:= $\$X5215H\ p_{\$T5215H}$
\# wff 1461 : $p_{\$T5215H}$:= $\$X5215H$
:= $\$A5215H\ p_{\$T5215H\,t}x_t$
\# wff 1394 : $p\,x_{\$T5215H}$:= $\$A5215H$
:= $\$H5215H$ %0
\# wff 1886 : $\supset \$HYP1\,\$C53X08A_o$:= $\$H5215H$
<< A5215H.r0t.txt
:= $\$T5215H$
:= $\$X5215H$
:= $\$A5215H$
:= $\$H5215H$
%0
\# $\supset \$HYP1\,(\supset(\exists\,u\,[\lambda x.(p\,x\,x)])\,(p\,x\,(j\,(p\,x))))$
\# $\supset_{ooo}\$HYP1_o(\supset_{ooo}(\exists_{o(o\backslash 3)\tau}u_\tau[\lambda x_u.(p_{out}x_t x_u)_o])(p_{out}x_t(j_{u(ou)}(p_{out}x_t))))$

.5

§r /27 y_u
\# $=[\lambda x.(p\,x\,x)]\,[\lambda y.(p\,x\,y)]$
§s %1 27 %0
\# $\supset \$HYP1\,(\supset(\exists\,u\,[\lambda y.(p\,x\,y)])\,(p\,x\,(j\,(p\,x))))$

.6

%$\$BTMP53X08A$
\# $\supset \$HYP1\,(\exists\,u\,[\lambda y.(p\,x\,y)])$:= $\$BTMP53X08A$
\# $\supset_{ooo}\$HYP1_o(\exists_{o(o\backslash 3)\tau}u_\tau[\lambda y_u.(p_{out}x_t y_u)_o])$:= $\$BTMP53X08A$
:= $\$BTMP53X08A$

use Proof Template A5224H (MP): H ⊃ A, H ⊃ (A ⊃ B) → H ⊃ B
:= $\$A5224H$ %0
\# wff 2010 : $\supset \$HYP1\,(\exists\,u\,[\lambda y.(p\,x\,y)])_o$:= $\$A5224H$
:= $\$AB5224H$ %1
\# wff 2089 : $\supset \$HYP1\,(\supset(\exists\,u\,[\lambda y.(p\,x\,y)])\,(p\,x\,(j\,(p\,x))))_o$:= $\$AB5224H$
<< A5224H.r0t.txt
:= $\$AB5224H$
:= $\$A5224H$
%0
\# $\supset \$HYP1\,(p\,x\,(j\,(p\,x)))$
\# $\supset_{ooo}\$HYP1_o(p_{out}x_t(j_{u(ou)}(p_{out}x_t)))$

use Proof Template A5220H (Gen): (H ⊃ A) → (H ⊃ ∀ x: A)
:= $\$T5220H\ t_\tau$
\# wff 4 : t_τ := $\$T5220H$

:= $\$X5220H\ \ x_{\$T5220H}$
wff 24 : $x_{\$T5220H}$:= $\$X5220H$
:= $\$A5220H$ %0
wff 2291 : $\supset \$HYP1\,(p\,\$X5220H\,(j\,(p\,\$X5220H)))_o$:= $\$A5220H$
<< A5220H.r0t.txt
:= $\$T5220H$
:= $\$X5220H$
:= $\$A5220H$
%0
$\supset \$HYP1\,(\forall t\,[\lambda x.(p\,x\,(j\,(p\,x)))])$
$\supset_{ooo}\$HYP1_o(\forall_{o(o\backslash 3)\tau}t_\tau[\lambda x_t.(p_{out}x_t(j_{u(ou)}(p_{out}x_t)))_o])$

reduce [\x.(j_(p_x))]_x
§\ $[\lambda x_t.(j_{u(ou)}(p_{out}x_t))_u]x_t$
$= ([\lambda x.(j\,(p\,x))]\,x)\,(j\,(p\,x))$
§= $[\lambda x_t.(j_{u(ou)}(p_{out}x_t))_u]x_t$
$= ([\lambda x.(j\,(p\,x))]\,x)\,([\lambda x.(j\,(p\,x))]\,x)$
§s %0 5 %1
$= (j\,(p\,x))\,([\lambda x.(j\,(p\,x))]\,x)$
§s %3 31 %0
$\supset \$HYP1\,(\forall t\,[\lambda x.(p\,x\,([\lambda x.(j\,(p\,x))]\,x))])$

§\ $[\lambda f_{ut}.(\forall_{o(o\backslash 3)\tau}t_\tau[\lambda x_t.(p_{out}x_t(f_{ut}x_t))_o])_o][\lambda x_t.(j_{u(ou)}(p_{out}x_t))_u]$
$= ([\lambda f.(\forall t\,[\lambda x.(p\,x\,(f\,x))])]\,[\lambda x.(j\,(p\,x))])\,(\forall t\,[\lambda x.(p\,x\,([\lambda x.(j\,(p\,x))]\,x))])$
§= $[\lambda f_{ut}.(\forall_{o(o\backslash 3)\tau}t_\tau[\lambda x_t.(p_{out}x_t(f_{ut}x_t))_o])_o][\lambda x_t.(j_{u(ou)}(p_{out}x_t))_u]$
$= ([\lambda f.(\forall t\,[\lambda x.(p\,x\,(f\,x))])]\,[\lambda x.(j\,(p\,x))])\,([\lambda f.(\forall t\,[\lambda x.(p\,x\,(f\,x))])]\,[\lambda x.(j\,(p\,x))])$
§s %0 5 %1
$= (\forall t\,[\lambda x.(p\,x\,([\lambda x.(j\,(p\,x))]\,x))])\,([\lambda f.(\forall t\,[\lambda x.(p\,x\,(f\,x))])]\,[\lambda x.(j\,(p\,x))])$
§s %3 3 %0
$\supset \$HYP1\,([\lambda f.(\forall t\,[\lambda x.(p\,x\,(f\,x))])]\,[\lambda x.(j\,(p\,x))])$

use Proof Template K8028 (∃ GenH): H ⊃ ([\x.B]A) → H ⊃ ∃ x: B
:= $\$H8028\ \wedge_{ooo}(\wedge_{ooo}\$AC53X08A_o\$A53X08A_o)\$C53X08A_o$
wff 1480 : $\wedge\,(\wedge\,\$AC53X08A\,\$A53X08A)\,\$C53X08A_{o,\dots}$:= $\$H8028$ $\$HYP1$
:= $\$T8028\ ut$
wff 1400 : ut_τ := $\$T8028$
:= $\$B8028$ %0/6
wff 1409 : $[\lambda f.(\forall t\,[\lambda x.(p\,x\,(f\,x))])]_{o\$T8028}$:= $\$B8028$
:= $\$A8028$ %0/7
wff 2425 : $[\lambda x.(j\,(p\,x))]_{\$T8028}$:= $\$A8028$
<< K8028.r0t.txt
:= $\$H8028$
:= $\$T8028$
:= $\$B8028$
:= $\$A8028$
%0
$\supset \$HYP1\,\$B53X08A$
$\supset_{ooo}\$HYP1_o\$B53X08A_o$

:= $\$ATMP53X08A$ %0
wff 6209 : $\supset \$HYP1\,\$B53X08A_o$:= $\$ATMP53X08A$

.7

%$K8005$
$\supset x\,x$:= $K8005$
$\supset_{ooo} x_o x_o$:= $K8005$

use Proof Template A5221 (Sub): B $\rightarrow$ B [x/A]
:= $\$B5221$ %0
wff 1357 : $\supset x\,x_{o,\ldots}$:= $\$B5221$ $K8005$
:= $\$T5221$ o
wff 2 : o_τ := $\$T5221$
:= $\$X5221$ x_o
wff 16 : x_o := $\$X5221$
:= $\$A5221$ $\wedge_{ooo}(\wedge_{ooo}\$AC53X08A_o\$A53X08A_o)\$C53X08A_o/5$
wff 1478 : $\wedge\,\$AC53X08A\,\$A53X08A_{o,\ldots}$:= $\$A5221$
<< A5221.r0t.txt
:= $\$B5221$
:= $\$T5221$
:= $\$X5221$
:= $\$A5221$
%0
$\supset(\wedge\,\$AC53X08A\,\$A53X08A)\,(\wedge\,\$AC53X08A\,\$A53X08A)$
$\supset_{ooo}(\wedge_{ooo}\$AC53X08A_o\$A53X08A_o)(\wedge_{ooo}\$AC53X08A_o\$A53X08A_o)$

use Proof Template K8019H: H $\supset$ (A $\wedge$ B) $\rightarrow$ H $\supset$ A, H $\supset$ B
:= $\$H8019H$ %0
wff 6219 : $\supset(\wedge\,\$AC53X08A\,\$A53X08A)\,(\wedge\,\$AC53X08A\,\$A53X08A)_{o,\ldots}$:= $\$H8019H$
<< K8019H.r0t.txt
:= $\$H8019H$
%$\$A8019H$
$\supset(\wedge\,\$AC53X08A\,\$A53X08A)\,\$AC53X08A$:= $\$A8019H$
$\supset_{ooo}(\wedge_{ooo}\$AC53X08A_o\$A53X08A_o)\$AC53X08A_o$:= $\$A8019H$
:= $\$A8019H$
:= $\$B8019H$

%$\$ATMP53X08A$
$\supset \$HYP1\,\$B53X08A$:= $\$ATMP53X08A$
$\supset_{ooo}\$HYP1_o\$B53X08A_o$:= $\$ATMP53X08A$
:= $\$ATMP53X08A$

use Proof Template A5245 (Rule C): H $\supset$ $\exists$ x: B, (H $\wedge$ (B [x/y])) $\supset$ A $\rightarrow$ H $\supset$ A
:= $\$T5245$ $u(ou)$
wff 1454 : $u(ou)_\tau$:= $\$T5245$
:= $\$X5245$ $j_{\$T5245}$

\# wff 1458 : $j_{\$T5245}$:= $\$X5245$
:= $\$Y5245$ $j_{\$T5245}$
\# wff 1458 : $j_{\$T5245}$:= $\$X5245$ $\$Y5245$
:= $\$B5245$ %1
\# wff 6282 : $\supset (\wedge \$AC53X08A \, \$A53X08A) \, \$AC53X08A_o$:= $\$B5245$
:= $\$A5245$ %0
\# wff 6209 : $\supset \$HYP1 \, \$B53X08A_o$:= $\$A5245$
<< A5245.r0t.txt
:= $\$T5245$
:= $\$X5245$
:= $\$Y5245$
:= $\$B5245$
:= $\$A5245$
%0
\# $\supset (\wedge \$AC53X08A \, \$A53X08A) \, \$B53X08A$
\# $\supset_{ooo}(\wedge_{ooo}\$AC53X08A_o\$A53X08A_o)\$B53X08A_o$

use Proof Template K8025 (Deduction Theorem): (H ∧ I) ⊃ A → H ⊃ (I ⊃ A)
<< K8025.r0t.txt
%0
\# $\supset \$AC53X08A \, (\supset \$A53X08A \, \$B53X08A)$
\# $\supset_{ooo}\$AC53X08A_o(\supset_{ooo}\$A53X08A_o\$B53X08A_o)$

##
Q.E.D.
##

%0
\# $\supset \$AC53X08A \, (\supset \$A53X08A \, \$B53X08A)$
\# $\supset_{ooo}\$AC53X08A_o(\supset_{ooo}\$A53X08A_o\$B53X08A_o)$

##
Undefine Syntactical Variables
##

:= $\$A53X08A$
:= $\$B53X08A$
:= $\$AC53X08A$
:= $\$C53X08A$
:= $\$HYP1$
:= $\$HYP2$

2.1.49 Results for File A53X08b.r0a.txt

##
Proof A53X08b (Part B ⊃ A): AC [t/b] ⊃ ∀ x: ∃ y: p_x_y = ∃ f: ∀ x: p_x_(f_x)
##

\#\#
\#\# Source: [cf. Andrews 2002 (ISBN 1-4020-0763-9), p. 237 (X5308)]
\#\#

\#\# Written by Ken Kubota (<mail@kenkubota.de>).
\#\#
\#\# This file is part of the publication of the mathematical logic $\mathcal{R}_0$.
\#\# For more information, visit: <http://doi.org/10.4444/100.10>
\#\#

<< K8005.r0.txt

\#\#
\#\# Axioms
\#\#

<< axiom_of_choice.r0a.txt

\#\#
\#\# Define Syntactical Variables
\#\#

\#\# left-hand side of the equation (equivalence)
:= $\$A53X08B$ $\forall_{o(o\backslash 3)\tau}t_\tau[\lambda x_t.(\exists_{o(o\backslash 3)\tau}u_\tau[\lambda y_u.(p_{out}x_ty_u)_o])_o]$
\# wff 1399 : $\forall t\,[\lambda x.(\exists u\,[\lambda y.(p\,x\,y)])]_o$:= $\$A53X08B$

\#\# right-hand side of the equation (equivalence)
:= $\$B53X08B$ $\exists_{o(o\backslash 3)\tau}(ut)_\tau[\lambda f_{ut}.(\forall_{o(o\backslash 3)\tau}t_\tau[\lambda x_t.(p_{out}x_t(f_{ut}x_t))_o])_o]$
\# wff 1410 : $\exists\,(ut)\,[\lambda f.(\forall t\,[\lambda x.(p\,x\,(f\,x))])]_o$:= $\$B53X08B$

\#\#
\#\# Proof
\#\#

\#\# .1

\#\# use Proof Template A5221 (Sub): B $\rightarrow$ B [x/A]
:= $\$B5221$ $\exists_{o(o\backslash 3)\tau}(t(ot))_\tau[\lambda j_{t(ot)}.(\forall_{o(o\backslash 3)\tau}(ot)_\tau[\lambda p_{ot}.(\supset_{ooo}(\exists_{o(o\backslash 3)\tau}t_\tau[\lambda x_t.(p_{ot}x_t)_o])(p_{ot}(j_{t(ot)}p_{ot})))_o])_o]$
\# wff 1386 : $\exists\,(t(ot))\,[\lambda j.(\forall\,(ot)\,[\lambda p.(\supset(\exists\,t\,[\lambda x.(p\,x)])\,(p\,(j\,p)))])]_o$:= $\$B5221$ AC
:= $\$T5221$ τ
\# wff 0 : τ_τ := $\$T5221$
:= $\$X5221$ t_τ
\# wff 4 : t_τ := $\$X5221$
:= $\$A5221$ u_τ
\# wff 1387 : u_τ := $\$A5221$
<< A5221.r0t.txt

:= $\$B5221$
:= $\$T5221$
:= $\$X5221$
:= $\$A5221$
:= $\$AC53X08B$ %0
wff 1474 : $\exists\,(u(ou))\,[\lambda j.(\forall\,(ou)\,[\lambda p.(\supset(\exists\,u\,[\lambda x.(p\,x)])\,(p\,(j\,p)))])]_{o,\dots}$:= $\$AC53X08B$

part of the Axiom of Choice (after the existential quantifier)
:= $\$C53X08B$ %0/7
wff 1472 : $\forall\,(ou)\,[\lambda p.(\supset(\exists\,u\,[\lambda x.(p\,x)])\,(p\,(j\,p)))]_o$:= $\$C53X08B$

hypotheses
:= $\$HYP1$ $\wedge_{ooo}(\wedge_{ooo}\$AC53X08B_o\$A53X08B_o)\$C53X08B_o$
wff 1480 : $\wedge\,(\wedge\,\$AC53X08B\,\$A53X08B)\,\$C53X08B_o$:= $\$HYP1$
:= $\$HYP2$ $\wedge_{ooo}(\wedge_{ooo}\$AC53X08B_o\$B53X08B_o)\$C53X08B_o$
wff 1483 : $\wedge\,(\wedge\,\$AC53X08B\,\$B53X08B)\,\$C53X08B_o$:= $\$HYP2$

.2

%$K8005$
$\supset x\,x$:= $K8005$
$\supset_{ooo}x_o x_o$:= $K8005$

use Proof Template A5221 (Sub): B → B [x/A]
:= $\$B5221$ %0
wff 1357 : $\supset x\,x_{o,\dots}$:= $\$B5221$ $K8005$
:= $\$T5221$ o
wff 2 : o_τ := $\$T5221$
:= $\$X5221$ x_o
wff 16 : x_o := $\$X5221$
:= $\$A5221$ $\wedge_{ooo}(\wedge_{ooo}\$AC53X08B_o\$B53X08B_o)\$C53X08B_o$
wff 1483 : $\wedge\,(\wedge\,\$AC53X08B\,\$B53X08B)\,\$C53X08B_o$:= $\$A5221$ $\$HYP2$
<< A5221.r0t.txt
:= $\$B5221$
:= $\$T5221$
:= $\$X5221$
:= $\$A5221$
%0
$\supset \$HYP2\,\$HYP2$
$\supset_{ooo}\$HYP2_o\$HYP2_o$

use Proof Template K8019H: H ⊃ (A ∧ B) → H ⊃ A, H ⊃ B
:= $\$H8019H$ %0
wff 1520 : $\supset \$HYP2\,\$HYP2_{o,\dots}$:= $\$H8019H$
<< K8019H.r0t.txt
:= $\$H8019H$
%$\$A8019H$
$\supset \$HYP2\,(\wedge\,\$AC53X08B\,\$B53X08B)$:= $\$A8019H$

\# $\supset_{ooo}\$HYP2_o(\wedge_{ooo}\$AC53X08B_o\$B53X08B_o)$:= $\$A8019H$
:= $\$A8019H$
:= $\$B8019H$

\#\# use Proof Template K8019H: H ⊃ (A ∧ B) → H ⊃ A, H ⊃ B
:= $\$H8019H$ %0
\# wff 1814 : $\supset \$HYP2\,(\wedge\,\$AC53X08B\,\$B53X08B)_o$:= $\$H8019H$
<< K8019H.r0t.txt
:= $\$H8019H$
%$\$B8019H$
\# $\supset \$HYP2\,\$B53X08B$:= $\$B8019H$
\# $\supset_{ooo}\$HYP2_o\$B53X08B_o$:= $\$B8019H$
:= $\$A8019H$
:= $\$B8019H$

:= $\$BTMP53X08B$ %0
\# wff 1951 : $\supset \$HYP2\,\$B53X08B_o$:= $\$BTMP53X08B$
:= $\$D53X08B$ $\forall_{o(o\backslash 3)\tau}t_\tau[\lambda x_t.(p_{out}x_t(f_{ut}x_t))_o]$
\# wff 1408 : $\forall t\,[\lambda x.(p\,x\,(f\,x))]_o$:= $\$D53X08B$

\#\# .3

%$K8005$
\# $\supset x\,x$:= $K8005$
\# $\supset_{ooo}x_o x_o$:= $K8005$

\#\# use Proof Template A5221 (Sub): B → B [x/A]
:= $\$B5221$ %0
\# wff 1357 : $\supset x\,x_{o,\ldots}$:= $\$B5221$ $K8005$
:= $\$T5221$ o
\# wff 2 : o_τ := $\$T5221$
:= $\$X5221$ x_o
\# wff 16 : x_o := $\$X5221$
:= $\$A5221$ $\wedge_{ooo}\$HYP2_o\$D53X08B_o$
\# wff 1952 : $\wedge\,\$HYP2\,\$D53X08B_o$:= $\$A5221$
<< A5221.r0t.txt
:= $\$B5221$
:= $\$T5221$
:= $\$X5221$
:= $\$A5221$
%0
\# $\supset (\wedge\,\$HYP2\,\$D53X08B)\,(\wedge\,\$HYP2\,\$D53X08B)$
\# $\supset_{ooo}(\wedge_{ooo}\$HYP2_o\$D53X08B_o)(\wedge_{ooo}\$HYP2_o\$D53X08B_o)$

\#\# use Proof Template K8019H: H ⊃ (A ∧ B) → H ⊃ A, H ⊃ B
:= $\$H8019H$ %0
\# wff 1962 : $\supset (\wedge\,\$HYP2\,\$D53X08B)\,(\wedge\,\$HYP2\,\$D53X08B)_{o,\ldots}$:= $\$H8019H$
<< K8019H.r0t.txt
:= $\$H8019H$

%$\$B8019H$
\# $\supset (\wedge\,\$HYP2\,\$D53X08B)\,\$D53X08B$:= $\$B8019H$
\# $\supset_{ooo}(\wedge_{ooo}\$HYP2_o\$D53X08B_o)\$D53X08B_o$:= $\$B8019H$
:= $\$A8019H$
:= $\$B8019H$
%0
\# $\supset (\wedge\,\$HYP2\,\$D53X08B)\,\$D53X08B$
\# $\supset_{ooo}(\wedge_{ooo}\$HYP2_o\$D53X08B_o)\$D53X08B_o$

\#\# .4

\#\# use Proof Template A5215H (∀ I): H ⊃ ∀ x: B → H ⊃ B [x/a]
:= $\$T5215H\ t_\tau$
\# wff 4 : t_τ := $\$T5215H$
:= $\$X5215H\ x_{\$T5215H}$
\# wff 24 : $x_{\$T5215H}$:= $\$X5215H$
:= $\$A5215H\ x_{\$T5215H}$
\# wff 24 : $x_{\$T5215H}$:= $\$A5215H$ $\$X5215H$
:= $\$H5215H$ %0
\# wff 2099 : $\supset (\wedge\,\$HYP2\,\$D53X08B)\,\$D53X08B_o$:= $\$H5215H$
<< A5215H.r0t.txt
:= $\$T5215H$
:= $\$X5215H$
:= $\$A5215H$
:= $\$H5215H$
%0
\# $\supset (\wedge\,\$HYP2\,\$D53X08B)\,(p\,x\,(f\,x))$
\# $\supset_{ooo}(\wedge_{ooo}\$HYP2_o\$D53X08B_o)(p_{out}x_t(f_{ut}x_t))$

\#\# .5

§\ $[\lambda y_u.(p_{out}x_ty_u)_o](f_{ut}x_t)$
\# $= ([\lambda y.(p\,x\,y)]\,(f\,x))\,(p\,x\,(f\,x))$
§= $[\lambda y_u.(p_{out}x_ty_u)_o](f_{ut}x_t)$
\# $= ([\lambda y.(p\,x\,y)]\,(f\,x))\,([\lambda y.(p\,x\,y)]\,(f\,x))$
§s %0 5 %1
\# $= (p\,x\,(f\,x))\,([\lambda y.(p\,x\,y)]\,(f\,x))$

§s %3 3 %0
\# $\supset (\wedge\,\$HYP2\,\$D53X08B)\,([\lambda y.(p\,x\,y)]\,(f\,x))$

\#\# use Proof Template K8028 (∃ GenH): H ⊃ ([\x.B]A) → H ⊃ ∃ x: B
:= $\$H8028\ \wedge_{ooo}\$HYP2_o\$D53X08B_o$
\# wff 1952 : $\wedge\,\$HYP2\,\$D53X08B_{o,\dots}$:= $\$H8028$
:= $\$T8028\ u_\tau$
\# wff 1387 : u_τ := $\$T8028$
:= $\$B8028$ %0/6
\# wff 1396 : $[\lambda y.(p\,x\,y)]_{o\,\$T8028}$:= $\$B8028$
:= $\$A8028$ %0/7

wff 1405 : $f\,x_{\$T8028}$:= $\$A8028$
<< K8028.r0t.txt
:= $\$H8028$
:= $\$T8028$
:= $\$B8028$
:= $\$A8028$
%0
$\supset (\wedge\, \$HYP2\, \$D53X08B)\, (\exists\, u\, [\lambda y.(p\, x\, y)])$
$\supset_{ooo}(\wedge_{ooo}\$HYP2_o\$D53X08B_o)(\exists_{o(o\backslash 3)\tau}u_\tau[\lambda y_u.(p_{out}x_t y_u)_o])$

.6

%$\$BTMP53X08B$
$\supset \$HYP2\, \$B53X08B$:= $\$BTMP53X08B$
$\supset_{ooo}\$HYP2_o\$B53X08B_o$:= $\$BTMP53X08B$
:= $\$BTMP53X08B$

use Proof Template A5245 (Rule C): H ⊃ ∃ x: B, (H ∧ (B [x/y])) ⊃ A → H ⊃ A
:= $\$T5245\ ut$
wff 1400 : ut_τ := $\$T5245$
:= $\$X5245\ f_{\$T5245}$
wff 1404 : $f_{\$T5245}$:= $\$X5245$
:= $\$Y5245\ f_{\$T5245}$
wff 1404 : $f_{\$T5245}$:= $\$X5245$ $\$Y5245$
:= $\$B5245$ %0
wff 1951 : $\supset \$HYP2\, \$B53X08B_o$:= $\$B5245$
:= $\$A5245$ %1
wff 6011 : $\supset (\wedge\, \$HYP2\, \$D53X08B)\, (\exists\, u\, [\lambda y.(p\, x\, y)])_o$:= $\$A5245$
<< A5245.r0t.txt
:= $\$T5245$
:= $\$X5245$
:= $\$Y5245$
:= $\$B5245$
:= $\$A5245$
%0
$\supset \$HYP2\, (\exists\, u\, [\lambda y.(p\, x\, y)])$
$\supset_{ooo}\$HYP2_o(\exists_{o(o\backslash 3)\tau}u_\tau[\lambda y_u.(p_{out}x_t y_u)_o])$

use Proof Template A5220H (Gen): (H ⊃ A) → (H ⊃ ∀ x: A)
:= $\$T5220H\ t_\tau$
wff 4 : t_τ := $\$T5220H$
:= $\$X5220H\ x_{\$T5220H}$
wff 24 : $x_{\$T5220H}$:= $\$X5220H$
:= $\$A5220H$ %0
wff 7269 : $\supset \$HYP2\, (\exists\, u\, [\lambda y.(p\, \$X5220H\, y)])_o$:= $\$A5220H$
<< A5220H.r0t.txt
:= $\$T5220H$
:= $\$X5220H$

:= $\$A5220H$
%0
$\supset \$HYP2\,\$A53X08B$
$\supset_{ooo}\$HYP2_o\$A53X08B_o$

:= $\$ATMP53X08B$ %0
wff 7400 : $\supset \$HYP2\,\$A53X08B_o$:= $\$ATMP53X08B$

.7

%$K8005$
$\supset x\,x$:= $K8005$
$\supset_{ooo}x_o x_o$:= $K8005$

use Proof Template A5221 (Sub): B $\rightarrow$ B [x/A]
:= $\$B5221$ %0
wff 1357 : $\supset x\,x_{o,\ldots}$:= $\$B5221$ $K8005$
:= $\$T5221$ o
wff 2 : o_τ := $\$T5221$
:= $\$X5221$ x_o
wff 16 : x_o := $\$X5221$
:= $\$A5221$ $\wedge_{ooo}(\wedge_{ooo}\$AC53X08B_o\$B53X08B_o)\$C53X08B_o/5$
wff 1481 : $\wedge\,\$AC53X08B\,\$B53X08B_{o,\ldots}$:= $\$A5221$
<< A5221.r0t.txt
:= $\$B5221$
:= $\$T5221$
:= $\$X5221$
:= $\$A5221$
%0
$\supset (\wedge\,\$AC53X08B\,\$B53X08B)\,(\wedge\,\$AC53X08B\,\$B53X08B)$
$\supset_{ooo}(\wedge_{ooo}\$AC53X08B_o\$B53X08B_o)(\wedge_{ooo}\$AC53X08B_o\$B53X08B_o)$

use Proof Template K8019H: H $\supset$ (A $\wedge$ B) $\rightarrow$ H $\supset$ A, H $\supset$ B
:= $\$H8019H$ %0
wff 7410 : $\supset (\wedge\,\$AC53X08B\,\$B53X08B)\,(\wedge\,\$AC53X08B\,\$B53X08B)_{o,\ldots}$:= $\$H8019H$
<< K8019H.r0t.txt
:= $\$H8019H$
%$\$A8019H$
$\supset (\wedge\,\$AC53X08B\,\$B53X08B)\,\$AC53X08B$:= $\$A8019H$
$\supset_{ooo}(\wedge_{ooo}\$AC53X08B_o\$B53X08B_o)\$AC53X08B_o$:= $\$A8019H$
:= $\$A8019H$
:= $\$B8019H$

%$\$ATMP53X08B$
$\supset \$HYP2\,\$A53X08B$:= $\$ATMP53X08B$
$\supset_{ooo}\$HYP2_o\$A53X08B_o$:= $\$ATMP53X08B$
:= $\$ATMP53X08B$

use Proof Template A5245 (Rule C): H ⊃ ∃ x: B, (H ∧ (B [x/y])) ⊃ A → H ⊃ A
:= $\$T5245\ u(ou)$
wff 1454 : $u(ou)_\tau$:= $\$T5245$
:= $\$X5245\ j_{\$T5245}$
wff 1458 : $j_{\$T5245}$:= $\$X5245$
:= $\$Y5245\ j_{\$T5245}$
wff 1458 : $j_{\$T5245}$:= $\$X5245$ $\$Y5245$
:= $\$B5245$ %1
wff 7473 : $\supset (\wedge \$AC53X08B\ \$B53X08B)\ \$AC53X08B_o$:= $\$B5245$
:= $\$A5245$ %0
wff 7400 : $\supset \$HYP2\ \$A53X08B_o$:= $\$A5245$
<< A5245.r0t.txt
:= $\$T5245$
:= $\$X5245$
:= $\$Y5245$
:= $\$B5245$
:= $\$A5245$
%0
$\supset (\wedge \$AC53X08B\ \$B53X08B)\ \$A53X08B$
$\supset_{ooo}(\wedge_{ooo}\$AC53X08B_o\$B53X08B_o)\$A53X08B_o$

use Proof Template K8025 (Deduction Theorem): (H ∧ I) ⊃ A → H ⊃ (I ⊃ A)
<< K8025.r0t.txt
%0
$\supset \$AC53X08B\ (\supset \$B53X08B\ \$A53X08B)$
$\supset_{ooo}\$AC53X08B_o(\supset_{ooo}\$B53X08B_o\$A53X08B_o)$

##
Q.E.D.
##

%0
$\supset \$AC53X08B\ (\supset \$B53X08B\ \$A53X08B)$
$\supset_{ooo}\$AC53X08B_o(\supset_{ooo}\$B53X08B_o\$A53X08B_o)$

##
Undefine Syntactical Variables
##

:= $\$A53X08B$
:= $\$B53X08B$
:= $\$AC53X08B$
:= $\$C53X08B$
:= $\$HYP1$
:= $\$HYP2$
:= $\$D53X08B$

2.1.50 Results for File A6100.r0.txt

##
Proof A6100: Peano's Postulate No. 1 for Andrews' Definition of Natural Numbers
##
##
Source: [Andrews 2002 (ISBN 1-4020-0763-9), p. 261]
##

Written by Ken Kubota (<mail@kenkubota.de>).
##
This file is part of the publication of the mathematical logic $\mathcal{R}_0$.
For more information, visit: <http://doi.org/10.4444/100.10>
##

<< natural_numbers_andrews.r0.txt
<< K8005.r0.txt

shorthands
:= $\$S\ \ o(ot)$
wff 22 : $o(ot)_\tau$:= $\$S\ \ SIGMA$
:= $\$ANSETZ\ \ p_{o\$S}ATZERO_{\$S}$
wff 312 : $p\,ATZERO_o$:= $\$ANSETZ$
:= $\$ANSETS\ \ \forall_{o(o\backslash 3)\tau}\$S_\tau[\lambda x_{\$S}.(\supset_{ooo}(p_{o\$S}x_{\$S})(p_{o\$S}(ATSUCC_{\$S\$S}x_{\$S})))_o]$
wff 322 : $\forall\$S\,[\lambda x.(\supset(p\,x)\,(p\,(ATSUCC\,x)))]_o$:= $\$ANSETS$
:= $\$ANBOTH\ \ \wedge_{ooo}\$ANSETZ_o\$ANSETS_o$
wff 326 : $\wedge\$ANSETZ\,\$ANSETS_o$:= $\$ANBOTH$
:= $\$P1APP\ \ P1_{o(o\backslash 5)(\backslash 4\backslash 4)\backslash 2\tau}\$S_\tau(AZERO_{o(o\backslash 3)\tau}t_\tau)(ASUCC_{o(o\backslash 4)(o(o\backslash 4))\tau}t_\tau)(ANSET_{o(o(o\backslash 4))\tau}t_\tau)$
wff 1541 : $P1\,\$S\,(AZERO\,t)\,(ASUCC\,t)\,(ANSET\,t)_o$:= $\$P1APP$

.0: expand Peano's postulate

§= $_o$ $\$P1APP$
$=\$P1APP\,\$P1APP$
§\ $P1_{o(o\backslash 5)(\backslash 4\backslash 4)\backslash 2\tau}\$S_\tau$
$=(P1\,\$S)\,[\lambda z.[\lambda s.[\lambda n.(n\,z)]]]$
§s %1 24 %0
$=\$P1APP\,([\lambda z.[\lambda s.[\lambda n.(n\,z)]]]\,(AZERO\,t)\,(ASUCC\,t)\,(ANSET\,t))$
§\ $[\lambda z_{\$S}.[\lambda s_{\$S\$S}.[\lambda n_{o\$S}.(n_{o\$S}z_{\$S})_o]_{(o(o\$S))}]_{(o(o\$S)(\$S\$S))}](AZERO_{o(o\backslash 3)\tau}t_\tau)$
$=([\lambda z.[\lambda s.[\lambda n.(n\,z)]]]\,(AZERO\,t))\,[\lambda s.[\lambda n.(n\,(AZERO\,t))]]$
§s %1 12 %0
$=\$P1APP\,([\lambda s.[\lambda n.(n\,(AZERO\,t))]]\,(ASUCC\,t)\,(ANSET\,t))$
§\ $[\lambda s_{\$S\$S}.[\lambda n_{o\$S}.(n_{o\$S}(AZERO_{o(o\backslash 3)\tau}t_\tau))_o]_{(o(o\$S))}](ASUCC_{o(o\backslash 4)(o(o\backslash 4))\tau}t_\tau)$
$=([\lambda s.[\lambda n.(n\,(AZERO\,t))]]\,(ASUCC\,t))\,[\lambda n.(n\,(AZERO\,t))]$
§s %1 6 %0
$=\$P1APP\,([\lambda n.(n\,(AZERO\,t))]\,(ANSET\,t))$
§\ $[\lambda n_{o\$S}.(n_{o\$S}(AZERO_{o(o\backslash 3)\tau}t_\tau))_o](ANSET_{o(o(o\backslash 4))\tau}t_\tau)$
$=([\lambda n.(n\,(AZERO\,t))]\,(ANSET\,t))\,(ANSET\,t\,(AZERO\,t))$

§s %1 3 %0
\# $=\$P1APP\,(ANSET\,t\,(AZERO\,t))$

:= $\$DTMP6100$ %0
\# wff 1571 : $=\$P1APP\,(ANSET\,t\,(AZERO\,t))_o$:= $\$DTMP6100$

\#\# .1

§= $_o$ /3
\# $=(ANSET\,t\,(AZERO\,t))\,(ANSET\,t\,(AZERO\,t))$

§\ $ANSET_{o(o(o\backslash 4))\tau}t_\tau$
\# $=(ANSET\,t)\,ATNSET$
§s %1 6 %0
\# $=(ANSET\,t\,(AZERO\,t))\,(ATNSET\,(AZERO\,t))$
§\ $ATNSET_{o\$S}(AZERO_{o(o\backslash 3)\tau}t_\tau)$
\# $=(ATNSET\,(AZERO\,t))\,(\forall\,(o\$S)\,[\lambda p.(\supset\$ANBOTH\,(p\,(AZERO\,t)))])$
§s %1 3 %0
\# $=(ANSET\,t\,(AZERO\,t))\,(\forall\,(o\$S)\,[\lambda p.(\supset\$ANBOTH\,(p\,(AZERO\,t)))])$

§\ $AZERO_{o(o\backslash 3)\tau}t_\tau$
\# $=(AZERO\,t)\,ATZERO$
§s %1 63 %0
\# $=(ANSET\,t\,(AZERO\,t))\,(\forall\,(o\$S)\,[\lambda p.(\supset\$ANBOTH\,\$ANSETZ)])$

\#\# use Proof Template A5201b (Swap): A = B → B = A
<< A5201b.r0t.txt
%0
\# $=(\forall\,(o\$S)\,[\lambda p.(\supset\$ANBOTH\,\$ANSETZ)])\,(ANSET\,t\,(AZERO\,t))$
\# $=_{ooo}(\forall_{o(o\backslash 3)\tau}(o\$S)_\tau[\lambda p_{o\$S}.(\supset_{ooo}\$ANBOTH_o\$ANSETZ_o)_o])\ldots$
$\ldots(ANSET_{o(o(o\backslash 4))\tau}t_\tau(AZERO_{o(o\backslash 3)\tau}t_\tau))$

:= $\$TMP6100$ %0
\# wff 1592 : $=(\forall\,(o\$S)\,[\lambda p.(\supset\$ANBOTH\,\$ANSETZ)])\,(ANSET\,t\,(AZERO\,t))_o$
:= $\$TMP6100$

%$K8005$
\# $\supset x\,x$:= $K8005$
\# $\supset_{ooo}x_ox_o$:= $K8005$

\#\# use Proof Template A5221 (Sub): B → B [x/A]
:= $\$B5221$ %0
\# wff 1520 : $\supset x\,x_{o,\ldots}$:= $\$B5221$ $K8005$
:= $\$T5221$ o
\# wff 2 : o_τ := $\$T5221$
:= $\$X5221$ x_o
\# wff 16 : x_o := $\$X5221$
:= $\$A5221$ %1/93
\# wff 326 : $\wedge\$ANSETZ\,\$ANSETS_o$:= $\$A5221$ $\$ANBOTH$

<< A5221.r0t.txt
:= $\$B5221$
:= $\$T5221$
:= $\$X5221$
:= $\$A5221$
%0
$\supset \$ANBOTH\,\$ANBOTH$
$\supset_{ooo}\$ANBOTH_o\$ANBOTH_o$

use Proof Template K8019H: H ⊃ (A ∧ B) → H ⊃ A, H ⊃ B
:= $\$H8019H$ %0
wff 1628 : $\supset \$ANBOTH\,\$ANBOTH_{o,\dots}$:= $\$H8019H$
<< K8019H.r0t.txt
:= $\$H8019H$
%$\$A8019H$
$\supset \$ANBOTH\,\$ANSETZ$:= $\$A8019H$
$\supset_{ooo}\$ANBOTH_o\$ANSETZ_o$:= $\$A8019H$
:= $\$A8019H$
:= $\$B8019H$
%0
$\supset \$ANBOTH\,\$ANSETZ$
$\supset_{ooo}\$ANBOTH_o\$ANSETZ_o$

use Proof Template A5220 (Gen): A → ∀ x: A
:= $\$T5220$ $o\,\$S$
wff 260 : $o\,\$S_\tau$:= $\$T5220$
:= $\$X5220$ $p_{\$T5220}$
wff 311 : $p_{\$T5220}$:= $\$X5220$
:= $\$A5220$ %0
wff 1587 : $\supset \$ANBOTH\,\$ANSETZ_o$:= $\$A5220$
<< A5220.r0t.txt
:= $\$T5220$
:= $\$X5220$
:= $\$A5220$
%0
$\forall\,(o\,\$S)\,[\lambda p.(\supset \$ANBOTH\,\$ANSETZ)]$
$\forall_{o(o\backslash 3)\tau}(o\,\$S)_\tau[\lambda p_{o\,\$S}.(\supset_{ooo}\$ANBOTH_o\$ANSETZ_o)_o]$

%$\$TMP6100$
$=(\forall\,(o\,\$S)\,[\lambda p.(\supset \$ANBOTH\,\$ANSETZ)])\,(ANSET\,t\,(AZERO\,t))$:= $\$TMP6100$
$=_{ooo}(\forall_{o(o\backslash 3)\tau}(o\,\$S)_\tau[\lambda p_{o\,\$S}.(\supset_{ooo}\$ANBOTH_o\$ANSETZ_o)_o])\dots$
$\dots(ANSET_{o(o(o\backslash 4))\tau}t_\tau(AZERO_{o(o\backslash 3)\tau}t_\tau))$:= $\$TMP6100$
:= $\$TMP6100$
§s %1 1 %0
$ANSET\,t\,(AZERO\,t)$

.2: match general definition

:= $\$TMP6100$ %0
wff 1569 : $ANSET\,t\,(AZERO\,t)_{o,\ldots}$:= $\$TMP6100$
%$\$DTMP6100$
$=\$P1APP\,\$TMP6100$:= $\$DTMP6100$
$=_{ooo}\$P1APP_o\$TMP6100_o$:= $\$DTMP6100$
:= $\$DTMP6100$

use Proof Template A5201b (Swap): A = B → B = A
<< A5201b.r0t.txt
%0
$=\$TMP6100\,\$P1APP$
$=_{ooo}\$TMP6100_o\$P1APP_o$

%$\$TMP6100$
$ANSET\,t\,(AZERO\,t)$:= $\$TMP6100$
$ANSET_{o(o(o\backslash 4))_\tau}t_\tau(AZERO_{o(o\backslash 3)_\tau}t_\tau)$:= $\$TMP6100$
:= $\$TMP6100$
§s %0 1 %1
$P1\,\$S\,(AZERO\,t)\,(ASUCC\,t)\,(ANSET\,t)$:= $\$P1APP$

:= $A6100$ %0
wff 1541 : $P1\,\$S\,(AZERO\,t)\,(ASUCC\,t)\,(ANSET\,t)_{o,\ldots}$:= $\$P1APP$ $A6100$

##
Q.E.D.
##

%0
$P1\,\$S\,(AZERO\,t)\,(ASUCC\,t)\,(ANSET\,t)$:= $\$P1APP$ $A6100$
$P1_{o(o\backslash 5)(\backslash 4\backslash 4)\backslash 2\tau}\$S_\tau(AZERO_{o(o\backslash 3)_\tau}t_\tau)(ASUCC_{o(o\backslash 4)(o(o\backslash 4))_\tau}t_\tau)\ldots$
$\ldots(ANSET_{o(o(o\backslash 4))_\tau}t_\tau)$:= $\$P1APP$ $A6100$

undefine local variables
:= $\$S$
:= $\$ANSETZ$
:= $\$ANSETS$
:= $\$ANBOTH$
:= $\$P1APP$

2.1.51 Results for File A6101.r0.txt

##
Proof A6101: Peano's Postulate No. 2 for Andrews' Definition of Natural Numbers
##
##
Source: [Andrews 2002 (ISBN 1-4020-0763-9), p. 261]
##

Written by Ken Kubota (<mail@kenkubota.de>).
##
This file is part of the publication of the mathematical logic $\mathcal{R}_0$.
For more information, visit: <http://doi.org/10.4444/100.10>
##

<< natural_numbers_andrews.r0.txt
<< K8005.r0.txt

shorthands
:= $\$S\ \ o(ot)$
wff 22 : $o(ot)_\tau$:= $\$S\ \ SIGMA$
:= $\$ANSETZ\ \ p_{o\$S}ATZERO_{\$S}$
wff 312 : $p\,ATZERO_o$:= $\$ANSETZ$
:= $\$ANSETS\ \ \forall_{o(o\backslash 3)\tau}\$S_\tau[\lambda x_{\$S}.(\supset_{ooo}(p_{o\$S}x_{\$S})(p_{o\$S}(ATSUCC_{\$S\$S}x_{\$S})))_o]$
wff 322 : $\forall\$S\,[\lambda x.(\supset(p\,x)\,(p\,(ATSUCC\,x)))]_o$:= $\$ANSETS$
:= $\$ANBOTH\ \ \wedge_{ooo}\$ANSETZ_o\$ANSETS_o$
wff 326 : $\wedge\,\$ANSETZ\,\$ANSETS_o$:= $\$ANBOTH$
:= $\$P2APP\ \ P2_{o(o\backslash 5)(\backslash 4\backslash 4)\backslash 2\tau}\$S_\tau(AZERO_{o(o\backslash 3)\tau}t_\tau)(ASUCC_{o(o\backslash 4)(o(o\backslash 4))\tau}t_\tau)(ANSET_{o(o(o\backslash 4))\tau}t_\tau)$
wff 1541 : $P2\,\$S\,(AZERO\,t)\,(ASUCC\,t)\,(ANSET\,t)_o$:= $\$P2APP$
:= $\$ANSETS2\ \ \forall_{o(o\backslash 3)\tau}\$S_\tau[\lambda x_{\$S}.(\supset_{ooo}(n_{o\$S}x_{\$S})(n_{o\$S}(s_{\$S\$S}x_{\$S})))_o]$
wff 1547 : $\forall\$S\,[\lambda x.(\supset(n\,x)\,(n\,(s\,x)))]_o$:= $\$ANSETS2$
:= $\$ANSETS3\ \ \forall_{o(o\backslash 3)\tau}\$S_\tau[\lambda x_{\$S}.(\supset_{ooo}(n_{o\$S}x_{\$S})(n_{o\$S}(ASUCC_{o(o\backslash 4)(o(o\backslash 4))\tau}t_\tau x_{\$S})))_o]$
wff 1552 : $\forall\$S\,[\lambda x.(\supset(n\,x)\,(n\,(ASUCC\,t\,x)))]_o$:= $\$ANSETS3$
:= $\$ANSETx\ \ \forall_{o(o\backslash 3)\tau}(o\$S)_\tau[\lambda p_{o\$S}.(\supset_{ooo}\$ANBOTH_o(p_{o\$S}x_{\$S}))_o]$
wff 1555 : $\forall\,(o\$S)\,[\lambda p.(\supset\$ANBOTH\,(p\,x))]_o$:= $\$ANSETx$

.0: expand Peano's Postulate

§= $\$P2APP$
$=\$P2APP\,\$P2APP$
§\ $P2_{o(o\backslash 5)(\backslash 4\backslash 4)\backslash 2\tau}\$S_\tau$
$=(P2\,\$S)\,[\lambda z.[\lambda s.[\lambda n.\$ANSETS2]]]$
§s %1 24 %0
$=\$P2APP\,([\lambda z.[\lambda s.[\lambda n.\$ANSETS2]]]\,(AZERO\,t)\,(ASUCC\,t)\,(ANSET\,t))$
§\ $[\lambda z_{\$S}.[\lambda s_{\$S\$S}.[\lambda n_{o\$S}.\$ANSETS2_o]_{(o(o\$S))}]_{(o(o\$S)(\$S\$S))}](AZERO_{o(o\backslash 3)\tau}t_\tau)$
$=([\lambda z.[\lambda s.[\lambda n.\$ANSETS2]]]\,(AZERO\,t))\,[\lambda s.[\lambda n.\$ANSETS2]]$
§s %1 12 %0
$=\$P2APP\,([\lambda s.[\lambda n.\$ANSETS2]]\,(ASUCC\,t)\,(ANSET\,t))$
§\ $[\lambda s_{\$S\$S}.[\lambda n_{o\$S}.\$ANSETS2_o]_{(o(o\$S))}](ASUCC_{o(o\backslash 4)(o(o\backslash 4))\tau}t_\tau)$
$=([\lambda s.[\lambda n.\$ANSETS2]]\,(ASUCC\,t))\,[\lambda n.\$ANSETS3]$
§s %1 6 %0
$=\$P2APP\,([\lambda n.\$ANSETS3]\,(ANSET\,t))$
§\ $[\lambda n_{o\$S}.\$ANSETS3_o](ANSET_{o(o(o\backslash 4))\tau}t_\tau)$
$=([\lambda n.\$ANSETS3]\,(ANSET\,t))\,(\forall\$S\,[\lambda x.(ADOT x\,(ANSET\,t\,(ASUCC\,t\,x)))])$
§s %1 3 %0
$=\$P2APP\,(\forall\$S\,[\lambda x.(ADOT x\,(ANSET\,t\,(ASUCC\,t\,x)))])$

:= $\$DTMP6101$ %0
\# wff 1584 : $=\$P2APP\,(\forall\,\$S\,[\lambda x.(ADOTx\,(ANSET\,t\,(ASUCC\,t\,x)))])_o$:= $\$DTMP6101$

\#\# .1

%$K8005$
\# $\supset x\,x$:= $K8005$
\# $\supset_{ooo} x_o x_o$:= $K8005$

\#\# use Proof Template A5221 (Sub): B → B [x/A]
:= $\$B5221$ %0
\# wff 1520 : $\supset x\,x_{o,\dots}$:= $\$B5221$ $K8005$
:= $\$T5221$ o
\# wff 2 : o_τ := $\$T5221$
:= $\$X5221$ x_o
\# wff 16 : x_o := $\$X5221$
:= $\$A5221$ $[\lambda n_{\$S}.(\forall_{o(o\backslash 3)\tau}(o\,\$S)_\tau[\lambda p_{o\,\$S}.(\supset_{ooo}\$ANBOTH_o(p_{o\,\$S}n_{\$S}))_o])_o]/61$
\# wff 326 : $\wedge\,\$ANSETZ\,\$ANSETS_o$:= $\$A5221$ $\$ANBOTH$
<< A5221.r0t.txt
:= $\$B5221$
:= $\$T5221$
:= $\$X5221$
:= $\$A5221$

:= $\$TMP6101$ %0
\# wff 1620 : $\supset\$ANBOTH\,\$ANBOTH_{o,\dots}$:= $\$TMP6101$

%$K8005$
\# $\supset x\,x$:= $K8005$
\# $\supset_{ooo} x_o x_o$:= $K8005$

\#\# use Proof Template A5221 (Sub): B → B [x/A]
:= $\$B5221$ %0
\# wff 1520 : $\supset x\,x_{o,\dots}$:= $\$B5221$ $K8005$
:= $\$T5221$ o
\# wff 2 : o_τ := $\$T5221$
:= $\$X5221$ x_o
\# wff 16 : x_o := $\$X5221$
:= $\$A5221$ $ANSET_{o(o(o\backslash 4))\tau}t_\tau x_{\$S}$
\# wff 363 : $ANSET\,t\,x_o$:= $\$A5221$
<< A5221.r0t.txt
:= $\$B5221$
:= $\$T5221$
:= $\$X5221$
:= $\$A5221$
%0
\# $ADOTx\,(ANSET\,t\,x)$
\# $ADOTx_{oo}(ANSET_{o(o(o\backslash 4))\tau}t_\tau x_{\$S})$

%$\$TMP6101$
$\supset \$ANBOTH\,\$ANBOTH$:= $\$TMP6101$
$\supset_{ooo}\$ANBOTH_o\$ANBOTH_o$:= $\$TMP6101$
:= $\$TMP6101$

use Proof Template K8004 (Trans): (H ⊕ A), B → H ⊃ B
:= $\$HA8004$ %1
wff 1631 : $ADOTx\,(ANSET\,t\,x)_{o,\ldots}$:= $\$HA8004$
:= $\$B8004$ %0
wff 1620 : $\supset \$ANBOTH\,\$ANBOTH_{o,\ldots}$:= $\$B8004$
<< K8004.r0t.txt
:= $\$HA8004$
:= $\$B8004$
%0
$ADOTx\,(\supset \$ANBOTH\,\$ANBOTH)$
$ADOTx_{oo}(\supset_{ooo}\$ANBOTH_o\$ANBOTH_o)$

use Proof Template K8026 (Deduction Theorem Reversed): H ⊃ (I ⊃ A) → (H ∧ I) ⊃ A
<< K8026.r0t.txt
%0
$\supset (\wedge\,(ANSET\,t\,x)\,\$ANBOTH)\,\$ANBOTH$
$\supset_{ooo}(\wedge_{ooo}(ANSET_{o(o(o\backslash 4))\tau}t_\tau x_{\$S})\$ANBOTH_o)\$ANBOTH_o$

:= $\$LTMP6101$ %0
wff 4384 : $\supset (\wedge\,(ANSET\,t\,x)\,\$ANBOTH)\,\$ANBOTH_{o,\ldots}$:= $\$LTMP6101$

.2

%$K8005$
$\supset x\,x$:= $K8005$
$\supset_{ooo}x_o x_o$:= $K8005$

use Proof Template A5221 (Sub): B → B [x/A]
:= $\$B5221$ %0
wff 1520 : $\supset x\,x_{o,\ldots}$:= $\$B5221$ $K8005$
:= $\$T5221$ o
wff 2 : o_τ := $\$T5221$
:= $\$X5221$ x_o
wff 16 : x_o := $\$X5221$
:= $\$A5221$ %1/5
wff 4344 : $\wedge\,(ANSET\,t\,x)\,\$ANBOTH_o$:= $\$A5221$
<< A5221.r0t.txt
:= $\$B5221$
:= $\$T5221$
:= $\$X5221$
:= $\$A5221$
%0

$\supset (\wedge\,(ANSET\,t\,x)\,\$ANBOTH)\,(\wedge\,(ANSET\,t\,x)\,\$ANBOTH)$
$\supset_{ooo}(\wedge_{ooo}(ANSET_{o(o(o\backslash 4))_\tau}t_\tau x_{\$S})\$ANBOTH_o)\ldots$
$\ldots(\wedge_{ooo}(ANSET_{o(o(o\backslash 4))_\tau}t_\tau x_{\$S})\$ANBOTH_o)$

use Proof Template K8019H: H ⊃ (A ∧ B) → H ⊃ A, H ⊃ B
:= $\$H8019H$ %0
wff 4400 : $\supset (\wedge\,(ANSET\,t\,x)\,\$ANBOTH)\,(\wedge\,(ANSET\,t\,x)\,\$ANBOTH)_{o,\ldots}$:=
$\$H8019H$
<< K8019H.r0t.txt
:= $\$H8019H$
%$\$A8019H$
$\supset (\wedge\,(ANSET\,t\,x)\,\$ANBOTH)\,(ANSET\,t\,x)$:= $\$A8019H$
$\supset_{ooo}(\wedge_{ooo}(ANSET_{o(o(o\backslash 4))_\tau}t_\tau x_{\$S})\$ANBOTH_o)(ANSET_{o(o(o\backslash 4))_\tau}t_\tau x_{\$S})$:=
$\$A8019H$
:= $\$A8019H$
:= $\$B8019H$
%0
$\supset (\wedge\,(ANSET\,t\,x)\,\$ANBOTH)\,(ANSET\,t\,x)$
$\supset_{ooo}(\wedge_{ooo}(ANSET_{o(o(o\backslash 4))_\tau}t_\tau x_{\$S})\$ANBOTH_o)(ANSET_{o(o(o\backslash 4))_\tau}t_\tau x_{\$S})$

§\ $ANSET_{o(o(o\backslash 4))_\tau}t_\tau$
$=(ANSET\,t)\,ATNSET$
§s %1 6 %0
$\supset (\wedge\,(ANSET\,t\,x)\,\$ANBOTH)\,(ATNSET\,x)$
§\ $ATNSET_{o\$S}x_{\$S}$
$=(ATNSET\,x)\,\$ANSETx$
§s %1 3 %0
$\supset (\wedge\,(ANSET\,t\,x)\,\$ANBOTH)\,\$ANSETx$

use Proof Template A5215H (∀ I): H ⊃ ∀ x: B → H ⊃ B [x/a]
:= $\$T5215H$ $o\$S$
wff 260 : $o\$S_\tau$:= $\$T5215H$
:= $\$X5215H$ $p_{\$T5215H}$
wff 311 : $p_{\$T5215H}$:= $\$X5215H$
:= $\$A5215H$ $p_{\$T5215H}$
wff 311 : $p_{\$T5215H}$:= $\$A5215H$ $\$X5215H$
:= $\$H5215H$ %0
wff 4601 : $\supset (\wedge\,(ANSET\,t\,x)\,\$ANBOTH)\,\$ANSETx_o$:= $\$H5215H$
<< A5215H.r0t.txt
:= $\$T5215H$
:= $\$X5215H$
:= $\$A5215H$
:= $\$H5215H$
%0
$\supset (\wedge\,(ANSET\,t\,x)\,\$ANBOTH)\,(\supset \$ANBOTH\,(p\,x))$
$\supset_{ooo}(\wedge_{ooo}(ANSET_{o(o(o\backslash 4))_\tau}t_\tau x_{\$S})\$ANBOTH_o)(\supset_{ooo}\$ANBOTH_o(p_{o\$S}x_{\$S}))$

use Proof Template A5224H (MP): H ⊃ A, H ⊃ (A ⊃ B) → H ⊃ B
:= $\$AB5224H$ %0

wff 4668 : $\supset(\wedge(ANSET\,t\,x)\,\$ANBOTH)\,(\supset \$ANBOTH\,(p\,x))_o$:= $\$AB5224H$
:= $\$A5224H$ $\supset_{ooo}(\wedge_{ooo}(ANSET_{o(o(o\backslash 4))\tau}t_\tau x_{\$S})\$ANBOTH_o)\$ANBOTH_o$
wff 4384 : $\supset(\wedge(ANSET\,t\,x)\,\$ANBOTH)\,\$ANBOTH_{o,\dots}$:= $\$A5224H$
$\$LTMP6101$
<< A5224H.r0t.txt
:= $\$AB5224H$
:= $\$A5224H$

:= $\$TMP6101$ %0
wff 4794 : $\supset(\wedge(ANSET\,t\,x)\,\$ANBOTH)\,(p\,x)_o$:= $\$TMP6101$

.3

%$\$LTMP6101$
$\supset(\wedge(ANSET\,t\,x)\,\$ANBOTH)\,\$ANBOTH$:= $\$LTMP6101$
$\supset_{ooo}(\wedge_{ooo}(ANSET_{o(o(o\backslash 4))\tau}t_\tau x_{\$S})\$ANBOTH_o)\$ANBOTH_o$:=
$\$LTMP6101$
:= $\$LTMP6101$

use Proof Template K8019H: H ⊃ (A ∧ B) → H ⊃ A, H ⊃ B
:= $\$H8019H$ %0
wff 4384 : $\supset(\wedge(ANSET\,t\,x)\,\$ANBOTH)\,\$ANBOTH_{o,\dots}$:= $\$H8019H$
<< K8019H.r0t.txt
:= $\$H8019H$
%$\$B8019H$
$\supset(\wedge(ANSET\,t\,x)\,\$ANBOTH)\,\$ANSETS$:= $\$B8019H$
$\supset_{ooo}(\wedge_{ooo}(ANSET_{o(o(o\backslash 4))\tau}t_\tau x_{\$S})\$ANBOTH_o)\$ANSETS_o$:= $\$B8019H$
:= $\$A8019H$
:= $\$B8019H$
%0
$\supset(\wedge(ANSET\,t\,x)\,\$ANBOTH)\,\$ANSETS$
$\supset_{ooo}(\wedge_{ooo}(ANSET_{o(o(o\backslash 4))\tau}t_\tau x_{\$S})\$ANBOTH_o)\$ANSETS_o$

use Proof Template A5215H (∀ I): H ⊃ ∀ x: B → H ⊃ B [x/a]
:= $\$T5215H$ $o(ot)$
wff 22 : $o(ot)_\tau$:= $\$S$ $\$T5215H$ $SIGMA$
:= $\$X5215H$ $x_{\$S}$
wff 315 : $x_{\$S}$:= $\$X5215H$
:= $\$A5215H$ $x_{\$S}$
wff 315 : $x_{\$S}$:= $\$A5215H$ $\$X5215H$
:= $\$H5215H$ %0
wff 4860 : $\supset(\wedge(ANSET\,t\,\$X5215H)\,\$ANBOTH)\,\$ANSETS_o$:= $\$H5215H$
<< A5215H.r0t.txt
:= $\$T5215H$
:= $\$X5215H$
:= $\$A5215H$
:= $\$H5215H$
%0
$\supset(\wedge(ANSET\,t\,x)\,\$ANBOTH)\,(\supset(p\,x)\,(p\,(ATSUCC\,x)))$

\# $\supset_{ooo}(\wedge_{ooo}(ANSET_{o(o(o\backslash 4))_\tau}t_\tau x_{\$S})\$ANBOTH_o)\ldots$
$\ldots(\supset_{ooo}(p_{o\$S}x_{\$S})(p_{o\$S}(ATSUCC_{\$S\$S}x_{\$S})))$

\#\# use Proof Template A5224H (MP): H ⊃ A, H ⊃ (A ⊃ B) → H ⊃ B
:= $\$AB5224H$ %0
\# wff 4923 : $\supset(\wedge(ANSET\,t\,x)\,\$ANBOTH)\,(\supset(p\,x)\,(p\,(ATSUCC\,x)))_o$:=
$\$AB5224H$
:= $\$A5224H$ $\supset_{ooo}(\wedge_{ooo}(ANSET_{o(o(o\backslash 4))_\tau}t_\tau x_{\$S})\$ANBOTH_o)(p_{o\$S}x_{\$S})$
\# wff 4794 : $\supset(\wedge(ANSET\,t\,x)\,\$ANBOTH)\,(p\,x)_o$:= $\$A5224H$ $\$TMP6101$
:= $\$TMP6101$
<< A5224H.r0t.txt
:= $\$AB5224H$
:= $\$A5224H$
%0
\# $\supset(\wedge(ANSET\,t\,x)\,\$ANBOTH)\,(p\,(ATSUCC\,x))$
\# $\supset_{ooo}(\wedge_{ooo}(ANSET_{o(o(o\backslash 4))_\tau}t_\tau x_{\$S})\$ANBOTH_o)(p_{o\$S}(ATSUCC_{\$S\$S}x_{\$S}))$

\#\# .4

\#\# use Proof Template K8025 (Deduction Theorem): (H ∧ I) ⊃ A → H ⊃ (I ⊃ A)
<< K8025.r0t.txt
%0
\# $ADOTx\,(\supset\$ANBOTH\,(p\,(ATSUCC\,x)))$
\# $ADOTx_{oo}(\supset_{ooo}\$ANBOTH_o(p_{o\$S}(ATSUCC_{\$S\$S}x_{\$S})))$

\#\# use Proof Template A5220H (Gen): (H ⊃ A) → (H ⊃ ∀ x: A)
:= $\$T5220H$ $o\$S$
\# wff 260 : $o\$S_\tau$:= $\$T5220H$
:= $\$X5220H$ $p_{\$T5220H}$
\# wff 311 : $p_{\$T5220H}$:= $\$X5220H$
:= $\$A5220H$ %0
\# wff 5028 : $ADOTx\,(\supset\$ANBOTH\,(\$X5220H\,(ATSUCC\,x)))_{o,\ldots}$:= $\$A5220H$
<< A5220H.r0t.txt
:= $\$T5220H$
:= $\$X5220H$
:= $\$A5220H$

:= $\$TMP6101$ %0
\# wff 5171 : $ADOTx\,(\forall\,(o\$S)\,[\lambda p.(\supset\$ANBOTH\,(p\,(ATSUCC\,x)))])_o$:=
$\$TMP6101$

§= $ANSET_{o(o(o\backslash 4))_\tau}t_\tau(ASUCC_{o(o\backslash 4)(o(o\backslash 4))_\tau}t_\tau x_{\$S})$
\# $=(ANSET\,t\,(ASUCC\,t\,x))\,(ANSET\,t\,(ASUCC\,t\,x))$
§\ $ANSET_{o(o(o\backslash 4))_\tau}t_\tau$
\# $=(ANSET\,t)\,ATNSET$
§s %1 10 %0
\# $=(ATNSET\,(ASUCC\,t\,x))\,(ANSET\,t\,(ASUCC\,t\,x))$
§\ $ATNSET_{o\$S}(ASUCC_{o(o\backslash 4)(o(o\backslash 4))_\tau}t_\tau x_{\$S})$
\# $=(ATNSET\,(ASUCC\,t\,x))\,(\forall\,(o\$S)\,[\lambda p.(\supset\$ANBOTH\,(p\,(ASUCC\,t\,x)))])$

§s %1 5 %0
$= (\forall (o\,\$S)\,[\lambda p.(\supset \$ANBOTH\,(p\,(ASUCC\,t\,x)))])\,(ANSET\,t\,(ASUCC\,t\,x))$
§\ $ASUCC_{o(o\backslash 4)(o(o\backslash 4))_\tau} t_\tau$
$= (ASUCC\,t)\,ATSUCC$
§s %1 190 %0
$= (\forall (o\,\$S)\,[\lambda p.(\supset \$ANBOTH\,(p\,(ATSUCC\,x)))])\,(ANSET\,t\,(ASUCC\,t\,x))$

%$\$TMP6101$
$ADOTx\,(\forall (o\,\$S)\,[\lambda p.(\supset \$ANBOTH\,(p\,(ATSUCC\,x)))])$:= $\$TMP6101$
$ADOTx_{oo} \ldots$
$\ldots (\forall_{o(o\backslash 3)\tau}(o\,\$S)_\tau[\lambda p_{o\,\$S}.(\supset_{ooo}\$ANBOTH_o(p_{o\,\$S}(ATSUCC_{\$S\,\$S}x_{\$S})))_o])$:= $\$TMP6101$
:= $\$TMP6101$
§s %0 3 %1
$ADOTx\,(ANSET\,t\,(ASUCC\,t\,x))$

.5: match general definition

use Proof Template A5220H (Gen): (H ⊃ A) → (H ⊃ ∀ x: A)
:= $\$T5220$ $o(ot)$
wff 22 : $o(ot)_\tau$:= $\$S$ $\$T5220$ $SIGMA$
:= $\$X5220$ $x_{\$S}$
wff 315 : $x_{\$S}$:= $\$X5220$
:= $\$A5220$ %0
wff 1580 : $ADOTx\,(ANSET\,t\,(ASUCC\,t\,\$X5220))_o$:= $\$A5220$
<< A5220.r0t.txt
:= $\$T5220$
:= $\$X5220$
:= $\$A5220$
%0
$\forall \$S\,[\lambda x.(ADOTx\,(ANSET\,t\,(ASUCC\,t\,x)))]$
$\forall_{o(o\backslash 3)\tau}\$S_\tau[\lambda x_{\$S}.(ADOTx_{oo}(ANSET_{o(o(o\backslash 4))_\tau}t_\tau(ASUCC_{o(o\backslash 4)(o(o\backslash 4))_\tau}t_\tau x_{\$S})))_o]$

:= $\$TMP6101$ %0
wff 1582 : $\forall \$S\,[\lambda x.(ADOTx\,(ANSET\,t\,(ASUCC\,t\,x)))]_{o,\ldots}$:= $\$TMP6101$
%$\$DTMP6101$
$= \$P2APP\,\$TMP6101$:= $\$DTMP6101$
$=_{o\omega\omega}\$P2APP_\omega\$TMP6101_\omega$:= $\$DTMP6101$
:= $\$DTMP6101$

use Proof Template A5201b (Swap): A = B → B = A
<< A5201b.r0t.txt
%0
$= \$TMP6101\,\$P2APP$
$=_{o\omega\omega}\$TMP6101_\omega\$P2APP_\omega$

%$\$TMP6101$
$\forall \$S\,[\lambda x.(ADOTx\,(ANSET\,t\,(ASUCC\,t\,x)))]$:= $\$TMP6101$
$\forall_{o(o\backslash 3)\tau}\$S_\tau \ldots$
$\ldots [\lambda x_{\$S}.(ADOTx_{oo}(ANSET_{o(o(o\backslash 4))_\tau}t_\tau(ASUCC_{o(o\backslash 4)(o(o\backslash 4))_\tau}t_\tau x_{\$S})))_o]$:= $\$TMP6101$

:= $\$TMP6101$
§s %0 1 %1
$P2\,\$S\,(AZERO\,t)\,(ASUCC\,t)\,(ANSET\,t)$:= $\$P2APP$

:= $A6101$ %0
wff 1541 : $P2\,\$S\,(AZERO\,t)\,(ASUCC\,t)\,(ANSET\,t)_{o,\ldots}$:= $\$P2APP$ $A6101$

##
Q.E.D.
##

%0
$P2\,\$S\,(AZERO\,t)\,(ASUCC\,t)\,(ANSET\,t)$:= $\$P2APP$ $A6101$
$P2_{o(o\backslash 5)(\backslash 4\backslash 4)\backslash 2\tau}\$S_{\tau}(AZERO_{o(o\backslash 3)\tau}t_{\tau})(ASUCC_{o(o\backslash 4)(o(o\backslash 4))\tau}t_{\tau})\ldots$
$\ldots(ANSET_{o(o(o\backslash 4))\tau}t_{\tau})$:= $\$P2APP$ $A6101$

undefine local variables
:= $\$S$
:= $\$ANSETZ$
:= $\$ANSETS$
:= $\$ANBOTH$
:= $\$P2APP$
:= $\$ANSETS2$
:= $\$ANSETS3$
:= $\$ANSETx$

2.1.52 Results for File A6102.r0.txt

##
Proof A6102: Peano's Postulate No. 5 for Andrews' Definition of Natural Numbers
##
##
Source: [Andrews 2002 (ISBN 1-4020-0763-9), p. 262]
##

Written by Ken Kubota (<mail@kenkubota.de>).
##
This file is part of the publication of the mathematical logic $\mathcal{R}_0$.
For more information, visit: <http://doi.org/10.4444/100.10>
##

<< natural_numbers_andrews.r0.txt
<< K8021.r0.txt
<< A6100.r0.txt
<< A6101.r0.txt

definition of P

:= $\$S \;\; o(ot)$
\# wff 22 : $o(ot)_{\tau}$:= $\$S \;\; SIGMA$
:= $\$P \;\; [\lambda t_{\$S}.(\wedge_{ooo}(ANSET_{o(o(o\backslash 4))\tau} t_{\tau} t_{\$S})(p_{o\$S} t_{\$S}))_{o}]$
\# wff 6719 : $[\lambda t.(\wedge\,(ANSET\,t\,t)\,(p\,t))]_{o\$S}$:= $\$P$

\#\# shorthands
:= $\$T3 \;\; o(o\backslash 3)\tau$
\# wff 33 : $o(o\backslash 3)\tau_{\tau}$:= $\$T3$
:= $\$T4 \;\; o(o(o\backslash 4))\tau$
\# wff 335 : $o(o(o\backslash 4))\tau_{\tau}$:= $\$T4$
:= $\$T44 \;\; o(o\backslash 4)(o(o\backslash 4))\tau$
\# wff 310 : $o(o\backslash 4)(o(o\backslash 4))\tau_{\tau}$:= $\$T44$
:= $\$To2S \;\; o(o\$S)$
\# wff 314 : $o(o\$S)_{\tau}$:= $\$To2S$
:= $\$To2S3 \;\; \$To2S\,(\$S\,\$S)$
\# wff 3246 : $\$To2S\,(\$S\,\$S)_{\tau}$:= $\$To2S3$
:= $\$P5APP \;\; P5_{o(o\backslash 5)(\backslash 4\backslash 4)\backslash 2\tau}\$S_{\tau}(AZERO_{\$T3}t_{\tau})(ASUCC_{\$T44}t_{\tau})(ANSET_{\$T4}t_{\tau})$
\# wff 6723 : $P5\,\$S\,(AZERO\,t)\,(ASUCC\,t)\,(ANSET\,t)_{o}$:= $\$P5APP$
:= $\$ANSETZ \;\; p_{o\$S} ATZERO_{\$S}$
\# wff 312 : $p\,ATZERO_{o,\ldots}$:= $\$ANSETZ$
:= $\$ANSETS \;\; \forall_{\$T3}\$S_{\tau}[\lambda x_{\$S}.(\supset_{ooo}(p_{o\$S}x_{\$S})(p_{o\$S}(ATSUCC_{\$S\$S}x_{\$S})))_{o}]$
\# wff 322 : $\forall\,\$S\,[\lambda x.(\supset(p\,x)\,(p\,(ATSUCC\,x)))]_{o,\ldots}$:= $\$ANSETS$
:= $\$ANBOTH \;\; \wedge_{ooo}\$ANSETZ_{o}\$ANSETS_{o}$
\# wff 326 : $\wedge\,\$ANSETZ\,\$ANSETS_{o,\ldots}$:= $\$ANBOTH$
:= $\$ANSETS2 \;\; \forall_{\$T3}\$S_{\tau}[\lambda x_{\$S}.(\supset_{ooo}(n_{o\$S}x_{\$S})(n_{o\$S}(s_{\$S\$S}x_{\$S})))_{o}]$
\# wff 3499 : $\forall\,\$S\,[\lambda x.(\supset(n\,x)\,(n\,(s\,x)))]_{o}$:= $\$ANSETS2$
:= $\$ANSETS3 \;\; \forall_{\$T3}\$S_{\tau}[\lambda x_{\$S}.(\supset_{ooo}(n_{o\$S}x_{\$S})(n_{o\$S}(ASUCC_{\$T44}t_{\tau}x_{\$S})))_{o}]$
\# wff 3504 : $\forall\,\$S\,[\lambda x.(\supset(n\,x)\,(n\,(ASUCC\,t\,x)))]_{o}$:= $\$ANSETS3$
:= $\$ANSETx \;\; \forall_{\$T3}(o\$S)_{\tau}[\lambda p_{o\$S}.(\supset_{ooo}\$ANBOTH_{o}(p_{o\$S}x_{\$S}))_{o}]$
\# wff 3507 : $\forall\,(o\$S)\,[\lambda p.(\supset\,\$ANBOTH\,(p\,x))]_{o,\ldots}$:= $\$ANSETx$
:= $\$ZRO \;\; p_{o\$S}z_{\$S}$
\# wff 6724 : $p\,z_{o}$:= $\$ZRO$
:= $\$SCC \;\; \forall_{\$T3}\$S_{\tau}[\lambda x_{\$S}.(\supset_{ooo}(n_{o\$S}x_{\$S})(\supset_{ooo}(p_{o\$S}x_{\$S})(p_{o\$S}(s_{\$S\$S}x_{\$S}))))_{o}]$
\# wff 6729 : $\forall\,\$S\,[\lambda x.(\supset(n\,x)\,(\supset(p\,x)\,(p\,(s\,x))))]_{o}$:= $\$SCC$
:= $\$ALL \;\; \forall_{\$T3}\$S_{\tau}[\lambda x_{\$S}.(\supset_{ooo}(n_{o\$S}x_{\$S})(p_{o\$S}x_{\$S}))_{o}]$
\# wff 6732 : $\forall\,\$S\,[\lambda x.(\supset(n\,x)\,(p\,x))]_{o}$:= $\$ALL$
:= $\$IDC \;\; \forall_{\$T3}(o\$S)_{\tau}[\lambda p_{o\$S}.(\supset_{ooo}(\wedge_{ooo}\$ZRO_{o}\$SCC_{o})\$ALL_{o})_{o}]$
\# wff 6738 : $\forall\,(o\$S)\,[\lambda p.(\supset(\wedge\,\$ZRO\,\$SCC)\,\$ALL)]_{o}$:= $\$IDC$
:= $\$P5S \;\; [\lambda z_{\$S}.[\lambda s_{\$S\$S}.[\lambda n_{o\$S}.\$IDC_{o}]_{\$To2S}]_{\$To2S3}](AZERO_{\$T3}t_{\tau})(ASUCC_{\$T44}t_{\tau})(ANSET_{\$T4}t_{\tau})$
\# wff 6744 : $[\lambda z.[\lambda s.[\lambda n.\$IDC]]]\,(AZERO\,t)\,(ASUCC\,t)\,(ANSET\,t)_{o}$:= $\$P5S$
:= $\$IDC0 \;\; \forall_{\$T3}(o\$S)_{\tau}[\lambda p_{o\$S}.(\supset_{ooo}(\wedge_{ooo}(p_{o\$S}(AZERO_{\$T3}t_{\tau}))\$SCC_{o})\$ALL_{o})_{o}]$
\# wff 6750 : $\forall\,(o\$S)\,[\lambda p.(\supset(\wedge\,(p\,(AZERO\,t))\,\$SCC)\,\$ALL)]_{o}$:= $\$IDC0$
:= $\$P5S0 \;\; [\lambda s_{\$S\$S}.[\lambda n_{o\$S}.\$IDC0_{o}]_{\$To2S}]$
\# wff 6752 : $[\lambda s.[\lambda n.\$IDC0]]_{\$To2S3}$:= $\$P5S0$
:= $\$ZRO2 \;\; p_{o\$S}(AZERO_{\$T3}t_{\tau})$
\# wff 3290 : $p\,(AZERO\,t)_{o}$:= $\$ZRO2$
:= $\$SCC2 \;\; \forall_{\$T3}\$S_{\tau}[\lambda x_{\$S}.(\supset_{ooo}(n_{o\$S}x_{\$S})(\supset_{ooo}(p_{o\$S}x_{\$S})(p_{o\$S}(ASUCC_{\$T44}t_{\tau}x_{\$S}))))_{o}]$
\# wff 6756 : $\forall\,\$S\,[\lambda x.(\supset(n\,x)\,(\supset(p\,x)\,(p\,(ASUCC\,t\,x))))]_{o}$:= $\$SCC2$
:= $\$P5S0SC \;\; [\lambda n_{o\$S}.(\forall_{\$T3}(o\$S)_{\tau}[\lambda p_{o\$S}.(\supset_{ooo}(\wedge_{ooo}\$ZRO2_{o}\$SCC2_{o})\$ALL_{o})_{o}])_{o}]$

wff 6762 : $[\lambda n.(\forall(o\$S)[\lambda p.(\supset(\wedge\$ZRO2\$SCC2)\$ALL)])]_{\$To2S}$:= $\$P5S0SC$
:= $\$SCC3$ $\forall_{\$T3}\$S_\tau[\lambda x_{\$S}.(ADOTx_{oo}(\supset_{ooo}(p_{o\$S}x_{\$S})(p_{o\$S}(ASUCC_{\$T44}t_\tau x_{\$S}))))_o]$
wff 6765 : $\forall\$S[\lambda x.(ADOTx(\supset(p\,x)(p(ASUCC\,t\,x))))]_o$:= $\$SCC3$
:= $\$ALL3$ $\forall_{\$T3}\$S_\tau[\lambda x_{\$S}.(ADOTx_{oo}(p_{o\$S}x_{\$S}))_o]$
wff 6768 : $\forall\$S[\lambda x.(ADOTx(p\,x))]_o$:= $\$ALL3$
:= $\$P5S0SCST$ $\forall_{\$T3}(o\$S)_\tau[\lambda p_{o\$S}.(\supset_{ooo}(\wedge_{ooo}\$ZRO2_o\$SCC3_o)\$ALL3_o)_o]$
wff 6773 : $\forall(o\$S)[\lambda p.(\supset(\wedge\$ZRO2\$SCC3)\$ALL3)]_o$:= $\$P5S0SCST$
:= $\$STSC$ $ANSET_{\$T4}t_\tau(ASUCC_{\$T44}t_\tau x_{\$S})$
wff 3530 : $ANSET\,t\,(ASUCC\,t\,x)_{o,\ldots}$:= $\$STSC$
:= $\$HPTMP$ $\supset_{ooo}(\wedge_{ooo}(\wedge_{ooo}\$ZRO2_o\$SCC3_o)(ANSET_{\$T4}t_\tau x_{\$S}))\$STSC_o$
wff 6777 : $\supset(\wedge(\wedge\$ZRO2\$SCC3)(ANSET\,t\,x))\$STSC_o$:= $\$HPTMP$
:= $\$SCCP$ $\forall_{\$T3}\$S_\tau[\lambda x_{\$S}.(\supset_{ooo}(\$P_{o\$S}x_{\$S})(\$P_{o\$S}(ASUCC_{\$T44}t_\tau x_{\$S})))_o]$
wff 6783 : $\forall\$S[\lambda x.(\supset(\$P\,x)(\$P(ASUCC\,t\,x)))]_o$:= $\$SCCP$
:= $\$SCCPT$ $\forall_{\$T3}\$S_\tau[\lambda x_{\$S}.(\supset_{ooo}(\$P_{o\$S}x_{\$S})(\$P_{o\$S}(ATSUCC_{\$S\$S}x_{\$S})))_o]$
wff 6787 : $\forall\$S[\lambda x.(\supset(\$P\,x)(\$P(ATSUCC\,x)))]_o$:= $\$SCCPT$
:= $\$ZROSCCT$ $\wedge_{ooo}(\$P_{o\$S}ATZERO_{\$S})\$SCCPT_o$
wff 6790 : $\wedge(\$P\,ATZERO)\$SCCPT_o$:= $\$ZROSCCT$
:= $\$HTMP2$ $\wedge_{ooo}(\wedge_{ooo}(\wedge_{ooo}\$ZRO2_o\$SCC3_o)(ANSET_{\$T4}t_\tau x_{\$S}))$
wff 6791 : $\wedge(\wedge(\wedge\$ZRO2\$SCC3)(ANSET\,t\,x))_{oo}$:= $\$HTMP2$

.0: expand Peano's postulate

§= $\$P5APP$
$=\$P5APP\,\$P5APP$
§\ $P5_{o(o\backslash5)(\backslash4\backslash4)\backslash2\tau}\$S_\tau$
$=(P5\,\$S)[\lambda z.[\lambda s.[\lambda n.\$IDC]]]$
§s %1 24 %0
$=\$P5APP\,\$P5S$
§\ $[\lambda z_{\$S}.[\lambda s_{\$S\$S}.[\lambda n_{o\$S}.\$IDC_o]_{\$To2S}]_{\$To2S3}](AZERO_{\$T3}t_\tau)$
$=([\lambda z.[\lambda s.[\lambda n.\$IDC]]](AZERO\,t))\,\$P5S0$
§s %1 12 %0
$=\$P5APP\,(\$P5S0\,(ASUCC\,t)\,(ANSET\,t))$
§\ $\$P5S0_{\$To2S3}(ASUCC_{\$T44}t_\tau)$
$=(\$P5S0\,(ASUCC\,t))\,\$P5S0SC$
§s %1 6 %0
$=\$P5APP\,(\$P5S0SC\,(ANSET\,t))$
§\ $\$P5S0SC_{\$To2S}(ANSET_{\$T4}t_\tau)$
$=(\$P5S0SC\,(ANSET\,t))\,\$P5S0SCST$
§s %1 3 %0
$=\$P5APP\,\$P5S0SCST$

:= $\$D0TMP$ %0
wff 6808 : $=\$P5APP\,\$P5S0SCST_o$:= $\$D0TMP$

.1

%$K8005$
$\supset x\,x$:= $K8005$
$\supset_{ooo}x_o x_o$:= $K8005$

use Proof Template A5221 (Sub): B $\rightarrow$ B [x/A]
:= $\$B5221$ %0
wff 1520 : $\supset x\, x_{o,\ldots}$:= $\$B5221$ $K8005$
:= $\$T5221$ o
wff 2 : o_τ := $\$T5221$
:= $\$X5221$ x_o
wff 16 : x_o := $\$X5221$
:= $\$A5221$ $\wedge_{ooo}\$ZRO2_o\$SCC3_o$
wff 6769 : $\wedge \$ZRO2\,\$SCC3_o$:= $\$A5221$
<< A5221.r0t.txt
:= $\$B5221$
:= $\$T5221$
:= $\$X5221$
:= $\$A5221$

:= $\$HTMP$ %0/5
wff 6769 : $\wedge \$ZRO2\,\$SCC3_o$:= $\$HTMP$
:= $\$D1TMP$ %0
wff 6817 : $\supset \$HTMP\,\$HTMP_{o,\ldots}$:= $\$D1TMP$

.2

%$K8005$
$\supset x\, x$:= $K8005$
$\supset_{ooo} x_o x_o$:= $K8005$

use Proof Template A5221 (Sub): B $\rightarrow$ B [x/A]
:= $\$B5221$ %0
wff 1520 : $\supset x\, x_{o,\ldots}$:= $\$B5221$ $K8005$
:= $\$T5221$ o
wff 2 : o_τ := $\$T5221$
:= $\$X5221$ x_o
wff 16 : x_o := $\$X5221$
:= $\$A5221$ $ANSET_{\$T4} t_\tau y_{\$S}$
wff 6821 : $ANSET\, t\, y_o$:= $\$A5221$
<< A5221.r0t.txt
:= $\$B5221$
:= $\$T5221$
:= $\$X5221$
:= $\$A5221$
%0
$\supset (ANSET\, t\, y)\, (ANSET\, t\, y)$
$\supset_{ooo}(ANSET_{\$T4} t_\tau y_{\$S})(ANSET_{\$T4} t_\tau y_{\$S})$

§\ $ANSET_{\$T4} t_\tau$
$= (ANSET\, t)\; ATNSET$
§s %1 6 %0
$\supset (ANSET\, t\, y)\, (ATNSET\, y)$

§\ $ATNSET_{o\$S}y_{\$S}$
\# $= (ATNSET\, y)\, (\forall\, (o\$S)\, [\lambda p.(\supset \$ANBOTH\, (p\, y))])$
§s %1 3 %0
\# $\supset (ANSET\, t\, y)\, (\forall\, (o\$S)\, [\lambda p.(\supset \$ANBOTH\, (p\, y))])$

\#\# use Proof Template A5215H (∀ I): H ⊃ ∀ x: B → H ⊃ B [x/a]
:= $\$T5215H$ $o\$S$
\# wff 260 : $o\$S_\tau$:= $\$T5215H$
:= $\$X5215H$ $p_{\$T5215H}$
\# wff 311 : $p_{\$T5215H}$:= $\$X5215H$
:= $\$A5215H$ $[\lambda t_{\$S}.(\wedge_{ooo}(ANSET_{\$T4}t_\tau t_{\$S})(\$X5215H_{\$T5215H}t_{\$S}))_o]$
\# wff 6719 : $[\lambda t.(\wedge\, (ANSET\, t\, t)\, (\$X5215H\, t))]_{\$T5215H}$:= $\$A5215H$ $\$P$
:= $\$H5215H$ %0
\# wff 6842 : $\supset (ANSET\, t\, y)\, (\forall \$T5215H\, [\lambda \$X5215H.(\supset \$ANBOTH\, (\$X5215H\, y))])_o$
:= $\$H5215H$
<< A5215H.r0t.txt
:= $\$T5215H$
:= $\$X5215H$
:= $\$A5215H$
:= $\$H5215H$

:= $\$D2TMP$ %0
\# wff 6934 : $\supset (ANSET\, t\, y)\, (\supset \$ZROSCCT\, (\$P\, y))_o$:= $\$D2TMP$

\#\# .3

%$\$D1TMP$
\# $\supset \$HTMP\, \$HTMP$:= $\$D1TMP$
\# $\supset_{ooo}\$HTMP_o\$HTMP_o$:= $\$D1TMP$
:= $\$D1TMP$

\#\# use Proof Template K8019H: H ⊃ (A ∧ B) → H ⊃ A, H ⊃ B
:= $\$H8019H$ %0
\# wff 6817 : $\supset \$HTMP\, \$HTMP_{o,\dots}$:= $\$H8019H$
<< K8019H.r0t.txt
:= $\$H8019H$
:= $\$ATMP$ $\supset_{ooo}\$HTMP_o\$ZRO2_o$
\# wff 7026 : $\supset \$HTMP\, \$ZRO2_o$:= $\$A8019H$ $\$ATMP$
:= $\$BTMP$ $\supset_{ooo}\$HTMP_o\$SCC3_o$
\# wff 7068 : $\supset \$HTMP\, \$SCC3_o$:= $\$B8019H$ $\$BTMP$
:= $\$A8019H$
:= $\$B8019H$

%$\$ATMP$
\# $\supset \$HTMP\, \$ZRO2$:= $\$ATMP$
\# $\supset_{ooo}\$HTMP_o\$ZRO2_o$:= $\$ATMP$
%$A6100$
\# $P1\, \$S\, (AZERO\, t)\, (ASUCC\, t)\, (ANSET\, t)$:= $A6100$
\# $P1_{o(o\backslash 5)(\backslash 4\backslash 4)\backslash 2\tau}\$S_\tau(AZERO_{\$T3}t_\tau)(ASUCC_{\$T44}t_\tau)(ANSET_{\$T4}t_\tau)$:= $A6100$

```
## use Proof Template K8004 (Trans):   (H ⊕ A), B   →   H ⊃ B
:=  $HA8004  %1
# wff      7026  :          ⊃ $HTMP $ZRO2_o          :=  $ATMP   $HA8004
:=  $B8004  %0
# wff      3252  :          P1 $S (AZERO t) (ASUCC t) (ANSET t)_{o,...}          :=  $B8004   A6100
<< K8004.r0t.txt
:=  $HA8004
:=  $B8004
%0
#                    ⊃ $HTMP A6100
#                    ⊃_{ooo}$HTMP_o A6100_o

§\  P1_{o(o\5)(\4\4)\2τ}$S_τ
#                    = (P1 $S) [λz.[λs.[λn.(n z)]]]
§s  %1  24  %0
#                    ⊃ $HTMP ([λz.[λs.[λn.(n z)]]] (AZERO t) (ASUCC t) (ANSET t))
§\  [λz_{$S}.[λs_{$S$S}.[λn_{o$S}.(n_{o$S}z_{$S})_o]_{$To2S}]_{$To2S3}](AZERO_{$T3}t_τ)
#                    = ([λz.[λs.[λn.(n z)]]] (AZERO t)) [λs.[λn.(n (AZERO t))]]
§s  %1  12  %0
#                    ⊃ $HTMP ([λs.[λn.(n (AZERO t))]] (ASUCC t) (ANSET t))
§\  [λs_{$S$S}.[λn_{o$S}.(n_{o$S}(AZERO_{$T3}t_τ))_o]_{$To2S}](ASUCC_{$T44}t_τ)
#                    = ([λs.[λn.(n (AZERO t))]] (ASUCC t)) [λn.(n (AZERO t))]
§s  %1  6  %0
#                    ⊃ $HTMP ([λn.(n (AZERO t))] (ANSET t))
§\  [λn_{o$S}.(n_{o$S}(AZERO_{$T3}t_τ))_o](ANSET_{$T4}t_τ)
#                    = ([λn.(n (AZERO t))] (ANSET t)) (ANSET t (AZERO t))
§s  %1  3  %0
#                    ⊃ $HTMP (ANSET t (AZERO t))

%$ATMP
#                    ⊃ $HTMP $ZRO2          :=  $ATMP
#                    ⊃_{ooo}$HTMP_o$ZRO2_o          :=  $ATMP
:=  $ATMP

## use Proof Template K8020H:   H ⊃ A, H ⊃ B   →   H ⊃ (A ∧ B)
:=  $A8020H  %1
# wff      7105  :          ⊃ $HTMP (ANSET t (AZERO t))_o          :=  $A8020H
:=  $B8020H  %0
# wff      7026  :          ⊃ $HTMP $ZRO2_o          :=  $B8020H
<< K8020H.r0t.txt
:=  $A8020H
:=  $B8020H
%0
#                    ⊃ $HTMP (∧ (ANSET t (AZERO t)) $ZRO2)
#                    ⊃_{ooo}$HTMP_o(∧_{ooo}(ANSET_{$T4}t_τ(AZERO_{$T3}t_τ))$ZRO2_o)

§\  AZERO_{$T3}t_τ
#                    = (AZERO t) ATZERO
```

§s %1 27 %0
$\supset \$HTMP\,(\wedge\,(ANSET\,t\,ATZERO)\,\$ZRO2)$
§\ $AZERO_{\$T3}t_\tau$
$=(AZERO\,t)\,ATZERO$
§s %1 15 %0
$\supset \$HTMP\,(\wedge\,(ANSET\,t\,ATZERO)\,\$ANSETZ)$

:= $\$TMP$ %0
wff 7234 : $\supset \$HTMP\,(\wedge\,(ANSET\,t\,ATZERO)\,\$ANSETZ)_o$:= $\$TMP$
§\ $\$P_{o\$S}ATZERO_{\$S}$
$=(\$P\,ATZERO)\,(\wedge\,(ANSET\,t\,ATZERO)\,\$ANSETZ)$

use Proof Template A5201b (Swap): A = B → B = A
<< A5201b.r0t.txt
%0
$=(\wedge\,(ANSET\,t\,ATZERO)\,\$ANSETZ)\,(\$P\,ATZERO)$
$=_{o\omega\omega}(\wedge_{ooo}(ANSET_{\$T4}t_\tau ATZERO_{\$S})\$ANSETZ_o)(\$P_{o\$S}ATZERO_{\$S})$

%$\$TMP$
$\supset \$HTMP\,(\wedge\,(ANSET\,t\,ATZERO)\,\$ANSETZ)$:= $\$TMP$
$\supset_{ooo}\$HTMP_o(\wedge_{ooo}(ANSET_{\$T4}t_\tau ATZERO_{\$S})\$ANSETZ_o)$:= $\$TMP$
:= $\$TMP$

§s %0 3 %1
$\supset \$HTMP\,(\$P\,ATZERO)$

:= $\$D3TMP$ %0
wff 7240 : $\supset \$HTMP\,(\$P\,ATZERO)_o$:= $\$D3TMP$

.4

%$\$BTMP$
$\supset \$HTMP\,\$SCC3$:= $\$BTMP$
$\supset_{ooo}\$HTMP_o\$SCC3_o$:= $\$BTMP$

%$A6101$
$P2\,\$S\,(AZERO\,t)\,(ASUCC\,t)\,(ANSET\,t)$:= $A6101$
$P2_{o(o\backslash 5)(\backslash 4\backslash 4)\backslash 2\tau}\$S_\tau(AZERO_{\$T3}t_\tau)(ASUCC_{\$T44}t_\tau)(ANSET_{\$T4}t_\tau)$:= $A6101$

use Proof Template K8004 (Trans): (H ⊕ A), B → H ⊃ B
:= $\$HA8004$ %1
wff 7068 : $\supset \$HTMP\,\$SCC3_o$:= $\$BTMP$ $\$HA8004$
:= $\$B8004$ %0
wff 3494 : $P2\,\$S\,(AZERO\,t)\,(ASUCC\,t)\,(ANSET\,t)_{o,\dots}$:= $\$B8004$ $A6101$
<< K8004.r0t.txt
:= $\$HA8004$
:= $\$B8004$
%0
$\supset \$HTMP\,A6101$

\# $\supset_{ooo}\$HTMP_oA6101_o$

§\ $P2_{o(o\backslash 5)(\backslash 4\backslash 4)\backslash 2\tau}\$S_\tau$
\# $= (P2\,\$S)\,[\lambda z.[\lambda s.[\lambda n.\$ANSETS2]]]$
§s %1 24 %0
\# $\supset \$HTMP\,([\lambda z.[\lambda s.[\lambda n.\$ANSETS2]]]\,(AZERO\,t)\,(ASUCC\,t)\,(ANSET\,t))$
§\ $[\lambda z_{\$S}.[\lambda s_{\$S\,\$S}.[\lambda n_{o\,\$S}.\$ANSETS2_o]_{\$To2S}]_{\$To2S3}](AZERO_{\$T3}t_\tau)$
\# $= ([\lambda z.[\lambda s.[\lambda n.\$ANSETS2]]]\,(AZERO\,t))\,[\lambda s.[\lambda n.\$ANSETS2]]$
§s %1 12 %0
\# $\supset \$HTMP\,([\lambda s.[\lambda n.\$ANSETS2]]\,(ASUCC\,t)\,(ANSET\,t))$
§\ $[\lambda s_{\$S\,\$S}.[\lambda n_{o\,\$S}.\$ANSETS2_o]_{\$To2S}](ASUCC_{\$T44}t_\tau)$
\# $= ([\lambda s.[\lambda n.\$ANSETS2]]\,(ASUCC\,t))\,[\lambda n.\$ANSETS3]$
§s %1 6 %0
\# $\supset \$HTMP\,([\lambda n.\$ANSETS3]\,(ANSET\,t))$
§\ $[\lambda n_{o\,\$S}.\$ANSETS3_o](ANSET_{\$T4}t_\tau)$
\# $= ([\lambda n.\$ANSETS3]\,(ANSET\,t))\,(\forall\,\$S\,[\lambda x.(ADOTx\,\$STSC)])$
§s %1 3 %0
\# $\supset \$HTMP\,(\forall\,\$S\,[\lambda x.(ADOTx\,\$STSC)])$

\## use Proof Template A5215H (∀ I): H ⊃ ∀ x: B → H ⊃ B [x/a]
:= $\$T5215H\ \ o(ot)$
\# wff 22 : $o(ot)_\tau$:= $\$S\ \ \$T5215H\ \ SIGMA$
:= $\$X5215H\ \ x_{\$S}$
\# wff 315 : $x_{\$S}$:= $\$X5215H$
:= $\$A5215H\ \ x_{\$S}$
\# wff 315 : $x_{\$S}$:= $\$A5215H\ \ \$X5215H$
:= $\$H5215H$ %0
\# wff 7278 : $\supset \$HTMP\,(\forall\,\$S\,[\lambda\$X5215H.(ADOTx\,\$STSC)])_o$:= $\$H5215H$
<< A5215H.r0t.txt
:= $\$T5215H$
:= $\$X5215H$
:= $\$A5215H$
:= $\$H5215H$
%0
\# $\supset \$HTMP\,(ADOTx\,\$STSC)$
\# $\supset_{ooo}\$HTMP_o(ADOTx_{oo}\$STSC_o)$

\## use Proof Template K8026 (Deduction Theorem Reversed): H ⊃ (I ⊃ A) → (H ∧ I) ⊃ A
<< K8026.r0t.txt
%0
\# $\supset (\wedge\,\$HTMP\,(ANSET\,t\,x))\,\$STSC$:= $\$HPTMP$
\# $\supset_{ooo}(\wedge_{ooo}\$HTMP_o(ANSET_{\$T4}t_\tau x_{\$S}))\$STSC_o$:= $\$HPTMP$

:= $\$TMP$ %0
\# wff 6777 : $\supset (\wedge\,\$HTMP\,(ANSET\,t\,x))\,\$STSC_{o,\ldots}$:= $\$HPTMP\ \ \$TMP$

$\%K8005$
\# $\supset x\,x$:= $K8005$

$\supset_{ooo}x_o x_o$:= $K8005$

use Proof Template A5221 (Sub): B → B [x/A]
:= $\$B5221$ %0
wff 1520 : $\supset x\, x_{o,\ldots}$:= $\$B5221$ $K8005$
:= $\$T5221$ o
wff 2 : o_τ := $\$T5221$
:= $\$X5221$ x_o
wff 16 : x_o := $\$X5221$
:= $\$A5221$ $p_{o\$S}x_{\$S}$
wff 316 : $p\, x_{o,\ldots}$:= $\$A5221$
<< A5221.r0t.txt
:= $\$B5221$
:= $\$T5221$
:= $\$X5221$
:= $\$A5221$
%0
$\supset (p\, x)\,(p\, x)$
$\supset_{ooo}(p_{o\$S}x_{\$S})(p_{o\$S}x_{\$S})$

%$\$TMP$
$\supset (\wedge\, \$HTMP\,(ANSET\, t\, x))\,\$STSC$:= $\$HPTMP$ $\$TMP$
$\supset_{ooo}(\wedge_{ooo}\$HTMP_o(ANSET_{\$T4}t_\tau x_{\$S}))\$STSC_o$:= $\$HPTMP$ $\$TMP$
:= $\$TMP$

use Proof Template K8004 (Trans): (H ⊕ A), B → H ⊃ B
:= $\$HA8004$ %1
wff 7404 : $\supset (p\, x)\,(p\, x)_{o,\ldots}$:= $\$HA8004$
:= $\$B8004$ %0
wff 6777 : $\supset (\wedge\, \$HTMP\,(ANSET\, t\, x))\,\$STSC_{o,\ldots}$:= $\$B8004$ $\$HPTMP$
<< K8004.r0t.txt
:= $\$HA8004$
:= $\$B8004$
%0
$\supset (p\, x)\,\$HPTMP$
$\supset_{ooo}(p_{o\$S}x_{\$S})\$HPTMP_o$

use Proof Template K8026 (Deduction Theorem Reversed): H ⊃ (I ⊃ A) → (H ∧ I) ⊃ A
<< K8026.r0t.txt
%0
$\supset (\wedge\,(p\, x)\,(\wedge\, \$HTMP\,(ANSET\, t\, x)))\,\$STSC$
$\supset_{ooo}(\wedge_{ooo}(p_{o\$S}x_{\$S})(\wedge_{ooo}\$HTMP_o(ANSET_{\$T4}t_\tau x_{\$S})))\$STSC_o$

use Proof Template K8027: (A ∧ B) ⊃ C → (B ∧ A) ⊃ C
<< K8027.r0t.txt
%0
$\supset (\$HTMP2\,(p\, x))\,\$STSC$
$\supset_{ooo}(\$HTMP2_{oo}(p_{o\$S}x_{\$S}))\$STSC_o$

:= $\$ATMP$ %0
wff 7612 : $\supset (\$HTMP2\,(p\,x))\,\$STSC_o$:= $\$ATMP$

%$\$BTMP$
$\supset \$HTMP\,\$SCC3$:= $\$BTMP$
$\supset_{ooo}\$HTMP_o\$SCC3_o$:= $\$BTMP$
:= $\$BTMP$

use Proof Template A5215H (∀ I): H ⊃ ∀ x: B → H ⊃ B [x/a]
:= $\$T5215H$ $o(ot)$
wff 22 : $o(ot)_\tau$:= $\$S$ $\$T5215H$ $SIGMA$
:= $\$X5215H$ $x_{\$S}$
wff 315 : $x_{\$S}$:= $\$X5215H$
:= $\$A5215H$ $x_{\$S}$
wff 315 : $x_{\$S}$:= $\$A5215H$ $\$X5215H$
:= $\$H5215H$ %0
wff 7068 : $\supset \$HTMP\,\$SCC3_o$:= $\$H5215H$
<< A5215H.r0t.txt
:= $\$T5215H$
:= $\$X5215H$
:= $\$A5215H$
:= $\$H5215H$
%0
$\supset \$HTMP\,(ADOTx\,(\supset (p\,x)\,(p\,(ASUCC\,t\,x))))$
$\supset_{ooo}\$HTMP_o(ADOTx_{oo}(\supset_{ooo}(p_{o\,\$S}x_{\$S})(p_{o\,\$S}(ASUCC_{\$T44}t_\tau x_{\$S}))))$

use Proof Template K8026 (Deduction Theorem Reversed): H ⊃ (I ⊃ A) → (H ∧ I) ⊃ A
<< K8026.r0t.txt
%0
$\supset (\wedge\,\$HTMP\,(ANSET\,t\,x))\,(\supset (p\,x)\,(p\,(ASUCC\,t\,x)))$
$\supset_{ooo}(\wedge_{ooo}\$HTMP_o(ANSET_{\$T4}t_\tau x_{\$S}))(\supset_{ooo}(p_{o\,\$S}x_{\$S})(p_{o\,\$S}(ASUCC_{\$T44}t_\tau x_{\$S})))$

use Proof Template K8026 (Deduction Theorem Reversed): H ⊃ (I ⊃ A) → (H ∧ I) ⊃ A
<< K8026.r0t.txt
%0
$\supset (\$HTMP2\,(p\,x))\,(p\,(ASUCC\,t\,x))$
$\supset_{ooo}(\$HTMP2_{oo}(p_{o\,\$S}x_{\$S}))(p_{o\,\$S}(ASUCC_{\$T44}t_\tau x_{\$S}))$

:= $\$BTMP$ %0
wff 7730 : $\supset (\$HTMP2\,(p\,x))\,(p\,(ASUCC\,t\,x))_{o,\ldots}$:= $\$BTMP$

%$\$ATMP$
$\supset (\$HTMP2\,(p\,x))\,\$STSC$:= $\$ATMP$
$\supset_{ooo}(\$HTMP2_{oo}(p_{o\,\$S}x_{\$S}))\$STSC_o$:= $\$ATMP$
:= $\$ATMP$
%$\$BTMP$

\# $\supset (\$HTMP2\,(p\,x))\,(p\,(ASUCC\,t\,x))$:= $\$BTMP$
\# $\supset_{ooo}(\$HTMP2_{oo}(p_{o\$S}x_{\$S}))(p_{o\$S}(ASUCC_{\$T44}t_{\tau}x_{\$S}))$:= $\$BTMP$
:= $\$BTMP$

\## use Proof Template K8020H: H ⊃ A, H ⊃ B → H ⊃ (A ∧ B)
:= $\$A8020H$ %1
\# wff 7612 : $\supset (\$HTMP2\,(p\,x))\,\$STSC_{o}$:= $\$A8020H$
:= $\$B8020H$ %0
\# wff 7730 : $\supset (\$HTMP2\,(p\,x))\,(p\,(ASUCC\,t\,x))_{o,\ldots}$:= $\$B8020H$
<< K8020H.r0t.txt
:= $\$A8020H$
:= $\$B8020H$

:= $\$TMP$ %0
\# wff 7877 : $\supset (\$HTMP2\,(p\,x))\,(\wedge\,\$STSC\,(p\,(ASUCC\,t\,x)))_{o}$:= $\$TMP$
§\ $\$P_{o\$S}(ASUCC_{\$T44}t_{\tau}x_{\$S})$
\# $= (\$P\,(ASUCC\,t\,x))\,(\wedge\,\$STSC\,(p\,(ASUCC\,t\,x)))$

\## use Proof Template A5201b (Swap): A = B → B = A
<< A5201b.r0t.txt
%0
\# $= (\wedge\,\$STSC\,(p\,(ASUCC\,t\,x)))\,(\$P\,(ASUCC\,t\,x))$
\# $=_{o\omega\omega}(\wedge_{ooo}\$STSC_{o}(p_{o\$S}(ASUCC_{\$T44}t_{\tau}x_{\$S})))(\$P_{o\$S}(ASUCC_{\$T44}t_{\tau}x_{\$S}))$

%$\$TMP$
\# $\supset (\$HTMP2\,(p\,x))\,(\wedge\,\$STSC\,(p\,(ASUCC\,t\,x)))$:= $\$TMP$
\# $\supset_{ooo}(\$HTMP2_{oo}(p_{o\$S}x_{\$S}))(\wedge_{ooo}\$STSC_{o}(p_{o\$S}(ASUCC_{\$T44}t_{\tau}x_{\$S})))$:=
$\$TMP$
:= $\$TMP$

§s %0 3 %1
\# $\supset (\$HTMP2\,(p\,x))\,(\$P\,(ASUCC\,t\,x))$

:= $\$HSWHYTMP$ %0
\# wff 7883 : $\supset (\$HTMP2\,(p\,x))\,(\$P\,(ASUCC\,t\,x))_{o}$:= $\$HSWHYTMP$

%$K8021$
\# $ASSOC\,o\,\wedge$:= $K8021$
\# $ASSOC_{o(\backslash 4\backslash 4\backslash 3)\tau}o_{\tau}\wedge_{ooo}$:= $K8021$
§\ $ASSOC_{o(\backslash 4\backslash 4\backslash 3)\tau}o_{\tau}$
\# $= (ASSOC\,o)\,[\lambda f.(= (f\,(f\,x\,y)\,z)\,(f\,x\,(f\,y\,z)))]$
§s %1 2 %0
\# $[\lambda f.(= (f\,(f\,x\,y)\,z)\,(f\,x\,(f\,y\,z)))]\,\wedge$
§\ $[\lambda f_{ooo}.(=_{ooo}(f_{ooo}(f_{ooo}x_{o}y_{o})z_{o})(f_{ooo}x_{o}(f_{ooo}y_{o}z_{o})))_{o}]\wedge_{ooo}$
\# $= ([\lambda f.(= (f\,(f\,x\,y)\,z)\,(f\,x\,(f\,y\,z)))]\,\wedge)\,(= (\wedge\,(\wedge\,x\,y)\,z)\,(\wedge\,x\,(\wedge\,y\,z)))$
§s %1 1 %0
\# $= (\wedge\,(\wedge\,x\,y)\,z)\,(\wedge\,x\,(\wedge\,y\,z))$

:= $\$SWHYTMP$ %0

wff 3222 : $= (\wedge (\wedge x\, y)\, z)\, (\wedge x\, (\wedge y\, z))_{o,\ldots}$:= $\$SWHYTMP$
%$\$HSWHYTMP$
$\supset (\$HTMP2\, (p\, x))\, (\$P\, (ASUCC\, t\, x))$:= $\$HSWHYTMP$
$\supset_{ooo}(\$HTMP2_{oo}(p_{o\$S}x_{\$S}))(\$P_{o\$S}(ASUCC_{\$T44}t_{\tau}x_{\$S}))$:= $\$HSWHYTMP$
%$\$SWHYTMP$
$= (\wedge (\wedge x\, y)\, z)\, (\wedge x\, (\wedge y\, z))$:= $\$SWHYTMP$
$=_{ooo}(\wedge_{ooo}(\wedge_{ooo}x_oy_o)z_o)(\wedge_{ooo}x_o(\wedge_{ooo}y_oz_o))$:= $\$SWHYTMP$
:= $\$SWHYTMP$

use Proof Template A5221 (Sub): B → B [x/A]
:= $\$B5221$ %0
wff 3222 : $= (\wedge (\wedge x\, y)\, z)\, (\wedge x\, (\wedge y\, z))_{o,\ldots}$:= $\$B5221$
:= $\$T5221$ o
wff 2 : o_{τ} := $\$T5221$
:= $\$X5221$ x_o
wff 16 : x_o := $\$X5221$
:= $\$A5221$ %1/85
wff 6769 : $\wedge \$ZRO2\, \$SCC3_{o,\ldots}$:= $\$A5221$ $\$HTMP$
<< A5221.r0t.txt
:= $\$B5221$
:= $\$T5221$
:= $\$X5221$
:= $\$A5221$

:= $\$SWHYTMP$ %0
wff 7920 : $= (\wedge (\wedge \$HTMP\, y)\, z)\, (\wedge \$HTMP\, (\wedge y\, z))_{o,\ldots}$:= $\$SWHYTMP$
%$\$HSWHYTMP$
$\supset (\$HTMP2\, (p\, x))\, (\$P\, (ASUCC\, t\, x))$:= $\$HSWHYTMP$
$\supset_{ooo}(\$HTMP2_{oo}(p_{o\$S}x_{\$S}))(\$P_{o\$S}(ASUCC_{\$T44}t_{\tau}x_{\$S}))$:= $\$HSWHYTMP$
%$\$SWHYTMP$
$= (\wedge (\wedge \$HTMP\, y)\, z)\, (\wedge \$HTMP\, (\wedge y\, z))$:= $\$SWHYTMP$
$=_{ooo}(\wedge_{ooo}(\wedge_{ooo}\$HTMP_oy_o)z_o)(\wedge_{ooo}\$HTMP_o(\wedge_{ooo}y_oz_o))$:= $\$SWHYTMP$
:= $\$SWHYTMP$

use Proof Template A5221 (Sub): B → B [x/A]
:= $\$B5221$ %0
wff 7920 : $= (\wedge (\wedge \$HTMP\, y)\, z)\, (\wedge \$HTMP\, (\wedge y\, z))_{o,\ldots}$:= $\$B5221$
:= $\$T5221$ o
wff 2 : o_{τ} := $\$T5221$
:= $\$X5221$ y_o
wff 34 : y_o := $\$X5221$
:= $\$A5221$ %1/43
wff 363 : $ANSET\, t\, x_{o,\ldots}$:= $\$A5221$
<< A5221.r0t.txt
:= $\$B5221$
:= $\$T5221$
:= $\$X5221$
:= $\$A5221$

:= $\$SWHYTMP$ %0
wff 7958 : $=(\$HTMP2\,z)\,(\wedge\,\$HTMP\,(\wedge\,(ANSET\,t\,x)\,z))_{o,\ldots}$:= $\$SWHYTMP$
%$\$HSWHYTMP$
$\supset(\$HTMP2\,(p\,x))\,(\$P\,(ASUCC\,t\,x))$:= $\$HSWHYTMP$
$\supset_{ooo}(\$HTMP2_{oo}(p_{o\$S}x_{\$S}))(\$P_{o\$S}(ASUCC_{\$T44}t_{\tau}x_{\$S}))$:= $\$HSWHYTMP$
%$\$SWHYTMP$
$=(\$HTMP2\,z)\,(\wedge\,\$HTMP\,(\wedge\,(ANSET\,t\,x)\,z))$:= $\$SWHYTMP$
$=_{ooo}(\$HTMP2_{oo}z_{o})(\wedge_{ooo}\$HTMP_{o}(\wedge_{ooo}(ANSET_{\$T4}t_{\tau}x_{\$S})z_{o}))$:=
$\$SWHYTMP$
:= $\$SWHYTMP$

use Proof Template A5221 (Sub): B $\rightarrow$ B [x/A]
:= $\$B5221$ %0
wff 7958 : $=(\$HTMP2\,z)\,(\wedge\,\$HTMP\,(\wedge\,(ANSET\,t\,x)\,z))_{o,\ldots}$:= $\$B5221$
:= $\$T5221$ o
wff 2 : o_{τ} := $\$T5221$
:= $\$X5221$ z_{o}
wff 3176 : z_{o} := $\$X5221$
:= $\$A5221$ %1/11
wff 316 : $p\,x_{o,\ldots}$:= $\$A5221$
<< A5221.r0t.txt
:= $\$B5221$
:= $\$T5221$
:= $\$X5221$
:= $\$A5221$

%$\$HSWHYTMP$
$\supset(\$HTMP2\,(p\,x))\,(\$P\,(ASUCC\,t\,x))$:= $\$HSWHYTMP$
$\supset_{ooo}(\$HTMP2_{oo}(p_{o\$S}x_{\$S}))(\$P_{o\$S}(ASUCC_{\$T44}t_{\tau}x_{\$S}))$:= $\$HSWHYTMP$
:= $\$HSWHYTMP$
§s %0 5 %1
$\supset(\wedge\,\$HTMP\,(\wedge\,(ANSET\,t\,x)\,(p\,x)))\,(\$P\,(ASUCC\,t\,x))$

:= $\$TMP$ %0
wff 7998 : $\supset(\wedge\,\$HTMP\,(\wedge\,(ANSET\,t\,x)\,(p\,x)))\,(\$P\,(ASUCC\,t\,x))_{o}$:= $\$TMP$
§\ $\$P_{o\$S}x_{\$S}$
$=(\$P\,x)\,(\wedge\,(ANSET\,t\,x)\,(p\,x))$

use Proof Template A5201b (Swap): A = B $\rightarrow$ B = A
<< A5201b.r0t.txt
%0
$=(\wedge\,(ANSET\,t\,x)\,(p\,x))\,(\$P\,x)$
$=_{o\omega\omega}(\wedge_{ooo}(ANSET_{\$T4}t_{\tau}x_{\$S})(p_{o\$S}x_{\$S}))(\$P_{o\$S}x_{\$S})$

%$\$TMP$
$\supset(\wedge\,\$HTMP\,(\wedge\,(ANSET\,t\,x)\,(p\,x)))\,(\$P\,(ASUCC\,t\,x))$:= $\$TMP$
...
$\ldots\supset_{ooo}(\wedge_{ooo}\$HTMP_{o}(\wedge_{ooo}(ANSET_{\$T4}t_{\tau}x_{\$S})(p_{o\$S}x_{\$S})))(\$P_{o\$S}(ASUCC_{\$T44}t_{\tau}x_{\$S}))$:=
$\$TMP$

:= $\$TMP$

§s %0 11 %1
$\supset (\wedge \$HTMP\,(\$P\,x))\,(\$P\,(ASUCC\,t\,x))$

use Proof Template K8025 (Deduction Theorem): (H ∧ I) ⊃ A → H ⊃ (I ⊃ A)
<< K8025.r0t.txt
%0
$\supset \$HTMP\,(\supset (\$P\,x)\,(\$P\,(ASUCC\,t\,x)))$
$\supset_{ooo}\$HTMP_o(\supset_{ooo}(\$P_{o\$S}x_{\$S})(\$P_{o\$S}(ASUCC_{\$T44}t_{\tau}x_{\$S})))$

use Proof Template A5220H (Gen): (H ⊃ A) → (H ⊃ ∀ x: A)
:= $\$T5220H$ $o(ot)$
wff 22 : $o(ot)_{\tau}$:= $\$S$ $\$T5220H$ $SIGMA$
:= $\$X5220H$ $x_{\$S}$
wff 315 : $x_{\$S}$:= $\$X5220H$
:= $\$A5220H$ %0
wff 8058 : $\supset \$HTMP\,(\supset (\$P\,\$X5220H)\,(\$P\,(ASUCC\,t\,\$X5220H)))_{o,\ldots}$:= $\$A5220H$
<< A5220H.r0t.txt
:= $\$T5220H$
:= $\$X5220H$
:= $\$A5220H$

:= $\$D4TMP$ %0
wff 8193 : $\supset \$HTMP\,\$SCCP_o$:= $\$D4TMP$

.5

%$\$D3TMP$
$\supset \$HTMP\,(\$P\,ATZERO)$:= $\$D3TMP$
$\supset_{ooo}\$HTMP_o(\$P_{o\$S}ATZERO_{\$S})$:= $\$D3TMP$
:= $\$D3TMP$
%$\$D4TMP$
$\supset \$HTMP\,\$SCCP$:= $\$D4TMP$
$\supset_{ooo}\$HTMP_o\$SCCP_o$:= $\$D4TMP$
:= $\$D4TMP$

use Proof Template K8020H: H ⊃ A, H ⊃ B → H ⊃ (A ∧ B)
:= $\$A8020H$ %1
wff 7240 : $\supset \$HTMP\,(\$P\,ATZERO)_o$:= $\$A8020H$
:= $\$B8020H$ %0
wff 8193 : $\supset \$HTMP\,\$SCCP_o$:= $\$B8020H$
<< K8020H.r0t.txt
:= $\$A8020H$
:= $\$B8020H$

:= $\$TMP$ %0
wff 8299 : $\supset \$HTMP\,(\wedge (\$P\,ATZERO)\,\$SCCP)_o$:= $\$TMP$

%$D2TMP
\# $\supset (ANSET\,t\,y)\,(\supset \$ZROSCCT\,(\$P\,y)) \quad := \quad \$D2TMP$
\# $\supset_{ooo}(ANSET_{\$T4}t_{\tau}y_{\$S})(\supset_{ooo}\$ZROSCCT_{o}(\$P_{o\$S}y_{\$S})) \quad := \quad \$D2TMP$
%$TMP
\# $\supset \$HTMP\,(\wedge\,(\$P\,ATZERO)\,\$SCCP) \quad := \quad \$TMP$
\# $\supset_{ooo}\$HTMP_{o}(\wedge_{ooo}(\$P_{o\$S}ATZERO_{\$S})\$SCCP_{o}) \quad := \quad \$TMP$

\#\# use Proof Template K8004 (Trans): (H ⊕ A), B → H ⊃ B
:= $HA8004 %1
\# wff 6934 : $\supset (ANSET\,t\,y)\,(\supset \$ZROSCCT\,(\$P\,y))_{o} \quad := \quad \$D2TMP \quad \$HA8004$
:= $B8004 %0
\# wff 8299 : $\supset \$HTMP\,(\wedge\,(\$P\,ATZERO)\,\$SCCP)_{o} \quad := \quad \$B8004 \quad \$TMP$
<< K8004.r0t.txt
:= $HA8004
:= $B8004
%0
\# $\supset (ANSET\,t\,y)\,\$TMP$
\# $\supset_{ooo}(ANSET_{\$T4}t_{\tau}y_{\$S})\$TMP_{o}$

\#\# use Proof Template K8026 (Deduction Theorem Reversed): H ⊃ (I ⊃ A) → (H ∧ I) ⊃ A
<< K8026.r0t.txt
%0
\# $\supset (\wedge\,(ANSET\,t\,y)\,\$HTMP)\,(\wedge\,(\$P\,ATZERO)\,\$SCCP)$
\# $\supset_{ooo}(\wedge_{ooo}(ANSET_{\$T4}t_{\tau}y_{\$S})\$HTMP_{o})(\wedge_{ooo}(\$P_{o\$S}ATZERO_{\$S})\$SCCP_{o})$

\#\# use Proof Template K8027: (A ∧ B) ⊃ C → (B ∧ A) ⊃ C
<< K8027.r0t.txt
%0
\# $\supset (\wedge\,\$HTMP\,(ANSET\,t\,y))\,(\wedge\,(\$P\,ATZERO)\,\$SCCP)$
\# $\supset_{ooo}(\wedge_{ooo}\$HTMP_{o}(ANSET_{\$T4}t_{\tau}y_{\$S}))(\wedge_{ooo}(\$P_{o\$S}ATZERO_{\$S})\$SCCP_{o})$

§\ $ASUCC_{\$T44}t_{\tau}$
\# $=(ASUCC\,t)\,ATSUCC$
§s %1 254 %0
\# $\supset (\wedge\,\$HTMP\,(ANSET\,t\,y))\,\$ZROSCCT$
:= $ATMP %0
\# wff 8477 : $\supset (\wedge\,\$HTMP\,(ANSET\,t\,y))\,\$ZROSCCT_{o} \quad := \quad \$ATMP$

%$TMP
\# $\supset \$HTMP\,(\wedge\,(\$P\,ATZERO)\,\$SCCP) \quad := \quad \$TMP$
\# $\supset_{ooo}\$HTMP_{o}(\wedge_{ooo}(\$P_{o\$S}ATZERO_{\$S})\$SCCP_{o}) \quad := \quad \$TMP$
:= $TMP
%$D2TMP
\# $\supset (ANSET\,t\,y)\,(\supset \$ZROSCCT\,(\$P\,y)) \quad := \quad \$D2TMP$
\# $\supset_{ooo}(ANSET_{\$T4}t_{\tau}y_{\$S})(\supset_{ooo}\$ZROSCCT_{o}(\$P_{o\$S}y_{\$S})) \quad := \quad \$D2TMP$
:= $D2TMP

\#\# use Proof Template K8004 (Trans): (H ⊕ A), B → H ⊃ B

:= $\$HA8004$ %1
wff 8299 : $\supset \$HTMP\,(\wedge\,(\$P\,ATZERO)\,\$SCCP)_{o,\ldots}$:= $\$HA8004$
:= $\$B8004$ %0
wff 6934 : $\supset (ANSET\,t\,y)\,(\supset \$ZROSCCT\,(\$P\,y))_{o}$:= $\$B8004$
<< K8004.r0t.txt
:= $\$HA8004$
:= $\$B8004$
%0
$\supset \$HTMP\,(\supset (ANSET\,t\,y)\,(\supset \$ZROSCCT\,(\$P\,y)))$
$\supset_{ooo}\$HTMP_{o}(\supset_{ooo}(ANSET_{\$T4}t_{\tau}y_{\$S})(\supset_{ooo}\$ZROSCCT_{o}(\$P_{o\$S}y_{\$S})))$

use Proof Template K8026 (Deduction Theorem Reversed): H ⊃ (I ⊃ A) → (H ∧ I) ⊃ A
<< K8026.r0t.txt
%0
$\supset (\wedge\,\$HTMP\,(ANSET\,t\,y))\,(\supset \$ZROSCCT\,(\$P\,y))$
$\supset_{ooo}(\wedge_{ooo}\$HTMP_{o}(ANSET_{\$T4}t_{\tau}y_{\$S}))(\supset_{ooo}\$ZROSCCT_{o}(\$P_{o\$S}y_{\$S}))$

:= $\$ABTMP$ %0
wff 8558 : $\supset (\wedge\,\$HTMP\,(ANSET\,t\,y))\,(\supset \$ZROSCCT\,(\$P\,y))_{o,\ldots}$:= $\$ABTMP$

%$\$ABTMP$
$\supset (\wedge\,\$HTMP\,(ANSET\,t\,y))\,(\supset \$ZROSCCT\,(\$P\,y))$:= $\$ABTMP$
$\supset_{ooo}(\wedge_{ooo}\$HTMP_{o}(ANSET_{\$T4}t_{\tau}y_{\$S}))(\supset_{ooo}\$ZROSCCT_{o}(\$P_{o\$S}y_{\$S}))$:= $\$ABTMP$
:= $\$ABTMP$
%$\$ATMP$
$\supset (\wedge\,\$HTMP\,(ANSET\,t\,y))\,\$ZROSCCT$:= $\$ATMP$
$\supset_{ooo}(\wedge_{ooo}\$HTMP_{o}(ANSET_{\$T4}t_{\tau}y_{\$S}))\$ZROSCCT_{o}$:= $\$ATMP$
:= $\$ATMP$

use Proof Template A5224H (MP): H ⊃ A, H ⊃ (A ⊃ B) → H ⊃ B
:= $\$AB5224H$ %1
wff 8558 : $\supset (\wedge\,\$HTMP\,(ANSET\,t\,y))\,(\supset \$ZROSCCT\,(\$P\,y))_{o,\ldots}$:= $\$AB5224H$
:= $\$A5224H$ %0
wff 8477 : $\supset (\wedge\,\$HTMP\,(ANSET\,t\,y))\,\$ZROSCCT_{o}$:= $\$A5224H$
<< A5224H.r0t.txt
:= $\$AB5224H$
:= $\$A5224H$
%0
$\supset (\wedge\,\$HTMP\,(ANSET\,t\,y))\,(\$P\,y)$
$\supset_{ooo}(\wedge_{ooo}\$HTMP_{o}(ANSET_{\$T4}t_{\tau}y_{\$S}))(\$P_{o\$S}y_{\$S})$

.6

§\ $\$P_{o\$S}y_{\$S}$
$= (\$P\,y)\,(\wedge\,(ANSET\,t\,y)\,(p\,y))$
§s %1 3 %0
$\supset (\wedge\,\$HTMP\,(ANSET\,t\,y))\,(\wedge\,(ANSET\,t\,y)\,(p\,y))$

use Proof Template K8019H: H ⊃ (A ∧ B) → H ⊃ A, H ⊃ B
:= $\$H8019H$ %0
wff 8713 : $\supset (\wedge \$HTMP\,(ANSET\,t\,y))\,(\wedge\,(ANSET\,t\,y)\,(p\,y))_o$:= $\$H8019H$
<< K8019H.r0t.txt
:= $\$H8019H$
%$\$B8019H$
$\supset (\wedge \$HTMP\,(ANSET\,t\,y))\,(p\,y)$:= $\$B8019H$
$\supset_{ooo}(\wedge_{ooo}\$HTMP_o(ANSET_{\$T4}t_\tau y_{\$S}))(p_{o\$S}y_{\$S})$:= $\$B8019H$
:= $\$A8019H$
:= $\$B8019H$

use Proof Template K8025 (Deduction Theorem): (H ∧ I) ⊃ A → H ⊃ (I ⊃ A)
<< K8025.r0t.txt
%0
$\supset \$HTMP\,(\supset (ANSET\,t\,y)\,(p\,y))$
$\supset_{ooo}\$HTMP_o(\supset_{ooo}(ANSET_{\$T4}t_\tau y_{\$S})(p_{o\$S}y_{\$S}))$

use Proof Template A5220H (Gen): (H ⊃ A) → (H ⊃ ∀ x: A)
:= $\$T5220H$ $o(ot)$
wff 22 : $o(ot)_\tau$:= $\$S$ $\$T5220H$ $SIGMA$
:= $\$X5220H$ $y_{\$S}$
wff 6820 : $y_{\$S}$:= $\$X5220H$
:= $\$A5220H$ %0
wff 8838 : $\supset \$HTMP\,(\supset (ANSET\,t\,\$X5220H)\,(p\,\$X5220H))_{o,\ldots}$:= $\$A5220H$
<< A5220H.r0t.txt
:= $\$T5220H$
:= $\$X5220H$
:= $\$A5220H$
%0
$\supset \$HTMP\,(\forall \$S\,[\lambda y.(\supset (ANSET\,t\,y)\,(p\,y))])$
$\supset_{ooo}\$HTMP_o(\forall_{\$T3}\$S_\tau[\lambda y_{\$S}.(\supset_{ooo}(ANSET_{\$T4}t_\tau y_{\$S})(p_{o\$S}y_{\$S}))_o])$

§r /7 $x_{\$S}$
$= [\lambda y.(\supset (ANSET\,t\,y)\,(p\,y))]\,[\lambda x.(ADOT x\,(p\,x))]$
§s %1 7 %0
$\supset \$HTMP\,\$ALL3$

use Proof Template A5220 (Gen): A → ∀ x: A
:= $\$T5220$ $o\$S$
wff 260 : $o\$S_\tau$:= $\$T5220$
:= $\$X5220$ $p_{\$T5220}$
wff 311 : $p_{\$T5220}$:= $\$X5220$
:= $\$A5220$ %0
wff 6771 : $\supset \$HTMP\,\$ALL3_o$:= $\$A5220$
<< A5220.r0t.txt
:= $\$T5220$
:= $\$X5220$
:= $\$A5220$

%0
\# $\forall\,(o\,\$S)\,[\lambda p.(\supset \$HTMP\,\$ALL3)]$:= $\$P5S0SCST$
\# $\forall_{\$T3}(o\,\$S)_\tau[\lambda p_{o\,\$S}.(\supset_{ooo}\$HTMP_o\$ALL3_o)_o]$:= $\$P5S0SCST$

\## .7: Match general definition

:= $\$TMP$ %0
\# wff 6773 : $\forall\,(o\,\$S)\,[\lambda p.(\supset \$HTMP\,\$ALL3)]_{o,\ldots}$:= $\$P5S0SCST$ $\$TMP$
%$\$D0TMP$
\# $=\$P5APP\,\$TMP$:= $\$D0TMP$
\# $=_{o\omega\omega}\$P5APP_\omega\$TMP_\omega$:= $\$D0TMP$
:= $\$D0TMP$

\## use Proof Template A5201b (Swap): A = B $\to$ B = A
<< A5201b.r0t.txt
%0
\# $=\$TMP\,\$P5APP$
\# $=_{o\omega\omega}\$TMP_\omega\$P5APP_\omega$

%$\$TMP$
\# $\forall\,(o\,\$S)\,[\lambda p.(\supset \$HTMP\,\$ALL3)]$:= $\$P5S0SCST$ $\$TMP$
\# $\forall_{\$T3}(o\,\$S)_\tau[\lambda p_{o\,\$S}.(\supset_{ooo}\$HTMP_o\$ALL3_o)_o]$:= $\$P5S0SCST$ $\$TMP$
:= $\$TMP$
§s %0 1 %1
\# $P5\,\$S\,(AZERO\,t)\,(ASUCC\,t)\,(ANSET\,t)$:= $\$P5APP$

:= $A6102$ %0
\# wff 6723 : $P5\,\$S\,(AZERO\,t)\,(ASUCC\,t)\,(ANSET\,t)_{o,\ldots}$:= $\$P5APP$ $A6102$

\##
\## Q.E.D.
\##

%0
\# $P5\,\$S\,(AZERO\,t)\,(ASUCC\,t)\,(ANSET\,t)$:= $\$P5APP$ $A6102$
\# $P5_{o(o\backslash 5)(\backslash 4\backslash 4)\backslash 2\tau}\$S_\tau(AZERO_{\$T3}t_\tau)(ASUCC_{\$T44}t_\tau)(ANSET_{\$T4}t_\tau)$:=
$\$P5APP$ $A6102$

\## undefine local variables
:= $\$S$
:= $\$T3$
:= $\$T4$
:= $\$T44$
:= $\$To2S$
:= $\$To2S3$
:= $\$P5APP$
:= $\$ANSETZ$
:= $\$ANSETS$

$:= \$ANBOTH$
$:= \$ANSETS2$
$:= \$ANSETS3$
$:= \$ANSETx$
$:= \$ZRO$
$:= \$SCC$
$:= \$ALL$
$:= \$IDC$
$:= \$P5S$
$:= \$IDC0$
$:= \$P5S0$
$:= \$ZRO2$
$:= \$SCC2$
$:= \$P5S0SC$
$:= \$SCC3$
$:= \$ALL3$
$:= \$P5S0SCST$
$:= \$STSC$
$:= \$HPTMP$
$:= \$SCCP$
$:= \$SCCPT$
$:= \$ZROSCCT$
$:= \$HTMP2$
$:= \$P$
$:= \$HTMP$

2.1.53 Results for File K8000.r0.txt

##
Proof K8000: (A ∧ T) = A; (T ∧ A) = A; (A ∧ T) = (T ∧ A)
##
##
Source: [Kubota 2017 (doi: 10.4444/100.10)]
##
Copyright (c) 2017 Owl of Minerva Press GmbH. All rights reserved.
Written by Ken Kubota (<mail@kenkubota.de>).
##
This file is part of the publication of the mathematical logic $\mathcal{R}_0$.
For more information, visit: <http://doi.org/10.4444/100.10>
##

<< basics.r0.txt
<< A5229.r0.txt

##
Proof
##

.a: (A ∧ T) = A

use Proof Template A5222 (Rule of Cases): [\x.A]T, [\x.A]F → A
:= $\$L5222$ $[\lambda x_o.(=_{ooo}(\wedge_{ooo}x_oT_o)x_o)_o]$
wff 1249 : $[\lambda x.(=(\wedge\, x\, T)\, x)]_{oo}$:= $\$L5222$
:= $\$X5222$ x_o
wff 16 : x_o := $\$X5222$
:= $\$T5222$ $\$L5222_{oo}T_o$
wff 1250 : $\$L5222\, T_o$:= $\$T5222$
:= $\$F5222$ $\$L5222_{oo}F_o$
wff 1251 : $\$L5222\, F_o$:= $\$F5222$

case T: (T ∧ T) = T
§\ $\$T5222$
$= \$T5222\, A5211$
use Proof Template A5201b (Swap): A = B → B = A
<< A5201b.r0t.txt
%0
$= A5211\, \$T5222$
$=_{o\omega\omega}A5211_\omega\$T5222_\omega$
%$A5211$
$= A5212\, T$:= $A5211$ $A5229a$
$=_{ooo}A5212_oT_o$:= $A5211$ $A5229a$
§s %0 1 %1
$\$L5222\, T$:= $\$T5222$

case F: (F ∧ T) = F
§\ $\$F5222$
$= \$F5222\, A5229c$
use Proof Template A5201b (Swap): A = B → B = A
<< A5201b.r0t.txt
%0
$= A5229c\, \$F5222$
$=_{o\omega\omega}A5229c_\omega\$F5222_\omega$
%$A5229c$
$= (\wedge\, F\, T)\, F$:= $A5229c$
$=_{ooo}(\wedge_{ooo}F_oT_o)F_o$:= $A5229c$
§s %0 1 %1
$\$L5222\, F$:= $\$F5222$

<< A5222.r0t.txt
:= $\$L5222$
:= $\$X5222$
:= $\$T5222$
:= $\$F5222$
%0
$= (\wedge\, x\, T)\, x$
$=_{ooo}(\wedge_{ooo}x_oT_o)x_o$

:= $K8000a$ %0
wff 1248 : $= (\wedge\, x\, T)\, x_{o,\dots}$:= $K8000a$

.b: (T ∧ A) = A [= A5216]

use Proof Template A5222 (Rule of Cases): [\x.A]T, [\x.A]F → A
:= $\$L5222$ $[\lambda x_o.(=_{ooo}(\wedge_{ooo}T_o x_o)x_o)_o]$
wff 597 : $[\lambda x.(= (\wedge\, T\, x)\, x)]_{oo,\dots}$:= $\$L5222$
:= $\$X5222$ x_o
wff 16 : x_o := $\$X5222$
:= $\$T5222$ $\$L5222_{oo}T_o$
wff 604 : $\$L5222\, T_{o,\dots}$:= $\$T5222$
:= $\$F5222$ $\$L5222_{oo}F_o$
wff 606 : $\$L5222\, F_{o,\dots}$:= $\$F5222$

case T: (T ∧ T) = T
§\ $\$T5222$
$= \$T5222\, A5211$
use Proof Template A5201b (Swap): A = B → B = A
<< A5201b.r0t.txt
%0
$= A5211\, \$T5222$
$=_{o\omega\omega}A5211_\omega \$T5222_\omega$
%$A5211$
$= A5212\, T$:= $A5211$ $A5229a$
$=_{ooo}A5212_o T_o$:= $A5211$ $A5229a$
§s %0 1 %1
$\$L5222\, T$:= $\$T5222$

case F: (T ∧ F) = F
§\ $\$F5222$
$= \$F5222\, A5214$
use Proof Template A5201b (Swap): A = B → B = A
<< A5201b.r0t.txt
%0
$= A5214\, \$F5222$
$=_{o\omega\omega}A5214_\omega \$F5222_\omega$
%$A5214$
$= (\wedge\, T\, F)\, F$:= $A5214$ $A5229b$
$=_{ooo}(\wedge_{ooo}T_o F_o)F_o$:= $A5214$ $A5229b$
§s %0 1 %1
$\$L5222\, F$:= $\$F5222$

<< A5222.r0t.txt
:= $\$L5222$
:= $\$X5222$
:= $\$T5222$
:= $\$F5222$
%0

\# $=(\wedge\, T\; x)\; x$
\# $=_{ooo}(\wedge_{ooo}T_o x_o)x_o$

:= $K8000b$ %0
\# wff 596 : $=(\wedge\, T\; x)\; x_{o,\ldots}$:= $K8000b$

\#\# .c: (A ∧ T) = (T ∧ A)

%$K8000b$
\# $=(\wedge\, T\; x)\; x$:= $K8000b$
\# $=_{ooo}(\wedge_{ooo}T_o x_o)x_o$:= $K8000b$
\#\# use Proof Template A5201b (Swap): A = B → B = A
<< A5201b.r0t.txt
%0
\# $=x\;(\wedge\, T\; x)$
\# $=_{ooo}x_o(\wedge_{ooo}T_o x_o)$
%$K8000a$
\# $=(\wedge\, x\; T)\; x$:= $K8000a$
\# $=_{ooo}(\wedge_{ooo}x_o T_o)x_o$:= $K8000a$
§s %0 3 %1
\# $=(\wedge\, x\; T)\;(\wedge\, T\; x)$
%0
\# $=(\wedge\, x\; T)\;(\wedge\, T\; x)$
\# $=_{ooo}(\wedge_{ooo}x_o T_o)(\wedge_{ooo}T_o x_o)$

:= $K8000c$ %0
\# wff 1410 : $=(\wedge\, x\; T)\;(\wedge\, T\; x)_o$:= $K8000c$

\#\#
\#\# Q.E.D.
\#\#

\#\# %K8000a
%$K8000a$
\# $=(\wedge\, x\; T)\; x$:= $K8000a$
\# $=_{ooo}(\wedge_{ooo}x_o T_o)x_o$:= $K8000a$

\#\# %K8000b
%$K8000b$
\# $=(\wedge\, T\; x)\; x$:= $K8000b$
\# $=_{ooo}(\wedge_{ooo}T_o x_o)x_o$:= $K8000b$

\#\# %K8000c
%$K8000c$
\# $=(\wedge\, x\; T)\;(\wedge\, T\; x)$:= $K8000c$
\# $=_{ooo}(\wedge_{ooo}x_o T_o)(\wedge_{ooo}T_o x_o)$:= $K8000c$

2.1.54 Results for File K8001.r0.txt

##
Proof K8001: (A ∧ F) = F; (F ∧ A) = F; (A ∧ F) = (F ∧ A)
##
##
Source: [Kubota 2017 (doi: 10.4444/100.10)]
##

Written by Ken Kubota (<mail@kenkubota.de>).
##
This file is part of the publication of the mathematical logic $\mathcal{R}_0$.
For more information, visit: <http://doi.org/10.4444/100.10>
##

<< basics.r0.txt
<< A5229.r0.txt

##
Proof
##

.a: (A ∧ F) = F

use Proof Template A5222 (Rule of Cases): [\x.A]T, [\x.A]F → A
:= $\$L5222\ [\lambda x_o.(=_{ooo}(\wedge_{ooo}x_oF_o)F_o)_o]$
wff 1249 : $[\lambda x.(=(\wedge\, x\, F)\, F)]_{oo}$:= $\$L5222$
:= $\$X5222\ x_o$
wff 16 : x_o := $\$X5222$
:= $\$T5222\ \$L5222_{oo}T_o$
wff 1250 : $\$L5222\, T_o$:= $\$T5222$
:= $\$F5222\ \$L5222_{oo}F_o$
wff 1251 : $\$L5222\, F_o$:= $\$F5222$

Case T: (T ∧ F) = F
§\ $\$T5222$
$= \$T5222\, A5214$
use Proof Template A5201b (Swap): A = B → B = A
<< A5201b.r0t.txt
%0
$= A5214\, \$T5222$
$=_{o\omega\omega}A5214_\omega \$T5222_\omega$
%$A5214$
$= (\wedge\, T\, F)\, F$:= $A5214$ $A5229b$
$=_{ooo}(\wedge_{ooo}T_oF_o)F_o$:= $A5214$ $A5229b$
§s %0 1 %1
$\$L5222\, T$:= $\$T5222$

```
## Case F:   (F ∧ F) = F
§\  $F5222
#                    = $F5222 A5229d
## use Proof Template A5201b (Swap):   A = B  →  B = A
<< A5201b.r0t.txt
%0
#                    = A5229d $F5222
#                    =_{oωω} A5229d_ω $F5222_ω
%A5229d
#                    = (∧ F F) F      :=   A5229d
#                    =_{ooo}(∧_{ooo} F_o F_o) F_o      :=   A5229d
§s  %0  1  %1
#                    $L5222 F      :=   $F5222

<< A5222.r0t.txt
:=   $L5222
:=   $X5222
:=   $T5222
:=   $F5222
%0
#                    = (∧ x F) F
#                    =_{ooo}(∧_{ooo} x_o F_o) F_o

:=   K8001a  %0
# wff      1248   :        = (∧ x F) F_{o,...}      :=   K8001a

## .b:   (F ∧ A) = F

## use Proof Template A5222 (Rule of Cases):   [\x.A]T, [\x.A]F   →   A
:=   $L5222  [λx_o.(=_{ooo}(∧_{ooo} F_o x_o) F_o)_o]
# wff      1359   :        [λx.(= (∧ F x) F)]_{oo}      :=   $L5222
:=   $X5222  x_o
# wff      16   :      x_o      :=   $X5222
:=   $T5222  $L5222_{oo} T_o
# wff      1360   :        $L5222 T_o      :=   $T5222
:=   $F5222  $L5222_{oo} F_o
# wff      1361   :        $L5222 F_o      :=   $F5222

## Case T:   (F ∧ T) = F
§\  $T5222
#                    = $T5222 A5229c
## use Proof Template A5201b (Swap):   A = B  →  B = A
<< A5201b.r0t.txt
%0
#                    = A5229c $T5222
#                    =_{oωω} A5229c_ω $T5222_ω
%A5229c
#                    = (∧ F T) F      :=   A5229c
#                    =_{ooo}(∧_{ooo} F_o T_o) F_o      :=   A5229c
```

§s %0 1 %1
$\$L5222\,T$:= $\$T5222$

Case F: (F ∧ F) = F
§\ $\$F5222$
$= \$F5222\,A5229d$
use Proof Template A5201b (Swap): A = B → B = A
<< A5201b.r0t.txt
%0
$= A5229d\,\$F5222$
$=_{o\omega\omega} A5229d_{\omega} \$F5222_{\omega}$
%$A5229d$
$= (\wedge\, F\, F)\, F$:= $A5229d$
$=_{ooo}(\wedge_{ooo} F_o F_o) F_o$:= $A5229d$
§s %0 1 %1
$\$L5222\,F$:= $\$F5222$

<< A5222.r0t.txt
:= $\$L5222$
:= $\$X5222$
:= $\$T5222$
:= $\$F5222$
%0
$= (\wedge\, F\, x)\, F$
$=_{ooo}(\wedge_{ooo} F_o x_o) F_o$

:= $K8001b$ %0
wff 1358 : $= (\wedge\, F\, x)\, F_{o,\ldots}$:= $K8001b$

.c: (A ∧ F) = (F ∧ A)

%$K8001b$
$= (\wedge\, F\, x)\, F$:= $K8001b$
$=_{ooo}(\wedge_{ooo} F_o x_o) F_o$:= $K8001b$
use Proof Template A5201b (Swap): A = B → B = A
<< A5201b.r0t.txt
%0
$= F\, (\wedge\, F\, x)$
$=_{ooo} F_o (\wedge_{ooo} F_o x_o)$
%$K8001a$
$= (\wedge\, x\, F)\, F$:= $K8001a$
$=_{ooo}(\wedge_{ooo} x_o F_o) F_o$:= $K8001a$
§s %0 3 %1
$= (\wedge\, x\, F)\, (\wedge\, F\, x)$
%0
$= (\wedge\, x\, F)\, (\wedge\, F\, x)$
$=_{ooo}(\wedge_{ooo} x_o F_o)(\wedge_{ooo} F_o x_o)$

:= $K8001c$ %0

wff 1434 : $=(\wedge\, x\, F)\,(\wedge\, F\, x)_o$:= $K8001c$

##
Q.E.D.
##

%K8001a
%$K8001a$
$=(\wedge\, x\, F)\, F$:= $K8001a$
$=_{ooo}(\wedge_{ooo} x_o F_o) F_o$:= $K8001a$

%K8001b
%$K8001b$
$=(\wedge\, F\, x)\, F$:= $K8001b$
$=_{ooo}(\wedge_{ooo} F_o x_o) F_o$:= $K8001b$

%K8001c
%$K8001c$
$=(\wedge\, x\, F)\,(\wedge\, F\, x)$:= $K8001c$
$=_{ooo}(\wedge_{ooo} x_o F_o)(\wedge_{ooo} F_o x_o)$:= $K8001c$

2.1.55 Results for File K8002.r0.txt

##
Proof K8002: (A ∧ A) = A
##
##
Source: [Kubota 2017 (doi: 10.4444/100.10)]
##

Written by Ken Kubota (<mail@kenkubota.de>).
##
This file is part of the publication of the mathematical logic $\mathcal{R}_0$.
For more information, visit: <http://doi.org/10.4444/100.10>
##

<< basics.r0.txt
<< A5229.r0.txt

##
Proof
##

use Proof Template A5222 (Rule of Cases): [\x.A]T, [\x.A]F → A
:= $\$L5222\ [\lambda x_o.(=_{ooo}(\wedge_{ooo} x_o x_o) x_o)_o]$
wff 1249 : $[\lambda x.(=(\wedge\, x\, x)\, x)]_{oo}$:= $\$L5222$
:= $\$X5222\ x_o$

```
# wff    16 :      x_o      := $X5222
:= $T5222 $L5222_oo T_o
# wff    1250 :      $L5222 T_o      := $T5222
:= $F5222 $L5222_oo F_o
# wff    1251 :      $L5222 F_o      := $F5222

## Case T:  (T ∧ T) = T
§\ $T5222
#                 = $T5222 A5211
## use Proof Template A5201b (Swap):  A = B  →  B = A
<< A5201b.r0t.txt
%0
#                 = A5211 $T5222
#                 =_oωω A5211_ω $T5222_ω
%A5211
#                 = A5212 T      :=  A5211  A5229a
#                 =_ooo A5212_o T_o      :=  A5211  A5229a
§s %0 1 %1
#                 $L5222 T      :=  $T5222

## Case F:  (F ∧ F) = F
§\ $F5222
#                 = $F5222 A5229d
## use Proof Template A5201b (Swap):  A = B  →  B = A
<< A5201b.r0t.txt
%0
#                 = A5229d $F5222
#                 =_oωω A5229d_ω $F5222_ω
%A5229d
#                 = (∧ F F) F      :=  A5229d
#                 =_ooo (∧_ooo F_o F_o) F_o      :=  A5229d
§s %0 1 %1
#                 $L5222 F      :=  $F5222

<< A5222.r0t.txt
:= $L5222
:= $X5222
:= $T5222
:= $F5222
%0
#                 = (∧ x x) x
#                 =_ooo (∧_ooo x_o x_o) x_o

:= K8002 %0
# wff    1248 :      = (∧ x x) x_{o,...}      := K8002


##
##  Q.E.D.
```

```
##

%0
#                 =(∧ x x) x        :=  K8002
#                 =ooo(∧oooxoxo)xo       :=  K8002
```

2.1.56 Results for File K8003.r0a.txt

```
##
##  Proof Template K8003 (Intro):  A  →  H ⊃ A
##       (Hypothesis Introduction)
##
##  Source: [Kubota 2017 (doi: 10.4444/100.10)]
##
##  Copyright (c) 2017 Owl of Minerva Press GmbH. All rights reserved.
##  Written by Ken Kubota (<mail@kenkubota.de>).
##
##  This file is part of the publication of the mathematical logic R0.
##  For more information, visit: <http://doi.org/10.4444/100.10>
##


##
##  Define Syntactical Variables
##

## the theorem A
:=  $A8003 ao
# wff    11  :      ao     :=  $A8003

## the hypotheses H
:=  $H8003 ho
# wff    12  :      ho     :=  $H8003


##
##  Assumptions and Resulting Syntactical Variables
##

§! $A8003
#               a      :=  $A8003


##
##  Include Proof Template
##

## <<< K8003.r0t.txt
## Include begin (K8003.r0t.txt) [oldfile=(K8003.r0a.txt)]
```

```
##
##   Proof Template K8003 (Intro):   A   →   H ⊃ A
##         (Hypothesis Introduction)
##
##   Source: [Kubota 2017 (doi: 10.4444/100.10)]
##
##   Copyright (c) 2017 Owl of Minerva Press GmbH. All rights reserved.
##   Written by Ken Kubota (<mail@kenkubota.de>).
##
##   This file is part of the publication of the mathematical logic R0.
##   For more information, visit: <http://doi.org/10.4444/100.10>
##

<< basics.r0.txt
<< K8000.r0.txt


##
##   Proof Template
##

## .1

§=  ⊃ooo$H8003o$A8003o
#                     = (⊃ $H8003 $A8003) (⊃ $H8003 $A8003)
§r  /12  xtmpo
#                     = ⊃ [λxtmp.[λy.(= xtmp (∧ xtmp y))]]
§s  %1  12  %0
#                     = (⊃ $H8003 $A8003) ([λxtmp.[λy.(= xtmp (∧ xtmp y))]] $H8003 $A8003)
§r  /25  ytmpo
#                     = [λy.(= xtmp (∧ xtmp y))] [λytmp.(= xtmp (∧ xtmp ytmp))]
§s  %1  25  %0
#                     = (⊃ $H8003 $A8003) ([λxtmp.[λytmp.(= xtmp (∧ xtmp ytmp))]] $H8003 $A8003)
§\  [λxtmpo.[λytmpo.(=oooxtmpo(∧oooxtmpoytmpo))o](oo)]$H8003o
#                     = ([λxtmp.[λytmp.(= xtmp (∧ xtmp ytmp))]] $H8003) ...
... [λytmp.(= $H8003 (∧ $H8003 ytmp))]
§s  %1  6  %0
#                     = (⊃ $H8003 $A8003) ([λytmp.(= $H8003 (∧ $H8003 ytmp))] $A8003)
§\  [λytmpo.(=ooo$H8003o(∧ooo$H8003oytmpo))o]$A8003o
#                     = ([λytmp.(= $H8003 (∧ $H8003 ytmp))] $A8003) (= $H8003 (∧ $H8003 $A8003))
§s  %1  3  %0
#                     = (⊃ $H8003 $A8003) (= $H8003 (∧ $H8003 $A8003))
:=  $TMP8003  %0
# wff     1451   :        = (⊃ $H8003 $A8003) (= $H8003 (∧ $H8003 $A8003))o        :=   $TMP8003

## .2

## use Proof Template A5219b (Rule T):   A   →   A = T
:=  $A5219b  ao
```

wff 11 : a_o := $\$A5219b$ $\$A8003$
<< A5219b.r0t.txt
:= $\$A5219b$
%0
$= \$A8003\,T$
$=_{ooo}\$A8003_o T_o$

.3

%$\$TMP8003$
$= (\supset \$H8003\,\$A8003)\,(= \$H8003\,(\wedge\,\$H8003\,\$A8003))$:= $\$TMP8003$
$=_{o\omega\omega}(\supset_{ooo}\$H8003_o\$A8003_o)(=_{ooo}\$H8003_o(\wedge_{ooo}\$H8003_o\$A8003_o))$:=
$\$TMP8003$
:= $\$TMP8003$
§s %0 15 %1
$= (\supset \$H8003\,\$A8003)\,(= \$H8003\,(\wedge\,\$H8003\,T))$
:= $\$TMP8003$ %0
wff 1471 : $= (\supset \$H8003\,\$A8003)\,(= \$H8003\,(\wedge\,\$H8003\,T))_o$:= $\$TMP8003$

.4

use Proof Template A5221 (Sub): B → B [x/A]
:= $\$B5221$ $=_{ooo}(\wedge_{ooo}x_oT_o)x_o$
wff 1249 : $= (\wedge\,x\,T)\,x_{o,\ldots}$:= $\$B5221$ $K8000a$
:= $\$T5221$ o
wff 2 : o_τ := $\$T5221$
:= $\$X5221$ x_o
wff 18 : x_o := $\$X5221$
:= $\$A5221$ h_o
wff 12 : h_o := $\$A5221$ $\$H8003$
<< A5221.r0t.txt
:= $\$B5221$
:= $\$T5221$
:= $\$X5221$
:= $\$A5221$
%0
$= (\wedge\,\$H8003\,T)\,\$H8003$
$=_{ooo}(\wedge_{ooo}\$H8003_oT_o)\$H8003_o$

.5

%$\$TMP8003$
$= (\supset \$H8003\,\$A8003)\,(= \$H8003\,(\wedge\,\$H8003\,T))$:= $\$TMP8003$
$=_{o\omega\omega}(\supset_{ooo}\$H8003_o\$A8003_o)(=_{ooo}\$H8003_o(\wedge_{ooo}\$H8003_oT_o))$:= $\$TMP8003$
:= $\$TMP8003$
§s %0 7 %1
$= (\supset \$H8003\,\$A8003)\,(= \$H8003\,\$H8003)$
use Proof Template A5201b (Swap): A = B → B = A
<< A5201b.r0t.txt

%0
$= (= \$H8003\ \$H8003)\ (\supset \$H8003\ \$A8003)$
$=_{o\omega\omega}(=_{ooo}\$H8003_o\$H8003_o)(\supset_{ooo}\$H8003_o\$A8003_o)$
§= $_o$ $\$H8003$
$= \$H8003\ \$H8003$
§s %0 1 %1
$\supset \$H8003\ \$A8003$
Include end (K8003.r0t.txt) [newfile=(K8003.r0a.txt)]
>>>

##
Undefine Syntactical Variables
##

:= $\$A8003$
:= $\$H8003$

##
Q.E.D.
##

%0
$\supset h\ a$
$\supset_{ooo}h_o a_o$

2.1.57 Results for File K8004.r0a.txt

##
Proof Template K8004 (Trans): (H $\oplus$ A), B $\rightarrow$ H $\supset$ B
for any operator $\oplus$, including "$\supset$" and "=" (Hypothesis Transfer)
##
Source: [Kubota 2017 (doi: 10.4444/100.10)]
##

Written by Ken Kubota (<mail@kenkubota.de>).
##
This file is part of the publication of the mathematical logic $\mathcal{R}_0$.
For more information, visit: <http://doi.org/10.4444/100.10>
##

<< basics.r0.txt

##
Define Syntactical Variables
##

hypothesis in theorem
:= $\$HA8004$ $\supset_{ooo} h_o a_o$
wff 210 : $\supset h\, a_o$:= $\$HA8004$

proposition
:= $\$B8004$ b_o
wff 58 : b_o := $\$B8004$

##
Assumptions and Resulting Syntactical Variables
##

§! $\$HA8004$
$\supset h\, a$:= $\$HA8004$
§! $\$B8004$
b := $\$B8004$

##
Include Proof Template
##

<<< K8004.r0t.txt
Include begin (K8004.r0t.txt) [oldfile=(K8004.r0a.txt)]
##
Proof Template K8004 (Trans): (H $\oplus$ A), B $\rightarrow$ H $\supset$ B
for any operator $\oplus$, including "$\supset$" and "$=$" (Hypothesis Transfer)
##
Source: [Kubota 2017 (doi: 10.4444/100.10)]
##

Written by Ken Kubota (<mail@kenkubota.de>).
##
This file is part of the publication of the mathematical logic $\mathcal{R}_0$.
For more information, visit: <http://doi.org/10.4444/100.10>
##

##
Proof Template
##

use Proof Template K8003 (Intro): A $\rightarrow$ H $\supset$ A
:= $\$A8003$ b_o
wff 58 : b_o := $\$A8003$ $\$B8004$
:= $\$H8003$ $\supset_{ooo} h_o a_o/5$
wff 208 : h_o := $\$H8003$
<< K8003.r0t.txt

:= $\$A8003$
:= $\$H8003$
%0
\# $\supset h\,\$B8004$
\# $\supset_{ooo} h_o \$B8004_o$
Include end (K8004.r0t.txt) [newfile=(K8004.r0a.txt)]
>>>

##
Undefine Syntactical Variables
##

:= $\$HA8004$
:= $\$B8004$

##
Q.E.D.
##

%0
\# $\supset h\,b$
\# $\supset_{ooo} h_o b_o$

2.1.58 Results for File K8005.r0.txt

##
Proof K8005: H ⊃ H
##
##
Source: [Kubota 2017 (doi: 10.4444/100.10)]
##

Written by Ken Kubota (<mail@kenkubota.de>).
##
This file is part of the publication of the mathematical logic $\mathcal{R}_0$.
For more information, visit: <http://doi.org/10.4444/100.10>
##

<< basics.r0.txt
<< K8002.r0.txt

##
Proof
##

.1

§= $\supset_{ooo} x_o x_o$
\# $= (\supset x\, x)\, (\supset x\, x)$
§\ $\supset_{ooo} x_o$
\# $= (\supset x)\, [\lambda y.(= x\, (\wedge\, x\, y))]$
§s %1 6 %0
\# $= (\supset x\, x)\, ([\lambda y.(= x\, (\wedge\, x\, y))]\, x)$
§\ $[\lambda y_o.(=_{ooo} x_o (\wedge_{ooo} x_o y_o))_o] x_o$
\# $= ([\lambda y.(= x\, (\wedge\, x\, y))]\, x)\, (= x\, (\wedge\, x\, x))$
§s %1 3 %0
\# $= (\supset x\, x)\, (= x\, (\wedge\, x\, x))$

\## .2

%$K8002$
\# $= (\wedge\, x\, x)\, x$:= $K8002$
\# $=_{ooo} (\wedge_{ooo} x_o x_o) x_o$:= $K8002$
§s %1 7 %0
\# $= (\supset x\, x)\, (= x\, x)$

\## .3

\## use Proof Template A5201b (Swap): A = B → B = A
<< A5201b.r0t.txt
%0
\# $= (= x\, x)\, (\supset x\, x)$
\# $=_{o\omega\omega} (=_{ooo} x_o x_o) (\supset_{ooo} x_o x_o)$

§= $_o\ x_o$
\# $= x\, x$
§s %0 1 %1
\# $\supset x\, x$

:= $K8005$ %0
\# wff 1357 : $\supset x\, x_{o,\ldots}$:= $K8005$

\##
\## Q.E.D.
\##

%0
\# $\supset x\, x$:= $K8005$
\# $\supset_{ooo} x_o x_o$:= $K8005$

2.1.59 Results for File K8006.r0.txt

\##
\## Proof Template K8006: (A * T) = (T * A), (A * F) = (F * A) → (A * B) = (B * A)

\## for any Boolean relation *
\##
\## Source: [Kubota 2017 (doi: 10.4444/100.10)]
\##

\## Written by Ken Kubota (<mail@kenkubota.de>).
\##
\## This file is part of the publication of the mathematical logic $\mathcal{R}_0$.
\## For more information, visit: <http://doi.org/10.4444/100.10>
\##

\##
\## Define Syntactical Variables
\##

<< K8000.r0.txt
<< K8001.r0.txt

\## the Boolean relation
:= $\$R8006\ \ [\lambda x_o.[\lambda y_o.(=_{o\omega\omega}[\lambda g_{ooo}.(g_{ooo}T_oT_o)_o][\lambda g_{ooo}.(g_{ooo}x_oy_o)_o])_o]_{(oo)}]$
\# wff 47 : $[\lambda x.[\lambda y.(=[\lambda g.(g\,T\,T)]\,[\lambda g.(g\,x\,y)])]]_{ooo}$:= $\$R8006$ $\wedge$

\## the theorem for case T (using variables x and y)
:= $\$T8006\ \ =_{ooo}(\wedge_{ooo}x_oT_o)(\wedge_{ooo}T_ox_o)$
\# wff 1410 : $=(\wedge\,x\,T)\,(\wedge\,T\,x)_o$:= $\$T8006$ $K8000c$

\## the theorem for case F (using variables x and y)
:= $\$F8006\ \ =_{ooo}(\wedge_{ooo}x_oF_o)(\wedge_{ooo}F_ox_o)$
\# wff 1564 : $=(\wedge\,x\,F)\,(\wedge\,F\,x)_o$:= $\$F8006$ $K8001c$

\##
\## Proof Template
\##

\## <<< K8006.r0t.txt
\## Include begin (K8006.r0t.txt) [oldfile=(K8006.r0.txt)]
\##
\## Proof Template K8006: (A * T) = (T * A), (A * F) = (F * A) → (A * B) = (B * A)
\## for any Boolean relation *
\##
\## Source: [Kubota 2017 (doi: 10.4444/100.10)]
\##

\## Written by Ken Kubota (<mail@kenkubota.de>).
\##
\## This file is part of the publication of the mathematical logic $\mathcal{R}_0$.
\## For more information, visit: <http://doi.org/10.4444/100.10>

##

Skipping file basics.r0.txt (already included)

##
Proof Template
##

use Proof Template A5222 (Rule of Cases): [\x.A]T, [\x.A]F $\rightarrow$ A
:= $\$L5222TMP$ $[\lambda y_o.(=_{ooo}(\wedge_{ooo}x_o y_o)(\wedge_{ooo}y_o x_o))_o]$
wff 1569 : $[\lambda y.(=(\wedge\, x\, y)\,(\wedge\, y\, x))]_{oo}$:= $\$L5222TMP$
:= $\$X5222TMP$ y_o
wff 34 : y_o := $\$X5222TMP$
:= $\$T5222TMP$ $\$L5222TMP_{oo}T_o$
wff 1570 : $\$L5222TMP\,T_o$:= $\$T5222TMP$
:= $\$F5222TMP$ $\$L5222TMP_{oo}F_o$
wff 1571 : $\$L5222TMP\,F_o$:= $\$F5222TMP$

case T: (A $\wedge$ T) = (T $\wedge$ A)
§\ $\$T5222TMP$
$= \$T5222TMP\,K8000c$
use Proof Template A5201b (Swap): A = B $\rightarrow$ B = A
<< A5201b.r0t.txt
%0
$= K8000c\,\$T5222TMP$
$=_{o\omega\omega}K8000c_\omega\$T5222TMP_\omega$
%$K8000c$
$= (\wedge\, x\, T)\,(\wedge\, T\, x)$:= $\$T8006$ $K8000c$
$=_{ooo}(\wedge_{ooo}x_o T_o)(\wedge_{ooo}T_o x_o)$:= $\$T8006$ $K8000c$
§s %0 1 %1
$\$L5222TMP\,T$:= $\$T5222TMP$

case F: (A $\wedge$ F) = (F $\wedge$ A)
§\ $\$F5222TMP$
$= \$F5222TMP\,K8001c$
use Proof Template A5201b (Swap): A = B $\rightarrow$ B = A
<< A5201b.r0t.txt
%0
$= K8001c\,\$F5222TMP$
$=_{o\omega\omega}K8001c_\omega\$F5222TMP_\omega$
%$K8001c$
$= (\wedge\, x\, F)\,(\wedge\, F\, x)$:= $\$F8006$ $K8001c$
$=_{ooo}(\wedge_{ooo}x_o F_o)(\wedge_{ooo}F_o x_o)$:= $\$F8006$ $K8001c$
§s %0 1 %1
$\$L5222TMP\,F$:= $\$F5222TMP$

replace free variable x by variable a avoiding a name collision

:= $\$L5222\ [\lambda \$X5222TMP_o.(=_{ooo}(\wedge_{ooo}a_o\$X5222TMP_o)(\wedge_{ooo}\$X5222TMP_oa_o))_o]$
wff 1587 : $[\lambda \$X5222TMP.(=(\wedge\, a\, \$X5222TMP)\,(\wedge\, \$X5222TMP\, a))]_{oo}$:= $\$L5222$

:= $\$X5222\ y_o$
wff 34 : y_o := $\$X5222$ $\$X5222TMP$

use Proof Template A5221 (Sub): B → B [x/A]
:= $\$B5221\ \$L5222TMP_{oo}T_o$
wff 1570 : $\$L5222TMP\,T_{o,\dots}$:= $\$B5221$ $\$T5222TMP$
:= $\$T5221\ o$
wff 2 : o_τ := $\$T5221$
:= $\$X5221\ x_o$
wff 16 : x_o := $\$X5221$
:= $\$A5221\ a_o$
wff 54 : a_o := $\$A5221$
<< A5221.r0t.txt
:= $\$B5221$
:= $\$T5221$
:= $\$X5221$
:= $\$A5221$
:= $\$T5222$ %0
wff 1623 : $\$L5222\,T_{o,\dots}$:= $\$T5222$
%0
$\$L5222\,T$:= $\$T5222$
$\$L5222_{oo}T_o$:= $\$T5222$

use Proof Template A5221 (Sub): B → B [x/A]
:= $\$B5221\ \$L5222TMP_{oo}F_o$
wff 1571 : $\$L5222TMP\,F_{o,\dots}$:= $\$B5221$ $\$F5222TMP$
:= $\$T5221\ o$
wff 2 : o_τ := $\$T5221$
:= $\$X5221\ x_o$
wff 16 : x_o := $\$X5221$
:= $\$A5221\ a_o$
wff 54 : a_o := $\$A5221$
<< A5221.r0t.txt
:= $\$B5221$
:= $\$T5221$
:= $\$X5221$
:= $\$A5221$
:= $\$F5222$ %0
wff 1657 : $\$L5222\,F_{o,\dots}$:= $\$F5222$
%0
$\$L5222\,F$:= $\$F5222$
$\$L5222_{oo}F_o$:= $\$F5222$

:= $\$L5222TMP$
:= $\$X5222TMP$
:= $\$T5222TMP$

:= $\$F5222TMP$

now actually use Proof Template A5222 (Rule of Cases): [\x.A]T, [\x.A]F → A
<< A5222.r0t.txt
:= $\$L5222$
:= $\$X5222$
:= $\$T5222$
:= $\$F5222$
%0
$= (\wedge\, a\, y)\, (\wedge\, y\, a)$
$=_{ooo}(\wedge_{ooo} a_o y_o)(\wedge_{ooo} y_o a_o)$

replace back

use Proof Template A5221 (Sub): B → B [x/A]
:= $\$B5221$ %0
wff 1586 : $= (\wedge\, a\, y)\, (\wedge\, y\, a)_{o,\ldots}$:= $\$B5221$
:= $\$T5221\ o$
wff 2 : o_τ := $\$T5221$
:= $\$X5221\ a_o$
wff 54 : a_o := $\$X5221$
:= $\$A5221\ x_o$
wff 16 : x_o := $\$A5221$
<< A5221.r0t.txt
:= $\$B5221$
:= $\$T5221$
:= $\$X5221$
:= $\$A5221$
%0
$= (\wedge\, x\, y)\, (\wedge\, y\, x)$
$=_{ooo}(\wedge_{ooo} x_o y_o)(\wedge_{ooo} y_o x_o)$

match general definition
§= $COMMT_{o(\backslash 4\backslash 4\backslash 3)\tau} o_\tau \wedge_{ooo}$
$= (COMMT\, o\, \wedge)\, (COMMT\, o\, \wedge)$
§\ $COMMT_{o(\backslash 4\backslash 4\backslash 3)\tau} o_\tau$
$= (COMMT\, o)\, [\lambda f.(= (f\, x\, y)\, (f\, y\, x))]$
§s %1 10 %0
$= ([\lambda f.(= (f\, x\, y)\, (f\, y\, x))]\, \wedge)\, (COMMT\, o\, \wedge)$
§\ $[\lambda f_{ooo}.(=_{ooo}(f_{ooo} x_o y_o)(f_{ooo} y_o x_o))_o]\wedge_{ooo}$
$= ([\lambda f.(= (f\, x\, y)\, (f\, y\, x))]\, \wedge)\, (= (\wedge\, x\, y)\, (\wedge\, y\, x))$
§s %1 5 %0
$= (= (\wedge\, x\, y)\, (\wedge\, y\, x))\, (COMMT\, o\, \wedge)$
§s %5 1 %0
$COMMT\, o\, \wedge$
Include end (K8006.r0t.txt) [newfile=(K8006.r0.txt)]
>>>

##
Undefine Syntactical Variables
##

:= $\$R8006$
:= $\$T8006$
:= $\$F8006$

##
Q.E.D.
##

%0
$COMMT\, o \wedge$
$COMMT_{o(\backslash 4\backslash 4\backslash 3)\tau}\, o_{\tau} \wedge_{ooo}$

2.1.60 Results for File K8007.r0.txt

##
Proof K8007: (A ∧ B) = (B ∧ A)
##
##
Source: [Kubota 2017 (doi: 10.4444/100.10)]
##

Written by Ken Kubota (<mail@kenkubota.de>).
##
This file is part of the publication of the mathematical logic $\mathcal{R}_0$.
For more information, visit: <http://doi.org/10.4444/100.10>
##

<< K8000.r0.txt
<< K8001.r0.txt

##
Proof
##

use Proof Template K8006: (A * T) = (T * A), (A * F) = (F * A) → (A * B) = (B * A)
:= $\$R8006\ [\lambda x_o.[\lambda y_o.(=_{o\omega\omega}[\lambda g_{ooo}.(g_{ooo}T_oT_o)_o][\lambda g_{ooo}.(g_{ooo}x_oy_o)_o])_o]_{(oo)}]$
wff 47 : $[\lambda x.[\lambda y.(=[\lambda g.(g\,T\,T)]\,[\lambda g.(g\,x\,y)])]]_{ooo}$:= $\$R8006$ ∧
:= $\$T8006\ =_{ooo}(\wedge_{ooo}x_oT_o)(\wedge_{ooo}T_ox_o)$
wff 1410 : $=(\wedge\,x\,T)\,(\wedge\,T\,x)_o$:= $\$T8006$ $K8000c$
:= $\$F8006\ =_{ooo}(\wedge_{ooo}x_oF_o)(\wedge_{ooo}F_ox_o)$
wff 1564 : $=(\wedge\,x\,F)\,(\wedge\,F\,x)_o$:= $\$F8006$ $K8001c$
<< K8006.r0t.txt

:= $\$R8006$
:= $\$T8006$
:= $\$F8006$

:= $K8007$ %0
wff 1773 : $COMMT\, o \wedge_{o,...}$:= $K8007$

##
Q.E.D.
##

%0
$COMMT\, o \wedge$:= $K8007$
$COMMT_{o(\backslash 4\backslash 4\backslash 3)\tau} o_\tau \wedge_{ooo}$:= $K8007$

2.1.61 Results for File K8008.r0.txt

##
Proof K8008: $\sim\sim$ A = A
##
##
Source: [Kubota 2017 (doi: 10.4444/100.10)]
##

Written by Ken Kubota (<mail@kenkubota.de>).
##
This file is part of the publication of the mathematical logic $\mathcal{R}_0$.
For more information, visit: <http://doi.org/10.4444/100.10>
##

<< basics.r0.txt
<< A5231.r0.txt

##
Proof
##

use Proof Template A5222 (Rule of Cases): [\x.A]T, [\x.A]F → A
:= $\$L5222$ $[\lambda a_o.(=_{ooo}(\sim_{oo}(\sim_{oo}a_o))a_o)_o]$
wff 1588 : $[\lambda a.(=(\sim(\sim a))\,a)]_{oo}$:= $\$L5222$
:= $\$X5222$ x_o
wff 16 : x_o := $\$X5222$
:= $\$T5222$ $\$L5222_{oo}T_o$
wff 1589 : $\$L5222\, T_o$:= $\$T5222$
:= $\$F5222$ $\$L5222_{oo}F_o$
wff 1590 : $\$L5222\, F_o$:= $\$F5222$

case T: $\sim\sim$ T = T
§= $\$T5222$
\# $=\$T5222\,\$T5222$
§\ $\$T5222$
\# $=\$T5222\,(=(\sim(\sim T))\,T)$
§s %1 3 %0
\# $=\$T5222\,(=(\sim(\sim T))\,T)$
%$A5231a$
\# $=(\sim T)\,F \quad := \quad A5231a$
\# $=_{ooo}(\sim_{oo}T_o)F_o \quad := \quad A5231a$
§s %1 27 %0
\# $=\$T5222\,A5231b$

use Proof Template A5201b (Swap): A = B $\rightarrow$ B = A
<< A5201b.r0t.txt
%0
\# $=A5231b\,\$T5222$
\# $=_{o\omega\omega}A5231b_\omega\$T5222_\omega$

%$A5231b$
\# $=(\sim F)\,T \quad := \quad A5231b$
\# $=_{ooo}(\sim_{oo}F_o)T_o \quad := \quad A5231b$
§s %0 1 %1
\# $\$L5222\,T \quad := \quad \$T5222$

case F: $\sim\sim$ F = F
§= $\$F5222$
\# $=\$F5222\,\$F5222$
§\ $\$F5222$
\# $=\$F5222\,(=(\sim(\sim F))\,F)$
§s %1 3 %0
\# $=\$F5222\,(=(\sim(\sim F))\,F)$
%$A5231b$
\# $=(\sim F)\,T \quad := \quad A5231b$
\# $=_{ooo}(\sim_{oo}F_o)T_o \quad := \quad A5231b$
§s %1 27 %0
\# $=\$F5222\,A5231a$

use Proof Template A5201b (Swap): A = B $\rightarrow$ B = A
<< A5201b.r0t.txt
%0
\# $=A5231a\,\$F5222$
\# $=_{o\omega\omega}A5231a_\omega\$F5222_\omega$

%$A5231a$
\# $=(\sim T)\,F \quad := \quad A5231a$
\# $=_{ooo}(\sim_{oo}T_o)F_o \quad := \quad A5231a$
§s %0 1 %1
\# $\$L5222\,F \quad := \quad \$F5222$

<< A5222.r0t.txt
:= $\$L5222$
:= $\$X5222$
:= $\$T5222$
:= $\$F5222$

:= $K8008$ %0
\# wff 1663 : $=(\sim(\sim x))\,x_{o,\ldots}$:= $K8008$

##
Q.E.D.
##

%0
\# $=(\sim(\sim x))\,x$:= $K8008$
\# $=_{ooo}(\sim_{oo}(\sim_{oo}x_o))x_o$:= $K8008$

2.1.62 Results for File K8009.r0.txt

##
Proof K8009: (A ∨ T) = T; (T ∨ A) = T; (A ∨ T) = (T ∨ A)
##
##
Source: [Kubota 2017 (doi: 10.4444/100.10)]
##

Written by Ken Kubota (<mail@kenkubota.de>).
##
This file is part of the publication of the mathematical logic $\mathcal{R}_0$.
For more information, visit: <http://doi.org/10.4444/100.10>
##

<< basics.r0.txt
<< A5231.r0.txt
<< K8001.r0.txt

##
Proof
##

.a: (A ∨ T) = T

§= $_o\ \vee_{ooo}x_oT_o$
\# $=(\vee\,x\,T)\,(\vee\,x\,T)$
§\ $\vee_{ooo}x_o$
\# $=(\vee\,x)\,[\lambda b.(\sim(\wedge\,(\sim x)\,(\sim b)))]$

§s %1 6 %0
\# $=(\vee\, x\, T)\, ([\lambda b.(\sim(\wedge\,(\sim x)\,(\sim b)))]\, T)$
§\ $[\lambda b_o.(\sim_{oo}(\wedge_{ooo}(\sim_{oo}x_o)(\sim_{oo}b_o)))_o]T_o$
\# $=([\lambda b.(\sim(\wedge\,(\sim x)\,(\sim b)))]\, T)\,(\sim(\wedge\,(\sim x)\,(\sim T)))$
§s %1 3 %0
\# $=(\vee\, x\, T)\,(\sim(\wedge\,(\sim x)\,(\sim T)))$
%$A5231a$
\# $=(\sim T)\, F$:= $A5231a$
\# $=_{ooo}(\sim_{oo}T_o)F_o$:= $A5231a$
§s %1 15 %0
\# $=(\vee\, x\, T)\,(\sim(\wedge\,(\sim x)\, F))$
:= $\$TMP8006$ %0
\# wff 1684 : $=(\vee\, x\, T)\,(\sim(\wedge\,(\sim x)\, F))_o$:= $\$TMP8006$

\## use Proof Template A5221 (Sub): B → B [x/A]
:= $\$B5221$ $=_{ooo}(\wedge_{ooo}x_oF_o)F_o$
\# wff 1433 : $=(\wedge\, x\, F)\, F_{o,\dots}$:= $\$B5221$ $K8001a$
:= $\$T5221$ o
\# wff 2 : o_τ := $\$T5221$
:= $\$X5221$ x_o
\# wff 16 : x_o := $\$X5221$
:= $\$A5221$ $\sim_{oo}\$X5221_o$
\# wff 1669 : $\sim\$X5221_o$:= $\$A5221$
<< A5221.r0t.txt
:= $\$B5221$
:= $\$T5221$
:= $\$X5221$
:= $\$A5221$
%0
\# $=(\wedge\,(\sim x)\, F)\, F$
\# $=_{ooo}(\wedge_{ooo}(\sim_{oo}x_o)F_o)F_o$

%$\$TMP8006$
\# $=(\vee\, x\, T)\,(\sim(\wedge\,(\sim x)\, F))$:= $\$TMP8006$
\# $=_{ooo}(\vee_{ooo}x_oT_o)(\sim_{oo}(\wedge_{ooo}(\sim_{oo}x_o)F_o))$:= $\$TMP8006$
:= $\$TMP8006$
§s %0 7 %1
\# $=(\vee\, x\, T)\,(\sim F)$

%$A5231b$
\# $=(\sim F)\, T$:= $A5231b$
\# $=_{ooo}(\sim_{oo}F_o)T_o$:= $A5231b$
§s %1 3 %0
\# $=(\vee\, x\, T)\, T$

:= $K8009a$ %0
\# wff 1698 : $=(\vee\, x\, T)\, T_o$:= $K8009a$

\## .b: (T ∨ A) = T

§= $_o\ \vee_{ooo}T_o x_o$
$=(\vee T\, x)\,(\vee T\, x)$
§\ $\vee_{ooo}T_o$
$=(\vee T)\,[\lambda b.(\sim(\wedge\,(\sim T)\,(\sim b)))]$
§s %1 6 %0
$=(\vee T\, x)\,([\lambda b.(\sim(\wedge\,(\sim T)\,(\sim b)))]\, x)$
§\ $[\lambda b_o.(\sim_{oo}(\wedge_{ooo}(\sim_{oo}T_o)(\sim_{oo}b_o)))_o]x_o$
$=([\lambda b.(\sim(\wedge\,(\sim T)\,(\sim b)))]\, x)\,(\sim(\wedge\,(\sim T)\,(\sim x)))$
§s %1 3 %0
$=(\vee T\, x)\,(\sim(\wedge\,(\sim T)\,(\sim x)))$
%$A5231a$
$=(\sim T)\, F$:= $A5231a$
$=_{ooo}(\sim_{oo}T_o)F_o$:= $A5231a$
§s %1 29 %0
$=(\vee T\, x)\,(\sim(\wedge\, F\,(\sim x)))$
:= $\$TMP8006$ %0
wff 1718 : $=(\vee T\, x)\,(\sim(\wedge\, F\,(\sim x)))_o$:= $\$TMP8006$

use Proof Template A5221 (Sub): B → B [x/A]
:= $\$B5221\ =_{ooo}(\wedge_{ooo}F_o x_o)F_o$
wff 1587 : $=(\wedge\, F\, x)\, F_{o,\dots}$:= $\$B5221$ $K8001b$
:= $\$T5221\ o$
wff 2 : o_τ := $\$T5221$
:= $\$X5221\ x_o$
wff 16 : x_o := $\$X5221$
:= $\$A5221\ \sim_{oo}\$X5221_o$
wff 1669 : $\sim\$X5221_o$:= $\$A5221$
<< A5221.r0t.txt
:= $\$B5221$
:= $\$T5221$
:= $\$X5221$
:= $\$A5221$
%0
$=(\wedge\, F\,(\sim x))\, F$
$=_{ooo}(\wedge_{ooo}F_o(\sim_{oo}x_o))F_o$

%$\$TMP8006$
$=(\vee T\, x)\,(\sim(\wedge\, F\,(\sim x)))$:= $\$TMP8006$
$=_{ooo}(\vee_{ooo}T_o x_o)(\sim_{oo}(\wedge_{ooo}F_o(\sim_{oo}x_o)))$:= $\$TMP8006$
:= $\$TMP8006$
§s %0 7 %1
$=(\vee T\, x)\,(\sim F)$

%$A5231b$
$=(\sim F)\, T$:= $A5231b$
$=_{ooo}(\sim_{oo}F_o)T_o$:= $A5231b$
§s %1 3 %0
$=(\vee T\, x)\, T$

:= $K8009b$ %0
wff 1749 : $=(\vee T\,x)\,T_o$:= $K8009b$

.c: (A ∨ T) = (T ∨ A)

%$K8009b$
$=(\vee T\,x)\,T$:= $K8009b$
$=_{ooo}(\vee_{ooo}T_o x_o)T_o$:= $K8009b$

use Proof Template A5201b (Swap): A = B → B = A
<< A5201b.r0t.txt
%0
$=T\,(\vee T\,x)$
$=_{ooo}T_o(\vee_{ooo}T_o x_o)$

%$K8009a$
$=(\vee x\,T)\,T$:= $K8009a$
$=_{ooo}(\vee_{ooo}x_o T_o)T_o$:= $K8009a$
§s %0 3 %1
$=(\vee x\,T)\,(\vee T\,x)$

:= $K8009c$ %0
wff 1751 : $=(\vee x\,T)\,(\vee T\,x)_o$:= $K8009c$

##
Q.E.D.
##

%K8009a
%$K8009a$
$=(\vee x\,T)\,T$:= $K8009a$
$=_{ooo}(\vee_{ooo}x_o T_o)T_o$:= $K8009a$

%K8009b
%$K8009b$
$=(\vee T\,x)\,T$:= $K8009b$
$=_{ooo}(\vee_{ooo}T_o x_o)T_o$:= $K8009b$

%K8009c
%$K8009c$
$=(\vee x\,T)\,(\vee T\,x)$:= $K8009c$
$=_{ooo}(\vee_{ooo}x_o T_o)(\vee_{ooo}T_o x_o)$:= $K8009c$

2.1.63 Results for File K8010.r0.txt

##
Proof K8010: (A ∨ F) = A; (F ∨ A) = A; (A ∨ F) = (F ∨ A)

##
##
Source: [Kubota 2017 (doi: 10.4444/100.10)]
##

Written by Ken Kubota (<mail@kenkubota.de>).
##
This file is part of the publication of the mathematical logic $\mathcal{R}_0$.
For more information, visit: <http://doi.org/10.4444/100.10>
##

<< A5231.r0.txt
<< K8000.r0.txt
<< K8008.r0.txt

##
Proof
##

.a: (A ∨ F) = A

§= $_o \ \vee_{ooo} x_o F_o$
$= (\vee\, x\, F)\, (\vee\, x\, F)$
§\ $\vee_{ooo} x_o$
$= (\vee\, x)\, [\lambda b.(\sim (\wedge\, (\sim x)\, (\sim b)))]$
§s %1 6 %0
$= (\vee\, x\, F)\, ([\lambda b.(\sim (\wedge\, (\sim x)\, (\sim b)))]\, F)$
§\ $[\lambda b_o.(\sim_{oo}(\wedge_{ooo}(\sim_{oo}x_o)(\sim_{oo}b_o)))_o]F_o$
$= ([\lambda b.(\sim (\wedge\, (\sim x)\, (\sim b)))]\, F)\, (\sim (\wedge\, (\sim x)\, (\sim F)))$
§s %1 3 %0
$= (\vee\, x\, F)\, (\sim (\wedge\, (\sim x)\, (\sim F)))$
%$A5231b$
$= (\sim F)\, T$:= $A5231b$
$=_{ooo}(\sim_{oo}F_o)T_o$:= $A5231b$
§s %1 15 %0
$= (\vee\, x\, F)\, (\sim (\wedge\, (\sim x)\, T))$
:= $\$TMP8010$ %0
wff 1828 : $= (\vee\, x\, F)\, (\sim (\wedge\, (\sim x)\, T))_o$:= $\$TMP8010$

use Proof Template A5221 (Sub): B → B [x/A]
:= $\$B5221$ $=_{ooo}(\wedge_{ooo}x_oT_o)x_o$
wff 1587 : $= (\wedge\, x\, T)\, x_{o,\ldots}$:= $\$B5221$ $K8000a$
:= $\$T5221$ o
wff 2 : o_τ := $\$T5221$
:= $\$X5221$ x_o
wff 16 : x_o := $\$X5221$
:= $\$A5221$ $\sim_{oo}\$X5221_o$
wff 1791 : $\sim \$X5221_o$:= $\$A5221$

<< A5221.r0t.txt
:= $\$B5221$
:= $\$T5221$
:= $\$X5221$
:= $\$A5221$
%0
\# $=(\wedge\,(\sim x)\,T)\,(\sim x)$
\# $=_{ooo}(\wedge_{ooo}(\sim_{oo}x_o)T_o)(\sim_{oo}x_o)$

%$\$TMP8010$
\# $=(\vee\,x\,F)\,(\sim(\wedge\,(\sim x)\,T))$:= $\$TMP8010$
\# $=_{ooo}(\vee_{ooo}x_oF_o)(\sim_{oo}(\wedge_{ooo}(\sim_{oo}x_o)T_o))$:= $\$TMP8010$
:= $\$TMP8010$
§s %0 7 %1
\# $=(\vee\,x\,F)\,(\sim(\sim x))$

%$K8008$
\# $=(\sim(\sim x))\,x$:= $K8008$
\# $=_{ooo}(\sim_{oo}(\sim_{oo}x_o))x_o$:= $K8008$
§s %1 3 %0
\# $=(\vee\,x\,F)\,x$

:= $K8010a$ %0
\# wff 1863 : $=(\vee\,x\,F)\,x_o$:= $K8010a$

\#\# .b: (F ∨ A) = A

§= $_o$ $\vee_{ooo}F_ox_o$
\# $=(\vee\,F\,x)\,(\vee\,F\,x)$
§\ $\vee_{ooo}F_o$
\# $=(\vee\,F)\,[\lambda b.(\sim(\wedge\,(\sim F)\,(\sim b)))]$
§s %1 6 %0
\# $=(\vee\,F\,x)\,([\lambda b.(\sim(\wedge\,(\sim F)\,(\sim b)))]\,x)$
§\ $[\lambda b_o.(\sim_{oo}(\wedge_{ooo}(\sim_{oo}F_o)(\sim_{oo}b_o)))_o]x_o$
\# $=([\lambda b.(\sim(\wedge\,(\sim F)\,(\sim b)))]\,x)\,(\sim(\wedge\,(\sim F)\,(\sim x)))$
§s %1 3 %0
\# $=(\vee\,F\,x)\,(\sim(\wedge\,(\sim F)\,(\sim x)))$
%$A5231b$
\# $=(\sim F)\,T$:= $A5231b$
\# $=_{ooo}(\sim_{oo}F_o)T_o$:= $A5231b$
§s %1 29 %0
\# $=(\vee\,F\,x)\,(\sim(\wedge\,T\,(\sim x)))$
:= $\$TMP8010$ %0
\# wff 1883 : $=(\vee\,F\,x)\,(\sim(\wedge\,T\,(\sim x)))_o$:= $\$TMP8010$

\#\# use Proof Template A5216: (T ∧ A) = A
:= $\$A5216$ $\sim_{oo}x_o$
\# wff 1791 : $\sim x_{o,\ldots}$:= $\$A5216$
<< A5216.r0t.txt

:= $\$A5216$
%0
$=(\wedge T\,(\sim x))\,(\sim x)$
$=_{ooo}(\wedge_{ooo}T_o(\sim_{oo}x_o))(\sim_{oo}x_o)$

%$\$TMP8010$
$=(\vee F\,x)\,(\sim(\wedge T\,(\sim x)))$:= $\$TMP8010$
$=_{ooo}(\vee_{ooo}F_ox_o)(\sim_{oo}(\wedge_{ooo}T_o(\sim_{oo}x_o)))$:= $\$TMP8010$
:= $\$TMP8010$
§s %0 7 %1
$=(\vee F\,x)\,(\sim(\sim x))$

%$K8008$
$=(\sim(\sim x))\,x$:= $K8008$
$=_{ooo}(\sim_{oo}(\sim_{oo}x_o))x_o$:= $K8008$
§s %1 3 %0
$=(\vee F\,x)\,x$

:= $K8010b$ %0
wff 1893 : $=(\vee F\,x)\,x_o$:= $K8010b$

.c: (A ∨ F) = (F ∨ A)

%$K8010b$
$=(\vee F\,x)\,x$:= $K8010b$
$=_{ooo}(\vee_{ooo}F_ox_o)x_o$:= $K8010b$

use Proof Template A5201b (Swap): A = B → B = A
<< A5201b.r0t.txt
%0
$=x\,(\vee F\,x)$
$=_{ooo}x_o(\vee_{ooo}F_ox_o)$

%$K8010a$
$=(\vee x\,F)\,x$:= $K8010a$
$=_{ooo}(\vee_{ooo}x_oF_o)x_o$:= $K8010a$
§s %0 3 %1
$=(\vee x\,F)\,(\vee F\,x)$

:= $K8010c$ %0
wff 1895 : $=(\vee x\,F)\,(\vee F\,x)_o$:= $K8010c$

##
Q.E.D.
##

%K8010a
%$K8010a$

$=(\vee\, x\, F)\, x$:= $K8010a$
$=_{ooo}(\vee_{ooo}x_oF_o)x_o$:= $K8010a$

%K8010b
%$K8010b$
$=(\vee\, F\, x)\, x$:= $K8010b$
$=_{ooo}(\vee_{ooo}F_ox_o)x_o$:= $K8010b$

%K8010c
%$K8010c$
$=(\vee\, x\, F)\,(\vee\, F\, x)$:= $K8010c$
$=_{ooo}(\vee_{ooo}x_oF_o)(\vee_{ooo}F_ox_o)$:= $K8010c$

2.1.64 Results for File K8011.r0.txt

##
Proof K8011: (A ∨ A) = A
##
##
Source: [Kubota 2017 (doi: 10.4444/100.10)]
##

Written by Ken Kubota (<mail@kenkubota.de>).
##
This file is part of the publication of the mathematical logic $\mathcal{R}_0$.
For more information, visit: <http://doi.org/10.4444/100.10>
##

<< A5232.r0.txt

##
Proof
##

use Proof Template A5222 (Rule of Cases): [\x.A]T, [\x.A]F → A
:= $\$L5222$ $[\lambda x_o.(=_{ooo}(\vee_{ooo}x_ox_o)x_o)_o]$
wff 1671 : $[\lambda x.(=(\vee\, x\, x)\, x)]_{oo}$:= $\$L5222$
:= $\$X5222$ x_o
wff 16 : x_o := $\$X5222$
:= $\$T5222$ $\$L5222_{oo}T_o$
wff 1672 : $\$L5222\, T_o$:= $\$T5222$
:= $\$F5222$ $\$L5222_{oo}F_o$
wff 1673 : $\$L5222\, F_o$:= $\$F5222$

case T: (T ∨ T) = T
§\ $\$T5222$
$=\$T5222\, A5232a$

use Proof Template A5201b (Swap): A = B $\rightarrow$ B = A
<< A5201b.r0t.txt
%0
$= A5232a \, \$T5222$
$=_{o\omega\omega} A5232a_\omega \$T5222_\omega$

%$A5232a$
$= (\vee \, T \, T) \, T$:= $A5232a$
$=_{ooo}(\vee_{ooo} T_o T_o) T_o$:= $A5232a$
§s %0 1 %1
$\$L5222 \, T$:= $\$T5222$

case F: (F $\vee$ F) = F
§\ $\$F5222$
$= \$F5222 \, A5232d$

use Proof Template A5201b (Swap): A = B $\rightarrow$ B = A
<< A5201b.r0t.txt
%0
$= A5232d \, \$F5222$
$=_{o\omega\omega} A5232d_\omega \$F5222_\omega$

%$A5232d$
$= (\vee \, F \, F) \, F$:= $A5232d$
$=_{ooo}(\vee_{ooo} F_o F_o) F_o$:= $A5232d$
§s %0 1 %1
$\$L5222 \, F$:= $\$F5222$

<< A5222.r0t.txt
:= $\$L5222$
:= $\$X5222$
:= $\$T5222$
:= $\$F5222$

:= $K8011$ %0
wff 1670 : $= (\vee \, x \, x) \, x_{o,\ldots}$:= $K8011$

##
Q.E.D.
##

%0
$= (\vee \, x \, x) \, x$:= $K8011$
$=_{ooo}(\vee_{ooo} x_o x_o) x_o$:= $K8011$

2.1.65 Results for File K8012.r0.txt

```
##
## Proof K8012: (A ∨ B) = (B ∨ A)
##
##
## Source: [Kubota 2017 (doi: 10.4444/100.10)]
##
```


```
## Written by Ken Kubota (<mail@kenkubota.de>).
##
## This file is part of the publication of the mathematical logic R0.
## For more information, visit: <http://doi.org/10.4444/100.10>
##

<< K8009.r0.txt
<< K8010.r0.txt


##
## Proof
##

## use Proof Template K8006: (A * T) = (T * A), (A * F) = (F * A) → (A * B) = (B * A)
```

:= $\$R8006$ $[\lambda a_o.[\lambda b_o.(\sim_{oo}(\wedge_{ooo}(\sim_{oo}a_o)(\sim_{oo}b_o)))_o]_{(oo)}]$

\# wff 65 : $[\lambda a.[\lambda b.(\sim(\wedge(\sim a)(\sim b)))]]_{ooo}$:= $\$R8006$ $\vee$

:= $\$T8006$ $=_{ooo}(\vee_{ooo}x_oT_o)(\vee_{ooo}T_ox_o)$

\# wff 1751 : $=(\vee\, x\, T)\,(\vee\, T\, x)_o$:= $\$T8006$ $K8009c$

:= $\$F8006$ $=_{ooo}(\vee_{ooo}x_oF_o)(\vee_{ooo}F_ox_o)$

\# wff 2049 : $=(\vee\, x\, F)\,(\vee\, F\, x)_o$:= $\$F8006$ $K8010c$

<< K8006.r0t.txt

:= $\$R8006$

:= $\$T8006$

:= $\$F8006$

:= $K8012$ %0

\# wff 2259 : $COMMT\, o \vee_{o,\ldots}$:= $K8012$

```
##
## Q.E.D.
##
```

%0

\# $COMMT\, o \vee$:= $K8012$

\# $COMMT_{o(\backslash 4\backslash 4\backslash 3)\tau}o_{\tau}\vee_{ooo}$:= $K8012$

2.1.66 Results for File K8013.r0a.txt

##
Proof Template K8013: $A \supset B, B \supset A \quad \rightarrow \quad A = B$
##
##
Source: [Kubota 2017 (doi: 10.4444/100.10)]
##

Written by Ken Kubota (<mail@kenkubota.de>).
##
This file is part of the publication of the mathematical logic $\mathcal{R}_0$.
For more information, visit: <http://doi.org/10.4444/100.10>
##

##
Define Syntactical Variables
##

proposition A
:= $\$A8013\ x_o$
wff 11 : x_o := $\$A8013$

proposition B
:= $\$B8013\ y_o$
wff 12 : y_o := $\$B8013$

##
Assumptions and Resulting Syntactical Variables
##

<< basics.r0.txt

§! $\supset_{ooo}\$A8013_o\$B8013_o$
$\supset \$A8013\, \$B8013$
§! $\supset_{ooo}\$B8013_o\$A8013_o$
$\supset \$B8013\, \$A8013$

##
Proof Template
##

<<< K8013.r0t.txt
Include begin (K8013.r0t.txt) [oldfile=(K8013.r0a.txt)]
##
Proof Template K8013: $A \supset B, B \supset A \quad \rightarrow \quad A = B$

##
##
Source: [Kubota 2017 (doi: 10.4444/100.10)]
##

Written by Ken Kubota (<mail@kenkubota.de>).
##
This file is part of the publication of the mathematical logic $\mathcal{R}_0$.
For more information, visit: <http://doi.org/10.4444/100.10>
##

<< K8007.r0.txt

##
Proof Template
##

.1

assumption 1
:= $\$HTMP8013\ \supset_{ooo}\$A8013_o\$B8013_o$
wff 209 : $\supset \$A8013\,\$B8013_o$:= $\$HTMP8013$

assumption 2
:= $\$ITMP8013\ \supset_{ooo}\$B8013_o\$A8013_o$
wff 211 : $\supset \$B8013\,\$A8013_o$:= $\$ITMP8013$

$\%K8007$
$COMMT\,o\,\wedge$:= $K8007$
$COMMT_{o(\backslash 4\backslash 4\backslash 3)\tau}o_\tau\wedge_{ooo}$:= $K8007$
§\ $COMMT_{o(\backslash 4\backslash 4\backslash 3)\tau}o_\tau$
$=(COMMT\,o)\,[\lambda f.(=(f\,\$A8013\,\$B8013)\,(f\,\$B8013\,\$A8013))]$
§s %1 2 %0
$[\lambda f.(=(f\,\$A8013\,\$B8013)\,(f\,\$B8013\,\$A8013))]\wedge$
§\ $[\lambda f_{ooo}.(=_{ooo}(f_{ooo}\$A8013_o\$B8013_o)(f_{ooo}\$B8013_o\$A8013_o))_o]\wedge_{ooo}$
$=([\lambda f.(=(f\,\$A8013\,\$B8013)\,(f\,\$B8013\,\$A8013))]\wedge)\ \ldots$
$\ldots(=(\wedge\,\$A8013\,\$B8013)\,(\wedge\,\$B8013\,\$A8013))$
§s %1 1 %0
$=(\wedge\,\$A8013\,\$B8013)\,(\wedge\,\$B8013\,\$A8013)$

.2

use Proof Template A5221 (Sub): B $\rightarrow$ B [x/A]
:= $\$B5221$ %0
wff 1572 : $=(\wedge\,\$A8013\,\$B8013)\,(\wedge\,\$B8013\,\$A8013)_{o,\ldots}$:= $\$B5221$
:= $\$T5221\ o$
wff 2 : o_τ := $\$T5221$
:= $\$X5221\ y_o$

```
# wff      12  :       y_o       :=  $B8013   $X5221
:=  $A5221  ytmp_o
# wff      1796  :       ytmp_o       :=  $A5221
<< A5221.r0t.txt
:=  $B5221
:=  $T5221
:=  $X5221
:=  $A5221
%0
#                   = (∧ $A8013 ytmp) (∧ ytmp $A8013)
#                   =_ooo(∧_ooo$A8013_o ytmp_o)(∧_ooo ytmp_o $A8013_o)

## use Proof Template A5221 (Sub):   B   →   B [x/A]
:=  $B5221  %0
# wff      1840  :       = (∧ $A8013 ytmp) (∧ ytmp $A8013)_o,...        :=  $B5221
:=  $T5221  o
# wff      2  :       o_τ       :=  $T5221
:=  $X5221  x_o
# wff      11  :       x_o       :=  $A8013   $X5221
:=  $A5221  y_o
# wff      12  :       y_o       :=  $A5221   $B8013
<< A5221.r0t.txt
:=  $B5221
:=  $T5221
:=  $X5221
:=  $A5221
%0
#                   = (∧ $B8013 ytmp) (∧ ytmp $B8013)
#                   =_ooo(∧_ooo$B8013_o ytmp_o)(∧_ooo ytmp_o $B8013_o)

## use Proof Template A5221 (Sub):   B   →   B [x/A]
:=  $B5221  %0
# wff      1877  :       = (∧ $B8013 ytmp) (∧ ytmp $B8013)_o,...        :=  $B5221
:=  $T5221  o
# wff      2  :       o_τ       :=  $T5221
:=  $X5221  ytmp_o
# wff      1796  :       ytmp_o       :=  $X5221
:=  $A5221  x_o
# wff      11  :       x_o       :=  $A5221   $A8013
<< A5221.r0t.txt
:=  $B5221
:=  $T5221
:=  $X5221
:=  $A5221
%0
#                   = (∧ $B8013 $A8013) (∧ $A8013 $B8013)
#                   =_ooo(∧_ooo$B8013_o $A8013_o)(∧_ooo$A8013_o $B8013_o)

## .3
```

$\%\$ITMP8013$
$\#\quad \supset \$B8013\,\$A8013 \quad := \quad \$ITMP8013$
$\#\quad \supset_{ooo}\$B8013_o\$A8013_o \quad := \quad \$ITMP8013$
$\S r\ \ /9\ \ ytmp_o$
$\#\quad = [\lambda\$B8013.(=\$A8013\,(\wedge\,\$A8013\,\$B8013))]\,[\lambda ytmp.(=\$A8013\,(\wedge\,\$A8013\,ytmp))]$
$\S s\ \ \%1\ \ 9\ \ \%0$
$\#\quad [\lambda\$A8013.[\lambda ytmp.(=\$A8013\,(\wedge\,\$A8013\,ytmp))]]\,\$B8013\,\$A8013$
$\S\backslash\ \ [\lambda\$A8013_o.[\lambda ytmp_o.(=_{ooo}\$A8013_o(\wedge_{ooo}\$A8013_o ytmp_o))_o]_{(oo)}]\$B8013_o$
$\#\quad = ([\lambda\$A8013.[\lambda ytmp.(=\$A8013\,(\wedge\,\$A8013\,ytmp))]]\,\$B8013)\ \ldots$
$\ldots[\lambda ytmp.(=\$B8013\,(\wedge\,\$B8013\,ytmp))]$
$\S s\ \ \%1\ \ 2\ \ \%0$
$\#\quad [\lambda ytmp.(=\$B8013\,(\wedge\,\$B8013\,ytmp))]\,\$A8013$
$\S\backslash\ \ [\lambda ytmp_o.(=_{ooo}\$B8013_o(\wedge_{ooo}\$B8013_o ytmp_o))_o]\$A8013_o$
$\#\quad = ([\lambda ytmp.(=\$B8013\,(\wedge\,\$B8013\,ytmp))]\,\$A8013)\,(=\$B8013\,(\wedge\,\$B8013\,\$A8013))$
$\S s\ \ \%1\ \ 1\ \ \%0$
$\#\quad = \$B8013\,(\wedge\,\$B8013\,\$A8013)$
$\S s\ \ \%0\ \ 3\ \ \%7$
$\#\quad = \$B8013\,(\wedge\,\$A8013\,\$B8013)$

use Proof Template A5201b (Swap): A = B → B = A
<< A5201b.r0t.txt
%0
$\#\quad = (\wedge\,\$A8013\,\$B8013)\,\$B8013$
$\#\quad =_{ooo}(\wedge_{ooo}\$A8013_o\$B8013_o)\$B8013_o$

.4

$\%\$HTMP8013$
$\#\quad \supset \$A8013\,\$B8013 \quad := \quad \$HTMP8013$
$\#\quad \supset_{ooo}\$A8013_o\$B8013_o \quad := \quad \$HTMP8013$
$\S r\ \ /9\ \ ytmp_o$
$\#\quad = [\lambda\$B8013.(=\$A8013\,(\wedge\,\$A8013\,\$B8013))]\,[\lambda ytmp.(=\$A8013\,(\wedge\,\$A8013\,ytmp))]$
$\S s\ \ \%1\ \ 9\ \ \%0$
$\#\quad [\lambda\$A8013.[\lambda ytmp.(=\$A8013\,(\wedge\,\$A8013\,ytmp))]]\,\$A8013\,\$B8013$
$\S\backslash\ \ [\lambda\$A8013_o.[\lambda ytmp_o.(=_{ooo}\$A8013_o(\wedge_{ooo}\$A8013_o ytmp_o))_o]_{(oo)}]\$A8013_o$
$\#\quad = ([\lambda\$A8013.[\lambda ytmp.(=\$A8013\,(\wedge\,\$A8013\,ytmp))]]\,\$A8013)\ \ldots$
$\ldots[\lambda ytmp.(=\$A8013\,(\wedge\,\$A8013\,ytmp))]$
$\S s\ \ \%1\ \ 2\ \ \%0$
$\#\quad [\lambda ytmp.(=\$A8013\,(\wedge\,\$A8013\,ytmp))]\,\$B8013$
$\S\backslash\ \ [\lambda ytmp_o.(=_{ooo}\$A8013_o(\wedge_{ooo}\$A8013_o ytmp_o))_o]\$B8013_o$
$\#\quad = ([\lambda ytmp.(=\$A8013\,(\wedge\,\$A8013\,ytmp))]\,\$B8013)\,(=\$A8013\,(\wedge\,\$A8013\,\$B8013))$
$\S s\ \ \%1\ \ 1\ \ \%0$
$\#\quad = \$A8013\,(\wedge\,\$A8013\,\$B8013)$
$\S s\ \ \%0\ \ 3\ \ \%7$
$\#\quad = \$A8013\,\$B8013$

undefine local variables
$:=\ \ \$HTMP8013$

:= $\$ITMP8013$
Include end (K8013.r0t.txt) [newfile=(K8013.r0a.txt)]
>>>

##
Undefine Syntactical Variables
##

:= $\$A8013$
:= $\$B8013$

##
Q.E.D.
##

%0
$=x\,y$
$=_{ooo}x_o y_o$

2.1.67 Results for File K8013H.r0a.txt

##
Proof Template K8013H: H ⊃ (A ⊃ B), H ⊃ (B ⊃ A) → H ⊃ (A = B)
##
##
Source: [Kubota 2017 (doi: 10.4444/100.10)]
##

Written by Ken Kubota (<mail@kenkubota.de>).
##
This file is part of the publication of the mathematical logic $\mathcal{R}_0$.
For more information, visit: <http://doi.org/10.4444/100.10>
##

##
Define Syntactical Variables
##

hypotheses H
:= $\$H8013H\ h_o$
wff 11 : h_o := $\$H8013H$

proposition A
:= $\$A8013H\ x_o$
wff 12 : x_o := $\$A8013H$

```
## proposition B
:=  $B8013H  y_o
# wff      13  :         y_o        :=  $B8013H


##
##   Assumptions and Resulting Syntactical Variables
##

<< basics.r0.txt

§! ⊃_ooo$H8013H_o(⊃_ooo$A8013H_o$B8013H_o)
#                       ⊃ $H8013H (⊃ $A8013H $B8013H)
§! ⊃_ooo$H8013H_o(⊃_ooo$B8013H_o$A8013H_o)
#                       ⊃ $H8013H (⊃ $B8013H $A8013H)


##
##   Proof Template
##

## <<< K8013H.r0t.txt
## Include begin (K8013H.r0t.txt) [oldfile=(K8013H.r0a.txt)]
##
##   Proof Template K8013H:   H ⊃ (A ⊃ B), H ⊃ (B ⊃ A)   →   H ⊃ (A = B)
##
##
##   Source: [Kubota 2017 (doi: 10.4444/100.10)]
##
##   Copyright (c) 2017 Owl of Minerva Press GmbH. All rights reserved.
##   Written by Ken Kubota (<mail@kenkubota.de>).
##
##   This file is part of the publication of the mathematical logic R0.
##   For more information, visit: <http://doi.org/10.4444/100.10>
##

<< K8007.r0.txt


##
##   Proof Template
##

## .1

## assumption 1
:=  $HTMP8013H  ⊃_ooo$H8013H_o(⊃_ooo$A8013H_o$B8013H_o)
# wff      212  :        ⊃ $H8013H (⊃ $A8013H $B8013H)_o        :=  $HTMP8013H
```

assumption 2
:= $\$ITMP8013H\ \supset_{ooo}\$H8013H_o(\supset_{ooo}\$B8013H_o\$A8013H_o)$
\# wff 215 : $\supset \$H8013H\,(\supset \$B8013H\,\$A8013H)_o$:= $\$ITMP8013H$

%$K8007$
\# $COMMT\,o\,\wedge$:= $K8007$
\# $COMMT_{o(\backslash 4\backslash 4\backslash 3)\tau}o_\tau\wedge_{ooo}$:= $K8007$
§\ $COMMT_{o(\backslash 4\backslash 4\backslash 3)\tau}o_\tau$
\# $=(COMMT\,o)\,[\lambda f.(=(f\,\$A8013H\,\$B8013H)\,(f\,\$B8013H\,\$A8013H))]$
§s %1 2 %0
\# $[\lambda f.(=(f\,\$A8013H\,\$B8013H)\,(f\,\$B8013H\,\$A8013H))]\wedge$
§\ $[\lambda f_{ooo}.(=_{ooo}(f_{ooo}\$A8013H_o\$B8013H_o)(f_{ooo}\$B8013H_o\$A8013H_o))_o]\wedge_{ooo}$
\# $=([\lambda f.(=(f\,\$A8013H\,\$B8013H)\,(f\,\$B8013H\,\$A8013H))]\wedge)\ldots$
$\ldots(=(\wedge\,\$A8013H\,\$B8013H)\,(\wedge\,\$B8013H\,\$A8013H))$
§s %1 1 %0
\# $=(\wedge\,\$A8013H\,\$B8013H)\,(\wedge\,\$B8013H\,\$A8013H)$

.2

use Proof Template A5221 (Sub): B → B [x/A]
:= $\$B5221$ %0
\# wff 1576 : $=(\wedge\,\$A8013H\,\$B8013H)\,(\wedge\,\$B8013H\,\$A8013H)_{o,\ldots}$:= $\$B5221$
:= $\$T5221\ o$
\# wff 2 : o_τ := $\$T5221$
:= $\$X5221\ y_o$
\# wff 13 : y_o := $\$B8013H$ $\$X5221$
:= $\$A5221\ ytmp_o$
\# wff 1800 : $ytmp_o$:= $\$A5221$
<< A5221.r0t.txt
:= $\$B5221$
:= $\$T5221$
:= $\$X5221$
:= $\$A5221$
%0
\# $=(\wedge\,\$A8013H\,ytmp)\,(\wedge\,ytmp\,\$A8013H)$
\# $=_{ooo}(\wedge_{ooo}\$A8013H_o ytmp_o)(\wedge_{ooo}ytmp_o\$A8013H_o)$

use Proof Template A5221 (Sub): B → B [x/A]
:= $\$B5221$ %0
\# wff 1844 : $=(\wedge\,\$A8013H\,ytmp)\,(\wedge\,ytmp\,\$A8013H)_{o,\ldots}$:= $\$B5221$
:= $\$T5221\ o$
\# wff 2 : o_τ := $\$T5221$
:= $\$X5221\ x_o$
\# wff 12 : x_o := $\$A8013H$ $\$X5221$
:= $\$A5221\ y_o$
\# wff 13 : y_o := $\$A5221$ $\$B8013H$
<< A5221.r0t.txt
:= $\$B5221$
:= $\$T5221$

:= $\$X5221$
:= $\$A5221$
%0
$= (\wedge\,\$B8013H\;ytmp)\;(\wedge\,ytmp\,\$B8013H)$
$=_{ooo}(\wedge_{ooo}\$B8013H_o ytmp_o)(\wedge_{ooo}ytmp_o\$B8013H_o)$

use Proof Template A5221 (Sub): B → B [x/A]
:= $\$B5221$ %0
wff 1881 : $= (\wedge\,\$B8013H\;ytmp)\;(\wedge\,ytmp\,\$B8013H)_{o,\ldots}$:= $\$B5221$
:= $\$T5221\;o$
wff 2 : o_τ := $\$T5221$
:= $\$X5221\;ytmp_o$
wff 1800 : $ytmp_o$:= $\$X5221$
:= $\$A5221\;x_o$
wff 12 : x_o := $\$A5221$ $\$A8013H$
<< A5221.r0t.txt
:= $\$B5221$
:= $\$T5221$
:= $\$X5221$
:= $\$A5221$
%0
$= (\wedge\,\$B8013H\,\$A8013H)\;(\wedge\,\$A8013H\,\$B8013H)$
$=_{ooo}(\wedge_{ooo}\$B8013H_o\$A8013H_o)(\wedge_{ooo}\$A8013H_o\$B8013H_o)$

.3

%$\$ITMP8013H$
$\supset \$H8013H\;(\supset \$B8013H\,\$A8013H)$:= $\$ITMP8013H$
$\supset_{ooo}\$H8013H_o(\supset_{ooo}\$B8013H_o\$A8013H_o)$:= $\$ITMP8013H$
§r /25 $ytmp_o$
$= [\lambda\$B8013H.(=\$A8013H\;(\wedge\,\$A8013H\,\$B8013H))]\ldots$
$\ldots$ $[\lambda ytmp.(=\$A8013H\;(\wedge\,\$A8013H\;ytmp))]$
§s %1 25 %0
$\supset \$H8013H\ldots$
$\ldots$ $([\lambda\$A8013H.[\lambda ytmp.(=\$A8013H\;(\wedge\,\$A8013H\;ytmp))]]\;\$B8013H\,\$A8013H)$
§\ $[\lambda\$A8013H_o.[\lambda ytmp_o.(=_{ooo}\$A8013H_o(\wedge_{ooo}\$A8013H_o ytmp_o))_o]_{(oo)}]\$B8013H_o$
$= ([\lambda\$A8013H.[\lambda ytmp.(=\$A8013H\;(\wedge\,\$A8013H\;ytmp))]]\;\$B8013H)\ldots$
$\ldots$ $[\lambda ytmp.(=\$B8013H\;(\wedge\,\$B8013H\;ytmp))]$
§s %1 6 %0
$\supset \$H8013H\;([\lambda ytmp.(=\$B8013H\;(\wedge\,\$B8013H\;ytmp))]\;\$A8013H)$
§\ $[\lambda ytmp_o.(=_{ooo}\$B8013H_o(\wedge_{ooo}\$B8013H_o ytmp_o))_o]\$A8013H_o$
$= ([\lambda ytmp.(=\$B8013H\;(\wedge\,\$B8013H\;ytmp))]\;\$A8013H)\ldots$
$\ldots$ $(=\$B8013H\;(\wedge\,\$B8013H\,\$A8013H))$
§s %1 3 %0
$\supset \$H8013H\;(=\$B8013H\;(\wedge\,\$B8013H\,\$A8013H))$
§s %0 7 %7
$\supset \$H8013H\;(=\$B8013H\;(\wedge\,\$A8013H\,\$B8013H))$

use Proof Template A5201bH (SwapH): H ⊃ (A = B) → H ⊃ (B = A)

```
<< A5201bH.r0t.txt
%0
#                    ⊃ $H8013H (= (∧ $A8013H $B8013H) $B8013H)
#                    ⊃_ooo$H8013H_o(=_ooo(∧_ooo$A8013H_o$B8013H_o)$B8013H_o)

## .4

%$HTMP8013H
#                    ⊃ $H8013H (⊃ $A8013H $B8013H)        :=  $HTMP8013H
#                    ⊃_ooo$H8013H_o(⊃_ooo$A8013H_o$B8013H_o)        :=  $HTMP8013H
§r  /25  ytmp_o
#                    = [λ$B8013H.(= $A8013H (∧ $A8013H $B8013H))] . . .
. . . [λytmp.(= $A8013H (∧ $A8013H ytmp))]
§s  %1  25  %0
#                    ⊃ $H8013H . . .
. . . ([λ$A8013H.[λytmp.(= $A8013H (∧ $A8013H ytmp))]] $A8013H $B8013H)
§\  [λ$A8013H_o.[λytmp_o.(=_ooo$A8013H_o(∧_ooo$A8013H_oytmp_o))_o]_(oo)]$A8013H_o
#                    = ([λ$A8013H.[λytmp.(= $A8013H (∧ $A8013H ytmp))]] $A8013H) . . .
. . . [λytmp.(= $A8013H (∧ $A8013H ytmp))]
§s  %1  6  %0
#                    ⊃ $H8013H ([λytmp.(= $A8013H (∧ $A8013H ytmp))] $B8013H)
§\  [λytmp_o.(=_ooo$A8013H_o(∧_ooo$A8013H_oytmp_o))_o]$B8013H_o
#                    = ([λytmp.(= $A8013H (∧ $A8013H ytmp))] $B8013H) . . .
. . . (= $A8013H (∧ $A8013H $B8013H))
§s  %1  3  %0
#                    ⊃ $H8013H (= $A8013H (∧ $A8013H $B8013H))
§s'  %0  3  %7
#                    ⊃ $H8013H (= $A8013H $B8013H)

## undefine local variables
:=  $HTMP8013H
:=  $ITMP8013H
## Include end (K8013H.r0t.txt) [newfile=(K8013H.r0a.txt)]
>>>


##
##   Undefine Syntactical Variables
##

:=  $H8013H
:=  $A8013H
:=  $B8013H


##
##   Q.E.D.
##
```

%0
$\supset h\,(=x\,y)$
$\supset_{ooo}h_o(=_{ooo}x_o y_o)$

2.1.68 Results for File K8014.r0.txt

```
##
##  Proof K8014:   (x = y)   =   (y = x)
##        for any x, y of any type
##
##  Source: [cf. Andrews 2002 (ISBN 1-4020-0763-9), pp. 232 f. (5302)]
##
##  Copyright (c) 2017 Owl of Minerva Press GmbH. All rights reserved.
##  Written by Ken Kubota (<mail@kenkubota.de>).
##
##  This file is part of the publication of the mathematical logic R0.
##  For more information, visit: <http://doi.org/10.4444/100.10>
##

<< basics.r0.txt
<< K8005.r0.txt



##
##  Proof
##

## .1
```

%K8005
$\supset x\,x$:= $K8005$
$\supset_{ooo}x_o x_o$:= $K8005$

use Proof Template A5221 (Sub): B → B [x/A]
:= $\$B5221$ %0
wff 1357 : $\supset x\,x_{o,\ldots}$:= $\$B5221$ $K8005$
:= $\$T5221$ o
wff 2 : o_τ := $\$T5221$
:= $\$X5221$ x_o
wff 16 : x_o := $\$X5221$
:= $\$A5221$ $=_{ott}x_t y_t$
wff 116 : $=x\,y_o$:= $\$A5221$
<< A5221.r0t.txt
:= $\$B5221$
:= $\$T5221$
:= $\$X5221$
:= $\$A5221$
%0
$\supset(=x\,y)\,(=x\,y)$

$\supset_{ooo}(=_{ott}x_t y_t)(=_{ott}x_t y_t)$

use Proof Template A5201bH (SwapH): H ⊃ (A = B) → H ⊃ (B = A)
<< A5201bH.r0t.txt
%0
$\supset (= x\, y)\, (= y\, x)$
$\supset_{ooo}(=_{ott}x_t y_t)(=_{ott}y_t x_t)$

.2

%$K8005$
$\supset x\, x$:= $K8005$
$\supset_{ooo}x_o x_o$:= $K8005$

use Proof Template A5221 (Sub): B → B [x/A]
:= $\$B5221$ %0
wff 1357 : $\supset x\, x_{o,\ldots}$:= $\$B5221$ $K8005$
:= $\$T5221$ o
wff 2 : o_τ := $\$T5221$
:= $\$X5221$ x_o
wff 16 : x_o := $\$X5221$
:= $\$A5221$ $=_{ott}y_t x_t$
wff 1634 : $= y\, x_o$:= $\$A5221$
<< A5221.r0t.txt
:= $\$B5221$
:= $\$T5221$
:= $\$X5221$
:= $\$A5221$
%0
$\supset (= y\, x)\, (= y\, x)$
$\supset_{ooo}(=_{ott}y_t x_t)(=_{ott}y_t x_t)$

use Proof Template A5201bH (SwapH): H ⊃ (A = B) → H ⊃ (B = A)
<< A5201bH.r0t.txt
%0
$\supset (= y\, x)\, (= x\, y)$
$\supset_{ooo}(=_{ott}y_t x_t)(=_{ott}x_t y_t)$

.3

use Proof Template K8013: A ⊃ B, B ⊃ A → A = B
:= $\$A8013$ $=_{ott}x_t y_t$
wff 116 : $= x\, y_{o,\ldots}$:= $\$A8013$
:= $\$B8013$ $=_{ott}y_t x_t$
wff 1634 : $= y\, x_{o,\ldots}$:= $\$B8013$
<< K8013.r0t.txt
:= $\$A8013$
:= $\$B8013$
%0

\# $= (= x\, y)\, (= y\, x)$
\# $=_{ooo}(=_{ott} x_t y_t)(=_{ott} y_t x_t)$

:= $K8014$ %0
\# wff 2234 : $= (= x\, y)\, (= y\, x)_o$:= $K8014$

\##
\## Q.E.D.
\##

%0
\# $= (= x\, y)\, (= y\, x)$:= $K8014$
\# $=_{ooo}(=_{ott} x_t y_t)(=_{ott} y_t x_t)$:= $K8014$

2.1.69 Results for File K8015.r0.txt

\##
\## Proof K8015: (A ⊃ F) = (∼ A)
\## (Proof by Contradiction)
\##
\## Source: [Kubota 2017 (doi: 10.4444/100.10)]
\##

\## Written by Ken Kubota (<mail@kenkubota.de>).
\##
\## This file is part of the publication of the mathematical logic $\mathcal{R}_0$.
\## For more information, visit: <http://doi.org/10.4444/100.10>
\##

<< K8014.r0.txt

\##
\## Proof
\##

§= $_o \supset_{ooo} x_o F_o$
\# $= (\supset x\, F)\, (\supset x\, F)$
§\ $\supset_{ooo} x_o$
\# $= (\supset x)\, [\lambda y.(= x\, (\wedge\, x\, y))]$
§s %1 6 %0
\# $= (\supset x\, F)\, ([\lambda y.(= x\, (\wedge\, x\, y))]\, F)$
§\ $[\lambda y_o.(=_{ooo} x_o(\wedge_{ooo} x_o y_o))_o] F_o$
\# $= ([\lambda y.(= x\, (\wedge\, x\, y))]\, F)\, (= x\, (\wedge\, x\, F))$
§s %1 3 %0
\# $= (\supset x\, F)\, (= x\, (\wedge\, x\, F))$

%$K8001a$

```
#                 = (∧ x F) F       :=  K8001a
#                 =ooo(∧oooxoFo)Fo       :=  K8001a
§s %1 7 %0
#                 = (⊃ x F) (= x F)
:= $TMP8015 %0
# wff   2245  :      = (⊃ x F) (= x F)o      :=  $TMP8015

%K8014
#                 = (= x y) (= y x)      :=  K8014
#                 =ooo(=ottxtyt)(=ottytxt)      :=  K8014

## use Proof Template A5221 (Sub):  B  →  B [x/A]
:= $B5221 %0
# wff   2234  :      = (= x y) (= y x)o      :=  $B5221  K8014
:= $T5221 τ
# wff   0  :      ττ      :=  $T5221
:= $X5221 tτ
# wff   4  :      tτ      :=  $X5221
:= $A5221 o
# wff   2  :      oτ      :=  $A5221
<< A5221.r0t.txt
:= $B5221
:= $T5221
:= $X5221
:= $A5221
%0
#                 = (= x y) (= y x)
#                 =ooo(=oooxoyo)(=oooyoxo)

## use Proof Template A5221 (Sub):  B  →  B [x/A]
:= $B5221 %0
# wff   2288  :      = (= x y) (= y x)o,...      :=  $B5221
:= $T5221 o
# wff   2  :      oτ      :=  $T5221
:= $X5221 yo
# wff   34  :      yo      :=  $X5221
:= $A5221 =o(oo)(oo)[λxo.To][λxo.xo]
# wff   20  :      = [λx.T] [λx.x]o,...      :=  $A5221  F
<< A5221.r0t.txt
:= $B5221
:= $T5221
:= $X5221
:= $A5221
%0
#                 = (= x F) (= F x)
#                 =ooo(=oooxoFo)(=oooFoxo)

%$TMP8015
#                 = (⊃ x F) (= x F)      :=  $TMP8015
```

\# $=_{ooo}(\supset_{ooo}x_oF_o)(=_{ooo}x_oF_o)$:= $\$TMP8015$
:= $\$TMP8015$
§s %0 3 %1
\# $=(\supset x\,F)\,(=F\,x)$

§\ $\sim_{oo}x_o$
\# $=(\sim x)\,(=F\,x)$

\## use Proof Template A5201b (Swap): A = B → B = A
<< A5201b.r0t.txt
%0
\# $=(=F\,x)\,(\sim x)$
\# $=_{o\omega\omega}(=_{ooo}F_ox_o)(\sim_{oo}x_o)$

§s %4 3 %0
\# $=(\supset x\,F)\,(\sim x)$

:= $K8015$ %0
\# wff 2334 : $=(\supset x\,F)\,(\sim x)_o$:= $K8015$

\##
\## Q.E.D.
\##

%0
\# $=(\supset x\,F)\,(\sim x)$:= $K8015$
\# $=_{ooo}(\supset_{ooo}x_oF_o)(\sim_{oo}x_o)$:= $K8015$

2.1.70 Results for File K8016.r0.txt

\##
\## Proof K8016: ∀ x: Px = ∼ ∃ x: ∼ Px
\##
\##
\## Source: [Kubota 2017 (doi: 10.4444/100.10)]
\##

\## Written by Ken Kubota (<mail@kenkubota.de>).
\##
\## This file is part of the publication of the mathematical logic $\mathcal{R}_0$.
\## For more information, visit: <http://doi.org/10.4444/100.10>
\##

<< basics.r0.txt
<< A5205.r0.txt
<< K8008.r0.txt

##
Proof
##

shorthands
:= $\$NE$ $[\lambda t_\tau.[\lambda p_{ot}.(\sim_{oo}(\exists_{o(o\backslash 3)\tau}t_\tau[\lambda x_t.(\sim_{oo}(p_{ot}x_t))_o]))_o]_{(o(ot))}]$
wff 1768 : $[\lambda t.[\lambda p.(\sim(\exists t\,[\lambda x.(\sim(p\,x))]))]]_{o(o\backslash 3)\tau}$:= $\$NE$
:= $\$DN$ $\sim_{oo}(\sim_{oo}(=_{o(ot)(ot)}[\lambda x_t.T_o][\lambda x_t.(\sim_{oo}([\lambda x_t.(\sim_{oo}(p_{ot}x_t))_o]x_t))_o]))$
wff 1774 : $\sim(\sim(=[\lambda x.T]\,[\lambda x.(\sim([\lambda x.(\sim(p\,x))]\,x))]))_o$:= $\$DN$
:= $\$LT$ $[\lambda t_\tau.[\lambda p_{ot}.(=_{o(ot)(ot)}[\lambda x_t.T_o][\lambda x_t.(\sim_{oo}(\sim_{oo}(p_{ot}x_t)))_o])_o]_{(o(ot))}]$
wff 1779 : $[\lambda t.[\lambda p.(=[\lambda x.T]\,[\lambda x.(\sim(\sim(p\,x)))])]]_{o(o\backslash 3)\tau}$:= $\$LT$

.1

§= $_{o(o\backslash 3)\tau}$ $\$NE$
$=\$NE\,\$NE$
§\ $\exists_{o(o\backslash 3)\tau}t_\tau$
$=(\exists\,t)\,[\lambda p.(\sim(=[\lambda x.T]\,[\lambda x.(\sim(p\,x))]))]$
§s %1 94 %0
$=[\lambda t.[\lambda p.(\sim([\lambda p.(\sim(=[\lambda x.T]\,[\lambda x.(\sim(p\,x))]))]\,[\lambda x.(\sim(p\,x))]))]]\,\$NE$
§\ $[\lambda p_{ot}.(\sim_{oo}(=_{o(ot)(ot)}[\lambda x_t.T_o][\lambda x_t.(\sim_{oo}(p_{ot}x_t))_o]))_o][\lambda x_t.(\sim_{oo}(p_{ot}x_t))_o]$
$=([\lambda p.(\sim(=[\lambda x.T]\,[\lambda x.(\sim(p\,x))]))]\,[\lambda x.(\sim(p\,x))])\ldots$
$\ldots(\sim(=[\lambda x.T]\,[\lambda x.(\sim([\lambda x.(\sim(p\,x))]\,x))]))$
§s %1 47 %0
$=[\lambda t.[\lambda p.\$DN]]\,\$NE$
:= $\$TMP8016$ %0
wff 1797 : $=[\lambda t.[\lambda p.\$DN]]\,\$NE_o$:= $\$TMP8016$

.2

%$K8008$
$=(\sim(\sim x))\,x$:= $K8008$
$=_{ooo}(\sim_{oo}(\sim_{oo}x_o))x_o$:= $K8008$

use Proof Template A5221 (Sub): B → B [x/A]
:= $\$B5221$ %0
wff 1750 : $=(\sim(\sim x))\,x_{o,\ldots}$:= $\$B5221$ $K8008$
:= $\$T5221$ o
wff 2 : o_τ := $\$T5221$
:= $\$X5221$ x_o
wff 16 : x_o := $\$X5221$
:= $\$A5221$ $=_{o(ot)(ot)}[\lambda x_t.T_o][\lambda x_t.(\sim_{oo}([\lambda x_t.(\sim_{oo}(p_{ot}x_t))_o]x_t))_o]$
wff 1772 : $=[\lambda x.T]\,[\lambda x.(\sim([\lambda x.(\sim(p\,x))]\,x))]_o$:= $\$A5221$
<< A5221.r0t.txt
:= $\$B5221$
:= $\$T5221$
:= $\$X5221$
:= $\$A5221$
%0

\# $= \$DN\,(= [\lambda x.T]\,[\lambda x.(\sim([\lambda x.(\sim(p\,x))]\,x))])$
\# $=_{ooo}\$DN_o(=_{o(ot)(ot)}[\lambda x_t.T_o][\lambda x_t.(\sim_{oo}([\lambda x_t.(\sim_{oo}(p_{ot}x_t))_o]x_t))_o])$

%$\$TMP8016$
\# $= [\lambda t.[\lambda p.\$DN]]\,\$NE$:= $\$TMP8016$
\# $=_{o(o(o\backslash 3)\tau)(o(o\backslash 3)\tau)}[\lambda t_\tau.[\lambda p_{ot}.\$DN_o]_{(o(ot))}]\$NE_{o(o\backslash 3)\tau}$:= $\$TMP8016$
:= $\$TMP8016$
§s %0 23 %1
\# $= [\lambda t.[\lambda p.(= [\lambda x.T]\,[\lambda x.(\sim([\lambda x.(\sim(p\,x))]\,x))])]]\,\$NE$
§\ $[\lambda x_t.(\sim_{oo}(p_{ot}x_t))_o]x_t$
\# $= ([\lambda x.(\sim(p\,x))]\,x)\,(\sim(p\,x))$
§s %1 191 %0
\# $= \$LT\,\$NE$
:= $\$TMP8016$ %0
\# wff 1838 : $= \$LT\,\$NE_o$:= $\$TMP8016$

\#\# .3

%$K8008$
\# $= (\sim(\sim x))\,x$:= $K8008$
\# $=_{ooo}(\sim_{oo}(\sim_{oo}x_o))x_o$:= $K8008$

\#\# use Proof Template A5221 (Sub): B $\rightarrow$ B [x/A]
:= $\$B5221$ %0
\# wff 1750 : $= (\sim(\sim x))\,x_{o,\ldots}$:= $\$B5221$ $K8008$
:= $\$T5221$ o
\# wff 2 : o_τ := $\$T5221$
:= $\$X5221$ x_o
\# wff 16 : x_o := $\$X5221$
:= $\$A5221$ $p_{ot}x_t$
\# wff 66 : $p\,x_o$:= $\$A5221$
<< A5221.r0t.txt
:= $\$B5221$
:= $\$T5221$
:= $\$X5221$
:= $\$A5221$
%0
\# $= (\sim(\sim(p\,x)))\,(p\,x)$
\# $=_{ooo}(\sim_{oo}(\sim_{oo}(p_{ot}x_t)))(p_{ot}x_t)$

%$\$TMP8016$
\# $= \$LT\,\$NE$:= $\$TMP8016$
\# $=_{o(o(o\backslash 3)\tau)(o(o\backslash 3)\tau)}\$LT_{o(o\backslash 3)\tau}\$NE_{o(o\backslash 3)\tau}$:= $\$TMP8016$
:= $\$TMP8016$
§s %0 95 %1
\# $= [\lambda t.[\lambda p.(= [\lambda x.T]\,[\lambda x.(p\,x)])]]\,\$NE$
:= $\$TMP8016$ %0
\# wff 1856 : $= [\lambda t.[\lambda p.(= [\lambda x.T]\,[\lambda x.(p\,x)])]]\,\$NE_o$:= $\$TMP8016$

.4

use Proof Template: A5205 Substitutions
:= $\$AA5205\ o$
wff 2 : o_τ := $\$AA5205$
:= $\$BA5205\ t_\tau$
wff 4 : t_τ := $\$BA5205$
:= $\$FA5205\ p_{o\$BA5205}$
wff 21 : $p_{o\$BA5205}$:= $\$FA5205$
<< a5205_substitutions.r0t.txt
:= $\$AA5205$
:= $\$BA5205$
:= $\$FA5205$
%0
$=p\,[\lambda y.(p\,y)]$
$=_{o(ot)(ot)}p_{ot}[\lambda y_t.(p_{ot}y_t)_o]$

.5

use Proof Template A5201b (Swap): A = B → B = A
<< A5201b.r0t.txt
%0
$=[\lambda y.(p\,y)]\,p$
$=_{o(ot)(ot)}[\lambda y_t.(p_{ot}y_t)_o]p_{ot}$

§r /5 x_t
$=[\lambda y.(p\,y)]\,[\lambda x.(p\,x)]$
§s %1 5 %0
$=[\lambda x.(p\,x)]\,p$

%$\$TMP8016$
$=[\lambda t.[\lambda p.(=[\lambda x.T]\,[\lambda x.(p\,x)])]]\,\$NE$:= $\$TMP8016$
$=_{o(o(o\backslash 3)\tau)(o(o\backslash 3)\tau)}[\lambda t_\tau.[\lambda p_{ot}.(=_{o(ot)(ot)}[\lambda x_t.T_o][\lambda x_t.(p_{ot}x_t)_o])_o]_{(o(ot))}]\$NE_{o(o\backslash 3)\tau}$
:= $\$TMP8016$
:= $\$TMP8016$
§s %0 47 %1
$=\forall\,\$NE$

:= $K8016$ %0
wff 2004 : $=\forall\,\$NE_o$:= $K8016$

undefine local variables
:= $\$NE$
:= $\$DN$
:= $\$LT$

##
Q.E.D.

##

%0
$=\forall\,[\lambda t.[\lambda p.(\sim(\exists t\,[\lambda x.(\sim(p\,x))]))]]$:= $K8016$
$=_{o(o(o\backslash 3)\tau)(o(o\backslash 3)\tau)}\forall_{o(o\backslash 3)\tau}[\lambda t_\tau.[\lambda p_{ot}.(\sim_{oo}(\exists_{o(o\backslash 3)\tau}t_\tau[\lambda x_t.(\sim_{oo}(p_{ot}x_t))_o]))_o]_{(o(ot))}]$
:= $K8016$

2.1.71 Results for File K8017.r0.txt

##
Proof K8017: ∃ x: Px = ∼ ∀ x: ∼ Px
##
##
Source: [Kubota 2017 (doi: 10.4444/100.10)]
##

Written by Ken Kubota (<mail@kenkubota.de>).
##
This file is part of the publication of the mathematical logic $\mathcal{R}_0$.
For more information, visit: <http://doi.org/10.4444/100.10>
##

<< basics.r0.txt

##
Proof
##

§= $_{o(o\backslash 3)\tau}$ $[\lambda t_\tau.[\lambda p_{ot}.(\sim_{oo}(\forall_{o(o\backslash 3)\tau}t_\tau[\lambda x_t.(\sim_{oo}(p_{ot}x_t))_o]))_o]_{(o(ot))}]$
$=[\lambda t.[\lambda p.(\sim(\forall t\,[\lambda x.(\sim(p\,x))]))]]\,[\lambda t.[\lambda p.(\sim(\forall t\,[\lambda x.(\sim(p\,x))]))]]$
§\ $\forall_{o(o\backslash 3)\tau}t_\tau$
$=(\forall t)\,[\lambda p.(=[\lambda x.T]\,p)]$
§s %1 94 %0
$=[\lambda t.[\lambda p.(\sim([\lambda p.(=[\lambda x.T]\,p)]\,[\lambda x.(\sim(p\,x))]))]]\,[\lambda t.[\lambda p.(\sim(\forall t\,[\lambda x.(\sim(p\,x))]))]]$
§\ $[\lambda p_{ot}.(=_{o(ot)(ot)}[\lambda x_t.T_o]p_{ot})_o][\lambda x_t.(\sim_{oo}(p_{ot}x_t))_o]$
$=([\lambda p.(=[\lambda x.T]\,p)]\,[\lambda x.(\sim(p\,x))])\,(=[\lambda x.T]\,[\lambda x.(\sim(p\,x))])$
§s %1 47 %0
$=\exists\,[\lambda t.[\lambda p.(\sim(\forall t\,[\lambda x.(\sim(p\,x))]))]]$

:= $K8017$ %0
wff 227 : $=\exists\,[\lambda t.[\lambda p.(\sim(\forall t\,[\lambda x.(\sim(p\,x))]))]]_o$:= $K8017$

##
Q.E.D.
##

%0

```
#          = ∃ [λt.[λp.(∼ (∀ t [λx.(∼ (p x))]))]]        :=   K8017
#          =_o(o(o\3)τ)(o(o\3)τ) ∃_o(o\3)τ [λt_τ.[λp_ot.(∼_oo(∀_o(o\3)τ t_τ [λx_t.(∼_oo(p_ot x_t))_o]))_o]_(o(ot))]
:=   K8017
```

2.1.72 Results for File K8018.r0.txt

```
##
##   Proof K8018:   (A ∧ B)   =   ∼ ( (∼ A) ∨ (∼ B) )
##
##
##   Source: [Kubota 2017 (doi: 10.4444/100.10)]
##
##   Copyright (c) 2017 Owl of Minerva Press GmbH. All rights reserved.
##   Written by Ken Kubota (<mail@kenkubota.de>).
##
##   This file is part of the publication of the mathematical logic R0.
##   For more information, visit: <http://doi.org/10.4444/100.10>
##

<< basics.r0.txt
<< A5205.r0.txt
<< K8008.r0.txt


##
##   Proof
##

## .1

§=  ooo  [λa_o.[λb_o.(∼_oo(∨_ooo(∼_oo a_o)(∼_oo b_o)))_o]_(oo)]
#          = [λa.[λb.(∼ (∨ (∼ a) (∼ b)))]] [λa.[λb.(∼ (∨ (∼ a) (∼ b)))]]
§\  ∨_ooo(∼_oo a_o)
#          = (∨ (∼ a)) [λb.(∼ (∧ (∼ (∼ a)) (∼ b)))]
§s  %1  62  %0
#          = [λa.[λb.(∼ (∨ (∼ a) (∼ b)))]] [λa.[λb.(∼ ([λb.(∼ (∧ (∼ (∼ a)) (∼ b)))] (∼ b)))]]
§\  [λb_o.(∼_oo(∧_ooo(∼_oo(∼_oo a_o))(∼_oo b_o)))_o](∼_oo b_o)
#          = ([λb.(∼ (∧ (∼ (∼ a)) (∼ b)))] (∼ b)) (∼ (∧ (∼ (∼ a)) (∼ (∼ b))))
§s  %1  31  %0
#          = [λa.[λb.(∼ (∨ (∼ a) (∼ b)))]] [λa.[λb.(∼ (∼ (∧ (∼ (∼ a)) (∼ (∼ b)))))]]
:=   $TMP8018  %0
# wff   1792   :   = [λa.[λb.(∼ (∨ (∼ a) (∼ b)))]] [λa.[λb.(∼ (∼ (∧ (∼ (∼ a)) (∼ (∼ b)))))]]_o
:=   $TMP8018

## .2

%K8008
#          = (∼ (∼ x)) x        :=   K8008
#          =_ooo(∼_oo(∼_oo x_o)) x_o        :=   K8008
```

use Proof Template A5221 (Sub): B $\rightarrow$ B [x/A]
:= $\$B5221$ %0
\# wff 1750 : $=(\sim(\sim x))\, x_{o,\ldots}$:= $\$B5221$ $K8008$
:= $\$T5221$ o
\# wff 2 : o_τ := $\$T5221$
:= $\$X5221$ x_o
\# wff 16 : x_o := $\$X5221$
:= $\$A5221$ a_o
\# wff 54 : a_o := $\$A5221$
<< A5221.r0t.txt
:= $\$B5221$
:= $\$T5221$
:= $\$X5221$
:= $\$A5221$
%0
\# $=(\sim(\sim a))\, a$
\# $=_{ooo}(\sim_{oo}(\sim_{oo}a_o))a_o$

%$\$TMP8018$
\# $=[\lambda a.[\lambda b.(\sim(\vee(\sim a)(\sim b)))]]\,[\lambda a.[\lambda b.(\sim(\sim(\wedge(\sim(\sim a))(\sim(\sim b)))))]]$:= $\$TMP8018$
\# $=_{o(ooo)(ooo)}[\lambda a_o.[\lambda b_o.(\sim_{oo}(\vee_{ooo}(\sim_{oo}a_o)(\sim_{oo}b_o)))_o]_{(oo)}]\ldots$
$\ldots[\lambda a_o.[\lambda b_o.(\sim_{oo}(\sim_{oo}(\wedge_{ooo}(\sim_{oo}(\sim_{oo}a_o))(\sim_{oo}(\sim_{oo}b_o)))))_o]_{(oo)}]$:= $\$TMP8018$
:= $\$TMP8018$
§s %0 253 %1
\# $=[\lambda a.[\lambda b.(\sim(\vee(\sim a)(\sim b)))]]\,[\lambda a.[\lambda b.(\sim(\sim(\wedge a(\sim(\sim b)))))]]$
:= $\$TMP8018$ %0
\# wff 1830 : $=[\lambda a.[\lambda b.(\sim(\vee(\sim a)(\sim b)))]]\,[\lambda a.[\lambda b.(\sim(\sim(\wedge a(\sim(\sim b)))))]]_o$:= $\$TMP8018$

.3

%$K8008$
\# $=(\sim(\sim x))\, x$:= $K8008$
\# $=_{ooo}(\sim_{oo}(\sim_{oo}x_o))x_o$:= $K8008$

use Proof Template A5221 (Sub): B $\rightarrow$ B [x/A]
:= $\$B5221$ %0
\# wff 1750 : $=(\sim(\sim x))\, x_{o,\ldots}$:= $\$B5221$ $K8008$
:= $\$T5221$ o
\# wff 2 : o_τ := $\$T5221$
:= $\$X5221$ x_o
\# wff 16 : x_o := $\$X5221$
:= $\$A5221$ b_o
\# wff 58 : b_o := $\$A5221$
<< A5221.r0t.txt
:= $\$B5221$
:= $\$T5221$

:= $\$X5221$
:= $\$A5221$
%0
$= (\sim (\sim b))\, b$
$=_{ooo}(\sim_{oo}(\sim_{oo}b_o))b_o$

%$\$TMP8018$
$= [\lambda a.[\lambda b.(\sim (\vee (\sim a)\, (\sim b)))]]\, [\lambda a.[\lambda b.(\sim (\sim (\wedge\, a\, (\sim (\sim b)))))]]$:= $\$TMP8018$
$=_{o(ooo)(ooo)}[\lambda a_o.[\lambda b_o.(\sim_{oo}(\vee_{ooo}(\sim_{oo}a_o)(\sim_{oo}b_o)))_o]_{(oo)}]\ldots$
$\ldots[\lambda a_o.[\lambda b_o.(\sim_{oo}(\sim_{oo}(\wedge_{ooo}a_o(\sim_{oo}(\sim_{oo}b_o)))))_o]_{(oo)}]$:= $\$TMP8018$
:= $\$TMP8018$
§s %0 127 %1
$= [\lambda a.[\lambda b.(\sim (\vee (\sim a)\, (\sim b)))]]\, [\lambda a.[\lambda b.(\sim (\sim (\wedge\, a\, b)))]]$
:= $\$TMP8018$ %0
wff 1848 : $= [\lambda a.[\lambda b.(\sim (\vee (\sim a)\, (\sim b)))]]\, [\lambda a.[\lambda b.(\sim (\sim (\wedge\, a\, b)))]]_o$:= $\$TMP8018$

.4

%$K8008$
$= (\sim (\sim x))\, x$:= $K8008$
$=_{ooo}(\sim_{oo}(\sim_{oo}x_o))x_o$:= $K8008$

use Proof Template A5221 (Sub): B $\rightarrow$ B [x/A]
:= $\$B5221$ %0
wff 1750 : $= (\sim (\sim x))\, x_{o,\ldots}$:= $\$B5221$ $K8008$
:= $\$T5221$ o
wff 2 : o_τ := $\$T5221$
:= $\$X5221$ x_o
wff 16 : x_o := $\$X5221$
:= $\$A5221$ $\wedge_{ooo}a_ob_o$
wff 1843 : $\wedge\, a\, b_o$:= $\$A5221$
<< A5221.r0t.txt
:= $\$B5221$
:= $\$T5221$
:= $\$X5221$
:= $\$A5221$
%0
$= (\sim (\sim (\wedge\, a\, b)))\, (\wedge\, a\, b)$
$=_{ooo}(\sim_{oo}(\sim_{oo}(\wedge_{ooo}a_ob_o)))(\wedge_{ooo}a_ob_o)$

%$\$TMP8018$
$= [\lambda a.[\lambda b.(\sim (\vee (\sim a)\, (\sim b)))]]\, [\lambda a.[\lambda b.(\sim (\sim (\wedge\, a\, b)))]]$:= $\$TMP8018$
$=_{o(ooo)(ooo)}[\lambda a_o.[\lambda b_o.(\sim_{oo}(\vee_{ooo}(\sim_{oo}a_o)(\sim_{oo}b_o)))_o]_{(oo)}]\ldots$
$\ldots[\lambda a_o.[\lambda b_o.(\sim_{oo}(\sim_{oo}(\wedge_{ooo}a_ob_o)))_o]_{(oo)}]$:= $\$TMP8018$
:= $\$TMP8018$
§s %0 15 %1
$= [\lambda a.[\lambda b.(\sim (\vee (\sim a)\, (\sim b)))]]\, [\lambda a.[\lambda b.(\wedge\, a\, b)]]$
:= $\$TMP8018$ %0
wff 1863 : $= [\lambda a.[\lambda b.(\sim (\vee (\sim a)\, (\sim b)))]]\, [\lambda a.[\lambda b.(\wedge\, a\, b)]]_o$:= $\$TMP8018$

.5

use Proof Template: A5205 Substitutions
:= $\$AA5205$ o
wff 2 : o_{τ} := $\$AA5205$
:= $\$BA5205$ o
wff 2 : o_{τ} := $\$AA5205$ $\$BA5205$
:= $\$FA5205$ $\wedge_{ooo} a_o$
wff 1824 : $\wedge\, a_{oo}$:= $\$FA5205$
<< a5205_substitutions.r0t.txt
:= $\$AA5205$
:= $\$BA5205$
:= $\$FA5205$
%0
$= (\wedge\, a)\, [\lambda y.(\wedge\, a\, y)]$
$=_{o(oo)(oo)} (\wedge_{ooo} a_o) [\lambda y_o.(\wedge_{ooo} a_o y_o)_o]$

§r /3 b_o
$= [\lambda y.(\wedge\, a\, y)]\, [\lambda b.(\wedge\, a\, b)]$
§s %1 3 %0
$= (\wedge\, a)\, [\lambda b.(\wedge\, a\, b)]$

use Proof Template A5201b (Swap): A = B → B = A
<< A5201b.r0t.txt
%0
$= [\lambda b.(\wedge\, a\, b)]\, (\wedge\, a)$
$=_{o(oo)(oo)} [\lambda b_o.(\wedge_{ooo} a_o b_o)_o] (\wedge_{ooo} a_o)$

%$\$TMP8018$
$= [\lambda a.[\lambda b.(\sim(\vee\, (\sim a)\, (\sim b)))]]\, [\lambda a.[\lambda b.(\wedge\, a\, b)]]$:= $\$TMP8018$
$=_{o(ooo)(ooo)} [\lambda a_o.[\lambda b_o.(\sim_{oo}(\vee_{ooo}(\sim_{oo} a_o)(\sim_{oo} b_o)))_o]_{(oo)}] \ldots$
$\ldots [\lambda a_o.[\lambda b_o.(\wedge_{ooo} a_o b_o)_o]_{(oo)}]$:= $\$TMP8018$
:= $\$TMP8018$
§s %0 7 %1
$= [\lambda a.[\lambda b.(\sim(\vee\, (\sim a)\, (\sim b)))]]\, [\lambda a.(\wedge\, a)]$
:= $\$TMP8018$ %0
wff 1989 : $= [\lambda a.[\lambda b.(\sim(\vee\, (\sim a)\, (\sim b)))]]\, [\lambda a.(\wedge\, a)]_o$:= $\$TMP8018$

.6

use Proof Template: A5205 Substitutions
:= $\$AA5205$ oo
wff 13 : oo_{τ} := $\$AA5205$
:= $\$BA5205$ o
wff 2 : o_{τ} := $\$BA5205$
:= $\$FA5205$ $[\lambda x_o.[\lambda y_o.(=_{o\omega\omega}[\lambda g_{\$AA5205\, o}.(g_{\$AA5205\, o} T_o T_o)_o][\lambda g_{\$AA5205\, o}.(g_{\$AA5205\, o} x_o y_o)_o])_o]_{\$AA5205}]$
wff 47 : $[\lambda x.[\lambda y.(=[\lambda g.(g\, T\, T)]\, [\lambda g.(g\, x\, y)])]]_{\$AA5205\, o}$:= $\$FA5205$ $\wedge$
<< a5205_substitutions.r0t.txt

:= $\$AA5205$
:= $\$BA5205$
:= $\$FA5205$
%0
$= \wedge [\lambda y.(\wedge y)]$
$=_{o(ooo)(ooo)} \wedge_{ooo}[\lambda y_o.(\wedge_{ooo} y_o)_{(oo)}]$

§r /3 a_o
$= [\lambda y.(\wedge y)]\, [\lambda a.(\wedge a)]$
§s %1 3 %0
$= \wedge [\lambda a.(\wedge a)]$

use Proof Template A5201b (Swap): A = B → B = A
<< A5201b.r0t.txt
%0
$= [\lambda a.(\wedge a)] \wedge$
$=_{o(ooo)(ooo)} [\lambda a_o.(\wedge_{ooo} a_o)_{(oo)}] \wedge_{ooo}$

%$\$TMP8018$
$= [\lambda a.[\lambda b.(\sim (\vee (\sim a)\, (\sim b)))]]\, [\lambda a.(\wedge a)]$:= $\$TMP8018$
$=_{o(ooo)(ooo)} [\lambda a_o.[\lambda b_o.(\sim_{oo}(\vee_{ooo}(\sim_{oo} a_o)(\sim_{oo} b_o)))_o]_{(oo)}][\lambda a_o.(\wedge_{ooo} a_o)_{(oo)}]$:= $\$TMP8018$
:= $\$TMP8018$
§s %0 3 %1
$= [\lambda a.[\lambda b.(\sim (\vee (\sim a)\, (\sim b)))]] \wedge$

use Proof Template A5201b (Swap): A = B → B = A
<< A5201b.r0t.txt
%0
$= \wedge [\lambda a.[\lambda b.(\sim (\vee (\sim a)\, (\sim b)))]]$
$=_{o(ooo)(ooo)} \wedge_{ooo}[\lambda a_o.[\lambda b_o.(\sim_{oo}(\vee_{ooo}(\sim_{oo} a_o)(\sim_{oo} b_o)))_o]_{(oo)}]$

:= $K8018$ %0
wff 2106 : $= \wedge [\lambda a.[\lambda b.(\sim (\vee (\sim a)\, (\sim b)))]]_o$:= $K8018$

##
Q.E.D.
##

%0
$= \wedge [\lambda a.[\lambda b.(\sim (\vee (\sim a)\, (\sim b)))]]$:= $K8018$
$=_{o(ooo)(ooo)} \wedge_{ooo}[\lambda a_o.[\lambda b_o.(\sim_{oo}(\vee_{ooo}(\sim_{oo} a_o)(\sim_{oo} b_o)))_o]_{(oo)}]$:= $K8018$

2.1.73 Results for File K8019.r0a.txt

##
Proof Template K8019: A ∧ B → A, B
##

```
##
##   Source: [Kubota 2017 (doi: 10.4444/100.10)]
##
##   Copyright (c) 2017 Owl of Minerva Press GmbH. All rights reserved.
##   Written by Ken Kubota (<mail@kenkubota.de>).
##
##   This file is part of the publication of the mathematical logic R0.
##   For more information, visit: <http://doi.org/10.4444/100.10>
##


##
##   Define Syntactical Variables
##

<< basics.r0.txt

## the assumption:   (A ∧ B)
:=  $H8019 ∧ooo xo yo
# wff      50  :        ∧ x yo       :=  $H8019


##
##   Assumptions and Resulting Syntactical Variables
##

§! $H8019
#                     ∧ x y      :=  $H8019


##
##   Include Proof Template
##

## <<< K8019.r0t.txt
## Include begin (K8019.r0t.txt) [oldfile=(K8019.r0a.txt)]
##
##   Proof Template K8019:   A ∧ B   →   A, B
##
##
##   Source: [Kubota 2017 (doi: 10.4444/100.10)]
##
##   Copyright (c) 2017 Owl of Minerva Press GmbH. All rights reserved.
##   Written by Ken Kubota (<mail@kenkubota.de>).
##
##   This file is part of the publication of the mathematical logic R0.
##   For more information, visit: <http://doi.org/10.4444/100.10>
##
```

<< A5200t.r0.txt

##
Proof Template
##

.1

%$\$H8019$
$\wedge\, x\, y$:= $\$H8019$
$\wedge_{ooo}x_o y_o$:= $\$H8019$
§\ $\wedge_{ooo}x_o$
$= (\wedge\, x)\, [\lambda y.(= [\lambda g.(g\, T\, T)]\, [\lambda g.(g\, x\, y)])]$
§s %1 2 %0
$[\lambda y.(= [\lambda g.(g\, T\, T)]\, [\lambda g.(g\, x\, y)])]\, y$
§\ $[\lambda y_o.(=_{o\omega\omega}[\lambda g_{ooo}.(g_{ooo}T_oT_o)_o][\lambda g_{ooo}.(g_{ooo}x_oy_o)_o])_o]y_o$
$= ([\lambda y.(= [\lambda g.(g\, T\, T)]\, [\lambda g.(g\, x\, y)])]\, y)\, (= [\lambda g.(g\, T\, T)]\, [\lambda g.(g\, x\, y)])$
§s %1 1 %0
$= [\lambda g.(g\, T\, T)]\, [\lambda g.(g\, x\, y)]$
§= $[\lambda g_{ooo}.(g_{ooo}T_oT_o)_o][\lambda x_o.[\lambda y_o.x_o]_{(oo)}]$
$= ([\lambda g.(g\, T\, T)]\, [\lambda x.[\lambda y.x]])\, ([\lambda g.(g\, T\, T)]\, [\lambda x.[\lambda y.x]])$
§s %0 6 %1
$= ([\lambda g.(g\, T\, T)]\, [\lambda x.[\lambda y.x]])\, ([\lambda g.(g\, x\, y)]\, [\lambda x.[\lambda y.x]])$
§\ $[\lambda g_{ooo}.(g_{ooo}T_oT_o)_o][\lambda x_o.[\lambda y_o.x_o]_{(oo)}]$
$= ([\lambda g.(g\, T\, T)]\, [\lambda x.[\lambda y.x]])\, ([\lambda x.[\lambda y.x]]\, T\, T)$
§s %1 5 %0
$= ([\lambda x.[\lambda y.x]]\, T\, T)\, ([\lambda g.(g\, x\, y)]\, [\lambda x.[\lambda y.x]])$
§\ $[\lambda g_{ooo}.(g_{ooo}x_oy_o)_o][\lambda x_o.[\lambda y_o.x_o]_{(oo)}]$
$= ([\lambda g.(g\, x\, y)]\, [\lambda x.[\lambda y.x]])\, ([\lambda x.[\lambda y.x]]\, x\, y)$
§s %1 3 %0
$= ([\lambda x.[\lambda y.x]]\, T\, T)\, ([\lambda x.[\lambda y.x]]\, x\, y)$
§\ $[\lambda x_o.[\lambda y_o.x_o]_{(oo)}]T_o$
$= ([\lambda x.[\lambda y.x]]\, T)\, [\lambda y.T]$
§s %1 10 %0
$= ([\lambda y.T]\, T)\, ([\lambda x.[\lambda y.x]]\, x\, y)$
§\ $[\lambda x_o.[\lambda y_o.x_o]_{(oo)}]x_o$
$= ([\lambda x.[\lambda y.x]]\, x)\, [\lambda y.x]$
§s %1 6 %0
$= ([\lambda y.T]\, T)\, ([\lambda y.x]\, y)$
§\ $[\lambda y_o.T_o]T_o$
$= ([\lambda y.T]\, T)\, T$
§s %1 5 %0
$= T\, ([\lambda y.x]\, y)$
§\ $[\lambda y_o.x_o]y_o$
$= ([\lambda y.x]\, y)\, x$
§s %1 3 %0
$= T\, x$
%T

\# $===$:= $A5200t$ T
\# $=_{o\omega\omega}=_{\omega}=_{\omega}$:= $A5200t$ T
§s %0 1 %1
\# x
:= \$$A8019$ %0
\# wff 16 : x_o := \$$A8019$

\#\# .2

%\$$H8019$
\# $\wedge\,\$A8019\,y$:= \$$H8019$
\# $\wedge_{ooo}\$A8019_o y_o$:= \$$H8019$
§\ $\wedge_{ooo}\$A8019_o$
\# $= (\wedge\,\$A8019)\,[\lambda y.(=[\lambda g.(g\,T\,T)]\,[\lambda g.(g\,\$A8019\,y)])]$
§s %1 2 %0
\# $[\lambda y.(=[\lambda g.(g\,T\,T)]\,[\lambda g.(g\,\$A8019\,y)])]\,y$
§\ $[\lambda y_o.(=_{o\omega\omega}[\lambda g_{ooo}.(g_{ooo}T_oT_o)_o][\lambda g_{ooo}.(g_{ooo}\$A8019_o y_o)_o])_o]y_o$
\# $= ([\lambda y.(=[\lambda g.(g\,T\,T)]\,[\lambda g.(g\,\$A8019\,y)])]\,y)\,(=[\lambda g.(g\,T\,T)]\,[\lambda g.(g\,\$A8019\,y)])$
§s %1 1 %0
\# $= [\lambda g.(g\,T\,T)]\,[\lambda g.(g\,\$A8019\,y)]$
§= $[\lambda g_{ooo}.(g_{ooo}T_oT_o)_o][\lambda \$A8019_o.[\lambda y_o.y_o]_{(oo)}]$
\# $= ([\lambda g.(g\,T\,T)]\,[\lambda\$A8019.[\lambda y.y]])\,([\lambda g.(g\,T\,T)]\,[\lambda\$A8019.[\lambda y.y]])$
§s %0 6 %1
\# $= ([\lambda g.(g\,T\,T)]\,[\lambda\$A8019.[\lambda y.y]])\,([\lambda g.(g\,\$A8019\,y)]\,[\lambda\$A8019.[\lambda y.y]])$
§r /15 z_o
\# $= [\lambda y.y]\,[\lambda z.z]$
§s %1 15 %0
\# $= ([\lambda g.(g\,T\,T)]\,[\lambda\$A8019.[\lambda y.y]])\,([\lambda g.(g\,\$A8019\,y)]\,[\lambda\$A8019.[\lambda z.z]])$
§\ $[\lambda g_{ooo}.(g_{ooo}T_oT_o)_o][\lambda\$A8019_o.[\lambda y_o.y_o]_{(oo)}]$
\# $= ([\lambda g.(g\,T\,T)]\,[\lambda\$A8019.[\lambda y.y]])\,([\lambda\$A8019.[\lambda y.y]]\,T\,T)$
§s %1 5 %0
\# $= ([\lambda\$A8019.[\lambda y.y]]\,T\,T)\,([\lambda g.(g\,\$A8019\,y)]\,[\lambda\$A8019.[\lambda z.z]])$
§\ $[\lambda g_{ooo}.(g_{ooo}\$A8019_o y_o)_o][\lambda\$A8019_o.[\lambda z_o.z_o]_{(oo)}]$
\# $= ([\lambda g.(g\,\$A8019\,y)]\,[\lambda\$A8019.[\lambda z.z]])\,([\lambda\$A8019.[\lambda z.z]]\,\$A8019\,y)$
§s %1 3 %0
\# $= ([\lambda\$A8019.[\lambda y.y]]\,T\,T)\,([\lambda\$A8019.[\lambda z.z]]\,\$A8019\,y)$
§\ $[\lambda\$A8019_o.[\lambda y_o.y_o]_{(oo)}]T_o$
\# $= ([\lambda\$A8019.[\lambda y.y]]\,T)\,[\lambda y.y]$
§s %1 10 %0
\# $= ([\lambda y.y]\,T)\,([\lambda\$A8019.[\lambda z.z]]\,\$A8019\,y)$
§\ $[\lambda\$A8019_o.[\lambda z_o.z_o]_{(oo)}]\$A8019_o$
\# $= ([\lambda\$A8019.[\lambda z.z]]\,\$A8019)\,[\lambda z.z]$
§s %1 6 %0
\# $= ([\lambda y.y]\,T)\,([\lambda z.z]\,y)$
§\ $[\lambda y_o.y_o]T_o$
\# $= ([\lambda y.y]\,T)\,T$
§s %1 5 %0
\# $= T\,([\lambda z.z]\,y)$
§\ $[\lambda z_o.z_o]y_o$

\# $= ([\lambda z.z]\, y)\, y$
§s %1 3 %0
\# $= T\, y$
%T
\# $===$:= *A5200t* T
\# $=_{o\omega\omega} =_{\omega} =_{\omega}$:= *A5200t* T
§s %0 1 %1
\# y
:= \$*B*8019 %0
\# wff 34 : y_o := \$*B*8019
\## Include end (K8019.r0t.txt) [newfile=(K8019.r0a.txt)]
>>>

\##
\## Undefine Syntactical Variables
\##

:= \$*H*8019

\##
\## Q.E.D.
\##

%\$*A*8019
\# x := \$*A*8019
\# x_o := \$*A*8019
%\$*B*8019
\# y := \$*B*8019
\# y_o := \$*B*8019

\##
\## Undefine Results
\##

:= \$*A*8019
:= \$*B*8019

2.1.74 Results for File K8019H.r0a.txt

\##
\## Proof Template K8019H: H ⊃ (A ∧ B) → H ⊃ A, H ⊃ B
\##
\##
\## Source: [Kubota 2017 (doi: 10.4444/100.10)]
\##

```
##  Written by Ken Kubota (<mail@kenkubota.de>).
##
##  This file is part of the publication of the mathematical logic R0.
##  For more information, visit: <http://doi.org/10.4444/100.10>
##


##
##  Define Syntactical Variables
##

<< basics.r0.txt

## the assumption:  H ⊃ (A ∧ B)
:=  $H8019H  ⊃_{ooo}h_o(∧_{ooo}x_o y_o)
# wff     210  :        ⊃ h (∧ x y)_o       :=  $H8019H


##
##  Assumptions and Resulting Syntactical Variables
##

§! $H8019H
#                  ⊃ h (∧ x y)      :=  $H8019H


##
##  Include Proof Template
##

## <<< K8019H.r0t.txt
## Include begin (K8019H.r0t.txt) [oldfile=(K8019H.r0a.txt)]
##
##  Proof Template K8019H:  H ⊃ (A ∧ B)  →  H ⊃ A, H ⊃ B
##
##
##  Source: [Kubota 2017 (doi: 10.4444/100.10)]
##
##  Copyright (c) 2017 Owl of Minerva Press GmbH. All rights reserved.
##  Written by Ken Kubota (<mail@kenkubota.de>).
##
##  This file is part of the publication of the mathematical logic R0.
##  For more information, visit: <http://doi.org/10.4444/100.10>
##

<< A5200t.r0.txt


##
```

Proof Template
##

.1: H ⊃ T

%T
$\quad = = = \quad := \quad A5200t \quad T$
$\quad =_{o\omega\omega} =_{\omega} =_{\omega} \quad := \quad A5200t \quad T$

use Proof Template K8003 (Intro): A → H ⊃ A
:= $\$A8003$ %0
wff 12 : $= = =_{o} \quad := \quad \$A8003 \quad A5200t \quad T$
:= $\$H8003 \ \supset_{ooo} h_o (\wedge_{ooo} x_o y_o)/5$
wff 208 : $h_o \quad := \quad \$H8003$
<< K8003.r0t.txt
:= $\$A8003$
:= $\$H8003$

:= $\$TTMP8019H$ %0
wff 1414 : $\supset h\, T_{o,\ldots} \quad := \quad \$TTMP8019H$

.2: H ⊃ A

%$\$H8019H$
$\quad \supset h\, (\wedge\, x\, y) \quad := \quad \$H8019H$
$\quad \supset_{ooo} h_o (\wedge_{ooo} x_o y_o) \quad := \quad \$H8019H$
§\ $\wedge_{ooo} x_o$
$\quad = (\wedge\, x)\, [\lambda y.(= [\lambda g.(g\, T\, T)]\, [\lambda g.(g\, x\, y)])]$
§s %1 6 %0
$\quad \supset h\, ([\lambda y.(= [\lambda g.(g\, T\, T)]\, [\lambda g.(g\, x\, y)])]\, y)$
§\ $[\lambda y_o.(=_{o\omega\omega} [\lambda g_{ooo}.(g_{ooo} T_o T_o)_o][\lambda g_{ooo}.(g_{ooo} x_o y_o)_o])_o] y_o$
$\quad = ([\lambda y.(= [\lambda g.(g\, T\, T)]\, [\lambda g.(g\, x\, y)])]\, y)\, (= [\lambda g.(g\, T\, T)]\, [\lambda g.(g\, x\, y)])$
§s %1 3 %0
$\quad \supset h\, (= [\lambda g.(g\, T\, T)]\, [\lambda g.(g\, x\, y)])$
:= $\$TMP8019H$ %0
wff 1499 : $\supset h\, (= [\lambda g.(g\, T\, T)]\, [\lambda g.(g\, x\, y)])_o \quad := \quad \$TMP8019H$

§= $[\lambda g_{ooo}.(g_{ooo} T_o T_o)_o][\lambda x_o.[\lambda y_o.x_o]_{(oo)}]$
$\quad = ([\lambda g.(g\, T\, T)]\, [\lambda x.[\lambda y.x]])\, ([\lambda g.(g\, T\, T)]\, [\lambda x.[\lambda y.x]])$

use Proof Template K8003 (Intro): A → H ⊃ A
:= $\$A8003$ %0
wff 1504 : $= ([\lambda g.(g\, T\, T)]\, [\lambda x.[\lambda y.x]])\, ([\lambda g.(g\, T\, T)]\, [\lambda x.[\lambda y.x]])_o \quad := \quad \$A8003$
:= $\$H8003 \ \supset_{ooo} h_o (\wedge_{ooo} x_o y_o)/5$
wff 208 : $h_o \quad := \quad \$H8003$
<< K8003.r0t.txt
:= $\$A8003$
:= $\$H8003$
%0

$\supset h\,(=([\lambda g.(g\,T\,T)]\,[\lambda x.[\lambda y.x]])\,([\lambda g.(g\,T\,T)]\,[\lambda x.[\lambda y.x]]))$
$\supset_{ooo} h_o \ldots$
$\ldots(=_{o\omega\omega}([\lambda g_{ooo}.(g_{ooo}T_oT_o)_o][\lambda x_o.[\lambda y_o.x_o]_{(oo)}])([\lambda g_{ooo}.(g_{ooo}T_oT_o)_o][\lambda x_o.[\lambda y_o.x_o]_{(oo)}]))$

%$TMP8019H
$\supset h\,(=[\lambda g.(g\,T\,T)]\,[\lambda g.(g\,x\,y)])$:= $TMP8019H
$\supset_{ooo} h_o(=_{o\omega\omega}[\lambda g_{ooo}.(g_{ooo}T_oT_o)_o][\lambda g_{ooo}.(g_{ooo}x_oy_o)_o])$:= $TMP8019H
:= $TMP8019H
§s' %1 6 %0
$\supset h\,(=([\lambda g.(g\,T\,T)]\,[\lambda x.[\lambda y.x]])\,([\lambda g.(g\,x\,y)]\,[\lambda x.[\lambda y.x]]))$
§\ $[\lambda g_{ooo}.(g_{ooo}T_oT_o)_o][\lambda x_o.[\lambda y_o.x_o]_{(oo)}]$
$=([\lambda g.(g\,T\,T)]\,[\lambda x.[\lambda y.x]])\,([\lambda x.[\lambda y.x]]\,T\,T)$
§s %1 13 %0
$\supset h\,(=([\lambda x.[\lambda y.x]]\,T\,T)\,([\lambda g.(g\,x\,y)]\,[\lambda x.[\lambda y.x]]))$
§\ $[\lambda g_{ooo}.(g_{ooo}x_oy_o)_o][\lambda x_o.[\lambda y_o.x_o]_{(oo)}]$
$=([\lambda g.(g\,x\,y)]\,[\lambda x.[\lambda y.x]])\,([\lambda x.[\lambda y.x]]\,x\,y)$
§s %1 7 %0
$\supset h\,(=([\lambda x.[\lambda y.x]]\,T\,T)\,([\lambda x.[\lambda y.x]]\,x\,y))$
§\ $[\lambda x_o.[\lambda y_o.x_o]_{(oo)}]T_o$
$=([\lambda x.[\lambda y.x]]\,T)\,[\lambda y.T]$
§s %1 26 %0
$\supset h\,(=([\lambda y.T]\,T)\,([\lambda x.[\lambda y.x]]\,x\,y))$
§\ $[\lambda x_o.[\lambda y_o.x_o]_{(oo)}]x_o$
$=([\lambda x.[\lambda y.x]]\,x)\,[\lambda y.x]$
§s %1 14 %0
$\supset h\,(=([\lambda y.T]\,T)\,([\lambda y.x]\,y))$
§\ $[\lambda y_o.T_o]T_o$
$=([\lambda y.T]\,T)\,T$
§s %1 13 %0
$\supset h\,(=T\,([\lambda y.x]\,y))$
§\ $[\lambda y_o.x_o]y_o$
$=([\lambda y.x]\,y)\,x$
§s %1 7 %0
$\supset h\,(=T\,x)$

%$TTMP8019H
$\supset h\,T$:= $TTMP8019H
$\supset_{ooo} h_oT_o$:= $TTMP8019H
§s' %0 1 %1
$\supset h\,x$

:= $A8019H %0
wff 1569 : $\supset h\,x_o$:= $A8019H

.3: H ⊃ B

%$H8019H
$\supset h\,(\wedge\,x\,y)$:= $H8019H
$\supset_{ooo} h_o(\wedge_{ooo}x_oy_o)$:= $H8019H

§\ $\wedge_{ooo}x_o$
$= (\wedge\, x)\, [\lambda y.(= [\lambda g.(g\,T\,T)]\, [\lambda g.(g\,x\,y)])]$
§s %1 6 %0
$\supset h\, ([\lambda y.(= [\lambda g.(g\,T\,T)]\, [\lambda g.(g\,x\,y)])]\, y)$
§\ $[\lambda y_o.(=_{o\omega\omega}[\lambda g_{ooo}.(g_{ooo}T_oT_o)_o][\lambda g_{ooo}.(g_{ooo}x_oy_o)_o])_o]y_o$
$= ([\lambda y.(= [\lambda g.(g\,T\,T)]\, [\lambda g.(g\,x\,y)])]\, y)\, (= [\lambda g.(g\,T\,T)]\, [\lambda g.(g\,x\,y)])$
§s %1 3 %0
$\supset h\, (= [\lambda g.(g\,T\,T)]\, [\lambda g.(g\,x\,y)])$
:= $\$TMP8019H$ %0
wff 1499 : $\supset h\, (= [\lambda g.(g\,T\,T)]\, [\lambda g.(g\,x\,y)])_o$:= $\$TMP8019H$

§= $[\lambda g_{ooo}.(g_{ooo}T_oT_o)_o][\lambda x_o.[\lambda y_o.y_o]_{(oo)}]$
$= ([\lambda g.(g\,T\,T)]\, [\lambda x.[\lambda y.y]])\, ([\lambda g.(g\,T\,T)]\, [\lambda x.[\lambda y.y]])$

use Proof Template K8003 (Intro): A $\rightarrow$ H $\supset$ A
:= $\$A8003$ %0
wff 1573 : $= ([\lambda g.(g\,T\,T)]\, [\lambda x.[\lambda y.y]])\, ([\lambda g.(g\,T\,T)]\, [\lambda x.[\lambda y.y]])_o$:= $\$A8003$
:= $\$H8003$ $\supset_{ooo}h_o(\wedge_{ooo}x_oy_o)/5$
wff 208 : h_o := $\$H8003$
<< K8003.r0t.txt
:= $\$A8003$
:= $\$H8003$
%0
$\supset h\, (= ([\lambda g.(g\,T\,T)]\, [\lambda x.[\lambda y.y]])\, ([\lambda g.(g\,T\,T)]\, [\lambda x.[\lambda y.y]]))$
$\supset_{ooo}h_o \ldots$
$\ldots (=_{o\omega\omega}([\lambda g_{ooo}.(g_{ooo}T_oT_o)_o][\lambda x_o.[\lambda y_o.y_o]_{(oo)}])([\lambda g_{ooo}.(g_{ooo}T_oT_o)_o][\lambda x_o.[\lambda y_o.y_o]_{(oo)}]))$

%$\$TMP8019H$
$\supset h\, (= [\lambda g.(g\,T\,T)]\, [\lambda g.(g\,x\,y)])$:= $\$TMP8019H$
$\supset_{ooo}h_o(=_{o\omega\omega}[\lambda g_{ooo}.(g_{ooo}T_oT_o)_o][\lambda g_{ooo}.(g_{ooo}x_oy_o)_o])$:= $\$TMP8019H$
:= $\$TMP8019H$
§s′ %1 6 %0
$\supset h\, (= ([\lambda g.(g\,T\,T)]\, [\lambda x.[\lambda y.y]])\, ([\lambda g.(g\,x\,y)]\, [\lambda x.[\lambda y.y]]))$
§\ $[\lambda g_{ooo}.(g_{ooo}T_oT_o)_o][\lambda x_o.[\lambda y_o.y_o]_{(oo)}]$
$= ([\lambda g.(g\,T\,T)]\, [\lambda x.[\lambda y.y]])\, ([\lambda x.[\lambda y.y]]\, T\,T)$
§s %1 13 %0
$\supset h\, (= ([\lambda x.[\lambda y.y]]\, T\,T)\, ([\lambda g.(g\,x\,y)]\, [\lambda x.[\lambda y.y]]))$
§\ $[\lambda g_{ooo}.(g_{ooo}x_oy_o)_o][\lambda x_o.[\lambda y_o.y_o]_{(oo)}]$
$= ([\lambda g.(g\,x\,y)]\, [\lambda x.[\lambda y.y]])\, ([\lambda x.[\lambda y.y]]\, x\,y)$
§s %1 7 %0
$\supset h\, (= ([\lambda x.[\lambda y.y]]\, T\,T)\, ([\lambda x.[\lambda y.y]]\, x\,y))$
§\ $[\lambda x_o.[\lambda y_o.y_o]_{(oo)}]T_o$
$= ([\lambda x.[\lambda y.y]]\, T)\, [\lambda y.y]$
§s %1 26 %0
$\supset h\, (= ([\lambda y.y]\, T)\, ([\lambda x.[\lambda y.y]]\, x\,y))$
§\ $[\lambda x_o.[\lambda y_o.y_o]_{(oo)}]x_o$
$= ([\lambda x.[\lambda y.y]]\, x)\, [\lambda y.y]$
§s %1 14 %0
$\supset h\, (= ([\lambda y.y]\, T)\, ([\lambda y.y]\, y))$

§\ $[\lambda y_o.y_o]T_o$
$= ([\lambda y.y]\, T)\, T$
§s %1 13 %0
$\supset h\, (= T\, ([\lambda y.y]\, y))$
§\ $[\lambda y_o.y_o]y_o$
$= ([\lambda y.y]\, y)\, y$
§s %1 7 %0
$\supset h\, (= T\, y)$

%$\$TTMP8019H$
$\supset h\, T$:= $\$TTMP8019H$
$\supset_{ooo} h_o T_o$:= $\$TTMP8019H$
§s' %0 1 %1
$\supset h\, y$

:= $\$B8019H$ %0
wff 1641 : $\supset h\, y_o$:= $\$B8019H$

undefine local variables
:= $\$TTMP8019H$
Include end (K8019H.r0t.txt) [newfile=(K8019H.r0a.txt)]
>>>

##
Undefine Syntactical Variables
##

:= $\$H8019H$

##
Q.E.D.
##

%$\$A8019H$
$\supset h\, x$:= $\$A8019H$
$\supset_{ooo} h_o x_o$:= $\$A8019H$
%$\$B8019H$
$\supset h\, y$:= $\$B8019H$
$\supset_{ooo} h_o y_o$:= $\$B8019H$

##
Undefine Results
##

:= $\$A8019H$
:= $\$B8019H$

2.1.75 Results for File K8020.r0a.txt

```
##
##  Proof Template K8020:  A, B  →  A ∧ B
##
##
##  Source: [Kubota 2017 (doi: 10.4444/100.10)]
##
##  Copyright (c) 2017 Owl of Minerva Press GmbH. All rights reserved.
##  Written by Ken Kubota (<mail@kenkubota.de>).
##
##  This file is part of the publication of the mathematical logic R0.
##  For more information, visit: <http://doi.org/10.4444/100.10>
##


##
##  Define Syntactical Variables
##

## assumption 1
:=  $A8020 x_o
# wff     11  :        x_o       :=  $A8020

## assumption 2
:=  $B8020 y_o
# wff     12  :        y_o       :=  $B8020


##
##  Assumptions and Resulting Syntactical Variables
##

§! $A8020
#                  x       :=  $A8020
§! $B8020
#                  y       :=  $B8020


##
##  Include Proof Template
##


## <<< K8020.r0t.txt
## Include begin (K8020.r0t.txt) [oldfile=(K8020.r0a.txt)]
##
##  Proof Template K8020:  A, B  →  A ∧ B
##
```

```
##
##   Source: [Kubota 2017 (doi: 10.4444/100.10)]
##
##   Copyright (c) 2017 Owl of Minerva Press GmbH. All rights reserved.
##   Written by Ken Kubota (<mail@kenkubota.de>).
##
##   This file is part of the publication of the mathematical logic R0.
##   For more information, visit: <http://doi.org/10.4444/100.10>
##

<< A5212.r0.txt



##
##   Proof Template
##

## .1

%A5212
#                    ∧TT          :=  A5212
#                    ∧_ooo T_o T_o    :=  A5212

:=  $TTMP8020  %0
# wff      160  :       ∧TT_o,...       :=  $TTMP8020   A5212

## .2

%$A8020
#                    x          :=  $A8020
#                    x_o          :=  $A8020

## use Proof Template A5219a (Rule T):   A   →   T = A
:=  $A5219a  %0
# wff      11  :       x_o       :=  $A5219a  $A8020
<< A5219a.r0t.txt
:=  $A5219a

:=  $ATMP8020  %0
# wff      321  :       =T$A8020_o,...       :=  $ATMP8020

## .3

%$B8020
#                    y          :=  $B8020
#                    y_o          :=  $B8020

## use Proof Template A5219a (Rule T):   A   →   T = A
:=  $A5219a  %0
```

```
# wff      12  :        y_o      :=  $A5219a  $B8020
<< A5219a.r0t.txt
:=  $A5219a

:=  $BTMP8020  %0
# wff      720  :         =T $B8020_o,...       :=  $BTMP8020

## .4

%A5212
#                     ∧TT        :=  $TTMP8020   A5212
#                     ∧_oooT_oT_o      :=  $TTMP8020   A5212
:=  $TTMP8020
%$ATMP8020
#                     =T $A8020      :=  $ATMP8020
#                     =_oooT_o$A8020_o      :=  $ATMP8020
:=  $ATMP8020
§s  %1  5  %0
#                     ∧ $A8020 T
%$BTMP8020
#                     =T $B8020      :=  $BTMP8020
#                     =_oooT_o$B8020_o      :=  $BTMP8020
:=  $BTMP8020
§s  %1  3  %0
#                     ∧ $A8020 $B8020
## Include end (K8020.r0t.txt) [newfile=(K8020.r0a.txt)]
>>>


##
##   Undefine Syntactical Variables
##

:=  $A8020
:=  $B8020



##
##   Q.E.D.
##

%0
#                     ∧ x y
#                     ∧_ooox_oy_o
```

2.1.76 Results for File K8020H.r0a.txt

```
##
##   Proof Template K8020H:   H ⊃ A, H ⊃ B   →   H ⊃ (A ∧ B)
```

##
##
Source: [Kubota 2017 (doi: 10.4444/100.10)]
##

Written by Ken Kubota (<mail@kenkubota.de>).
##
This file is part of the publication of the mathematical logic $\mathcal{R}_0$.
For more information, visit: <http://doi.org/10.4444/100.10>
##

##
Define Syntactical Variables
##

<< basics.r0.txt

assumption 1
:= $\$A8020H\ \supset_{ooo} h_o x_o$
\# wff 210 : $\supset h\, x_o$:= $\$A8020H$

assumption 2
:= $\$B8020H\ \supset_{ooo} h_o y_o$
\# wff 211 : $\supset h\, y_o$:= $\$B8020H$

##
Assumptions and Resulting Syntactical Variables
##

§! $\$A8020H$
\# $\supset h\, x$:= $\$A8020H$
§! $\$B8020H$
\# $\supset h\, y$:= $\$B8020H$

##
Include Proof Template
##

<<< K8020H.r0t.txt
Include begin (K8020H.r0t.txt) [oldfile=(K8020H.r0a.txt)]
##
Proof Template K8020H: H ⊃ A, H ⊃ B → H ⊃ (A ∧ B)
##
##
Source: [Kubota 2017 (doi: 10.4444/100.10)]
##

##
This file is part of the publication of the mathematical logic $\mathcal{R}_0$.
For more information, visit: <http://doi.org/10.4444/100.10>
##

<< A5212.r0.txt

##
Proof Template
##

.1

%$A5212$
$\wedge T T$:= $A5212$
$\wedge_{ooo} T_o T_o$:= $A5212$

use Proof Template K8003 (Intro): A → H ⊃ A
:= $\$A8003$ %0
wff 242 : $\wedge T T_{o,...}$:= $\$A8003$ $A5212$
:= $\$H8003$ $\supset_{ooo} h_o x_o/5$
wff 208 : h_o := $\$H8003$
<< K8003.r0t.txt
:= $\$A8003$
:= $\$H8003$

:= $\$TTMP8020H$ %0
wff 1415 : $\supset h\, A5212_{o,...}$:= $\$TTMP8020H$

.2

%$\$A8020H$
$\supset h\, x$:= $\$A8020H$
$\supset_{ooo} h_o x_o$:= $\$A8020H$

use Proof Template A5219aH (Rule T): H ⊃ A → H ⊃ (T = A)
:= $\$A5219aH$ %0
wff 210 : $\supset h\, x_o$:= $\$A5219aH$ $\$A8020H$
<< A5219aH.r0t.txt
:= $\$A5219aH$

:= $\$ATMP8020H$ %0
wff 1544 : $\supset h\,(=T\,x)_o$:= $\$ATMP8020H$

.3

%$\$B8020H$
$\supset h\, y$:= $\$B8020H$
$\supset_{ooo} h_o y_o$:= $\$B8020H$

use Proof Template A5219aH (Rule T): H ⊃ A → H ⊃ (T = A)
:= $\$A5219aH$ %0
wff 211 : $\supset h\, y_o$:= $\$A5219aH$ $\$B8020H$
<< A5219aH.r0t.txt
:= $\$A5219aH$

:= $\$BTMP8020H$ %0
wff 1595 : $\supset h\, (=T\, y)_o$:= $\$BTMP8020H$

.4

%$\$TTMP8020H$
$\supset h\, A5212$:= $\$TTMP8020H$
$\supset_{ooo} h_o A5212_o$:= $\$TTMP8020H$
:= $\$TTMP8020H$
%$\$ATMP8020H$
$\supset h\, (=T\, x)$:= $\$ATMP8020H$
$\supset_{ooo} h_o (=_{ooo} T_o x_o)$:= $\$ATMP8020H$
:= $\$ATMP8020H$
§s' %1 5 %0
$\supset h\, (\wedge\, x\, T)$
%$\$BTMP8020H$
$\supset h\, (=T\, y)$:= $\$BTMP8020H$
$\supset_{ooo} h_o (=_{ooo} T_o y_o)$:= $\$BTMP8020H$
:= $\$BTMP8020H$
§s' %1 3 %0
$\supset h\, (\wedge\, x\, y)$
Include end (K8020H.r0t.txt) [newfile=(K8020H.r0a.txt)]
>>>

##
Undefine Syntactical Variables
##

:= $\$A8020H$
:= $\$B8020H$

##
Q.E.D.
##

%0
$\supset h\, (\wedge\, x\, y)$

\# $\supset_{ooo} h_o (\wedge_{ooo} x_o y_o)$

2.1.77 Results for File K8021.r0.txt

\##
\## Proof K8021: (A ∧ B) ∧ C = A ∧ (B ∧ C)
\##
\##
\## Source: [Kubota 2017 (doi: 10.4444/100.10)]
\##
\## Copyright (c) 2017 Owl of Minerva Press GmbH. All rights reserved.
\## Written by Ken Kubota (<mail@kenkubota.de>).
\##
\## This file is part of the publication of the mathematical logic $\mathcal{R}_0$.
\## For more information, visit: <http://doi.org/10.4444/100.10>
\##

<< basics.r0.txt
<< K8005.r0.txt

\##
\## Proof
\##

\## .1a

%$K8005$
\# $\supset x\, x$:= $K8005$
\# $\supset_{ooo} x_o x_o$:= $K8005$

\## use Proof Template A5221 (Sub): B → B [x/A]
:= $\$B5221$ %0
\# wff 1357 : $\supset x\, x_{o,\ldots}$:= $\$B5221$ $K8005$
:= $\$T5221$ o
\# wff 2 : o_τ := $\$T5221$
:= $\$X5221$ x_o
\# wff 16 : x_o := $\$X5221$
:= $\$A5221$ $\wedge_{ooo}(\wedge_{ooo} a_o b_o) c_o$
\# wff 1375 : $\wedge (\wedge\, a\, b)\, c_o$:= $\$A5221$
<< A5221.r0t.txt
:= $\$B5221$
:= $\$T5221$
:= $\$X5221$
:= $\$A5221$
%0
\# $\supset (\wedge (\wedge\, a\, b)\, c)\, (\wedge (\wedge\, a\, b)\, c)$
\# $\supset_{ooo}(\wedge_{ooo}(\wedge_{ooo} a_o b_o) c_o)(\wedge_{ooo}(\wedge_{ooo} a_o b_o) c_o)$

.1b

use Proof Template K8019H: H $\supset$ (A $\wedge$ B) $\rightarrow$ H $\supset$ A, H $\supset$ B
:= $\$H8019H$ %0
wff 1412 : $\supset(\wedge(\wedge\, a\, b)\, c)\,(\wedge(\wedge\, a\, b)\, c)_{o,\ldots}$:= $\$H8019H$
<< K8019H.r0t.txt
:= $\$H8019H$
:= $\$ABTMP8021$ $\supset_{ooo}(\wedge_{ooo}(\wedge_{ooo}a_o b_o)c_o)(\wedge_{ooo}a_o b_o)$
wff 1706 : $\supset(\wedge(\wedge\, a\, b)\, c)\,(\wedge\, a\, b)_o$:= $\$A8019H$ $\$ABTMP8021$
:= $\$CTMP8021$ $\supset_{ooo}(\wedge_{ooo}(\wedge_{ooo}a_o b_o)c_o)c_o$
wff 1778 : $\supset(\wedge(\wedge\, a\, b)\, c)\, c_o$:= $\$B8019H$ $\$CTMP8021$
:= $\$A8019H$
:= $\$B8019H$
%0
$\supset(\wedge(\wedge\, a\, b)\, c)\, c$:= $\$CTMP8021$
$\supset_{ooo}(\wedge_{ooo}(\wedge_{ooo}a_o b_o)c_o)c_o$:= $\$CTMP8021$

.1c

%$\$ABTMP8021$
$\supset(\wedge(\wedge\, a\, b)\, c)\,(\wedge\, a\, b)$:= $\$ABTMP8021$
$\supset_{ooo}(\wedge_{ooo}(\wedge_{ooo}a_o b_o)c_o)(\wedge_{ooo}a_o b_o)$:= $\$ABTMP8021$

use Proof Template K8019H: H $\supset$ (A $\wedge$ B) $\rightarrow$ H $\supset$ A, H $\supset$ B
:= $\$H8019H$ %0
wff 1706 : $\supset(\wedge(\wedge\, a\, b)\, c)\,(\wedge\, a\, b)_o$:= $\$ABTMP8021$ $\$H8019H$
<< K8019H.r0t.txt
:= $\$H8019H$
:= $\$ATMP8021$ $\supset_{ooo}(\wedge_{ooo}(\wedge_{ooo}a_o b_o)c_o)a_o$
wff 1819 : $\supset(\wedge(\wedge\, a\, b)\, c)\, a_o$:= $\$A8019H$ $\$ATMP8021$
:= $\$BTMP8021$ $\supset_{ooo}(\wedge_{ooo}(\wedge_{ooo}a_o b_o)c_o)b_o$
wff 1844 : $\supset(\wedge(\wedge\, a\, b)\, c)\, b_o$:= $\$B8019H$ $\$BTMP8021$
:= $\$A8019H$
:= $\$B8019H$
%0
$\supset(\wedge(\wedge\, a\, b)\, c)\, b$:= $\$BTMP8021$
$\supset_{ooo}(\wedge_{ooo}(\wedge_{ooo}a_o b_o)c_o)b_o$:= $\$BTMP8021$

:= $\$ABTMP8021$

.1d

%$\$BTMP8021$
$\supset(\wedge(\wedge\, a\, b)\, c)\, b$:= $\$BTMP8021$
$\supset_{ooo}(\wedge_{ooo}(\wedge_{ooo}a_o b_o)c_o)b_o$:= $\$BTMP8021$
:= $\$BTMP8021$
%$\$CTMP8021$
$\supset(\wedge(\wedge\, a\, b)\, c)\, c$:= $\$CTMP8021$
$\supset_{ooo}(\wedge_{ooo}(\wedge_{ooo}a_o b_o)c_o)c_o$:= $\$CTMP8021$

:= $\$CTMP8021$

use Proof Template K8020H: H ⊃ A, H ⊃ B → H ⊃ (A ∧ B)
:= $\$A8020H$ %1
wff 1844 : $\supset (\wedge (\wedge\, a\, b)\, c)\, b_o$:= $\$A8020H$
:= $\$B8020H$ %0
wff 1778 : $\supset (\wedge (\wedge\, a\, b)\, c)\, c_o$:= $\$B8020H$
<< K8020H.r0t.txt
:= $\$A8020H$
:= $\$B8020H$

:= $\$BCTMP8020$ %0
wff 1979 : $\supset (\wedge (\wedge\, a\, b)\, c)\, (\wedge\, b\, c)_o$:= $\$BCTMP8020$

%$\$ATMP8021$
$\supset (\wedge (\wedge\, a\, b)\, c)\, a$:= $\$ATMP8021$
$\supset_{ooo}(\wedge_{ooo}(\wedge_{ooo}a_o b_o)c_o)a_o$:= $\$ATMP8021$
:= $\$ATMP8021$
%$\$BCTMP8020$
$\supset (\wedge (\wedge\, a\, b)\, c)\, (\wedge\, b\, c)$:= $\$BCTMP8020$
$\supset_{ooo}(\wedge_{ooo}(\wedge_{ooo}a_o b_o)c_o)(\wedge_{ooo}b_o c_o)$:= $\$BCTMP8020$
:= $\$BCTMP8020$

use Proof Template K8020H: H ⊃ A, H ⊃ B → H ⊃ (A ∧ B)
:= $\$A8020H$ %1
wff 1819 : $\supset (\wedge (\wedge\, a\, b)\, c)\, a_o$:= $\$A8020H$
:= $\$B8020H$ %0
wff 1979 : $\supset (\wedge (\wedge\, a\, b)\, c)\, (\wedge\, b\, c)_o$:= $\$B8020H$
<< K8020H.r0t.txt
:= $\$A8020H$
:= $\$B8020H$

:= $\$ABC1TMP8020$ %0
wff 2085 : $\supset (\wedge (\wedge\, a\, b)\, c)\, (\wedge\, a\, (\wedge\, b\, c))_o$:= $\$ABC1TMP8020$

.2a

%$K8005$
$\supset x\, x$:= $K8005$
$\supset_{ooo}x_o x_o$:= $K8005$

use Proof Template A5221 (Sub): B → B [x/A]
:= $\$B5221$ %0
wff 1357 : $\supset x\, x_{o',\dots}$:= $\$B5221$ $K8005$
:= $\$T5221$ o
wff 2 : o_τ := $\$T5221$
:= $\$X5221$ x_o
wff 16 : x_o := $\$X5221$
:= $\$A5221$ $\wedge_{ooo}a_o(\wedge_{ooo}b_o c_o)$

wff 2084 : $\wedge a\,(\wedge b\,c)_o$:= $\$A5221$
<< A5221.r0t.txt
:= $\$B5221$
:= $\$T5221$
:= $\$X5221$
:= $\$A5221$
%0
$\supset (\wedge a\,(\wedge b\,c))\,(\wedge a\,(\wedge b\,c))$
$\supset_{ooo}(\wedge_{ooo}a_o(\wedge_{ooo}b_oc_o))(\wedge_{ooo}a_o(\wedge_{ooo}b_oc_o))$

.2b

use Proof Template K8019H: H ⊃ (A ∧ B) → H ⊃ A, H ⊃ B
:= $\$H8019H$ %0
wff 2095 : $\supset (\wedge a\,(\wedge b\,c))\,(\wedge a\,(\wedge b\,c))_{o,\dots}$:= $\$H8019H$
<< K8019H.r0t.txt
:= $\$H8019H$
:= $\$ATMP8021$ $\supset_{ooo}(\wedge_{ooo}a_o(\wedge_{ooo}b_oc_o))a_o$
wff 2178 : $\supset (\wedge a\,(\wedge b\,c))\,a_o$:= $\$A8019H$ $\$ATMP8021$
:= $\$BCTMP8020$ $\supset_{ooo}(\wedge_{ooo}a_o(\wedge_{ooo}b_oc_o))(\wedge_{ooo}b_oc_o)$
wff 2217 : $\supset (\wedge a\,(\wedge b\,c))\,(\wedge b\,c)_o$:= $\$B8019H$ $\$BCTMP8020$
:= $\$A8019H$
:= $\$B8019H$
%0
$\supset (\wedge a\,(\wedge b\,c))\,(\wedge b\,c)$:= $\$BCTMP8020$
$\supset_{ooo}(\wedge_{ooo}a_o(\wedge_{ooo}b_oc_o))(\wedge_{ooo}b_oc_o)$:= $\$BCTMP8020$

.2c

%$\$BCTMP8020$
$\supset (\wedge a\,(\wedge b\,c))\,(\wedge b\,c)$:= $\$BCTMP8020$
$\supset_{ooo}(\wedge_{ooo}a_o(\wedge_{ooo}b_oc_o))(\wedge_{ooo}b_oc_o)$:= $\$BCTMP8020$

use Proof Template K8019H: H ⊃ (A ∧ B) → H ⊃ A, H ⊃ B
:= $\$H8019H$ %0
wff 2217 : $\supset (\wedge a\,(\wedge b\,c))\,(\wedge b\,c)_o$:= $\$BCTMP8020$ $\$H8019H$
<< K8019H.r0t.txt
:= $\$H8019H$
:= $\$BTMP8021$ $\supset_{ooo}(\wedge_{ooo}a_o(\wedge_{ooo}b_oc_o))b_o$
wff 2257 : $\supset (\wedge a\,(\wedge b\,c))\,b_o$:= $\$A8019H$ $\$BTMP8021$
:= $\$CTMP8021$ $\supset_{ooo}(\wedge_{ooo}a_o(\wedge_{ooo}b_oc_o))c_o$
wff 2276 : $\supset (\wedge a\,(\wedge b\,c))\,c_o$:= $\$B8019H$ $\$CTMP8021$
:= $\$A8019H$
:= $\$B8019H$
%0
$\supset (\wedge a\,(\wedge b\,c))\,c$:= $\$CTMP8021$
$\supset_{ooo}(\wedge_{ooo}a_o(\wedge_{ooo}b_oc_o))c_o$:= $\$CTMP8021$

:= $\$BCTMP8020$

.2d

%$\$ATMP8021$
$\supset (\wedge\, a\, (\wedge\, b\, c))\, a$:= $\$ATMP8021$
$\supset_{ooo}(\wedge_{ooo} a_o (\wedge_{ooo} b_o c_o)) a_o$:= $\$ATMP8021$
:= $\$ATMP8021$
%$\$BTMP8021$
$\supset (\wedge\, a\, (\wedge\, b\, c))\, b$:= $\$BTMP8021$
$\supset_{ooo}(\wedge_{ooo} a_o (\wedge_{ooo} b_o c_o)) b_o$:= $\$BTMP8021$
:= $\$BTMP8021$

use Proof Template K8020H: H ⊃ A, H ⊃ B → H ⊃ (A ∧ B)
:= $\$A8020H$ %1
wff 2178 : $\supset (\wedge\, a\, (\wedge\, b\, c))\, a_o$:= $\$A8020H$
:= $\$B8020H$ %0
wff 2257 : $\supset (\wedge\, a\, (\wedge\, b\, c))\, b_o$:= $\$B8020H$
<< K8020H.r0t.txt
:= $\$A8020H$
:= $\$B8020H$

:= $\$ABTMP8021$ %0
wff 2331 : $\supset (\wedge\, a\, (\wedge\, b\, c))\, (\wedge\, a\, b)_o$:= $\$ABTMP8021$

%$\$ABTMP8021$
$\supset (\wedge\, a\, (\wedge\, b\, c))\, (\wedge\, a\, b)$:= $\$ABTMP8021$
$\supset_{ooo}(\wedge_{ooo} a_o (\wedge_{ooo} b_o c_o))(\wedge_{ooo} a_o b_o)$:= $\$ABTMP8021$
:= $\$ABTMP8021$
%$\$CTMP8021$
$\supset (\wedge\, a\, (\wedge\, b\, c))\, c$:= $\$CTMP8021$
$\supset_{ooo}(\wedge_{ooo} a_o (\wedge_{ooo} b_o c_o)) c_o$:= $\$CTMP8021$
:= $\$CTMP8021$

use Proof Template K8020H: H ⊃ A, H ⊃ B → H ⊃ (A ∧ B)
:= $\$A8020H$ %1
wff 2331 : $\supset (\wedge\, a\, (\wedge\, b\, c))\, (\wedge\, a\, b)_o$:= $\$A8020H$
:= $\$B8020H$ %0
wff 2276 : $\supset (\wedge\, a\, (\wedge\, b\, c))\, c_o$:= $\$B8020H$
<< K8020H.r0t.txt
:= $\$A8020H$
:= $\$B8020H$

:= $\$ABC2TMP8020$ %0
wff 2403 : $\supset (\wedge\, a\, (\wedge\, b\, c))\, (\wedge\, (\wedge\, a\, b)\, c)_o$:= $\$ABC2TMP8020$

.3

use Proof Template K8013: A ⊃ B, B ⊃ A → A = B
:= $\$A8013$ $\supset_{ooo}(\wedge_{ooo}(\wedge_{ooo} a_o b_o) c_o)(\wedge_{ooo} a_o (\wedge_{ooo} b_o c_o))/5$

wff 1375 : $\wedge(\wedge\, a\, b)\, c_{o,\ldots}$:= $\$A8013$
:= $\$B8013$ $\supset_{ooo}(\wedge_{ooo}a_o(\wedge_{ooo}b_oc_o))\$A8013_o/5$
wff 2084 : $\wedge\, a\,(\wedge\, b\, c)_{o,\ldots}$:= $\$B8013$
<< K8013.r0t.txt
:= $\$A8013$
:= $\$B8013$
%0
$=(\wedge\,(\wedge\, a\, b)\, c)\,(\wedge\, a\,(\wedge\, b\, c))$
$=_{ooo}(\wedge_{ooo}(\wedge_{ooo}a_ob_o)c_o)(\wedge_{ooo}a_o(\wedge_{ooo}b_oc_o))$

:= $\$ABC1TMP8020$
:= $\$ABC2TMP8020$

.4: Rename variables

use Proof Template A5221 (Sub): B → B [x/A]
:= $\$B5221$ %0
wff 2928 : $=(\wedge\,(\wedge\, a\, b)\, c)\,(\wedge\, a\,(\wedge\, b\, c))_o$:= $\$B5221$
:= $\$T5221$ o
wff 2 : o_τ := $\$T5221$
:= $\$X5221$ a_o
wff 54 : a_o := $\$X5221$
:= $\$A5221$ x_o
wff 16 : x_o := $\$A5221$
<< A5221.r0t.txt
:= $\$B5221$
:= $\$T5221$
:= $\$X5221$
:= $\$A5221$
%0
$=(\wedge\,(\wedge\, x\, b)\, c)\,(\wedge\, x\,(\wedge\, b\, c))$
$=_{ooo}(\wedge_{ooo}(\wedge_{ooo}x_ob_o)c_o)(\wedge_{ooo}x_o(\wedge_{ooo}b_oc_o))$

use Proof Template A5221 (Sub): B → B [x/A]
:= $\$B5221$ %0
wff 2965 : $=(\wedge\,(\wedge\, x\, b)\, c)\,(\wedge\, x\,(\wedge\, b\, c))_{o,\ldots}$:= $\$B5221$
:= $\$T5221$ o
wff 2 : o_τ := $\$T5221$
:= $\$X5221$ b_o
wff 58 : b_o := $\$X5221$
:= $\$A5221$ y_o
wff 34 : y_o := $\$A5221$
<< A5221.r0t.txt
:= $\$B5221$
:= $\$T5221$
:= $\$X5221$
:= $\$A5221$
%0
$=(\wedge\,(\wedge\, x\, y)\, c)\,(\wedge\, x\,(\wedge\, y\, c))$

$=_{ooo}(\wedge_{ooo}(\wedge_{ooo}x_o y_o)c_o)(\wedge_{ooo}x_o(\wedge_{ooo}y_o c_o))$

use Proof Template A5221 (Sub): B $\rightarrow$ B [x/A]
:= $\$B5221$ %0
wff 3010 : $=(\wedge\,(\wedge\,x\,y)\,c)\,(\wedge\,x\,(\wedge\,y\,c))_{o,\ldots}$:= $\$B5221$
:= $\$T5221$ o
wff 2 : o_τ := $\$T5221$
:= $\$X5221$ c_o
wff 1374 : c_o := $\$X5221$
:= $\$A5221$ z_o
wff 3013 : z_o := $\$A5221$
<< A5221.r0t.txt
:= $\$B5221$
:= $\$T5221$
:= $\$X5221$
:= $\$A5221$
%0
$=(\wedge\,(\wedge\,x\,y)\,z)\,(\wedge\,x\,(\wedge\,y\,z))$
$=_{ooo}(\wedge_{ooo}(\wedge_{ooo}x_o y_o)z_o)(\wedge_{ooo}x_o(\wedge_{ooo}y_o z_o))$

:= $\$TMP8020$ %0
wff 3059 : $=(\wedge\,(\wedge\,x\,y)\,z)\,(\wedge\,x\,(\wedge\,y\,z))_{o,\ldots}$:= $\$TMP8020$

.5: Match general definition

§= $ASSOC_{o(\backslash 4\backslash 4\backslash 3)\tau}o_\tau\wedge_{ooo}$
$=(ASSOC\,o\,\wedge)\,(ASSOC\,o\,\wedge)$
§\ $ASSOC_{o(\backslash 4\backslash 4\backslash 3)\tau}o_\tau$
$=(ASSOC\,o)\,[\lambda f.(=(f\,(f\,x\,y)\,z)\,(f\,x\,(f\,y\,z)))]$
§s %1 6 %0
$=(ASSOC\,o\,\wedge)\,([\lambda f.(=(f\,(f\,x\,y)\,z)\,(f\,x\,(f\,y\,z)))]\,\wedge)$
§\ $[\lambda f_{ooo}.(=_{ooo}(f_{ooo}(f_{ooo}x_o y_o)z_o)(f_{ooo}x_o(f_{ooo}y_o z_o)))_o]\wedge_{ooo}$
$=([\lambda f.(=(f\,(f\,x\,y)\,z)\,(f\,x\,(f\,y\,z)))]\,\wedge)\,\$TMP8020$
§s %1 3 %0
$=(ASSOC\,o\,\wedge)\,\$TMP8020$

use Proof Template A5201b (Swap): A = B $\rightarrow$ B = A
<< A5201b.r0t.txt
%0
$=\$TMP8020\,(ASSOC\,o\,\wedge)$
$=_{o\omega\omega}\$TMP8020_\omega(ASSOC_{o(\backslash 4\backslash 4\backslash 3)\tau}o_\tau\wedge_{ooo})$

%$\$TMP8020$
$=(\wedge\,(\wedge\,x\,y)\,z)\,(\wedge\,x\,(\wedge\,y\,z))$:= $\$TMP8020$
$=_{ooo}(\wedge_{ooo}(\wedge_{ooo}x_o y_o)z_o)(\wedge_{ooo}x_o(\wedge_{ooo}y_o z_o))$:= $\$TMP8020$
:= $\$TMP8020$
§s %0 1 %1
$ASSOC\,o\,\wedge$

:= $K8021$ %0
wff 3063 : $ASSOC\, o \wedge_{o, \ldots}$:= $K8021$

##
Q.E.D.
##

%0
$ASSOC\, o \wedge$:= $K8021$
$ASSOC_{o(\backslash 4\backslash 4\backslash 3)\tau} o_\tau \wedge_{ooo}$:= $K8021$

2.1.78 Results for File K8022.r0.txt

##
Proof K8022: A ⊃ B = (∼ A) ∨ B
##
##
Source: [Kubota 2017 (doi: 10.4444/100.10)]
##

Written by Ken Kubota (<mail@kenkubota.de>).
##
This file is part of the publication of the mathematical logic $\mathcal{R}_0$.
For more information, visit: <http://doi.org/10.4444/100.10>
##

<< basics.r0.txt
<< A5200t.r0.txt
<< A5205.r0.txt
<< A5228.r0.txt
<< A5230.r0.txt
<< A5231.r0.txt
<< A5232.r0.txt

##
Proof
##

.1: main case T

use Proof Template A5222 (Rule of Cases): [\x.A]T, [\x.A]F → A
§\ $[\lambda x_o.[\lambda y_o.(=_{ooo}(\supset_{ooo}x_o y_o)(\vee_{ooo}(\sim_{oo}x_o)y_o))_o]_{(oo)}]T_o$
$= ([\lambda x.[\lambda y.(= (\supset x\, y)\, (\vee\, (\sim x)\, y))]]\, T)\, [\lambda y.(= (\supset T\, y)\, (\vee\, (\sim T)\, y))]$
:= $\$L5222$ %0/3
wff 1887 : $[\lambda y.(= (\supset T\, y)\, (\vee\, (\sim T)\, y))]_{oo, \ldots}$:= $\$L5222$
:= $\$X5222\ y_o$
wff 34 : y_o := $\$X5222$

:= $\$T5222$ $\$L5222_{oo}T_o$
wff 1889 : $\$L5222\,T_o$:= $\$T5222$
:= $\$F5222$ $\$L5222_{oo}F_o$
wff 1890 : $\$L5222\,F_o$:= $\$F5222$

case T
§= $\$T5222$
$= \$T5222\,\$T5222$
§\ $\$T5222$
$= \$T5222\,(=(\supset T\,T)\,(\vee\,(\sim T)\,T))$
§s %1 3 %0
$= \$T5222\,(=(\supset T\,T)\,(\vee\,(\sim T)\,T))$
%$A5231a$
$=(\sim T)\,F$:= $A5231a$
$=_{ooo}(\sim_{oo}T_o)F_o$:= $A5231a$
§s %1 29 %0
$= \$T5222\,(=(\supset T\,T)\,(\vee\,F\,T))$
%$A5228a$
$=(\supset T\,T)\,T$:= $A5228a$
$=_{ooo}(\supset_{ooo}T_oT_o)T_o$:= $A5228a$
§s %1 13 %0
$= \$T5222\,(=T\,(\vee\,F\,T))$
%$A5232c$
$=(\vee\,F\,T)\,T$:= $A5232c$
$=_{ooo}(\vee_{ooo}F_oT_o)T_o$:= $A5232c$
§s %1 7 %0
$= \$T5222\,(=T\,T)$
%$A5230a$
$=(=T\,T)\,T$:= $A5230a$
$=_{ooo}(=_{ooo}T_oT_o)T_o$:= $A5230a$
§s %1 3 %0
$= \$T5222\,T$
use Proof Template A5201b (Swap): A = B → B = A
<< A5201b.r0t.txt
%0
$=T\,\$T5222$
$=_{o\omega\omega}T_\omega\$T5222_\omega$
%T
$===$:= $A5200t$ T
$=_{o\omega\omega}=_\omega=_\omega$:= $A5200t$ T
§s %0 1 %1
$\$L5222\,T$:= $\$T5222$

case F
§= $\$F5222$
$= \$F5222\,\$F5222$
§\ $\$F5222$
$= \$F5222\,(=(\supset T\,F)\,(\vee\,(\sim T)\,F))$
§s %1 3 %0

\# $= \$F5222\,(=(\supset T\,F)\,(\vee\,(\sim T)\,F))$
$\%A5231a$
\# $=(\sim T)\,F$:= $A5231a$
\# $=_{ooo}(\sim_{oo}T_o)F_o$:= $A5231a$
§s %1 29 %0
\# $= \$F5222\,(=(\supset T\,F)\,(\vee\,F\,F))$
$\%A5228b$
\# $=(\supset T\,F)\,F$:= $A5228b$
\# $=_{ooo}(\supset_{ooo}T_oF_o)F_o$:= $A5228b$
§s %1 13 %0
\# $= \$F5222\,(=F\,(\vee\,F\,F))$
$\%A5232d$
\# $=(\vee\,F\,F)\,F$:= $A5232d$
\# $=_{ooo}(\vee_{ooo}F_oF_o)F_o$:= $A5232d$
§s %1 7 %0
\# $= \$F5222\,(=F\,F)$
$\%A5230d$
\# $=(=F\,F)\,T$:= $A5230d$
\# $=_{ooo}(=_{ooo}F_oF_o)T_o$:= $A5230d$
§s %1 3 %0
\# $= \$F5222\,T$
\#\# use Proof Template A5201b (Swap): A = B → B = A
<< A5201b.r0t.txt
%0
\# $=T\,\$F5222$
\# $=_{o\omega\omega}T_\omega\$F5222_\omega$
$\%T$
\# $===$:= $A5200t\quad T$
\# $=_{o\omega\omega}=_\omega=_\omega$:= $A5200t\quad T$
§s %0 1 %1
\# $\$L5222\,F$:= $\$F5222$

<< A5222.r0t.txt
:= $\$L5222$
:= $\$X5222$
:= $\$T5222$
:= $\$F5222$

:= $\$TTMP8022$ %0
\# wff 1886 : $=(\supset T\,y)\,(\vee\,(\sim T)\,y)_{o,\ldots}$:= $\$TTMP8022$

\#\# .2: main case F

\#\# use Proof Template A5222 (Rule of Cases): [\x.A]T, [\x.A]F → A
§\ $[\lambda x_o.[\lambda y_o.(=_{ooo}(\supset_{ooo}x_oy_o)(\vee_{ooo}(\sim_{oo}x_o)y_o))_o]_{(oo)}]F_o$
\# $=([\lambda x.[\lambda y.(=(\supset x\,y)\,(\vee\,(\sim x)\,y))]]\,F)\,[\lambda y.(=(\supset F\,y)\,(\vee\,(\sim F)\,y))]$
:= $\$L5222$ %0/3
\# wff 1991 : $[\lambda y.(=(\supset F\,y)\,(\vee\,(\sim F)\,y))]_{oo,\ldots}$:= $\$L5222$
:= $\$X5222\ y_o$

\# wff 34 : y_o := $\$X5222$
:= $\$T5222$ $\$L5222_{oo}T_o$
\# wff 1993 : $\$L5222\,T_o$:= $\$T5222$
:= $\$F5222$ $\$L5222_{oo}F_o$
\# wff 1994 : $\$L5222\,F_o$:= $\$F5222$

\## case T
§= $\$T5222$
\# $=\$T5222\,\$T5222$
§\ $\$T5222$
\# $=\$T5222\,(=(\supset F\,T)\,(\vee\,(\sim F)\,T))$
§s %1 3 %0
\# $=\$T5222\,(=(\supset F\,T)\,(\vee\,(\sim F)\,T))$
%$A5231b$
\# $=(\sim F)\,T$:= $A5231b$
\# $=_{ooo}(\sim_{oo}F_o)T_o$:= $A5231b$
§s %1 29 %0
\# $=\$T5222\,(=(\supset F\,T)\,(\vee\,T\,T))$
%$A5228c$
\# $=(\supset F\,T)\,T$:= $A5228c$
\# $=_{ooo}(\supset_{ooo}F_oT_o)T_o$:= $A5228c$
§s %1 13 %0
\# $=\$T5222\,(=T\,(\vee\,T\,T))$
%$A5232a$
\# $=(\vee\,T\,T)\,T$:= $A5232a$
\# $=_{ooo}(\vee_{ooo}T_oT_o)T_o$:= $A5232a$
§s %1 7 %0
\# $=\$T5222\,(=T\,T)$
%$A5230a$
\# $=(=T\,T)\,T$:= $A5230a$
\# $=_{ooo}(=_{ooo}T_oT_o)T_o$:= $A5230a$
§s %1 3 %0
\# $=\$T5222\,T$
\## use Proof Template A5201b (Swap): A = B → B = A
<< A5201b.r0t.txt
%0
\# $=T\,\$T5222$
\# $=_{o\omega\omega}T_\omega\$T5222_\omega$
%T
\# $===$:= $A5200t$ T
\# $=_{o\omega\omega}=_\omega=_\omega$:= $A5200t$ T
§s %0 1 %1
\# $\$L5222\,T$:= $\$T5222$

\## case F
§= $\$F5222$
\# $=\$F5222\,\$F5222$
§\ $\$F5222$
\# $=\$F5222\,(=(\supset F\,F)\,(\vee\,(\sim F)\,F))$

§s %1 3 %0
$= \$F5222\,(=(\supset F\,F)\,(\vee\,(\sim F)\,F))$
%$A5231b$
$= (\sim F)\,T$:= $A5231b$
$=_{ooo}(\sim_{oo}F_o)T_o$:= $A5231b$
§s %1 29 %0
$= \$F5222\,(=(\supset F\,F)\,(\vee\,T\,F))$
%$A5228d$
$= (\supset F\,F)\,T$:= $A5228d$
$=_{ooo}(\supset_{ooo}F_oF_o)T_o$:= $A5228d$
§s %1 13 %0
$= \$F5222\,(=T\,(\vee\,T\,F))$
%$A5232b$
$= (\vee\,T\,F)\,T$:= $A5232b$
$=_{ooo}(\vee_{ooo}T_oF_o)T_o$:= $A5232b$
§s %1 7 %0
$= \$F5222\,(=T\,T)$
%$A5230a$
$= (=T\,T)\,T$:= $A5230a$
$=_{ooo}(=_{ooo}T_oT_o)T_o$:= $A5230a$
§s %1 3 %0
$= \$F5222\,T$
use Proof Template A5201b (Swap): A = B → B = A
<< A5201b.r0t.txt
%0
$= T\,\$F5222$
$=_{o\omega\omega}T_\omega\$F5222_\omega$
%T
$===$:= $A5200t$ T
$=_{o\omega\omega}=_\omega=_\omega$:= $A5200t$ T
§s %0 1 %1
$\$L5222\,F$:= $\$F5222$

<< A5222.r0t.txt
:= $\$L5222$
:= $\$X5222$
:= $\$T5222$
:= $\$F5222$

:= $\$FTMP8022$ %0
wff 1990 : $= (\supset F\,y)\,(\vee\,(\sim F)\,y)_{o,\dots}$:= $\$FTMP8022$

.3

use Proof Template A5222 (Rule of Cases): [\x.A]T, [\x.A]F → A
:= $\$L5222$ $[\lambda x_o.(=_{ooo}(\supset_{ooo}x_oy_o)(\vee_{ooo}(\sim_{oo}x_o)y_o))_o]$
wff 2086 : $[\lambda x.(=(\supset x\,y)\,(\vee\,(\sim x)\,y))]_{oo}$:= $\$L5222$
:= $\$X5222$ x_o
wff 16 : x_o := $\$X5222$

```
:=  $T5222  $L5222_oo T_o
 # wff    2087  :    $L5222 T_o    :=  $T5222
:=  $F5222  $L5222_oo F_o
 # wff    2088  :    $L5222 F_o    :=  $F5222

## case T
§=  $T5222
#               = $T5222 $T5222
§\  $T5222
#               = $T5222 $TTMP8022
§s  %1  3  %0
#               = $T5222 $TTMP8022
## use Proof Template A5201b (Swap):   A = B   →   B = A
<< A5201b.r0t.txt
%0
#               = $TTMP8022 $T5222
#               =_oωω $TTMP8022_ω $T5222_ω
%$TTMP8022
#               = (⊃ T y) (∨ (∼ T) y)     :=  $TTMP8022
#               =_ooo(⊃_ooo T_o y_o)(∨_ooo(∼_oo T_o) y_o)     :=  $TTMP8022
:=  $TTMP8022
§s  %0  1  %1
#               $L5222 T     :=  $T5222

## case F
§=  $F5222
#               = $F5222 $F5222
§\  $F5222
#               = $F5222 $FTMP8022
§s  %1  3  %0
#               = $F5222 $FTMP8022
## use Proof Template A5201b (Swap):   A = B   →   B = A
<< A5201b.r0t.txt
%0
#               = $FTMP8022 $F5222
#               =_oωω $FTMP8022_ω $F5222_ω
%$FTMP8022
#               = (⊃ F y) (∨ (∼ F) y)     :=  $FTMP8022
#               =_ooo(⊃_ooo F_o y_o)(∨_ooo(∼_oo F_o) y_o)     :=  $FTMP8022
:=  $FTMP8022
§s  %0  1  %1
#               $L5222 F     :=  $F5222

<< A5222.r0t.txt
:=  $L5222
:=  $X5222
:=  $T5222
:=  $F5222
%0
```

\# $= (\supset x\, y)\, (\vee\, (\sim x)\, y)$
\# $=_{ooo}(\supset_{ooo} x_o y_o)(\vee_{ooo}(\sim_{oo} x_o) y_o)$

\#\# .4

§= $_{ooo}$ $[\lambda x_o.[\lambda y_o.(\supset_{ooo} x_o y_o)_o]_{(oo)}]$
\# $= [\lambda x.[\lambda y.(\supset x\, y)]]\, [\lambda x.[\lambda y.(\supset x\, y)]]$
§s %0 15 %1
\# $= [\lambda x.[\lambda y.(\supset x\, y)]]\, [\lambda x.[\lambda y.(\vee\, (\sim x)\, y)]]$
:= $\$TMP8022$ %0
\# wff 2167 : $= [\lambda x.[\lambda y.(\supset x\, y)]]\, [\lambda x.[\lambda y.(\vee\, (\sim x)\, y)]]_o$:= $\$TMP8022$

\#\# use Proof Template: A5205 Substitutions
:= $\$AA5205$ o
\# wff 2 : o_τ := $\$AA5205$
:= $\$BA5205$ o
\# wff 2 : o_τ := $\$AA5205$ $\$BA5205$
:= $\$FA5205$ $\supset_{ooo} x_o$
\# wff 1873 : $\supset x_{oo}$:= $\$FA5205$
<< a5205_substitutions.r0t.txt
:= $\$AA5205$
:= $\$BA5205$
:= $\$FA5205$
%0
\# $= (\supset x)\, [\lambda y.(\supset x\, y)]$
\# $=_{o(oo)(oo)}(\supset_{ooo} x_o)[\lambda y_o.(\supset_{ooo} x_o y_o)_o]$

\#\# use Proof Template A5201b (Swap): A = B → B = A
<< A5201b.r0t.txt
%0
\# $= [\lambda y.(\supset x\, y)]\, (\supset x)$
\# $=_{o(oo)(oo)}[\lambda y_o.(\supset_{ooo} x_o y_o)_o](\supset_{ooo} x_o)$
%$\$TMP8022$
\# $= [\lambda x.[\lambda y.(\supset x\, y)]]\, [\lambda x.[\lambda y.(\vee\, (\sim x)\, y)]]$:= $\$TMP8022$
\# $=_{o(ooo)(ooo)}[\lambda x_o.[\lambda y_o.(\supset_{ooo} x_o y_o)_o]_{(oo)}][\lambda x_o.[\lambda y_o.(\vee_{ooo}(\sim_{oo} x_o) y_o)_o]_{(oo)}]$:=
$\$TMP8022$
:= $\$TMP8022$
§s %0 11 %1
\# $= [\lambda x.(\supset x)]\, [\lambda x.[\lambda y.(\vee\, (\sim x)\, y)]]$
:= $\$TMP8022$ %0
\# wff 2289 : $= [\lambda x.(\supset x)]\, [\lambda x.[\lambda y.(\vee\, (\sim x)\, y)]]_o$:= $\$TMP8022$

\#\# use Proof Template: A5205 Substitutions
:= $\$AA5205$ oo
\# wff 13 : oo_τ := $\$AA5205$
:= $\$BA5205$ o
\# wff 2 : o_τ := $\$BA5205$
:= $\$FA5205$ $[\lambda x_o.[\lambda y_o.(=_{\$AA5205\, o} x_o(\wedge_{\$AA5205\, o} x_o y_o))_o]_{\$AA5205}]$
\# wff 53 : $[\lambda x.[\lambda y.(= x\, (\wedge\, x\, y))]]_{\$AA5205\, o}$:= $\$FA5205$ $\supset$

<< a5205_substitutions.r0t.txt
:= $\$AA5205$
:= $\$BA5205$
:= $\$FA5205$
%0
\# $= \supset [\lambda y.(\supset y)]$
\# $=_{o(ooo)(ooo)} \supset_{ooo} [\lambda y_o.(\supset_{ooo} y_o)_{(oo)}]$

§r /3 x_o
\# $= [\lambda y.(\supset y)]\, [\lambda x.(\supset x)]$
§s %1 3 %0
\# $= \supset [\lambda x.(\supset x)]$
\## use Proof Template A5201b (Swap): A = B → B = A
<< A5201b.r0t.txt
%0
\# $= [\lambda x.(\supset x)] \supset$
\# $=_{o(ooo)(ooo)} [\lambda x_o.(\supset_{ooo} x_o)_{(oo)}] \supset_{ooo}$
%$\$TMP8022$
\# $= [\lambda x.(\supset x)]\, [\lambda x.[\lambda y.(\vee (\sim x)\, y)]]$:= $\$TMP8022$
\# $=_{o(ooo)(ooo)} [\lambda x_o.(\supset_{ooo} x_o)_{(oo)}] [\lambda x_o.[\lambda y_o.(\vee_{ooo} (\sim_{oo} x_o) y_o)_o]_{(oo)}]$:= $\$TMP8022$
:= $\$TMP8022$
§s %0 5 %1
\# $= \supset [\lambda x.[\lambda y.(\vee (\sim x)\, y)]]$

:= $K8022$ %0
\# wff 2404 : $= \supset [\lambda x.[\lambda y.(\vee (\sim x)\, y)]]_o$:= $K8022$

\##
\## Q.E.D.
\##

%0
\# $= \supset [\lambda x.[\lambda y.(\vee (\sim x)\, y)]]$:= $K8022$
\# $=_{o(ooo)(ooo)} \supset_{ooo} [\lambda x_o.[\lambda y_o.(\vee_{ooo} (\sim_{oo} x_o) y_o)_o]_{(oo)}]$:= $K8022$

2.1.79 Results for File K8023.r0.txt

\##
\## Proof K8023: (A ∨ B) ∨ C = A ∨ (B ∨ C)
\##
\##
\## Source: [Kubota 2017 (doi: 10.4444/100.10)]
\##

\## Written by Ken Kubota (<mail@kenkubota.de>).
\##
\## This file is part of the publication of the mathematical logic $\mathcal{R}_0$.
\## For more information, visit: <http://doi.org/10.4444/100.10>

```
##

<< basics.r0.txt
<< A5200t.r0.txt
<< A5230.r0.txt
<< A5232.r0.txt


##
##  Proof
##
```

:= $\$LTMP8023$ $[\lambda x_o.[\lambda y_o.[\lambda z_o.(=_{ooo}(\vee_{ooo}(\vee_{ooo}x_o y_o)z_o)(\vee_{ooo}x_o(\vee_{ooo}y_o z_o)))_o]_{(oo)}]_{(ooo)}]$
wff 1679 : $[\lambda x.[\lambda y.[\lambda z.(=(\vee(\vee x\, y)\, z)\,(\vee x\,(\vee y\, z)))]]]_{oooo}$:= $\$LTMP8023$

.1: Subcase TT

:= $\$TTTMP8023$ $\$LTMP8023_{oooo}T_o T_o$
wff 1682 : $\$LTMP8023\, T\, T_{oo}$:= $\$TTTMP8023$
§= $\$TTTMP8023$
$= \$TTTMP8023\, \$TTTMP8023$
§\ $\$LTMP8023_{oooo}T_o$
$= (\$LTMP8023\, T)\, [\lambda y.[\lambda z.(=(\vee(\vee T\, y)\, z)\,(\vee T\,(\vee y\, z)))]]$
§s %1 6 %0
$= \$TTTMP8023\, ([\lambda y.[\lambda z.(=(\vee(\vee T\, y)\, z)\,(\vee T\,(\vee y\, z)))]]\, T)$
§\ $[\lambda y_o.[\lambda z_o.(=_{ooo}(\vee_{ooo}(\vee_{ooo}T_o y_o)z_o)(\vee_{ooo}T_o(\vee_{ooo}y_o z_o)))_o]_{(oo)}]T_o$
$= ([\lambda y.[\lambda z.(=(\vee(\vee T\, y)\, z)\,(\vee T\,(\vee y\, z)))]]\, T)\, [\lambda z.(=(\vee(\vee T\, T)\, z)\,(\vee T\,(\vee T\, z)))]$
§s %1 3 %0
$= \$TTTMP8023\, [\lambda z.(=(\vee(\vee T\, T)\, z)\,(\vee T\,(\vee T\, z)))]$

use Proof Template A5222 (Rule of Cases): [\x.A]T, [\x.A]F → A
:= $\$L5222$ %0/3
wff 1704 : $[\lambda z.(=(\vee(\vee T\, T)\, z)\,(\vee T\,(\vee T\, z)))]_{oo,\ldots}$:= $\$L5222$
:= $\$X5222$ z_o
wff 1667 : z_o := $\$X5222$
:= $\$T5222$ $\$L5222_{oo}T_o$
wff 1707 : $\$L5222\, T_o$:= $\$T5222$
:= $\$F5222$ $\$L5222_{oo}F_o$
wff 1708 : $\$L5222\, F_o$:= $\$F5222$

case T
§= $\$T5222$
$= \$T5222\, \$T5222$
§\ $\$T5222$
$= \$T5222\, (=(\vee(\vee T\, T)\, T)\,(\vee T\,(\vee T\, T)))$
§s %1 3 %0
$= \$T5222\, (=(\vee(\vee T\, T)\, T)\,(\vee T\,(\vee T\, T)))$
%$A5232a$
$= (\vee T\, T)\, T$:= $A5232a$

\# $=_{ooo}(\vee_{ooo}T_oT_o)T_o$:= $A5232a$
§s %1 53 %0
\# $= \$T5222\,(=(\vee\,T\,T)\,(\vee\,T\,(\vee\,T\,T)))$
%$A5232a$
\# $=(\vee\,T\,T)\,T$:= $A5232a$
\# $=_{ooo}(\vee_{ooo}T_oT_o)T_o$:= $A5232a$
§s %1 13 %0
\# $= \$T5222\,(=T\,(\vee\,T\,(\vee\,T\,T)))$
%$A5232a$
\# $=(\vee\,T\,T)\,T$:= $A5232a$
\# $=_{ooo}(\vee_{ooo}T_oT_o)T_o$:= $A5232a$
§s %1 15 %0
\# $= \$T5222\,(=T\,(\vee\,T\,T))$
%$A5232a$
\# $=(\vee\,T\,T)\,T$:= $A5232a$
\# $=_{ooo}(\vee_{ooo}T_oT_o)T_o$:= $A5232a$
§s %1 7 %0
\# $= \$T5222\,(=T\,T)$
%$A5230a$
\# $=(=T\,T)\,T$:= $A5230a$
\# $=_{ooo}(=_{ooo}T_oT_o)T_o$:= $A5230a$
§s %1 3 %0
\# $= \$T5222\,T$
\## use Proof Template A5201b (Swap): A = B → B = A
<< A5201b.r0t.txt
%0
\# $=T\,\$T5222$
\# $=_{o\omega\omega}T_\omega \$T5222_\omega$
%T
\# $===$:= $A5200t$ T
\# $=_{o\omega\omega}=_\omega=_\omega$:= $A5200t$ T
§s %0 1 %1
\# $\$L5222\,T$:= $\$T5222$

\## case F
§= $\$F5222$
\# $= \$F5222\,\$F5222$
§\ $\$F5222$
\# $= \$F5222\,(=(\vee\,(\vee\,T\,T)\,F)\,(\vee\,T\,(\vee\,T\,F)))$
§s %1 3 %0
\# $= \$F5222\,(=(\vee\,(\vee\,T\,T)\,F)\,(\vee\,T\,(\vee\,T\,F)))$
%$A5232a$
\# $=(\vee\,T\,T)\,T$:= $A5232a$
\# $=_{ooo}(\vee_{ooo}T_oT_o)T_o$:= $A5232a$
§s %1 53 %0
\# $= \$F5222\,(=(\vee\,T\,F)\,(\vee\,T\,(\vee\,T\,F)))$
%$A5232b$
\# $=(\vee\,T\,F)\,T$:= $A5232b$
\# $=_{ooo}(\vee_{ooo}T_oF_o)T_o$:= $A5232b$

§s %1 13 %0
$= \$F5222\,(= T\,(\vee T\,(\vee T\,F)))$
%$A5232b$
$= (\vee T\,F)\,T$:= $A5232b$
$=_{ooo}(\vee_{ooo}T_oF_o)T_o$:= $A5232b$
§s %1 15 %0
$= \$F5222\,(= T\,(\vee T\,T))$
%$A5232a$
$= (\vee T\,T)\,T$:= $A5232a$
$=_{ooo}(\vee_{ooo}T_oT_o)T_o$:= $A5232a$
§s %1 7 %0
$= \$F5222\,(= T\,T)$
%$A5230a$
$= (= T\,T)\,T$:= $A5230a$
$=_{ooo}(=_{ooo}T_oT_o)T_o$:= $A5230a$
§s %1 3 %0
$= \$F5222\,T$
use Proof Template A5201b (Swap): A = B → B = A
<< A5201b.r0t.txt
%0
$= T\,\$F5222$
$=_{o\omega\omega}T_\omega\$F5222_\omega$
%T
$= = =$:= $A5200t$ T
$=_{o\omega\omega}=_\omega=_\omega$:= $A5200t$ T
§s %0 1 %1
$\$L5222\,F$:= $\$F5222$

<< A5222.r0t.txt
:= $\$L5222$
:= $\$X5222$
:= $\$T5222$
:= $\$F5222$

:= $\$TTTMP8023$
:= $\$TTTMP8023$ %0
wff 1703 : $= (\vee\,(\vee T\,T)\,z)\,(\vee T\,(\vee T\,z))_{o,\ldots}$:= $\$TTTMP8023$

.2: Subcase TF

:= $\$TFTMP8023$ $\$LTMP8023_{oooo}T_oF_o$
wff 1814 : $\$LTMP8023\,T\,F_{oo}$:= $\$TFTMP8023$
§= $\$TFTMP8023$
$= \$TFTMP8023\,\$TFTMP8023$
§\ $\$LTMP8023_{oooo}T_o$
$= (\$LTMP8023\,T)\,[\lambda y.[\lambda z.(= (\vee\,(\vee T\,y)\,z)\,(\vee T\,(\vee y\,z)))]]$
§s %1 6 %0
$= \$TFTMP8023\,([\lambda y.[\lambda z.(= (\vee\,(\vee T\,y)\,z)\,(\vee T\,(\vee y\,z)))]]\,F)$
§\ $[\lambda y_o.[\lambda z_o.(=_{ooo}(\vee_{ooo}(\vee_{ooo}T_oy_o)z_o)(\vee_{ooo}T_o(\vee_{ooo}y_oz_o)))_o]_{(oo)}]F_o$

\# $= ([\lambda y.[\lambda z.(= (\vee (\vee T\, y)\, z)\, (\vee T\, (\vee y\, z)))]]\, F)\, [\lambda z.(= (\vee (\vee T\, F)\, z)\, (\vee T\, (\vee F\, z)))]$
§s %1 3 %0
\# $= \$TFTMP8023\, [\lambda z.(= (\vee (\vee T\, F)\, z)\, (\vee T\, (\vee F\, z)))]$

\## use Proof Template A5222 (Rule of Cases): [\x.A]T, [\x.A]F → A
:= $\$L5222$ %0/3
\# wff 1826 : $[\lambda z.(= (\vee (\vee T\, F)\, z)\, (\vee T\, (\vee F\, z)))]_{oo,\ldots}$:= $\$L5222$
:= $\$X5222$ z_o
\# wff 1667 : z_o := $\$X5222$
:= $\$T5222$ $\$L5222_{oo}T_o$
\# wff 1829 : $\$L5222\, T_o$:= $\$T5222$
:= $\$F5222$ $\$L5222_{oo}F_o$
\# wff 1830 : $\$L5222\, F_o$:= $\$F5222$

\## case T
§= $\$T5222$
\# $= \$T5222\, \$T5222$
§\ $\$T5222$
\# $= \$T5222\, (= (\vee (\vee T\, F)\, T)\, (\vee T\, (\vee F\, T)))$
§s %1 3 %0
\# $= \$T5222\, (= (\vee (\vee T\, F)\, T)\, (\vee T\, (\vee F\, T)))$
%$A5232b$
\# $= (\vee T\, F)\, T$:= $A5232b$
\# $=_{ooo}(\vee_{ooo}T_oF_o)T_o$:= $A5232b$
§s %1 53 %0
\# $= \$T5222\, (= (\vee T\, T)\, (\vee T\, (\vee F\, T)))$
%$A5232a$
\# $= (\vee T\, T)\, T$:= $A5232a$
\# $=_{ooo}(\vee_{ooo}T_oT_o)T_o$:= $A5232a$
§s %1 13 %0
\# $= \$T5222\, (= T\, (\vee T\, (\vee F\, T)))$
%$A5232c$
\# $= (\vee F\, T)\, T$:= $A5232c$
\# $=_{ooo}(\vee_{ooo}F_oT_o)T_o$:= $A5232c$
§s %1 15 %0
\# $= \$T5222\, (= T\, (\vee T\, T))$
%$A5232a$
\# $= (\vee T\, T)\, T$:= $A5232a$
\# $=_{ooo}(\vee_{ooo}T_oT_o)T_o$:= $A5232a$
§s %1 7 %0
\# $= \$T5222\, (= T\, T)$
%$A5230a$
\# $= (= T\, T)\, T$:= $A5230a$
\# $=_{ooo}(=_{ooo}T_oT_o)T_o$:= $A5230a$
§s %1 3 %0
\# $= \$T5222\, T$
\## use Proof Template A5201b (Swap): A = B → B = A
<< A5201b.r0t.txt
%0

```
#           = T $T5222
#           =_{oωω} T_ω $T5222_ω
%T
#           = = =          :=   A5200t   T
#           =_{oωω} =_ω =_ω     :=   A5200t   T
§s %0 1 %1
#           $L5222 T        :=   $T5222

## case F
§= $F5222
#           = $F5222 $F5222
§\ $F5222
#           = $F5222 (= (∨ (∨ T F) F) (∨ T (∨ F F)))
§s %1 3 %0
#           = $F5222 (= (∨ (∨ T F) F) (∨ T (∨ F F)))
%A5232b
#           = (∨ T F) T        :=   A5232b
#           =_{ooo}(∨_{ooo} T_o F_o) T_o     :=   A5232b
§s %1 53 %0
#           = $F5222 (= (∨ T F) (∨ T (∨ F F)))
%A5232b
#           = (∨ T F) T        :=   A5232b
#           =_{ooo}(∨_{ooo} T_o F_o) T_o     :=   A5232b
§s %1 13 %0
#           = $F5222 (= T (∨ T (∨ F F)))
%A5232d
#           = (∨ F F) F        :=   A5232d
#           =_{ooo}(∨_{ooo} F_o F_o) F_o     :=   A5232d
§s %1 15 %0
#           = $F5222 (= T (∨ T F))
%A5232b
#           = (∨ T F) T        :=   A5232b
#           =_{ooo}(∨_{ooo} T_o F_o) T_o     :=   A5232b
§s %1 7 %0
#           = $F5222 (= T T)
%A5230a
#           = (= T T) T        :=   A5230a
#           =_{ooo}(=_{ooo} T_o T_o) T_o     :=   A5230a
§s %1 3 %0
#           = $F5222 T
## use Proof Template A5201b (Swap):   A = B   →   B = A
<< A5201b.r0t.txt
%0
#           = T $F5222
#           =_{oωω} T_ω $F5222_ω
%T
#           = = =          :=   A5200t   T
#           =_{oωω} =_ω =_ω     :=   A5200t   T
§s %0 1 %1
```

$\$L5222\,F \quad := \quad \$F5222$

<< A5222.r0t.txt
:= $\$L5222$
:= $\$X5222$
:= $\$T5222$
:= $\$F5222$

:= $\$TFTMP8023$
:= $\$TFTMP8023$ %0
wff 1825 : $=(\vee\,(\vee\,T\,F)\,z)\,(\vee\,T\,(\vee\,F\,z))_{o,\ldots}$:= $\$TFTMP8023$

.3: Subcase FT

:= $\$FTTMP8023$ $\$LTMP8023_{oooo}F_oT_o$
wff 1933 : $\$LTMP8023\,F\,T_{oo}$:= $\$FTTMP8023$
§= $\$FTTMP8023$
$=\$FTTMP8023\,\$FTTMP8023$
§\ $\$LTMP8023_{oooo}F_o$
$=(\$LTMP8023\,F)\,[\lambda y.[\lambda z.(=(\vee\,(\vee\,F\,y)\,z)\,(\vee\,F\,(\vee\,y\,z)))]]$
§s %1 6 %0
$=\$FTTMP8023\,([\lambda y.[\lambda z.(=(\vee\,(\vee\,F\,y)\,z)\,(\vee\,F\,(\vee\,y\,z)))]]\,T)$
§\ $[\lambda y_o.[\lambda z_o.(=_{ooo}(\vee_{ooo}(\vee_{ooo}F_oy_o)z_o)(\vee_{ooo}F_o(\vee_{ooo}y_oz_o)))_o]_{(oo)}]T_o$
$=([\lambda y.[\lambda z.(=(\vee\,(\vee\,F\,y)\,z)\,(\vee\,F\,(\vee\,y\,z)))]]\,T)\,[\lambda z.(=(\vee\,(\vee\,F\,T)\,z)\,(\vee\,F\,(\vee\,T\,z)))]$
§s %1 3 %0
$=\$FTTMP8023\,[\lambda z.(=(\vee\,(\vee\,F\,T)\,z)\,(\vee\,F\,(\vee\,T\,z)))]$

use Proof Template A5222 (Rule of Cases): [\x.A]T, [\x.A]F → A
:= $\$L5222$ %0/3
wff 1954 : $[\lambda z.(=(\vee\,(\vee\,F\,T)\,z)\,(\vee\,F\,(\vee\,T\,z)))]_{oo,\ldots}$:= $\$L5222$
:= $\$X5222$ z_o
wff 1667 : z_o := $\$X5222$
:= $\$T5222$ $\$L5222_{oo}T_o$
wff 1957 : $\$L5222\,T_o$:= $\$T5222$
:= $\$F5222$ $\$L5222_{oo}F_o$
wff 1958 : $\$L5222\,F_o$:= $\$F5222$

case T
§= $\$T5222$
$=\$T5222\,\$T5222$
§\ $\$T5222$
$=\$T5222\,(=(\vee\,(\vee\,F\,T)\,T)\,(\vee\,F\,(\vee\,T\,T)))$
§s %1 3 %0
$=\$T5222\,(=(\vee\,(\vee\,F\,T)\,T)\,(\vee\,F\,(\vee\,T\,T)))$
%$A5232c$
$=(\vee\,F\,T)\,T$:= $A5232c$
$=_{ooo}(\vee_{ooo}F_oT_o)T_o$:= $A5232c$
§s %1 53 %0
$=\$T5222\,(=(\vee\,T\,T)\,(\vee\,F\,(\vee\,T\,T)))$

%$A5232a$
$= (\vee\, T\, T)\, T \quad := \quad A5232a$
$=_{ooo}(\vee_{ooo} T_o T_o) T_o \quad := \quad A5232a$
§s %1 13 %0
$= \$T5222\, (= T\, (\vee\, F\, (\vee\, T\, T)))$
%$A5232a$
$= (\vee\, T\, T)\, T \quad := \quad A5232a$
$=_{ooo}(\vee_{ooo} T_o T_o) T_o \quad := \quad A5232a$
§s %1 15 %0
$= \$T5222\, (= T\, (\vee\, F\, T))$
%$A5232c$
$= (\vee\, F\, T)\, T \quad := \quad A5232c$
$=_{ooo}(\vee_{ooo} F_o T_o) T_o \quad := \quad A5232c$
§s %1 7 %0
$= \$T5222\, (= T\, T)$
%$A5230a$
$= (= T\, T)\, T \quad := \quad A5230a$
$=_{ooo}(=_{ooo} T_o T_o) T_o \quad := \quad A5230a$
§s %1 3 %0
$= \$T5222\, T$
use Proof Template A5201b (Swap): A = B $\to$ B = A
<< A5201b.r0t.txt
%0
$= T\, \$T5222$
$=_{o\omega\omega} T_\omega \$T5222_\omega$
%T
$= = = \quad := \quad A5200t \quad T$
$=_{o\omega\omega} =_\omega =_\omega \quad := \quad A5200t \quad T$
§s %0 1 %1
$\$L5222\, T \quad := \quad \$T5222$

case F
§= $\$F5222$
$= \$F5222\, \$F5222$
§\ $\$F5222$
$= \$F5222\, (= (\vee\, (\vee\, F\, T)\, F)\, (\vee\, F\, (\vee\, T\, F)))$
§s %1 3 %0
$= \$F5222\, (= (\vee\, (\vee\, F\, T)\, F)\, (\vee\, F\, (\vee\, T\, F)))$
%$A5232c$
$= (\vee\, F\, T)\, T \quad := \quad A5232c$
$=_{ooo}(\vee_{ooo} F_o T_o) T_o \quad := \quad A5232c$
§s %1 53 %0
$= \$F5222\, (= (\vee\, T\, F)\, (\vee\, F\, (\vee\, T\, F)))$
%$A5232b$
$= (\vee\, T\, F)\, T \quad := \quad A5232b$
$=_{ooo}(\vee_{ooo} T_o F_o) T_o \quad := \quad A5232b$
§s %1 13 %0
$= \$F5222\, (= T\, (\vee\, F\, (\vee\, T\, F)))$
%$A5232b$

\# $= (\vee\, T\, F)\, T$:= $A5232b$
\# $=_{ooo}(\vee_{ooo}T_oF_o)T_o$:= $A5232b$
§s %1 15 %0
\# $= \$F5222\, (= T\, (\vee\, F\, T))$
%$A5232c$
\# $= (\vee\, F\, T)\, T$:= $A5232c$
\# $=_{ooo}(\vee_{ooo}F_oT_o)T_o$:= $A5232c$
§s %1 7 %0
\# $= \$F5222\, (= T\, T)$
%$A5230a$
\# $= (= T\, T)\, T$:= $A5230a$
\# $=_{ooo}(=_{ooo}T_oT_o)T_o$:= $A5230a$
§s %1 3 %0
\# $= \$F5222\, T$
\## use Proof Template A5201b (Swap): A = B → B = A
<< A5201b.r0t.txt
%0
\# $= T\, \$F5222$
\# $=_{o\omega\omega}T_\omega \$F5222_\omega$
%T
\# $= = =$:= $A5200t$ T
\# $=_{o\omega\omega}=_\omega=_\omega$:= $A5200t$ T
§s %0 1 %1
\# $\$L5222\, F$:= $\$F5222$

<< A5222.r0t.txt
:= $\$L5222$
:= $\$X5222$
:= $\$T5222$
:= $\$F5222$

:= $\$FTTMP8023$
:= $\$FTTMP8023$ %0
\# wff 1953 : $= (\vee\, (\vee\, F\, T)\, z)\, (\vee\, F\, (\vee\, T\, z))_{o,\ldots}$:= $\$FTTMP8023$

\## .4: Subcase FF

:= $\$FFTMP8023$ $\$LTMP8023_{oooo}F_oF_o$
\# wff 2059 : $\$LTMP8023\, F\, F_{oo}$:= $\$FFTMP8023$
§= $\$FFTMP8023$
\# $= \$FFTMP8023\, \$FFTMP8023$
§\ $\$LTMP8023_{oooo}F_o$
\# $= (\$LTMP8023\, F)\, [\lambda y.[\lambda z.(= (\vee\, (\vee\, F\, y)\, z)\, (\vee\, F\, (\vee\, y\, z)))]]$
§s %1 6 %0
\# $= \$FFTMP8023\, ([\lambda y.[\lambda z.(= (\vee\, (\vee\, F\, y)\, z)\, (\vee\, F\, (\vee\, y\, z)))]]\, F)$
§\ $[\lambda y_o.[\lambda z_o.(=_{ooo}(\vee_{ooo}(\vee_{ooo}F_oy_o)z_o)(\vee_{ooo}F_o(\vee_{ooo}y_oz_o)))_o]_{(oo)}]F_o$
\# $= ([\lambda y.[\lambda z.(= (\vee\, (\vee\, F\, y)\, z)\, (\vee\, F\, (\vee\, y\, z)))]]\, F)\, [\lambda z.(= (\vee\, (\vee\, F\, F)\, z)\, (\vee\, F\, (\vee\, F\, z)))]$
§s %1 3 %0
\# $= \$FFTMP8023\, [\lambda z.(= (\vee\, (\vee\, F\, F)\, z)\, (\vee\, F\, (\vee\, F\, z)))]$

use Proof Template A5222 (Rule of Cases): [\x.A]T, [\x.A]F $\rightarrow$ A
:= $\$L5222$ %0/3
wff 2070 : $[\lambda z.(=(\vee(\vee F\,F)\,z)\,(\vee F\,(\vee F\,z)))]_{oo,\ldots}$:= $\$L5222$
:= $\$X5222$ z_o
wff 1667 : z_o := $\$X5222$
:= $\$T5222$ $\$L5222_{oo}T_o$
wff 2073 : $\$L5222\,T_o$:= $\$T5222$
:= $\$F5222$ $\$L5222_{oo}F_o$
wff 2074 : $\$L5222\,F_o$:= $\$F5222$

case T
§= $\$T5222$
$=\$T5222\,\$T5222$
§\ $\$T5222$
$=\$T5222\,(=(\vee(\vee F\,F)\,T)\,(\vee F\,(\vee F\,T)))$
§s %1 3 %0
$=\$T5222\,(=(\vee(\vee F\,F)\,T)\,(\vee F\,(\vee F\,T)))$
%$A5232d$
$=(\vee F\,F)\,F$:= $A5232d$
$=_{ooo}(\vee_{ooo}F_oF_o)F_o$:= $A5232d$
§s %1 53 %0
$=\$T5222\,(=(\vee F\,T)\,(\vee F\,(\vee F\,T)))$
%$A5232c$
$=(\vee F\,T)\,T$:= $A5232c$
$=_{ooo}(\vee_{ooo}F_oT_o)T_o$:= $A5232c$
§s %1 13 %0
$=\$T5222\,(=T\,(\vee F\,(\vee F\,T)))$
%$A5232c$
$=(\vee F\,T)\,T$:= $A5232c$
$=_{ooo}(\vee_{ooo}F_oT_o)T_o$:= $A5232c$
§s %1 15 %0
$=\$T5222\,(=T\,(\vee F\,T))$
%$A5232c$
$=(\vee F\,T)\,T$:= $A5232c$
$=_{ooo}(\vee_{ooo}F_oT_o)T_o$:= $A5232c$
§s %1 7 %0
$=\$T5222\,(=T\,T)$
%$A5230a$
$=(=T\,T)\,T$:= $A5230a$
$=_{ooo}(=_{ooo}T_oT_o)T_o$:= $A5230a$
§s %1 3 %0
$=\$T5222\,T$
use Proof Template A5201b (Swap): A = B $\rightarrow$ B = A
<< A5201b.r0t.txt
%0
$=T\,\$T5222$
$=_{o\omega\omega}T_\omega\$T5222_\omega$
%T

$===$:= $A5200t$ T
$=_{o\omega\omega}=_{\omega}=_{\omega}$:= $A5200t$ T
§s %0 1 %1
$\$L5222\,T$:= $\$T5222$

case F
§= $\$F5222$
$=\$F5222\,\$F5222$
§\ $\$F5222$
$=\$F5222\,(=(\vee\,(\vee\,F\,F)\,F)\,(\vee\,F\,(\vee\,F\,F)))$
§s %1 3 %0
$=\$F5222\,(=(\vee\,(\vee\,F\,F)\,F)\,(\vee\,F\,(\vee\,F\,F)))$
%$A5232d$
$=(\vee\,F\,F)\,F$:= $A5232d$
$=_{ooo}(\vee_{ooo}F_oF_o)F_o$:= $A5232d$
§s %1 53 %0
$=\$F5222\,(=(\vee\,F\,F)\,(\vee\,F\,(\vee\,F\,F)))$
%$A5232d$
$=(\vee\,F\,F)\,F$:= $A5232d$
$=_{ooo}(\vee_{ooo}F_oF_o)F_o$:= $A5232d$
§s %1 13 %0
$=\$F5222\,(=F\,(\vee\,F\,(\vee\,F\,F)))$
%$A5232d$
$=(\vee\,F\,F)\,F$:= $A5232d$
$=_{ooo}(\vee_{ooo}F_oF_o)F_o$:= $A5232d$
§s %1 15 %0
$=\$F5222\,(=F\,(\vee\,F\,F))$
%$A5232d$
$=(\vee\,F\,F)\,F$:= $A5232d$
$=_{ooo}(\vee_{ooo}F_oF_o)F_o$:= $A5232d$
§s %1 7 %0
$=\$F5222\,(=F\,F)$
%$A5230d$
$=(=F\,F)\,T$:= $A5230d$
$=_{ooo}(=_{ooo}F_oF_o)T_o$:= $A5230d$
§s %1 3 %0
$=\$F5222\,T$
use Proof Template A5201b (Swap): A = B → B = A
<< A5201b.r0t.txt
%0
$=T\,\$F5222$
$=_{o\omega\omega}T_{\omega}\$F5222_{\omega}$
%T
$===$:= $A5200t$ T
$=_{o\omega\omega}=_{\omega}=_{\omega}$:= $A5200t$ T
§s %0 1 %1
$\$L5222\,F$:= $\$F5222$

<< A5222.r0t.txt

:= $\$L5222$
:= $\$X5222$
:= $\$T5222$
:= $\$F5222$

:= $\$FFTMP8023$
:= $\$FFTMP8023$ %0
wff 2069 : $=(\lor(\lor F\,F)\,z)\,(\lor F\,(\lor F\,z))_{o,\dots}$:= $\$FFTMP8023$

.5: Case T

:= $\$TTMP8023$ $[\lambda y_o.(\$LTMP8023_{oooo}T_o y_o z_o)_o]$
wff 2177 : $[\lambda y.(\$LTMP8023\,T\,y\,z)]_{oo}$:= $\$TTMP8023$
§= $\$TTMP8023$
$=\$TTMP8023\,\$TTMP8023$
§\ $\$LTMP8023_{oooo}T_o$
$=(\$LTMP8023\,T)\,[\lambda y.[\lambda z.(=(\lor(\lor T\,y)\,z)\,(\lor T\,(\lor y\,z)))]]$
§s %1 28 %0
$=\$TTMP8023\,[\lambda y.([\lambda y.[\lambda z.(=(\lor(\lor T\,y)\,z)\,(\lor T\,(\lor y\,z)))]]\,y\,z)]$
§\ $[\lambda y_o.[\lambda z_o.(=_{ooo}(\lor_{ooo}(\lor_{ooo}T_o y_o)z_o)(\lor_{ooo}T_o(\lor_{ooo}y_o z_o)))_o]_{(oo)}]y_o$
$=([\lambda y.[\lambda z.(=(\lor(\lor T\,y)\,z)\,(\lor T\,(\lor y\,z)))]]\,y)\,[\lambda z.(=(\lor(\lor T\,y)\,z)\,(\lor T\,(\lor y\,z)))]$
§s %1 14 %0
$=\$TTMP8023\,[\lambda y.([\lambda z.(=(\lor(\lor T\,y)\,z)\,(\lor T\,(\lor y\,z)))]\,z)]$
§\ $[\lambda z_o.(=_{ooo}(\lor_{ooo}(\lor_{ooo}T_o y_o)z_o)(\lor_{ooo}T_o(\lor_{ooo}y_o z_o)))_o]z_o$
$=([\lambda z.(=(\lor(\lor T\,y)\,z)\,(\lor T\,(\lor y\,z)))]\,z)\,(=(\lor(\lor T\,y)\,z)\,(\lor T\,(\lor y\,z)))$
§s %1 7 %0
$=\$TTMP8023\,[\lambda y.(=(\lor(\lor T\,y)\,z)\,(\lor T\,(\lor y\,z)))]$

use Proof Template A5222 (Rule of Cases): [\x.A]T, [\x.A]F → A
:= $\$L5222$ %0/3
wff 2191 : $[\lambda y.(=(\lor(\lor T\,y)\,z)\,(\lor T\,(\lor y\,z)))]_{oo,\dots}$:= $\$L5222$
:= $\$X5222$ y_o
wff 34 : y_o := $\$X5222$
:= $\$T5222$ $\$L5222_{oo}T_o$
wff 2193 : $\$L5222\,T_o$:= $\$T5222$
:= $\$F5222$ $\$L5222_{oo}F_o$
wff 2194 : $\$L5222\,F_o$:= $\$F5222$

case T
§= $\$T5222$
$=\$T5222\,\$T5222$
§\ $\$T5222$
$=\$T5222\,\$TTTMP8023$
§s %1 3 %0
$=\$T5222\,\$TTTMP8023$
use Proof Template A5201b (Swap): A = B → B = A
<< A5201b.r0t.txt
%0
$=\$TTTMP8023\,\$T5222$

\# $=_{o\omega\omega} \$TTTMP8023_{\omega} \$T5222_{\omega}$
%\$TTTMP8023
\# $= (\vee\, (\vee\, T\, T)\, z)\, (\vee\, T\, (\vee\, T\, z))$:= \$TTTMP8023
\# $=_{ooo}(\vee_{ooo}(\vee_{ooo}T_oT_o)z_o)(\vee_{ooo}T_o(\vee_{ooo}T_oz_o))$:= \$TTTMP8023
:= \$TTTMP8023
§s %0 1 %1
\# \$L5222 T := \$T5222

\## case F
§= \$F5222
\# = \$F5222 \$F5222
§\ \$F5222
\# = \$F5222 \$TFTMP8023
§s %1 3 %0
\# = \$F5222 \$TFTMP8023
\## use Proof Template A5201b (Swap): A = B → B = A
<< A5201b.r0t.txt
%0
\# = \$TFTMP8023 \$F5222
\# $=_{o\omega\omega} \$TFTMP8023_{\omega} \$F5222_{\omega}$
%\$TFTMP8023
\# $= (\vee\, (\vee\, T\, F)\, z)\, (\vee\, T\, (\vee\, F\, z))$:= \$TFTMP8023
\# $=_{ooo}(\vee_{ooo}(\vee_{ooo}T_oF_o)z_o)(\vee_{ooo}T_o(\vee_{ooo}F_oz_o))$:= \$TFTMP8023
:= \$TFTMP8023
§s %0 1 %1
\# \$L5222 F := \$F5222

<< A5222.r0t.txt
:= \$L5222
:= \$X5222
:= \$T5222
:= \$F5222

:= \$TTMP8023
:= \$TTMP8023 %0
\# wff 1691 : $= (\vee\, (\vee\, T\, y)\, z)\, (\vee\, T\, (\vee\, y\, z))_{o,\dots}$:= \$TTMP8023

\## .6: Case F

:= \$FTMP8023 $[\lambda y_o.(\$LTMP8023_{oooo}F_oy_oz_o)_o]$
\# wff 2282 : $[\lambda y.(\$LTMP8023\, F\, y\, z)]_{oo}$:= \$FTMP8023
§= \$FTMP8023
\# = \$FTMP8023 \$FTMP8023
§\ $\$LTMP8023_{oooo}F_o$
\# $= (\$LTMP8023\, F)\, [\lambda y.[\lambda z.(= (\vee\, (\vee\, F\, y)\, z)\, (\vee\, F\, (\vee\, y\, z)))]]$
§s %1 28 %0
\# $= \$FTMP8023\, [\lambda y.([\lambda y.[\lambda z.(= (\vee\, (\vee\, F\, y)\, z)\, (\vee\, F\, (\vee\, y\, z)))]]\, y\, z)]$
§\ $[\lambda y_o.[\lambda z_o.(=_{ooo}(\vee_{ooo}(\vee_{ooo}F_oy_o)z_o)(\vee_{ooo}F_o(\vee_{ooo}y_oz_o)))_o]_{(oo)}]y_o$
\# $= ([\lambda y.[\lambda z.(= (\vee\, (\vee\, F\, y)\, z)\, (\vee\, F\, (\vee\, y\, z)))]]\, y)\, [\lambda z.(= (\vee\, (\vee\, F\, y)\, z)\, (\vee\, F\, (\vee\, y\, z)))]$

§s %1 14 %0
$= \$FTMP8023\,[\lambda y.([\lambda z.(=(\vee\,(\vee\,F\,y)\,z)\,(\vee\,F\,(\vee\,y\,z)))]\,z)]$
§\ $[\lambda z_o.(=_{ooo}(\vee_{ooo}(\vee_{ooo}F_o y_o)z_o)(\vee_{ooo}F_o(\vee_{ooo}y_o z_o)))_o]z_o$
$= ([\lambda z.(=(\vee\,(\vee\,F\,y)\,z)\,(\vee\,F\,(\vee\,y\,z)))]\,z)\,(=(\vee\,(\vee\,F\,y)\,z)\,(\vee\,F\,(\vee\,y\,z)))$
§s %1 7 %0
$= \$FTMP8023\,[\lambda y.(=(\vee\,(\vee\,F\,y)\,z)\,(\vee\,F\,(\vee\,y\,z)))]$

use Proof Template A5222 (Rule of Cases): [\x.A]T, [\x.A]F $\rightarrow$ A
:= $\$L5222$ %0/3
wff 2296 : $[\lambda y.(=(\vee\,(\vee\,F\,y)\,z)\,(\vee\,F\,(\vee\,y\,z)))]_{oo,\ldots}$:= $\$L5222$
:= $\$X5222\ y_o$
wff 34 : y_o := $\$X5222$
:= $\$T5222\ \$L5222_{oo}T_o$
wff 2298 : $\$L5222\,T_o$:= $\$T5222$
:= $\$F5222\ \$L5222_{oo}F_o$
wff 2299 : $\$L5222\,F_o$:= $\$F5222$

case T
§= $\$T5222$
$= \$T5222\,\$T5222$
§\ $\$T5222$
$= \$T5222\,\$FTTMP8023$
§s %1 3 %0
$= \$T5222\,\$FTTMP8023$
use Proof Template A5201b (Swap): A = B $\rightarrow$ B = A
<< A5201b.r0t.txt
%0
$= \$FTTMP8023\,\$T5222$
$=_{o\omega\omega}\$FTTMP8023_\omega \$T5222_\omega$
%$\$FTTMP8023$
$= (\vee\,(\vee\,F\,T)\,z)\,(\vee\,F\,(\vee\,T\,z))$:= $\$FTTMP8023$
$=_{ooo}(\vee_{ooo}(\vee_{ooo}F_o T_o)z_o)(\vee_{ooo}F_o(\vee_{ooo}T_o z_o))$:= $\$FTTMP8023$
:= $\$FTTMP8023$
§s %0 1 %1
$\$L5222\,T$:= $\$T5222$

case F
§= $\$F5222$
$= \$F5222\,\$F5222$
§\ $\$F5222$
$= \$F5222\,\$FFTMP8023$
§s %1 3 %0
$= \$F5222\,\$FFTMP8023$
use Proof Template A5201b (Swap): A = B $\rightarrow$ B = A
<< A5201b.r0t.txt
%0
$= \$FFTMP8023\,\$F5222$
$=_{o\omega\omega}\$FFTMP8023_\omega \$F5222_\omega$
%$\$FFTMP8023$

\# $= (\vee\,(\vee\,F\,F)\,z)\,(\vee\,F\,(\vee\,F\,z))$:= $\$FFTMP8023$
\# $=_{ooo}(\vee_{ooo}(\vee_{ooo}F_oF_o)z_o)(\vee_{ooo}F_o(\vee_{ooo}F_oz_o))$:= $\$FFTMP8023$
:= $\$FFTMP8023$
§s %0 1 %1
\# $\$L5222\,F$:= $\$F5222$

<< A5222.r0t.txt
:= $\$L5222$
:= $\$X5222$
:= $\$T5222$
:= $\$F5222$

:= $\$FTMP8023$
:= $\$FTMP8023$ %0
\# wff 1942 : $= (\vee\,(\vee\,F\,y)\,z)\,(\vee\,F\,(\vee\,y\,z))_{o,\ldots}$:= $\$FTMP8023$

.7: General case

:= $\$TMP8023$ $[\lambda x_o.(\$LTMP8023_{oooo}x_oy_oz_o)_o]$
\# wff 2383 : $[\lambda x.(\$LTMP8023\,x\,y\,z)]_{oo}$:= $\$TMP8023$
§= $\$TMP8023$
\# $= \$TMP8023\,\$TMP8023$
§\ $\$LTMP8023_{oooo}x_o$
\# $= (\$LTMP8023\,x)\,[\lambda y.[\lambda z.(= (\vee\,(\vee\,x\,y)\,z)\,(\vee\,x\,(\vee\,y\,z)))]]$
§s %1 28 %0
\# $= \$TMP8023\,[\lambda x.([\lambda y.[\lambda z.(= (\vee\,(\vee\,x\,y)\,z)\,(\vee\,x\,(\vee\,y\,z)))]]\,y\,z)]$
§\ $[\lambda y_o.[\lambda z_o.(=_{ooo}(\vee_{ooo}(\vee_{ooo}x_oy_o)z_o)(\vee_{ooo}x_o(\vee_{ooo}y_oz_o)))_o]_{(oo)}]y_o$
\# $= ([\lambda y.[\lambda z.(= (\vee\,(\vee\,x\,y)\,z)\,(\vee\,x\,(\vee\,y\,z)))]]\,y)\,[\lambda z.(= (\vee\,(\vee\,x\,y)\,z)\,(\vee\,x\,(\vee\,y\,z)))]$
§s %1 14 %0
\# $= \$TMP8023\,[\lambda x.([\lambda z.(= (\vee\,(\vee\,x\,y)\,z)\,(\vee\,x\,(\vee\,y\,z)))]\,z)]$
§\ $[\lambda z_o.(=_{ooo}(\vee_{ooo}(\vee_{ooo}x_oy_o)z_o)(\vee_{ooo}x_o(\vee_{ooo}y_oz_o)))_o]z_o$
\# $= ([\lambda z.(= (\vee\,(\vee\,x\,y)\,z)\,(\vee\,x\,(\vee\,y\,z)))]\,z)\,(= (\vee\,(\vee\,x\,y)\,z)\,(\vee\,x\,(\vee\,y\,z)))$
§s %1 7 %0
\# $= \$TMP8023\,[\lambda x.(= (\vee\,(\vee\,x\,y)\,z)\,(\vee\,x\,(\vee\,y\,z)))]$

use Proof Template A5222 (Rule of Cases): [\x.A]T, [\x.A]F → A
:= $\$L5222$ %0/3
\# wff 2399 : $[\lambda x.(= (\vee\,(\vee\,x\,y)\,z)\,(\vee\,x\,(\vee\,y\,z)))]_{oo,\ldots}$:= $\$L5222$
:= $\$X5222$ x_o
\# wff 16 : x_o := $\$X5222$
:= $\$T5222$ $\$L5222_{oo}T_o$
\# wff 2401 : $\$L5222\,T_o$:= $\$T5222$
:= $\$F5222$ $\$L5222_{oo}F_o$
\# wff 2402 : $\$L5222\,F_o$:= $\$F5222$

case T
§= $\$T5222$
\# $= \$T5222\,\$T5222$
§\ $\$T5222$

\# $= \$T5222\, \$TTMP8023$
§s %1 3 %0
\# $= \$T5222\, \$TTMP8023$
\#\# use Proof Template A5201b (Swap): A = B → B = A
<< A5201b.r0t.txt
%0
\# $= \$TTMP8023\, \$T5222$
\# $=_{o\omega\omega} \$TTMP8023_{\omega} \$T5222_{\omega}$
%$\$TTMP8023$
\# $= (\vee\, (\vee\, T\, y)\, z)\, (\vee\, T\, (\vee\, y\, z))$:= $\$TTMP8023$
\# $=_{ooo}(\vee_{ooo}(\vee_{ooo}T_{o}y_{o})z_{o})(\vee_{ooo}T_{o}(\vee_{ooo}y_{o}z_{o}))$:= $\$TTMP8023$
:= $\$TTMP8023$
§s %0 1 %1
\# $\$L5222\, T$:= $\$T5222$

\#\# case F
§= $\$F5222$
\# $= \$F5222\, \$F5222$
§\ $\$F5222$
\# $= \$F5222\, \$FTMP8023$
§s %1 3 %0
\# $= \$F5222\, \$FTMP8023$
\#\# use Proof Template A5201b (Swap): A = B → B = A
<< A5201b.r0t.txt
%0
\# $= \$FTMP8023\, \$F5222$
\# $=_{o\omega\omega} \$FTMP8023_{\omega} \$F5222_{\omega}$
%$\$FTMP8023$
\# $= (\vee\, (\vee\, F\, y)\, z)\, (\vee\, F\, (\vee\, y\, z))$:= $\$FTMP8023$
\# $=_{ooo}(\vee_{ooo}(\vee_{ooo}F_{o}y_{o})z_{o})(\vee_{ooo}F_{o}(\vee_{ooo}y_{o}z_{o}))$:= $\$FTMP8023$
:= $\$FTMP8023$
§s %0 1 %1
\# $\$L5222\, F$:= $\$F5222$

<< A5222.r0t.txt
:= $\$L5222$
:= $\$X5222$
:= $\$T5222$
:= $\$F5222$

:= $\$TMP8023$
:= $\$TMP8023$ %0
\# wff 1676 : $= (\vee\, (\vee\, x\, y)\, z)\, (\vee\, x\, (\vee\, y\, z))_{o,\dots}$:= $\$TMP8023$

\#\# .8: Match general definition

§= $ASSOC_{o(\backslash 4 \backslash 4 \backslash 3)\tau} o_{\tau} \vee_{ooo}$
\# $= (ASSOC\, o \vee)\, (ASSOC\, o \vee)$
§\ $ASSOC_{o(\backslash 4 \backslash 4 \backslash 3)\tau} o_{\tau}$

$= (ASSOC\, o)\, [\lambda f.(= (f\, (f\, x\, y)\, z)\, (f\, x\, (f\, y\, z)))]$
§s %1 6 %0
$= (ASSOC\, o \vee)\, ([\lambda f.(= (f\, (f\, x\, y)\, z)\, (f\, x\, (f\, y\, z)))] \vee)$
§\ $[\lambda f_{ooo}.(=_{ooo}(f_{ooo}(f_{ooo}x_o y_o)z_o)(f_{ooo}x_o(f_{ooo}y_o z_o)))_o]\vee_{ooo}$
$= ([\lambda f.(= (f\, (f\, x\, y)\, z)\, (f\, x\, (f\, y\, z)))] \vee)\, \$TMP8023$
§s %1 3 %0
$= (ASSOC\, o \vee)\, \$TMP8023$
use Proof Template A5201b (Swap): A = B → B = A
<< A5201b.r0t.txt
%0
$= \$TMP8023\, (ASSOC\, o \vee)$
$=_{o\omega\omega} \$TMP8023_{\omega} (ASSOC_{o(\backslash 4 \backslash 4 \backslash 3)\tau} o_{\tau} \vee_{ooo})$
%$TMP8023
$= (\vee\, (\vee\, x\, y)\, z)\, (\vee\, x\, (\vee\, y\, z))$:= $TMP8023
$=_{ooo}(\vee_{ooo}(\vee_{ooo}x_o y_o)z_o)(\vee_{ooo}x_o(\vee_{ooo}y_o z_o))$:= $TMP8023
:= $TMP8023
§s %0 1 %1
$ASSOC\, o \vee$

:= K8023 %0
wff 2475 : $ASSOC\, o \vee_{o,\dots}$:= K8023

:= $LTMP8023

##
Q.E.D.
##

%0
$ASSOC\, o \vee$:= K8023
$ASSOC_{o(\backslash 4 \backslash 4 \backslash 3)\tau} o_{\tau} \vee_{ooo}$:= K8023

2.1.80 Results for File K8024.r0.txt

##
Proof K8024 (Generalized Deduction Theorem): (H ∧ I) ⊃ A = H ⊃ (I ⊃ A)
##
##
Source: [Kubota 2017 (doi: 10.4444/100.10)]
##

Written by Ken Kubota (<mail@kenkubota.de>).
##
This file is part of the publication of the mathematical logic $\mathcal{R}_0$.
For more information, visit: <http://doi.org/10.4444/100.10>
##

<< basics.r0.txt
<< K8008.r0.txt
<< K8018.r0.txt
<< K8022.r0.txt
<< K8023.r0.txt

##
Proof
##

.1

%$K8022$
$=\supset[\lambda x.[\lambda y.(\vee\,(\sim x)\,y)]]$:= $K8022$
$=_{o(ooo)(ooo)}\supset_{ooo}[\lambda x_o.[\lambda y_o.(\vee_{ooo}(\sim_{oo}x_o)y_o)_o]_{(oo)}]$:= $K8022$
§= $_o\ \supset_{ooo}(\wedge_{ooo}h_oj_o)x_o$
$=(\supset(\wedge\,h\,j)\,x)\,(\supset(\wedge\,h\,j)\,x)$
§s %0 12 %1
$=(\supset(\wedge\,h\,j)\,x)\,([\lambda x.[\lambda y.(\vee\,(\sim x)\,y)]]\,(\wedge\,h\,j)\,x)$
§\ $[\lambda x_o.[\lambda y_o.(\vee_{ooo}(\sim_{oo}x_o)y_o)_o]_{(oo)}](\wedge_{ooo}h_oj_o)$
$=([\lambda x.[\lambda y.(\vee\,(\sim x)\,y)]]\,(\wedge\,h\,j))\,[\lambda y.(\vee\,(\sim(\wedge\,h\,j))\,y)]$
§s %1 6 %0
$=(\supset(\wedge\,h\,j)\,x)\,([\lambda y.(\vee\,(\sim(\wedge\,h\,j))\,y)]\,x)$
§\ $[\lambda y_o.(\vee_{ooo}(\sim_{oo}(\wedge_{ooo}h_oj_o))y_o)_o]x_o$
$=([\lambda y.(\vee\,(\sim(\wedge\,h\,j))\,y)]\,x)\,(\vee\,(\sim(\wedge\,h\,j))\,x)$
§s %1 3 %0
$=(\supset(\wedge\,h\,j)\,x)\,(\vee\,(\sim(\wedge\,h\,j))\,x)$

.2

%$K8018$
$=\wedge[\lambda a.[\lambda b.(\sim(\vee\,(\sim a)\,(\sim b)))]]$:= $K8018$
$=_{o(ooo)(ooo)}\wedge_{ooo}[\lambda a_o.[\lambda b_o.(\sim_{oo}(\vee_{ooo}(\sim_{oo}a_o)(\sim_{oo}b_o)))_o]_{(oo)}]$:= $K8018$
§s %1 108 %0
$=(\supset(\wedge\,h\,j)\,x)\,(\vee\,(\sim([\lambda a.[\lambda b.(\sim(\vee\,(\sim a)\,(\sim b)))]]\,h\,j))\,x)$
§\ $[\lambda a_o.[\lambda b_o.(\sim_{oo}(\vee_{ooo}(\sim_{oo}a_o)(\sim_{oo}b_o)))_o]_{(oo)}]h_o$
$=([\lambda a.[\lambda b.(\sim(\vee\,(\sim a)\,(\sim b)))]]\,h)\,[\lambda b.(\sim(\vee\,(\sim h)\,(\sim b)))]$
§s %1 54 %0
$=(\supset(\wedge\,h\,j)\,x)\,(\vee\,(\sim([\lambda b.(\sim(\vee\,(\sim h)\,(\sim b)))]\,j))\,x)$
§\ $[\lambda b_o.(\sim_{oo}(\vee_{ooo}(\sim_{oo}h_o)(\sim_{oo}b_o)))_o]j_o$
$=([\lambda b.(\sim(\vee\,(\sim h)\,(\sim b)))]\,j)\,(\sim(\vee\,(\sim h)\,(\sim j)))$
§s %1 27 %0
$=(\supset(\wedge\,h\,j)\,x)\,(\vee\,(\sim(\sim(\vee\,(\sim h)\,(\sim j))))\,x)$

:= $\$TMP8025$ %0
wff 3509 : $=(\supset(\wedge\,h\,j)\,x)\,(\vee\,(\sim(\sim(\vee\,(\sim h)\,(\sim j))))\,x)_o$:= $\$TMP8025$

.3

$\%K8008$
$= (\sim (\sim x))\, x$:= $K8008$
$=_{ooo}(\sim_{oo}(\sim_{oo}x_o))x_o$:= $K8008$

use Proof Template A5221 (Sub): B $\rightarrow$ B [x/A]
:= $\$B5221$ %0
wff 1663 : $= (\sim (\sim x))\, x_{o,\ldots}$:= $\$B5221$ $K8008$
:= $\$T5221$ o
wff 2 : o_τ := $\$T5221$
:= $\$X5221$ x_o
wff 16 : x_o := $\$X5221$
:= $\$A5221$ $\vee_{ooo}(\sim_{oo}h_o)(\sim_{oo}j_o)$
wff 3503 : $\vee (\sim h)\, (\sim j)_o$:= $\$A5221$
<< A5221.r0t.txt
:= $\$B5221$
:= $\$T5221$
:= $\$X5221$
:= $\$A5221$
%0
$= (\sim (\sim (\vee (\sim h)\, (\sim j))))\, (\vee (\sim h)\, (\sim j))$
$=_{ooo}(\sim_{oo}(\sim_{oo}(\vee_{ooo}(\sim_{oo}h_o)(\sim_{oo}j_o))))(\vee_{ooo}(\sim_{oo}h_o)(\sim_{oo}j_o))$

$\%\$TMP8025$
$= (\supset (\wedge\, h\, j)\, x)\, (\vee (\sim (\sim (\vee (\sim h)\, (\sim j))))\, x)$:= $\$TMP8025$
$=_{ooo}(\supset_{ooo}(\wedge_{ooo}h_oj_o)x_o)(\vee_{ooo}(\sim_{oo}(\sim_{oo}(\vee_{ooo}(\sim_{oo}h_o)(\sim_{oo}j_o))))x_o)$:=
$\$TMP8025$
:= $\$TMP8025$
§s %0 13 %1
$= (\supset (\wedge\, h\, j)\, x)\, (\vee (\vee (\sim h)\, (\sim j))\, x)$
:= $\$TMP8025$ %0
wff 3524 : $= (\supset (\wedge\, h\, j)\, x)\, (\vee (\vee (\sim h)\, (\sim j))\, x)_o$:= $\$TMP8025$

.4

$\%K8023$
$ASSOC\, o \vee$:= $K8023$
$ASSOC_{o(\backslash 4\backslash 4\backslash 3)\tau}o_\tau\vee_{ooo}$:= $K8023$
§\ $ASSOC_{o(\backslash 4\backslash 4\backslash 3)\tau}o_\tau$
$= (ASSOC\, o)\, [\lambda f.(= (f\, (f\, x\, y)\, z)\, (f\, x\, (f\, y\, z)))]$
§s %1 2 %0
$[\lambda f.(= (f\, (f\, x\, y)\, z)\, (f\, x\, (f\, y\, z)))] \vee$
§\ $[\lambda f_{ooo}.(=_{ooo}(f_{ooo}(f_{ooo}x_oy_o)z_o)(f_{ooo}x_o(f_{ooo}y_oz_o)))_o]\vee_{ooo}$
$= ([\lambda f.(= (f\, (f\, x\, y)\, z)\, (f\, x\, (f\, y\, z)))] \vee)\, (= (\vee (\vee\, x\, y)\, z)\, (\vee\, x\, (\vee\, y\, z)))$
§s %1 1 %0
$= (\vee (\vee\, x\, y)\, z)\, (\vee\, x\, (\vee\, y\, z))$

use Proof Template A5221 (Sub): B $\rightarrow$ B [x/A]
:= $\$B5221$ %0

\# wff 2648 : $=(\vee\,(\vee\,x\,y)\,z)\,(\vee\,x\,(\vee\,y\,z))_{o,\ldots}$:= $\$B5221$
:= $\$T5221$ o
\# wff 2 : o_{τ} := $\$T5221$
:= $\$X5221$ x_o
\# wff 16 : x_o := $\$X5221$
:= $\$A5221$ $\sim_{oo}h_o$
\# wff 3490 : $\sim h_o$:= $\$A5221$
<< A5221.r0t.txt
:= $\$B5221$
:= $\$T5221$
:= $\$X5221$
:= $\$A5221$
%0
\# $=(\vee\,(\vee\,(\sim h)\,y)\,z)\,(\vee\,(\sim h)\,(\vee\,y\,z))$
\# $=_{ooo}(\vee_{ooo}(\vee_{ooo}(\sim_{oo}h_o)y_o)z_o)(\vee_{ooo}(\sim_{oo}h_o)(\vee_{ooo}y_oz_o))$

\#\# use Proof Template A5221 (Sub): B $\rightarrow$ B [x/A]
:= $\$B5221$ %0
\# wff 3559 : $=(\vee\,(\vee\,(\sim h)\,y)\,z)\,(\vee\,(\sim h)\,(\vee\,y\,z))_{o,\ldots}$:= $\$B5221$
:= $\$T5221$ o
\# wff 2 : o_{τ} := $\$T5221$
:= $\$X5221$ y_o
\# wff 34 : y_o := $\$X5221$
:= $\$A5221$ $\sim_{oo}j_o$
\# wff 3502 : $\sim j_o$:= $\$A5221$
<< A5221.r0t.txt
:= $\$B5221$
:= $\$T5221$
:= $\$X5221$
:= $\$A5221$
%0
\# $=(\vee\,(\vee\,(\sim h)\,(\sim j))\,z)\,(\vee\,(\sim h)\,(\vee\,(\sim j)\,z))$
\# $=_{ooo}(\vee_{ooo}(\vee_{ooo}(\sim_{oo}h_o)(\sim_{oo}j_o))z_o)(\vee_{ooo}(\sim_{oo}h_o)(\vee_{ooo}(\sim_{oo}j_o)z_o))$

\#\# use Proof Template A5221 (Sub): B $\rightarrow$ B [x/A]
:= $\$B5221$ %0
\# wff 3602 : $=(\vee\,(\vee\,(\sim h)\,(\sim j))\,z)\,(\vee\,(\sim h)\,(\vee\,(\sim j)\,z))_{o,\ldots}$:= $\$B5221$
:= $\$T5221$ o
\# wff 2 : o_{τ} := $\$T5221$
:= $\$X5221$ z_o
\# wff 2639 : z_o := $\$X5221$
:= $\$A5221$ x_o
\# wff 16 : x_o := $\$A5221$
<< A5221.r0t.txt
:= $\$B5221$
:= $\$T5221$
:= $\$X5221$
:= $\$A5221$
%0

\# $=(\vee\,(\vee\,(\sim h)\,(\sim j))\,x)\,(\vee\,(\sim h)\,(\vee\,(\sim j)\,x))$
\# $=_{ooo}(\vee_{ooo}(\vee_{ooo}(\sim_{oo}h_o)(\sim_{oo}j_o))x_o)(\vee_{ooo}(\sim_{oo}h_o)(\vee_{ooo}(\sim_{oo}j_o)x_o))$

%$\$TMP8025$
\# $=(\supset(\wedge\,h\,j)\,x)\,(\vee\,(\vee\,(\sim h)\,(\sim j))\,x)$:= $\$TMP8025$
\# $=_{ooo}(\supset_{ooo}(\wedge_{ooo}h_oj_o)x_o)(\vee_{ooo}(\vee_{ooo}(\sim_{oo}h_o)(\sim_{oo}j_o))x_o)$:= $\$TMP8025$
:= $\$TMP8025$
§s %0 3 %1
\# $=(\supset(\wedge\,h\,j)\,x)\,(\vee\,(\sim h)\,(\vee\,(\sim j)\,x))$
:= $\$TMP8025$ %0
\# wff 3648 : $=(\supset(\wedge\,h\,j)\,x)\,(\vee\,(\sim h)\,(\vee\,(\sim j)\,x))_o$:= $\$TMP8025$

\## .5

%$K8022$
\# $=\supset[\lambda x.[\lambda y.(\vee\,(\sim x)\,y)]]$:= $K8022$
\# $=_{o(ooo)(ooo)}\supset_{ooo}[\lambda x_o.[\lambda y_o.(\vee_{ooo}(\sim_{oo}x_o)y_o)_o]_{(oo)}]$:= $K8022$
§= $\supset_{ooo}j_ox_o$
\# $=(\supset j\,x)\,(\supset j\,x)$
§s %0 12 %1
\# $=(\supset j\,x)\,([\lambda x.[\lambda y.(\vee\,(\sim x)\,y)]]\,j\,x)$
§\ $[\lambda x_o.[\lambda y_o.(\vee_{ooo}(\sim_{oo}x_o)y_o)_o]_{(oo)}]j_o$
\# $=([\lambda x.[\lambda y.(\vee\,(\sim x)\,y)]]\,j)\,[\lambda y.(\vee\,(\sim j)\,y)]$
§s %1 6 %0
\# $=(\supset j\,x)\,([\lambda y.(\vee\,(\sim j)\,y)]\,x)$
§\ $[\lambda y_o.(\vee_{ooo}(\sim_{oo}j_o)y_o)_o]x_o$
\# $=([\lambda y.(\vee\,(\sim j)\,y)]\,x)\,(\vee\,(\sim j)\,x)$
§s %1 3 %0
\# $=(\supset j\,x)\,(\vee\,(\sim j)\,x)$

\## use Proof Template A5201b (Swap): A = B → B = A
<< A5201b.r0t.txt
%0
\# $=(\vee\,(\sim j)\,x)\,(\supset j\,x)$
\# $=_{o\omega\omega}(\vee_{ooo}(\sim_{oo}j_o)x_o)(\supset_{ooo}j_ox_o)$

%$\$TMP8025$
\# $=(\supset(\wedge\,h\,j)\,x)\,(\vee\,(\sim h)\,(\vee\,(\sim j)\,x))$:= $\$TMP8025$
\# $=_{ooo}(\supset_{ooo}(\wedge_{ooo}h_oj_o)x_o)(\vee_{ooo}(\sim_{oo}h_o)(\vee_{ooo}(\sim_{oo}j_o)x_o))$:= $\$TMP8025$
:= $\$TMP8025$
§s %0 7 %1
\# $=(\supset(\wedge\,h\,j)\,x)\,(\vee\,(\sim h)\,(\supset j\,x))$
:= $\$TMP8025$ %0
\# wff 3668 : $=(\supset(\wedge\,h\,j)\,x)\,(\vee\,(\sim h)\,(\supset j\,x))_o$:= $\$TMP8025$

\## .6

%$K8022$
\# $=\supset[\lambda x.[\lambda y.(\vee\,(\sim x)\,y)]]$:= $K8022$

$=_{o(ooo)(ooo)} \supset_{ooo} [\lambda x_o.[\lambda y_o.(\vee_{ooo}(\sim_{oo} x_o) y_o)_o]_{(oo)}]$:= $K8022$
§= $\supset_{ooo} h_o (\supset_{ooo} j_o x_o)$
$= (\supset h\,(\supset j\,x))\,(\supset h\,(\supset j\,x))$
§s %0 12 %1
$= (\supset h\,(\supset j\,x))\,([\lambda x.[\lambda y.(\vee\,(\sim x)\,y)]]\,h\,(\supset j\,x))$
§\ $[\lambda x_o.[\lambda y_o.(\vee_{ooo}(\sim_{oo} x_o) y_o)_o]_{(oo)}] h_o$
$= ([\lambda x.[\lambda y.(\vee\,(\sim x)\,y)]]\,h)\,[\lambda y.(\vee\,(\sim h)\,y)]$
§s %1 6 %0
$= (\supset h\,(\supset j\,x))\,([\lambda y.(\vee\,(\sim h)\,y)]\,(\supset j\,x))$
§\ $[\lambda y_o.(\vee_{ooo}(\sim_{oo} h_o) y_o)_o](\supset_{ooo} j_o x_o)$
$= ([\lambda y.(\vee\,(\sim h)\,y)]\,(\supset j\,x))\,(\vee\,(\sim h)\,(\supset j\,x))$
§s %1 3 %0
$= (\supset h\,(\supset j\,x))\,(\vee\,(\sim h)\,(\supset j\,x))$

use Proof Template A5201b (Swap): A = B → B = A
<< A5201b.r0t.txt
%0
$= (\vee\,(\sim h)\,(\supset j\,x))\,(\supset h\,(\supset j\,x))$
$=_{o\omega\omega} (\vee_{ooo}(\sim_{oo} h_o)(\supset_{ooo} j_o x_o))(\supset_{ooo} h_o (\supset_{ooo} j_o x_o))$

%$TMP8025
$= (\supset (\wedge\,h\,j)\,x)\,(\vee\,(\sim h)\,(\supset j\,x))$:= $\$TMP8025$
$=_{ooo} (\supset_{ooo}(\wedge_{ooo} h_o j_o) x_o)(\vee_{ooo}(\sim_{oo} h_o)(\supset_{ooo} j_o x_o))$:= $\$TMP8025$
:= $\$TMP8025$
§s %0 3 %1
$= (\supset (\wedge\,h\,j)\,x)\,(\supset h\,(\supset j\,x))$

:= $K8024$ %0
wff 3686 : $= (\supset (\wedge\,h\,j)\,x)\,(\supset h\,(\supset j\,x))_o$:= $K8024$

##
Q.E.D.
##

%0
$= (\supset (\wedge\,h\,j)\,x)\,(\supset h\,(\supset j\,x))$:= $K8024$
$=_{ooo} (\supset_{ooo}(\wedge_{ooo} h_o j_o) x_o)(\supset_{ooo} h_o (\supset_{ooo} j_o x_o))$:= $K8024$

2.1.81 Results for File K8025.r0a.txt

##
Proof Template K8025 (Deduction Theorem): (H ∧ I) ⊃ A → H ⊃ (I ⊃ A)
##
##
Source: [cf. Andrews 2002 (ISBN 1-4020-0763-9), pp. 228 f. (5240)]
##

Written by Ken Kubota (<mail@kenkubota.de>).

```
##
##   This file is part of the publication of the mathematical logic R0.
##   For more information, visit: <http://doi.org/10.4444/100.10>
##


##
##   Assumptions and Resulting Syntactical Variables
##

<< basics.r0.txt

## the assumption as last theorem on stack (%0)
§! ⊃ooo(∧ooo ho jo) xo
#                ⊃ (∧ h j) x


##
##   Include Proof Template
##

## <<< K8025.r0t.txt
## Include begin (K8025.r0t.txt) [oldfile=(K8025.r0a.txt)]
##
##   Proof Template K8025 (Deduction Theorem):  (H ∧ I) ⊃ A   →   H ⊃ (I ⊃ A)
##
##
##   Source: [cf. Andrews 2002 (ISBN 1-4020-0763-9), pp. 228 f. (5240)]
##
##   Copyright (c) 2017 Owl of Minerva Press GmbH. All rights reserved.
##   Written by Ken Kubota (<mail@kenkubota.de>).
##
##   This file is part of the publication of the mathematical logic R0.
##   For more information, visit: <http://doi.org/10.4444/100.10>
##


## define variable first (before inclusion of file)
:=  $STMPDED8025  %0
# wff     213  :      ⊃ (∧ h j) xo       :=   $STMPDED8025


##
##   Proof Template
##

<< K8024.r0.txt
%K8024
#                 = $STMPDED8025 (⊃ h (⊃ j x))       :=   K8024
```

```
#            =_ooo$STMPDED8025_o(⊃_ooo h_o(⊃_ooo j_o x_o))       :=  K8024

:=  $TMPDED8025  %0
# wff     3686  :       =$STMPDED8025 (⊃ h (⊃ j x))_o       :=  $TMPDED8025   K8024
%$STMPDED8025
#             ⊃ (∧ h j) x       :=  $STMPDED8025
#             ⊃_ooo(∧_ooo h_o j_o)x_o       :=  $STMPDED8025
%K8024
#             =$STMPDED8025 (⊃ h (⊃ j x))       :=  $TMPDED8025  K8024
#             =_ooo$STMPDED8025_o(⊃_ooo h_o(⊃_ooo j_o x_o))       :=  $TMPDED8025   K8024
:=  $TMPDED8025

## use Proof Template A5221 (Sub):   B   →   B [x/A]
:=  $B5221  %0
# wff     3686  :       =$STMPDED8025 (⊃ h (⊃ j x))_o       :=  $B5221   K8024
:=  $T5221  o
# wff     2  :       o_τ       :=  $T5221
:=  $X5221  h_o
# wff     208  :       h_o       :=  $X5221
:=  $A5221  %1/21
# wff     208  :       h_o       :=  $A5221   $X5221
<< A5221.r0t.txt
:=  $B5221
:=  $T5221
:=  $X5221
:=  $A5221

:=  $TMPDED8025  %0
# wff     3686  :       =$STMPDED8025 (⊃ h (⊃ j x))_o,...       :=  $TMPDED8025   K8024
%$STMPDED8025
#             ⊃ (∧ h j) x       :=  $STMPDED8025
#             ⊃_ooo(∧_ooo h_o j_o)x_o       :=  $STMPDED8025
%K8024
#             =$STMPDED8025 (⊃ h (⊃ j x))       :=  $TMPDED8025  K8024
#             =_ooo$STMPDED8025_o(⊃_ooo h_o(⊃_ooo j_o x_o))       :=  $TMPDED8025   K8024
:=  $TMPDED8025

## use Proof Template A5221 (Sub):   B   →   B [x/A]
:=  $B5221  %0
# wff     3686  :       =$STMPDED8025 (⊃ h (⊃ j x))_o,...       :=  $B5221   K8024
:=  $T5221  o
# wff     2  :       o_τ       :=  $T5221
:=  $X5221  j_o
# wff     210  :       j_o       :=  $X5221
:=  $A5221  %1/11
# wff     210  :       j_o       :=  $A5221   $X5221
<< A5221.r0t.txt
:=  $B5221
:=  $T5221
```

:= $\$X5221$
:= $\$A5221$

:= $\$TMPDED8025$ %0
wff 3686 : $=\$STMPDED8025\,(\supset h\,(\supset j\,x))_{o,\ldots}$:= $\$TMPDED8025$ $K8024$
%$\$STMPDED8025$
$\supset(\wedge\,h\,j)\,x$:= $\$STMPDED8025$
$\supset_{ooo}(\wedge_{ooo}h_oj_o)x_o$:= $\$STMPDED8025$
%$K8024$
$=\$STMPDED8025\,(\supset h\,(\supset j\,x))$:= $\$TMPDED8025$ $K8024$
$=_{ooo}\$STMPDED8025_o(\supset_{ooo}h_o(\supset_{ooo}j_ox_o))$:= $\$TMPDED8025$ $K8024$
:= $\$TMPDED8025$

use Proof Template A5221 (Sub): B → B [x/A]
:= $\$B5221$ %0
wff 3686 : $=\$STMPDED8025\,(\supset h\,(\supset j\,x))_{o,\ldots}$:= $\$B5221$ $K8024$
:= $\$T5221$ o
wff 2 : o_τ := $\$T5221$
:= $\$X5221$ x_o
wff 16 : x_o := $\$X5221$
:= $\$A5221$ %1/3
wff 16 : x_o := $\$A5221$ $\$X5221$
<< A5221.r0t.txt
:= $\$B5221$
:= $\$T5221$
:= $\$X5221$
:= $\$A5221$
%0
$=\$STMPDED8025\,(\supset h\,(\supset j\,x))$:= $K8024$
$=_{ooo}\$STMPDED8025_o(\supset_{ooo}h_o(\supset_{ooo}j_ox_o))$:= $K8024$

%$\$STMPDED8025$
$\supset(\wedge\,h\,j)\,x$:= $\$STMPDED8025$
$\supset_{ooo}(\wedge_{ooo}h_oj_o)x_o$:= $\$STMPDED8025$
:= $\$STMPDED8025$
§s %0 1 %1
$\supset h\,(\supset j\,x)$
Include end (K8025.r0t.txt) [newfile=(K8025.r0a.txt)]
>>>

##
Q.E.D.
##

%0
$\supset h\,(\supset j\,x)$
$\supset_{ooo}h_o(\supset_{ooo}j_ox_o)$

2.1.82 Results for File K8026.r0a.txt

```
##
## Proof Template K8026 (Deduction Theorem Reversed): H ⊃ (I ⊃ A) → (H ∧ I) ⊃ A
##
##
## Source: [Kubota 2017 (doi: 10.4444/100.10)]
##
```

```
## Written by Ken Kubota (<mail@kenkubota.de>).
##
## This file is part of the publication of the mathematical logic R0.
## For more information, visit: <http://doi.org/10.4444/100.10>
##
```

```
##
## Assumptions and Resulting Syntactical Variables
##

<< basics.r0.txt

## the assumption as last theorem on stack (%0)
```
§! $\supset_{ooo} h_o (\supset_{ooo} j_o x_o)$
\# $\supset h\, (\supset j\, x)$

```
##
## Include Proof Template
##

## <<< K8026.r0t.txt
## Include begin (K8026.r0t.txt) [oldfile=(K8026.r0a.txt)]
##
## Proof Template K8026 (Deduction Theorem Reversed): H ⊃ (I ⊃ A) → (H ∧ I) ⊃ A
##
##
## Source: [Kubota 2017 (doi: 10.4444/100.10)]
##
```
Copyright (c) 2017 Owl of Minerva Press GmbH. All rights reserved.
```
## Written by Ken Kubota (<mail@kenkubota.de>).
##
## This file is part of the publication of the mathematical logic R0.
## For more information, visit: <http://doi.org/10.4444/100.10>
##

## define variable first (before inclusion of file)
```
:= $\$STMPDED$ %0

wff 213 : $\supset h\,(\supset j\,x)_o$:= $\$STMPDED$

##
Proof Template
##

<< K8024.r0.txt
%$K8024$
$=(\supset(\wedge\,h\,j)\,x)\,\$STMPDED$:= $K8024$
$=_{ooo}(\supset_{ooo}(\wedge_{ooo}h_o j_o)x_o)\$STMPDED_o$:= $K8024$

:= $\$TMPDED$ %0
wff 3686 : $=(\supset(\wedge\,h\,j)\,x)\,\$STMPDED_o$:= $\$TMPDED$ $K8024$
%$\$STMPDED$
$\supset h\,(\supset j\,x)$:= $\$STMPDED$
$\supset_{ooo}h_o(\supset_{ooo}j_o x_o)$:= $\$STMPDED$
%$K8024$
$=(\supset(\wedge\,h\,j)\,x)\,\$STMPDED$:= $\$TMPDED$ $K8024$
$=_{ooo}(\supset_{ooo}(\wedge_{ooo}h_o j_o)x_o)\$STMPDED_o$:= $\$TMPDED$ $K8024$
:= $\$TMPDED$

use Proof Template A5221 (Sub): B $\rightarrow$ B [x/A]
:= $\$B5221$ %0
wff 3686 : $=(\supset(\wedge\,h\,j)\,x)\,\$STMPDED_o$:= $\$B5221$ $K8024$
:= $\$T5221$ o
wff 2 : o_τ := $\$T5221$
:= $\$X5221$ h_o
wff 208 : h_o := $\$X5221$
:= $\$A5221$ %1/5
wff 208 : h_o := $\$A5221$ $\$X5221$
<< A5221.r0t.txt
:= $\$B5221$
:= $\$T5221$
:= $\$X5221$
:= $\$A5221$

:= $\$TMPDED$ %0
wff 3686 : $=(\supset(\wedge\,h\,j)\,x)\,\$STMPDED_{o,\ldots}$:= $\$TMPDED$ $K8024$
%$\$STMPDED$
$\supset h\,(\supset j\,x)$:= $\$STMPDED$
$\supset_{ooo}h_o(\supset_{ooo}j_o x_o)$:= $\$STMPDED$
%$K8024$
$=(\supset(\wedge\,h\,j)\,x)\,\$STMPDED$:= $\$TMPDED$ $K8024$
$=_{ooo}(\supset_{ooo}(\wedge_{ooo}h_o j_o)x_o)\$STMPDED_o$:= $\$TMPDED$ $K8024$
:= $\$TMPDED$

use Proof Template A5221 (Sub): B $\rightarrow$ B [x/A]
:= $\$B5221$ %0

\# wff 3686 : $=(\supset(\wedge\, h\, j)\, x)\, \$STMPDED_{o,\ldots}$:= $\$B5221$ $K8024$
:= $\$T5221$ o
\# wff 2 : o_τ := $\$T5221$
:= $\$X5221$ j_o
\# wff 210 : j_o := $\$X5221$
:= $\$A5221$ %1/13
\# wff 210 : j_o := $\$A5221$ $\$X5221$
<< A5221.r0t.txt
:= $\$B5221$
:= $\$T5221$
:= $\$X5221$
:= $\$A5221$

:= $\$TMPDED$ %0
\# wff 3686 : $=(\supset(\wedge\, h\, j)\, x)\, \$STMPDED_{o,\ldots}$:= $\$TMPDED$ $K8024$
%$\$STMPDED$
\# $\supset h\,(\supset j\, x)$:= $\$STMPDED$
\# $\supset_{ooo} h_o(\supset_{ooo} j_o x_o)$:= $\$STMPDED$
%$K8024$
\# $=(\supset(\wedge\, h\, j)\, x)\, \$STMPDED$:= $\$TMPDED$ $K8024$
\# $=_{ooo}(\supset_{ooo}(\wedge_{ooo} h_o j_o) x_o)\$STMPDED_o$:= $\$TMPDED$ $K8024$
:= $\$TMPDED$

\#\# use Proof Template A5221 (Sub): B → B [x/A]
:= $\$B5221$ %0
\# wff 3686 : $=(\supset(\wedge\, h\, j)\, x)\, \$STMPDED_{o,\ldots}$:= $\$B5221$ $K8024$
:= $\$T5221$ o
\# wff 2 : o_τ := $\$T5221$
:= $\$X5221$ x_o
\# wff 16 : x_o := $\$X5221$
:= $\$A5221$ %1/7
\# wff 16 : x_o := $\$A5221$ $\$X5221$
<< A5221.r0t.txt
:= $\$B5221$
:= $\$T5221$
:= $\$X5221$
:= $\$A5221$
%0
\# $=(\supset(\wedge\, h\, j)\, x)\, \$STMPDED$:= $K8024$
\# $=_{ooo}(\supset_{ooo}(\wedge_{ooo} h_o j_o) x_o)\$STMPDED_o$:= $K8024$

\#\# use Proof Template A5201b (Swap): A = B → B = A
<< A5201b.r0t.txt
%0
\# $=\$STMPDED\,(\supset(\wedge\, h\, j)\, x)$
\# $=_{ooo}\$STMPDED_o(\supset_{ooo}(\wedge_{ooo} h_o j_o) x_o)$

%$\$STMPDED$
\# $\supset h\,(\supset j\, x)$:= $\$STMPDED$

\# $\supset_{ooo} h_o (\supset_{ooo} j_o x_o)$:= $\$STMPDED$
:= $\$STMPDED$
§s %0 1 %1
\# $\supset (\wedge\, h\, j)\, x$
Include end (K8026.r0t.txt) [newfile=(K8026.r0a.txt)]
>>>

##
Q.E.D.
##

%0
\# $\supset (\wedge\, h\, j)\, x$
\# $\supset_{ooo} (\wedge_{ooo} h_o j_o) x_o$

2.1.83 Results for File K8027.r0a.txt

##
Proof Template K8027: (A ∧ B) ⊃ C → (B ∧ A) ⊃ C
(Hypotheses Swap)
##
Source: [Kubota 2017 (doi: 10.4444/100.10)]
##
Copyright (c) 2017 Owl of Minerva Press GmbH. All rights reserved.
Written by Ken Kubota (<mail@kenkubota.de>).
##
This file is part of the publication of the mathematical logic $\mathcal{R}_0$.
For more information, visit: <http://doi.org/10.4444/100.10>
##

##
Assumptions and Resulting Syntactical Variables
##

<< basics.r0.txt

the assumption as last theorem on stack (%0)
§! $\supset_{ooo} (\wedge_{ooo} a_o b_o) c_o$
\# $\supset (\wedge\, a\, b)\, c$

##
Include Proof Template
##

<<< K8027.r0t.txt
Include begin (K8027.r0t.txt) [oldfile=(K8027.r0a.txt)]

##
Proof Template K8027: (A ∧ B) ⊃ C → (B ∧ A) ⊃ C
(Hypotheses Swap)
##
Source: [Kubota 2017 (doi: 10.4444/100.10)]
##

Written by Ken Kubota (<mail@kenkubota.de>).
##
This file is part of the publication of the mathematical logic $\mathcal{R}_0$.
For more information, visit: <http://doi.org/10.4444/100.10>
##

define variable first (before inclusion of file)
:= $\$HTMPSWPHYP$ %0
wff 212 : $\supset (\wedge\, a\, b)\, c_o$:= $\$HTMPSWPHYP$

##
Proof Template
##

<< K8007.r0.txt
$\%K8007$
$COMMT\, o\, \wedge$:= $K8007$
$COMMT_{o(\backslash 4\backslash 4\backslash 3)\tau}o_\tau \wedge_{ooo}$:= $K8007$
§\ $COMMT_{o(\backslash 4\backslash 4\backslash 3)\tau}o_\tau$
$= (COMMT\, o)\, [\lambda f.(= (f\, x\, y)\, (f\, y\, x))]$
§s %1 2 %0
$[\lambda f.(= (f\, x\, y)\, (f\, y\, x))]\, \wedge$
§\ $[\lambda f_{ooo}.(=_{ooo}(f_{ooo}x_o y_o)(f_{ooo}y_o x_o))_o]\wedge_{ooo}$
$= ([\lambda f.(= (f\, x\, y)\, (f\, y\, x))]\, \wedge)\, (= (\wedge\, x\, y)\, (\wedge\, y\, x))$
§s %1 1 %0
$= (\wedge\, x\, y)\, (\wedge\, y\, x)$

:= $\$TMPSWPHYP$ %0
wff 1573 : $= (\wedge\, x\, y)\, (\wedge\, y\, x)_{o,\ldots}$:= $\$TMPSWPHYP$
$\%\$HTMPSWPHYP$
$\supset (\wedge\, a\, b)\, c$:= $\$HTMPSWPHYP$
$\supset_{ooo}(\wedge_{ooo}a_o b_o)c_o$:= $\$HTMPSWPHYP$
$\%\$TMPSWPHYP$
$= (\wedge\, x\, y)\, (\wedge\, y\, x)$:= $\$TMPSWPHYP$
$=_{ooo}(\wedge_{ooo}x_o y_o)(\wedge_{ooo}y_o x_o)$:= $\$TMPSWPHYP$
:= $\$TMPSWPHYP$

use Proof Template A5221 (Sub): B → B [x/A]
:= $\$B5221$ %0
wff 1573 : $= (\wedge\, x\, y)\, (\wedge\, y\, x)_{o,\ldots}$:= $\$B5221$

:= $\$T5221$ o
wff 2 : o_τ := $\$T5221$
:= $\$X5221$ x_o
wff 16 : x_o := $\$X5221$
:= $\$A5221$ %1/21
wff 54 : a_o := $\$A5221$
<< A5221.r0t.txt
:= $\$B5221$
:= $\$T5221$
:= $\$X5221$
:= $\$A5221$

:= $\$TMPSWPHYP$ %0
wff 1590 : $=(\wedge\, a\, y)\,(\wedge\, y\, a)_{o,\ldots}$:= $\$TMPSWPHYP$
%$\$HTMPSWPHYP$
$\supset(\wedge\, a\, b)\, c$:= $\$HTMPSWPHYP$
$\supset_{ooo}(\wedge_{ooo} a_o b_o) c_o$:= $\$HTMPSWPHYP$
%$\$TMPSWPHYP$
$=(\wedge\, a\, y)\,(\wedge\, y\, a)$:= $\$TMPSWPHYP$
$=_{ooo}(\wedge_{ooo} a_o y_o)(\wedge_{ooo} y_o a_o)$:= $\$TMPSWPHYP$
:= $\$TMPSWPHYP$

use Proof Template A5221 (Sub): B → B [x/A]
:= $\$B5221$ %0
wff 1590 : $=(\wedge\, a\, y)\,(\wedge\, y\, a)_{o,\ldots}$:= $\$B5221$
:= $\$T5221$ o
wff 2 : o_τ := $\$T5221$
:= $\$X5221$ y_o
wff 34 : y_o := $\$X5221$
:= $\$A5221$ %1/11
wff 58 : b_o := $\$A5221$
<< A5221.r0t.txt
:= $\$B5221$
:= $\$T5221$
:= $\$X5221$
:= $\$A5221$
%0
$=(\wedge\, a\, b)\,(\wedge\, b\, a)$
$=_{ooo}(\wedge_{ooo} a_o b_o)(\wedge_{ooo} b_o a_o)$

%$\$HTMPSWPHYP$
$\supset(\wedge\, a\, b)\, c$:= $\$HTMPSWPHYP$
$\supset_{ooo}(\wedge_{ooo} a_o b_o) c_o$:= $\$HTMPSWPHYP$
:= $\$HTMPSWPHYP$
§s %0 5 %1
$\supset(\wedge\, b\, a)\, c$
Include end (K8027.r0t.txt) [newfile=(K8027.r0a.txt)]
>>>

##
Q.E.D.
##

%0
$\supset (\wedge\, b\, a)\, c$
$\supset_{ooo}(\wedge_{ooo} b_o a_o) c_o$

2.1.84 Results for File K8028.r0a.txt

##
Proof Template K8028 (∃ GenH): H ⊃ ([\x.B]A) → H ⊃ ∃ x: B
for any x of any type (Rule of Existential Generalization – with hypothesis)
##
Source: [cf. Andrews 2002 (ISBN 1-4020-0763-9), p. 229 (5242)]
##

Written by Ken Kubota (<mail@kenkubota.de>).
##
This file is part of the publication of the mathematical logic $\mathcal{R}_0$.
For more information, visit: <http://doi.org/10.4444/100.10>
##

##
Define Syntactical Variables
##

hypothesis: H
:= $\$H8028\ h_o$
wff 11 : h_o := $\$H8028$

type of substitute
:= $\$T8028\ t_\tau$
wff 4 : t_τ := $\$T8028$

proposition: [\x.B]
:= $\$B8028\ b_{o\$T8028}$
wff 12 : $b_{o\$T8028}$:= $\$B8028$

substitute: A
:= $\$A8028\ a_{\$T8028}$
wff 13 : $a_{\$T8028}$:= $\$A8028$

##
Assumptions and Resulting Syntactical Variables
##

<< basics.r0.txt

given proposition
§! $\supset_{ooo}\$H8028_o(\$B8028_{o\,\$T8028}\$A8028_{\$T8028})$
\# $\supset \$H8028\,(\$B8028\,\$A8028)$

\##
\##   Proof Template
\##

\## <<< K8028.r0t.txt
\## Include begin (K8028.r0t.txt) [oldfile=(K8028.r0a.txt)]
\##
\##   Proof Template K8028 (∃ GenH):   H ⊃ ([\x.B]A)   →   H ⊃ ∃ x: B
\##   for any x of any type (Rule of Existential Generalization – with hypothesis)
\##
\##   Source: [cf. Andrews 2002 (ISBN 1-4020-0763-9), p. 229 (5242)]
\##

\##   Written by Ken Kubota (<mail@kenkubota.de>).
\##
\##   This file is part of the publication of the mathematical logic $\mathcal{R}_0$.
\##   For more information, visit: <http://doi.org/10.4444/100.10>
\##

\## given proposition
:= $\$PTMP8028\ \supset_{ooo}\$H8028_o(\$B8028_{o\,\$T8028}\$A8028_{\$T8028})$
\# wff   213 : $\supset \$H8028\,(\$B8028\,\$A8028)_o$:= $\$PTMP8028$

\## Skipping file basics.r0.txt (already included)
<< A5205.r0.txt
<< A5231.r0.txt
<< K8005.r0.txt
<< K8008.r0.txt
<< K8015.r0.txt
<< K8017.r0.txt

\##
\##   Proof Template
\##

\## .1

$\%K8005$
\# $\supset x\,x$:= $K8005$
\# $\supset_{ooo} x_o x_o$:= $K8005$

use Proof Template A5221 (Sub): B $\rightarrow$ B [x/A]
:= $\$B5221$ %0
wff 1755 : $\supset x\, x_{o,\ldots}$:= $\$B5221$ $K8005$
:= $\$T5221$ o
wff 2 : o_τ := $\$T5221$
:= $\$X5221$ x_o
wff 19 : x_o := $\$X5221$
:= $\$A5221$ $\sim_{oo}(\exists_{o(o\backslash 3)\tau}\$T8028_\tau\$B8028_{o\,\$T8028})$
wff 2764 : $\sim(\exists\,\$T8028\,\$B8028)_o$:= $\$A5221$
<< A5221.r0t.txt
:= $\$B5221$
:= $\$T5221$
:= $\$X5221$
:= $\$A5221$
%0
$\supset(\sim(\exists\,\$T8028\,\$B8028))\,(\sim(\exists\,\$T8028\,\$B8028))$
$\supset_{ooo}(\sim_{oo}(\exists_{o(o\backslash 3)\tau}\$T8028_\tau\$B8028_{o\,\$T8028}))(\sim_{oo}(\exists_{o(o\backslash 3)\tau}\$T8028_\tau\$B8028_{o\,\$T8028}))$

.2

%$K8017$
$=\exists\,[\lambda\$T8028.[\lambda p.(\sim(\forall\,\$T8028\,[\lambda x.(\sim(p\,x))]))]]$:= $K8017$
$=_{o(o(o\backslash 3)\tau)(o(o\backslash 3)\tau)}\exists_{o(o\backslash 3)\tau}\ldots$
$\ldots[\lambda\$T8028_\tau.[\lambda p_{o\,\$T8028}.(\sim_{oo}(\forall_{o(o\backslash 3)\tau}\$T8028_\tau[\lambda x_{\$T8028}.(\sim_{oo}(p_{o\,\$T8028}x_{\$T8028}))_o]))_o]_{(o(o\,\$T8028))}]$
:= $K8017$
§s %1 28 %0
$\supset(\sim(\exists\,\$T8028\,\$B8028))\ \ldots$
$\ldots(\sim([\lambda\$T8028.[\lambda p.(\sim(\forall\,\$T8028\,[\lambda x.(\sim(p\,x))]))]]\,\$T8028\,\$B8028))$
§\ $[\lambda\$T8028_\tau.[\lambda p_{o\,\$T8028}.(\sim_{oo}(\forall_{o(o\backslash 3)\tau}\$T8028_\tau[\lambda x_{\$T8028}.(\sim_{oo}(p_{o\,\$T8028}x_{\$T8028}))_o]))_o]_{(o(o\,\$T8028))}]\ldots$
$\ldots\$T8028_\tau$
$=([\lambda\$T8028.[\lambda p.(\sim(\forall\,\$T8028\,[\lambda x.(\sim(p\,x))]))]]\,\$T8028)\ \ldots$
$\ldots[\lambda p.(\sim(\forall\,\$T8028\,[\lambda x.(\sim(p\,x))]))]$
§s %1 14 %0
$\supset(\sim(\exists\,\$T8028\,\$B8028))\,(\sim([\lambda p.(\sim(\forall\,\$T8028\,[\lambda x.(\sim(p\,x))]))]\,\$B8028))$
§\ $[\lambda p_{o\,\$T8028}.(\sim_{oo}(\forall_{o(o\backslash 3)\tau}\$T8028_\tau[\lambda x_{\$T8028}.(\sim_{oo}(p_{o\,\$T8028}x_{\$T8028}))_o]))_o]\$B8028_{o\,\$T8028}$
$=([\lambda p.(\sim(\forall\,\$T8028\,[\lambda x.(\sim(p\,x))]))]\,\$B8028)\,(\sim(\forall\,\$T8028\,[\lambda x.(\sim(\$B8028\,x))]))$
§s %1 7 %0
$\supset(\sim(\exists\,\$T8028\,\$B8028))\,(\sim(\sim(\forall\,\$T8028\,[\lambda x.(\sim(\$B8028\,x))])))$
:= $\$TTMP8028$ %0
wff 2794 : $\supset(\sim(\exists\,\$T8028\,\$B8028))\,(\sim(\sim(\forall\,\$T8028\,[\lambda x.(\sim(\$B8028\,x))])))_o$:=
$\$TTMP8028$

.3

%$K8008$
$=(\sim(\sim x))\,x$:= $K8008$
$=_{ooo}(\sim_{oo}(\sim_{oo}x_o))x_o$:= $K8008$

use Proof Template A5221 (Sub): B $\rightarrow$ B [x/A]
:= $\$B5221$ %0
wff 1847 : $=(\sim(\sim x))\,x_{o,\ldots}$:= $\$B5221$ $K8008$
:= $\$T5221$ o
wff 2 : o_τ := $\$T5221$
:= $\$X5221$ x_o
wff 19 : x_o := $\$X5221$
:= $\$A5221$ %1/15
wff 2790 : $\forall \$T8028\,[\lambda x.(\sim(\$B8028\,x))]_o$:= $\$A5221$
<< A5221.r0t.txt
:= $\$B5221$
:= $\$T5221$
:= $\$X5221$
:= $\$A5221$
%0
$=(\sim(\sim(\forall \$T8028\,[\lambda x.(\sim(\$B8028\,x))])))\,(\forall \$T8028\,[\lambda x.(\sim(\$B8028\,x))])$
$=_{ooo}(\sim_{oo}(\sim_{oo}(\forall_{o(o\backslash 3)\tau}\$T8028_\tau[\lambda x_{\$T8028}.(\sim_{oo}(\$B8028_{o\,\$T8028}x_{\$T8028}))_o])))\ldots$
$\ldots(\forall_{o(o\backslash 3)\tau}\$T8028_\tau[\lambda x_{\$T8028}.(\sim_{oo}(\$B8028_{o\,\$T8028}x_{\$T8028}))_o])$

%$\$TTMP8028$
$\supset(\sim(\exists \$T8028\,\$B8028))\,(\sim(\sim(\forall \$T8028\,[\lambda x.(\sim(\$B8028\,x))])))$:=
$\$TTMP8028$
$\supset_{ooo}(\sim_{oo}(\exists_{o(o\backslash 3)\tau}\$T8028_\tau\$B8028_{o\,\$T8028}))\ldots$
$\ldots(\sim_{oo}(\sim_{oo}(\forall_{o(o\backslash 3)\tau}\$T8028_\tau[\lambda x_{\$T8028}.(\sim_{oo}(\$B8028_{o\,\$T8028}x_{\$T8028}))_o])))$:= $\$TTMP8028$
:= $\$TTMP8028$
§s %0 3 %1
$\supset(\sim(\exists \$T8028\,\$B8028))\,(\forall \$T8028\,[\lambda x.(\sim(\$B8028\,x))])$
§\ $\forall_{o(o\backslash 3)\tau}\$T8028_\tau$
$=(\forall \$T8028)\,[\lambda p.(=[\lambda x.T]\,p)]$
§s %1 6 %0
$\supset(\sim(\exists \$T8028\,\$B8028))\,([\lambda p.(=[\lambda x.T]\,p)]\,[\lambda x.(\sim(\$B8028\,x))])$
§\ $[\lambda p_{o\,\$T8028}.(=_{o(o\,\$T8028)(o\,\$T8028)}[\lambda x_{\$T8028}.T_o]p_{o\,\$T8028})_o][\lambda x_{\$T8028}.(\sim_{oo}(\$B8028_{o\,\$T8028}x_{\$T8028}))_o]$
$=([\lambda p.(=[\lambda x.T]\,p)]\,[\lambda x.(\sim(\$B8028\,x))])\,(=[\lambda x.T]\,[\lambda x.(\sim(\$B8028\,x))])$
§s %1 3 %0
$\supset(\sim(\exists \$T8028\,\$B8028))\,(=[\lambda x.T]\,[\lambda x.(\sim(\$B8028\,x))])$
:= $\$TTMP8028$ %0
wff 2834 : $\supset(\sim(\exists \$T8028\,\$B8028))\,(=[\lambda x.T]\,[\lambda x.(\sim(\$B8028\,x))])_o$:=
$\$TTMP8028$

.4

§= $_o$ $[\lambda x_{\$T8028}.T_o]\$A8028_{\$T8028}$
$=([\lambda x.T]\,\$A8028)\,([\lambda x.T]\,\$A8028)$

use Proof Template K8003 (Intro): A $\rightarrow$ H $\supset$ A
:= $\$A8003$ %0
wff 2837 : $=([\lambda x.T]\,\$A8028)\,([\lambda x.T]\,\$A8028)_o$:= $\$A8003$
:= $\$H8003$ $\sim_{oo}(\exists_{o(o\backslash 3)\tau}\$T8028_\tau\$B8028_{o\,\$T8028})$
wff 2764 : $\sim(\exists \$T8028\,\$B8028)_o$:= $\$H8003$

<< K8003.r0t.txt
:= $\$A8003$
:= $\$H8003$

:= $\$HTMP8028$ %0
\# wff 2838 : $\supset (\sim (\exists\, \$T8028\, \$B8028))\, (= ([\lambda x.T]\, \$A8028)\, ([\lambda x.T]\, \$A8028))_{o,\ldots}$:=
$\$HTMP8028$

%$\$TTMP8028$
\# $\supset (\sim (\exists\, \$T8028\, \$B8028))\, (= [\lambda x.T]\, [\lambda x.(\sim (\$B8028\, x))])$:= $\$TTMP8028$
\# $\supset_{ooo}(\sim_{oo}(\exists_{o(o\backslash 3)\tau}\$T8028_{\tau}\$B8028_{o\,\$T8028}))\ldots$
$\ldots(=_{o(o\,\$T8028)(o\,\$T8028)}[\lambda x_{\$T8028}.T_o][\lambda x_{\$T8028}.(\sim_{oo}(\$B8028_{o\,\$T8028}x_{\$T8028}))_o])$:= $\$TTMP8028$
:= $\$TTMP8028$
§s' %1 6 %0
\# $\supset (\sim (\exists\, \$T8028\, \$B8028))\, (= ([\lambda x.T]\, \$A8028)\, ([\lambda x.(\sim (\$B8028\, x))]\, \$A8028))$
§\ $[\lambda x_{\$T8028}.T_o]\$A8028_{\$T8028}$
\# $= ([\lambda x.T]\, \$A8028)\, T$
§s %1 13 %0
\# $\supset (\sim (\exists\, \$T8028\, \$B8028))\, (= T\, ([\lambda x.(\sim (\$B8028\, x))]\, \$A8028))$
§\ $[\lambda x_{\$T8028}.(\sim_{oo}(\$B8028_{o\,\$T8028}x_{\$T8028}))_o]\$A8028_{\$T8028}$
\# $= ([\lambda x.(\sim (\$B8028\, x))]\, \$A8028)\, (\sim (\$B8028\, \$A8028))$
§s %1 7 %0
\# $\supset (\sim (\exists\, \$T8028\, \$B8028))\, (= T\, (\sim (\$B8028\, \$A8028)))$

\#\# use Proof Template A5219cH (Rule T): H ⊃ (T = A) → H ⊃ A
:= $\$A5219cH$ %0
\# wff 2904 : $\supset (\sim (\exists\, \$T8028\, \$B8028))\, (= T\, (\sim (\$B8028\, \$A8028)))_o$:= $\$A5219cH$
<< A5219cH.r0t.txt
:= $\$A5219cH$

:= $\$TTMP8028$ %0
\# wff 2951 : $\supset (\sim (\exists\, \$T8028\, \$B8028))\, (\sim (\$B8028\, \$A8028))_o$:= $\$TTMP8028$
%$\$PTMP8028$
\# $\supset \$H8028\, (\$B8028\, \$A8028)$:= $\$PTMP8028$
\# $\supset_{ooo}\$H8028_o(\$B8028_{o\,\$T8028}\$A8028_{\$T8028})$:= $\$PTMP8028$
%$\$TTMP8028$
\# $\supset (\sim (\exists\, \$T8028\, \$B8028))\, (\sim (\$B8028\, \$A8028))$:= $\$TTMP8028$
\# $\supset_{ooo}(\sim_{oo}(\exists_{o(o\backslash 3)\tau}\$T8028_{\tau}\$B8028_{o\,\$T8028}))(\sim_{oo}(\$B8028_{o\,\$T8028}\$A8028_{\$T8028}))$
:= $\$TTMP8028$
:= $\$TTMP8028$

\#\# use Proof Template K8004 (Trans): (H ⊕ A), B → H ⊃ B
:= $\$HA8004$ %1
\# wff 213 : $\supset \$H8028\, (\$B8028\, \$A8028)_o$:= $\$HA8004$ $\$PTMP8028$
:= $\$B8004$ %0
\# wff 2951 : $\supset (\sim (\exists\, \$T8028\, \$B8028))\, (\sim (\$B8028\, \$A8028))_o$:= $\$B8004$
<< K8004.r0t.txt
:= $\$HA8004$
:= $\$B8004$

%0
$\supset \$H8028\,(\supset(\sim(\exists\,\$T8028\,\$B8028))\,(\sim(\$B8028\,\$A8028)))$
$\supset_{ooo}\$H8028_o\ldots$
$\ldots(\supset_{ooo}(\sim_{oo}(\exists_{o(o\backslash 3)\tau}\$T8028_\tau\$B8028_{o\,\$T8028}))(\sim_{oo}(\$B8028_{o\,\$T8028}\$A8028_{\$T8028})))$

use Proof Template K8026 (Deduction Theorem Reversed): H ⊃ (I ⊃ A) → (H ∧ I) ⊃ A
<< K8026.r0t.txt
:= $\$NTMP8028$ %0
wff 4993 : $\supset(\wedge\,\$H8028\,(\sim(\exists\,\$T8028\,\$B8028)))\,(\sim(\$B8028\,\$A8028))_{o,\ldots}$:= $\$NTMP8028$

.5

%$\$HTMP8028$
$\supset(\sim(\exists\,\$T8028\,\$B8028))\,(=([\lambda x.T]\,\$A8028)\,([\lambda x.T]\,\$A8028))$:= $\$HTMP8028$
$\supset_{ooo}(\sim_{oo}(\exists_{o(o\backslash 3)\tau}\$T8028_\tau\$B8028_{o\,\$T8028}))\ldots$
$\ldots(=_{ooo}([\lambda x_{\$T8028}.T_o]\$A8028_{\$T8028})([\lambda x_{\$T8028}.T_o]\$A8028_{\$T8028}))$:= $\$HTMP8028$
:= $\$HTMP8028$
%$\$PTMP8028$
$\supset \$H8028\,(\$B8028\,\$A8028)$:= $\$PTMP8028$
$\supset_{ooo}\$H8028_o(\$B8028_{o\,\$T8028}\$A8028_{\$T8028})$:= $\$PTMP8028$
:= $\$PTMP8028$

use Proof Template K8004 (Trans): (H ⊕ A), B → H ⊃ B
:= $\$HA8004$ %1
wff 2838 : $\supset(\sim(\exists\,\$T8028\,\$B8028))\,(=([\lambda x.T]\,\$A8028)\,([\lambda x.T]\,\$A8028))_{o,\ldots}$:= $\$HA8004$
:= $\$B8004$ %0
wff 213 : $\supset \$H8028\,(\$B8028\,\$A8028)_o$:= $\$B8004$
<< K8004.r0t.txt
:= $\$HA8004$
:= $\$B8004$
%0
$\supset(\sim(\exists\,\$T8028\,\$B8028))\,(\supset \$H8028\,(\$B8028\,\$A8028))$
$\supset_{ooo}(\sim_{oo}(\exists_{o(o\backslash 3)\tau}\$T8028_\tau\$B8028_{o\,\$T8028}))\ldots$
$\ldots(\supset_{ooo}\$H8028_o(\$B8028_{o\,\$T8028}\$A8028_{\$T8028}))$

use Proof Template K8026 (Deduction Theorem Reversed): H ⊃ (I ⊃ A) → (H ∧ I) ⊃ A
<< K8026.r0t.txt
%0
$\supset(\wedge(\sim(\exists\,\$T8028\,\$B8028))\,\$H8028)\,(\$B8028\,\$A8028)$
$\supset_{ooo}(\wedge_{ooo}(\sim_{oo}(\exists_{o(o\backslash 3)\tau}\$T8028_\tau\$B8028_{o\,\$T8028}))\$H8028_o)\ldots$
$\ldots(\$B8028_{o\,\$T8028}\$A8028_{\$T8028})$

.6

use Proof Template K8027: $(\mathrm{A} \wedge \mathrm{B}) \supset \mathrm{C} \quad \rightarrow \quad (\mathrm{B} \wedge \mathrm{A}) \supset \mathrm{C}$
<< K8027.r0t.txt
%0
$\supset (\wedge \, \$H8028 \, (\sim (\exists \, \$T8028 \, \$B8028))) \, (\$B8028 \, \$A8028)$
$\supset_{ooo}(\wedge_{ooo}\$H8028_{o}(\sim_{oo}(\exists_{o(o\backslash 3)\tau}\$T8028_{\tau}\$B8028_{o\,\$T8028}))) \ldots$
$\ldots (\$B8028_{o\,\$T8028}\$A8028_{\$T8028})$

.7

use Proof Template A5219bH (Rule T): $\mathrm{H} \supset \mathrm{A} \quad \rightarrow \quad \mathrm{H} \supset (\mathrm{A} = \mathrm{T})$
:= $\$A5219bH$ %0
wff 5186 : $\supset (\wedge \, \$H8028 \, (\sim (\exists \, \$T8028 \, \$B8028))) \, (\$B8028 \, \$A8028)_{o}$:= $\$A5219bH$
<< A5219bH.r0t.txt
:= $\$A5219bH$
%0
$\supset (\wedge \, \$H8028 \, (\sim (\exists \, \$T8028 \, \$B8028))) \, (= (\$B8028 \, \$A8028) \, T)$
$\supset_{ooo}(\wedge_{ooo}\$H8028_{o}(\sim_{oo}(\exists_{o(o\backslash 3)\tau}\$T8028_{\tau}\$B8028_{o\,\$T8028}))) \ldots$
$\ldots (=_{ooo}(\$B8028_{o\,\$T8028}\$A8028_{\$T8028})T_{o})$

%$\$NTMP8028$
$\supset (\wedge \, \$H8028 \, (\sim (\exists \, \$T8028 \, \$B8028))) \, (\sim (\$B8028 \, \$A8028))$:= $\$NTMP8028$
$\supset_{ooo}(\wedge_{ooo}\$H8028_{o}(\sim_{oo}(\exists_{o(o\backslash 3)\tau}\$T8028_{\tau}\$B8028_{o\,\$T8028}))) \ldots$
$\ldots (\sim_{oo}(\$B8028_{o\,\$T8028}\$A8028_{\$T8028}))$:= $\$NTMP8028$
:= $\$NTMP8028$
§s' %0 3 %1
$\supset (\wedge \, \$H8028 \, (\sim (\exists \, \$T8028 \, \$B8028))) \, (\sim T)$
:= $\$NTMP8028$ %0
wff 5289 : $\supset (\wedge \, \$H8028 \, (\sim (\exists \, \$T8028 \, \$B8028))) \, (\sim T)_{o}$:= $\$NTMP8028$

%$A5231a$
$= (\sim T) \, F$:= $A5231a$
$=_{ooo}(\sim_{oo}T_{o})F_{o}$:= $A5231a$

use Proof Template K8004 (Trans): $(\mathrm{H} \oplus \mathrm{A}), \mathrm{B} \quad \rightarrow \quad \mathrm{H} \supset \mathrm{B}$
:= $\$HA8004$ %1
wff 5289 : $\supset (\wedge \, \$H8028 \, (\sim (\exists \, \$T8028 \, \$B8028))) \, (\sim T)_{o}$:= $\$HA8004$
$\$NTMP8028$
:= $\$B8004$ %0
wff 1673 : $= (\sim T) \, F_{o,\ldots}$:= $\$B8004$ $A5231a$
<< K8004.r0t.txt
:= $\$HA8004$
:= $\$B8004$
%0
$\supset (\wedge \, \$H8028 \, (\sim (\exists \, \$T8028 \, \$B8028))) \, A5231a$
$\supset_{ooo}(\wedge_{ooo}\$H8028_{o}(\sim_{oo}(\exists_{o(o\backslash 3)\tau}\$T8028_{\tau}\$B8028_{o\,\$T8028})))A5231a_{o}$

%$\$NTMP8028$
$\supset (\wedge \, \$H8028 \, (\sim (\exists \, \$T8028 \, \$B8028))) \, (\sim T)$:= $\$NTMP8028$
$\supset_{ooo}(\wedge_{ooo}\$H8028_{o}(\sim_{oo}(\exists_{o(o\backslash 3)\tau}\$T8028_{\tau}\$B8028_{o\,\$T8028})))(\sim_{oo}T_{o})$:=

$\$NTMP8028$
:= $\$NTMP8028$
§s' %0 1 %1
\# $\supset (\wedge \$H8028 (\sim (\exists \$T8028 \$B8028))) F$

\## .8

<< K8025.r0t.txt
:= $\$DTMP8028$ %0
\# wff 5331 : $\supset \$H8028 (\supset (\sim (\exists \$T8028 \$B8028)) F)_{o,\dots}$:= $\$DTMP8028$

%$K8015$
\# $= (\supset x\, F) (\sim x)$:= $K8015$
\# $=_{ooo}(\supset_{ooo} x_o F_o)(\sim_{oo} x_o)$:= $K8015$

\## use Proof Template A5221 (Sub): B → B [x/A]
:= $\$B5221$ %0
\# wff 2742 : $= (\supset x\, F) (\sim x)_o$:= $\$B5221$ $K8015$
:= $\$T5221$ o
\# wff 2 : o_τ := $\$T5221$
:= $\$X5221$ x_o
\# wff 19 : x_o := $\$X5221$
:= $\$A5221$ %1/13
\# wff 2764 : $\sim (\exists \$T8028 \$B8028)_{o,\dots}$:= $\$A5221$
<< A5221.r0t.txt
:= $\$B5221$
:= $\$T5221$
:= $\$X5221$
:= $\$A5221$
%0
\# $= (\supset (\sim (\exists \$T8028 \$B8028)) F) (\sim (\sim (\exists \$T8028 \$B8028)))$
\# $=_{ooo}(\supset_{ooo}(\sim_{oo}(\exists_{o(o\backslash 3)\tau} \$T8028_\tau \$B8028_{o\,\$T8028}))F_o)\dots$
$\dots(\sim_{oo}(\sim_{oo}(\exists_{o(o\backslash 3)\tau} \$T8028_\tau \$B8028_{o\,\$T8028})))$

:= $\$TTMP8028$ %0
\# wff 5368 : $= (\supset (\sim (\exists \$T8028 \$B8028)) F) (\sim (\sim (\exists \$T8028 \$B8028)))_{o,\dots}$:=
$\$TTMP8028$

\## .9

%$K8008$
\# $= (\sim (\sim x))\, x$:= $K8008$
\# $=_{ooo}(\sim_{oo}(\sim_{oo} x_o))x_o$:= $K8008$

\## use Proof Template A5221 (Sub): B → B [x/A]
:= $\$B5221$ %0
\# wff 1847 : $= (\sim (\sim x))\, x_{o,\dots}$:= $\$B5221$ $K8008$
:= $\$T5221$ o
\# wff 2 : o_τ := $\$T5221$

:= $\$X5221\ x_o$
\# wff 19 : x_o := $\$X5221$
:= $\$A5221$ %1/15
\# wff 2763 : $\exists\,\$T8028\,\$B8028_o$:= $\$A5221$
<< A5221.r0t.txt
:= $\$B5221$
:= $\$T5221$
:= $\$X5221$
:= $\$A5221$
%0
\# $=(\sim(\sim(\exists\,\$T8028\,\$B8028)))\,(\exists\,\$T8028\,\$B8028)$
\# $=_{ooo}(\sim_{oo}(\sim_{oo}(\exists_{o(o\backslash 3)\tau}\$T8028_\tau\$B8028_{o\,\$T8028})))(\exists_{o(o\backslash 3)\tau}\$T8028_\tau\$B8028_{o\,\$T8028})$

%$\$TTMP8028$
\# $=(\supset(\sim(\exists\,\$T8028\,\$B8028))\,F)\,(\sim(\sim(\exists\,\$T8028\,\$B8028)))$:= $\$TTMP8028$
\# $=_{ooo}(\supset_{ooo}(\sim_{oo}(\exists_{o(o\backslash 3)\tau}\$T8028_\tau\$B8028_{o\,\$T8028}))F_o)\ldots$
$\ldots(\sim_{oo}(\sim_{oo}(\exists_{o(o\backslash 3)\tau}\$T8028_\tau\$B8028_{o\,\$T8028})))$:= $\$TTMP8028$
:= $\$TTMP8028$
§s %0 3 %1
\# $=(\supset(\sim(\exists\,\$T8028\,\$B8028))\,F)\,(\exists\,\$T8028\,\$B8028)$
:= $\$TTMP8028$ %0
\# wff 5383 : $=(\supset(\sim(\exists\,\$T8028\,\$B8028))\,F)\,(\exists\,\$T8028\,\$B8028)_o$:= $\$TTMP8028$

%$\$DTMP8028$
\# $\supset\$H8028\,(\supset(\sim(\exists\,\$T8028\,\$B8028))\,F)$:= $\$DTMP8028$
\# $\supset_{ooo}\$H8028_o(\supset_{ooo}(\sim_{oo}(\exists_{o(o\backslash 3)\tau}\$T8028_\tau\$B8028_{o\,\$T8028}))F_o)$:= $\$DTMP8028$
%$\$TTMP8028$
\# $=(\supset(\sim(\exists\,\$T8028\,\$B8028))\,F)\,(\exists\,\$T8028\,\$B8028)$:= $\$TTMP8028$
\# $=_{ooo}(\supset_{ooo}(\sim_{oo}(\exists_{o(o\backslash 3)\tau}\$T8028_\tau\$B8028_{o\,\$T8028}))F_o)\ldots$
$\ldots(\exists_{o(o\backslash 3)\tau}\$T8028_\tau\$B8028_{o\,\$T8028})$:= $\$TTMP8028$
:= $\$TTMP8028$

\#\# use Proof Template K8004 (Trans): (H ⊕ A), B → H ⊃ B
:= $\$HA8004$ %1
\# wff 5331 : $\supset\$H8028\,(\supset(\sim(\exists\,\$T8028\,\$B8028))\,F)_{o,\ldots}$ ...
... := $\$DTMP8028$ $\$HA8004$
:= $\$B8004$ %0
\# wff 5383 : $=(\supset(\sim(\exists\,\$T8028\,\$B8028))\,F)\,(\exists\,\$T8028\,\$B8028)_o$:= $\$B8004$
<< K8004.r0t.txt
:= $\$HA8004$
:= $\$B8004$
%0
\# $\supset\$H8028\,(=(\supset(\sim(\exists\,\$T8028\,\$B8028))\,F)\,(\exists\,\$T8028\,\$B8028))$
\# $\supset_{ooo}\$H8028_o\ldots$
$\ldots(=_{ooo}(\supset_{ooo}(\sim_{oo}(\exists_{o(o\backslash 3)\tau}\$T8028_\tau\$B8028_{o\,\$T8028}))F_o)(\exists_{o(o\backslash 3)\tau}\$T8028_\tau\$B8028_{o\,\$T8028}))$

%$\$DTMP8028$
\# $\supset\$H8028\,(\supset(\sim(\exists\,\$T8028\,\$B8028))\,F)$:= $\$DTMP8028$
\# $\supset_{ooo}\$H8028_o(\supset_{ooo}(\sim_{oo}(\exists_{o(o\backslash 3)\tau}\$T8028_\tau\$B8028_{o\,\$T8028}))F_o)$:= $\$DTMP8028$

:= $\$DTMP8028$
§s' %0 1 %1
\# $\supset \$H8028\,(\exists\,\$T8028\,\$B8028)$
\## Include end (K8028.r0t.txt) [newfile=(K8028.r0a.txt)]
>>>

\##
\## Undefine Syntactical Variables
\##

:= $\$H8028$
:= $\$T8028$
:= $\$B8028$
:= $\$A8028$

\##
\## Q.E.D.
\##

%0
\# $\supset h\,(\exists\,t\,b)$
\# $\supset_{ooo} h_o (\exists_{o(o\backslash 3)\tau} t_\tau b_{ot})$

2.1.85 Results for File K8029.r0.txt

\##
\## Proof K8029: A ⊃ B = (∼ B) ⊃ (∼ A)
\##
\##
\## Source: [Kubota 2017 (doi: 10.4444/100.10)]
\##

\## Written by Ken Kubota (<mail@kenkubota.de>).
\##
\## This file is part of the publication of the mathematical logic $\mathcal{R}_0$.
\## For more information, visit: <http://doi.org/10.4444/100.10>
\##

<< basics.r0.txt
<< K8008.r0.txt
<< K8012.r0.txt
<< K8022.r0.txt

\##
\## Proof
\##

.1

§= $_o \supset_{ooo} x_o y_o$
$= (\supset x\, y)\, (\supset x\, y)$
%$K8022$
$= \supset [\lambda x.[\lambda y.(\vee\, (\sim x)\, y)]]$:= $K8022$
$=_{o(ooo)(ooo)} \supset_{ooo} [\lambda x_o.[\lambda y_o.(\vee_{ooo} (\sim_{oo} x_o) y_o)_o]_{(oo)}]$:= $K8022$
§s %1 12 %0
$= (\supset x\, y)\, ([\lambda x.[\lambda y.(\vee\, (\sim x)\, y)]]\, x\, y)$
§\ $[\lambda x_o.[\lambda y_o.(\vee_{ooo} (\sim_{oo} x_o) y_o)_o]_{(oo)}] x_o$
$= ([\lambda x.[\lambda y.(\vee\, (\sim x)\, y)]]\, x)\, [\lambda y.(\vee\, (\sim x)\, y)]$
§s %1 6 %0
$= (\supset x\, y)\, ([\lambda y.(\vee\, (\sim x)\, y)]\, y)$
§\ $[\lambda y_o.(\vee_{ooo} (\sim_{oo} x_o) y_o)_o] y_o$
$= ([\lambda y.(\vee\, (\sim x)\, y)]\, y)\, (\vee\, (\sim x)\, y)$
§s %1 3 %0
$= (\supset x\, y)\, (\vee\, (\sim x)\, y)$
:= $\$TMP8029$ %0
wff 2553 : $= (\supset x\, y)\, (\vee\, (\sim x)\, y)_{o,\dots}$:= $\$TMP8029$

.2

%$K8012$
$COMMT\, o \vee$:= $K8012$
$COMMT_{o(\backslash 4 \backslash 4 \backslash 3)\tau} o_\tau \vee_{ooo}$:= $K8012$
§\ $COMMT_{o(\backslash 4 \backslash 4 \backslash 3)\tau} o_\tau$
$= (COMMT\, o)\, [\lambda f.(= (f\, x\, y)\, (f\, y\, x))]$
§s %1 2 %0
$[\lambda f.(= (f\, x\, y)\, (f\, y\, x))]\, \vee$
§\ $[\lambda f_{ooo}.(=_{ooo} (f_{ooo} x_o y_o)(f_{ooo} y_o x_o))_o] \vee_{ooo}$
$= ([\lambda f.(= (f\, x\, y)\, (f\, y\, x))]\, \vee)\, (= (\vee\, x\, y)\, (\vee\, y\, x))$
§s %1 1 %0
$= (\vee\, x\, y)\, (\vee\, y\, x)$

use Proof Template A5221 (Sub): B → B [x/A]
:= $\$B5221$ %0
wff 2054 : $= (\vee\, x\, y)\, (\vee\, y\, x)_{o,\dots}$:= $\$B5221$
:= $\$T5221$ o
wff 2 : o_τ := $\$T5221$
:= $\$X5221$ x_o
wff 16 : x_o := $\$X5221$
:= $\$A5221$ $\sim_{oo} \$X5221_o$
wff 1660 : $\sim \$X5221_{o,\dots}$:= $\$A5221$
<< A5221.r0t.txt
:= $\$B5221$
:= $\$T5221$
:= $\$X5221$
:= $\$A5221$

%0
$= (\vee\,(\sim x)\,y)\,(\vee\,y\,(\sim x))$
$=_{ooo}(\vee_{ooo}(\sim_{oo}x_o)y_o)(\vee_{ooo}y_o(\sim_{oo}x_o))$

%$\$TMP8029$
$= (\supset x\,y)\,(\vee\,(\sim x)\,y)$:= $\$TMP8029$
$=_{ooo}(\supset_{ooo}x_oy_o)(\vee_{ooo}(\sim_{oo}x_o)y_o)$:= $\$TMP8029$
:= $\$TMP8029$
§s %0 3 %1
$= (\supset x\,y)\,(\vee\,y\,(\sim x))$
:= $\$LTMP8029$ %0
wff 3123 : $= (\supset x\,y)\,(\vee\,y\,(\sim x))_o$:= $\$LTMP8029$

.3

§= $_o\ \supset_{ooo}(\sim_{oo}y_o)(\sim_{oo}x_o)$
$= (\supset (\sim y)\,(\sim x))\,(\supset (\sim y)\,(\sim x))$
%$K8022$
$= \supset [\lambda x.[\lambda y.(\vee\,(\sim x)\,y)]]$:= $K8022$
$=_{o(ooo)(ooo)}\supset_{ooo}[\lambda x_o.[\lambda y_o.(\vee_{ooo}(\sim_{oo}x_o)y_o)_o]_{(oo)}]$:= $K8022$
§s %1 12 %0
$= (\supset (\sim y)\,(\sim x))\,([\lambda x.[\lambda y.(\vee\,(\sim x)\,y)]]\,(\sim y)\,(\sim x))$
§r /25 z_o
$= [\lambda y.(\vee\,(\sim x)\,y)]\,[\lambda z.(\vee\,(\sim x)\,z)]$
§s %1 25 %0
$= (\supset (\sim y)\,(\sim x))\,([\lambda x.[\lambda z.(\vee\,(\sim x)\,z)]]\,(\sim y)\,(\sim x))$
§\ $[\lambda x_o.[\lambda z_o.(\vee_{ooo}(\sim_{oo}x_o)z_o)_o]_{(oo)}](\sim_{oo}y_o)$
$= ([\lambda x.[\lambda z.(\vee\,(\sim x)\,z)]]\,(\sim y))\,[\lambda z.(\vee\,(\sim (\sim y))\,z)]$
§s %1 6 %0
$= (\supset (\sim y)\,(\sim x))\,([\lambda z.(\vee\,(\sim (\sim y))\,z)]\,(\sim x))$
§\ $[\lambda z_o.(\vee_{ooo}(\sim_{oo}(\sim_{oo}y_o))z_o)_o](\sim_{oo}x_o)$
$= ([\lambda z.(\vee\,(\sim (\sim y))\,z)]\,(\sim x))\,(\vee\,(\sim (\sim y))\,(\sim x))$
§s %1 3 %0
$= (\supset (\sim y)\,(\sim x))\,(\vee\,(\sim (\sim y))\,(\sim x))$
:= $\$TMP8029$ %0
wff 3152 : $= (\supset (\sim y)\,(\sim x))\,(\vee\,(\sim (\sim y))\,(\sim x))_o$:= $\$TMP8029$

.4

%$K8008$
$= (\sim (\sim x))\,x$:= $K8008$
$=_{ooo}(\sim_{oo}(\sim_{oo}x_o))x_o$:= $K8008$

use Proof Template A5221 (Sub): B → B [x/A]
:= $\$B5221$ %0
wff 1663 : $= (\sim (\sim x))\,x_{o,\ldots}$:= $\$B5221$ $K8008$
:= $\$T5221$ o
wff 2 : o_τ := $\$T5221$
:= $\$X5221$ x_o

\# wff 16 : x_o := $\$X5221$
:= $\$A5221$ y_o
\# wff 34 : y_o := $\$A5221$
<< A5221.r0t.txt
:= $\$B5221$
:= $\$T5221$
:= $\$X5221$
:= $\$A5221$
%0
\# $= (\sim (\sim y))\, y$
\# $=_{ooo}(\sim_{oo}(\sim_{oo}y_o))y_o$

%$\$TMP8029$
\# $= (\supset (\sim y)\, (\sim x))\, (\vee\, (\sim (\sim y))\, (\sim x))$:= $\$TMP8029$
\# $=_{ooo}(\supset_{ooo}(\sim_{oo}y_o)(\sim_{oo}x_o))(\vee_{ooo}(\sim_{oo}(\sim_{oo}y_o))(\sim_{oo}x_o))$:= $\$TMP8029$
:= $\$TMP8029$
§s %0 13 %1
\# $= (\supset (\sim y)\, (\sim x))\, (\vee\, y\, (\sim x))$

\## use Proof Template A5201b (Swap): A = B → B = A
<< A5201b.r0t.txt
%0
\# $= (\vee\, y\, (\sim x))\, (\supset (\sim y)\, (\sim x))$
\# $=_{ooo}(\vee_{ooo}y_o(\sim_{oo}x_o))(\supset_{ooo}(\sim_{oo}y_o)(\sim_{oo}x_o))$

%$\$LTMP8029$
\# $= (\supset x\, y)\, (\vee\, y\, (\sim x))$:= $\$LTMP8029$
\# $=_{ooo}(\supset_{ooo}x_oy_o)(\vee_{ooo}y_o(\sim_{oo}x_o))$:= $\$LTMP8029$
:= $\$LTMP8029$
§s %0 3 %1
\# $= (\supset x\, y)\, (\supset (\sim y)\, (\sim x))$

:= $K8029$ %0
\# wff 3185 : $= (\supset x\, y)\, (\supset (\sim y)\, (\sim x))_o$:= $K8029$

\##
\## Q.E.D.
\##

%0
\# $= (\supset x\, y)\, (\supset (\sim y)\, (\sim x))$:= $K8029$
\# $=_{ooo}(\supset_{ooo}x_oy_o)(\supset_{ooo}(\sim_{oo}y_o)(\sim_{oo}x_o))$:= $K8029$

2.1.86 Results for File K8030.r0a.txt

\##
\## Proof Template K8030 (∃ Rule): (H ∧ B) ⊃ A → (H ∧ ∃ x: B) ⊃ A
\## for any x of any type, provided x is not free in H or in A (Existential Rule)

```
##
##   Source: [cf. Andrews 2002 (ISBN 1-4020-0763-9), p. 230 (5244)]
##
##   Copyright (c) 2017 Owl of Minerva Press GmbH. All rights reserved.
##   Written by Ken Kubota (<mail@kenkubota.de>).
##
##   This file is part of the publication of the mathematical logic R0.
##   For more information, visit: <http://doi.org/10.4444/100.10>
##


##
##   Define Syntactical Variables
##

<< basics.r0.txt


## type of variable
:=  $T8030 u_τ
# wff    208  :        u_τ       :=  $T8030

## the variable
:=  $X8030 x_$T8030
# wff    209  :        x_$T8030       :=  $X8030

## the proposition
:=  $A8030 ⊃_ooo(∧_ooo h_o b_o)a_o
# wff    214  :        ⊃ (∧ h b) a_o      :=  $A8030


##
##   Assumptions and Resulting Syntactical Variables
##

§! $A8030
#                 ⊃ (∧ h b) a      :=  $A8030


##
##   Proof Template
##

## <<< K8030.r0t.txt
## Include begin (K8030.r0t.txt) [oldfile=(K8030.r0a.txt)]
##
##   Proof Template K8030 (∃ Rule):   (H ∧ B) ⊃ A   →   (H ∧ ∃ x: B) ⊃ A
##         for any x of any type, provided x is not free in H or in A (Existential Rule)
##
##   Source: [cf. Andrews 2002 (ISBN 1-4020-0763-9), p. 230 (5244)]
```

##

Written by Ken Kubota (<mail@kenkubota.de>).
##
This file is part of the publication of the mathematical logic $\mathcal{R}_0$.
For more information, visit: <http://doi.org/10.4444/100.10>
##

define variable first (save before inclusion of files)
:= $\$TMP8030$ %0
wff 214 : $\supset (\wedge\, h\, b)\, a_o$:= $\$A8030$ $\$TMP8030$

<< K8008.r0.txt
<< K8016.r0.txt
<< K8017.r0.txt
<< K8029.r0.txt

shorthands
:= $\$PTMP8030$ $[\lambda p_{o\,\$T8030}.(\sim_{oo}(\exists_{o(o\backslash 3)\tau}\$T8030_\tau[\lambda \$X8030_{\$T8030}.(\sim_{oo}(p_{o\,\$T8030}\$X8030_{\$T8030}))_o]))_o]$
wff 3371 : $[\lambda p.(\sim(\exists\,\$T8030\,[\lambda \$X8030.(\sim(p\,\$X8030))]))]_{o(o\,\$T8030)}$:= $\$PTMP8030$
:= $\$PBTMP8030$ $[\lambda \$X8030_{\$T8030}.(\sim_{oo}b_o)_o]\$X8030_{\$T8030}$
wff 3373 : $[\lambda \$X8030.(\sim b)]\,\$X8030_o$:= $\$PBTMP8030$
:= $\$ETMP8030$ $\exists_{o(o\backslash 3)\tau}\$T8030_\tau[\lambda \$X8030_{\$T8030}.b_o]$
wff 3375 : $\exists\,\$T8030\,[\lambda \$X8030.b]_o$:= $\$ETMP8030$

%$\$A8030$
$\supset (\wedge\, h\, b)\, a$:= $\$A8030$ $\$TMP8030$
$\supset_{ooo}(\wedge_{ooo}h_o b_o)a_o$:= $\$A8030$ $\$TMP8030$
:= $\$TMP8030$

##
Proof Template
##

.1

use Proof Template K8025 (Deduction Theorem): (H ∧ I) ⊃ A → H ⊃ (I ⊃ A)
<< K8025.r0t.txt
:= $\$TMP8030$ %0
wff 4636 : $\supset h\,(\supset b\, a)_{o,\ldots}$:= $\$TMP8030$

.2

%$K8029$
$=(\supset x\, y)\,(\supset(\sim y)\,(\sim x))$:= $K8029$
$=_{ooo}(\supset_{ooo}x_o y_o)(\supset_{ooo}(\sim_{oo}y_o)(\sim_{oo}x_o))$:= $K8029$

use Proof Template A5221 (Sub): B → B [x/A]
:= $\$B5221$ %0
wff 3361 : $=(\supset x\, y)\,(\supset(\sim y)\,(\sim x))_o$:= $\$B5221$ $K8029$
:= $\$T5221$ o
wff 2 : o_τ := $\$T5221$
:= $\$X5221$ x_o
wff 16 : x_o := $\$X5221$
:= $\$A5221$ $\supset_{ooo}h_o(\supset_{ooo}b_oa_o)/13$
wff 58 : b_o := $\$A5221$
<< A5221.r0t.txt
:= $\$B5221$
:= $\$T5221$
:= $\$X5221$
:= $\$A5221$
%0
$=(\supset b\, y)\,(\supset(\sim y)\,(\sim b))$
$=_{ooo}(\supset_{ooo}b_oy_o)(\supset_{ooo}(\sim_{oo}y_o)(\sim_{oo}b_o))$

use Proof Template A5221 (Sub): B → B [x/A]
:= $\$B5221$ %0
wff 4674 : $=(\supset b\, y)\,(\supset(\sim y)\,(\sim b))_{o,\dots}$:= $\$B5221$
:= $\$T5221$ o
wff 2 : o_τ := $\$T5221$
:= $\$X5221$ y_o
wff 34 : y_o := $\$X5221$
:= $\$A5221$ $\supset_{ooo}h_o(\supset_{ooo}b_oa_o)/7$
wff 54 : a_o := $\$A5221$
<< A5221.r0t.txt
:= $\$B5221$
:= $\$T5221$
:= $\$X5221$
:= $\$A5221$
%0
$=(\supset b\, a)\,(\supset(\sim a)\,(\sim b))$
$=_{ooo}(\supset_{ooo}b_oa_o)(\supset_{ooo}(\sim_{oo}a_o)(\sim_{oo}b_o))$

%$\$TMP8030$
$\supset h\,(\supset b\, a)$:= $\$TMP8030$
$\supset_{ooo}h_o(\supset_{ooo}b_oa_o)$:= $\$TMP8030$
:= $\$TMP8030$
§s %0 3 %1
$\supset h\,(\supset(\sim a)\,(\sim b))$

use Proof Template K8026 (Deduction Theorem Reversed): H ⊃ (I ⊃ A) → (H ∧ I) ⊃ A
<< K8026.r0t.txt
%0
$\supset(\wedge\, h\,(\sim a))\,(\sim b)$
$\supset_{ooo}(\wedge_{ooo}h_o(\sim_{oo}a_o))(\sim_{oo}b_o)$

.3

use Proof Template A5220H (Gen): (H ⊃ A) → (H ⊃ ∀ x: A)
:= $\$T5220H\ u_\tau$
wff 208 : u_τ := $\$T5220H\ \$T8030$
:= $\$X5220H\ x_{\$T8030}$
wff 209 : $x_{\$T8030}$:= $\$X5220H\ \$X8030$
:= $\$A5220H$ %0
wff 4767 : $\supset (\wedge\, h\, (\sim a))\, (\sim b)_{o,\dots}$:= $\$A5220H$
<< A5220H.r0t.txt
:= $\$T5220H$
:= $\$X5220H$
:= $\$A5220H$
%0
$\supset (\wedge\, h\, (\sim a))\, (\forall\, \$T8030\, [\lambda \$X8030.(\sim b)])$
$\supset_{ooo}(\wedge_{ooo} h_o(\sim_{oo} a_o))(\forall_{o(o\backslash 3)\tau}\$T8030_\tau[\lambda\$X8030_{\$T8030}.(\sim_{oo} b_o)_o])$

.4

%$K8016$
$= \forall\, [\lambda t.[\lambda p.(\sim (\exists\, t\, [\lambda x.(\sim (p\, x))]))]]$:= $K8016$
$=_{o(o(o\backslash 3)\tau)(o(o\backslash 3)\tau)}\forall_{o(o\backslash 3)\tau}[\lambda t_\tau.[\lambda p_{ot}.(\sim_{oo}(\exists_{o(o\backslash 3)\tau} t_\tau[\lambda x_t.(\sim_{oo}(p_{ot} x_t))_o]))_o]_{(o(ot))}]$
:= $K8016$
§= $\forall_{o(o\backslash 3)\tau}\$T8030_\tau$
$= (\forall\, \$T8030)\, (\forall\, \$T8030)$
§s %0 6 %1
$= (\forall\, \$T8030)\, ([\lambda t.[\lambda p.(\sim (\exists\, t\, [\lambda x.(\sim (p\, x))]))]]\, \$T8030)$
§\ $[\lambda t_\tau.[\lambda p_{ot}.(\sim_{oo}(\exists_{o(o\backslash 3)\tau} t_\tau[\lambda x_t.(\sim_{oo}(p_{ot} x_t))_o]))_o]_{(o(ot))}]\$T8030_\tau$
$= ([\lambda t.[\lambda p.(\sim (\exists\, t\, [\lambda x.(\sim (p\, x))]))]]\, \$T8030)\, \$PTMP8030$
§s %1 3 %0
$= (\forall\, \$T8030)\, \$PTMP8030$
§s %5 6 %0
$\supset (\wedge\, h\, (\sim a))\, (\$PTMP8030\, [\lambda \$X8030.(\sim b)])$
§\ $\$PTMP8030_{o(o\,\$T8030)}[\lambda\$X8030_{\$T8030}.(\sim_{oo} b_o)_o]$
$= (\$PTMP8030\, [\lambda \$X8030.(\sim b)])\, (\sim (\exists\, \$T8030\, [\lambda \$X8030.(\sim \$PBTMP8030)]))$
§s %1 3 %0
$\supset (\wedge\, h\, (\sim a))\, (\sim (\exists\, \$T8030\, [\lambda \$X8030.(\sim \$PBTMP8030)]))$
§\ $\$PBTMP8030$
$= \$PBTMP8030\, (\sim b)$
§s %1 63 %0
$\supset (\wedge\, h\, (\sim a))\, (\sim (\exists\, \$T8030\, [\lambda \$X8030.(\sim (\sim b))]))$
§r /15 $\$X8030$
$= [\lambda \$X8030.(\sim (\sim b))]\, [\lambda \$X8030.(\sim (\sim b))]$
§s %1 15 %0
$\supset (\wedge\, h\, (\sim a))\, (\sim (\exists\, \$T8030\, [\lambda \$X8030.(\sim (\sim b))]))$
:= $\$TMP8030$ %0
wff 4977 : $\supset (\wedge\, h\, (\sim a))\, (\sim (\exists\, \$T8030\, [\lambda \$X8030.(\sim (\sim b))]))_o$:= $\$TMP8030$

.5

%$K8008$
$=(\sim(\sim x))\,x$:= $K8008$
$=_{ooo}(\sim_{oo}(\sim_{oo}x_o))x_o$:= $K8008$

use Proof Template A5221 (Sub): B $\to$ B [x/A]
:= $\$B5221$ %0
wff 1670 : $=(\sim(\sim x))\,x_{o,\ldots}$:= $\$B5221$ $K8008$
:= $\$T5221$ o
wff 2 : o_τ := $\$T5221$
:= $\$X5221$ x_o
wff 16 : x_o := $\$X5221$
:= $\$A5221$ %1/127
wff 58 : b_o := $\$A5221$
<< A5221.r0t.txt
:= $\$B5221$
:= $\$T5221$
:= $\$X5221$
:= $\$A5221$
%0
$=(\sim(\sim b))\,b$
$=_{ooo}(\sim_{oo}(\sim_{oo}b_o))b_o$

%$\$TMP8030$
$\supset(\wedge\,h\,(\sim a))\,(\sim(\exists\,\$T8030\,[\lambda\$X8030.(\sim(\sim b))]))$:= $\$TMP8030$
$\supset_{ooo}(\wedge_{ooo}h_o(\sim_{oo}a_o))(\sim_{oo}(\exists_{o(o\backslash 3)\tau}\$T8030_\tau[\lambda\$X8030_{\$T8030}.(\sim_{oo}(\sim_{oo}b_o))_o]))$:= $\$TMP8030$
:= $\$TMP8030$
§s %0 31 %1
$\supset(\wedge\,h\,(\sim a))\,(\sim\$ETMP8030)$
<< K8025.r0t.txt
:= $\$TMP8030$ %0
wff 4992 : $\supset h\,(\supset(\sim a)\,(\sim\$ETMP8030))_{o,\ldots}$:= $\$TMP8030$

.6

%$K8029$
$=(\supset x\,y)\,(\supset(\sim y)\,(\sim x))$:= $K8029$
$=_{ooo}(\supset_{ooo}x_oy_o)(\supset_{ooo}(\sim_{oo}y_o)(\sim_{oo}x_o))$:= $K8029$

use Proof Template A5221 (Sub): B $\to$ B [x/A]
:= $\$B5221$ %0
wff 3361 : $=(\supset x\,y)\,(\supset(\sim y)\,(\sim x))_{o,\ldots}$:= $\$B5221$ $K8029$
:= $\$T5221$ o
wff 2 : o_τ := $\$T5221$
:= $\$X5221$ y_o
wff 34 : y_o := $\$X5221$
:= $\$A5221$ $\supset_{ooo}h_o(\supset_{ooo}(\sim_{oo}a_o)(\sim_{oo}\$ETMP8030_o))$/27

\# wff 54 : a_o := $\$A5221$
<< A5221.r0t.txt
:= $\$B5221$
:= $\$T5221$
:= $\$X5221$
:= $\$A5221$
%0
\# $= (\supset x\, a)\, (\supset (\sim a)\, (\sim x))$
\# $=_{ooo}(\supset_{ooo} x_o a_o)(\supset_{ooo}(\sim_{oo} a_o)(\sim_{oo} x_o))$

\## use Proof Template A5221 (Sub): B → B [x/A]
:= $\$B5221$ %0
\# wff 5014 : $= (\supset x\, a)\, (\supset (\sim a)\, (\sim x))_{o,\ldots}$:= $\$B5221$
:= $\$T5221$ o
\# wff 2 : o_τ := $\$T5221$
:= $\$X5221$ x_o
\# wff 16 : x_o := $\$X5221$
:= $\$A5221$ $\supset_{ooo} h_o(\supset_{ooo}(\sim_{oo} a_o)(\sim_{oo}\$ETMP8030_o))/15$
\# wff 3375 : $\exists\, \$T8030\, [\lambda \$X8030.b]_o$:= $\$A5221$ $\$ETMP8030$
<< A5221.r0t.txt
:= $\$B5221$
:= $\$T5221$
:= $\$X5221$
:= $\$A5221$
%0
\# $= (\supset \$ETMP8030\, a)\, (\supset (\sim a)\, (\sim \$ETMP8030))$
\# $=_{ooo}(\supset_{ooo}\$ETMP8030_o a_o)(\supset_{ooo}(\sim_{oo} a_o)(\sim_{oo}\$ETMP8030_o))$

\## use Proof Template A5201b (Swap): A = B → B = A
<< A5201b.r0t.txt
%0
\# $= (\supset (\sim a)\, (\sim \$ETMP8030))\, (\supset \$ETMP8030\, a)$
\# $=_{ooo}(\supset_{ooo}(\sim_{oo} a_o)(\sim_{oo}\$ETMP8030_o))(\supset_{ooo}\$ETMP8030_o a_o)$

%$\$TMP8030$
\# $\supset h\, (\supset (\sim a)\, (\sim \$ETMP8030))$:= $\$TMP8030$
\# $\supset_{ooo} h_o(\supset_{ooo}(\sim_{oo} a_o)(\sim_{oo}\$ETMP8030_o))$:= $\$TMP8030$
:= $\$TMP8030$

§s %0 3 %1
\# $\supset h\, (\supset \$ETMP8030\, a)$

\## use Proof Template K8026 (Deduction Theorem Reversed): H ⊃ (I ⊃ A) → (H ∧ I) ⊃ A
<< K8026.r0t.txt
%0
\# $\supset (\wedge\, h\, \$ETMP8030)\, a$
\# $\supset_{ooo}(\wedge_{ooo} h_o \$ETMP8030_o) a_o$

undefine local variables
:= $\$PTMP8030$
:= $\$PBTMP8030$
:= $\$ETMP8030$
Include end (K8030.r0t.txt) [newfile=(K8030.r0a.txt)]
>>>

##
Undefine Syntactical Variables
##

:= $\$T8030$
:= $\$X8030$
:= $\$A8030$

##
Q.E.D.
##

%0
$\supset (\wedge\, h\, (\exists\, u\, [\lambda x.b]))\, a$
$\supset_{ooo}(\wedge_{ooo}h_o(\exists_{o(o\backslash 3)\tau}u_\tau[\lambda x_u.b_o]))a_o$

2.1.87 Results for File K8031.r0a.txt

##
Proof Template K8031 (∃ Gen): ([\x.B]A) → ∃ x: B
for any x of any type (Rule of Existential Generalization – without hypothesis)
##
Source: [cf. Andrews 2002 (ISBN 1-4020-0763-9), p. 229 (5242)]
##
Copyright (c) 2017 Owl of Minerva Press GmbH. All rights reserved.
Written by Ken Kubota (<mail@kenkubota.de>).
##
This file is part of the publication of the mathematical logic $\mathcal{R}_0$.
For more information, visit: <http://doi.org/10.4444/100.10>
##

##
Define Syntactical Variables
##

type of the variable and the substitute
:= $\$T8031\ t_\tau$
wff 4 : t_τ := $\$T8031$

proposition: [\x.B]
:= $\$B8031$ $b_{o\$T8031}$
wff 11 : $b_{o\$T8031}$:= $\$B8031$

substitute: A
:= $\$A8031$ $a_{\$T8031}$
wff 12 : $a_{\$T8031}$:= $\$A8031$

##
Assumptions and Resulting Syntactical Variables
##

<< basics.r0.txt

given proposition
§! $\$B8031_{o\$T8031}\$A8031_{\$T8031}$
$\$B8031\,\$A8031$

##
Proof Template
##

<<< K8031.r0t.txt
Include begin (K8031.r0t.txt) [oldfile=(K8031.r0a.txt)]
##
Proof Template K8031 (∃ Gen): ([\x.B]A) → ∃ x: B
for any x of any type (Rule of Existential Generalization – without hypothesis)
##
Source: [cf. Andrews 2002 (ISBN 1-4020-0763-9), p. 229 (5242)]
##

Written by Ken Kubota (<mail@kenkubota.de>).
##
This file is part of the publication of the mathematical logic $\mathcal{R}_0$.
For more information, visit: <http://doi.org/10.4444/100.10>
##

<< A5223.r0.txt

##
Proof Template
##

:= $\$P8031$ $\$B8031_{o\$T8031}\$A8031_{\$T8031}$
wff 210 : $\$B8031\,\$A8031_{o}$:= $\$P8031$

```
## .1

## use Proof Template K8003 (Intro):   A   →   H ⊃ A
:=  $A8003  $B8031_o $T8031 $A8031_$T8031
# wff     210  :        $B8031 $A8031_o        :=  $A8003   $P8031
:=  $H8003  =_oωω=_ω=_ω
# wff     14  :        = = =_o,...        :=  $H8003   A5200t   T
<< K8003.r0t.txt
:=  $A8003
:=  $H8003
%0
#                ⊃ T $P8031
#                ⊃_ooo T_o $P8031_o

## .2

## use Proof Template K8028 (∃ GenH):   H ⊃ ([\x.B]A)   →   H ⊃ ∃ x: B
:=  $H8028  =_oωω=_ω=_ω
# wff     14  :        = = =_o,...        :=  $H8028   A5200t   T
:=  $T8028  t_τ
# wff     4  :        t_τ        :=  $T8028   $T8031
:=  $B8028  b_o $T8031
# wff     11  :        b_o $T8031        :=  $B8028   $B8031
:=  $A8028  a_$T8031
# wff     12  :        a_$T8031        :=  $A8028   $A8031
<< K8028.r0t.txt
:=  $H8028
:=  $T8028
:=  $B8028
:=  $A8028

:=  $TTMP8031  %0
# wff     5438  :        ⊃ T (∃ $T8031 $B8031)_o        :=  $TTMP8031

## .3

## use Proof Template A5221 (Sub):   B   →   B [x/A]
:=  $B5221  =_ooo(⊃_ooo T_o y_o) y_o
# wff     826  :        = (⊃ T y) y_o,...        :=  $B5221   A5223
:=  $T5221  o
# wff     2  :        o_τ        :=  $T5221
:=  $X5221  y_o
# wff     36  :        y_o        :=  $X5221
:=  $A5221  %0/3
# wff     2839  :        ∃ $T8031 $B8031_o,...        :=  $A5221
<< A5221.r0t.txt
:=  $B5221
:=  $T5221
```

:= $\$X5221$
:= $\$A5221$
%0
$= \$TTMP8031\,(\exists\,\$T8031\,\$B8031)$
$=_{ooo}\$TTMP8031_{o}(\exists_{o(o\backslash 3)\tau}\$T8031_{\tau}\$B8031_{o\,\$T8031})$

%$\$TTMP8031$
$\supset T\,(\exists\,\$T8031\,\$B8031)$:= $\$TTMP8031$
$\supset_{ooo}T_{o}(\exists_{o(o\backslash 3)\tau}\$T8031_{\tau}\$B8031_{o\,\$T8031})$:= $\$TTMP8031$
:= $\$TTMP8031$
§s %0 1 %1
$\exists\,\$T8031\,\$B8031$

undefine local variables
:= $\$P8031$
Include end (K8031.r0t.txt) [newfile=(K8031.r0a.txt)]
>>>

##
Undefine Syntactical Variables
##

:= $\$T8031$
:= $\$B8031$
:= $\$A8031$

##
Q.E.D.
##

%0
$\exists\,t\,b$
$\exists_{o(o\backslash 3)\tau}t_{\tau}b_{ot}$

2.1.88 Results for File K8032.r0a.txt

##
Proof Template K8032 (⊃ ∀ Rule): H ⊃ (A ⊃ B) → H ⊃ (A ⊃ ∀ x: B)
##
##
Source: [cf. Andrews 2002 (ISBN 1-4020-0763-9), p. 227 (5237)]
##

Written by Ken Kubota (<mail@kenkubota.de>).
##
This file is part of the publication of the mathematical logic $\mathcal{R}_0$.
For more information, visit: <http://doi.org/10.4444/100.10>

##

##
Define Syntactical Variables
##

<< basics.r0.txt

proposition: H ⊃ (A ⊃ B)
:= $P8032 $\supset_{ooo} h_o (\supset_{ooo} a_o b_o)$
wff 212 : $\supset h\, (\supset a\, b)_o$:= $P8032

type of variable
:= $T8032 o
wff 2 : o_τ := $T8032

the variable
:= $X8032 x_o
wff 16 : x_o := $X8032

##
Assumptions and Resulting Syntactical Variables
##

given proposition
§! $P8032
$\supset h\, (\supset a\, b)$:= $P8032

##
Proof Template
##

<<< K8032.r0t.txt
Include begin (K8032.r0t.txt) [oldfile=(K8032.r0a.txt)]
##
Proof Template K8032 (⊃ ∀ Rule): H ⊃ (A ⊃ B) → H ⊃ (A ⊃ ∀ x: B)
##
##
Source: [cf. Andrews 2002 (ISBN 1-4020-0763-9), p. 227 (5237)]
##

Written by Ken Kubota (<mail@kenkubota.de>).
##
This file is part of the publication of the mathematical logic $\mathcal{R}_0$.
For more information, visit: <http://doi.org/10.4444/100.10>
##

```
##
## Proof Template
##

## the assumption as last theorem on stack (%0)
%$P8032
#                ⊃ h (⊃ a b)         :=  $P8032
#                ⊃ooo ho(⊃ooo ao bo)     :=  $P8032

## use Proof Template K8026 (Deduction Theorem Reversed): H ⊃ (I ⊃ A) → (H ∧ I)
⊃ A
<< K8026.r0t.txt
%0
#                ⊃ (∧ h a) b
#                ⊃ooo(∧ooo ho ao)bo

## use Proof Template A5220H (Gen): (H ⊃ A) → (H ⊃ ∀ x: A)
:= $T5220H o
# wff    2 :        oτ      := $T5220H $T8032
:= $X5220H xo
# wff    16 :        xo      := $X5220H $X8032
:= $A5220H %0
# wff    3794 :      ⊃ (∧ h a) bo,...      := $A5220H
<< A5220H.r0t.txt
:= $T5220H
:= $X5220H
:= $A5220H
%0
#                ⊃ (∧ h a) (∀ o [λ$X8032.b])
#                ⊃ooo(∧ooo ho ao)(∀o(o\3)τ oτ[λ$X8032o.bo])

## use Proof Template K8025 (Deduction Theorem): (H ∧ I) ⊃ A → H ⊃ (I ⊃ A)
<< K8025.r0t.txt
%0
#                ⊃ h (⊃ a (∀ o [λ$X8032.b]))
#                ⊃ooo ho(⊃ooo ao(∀o(o\3)τ oτ[λ$X8032o.bo]))
## Include end (K8032.r0t.txt) [newfile=(K8032.r0a.txt)]
>>>


##
## Undefine Syntactical Variables
##

:= $P8032
:= $T8032
:= $X8032
```

```
##
##   Q.E.D.
##
```

%0

\# $\supset h\,(\supset a\,(\forall o\,[\lambda x.b]))$

\# $\supset_{ooo} h_o (\supset_{ooo} a_o (\forall_{o(o\backslash 3)\tau} o_\tau [\lambda x_o.b_o]))$

2.1.89 Results for File K8033.r0.txt

```
##
##   Proof K8033:  ∀ x: ∃1 y: P x y  ⊃  ∃ f: ∀ x: P x (f x)
##
##
##   Source: [cf. https://sourceforge.net/p/hol/mailman/message/35361865/ (Sep. 11, 2016)]
##
##   Copyright (c) 2017 Owl of Minerva Press GmbH. All rights reserved.
##   Written by Ken Kubota (<mail@kenkubota.de>).
##
##   This file is part of the publication of the mathematical logic R0.
##   For more information, visit: <http://doi.org/10.4444/100.10>
##
```

```
<< basics.r0.txt
<< K8005.r0.txt
<< A5311.r0.txt
```

```
##
##   Proof
##
```

\#\# .1

:= $\$HYP8033\ \forall_{o(o\backslash 3)\tau} t_\tau [\lambda x_t.(\exists_{1\,o(o\backslash 3)\tau} u_\tau [\lambda y_u.(p_{out} x_t y_u)_o])_o]$

\# wff 5480 : $\forall t\,[\lambda x.(\exists_1 u\,[\lambda y.(p\,x\,y)])]_o$:= $\$HYP8033$

$\%K8005$

\# $\supset x\,x$:= $K8005$

\# $\supset_{ooo} x_o x_o$:= $K8005$

\#\# use Proof Template A5221 (Sub): B → B [x/A]

:= $\$B5221$ %0

\# wff 1357 : $\supset x\,x_{o,\ldots}$:= $\$B5221\ \ K8005$

:= $\$T5221\ o$

\# wff 2 : o_τ := $\$T5221$

:= $\$X5221\ x_o$

wff 16 : x_o := $\$X5221$
:= $\$A5221$ $\forall_{o(o\backslash 3)\tau} t_\tau[\lambda x_t.(\exists_{1\,o(o\backslash 3)\tau} u_\tau[\lambda y_u.(p_{out} x_t y_u)_o])_o]$
wff 5480 : $\forall t\,[\lambda x.(\exists_1\, u\,[\lambda y.(p\,x\,y)])]_o$:= $\$A5221$ $\$HYP8033$
<< A5221.r0t.txt
:= $\$B5221$
:= $\$T5221$
:= $\$X5221$
:= $\$A5221$
%0
$\supset \$HYP8033\,\$HYP8033$
$\supset_{ooo}\$HYP8033_o\$HYP8033_o$

.2

use Proof Template A5215H (∀ I): H ⊃ ∀ x: B → H ⊃ B [x/a]
:= $\$T5215H$ t_τ
wff 4 : t_τ := $\$T5215H$
:= $\$X5215H$ $x_{\$T5215H}$
wff 24 : $x_{\$T5215H}$:= $\$X5215H$
:= $\$A5215H$ $x_{\$T5215H}$
wff 24 : $x_{\$T5215H}$:= $\$A5215H$ $\$X5215H$
:= $\$H5215H$ %0
wff 5490 : $\supset \$HYP8033\,\$HYP8033_{o,\ldots}$:= $\$H5215H$
<< A5215H.r0t.txt
:= $\$T5215H$
:= $\$X5215H$
:= $\$A5215H$
:= $\$H5215H$
%0
$\supset \$HYP8033\,(\exists_1\, u\,[\lambda y.(p\,x\,y)])$
$\supset_{ooo}\$HYP8033_o(\exists_{1\,o(o\backslash 3)\tau} u_\tau[\lambda y_u.(p_{out} x_t y_u)_o])$

:= $\$LTMP8033$ %0
wff 5584 : $\supset \$HYP8033\,(\exists_1\, u\,[\lambda y.(p\,x\,y)])_o$:= $\$LTMP8033$

.3

%$A5311$
$\supset (\exists_1\, t\,[\lambda y.(p\,y)])\,(p\,(\iota\, p))$:= $A5311$
$\supset_{ooo}(\exists_{1\,o(o\backslash 3)\tau} t_\tau[\lambda y_t.(p_{ot} y_t)_o])(p_{ot}(\iota_{t(ot)} p_{ot}))$:= $A5311$

use Proof Template A5221 (Sub): B → B [x/A]
:= $\$B5221$ %0
wff 5467 : $\supset (\exists_1\, t\,[\lambda y.(p\,y)])\,(p\,(\iota\, p))_o$:= $\$B5221$ $A5311$
:= $\$T5221$ τ
wff 0 : τ_τ := $\$T5221$
:= $\$X5221$ t_τ
wff 4 : t_τ := $\$X5221$
:= $\$A5221$ u_τ

wff 5468 : u_τ := $\$A5221$
<< A5221.r0t.txt
:= $\$B5221$
:= $\$T5221$
:= $\$X5221$
:= $\$A5221$
%0
$\supset (\exists_1 u\, [\lambda y.(p\, y)])\, (p\, (\iota\, p))$
$\supset_{ooo}(\exists_{1o(o\backslash 3)\tau} u_\tau[\lambda y_u.(p_{ou} y_u)_o])(p_{ou}(\iota_{u(ou)} p_{ou}))$

use Proof Template A5221 (Sub): B → B [x/A]
:= $\$B5221$ %0
wff 5637 : $\supset (\exists_1 u\, [\lambda y.(p\, y)])\, (p\, (\iota\, p))_{o,\ldots}$:= $\$B5221$
:= $\$T5221$ ou
wff 5470 : ou_τ := $\$T5221$
:= $\$X5221$ $p_{\$T5221}$
wff 5629 : $p_{\$T5221}$:= $\$X5221$
:= $\$A5221$ $p_{\$T5221\, t} x_t$
wff 5475 : $p\, x_{\$T5221}$:= $\$A5221$
<< A5221.r0t.txt
:= $\$B5221$
:= $\$T5221$
:= $\$X5221$
:= $\$A5221$
%0
$\supset (\exists_1 u\, [\lambda y.(p\, x\, y)])\, (p\, x\, (\iota\, (p\, x)))$
$\supset_{ooo}(\exists_{1o(o\backslash 3)\tau} u_\tau[\lambda y_u.(p_{out} x_t y_u)_o])(p_{out} x_t(\iota_{u(ou)}(p_{out} x_t)))$

use Proof Template K8003 (Intro): A → H ⊃ A
:= $\$A8003$ %0
wff 5696 : $\supset (\exists_1 u\, [\lambda y.(p\, x\, y)])\, (p\, x\, (\iota\, (p\, x)))_{o,\ldots}$:= $\$A8003$
:= $\$H8003$ $\supset_{ooo}\$HYP8033_o(\exists_{1o(o\backslash 3)\tau} u_\tau[\lambda y_u.(p_{out} x_t y_u)_o])/5$
wff 5480 : $\forall t\, [\lambda x.(\exists_1 u\, [\lambda y.(p\, x\, y)])]_{o,\ldots}$:= $\$H8003$ $\$HYP8033$
<< K8003.r0t.txt
:= $\$A8003$
:= $\$H8003$
%0
$\supset \$HYP8033\, (\supset (\exists_1 u\, [\lambda y.(p\, x\, y)])\, (p\, x\, (\iota\, (p\, x))))$
$\supset_{ooo}\$HYP8033_o(\supset_{ooo}(\exists_{1o(o\backslash 3)\tau} u_\tau[\lambda y_u.(p_{out} x_t y_u)_o])(p_{out} x_t(\iota_{u(ou)}(p_{out} x_t))))$

.4

%$\$LTMP8033$
$\supset \$HYP8033\, (\exists_1 u\, [\lambda y.(p\, x\, y)])$:= $\$LTMP8033$
$\supset_{ooo}\$HYP8033_o(\exists_{1o(o\backslash 3)\tau} u_\tau[\lambda y_u.(p_{out} x_t y_u)_o])$:= $\$LTMP8033$
:= $\$LTMP8033$

use Proof Template A5224H (MP): H ⊃ A, H ⊃ (A ⊃ B) → H ⊃ B
:= $\$A5224H$ %0

\# wff 5584 : $\supset \$HYP8033\,(\exists_1 u\,[\lambda y.(p\,x\,y)])_o$:= $\$A5224H$
:= $\$AB5224H$ %1
\# wff 5699 : $\supset \$HYP8033\,(\supset (\exists_1 u\,[\lambda y.(p\,x\,y)])\,(p\,x\,(\iota\,(p\,x))))_{o,\ldots}$:= $\$AB5224H$
<< A5224H.r0t.txt
:= $\$AB5224H$
:= $\$A5224H$
%0
\# $\supset \$HYP8033\,(p\,x\,(\iota\,(p\,x)))$
\# $\supset_{ooo}\$HYP8033_o(p_{out}x_t(\iota_{u(ou)}(p_{out}x_t)))$

\#\# .5

§\ $[\lambda x_t.(\iota_{u(ou)}(p_{out}x_t))_u]x_t$
\# $= ([\lambda x.(\iota\,(p\,x))]\,x)\,(\iota\,(p\,x))$

§= $[\lambda x_t.(\iota_{u(ou)}(p_{out}x_t))_u]x_t$
\# $= ([\lambda x.(\iota\,(p\,x))]\,x)\,([\lambda x.(\iota\,(p\,x))]\,x)$
§s %0 5 %1
\# $= (\iota\,(p\,x))\,([\lambda x.(\iota\,(p\,x))]\,x)$

§s %3 7 %0
\# $\supset \$HYP8033\,(p\,x\,([\lambda x.(\iota\,(p\,x))]\,x))$

\#\# .6

\#\# use Proof Template A5220H (Gen): (H ⊃ A) → (H ⊃ ∀ x: A)
:= $\$T5220H\ t_\tau$
\# wff 4 : t_τ := $\$T5220H$
:= $\$X5220H\ x_{\$T5220H}$
\# wff 24 : $x_{\$T5220H}$:= $\$X5220H$
:= $\$A5220H$ %0
\# wff 5872 : $\supset \$HYP8033\,(p\,\$X5220H\,([\lambda\$X5220H.(\iota\,(p\,\$X5220H))]\,\$X5220H))_o$
:= $\$A5220H$
<< A5220H.r0t.txt
:= $\$T5220H$
:= $\$X5220H$
:= $\$A5220H$
%0
\# $\supset \$HYP8033\,(\forall t\,[\lambda x.(p\,x\,([\lambda x.(\iota\,(p\,x))]\,x))])$
\# $\supset_{ooo}\$HYP8033_o(\forall_{o(o\backslash 3)\tau}t_\tau[\lambda x_t.(p_{out}x_t([\lambda x_t.(\iota_{u(ou)}(p_{out}x_t))_u]x_t))_o])$

\#\# .7

§\ $[\lambda f_{ut}.(\forall_{o(o\backslash 3)\tau}t_\tau[\lambda x_t.(p_{out}x_t(f_{ut}x_t))_o])_o][\lambda x_t.(\iota_{u(ou)}(p_{out}x_t))_u]$
\# $= ([\lambda f.(\forall t\,[\lambda x.(p\,x\,(f\,x))])]\,[\lambda x.(\iota\,(p\,x))])\,(\forall t\,[\lambda x.(p\,x\,([\lambda x.(\iota\,(p\,x))]\,x))])$

§= $[\lambda f_{ut}.(\forall_{o(o\backslash 3)\tau}t_\tau[\lambda x_t.(p_{out}x_t(f_{ut}x_t))_o])_o][\lambda x_t.(\iota_{u(ou)}(p_{out}x_t))_u]$
\# $= ([\lambda f.(\forall t\,[\lambda x.(p\,x\,(f\,x))])]\,[\lambda x.(\iota\,(p\,x))])\,([\lambda f.(\forall t\,[\lambda x.(p\,x\,(f\,x))])]\,[\lambda x.(\iota\,(p\,x))])$
§s %0 5 %1

\# $= (\forall t\,[\lambda x.(p\,x\,([\lambda x.(\iota\,(p\,x))]\,x))])\,([\lambda f.(\forall t\,[\lambda x.(p\,x\,(f\,x))])]\,[\lambda x.(\iota\,(p\,x))])$

§s %3 3 %0
\# $\supset \$HYP8033\,([\lambda f.(\forall t\,[\lambda x.(p\,x\,(f\,x))])]\,[\lambda x.(\iota\,(p\,x))])$

\## .8

\## use Proof Template K8028 (∃ GenH): H ⊃ ([\x.B]A) → H ⊃ ∃ x: B
:= $\$H8028\ \forall_{o(o\backslash 3)\tau}t_\tau[\lambda x_t.(\exists 1_{o(o\backslash 3)\tau}u_\tau[\lambda y_u.(p_{out}x_ty_u)_o])_o]$
\# wff 5480 : $\forall t\,[\lambda x.(\exists_1\,u\,[\lambda y.(p\,x\,y)])]_{o,\ldots}$:= $\$H8028$ $\$HYP8033$
:= $\$T8028\ ut$
\# wff 5864 : ut_τ := $\$T8028$
:= $\$B8028$ %0/6
\# wff 6015 : $[\lambda f.(\forall t\,[\lambda x.(p\,x\,(f\,x))])]_{o\$T8028}$:= $\$B8028$
:= $\$A8028$ %0/7
\# wff 5863 : $[\lambda x.(\iota\,(p\,x))]_{\$T8028}$:= $\$A8028$
<< K8028.r0t.txt
:= $\$H8028$
:= $\$T8028$
:= $\$B8028$
:= $\$A8028$
%0
\# $\supset \$HYP8033\,(\exists\,(ut)\,[\lambda f.(\forall t\,[\lambda x.(p\,x\,(f\,x))])])$
\# $\supset_{ooo}\$HYP8033_o(\exists_{o(o\backslash 3)\tau}(ut)_\tau[\lambda f_{ut}.(\forall_{o(o\backslash 3)\tau}t_\tau[\lambda x_t.(p_{out}x_t(f_{ut}x_t))_o])_o])$

:= $K8033$ %0
\# wff 7296 : $\supset \$HYP8033\,(\exists\,(ut)\,[\lambda f.(\forall t\,[\lambda x.(p\,x\,(f\,x))])])_o$:= $K8033$

\## undefine local variables
:= $\$HYP8033$

\##
\## Q.E.D.
\##

%0
\# $\supset (\forall t\,[\lambda x.(\exists_1\,u\,[\lambda y.(p\,x\,y)])])\,(\exists\,(ut)\,[\lambda f.(\forall t\,[\lambda x.(p\,x\,(f\,x))])])$:= $K8033$
\# $\supset_{ooo}(\forall_{o(o\backslash 3)\tau}t_\tau[\lambda x_t.(\exists 1_{o(o\backslash 3)\tau}u_\tau[\lambda y_u.(p_{out}x_ty_u)_o])_o])\ldots$
$\ldots(\exists_{o(o\backslash 3)\tau}(ut)_\tau[\lambda f_{ut}.(\forall_{o(o\backslash 3)\tau}t_\tau[\lambda x_t.(p_{out}x_t(f_{ut}x_t))_o])_o])$:= $K8033$

2.1.90 Results for File a5205_substitutions.r0.txt

\##
\## Proof Template: A5205 Substitutions
\##
\##
\## Source: [Kubota 2017 (doi: 10.4444/100.10)]
\##

Written by Ken Kubota (<mail@kenkubota.de>).
##
This file is part of the publication of the mathematical logic $\mathcal{R}_0$.
For more information, visit: <http://doi.org/10.4444/100.10>
##

##
Define Syntactical Variables
##

replacement for type a (alpha) in A5205
:= $\$AA5205\ o$
wff 2 : o_τ := $\$AA5205$

replacement for type b (beta) in A5205
:= $\$BA5205\ t_\tau$
wff 4 : t_τ := $\$BA5205$

replacement for f() in A5205
:= $\$FA5205\ p_{o\$BA5205}$
wff 11 : $p_{o\$BA5205}$:= $\$FA5205$

##
Include Proof Template
##

<<< a5205_substitutions.r0t.txt
Include begin (a5205_substitutions.r0t.txt) [oldfile=(a5205_substitutions.r0.txt)]
##
Proof Template: Axiom 2 Substitutions
##
##
Source: [Kubota 2017 (doi: 10.4444/100.10)]
##

Written by Ken Kubota (<mail@kenkubota.de>).
##
This file is part of the publication of the mathematical logic $\mathcal{R}_0$.
For more information, visit: <http://doi.org/10.4444/100.10>
##

<< A5205.r0.txt

##
Proof Template

```
##

## .1

%A5205
#                    = f [λy.(f y)]        :=   A5205
#                    =o(ab)(ab) fab[λyb.(fab yb)a]       :=   A5205

## .1a Replace type a (alpha) in A5205

## use Proof Template A5221 (Sub):   B   →   B [x/A]
:=  $B5221  %0
# wff      761   :        = f [λy.(f y)]o,...       :=   $B5221   A5205
:=  $T5221  τ
# wff       0   :         ττ        :=   $T5221
:=  $X5221  aτ
# wff       93   :         aτ        :=   $X5221
:=  $A5221  o
# wff       2   :         oτ        :=   $A5221   $AA5205
<< A5221.r0t.txt
:=  $B5221
:=  $T5221
:=  $X5221
:=  $A5221
%0
#                    = f [λy.(f y)]
#                    =o(ob)(ob) fob[λyb.(fob yb)o]

## .1b Replace type b (beta) in A5205

## use Proof Template A5221 (Sub):   B   →   B [x/A]
:=  $B5221  %0
# wff      847   :        = f [λy.(f y)]o,...       :=   $B5221
:=  $T5221  τ
# wff       0   :         ττ        :=   $T5221
:=  $X5221  bτ
# wff       12   :         bτ        :=   $X5221
:=  $A5221  tτ
# wff       4   :         tτ        :=   $A5221   $BA5205
<< A5221.r0t.txt
:=  $B5221
:=  $T5221
:=  $X5221
:=  $A5221
%0
#                    = f [λy.(f y)]
#                    =o(o$BA5205)(o$BA5205) fo$BA5205[λy$BA5205.(fo$BA5205 y$BA5205)o]

## .1c Replace f() in A5205
```

use Proof Template A5221 (Sub): B $\rightarrow$ B [x/A]
:= $\$B5221$ %0
\# wff 895 : $= f\,[\lambda y.(f\,y)]_{o,\ldots}$:= $\$B5221$
:= $\$T5221$ $o\$BA5205$
\# wff 5 : $o\$BA5205_{\tau}$:= $\$T5221$
:= $\$X5221$ $f_{\$T5221}$
\# wff 891 : $f_{\$T5221}$:= $\$X5221$
:= $\$A5221$ $p_{\$T5221}$
\# wff 11 : $p_{\$T5221}$:= $\$A5221$ $\$FA5205$
<< A5221.r0t.txt
:= $\$B5221$
:= $\$T5221$
:= $\$X5221$
:= $\$A5221$
%0
\# $= \$FA5205\,[\lambda y.(\$FA5205\,y)]$
\# $=_{o(o\$BA5205)(o\$BA5205)}\$FA5205_{o\$BA5205}[\lambda y_{\$BA5205}.(\$FA5205_{o\$BA5205}y_{\$BA5205})_o]$
Include end (a5205_substitutions.r0t.txt) [newfile=(a5205_substitutions.r0.txt)]
>>>

##
Undefine Syntactical Variables
##

:= $\$AA5205$
:= $\$BA5205$
:= $\$FA5205$

##
Q.E.D.
##

%0
\# $= p\,[\lambda y.(p\,y)]$
\# $=_{o(ot)(ot)}p_{ot}[\lambda y_t.(p_{ot}y_t)_o]$

2.1.91 Results for File axiom2_substitutions.r0.txt

##
Proof Template: Axiom 2 Substitutions
##
##
Source: [Kubota 2017 (doi: 10.4444/100.10)]
##

Written by Ken Kubota (<mail@kenkubota.de>).

##
This file is part of the publication of the mathematical logic $\mathcal{R}_0$.
For more information, visit: <http://doi.org/10.4444/100.10>
##

##
Define Syntactical Variables
##

<< basics.r0.txt

replacement for type a (alpha) in Axiom 2
:= $\$AA2$ oa
wff 172 : oa_{τ} := $\$AA2$

replacement for h() in Axiom 2
:= $\$HA2$ $[\lambda f_{\$AA2}.(f_{\$AA2}x_a)_o]$
wff 210 : $[\lambda f.(f\ x)]_{o\$AA2}$:= $\$HA2$

replacement for x in Axiom 2
:= $\$XA2$ $[\lambda x_a.T_o]$
wff 212 : $[\lambda x.T]_{\$AA2}$:= $\$XA2$

replacement for y in Axiom 2
:= $\$YA2$ $f_{\$AA2}$
wff 208 : $f_{\$AA2}$:= $\$YA2$

##
Include Proof Template
##

<<< axiom2_substitutions.r0t.txt
Include begin (axiom2_substitutions.r0t.txt) [oldfile=(axiom2_substitutions.r0.txt)]
##
Proof Template: Axiom 2 Substitutions
##
##
Source: [Kubota 2017 (doi: 10.4444/100.10)]
##

Written by Ken Kubota (<mail@kenkubota.de>).
##
This file is part of the publication of the mathematical logic $\mathcal{R}_0$.
For more information, visit: <http://doi.org/10.4444/100.10>
##

Skipping file axioms.r0.txt (already included)

##
Proof Template
##

.1

Axiom 2: One of the Basic Properties of Equality
%A2
$\supset (= x\, y)\, (= (h\, x)\, (h\, y))$:= A2
$\supset_{ooo}(=_{\$AA2\, a} x_a y_a)(=_{ooo}(h_{\$AA2} x_a)(h_{\$AA2} y_a))$:= A2

.1a Replace type a (alpha) in Axiom 2

use Proof Template A5221 (Sub): B $\rightarrow$ B [x/A]
:= $\$B5221$ %0
wff 184 : $\supset (= x\, y)\, (= (h\, x)\, (h\, y))_o$:= $\$B5221$ $A2$
:= $\$T5221$ τ
wff 0 : τ_τ := $\$T5221$
:= $\$X5221$ a_τ
wff 171 : a_τ := $\$X5221$
:= $\$A5221$ $o\$X5221$
wff 172 : $o\$X5221_\tau$:= $\$A5221$ $\$AA2$
<< A5221.r0t.txt
:= $\$B5221$
:= $\$T5221$
:= $\$X5221$
:= $\$A5221$
%0
$\supset (= x\, y)\, (= (h\, x)\, (h\, y))$
$\supset_{ooo}(=_{o\,\$AA2\,\$AA2} x_{\$AA2} y_{\$AA2})(=_{ooo}(h_{o\,\$AA2} x_{\$AA2})(h_{o\,\$AA2} y_{\$AA2}))$

.1b Replace h() in Axiom 2

use Proof Template A5221 (Sub): B $\rightarrow$ B [x/A]
:= $\$B5221$ %0
wff 853 : $\supset (= x\, y)\, (= (h\, x)\, (h\, y))_{o,\dots}$:= $\$B5221$
:= $\$T5221$ $o\$AA2$
wff 211 : $o\$AA2_\tau$:= $\$T5221$
:= $\$X5221$ $h_{\$T5221}$
wff 848 : $h_{\$T5221}$:= $\$X5221$
:= $\$A5221$ $[\lambda \$YA2_{\$AA2}.(\$YA2_{\$AA2} x_a)_o]$
wff 210 : $[\lambda \$YA2.(\$YA2\, x)]_{\$T5221}$:= $\$A5221$ $\$HA2$
<< A5221.r0t.txt
:= $\$B5221$
:= $\$T5221$
:= $\$X5221$
:= $\$A5221$

%0
$\supset (= x\, y)\, (= (\$HA2\, x)\, (\$HA2\, y))$
$\supset_{ooo}(=_{o\$AA2\$AA2} x_{\$AA2} y_{\$AA2})(=_{ooo}(\$HA2_{o\$AA2} x_{\$AA2})(\$HA2_{o\$AA2} y_{\$AA2}))$

.1c Replace x in Axiom 2

use Proof Template A5221 (Sub): B → B [x/A]
:= $\$B5221$ %0
wff 914 : $\supset (= x\, y)\, (= (\$HA2\, x)\, (\$HA2\, y))_{o,\ldots}$:= $\$B5221$
:= $\$T5221$ oa
wff 172 : oa_{τ} := $\$AA2$ $\$T5221$
:= $\$X5221$ $x_{\$AA2}$
wff 843 : $x_{\$AA2}$:= $\$X5221$
:= $\$A5221$ $[\lambda x_a.T_o]$
wff 212 : $[\lambda x.T]_{\$AA2}$:= $\$A5221$ $\$XA2$
<< A5221.r0t.txt
:= $\$B5221$
:= $\$T5221$
:= $\$X5221$
:= $\$A5221$
%0
$\supset (= \$XA2\, y)\, (= (\$HA2\, \$XA2)\, (\$HA2\, y))$
$\supset_{ooo}(=_{o\$AA2\$AA2} \$XA2_{\$AA2} y_{\$AA2}) \ldots$
$\ldots (=_{ooo}(\$HA2_{o\$AA2} \$XA2_{\$AA2})(\$HA2_{o\$AA2} y_{\$AA2}))$

.1d Replace y in Axiom 2

use Proof Template A5221 (Sub): B → B [x/A]
:= $\$B5221$ %0
wff 969 : $\supset (= \$XA2\, y)\, (= (\$HA2\, \$XA2)\, (\$HA2\, y))_{o,\ldots}$:= $\$B5221$
:= $\$T5221$ oa
wff 172 : oa_{τ} := $\$AA2$ $\$T5221$
:= $\$X5221$ $y_{\$AA2}$
wff 845 : $y_{\$AA2}$:= $\$X5221$
:= $\$A5221$ $f_{\$AA2}$
wff 208 : $f_{\$AA2}$:= $\$A5221$ $\$YA2$
<< A5221.r0t.txt
:= $\$B5221$
:= $\$T5221$
:= $\$X5221$
:= $\$A5221$
%0
$\supset (= \$XA2\, \$YA2)\, (= (\$HA2\, \$XA2)\, (\$HA2\, \$YA2))$
$\supset_{ooo}(=_{o\$AA2\$AA2} \$XA2_{\$AA2} \$YA2_{\$AA2}) \ldots$
$\ldots (=_{ooo}(\$HA2_{o\$AA2} \$XA2_{\$AA2})(\$HA2_{o\$AA2} \$YA2_{\$AA2}))$
Include end (axiom2_substitutions.r0t.txt) [newfile=(axiom2_substitutions.r0.txt)]
>>>

##
Undefine Syntactical Variables
##

:= $\$AA2$
:= $\$HA2$
:= $\$XA2$
:= $\$YA2$

##
Q.E.D.
##

%0
$\supset (= [\lambda x.T] \, f) \, (= ([\lambda f.(f \; x)] \, [\lambda x.T]) \, ([\lambda f.(f \; x)] \, f))$
$\supset_{ooo}(=_{o(oa)(oa)}[\lambda x_a.T_o]f_{oa})(=_{ooo}([\lambda f_{oa}.(f_{oa}x_a)_o][\lambda x_a.T_o])([\lambda f_{oa}.(f_{oa}x_a)_o]f_{oa}))$

2.1.92 Results for File axiom3_substitutions.r0.txt

##
Proof Template: Axiom 3 Substitutions
##
##
Source: [Kubota 2017 (doi: 10.4444/100.10)]
##

Written by Ken Kubota (<mail@kenkubota.de>).
##
This file is part of the publication of the mathematical logic $\mathcal{R}_0$.
For more information, visit: <http://doi.org/10.4444/100.10>
##

##
Define Syntactical Variables
##

replacement for type a (alpha) in Axiom 3
:= $\$AA3 \; t_\tau$
wff 4 : t_τ := $\$AA3$

replacement for type b (beta) in Axiom 3
:= $\$BA3 \; u_\tau$
wff 11 : u_τ := $\$BA3$

replacement for f() in Axiom 3
:= $\$FA3 \; y_{\$AA3\,\$BA3}$
wff 13 : $y_{\$AA3\,\$BA3}$:= $\$FA3$

replacement for g() in Axiom 3
:= $\$GA3$ $z_{\$AA3\,\$BA3}$
wff 14 : $z_{\$AA3\,\$BA3}$:= $\$GA3$

##
Include Proof Template
##

<<< axiom3_substitutions.r0t.txt
Include begin (axiom3_substitutions.r0t.txt) [oldfile=(axiom3_substitutions.r0.txt)]
##
Proof Template: Axiom 3 Substitutions
##
##
Source: [Kubota 2017 (doi: 10.4444/100.10)]
##

Written by Ken Kubota (<mail@kenkubota.de>).
##
This file is part of the publication of the mathematical logic $\mathcal{R}_0$.
For more information, visit: <http://doi.org/10.4444/100.10>
##

<< axioms.r0.txt

##
Proof Template
##

.1

Axiom 3: Axiom of Extensionality
%$A3$
$=(= f\,g)\,(\forall\, b\,[\lambda x.(=(f\,x)\,(g\,x))])$:= $A3$
$=_{ooo}(=_{o(ab)(ab)} f_{ab} g_{ab})(\forall_{o(o\backslash 3)\tau} b_\tau [\lambda x_b.(=_{oaa}(f_{ab} x_b)(g_{ab} x_b))_o])$:= $A3$

.1a Replace type a (alpha) in Axiom 3

use Proof Template A5209 (incl. A5204): B = C → (B = C) [x/A]
:= $\$M5209$ o
wff 2 : o_τ := $\$M5209$
:= $\$E5209$ %0
wff 128 : $=(= f\,g)\,(\forall\, b\,[\lambda x.(=(f\,x)\,(g\,x))])_o$:= $\$E5209$ $A3$
:= $\$T5209$ τ
wff 0 : τ_τ := $\$T5209$
:= $\$X5209$ a_τ

wff 95 : a_τ := $\$X5209$
:= $\$A5209\ t_\tau$
wff 4 : t_τ := $\$A5209$ $\$AA3$
<< A5209.r0t.txt
:= $\$M5209$
:= $\$E5209$
:= $\$T5209$
:= $\$X5209$
:= $\$A5209$
%0
$=(=f\,g)\,(\forall\, b\,[\lambda x.(=(f\,x)\,(g\,x))])$
$=_{ooo}(=_{o(\$AA3\,b)(\$AA3\,b)}f_{\$AA3\,b}g_{\$AA3\,b})\cdots$
$\ldots(\forall_{o(o\backslash 3)\tau}b_\tau[\lambda x_b.(=_{o\,\$AA3\,\$AA3}(f_{\$AA3\,b}x_b)(g_{\$AA3\,b}x_b))_o])$

.1b Replace type b (beta) in Axiom 3

use Proof Template A5209 (incl. A5204): B = C → (B = C) [x/A]
:= $\$M5209\ o$
wff 2 : o_τ := $\$M5209$
:= $\$E5209$ %0
wff 161 : $=(=f\,g)\,(\forall\, b\,[\lambda x.(=(f\,x)\,(g\,x))])_o$:= $\$E5209$
:= $\$T5209\ \tau$
wff 0 : τ_τ := $\$T5209$
:= $\$X5209\ b_\tau$
wff 109 : b_τ := $\$X5209$
:= $\$A5209\ u_\tau$
wff 11 : u_τ := $\$A5209$ $\$BA3$
<< A5209.r0t.txt
:= $\$M5209$
:= $\$E5209$
:= $\$T5209$
:= $\$X5209$
:= $\$A5209$
%0
$=(=f\,g)\,(\forall\,\$BA3\,[\lambda x.(=(f\,x)\,(g\,x))])$
$=_{ooo}(=_{o(\$AA3\,\$BA3)(\$AA3\,\$BA3)}f_{\$AA3\,\$BA3}g_{\$AA3\,\$BA3})\cdots$
$\ldots(\forall_{o(o\backslash 3)\tau}\$BA3_\tau[\lambda x_{\$BA3}.(=_{o\,\$AA3\,\$AA3}(f_{\$AA3\,\$BA3}x_{\$BA3})(g_{\$AA3\,\$BA3}x_{\$BA3}))_o])$

.1c Replace f() in Axiom 3

use Proof Template A5209 (incl. A5204): B = C → (B = C) [x/A]
:= $\$M5209\ o$
wff 2 : o_τ := $\$M5209$
:= $\$E5209$ %0
wff 191 : $=(=f\,g)\,(\forall\,\$BA3\,[\lambda x.(=(f\,x)\,(g\,x))])_o$:= $\$E5209$
:= $\$T5209\ \$AA3\,\$BA3$
wff 12 : $\$AA3\,\$BA3_\tau$:= $\$T5209$
:= $\$X5209\ f_{\$T5209}$
wff 172 : $f_{\$T5209}$:= $\$X5209$

:= $\$A5209 \quad y_{\$T5209}$
wff 13 : $y_{\$T5209}$:= $\$A5209 \quad \$FA3$
<< A5209.r0t.txt
:= $\$M5209$
:= $\$E5209$
:= $\$T5209$
:= $\$X5209$
:= $\$A5209$
%0
$= (= \$FA3\, g)\, (\forall\, \$BA3\, [\lambda x.(= (\$FA3\, x)\, (g\, x))])$
$=_{ooo}(=_{o(\$AA3\,\$BA3)(\$AA3\,\$BA3)}\$FA3_{\$AA3\,\$BA3}g_{\$AA3\,\$BA3})\ldots$
$\ldots(\forall_{o(o\backslash 3)\tau}\$BA3_{\tau}[\lambda x_{\$BA3}.(=_{o\,\$AA3\,\$AA3}(\$FA3_{\$AA3\,\$BA3}x_{\$BA3})(g_{\$AA3\,\$BA3}x_{\$BA3}))_o])$

.1d Replace g() in Axiom 3

use Proof Template A5209 (incl. A5204): B = C → (B = C) [x/A]
:= $\$M5209\ o$
wff 2 : o_{τ} := $\$M5209$
:= $\$E5209$ %0
wff 212 : $= (= \$FA3\, g)\, (\forall\, \$BA3\, [\lambda x.(= (\$FA3\, x)\, (g\, x))])_o$:= $\$E5209$
:= $\$T5209 \quad \$AA3\,\$BA3$
wff 12 : $\$AA3\,\$BA3_{\tau}$:= $\$T5209$
:= $\$X5209 \quad g_{\$T5209}$
wff 174 : $g_{\$T5209}$:= $\$X5209$
:= $\$A5209 \quad z_{\$T5209}$
wff 14 : $z_{\$T5209}$:= $\$A5209 \quad \$GA3$
<< A5209.r0t.txt
:= $\$M5209$
:= $\$E5209$
:= $\$T5209$
:= $\$X5209$
:= $\$A5209$
%0
$= (= \$FA3\, \$GA3)\, (\forall\, \$BA3\, [\lambda x.(= (\$FA3\, x)\, (\$GA3\, x))])$
$=_{ooo}(=_{o(\$AA3\,\$BA3)(\$AA3\,\$BA3)}\$FA3_{\$AA3\,\$BA3}\$GA3_{\$AA3\,\$BA3})\ldots$
$\ldots(\forall_{o(o\backslash 3)\tau}\$BA3_{\tau}[\lambda x_{\$BA3}.(=_{o\,\$AA3\,\$AA3}(\$FA3_{\$AA3\,\$BA3}x_{\$BA3})(\$GA3_{\$AA3\,\$BA3}x_{\$BA3}))_o])$
Include end (axiom3_substitutions.r0t.txt) [newfile=(axiom3_substitutions.r0.txt)]
>>>

##
Undefine Syntactical Variables
##

:= $\$AA3$
:= $\$BA3$
:= $\$FA3$
:= $\$GA3$

##
Q.E.D.
##

%0
$= (= y\, z)\, (\forall\, u\, [\lambda x.(= (y\, x)\, (z\, x))])$
$=_{ooo}(=_{o(tu)(tu)} y_{tu} z_{tu})(\forall_{o(o\backslash 3)\tau} u_\tau [\lambda x_u.(=_{ott}(y_{tu} x_u)(z_{tu} x_u))_o])$

2.1.93 Results for File axiom_of_choice.r0a.txt

##
Axiom of Choice
##
##
Source: [Andrews 2002 (ISBN 1-4020-0763-9), p. 236]
##
Copyright (c) 2017 Owl of Minerva Press GmbH. All rights reserved.
Written by Ken Kubota (<mail@kenkubota.de>).
##
This file is part of the publication of the mathematical logic $\mathcal{R}_0$.
For more information, visit: <http://doi.org/10.4444/100.10>
##

<< definitions1.r0.txt

##
Axiom of Choice
##

:= AC $\exists_{o(o\backslash 3)\tau}(t(ot))_\tau[\lambda j_{t(ot)}.(\forall_{o(o\backslash 3)\tau}(ot)_\tau[\lambda p_{ot}.(\supset_{ooo}(\exists_{o(o\backslash 3)\tau} t_\tau [\lambda x_t.(p_{ot} x_t)_o])(p_{ot}(j_{t(ot)} p_{ot})))_o])_o]$
wff 96 : $\exists\, (t(ot))\, [\lambda j.(\forall\, (ot)\, [\lambda p.(\supset (\exists\, t\, [\lambda x.(p\, x)])\, (p\, (j\, p)))])]_o$:= AC
§! AC
$\exists\, (t(ot))\, [\lambda j.(\forall\, (ot)\, [\lambda p.(\supset (\exists\, t\, [\lambda x.(p\, x)])\, (p\, (j\, p)))])]$:= AC

2.1.94 Results for File axioms.r0.txt

##
Axioms
##
##
Source: [Andrews 2002 (ISBN 1-4020-0763-9), p. 213]
##
Copyright (c) 2017 Owl of Minerva Press GmbH. All rights reserved.
Written by Ken Kubota (<mail@kenkubota.de>).
##
This file is part of the publication of the mathematical logic $\mathcal{R}_0$.

For more information, visit: <http://doi.org/10.4444/100.10>
##

<< definitions1.r0.txt

##
Axiom 1: Truth and Falsehood Are the Only Truth Values
##

:= $A1$ $=_{ooo}(\wedge_{ooo}(g_{oo}T_o)(g_{oo}F_o))(\forall_{o(o\backslash 3)_\tau}o_\tau[\lambda x_o.(g_{oo}x_o)_o])$
\# wff 90 : $=(\wedge\,(g\,T)\,(g\,F))\,(\forall\,o\,[\lambda x.(g\,x)])_o$:= $A1$
§! $A1$
\# $=(\wedge\,(g\,T)\,(g\,F))\,(\forall\,o\,[\lambda x.(g\,x)])$:= $A1$

##
Axiom 2: One of the Basic Properties of Equality
##

:= $A2$ $\supset_{ooo}(=_{oaa}x_ay_a)(=_{ooo}(h_{oa}x_a)(h_{oa}y_a))$
\# wff 104 : $\supset(=x\,y)\,(=(h\,x)\,(h\,y))_o$:= $A2$
§! $A2$
\# $\supset(=x\,y)\,(=(h\,x)\,(h\,y))$:= $A2$

##
Axiom 3: Axiom of Extensionality
##

:= $A3$ $=_{ooo}(=_{o(ab)(ab)}f_{ab}g_{ab})(\forall_{o(o\backslash 3)_\tau}b_\tau[\lambda x_b.(=_{oaa}(f_{ab}x_b)(g_{ab}x_b))_o])$
\# wff 124 : $=(=f\,g)\,(\forall\,b\,[\lambda x.(=(f\,x)\,(g\,x))])_o$:= $A3$
§! $A3$
\# $=(=f\,g)\,(\forall\,b\,[\lambda x.(=(f\,x)\,(g\,x))])$:= $A3$

##
Axiom 4: Axiom of Lambda Conversion
##

Replaced by Rule 2 (Lambda Conversion)
[cf. Andrews 2002 (ISBN 1-4020-0763-9), pp. 218 f. (5207)]
##
"5207 could be taken as an axiom schema in place of 4_1 - 4_5,
and for some purposes this would be desirable,
since 5207 has a conceptual simplicity and unity
which is not apparent in 4_1 - 4_5." [Andrews 2002, p. 214]

```
##
## Axiom 5: Axiom of Descriptions
##
```

$:=\ A5\ =_{ott}(\iota_{t(ot)}(=_{ott}y_t))y_t$

$\#$ wff 129 : $=(\iota\,(=y))\,y_o$ $:=$ $A5$

§! $A5$

$\#$ $=(\iota\,(=y))\,y$ $:=$ $A5$

2.1.95 Results for File basics.r0.txt

```
##
## Basics
##
##
## Source: [Kubota 2017 (doi: 10.4444/100.10)]
##
## Copyright (c) 2017 Owl of Minerva Press GmbH. All rights reserved.
## Written by Ken Kubota (<mail@kenkubota.de>).
##
## This file is part of the publication of the mathematical logic R0.
## For more information, visit: <http://doi.org/10.4444/100.10>
##
```

```
<< definitions1.r0.txt
<< definitions2.r0.txt
<< definitions3.r0.txt
<< axioms.r0.txt
```

2.1.96 Results for File composition.r0.txt

```
##
## Associativity of the Composition of Functions
##
##
## Source: [Kubota 2017 (doi: 10.4444/100.10)]
##
## Copyright (c) 2017 Owl of Minerva Press GmbH. All rights reserved.
## Written by Ken Kubota (<mail@kenkubota.de>).
##
## This file is part of the publication of the mathematical logic R0.
## For more information, visit: <http://doi.org/10.4444/100.10>
##
```

```
<< basics.r0.txt
```

$:=\ COMPS\ \ldots$

$\ldots [\lambda a_\tau.[\lambda b_\tau.[\lambda c_\tau.[\lambda g_{ab}.[\lambda f_{bc}.[\lambda x_c.(g_{ab}(f_{bc}x_c))_a]_{(ac)}]_{(ac(bc))}]_{(ac(bc)(ab))}]_{(a\backslash 4(b\backslash 4)(ab)\tau)}]_{(a\backslash 4(\backslash 5\backslash 4)(a\backslash 4)\tau\tau)}]$
wff 233 : $[\lambda a.[\lambda b.[\lambda c.[\lambda g.[\lambda f.[\lambda x.(g\,(f\,x))]]]]]]_{\backslash 6\backslash 4(\backslash 5\backslash 4)(\backslash 5\backslash 4)\tau\tau\tau}$:= $COMPS$

.1

:= $\$GF$ $COMPS_{\backslash 6\backslash 4(\backslash 5\backslash 4)(\backslash 5\backslash 4)\tau\tau\tau}u_\tau v_\tau w_\tau g_{uv} f_{vw}$
wff 264 : $COMPS\,u\,v\,w\,g\,f_{uw}$:= $\$GF$
§= $\$GF$
$= \$GF\,\$GF$
§\ $COMPS_{\backslash 6\backslash 4(\backslash 5\backslash 4)(\backslash 5\backslash 4)\tau\tau\tau}u_\tau$
$= (COMPS\,u)\,[\lambda b.[\lambda c.[\lambda g.[\lambda f.[\lambda x.(g\,(f\,x))]]]]]$
§s %1 48 %0
$= \$GF\,([\lambda b.[\lambda c.[\lambda g.[\lambda f.[\lambda x.(g\,(f\,x))]]]]]\,v\,w\,g\,f)$
§\ $[\lambda b_\tau.[\lambda c_\tau.[\lambda g_{ub}.[\lambda f_{bc}.[\lambda x_c.(g_{ub}(f_{bc}x_c))_u]_{(uc)}]_{(uc(bc))}]_{(uc(bc)(ub))}]_{(u\backslash 4(b\backslash 4)(ub)\tau)}]v_\tau$
$= ([\lambda b.[\lambda c.[\lambda g.[\lambda f.[\lambda x.(g\,(f\,x))]]]]]\,v)\,[\lambda c.[\lambda g.[\lambda f.[\lambda x.(g\,(f\,x))]]]]$
§s %1 24 %0
$= \$GF\,([\lambda c.[\lambda g.[\lambda f.[\lambda x.(g\,(f\,x))]]]]\,w\,g\,f)$
§\ $[\lambda c_\tau.[\lambda g_{uv}.[\lambda f_{vc}.[\lambda x_c.(g_{uv}(f_{vc}x_c))_u]_{(uc)}]_{(uc(vc))}]_{(uc(vc)(uv))}]w_\tau$
$= ([\lambda c.[\lambda g.[\lambda f.[\lambda x.(g\,(f\,x))]]]]\,w)\,[\lambda g.[\lambda f.[\lambda x.(g\,(f\,x))]]]$
§s %1 12 %0
$= \$GF\,([\lambda g.[\lambda f.[\lambda x.(g\,(f\,x))]]]\,g\,f)$
§\ $[\lambda g_{uv}.[\lambda f_{vw}.[\lambda x_w.(g_{uv}(f_{vw}x_w))_u]_{(uw)}]_{(uw(vw))}]g_{uv}$
$= ([\lambda g.[\lambda f.[\lambda x.(g\,(f\,x))]]]\,g)\,[\lambda f.[\lambda x.(g\,(f\,x))]]$
§s %1 6 %0
$= \$GF\,([\lambda f.[\lambda x.(g\,(f\,x))]]\,f)$
§\ $[\lambda f_{vw}.[\lambda x_w.(g_{uv}(f_{vw}x_w))_u]_{(uw)}]f_{vw}$
$= ([\lambda f.[\lambda x.(g\,(f\,x))]]\,f)\,[\lambda x.(g\,(f\,x))]$
§s %1 3 %0
$= \$GF\,[\lambda x.(g\,(f\,x))]$

:= $\$HxGF$ $COMPS_{\backslash 6\backslash 4(\backslash 5\backslash 4)(\backslash 5\backslash 4)\tau\tau\tau}t_\tau u_\tau w_\tau h_{tu}\$GF_{uw}$
wff 339 : $COMPS\,t\,u\,w\,h\,\$GF_{tw}$:= $\$HxGF$
§= $\$HxGF$
$= \$HxGF\,\$HxGF$
§s %0 7 %1
$= \$HxGF\,(COMPS\,t\,u\,w\,h\,[\lambda x.(g\,(f\,x))])$
§\ $COMPS_{\backslash 6\backslash 4(\backslash 5\backslash 4)(\backslash 5\backslash 4)\tau\tau\tau}t_\tau$
$= (COMPS\,t)\,[\lambda b.[\lambda c.[\lambda g.[\lambda f.[\lambda x.(g\,(f\,x))]]]]]$
§s %1 48 %0
$= \$HxGF\,([\lambda b.[\lambda c.[\lambda g.[\lambda f.[\lambda x.(g\,(f\,x))]]]]]\,u\,w\,h\,[\lambda x.(g\,(f\,x))])$
§\ $[\lambda b_\tau.[\lambda c_\tau.[\lambda g_{tb}.[\lambda f_{bc}.[\lambda x_c.(g_{tb}(f_{bc}x_c))_t]_{(tc)}]_{(tc(bc))}]_{(tc(bc)(tb))}]_{(t\backslash 4(b\backslash 4)(tb)\tau)}]u_\tau$
$= ([\lambda b.[\lambda c.[\lambda g.[\lambda f.[\lambda x.(g\,(f\,x))]]]]]\,u)\,[\lambda c.[\lambda g.[\lambda f.[\lambda x.(g\,(f\,x))]]]]$
§s %1 24 %0
$= \$HxGF\,([\lambda c.[\lambda g.[\lambda f.[\lambda x.(g\,(f\,x))]]]]\,w\,h\,[\lambda x.(g\,(f\,x))])$
§\ $[\lambda c_\tau.[\lambda g_{tu}.[\lambda f_{uc}.[\lambda x_c.(g_{tu}(f_{uc}x_c))_t]_{(tc)}]_{(tc(uc))}]_{(tc(uc)(tu))}]w_\tau$
$= ([\lambda c.[\lambda g.[\lambda f.[\lambda x.(g\,(f\,x))]]]]\,w)\,[\lambda g.[\lambda f.[\lambda x.(g\,(f\,x))]]]$
§s %1 12 %0
$= \$HxGF\,([\lambda g.[\lambda f.[\lambda x.(g\,(f\,x))]]]\,h\,[\lambda x.(g\,(f\,x))])$
§\ $[\lambda g_{tu}.[\lambda f_{uw}.[\lambda x_w.(g_{tu}(f_{uw}x_w))_t]_{(tw)}]_{(tw(uw))}]h_{tu}$

$\#\qquad = ([\lambda g.[\lambda f.[\lambda x.(g\,(f\,x))]]]\,h)\,[\lambda f.[\lambda x.(h\,(f\,x))]]$
$\S s\ \ \%1\ \ 6\ \ \%0$
$\#\qquad = \$HxGF\,([\lambda f.[\lambda x.(h\,(f\,x))]]\,[\lambda x.(g\,(f\,x))])$
$\S\backslash\ \ [\lambda f_{uw}.[\lambda x_w.(h_{tu}(f_{uw}x_w))_t]_{(tw)}][\lambda x_w.(g_{uv}(f_{vw}x_w))_u]$
$\#\qquad = ([\lambda f.[\lambda x.(h\,(f\,x))]]\,[\lambda x.(g\,(f\,x))])\,[\lambda x.(h\,([\lambda x.(g\,(f\,x))]\,x))]$
$\S s\ \ \%1\ \ 3\ \ \%0$
$\#\qquad = \$HxGF\,[\lambda x.(h\,([\lambda x.(g\,(f\,x))]\,x))]$
$\S\backslash\ \ [\lambda x_w.(g_{uv}(f_{vw}x_w))_u]x_w$
$\#\qquad = ([\lambda x.(g\,(f\,x))]\,x)\,(g\,(f\,x))$
$\S s\ \ \%1\ \ 15\ \ \%0$
$\#\qquad = \$HxGF\,[\lambda x.(h\,(g\,(f\,x)))]$

$:=\ \ \$TMP1\ \ \%0$
$\#$ wff 409 : $= \$HxGF\,[\lambda x.(h\,(g\,(f\,x)))]_o \qquad :=\ \ \$TMP1$

$\#\#\ .2$

$:=\ \ \$HG\ \ COMPS_{\backslash 6\backslash 4(\backslash 5\backslash 4)(\backslash 5\backslash 4)\tau\tau\tau}t_\tau u_\tau v_\tau h_{tu}g_{uv}$
$\#$ wff 415 : $COMPS\,t\,u\,v\,h\,g_{tv} \qquad :=\ \ \$HG$
$\S{=}\ \ \$HG$
$\#\qquad = \$HG\,\$HG$
$\S\backslash\ \ COMPS_{\backslash 6\backslash 4(\backslash 5\backslash 4)(\backslash 5\backslash 4)\tau\tau\tau}t_\tau$
$\#\qquad = (COMPS\,t)\,[\lambda b.[\lambda c.[\lambda g.[\lambda f.[\lambda x.(g\,(f\,x))]]]]]$
$\S s\ \ \%1\ \ 48\ \ \%0$
$\#\qquad = \$HG\,([\lambda b.[\lambda c.[\lambda g.[\lambda f.[\lambda x.(g\,(f\,x))]]]]]\,u\,v\,h\,g)$
$\S\backslash\ \ [\lambda b_\tau.[\lambda c_\tau.[\lambda g_{tb}.[\lambda f_{bc}.[\lambda x_c.(g_{tb}(f_{bc}x_c))_t]_{(tc)}]_{(tc(bc))}]_{(tc(bc)(tb))}]_{(t\backslash 4(b\backslash 4)(tb)\tau)}]u_\tau$
$\#\qquad = ([\lambda b.[\lambda c.[\lambda g.[\lambda f.[\lambda x.(g\,(f\,x))]]]]]\,u)\,[\lambda c.[\lambda g.[\lambda f.[\lambda x.(g\,(f\,x))]]]]$
$\S s\ \ \%1\ \ 24\ \ \%0$
$\#\qquad = \$HG\,([\lambda c.[\lambda g.[\lambda f.[\lambda x.(g\,(f\,x))]]]]\,v\,h\,g)$
$\S\backslash\ \ [\lambda c_\tau.[\lambda g_{tu}.[\lambda f_{uc}.[\lambda x_c.(g_{tu}(f_{uc}x_c))_t]_{(tc)}]_{(tc(uc))}]_{(tc(uc)(tu))}]v_\tau$
$\#\qquad = ([\lambda c.[\lambda g.[\lambda f.[\lambda x.(g\,(f\,x))]]]]\,v)\,[\lambda g.[\lambda f.[\lambda x.(g\,(f\,x))]]]$
$\S s\ \ \%1\ \ 12\ \ \%0$
$\#\qquad = \$HG\,([\lambda g.[\lambda f.[\lambda x.(g\,(f\,x))]]]\,h\,g)$
$\S\backslash\ \ [\lambda g_{tu}.[\lambda f_{uv}.[\lambda x_v.(g_{tu}(f_{uv}x_v))_t]_{(tv)}]_{(tv(uv))}]h_{tu}$
$\#\qquad = ([\lambda g.[\lambda f.[\lambda x.(g\,(f\,x))]]]\,h)\,[\lambda f.[\lambda x.(h\,(f\,x))]]$
$\S s\ \ \%1\ \ 6\ \ \%0$
$\#\qquad = \$HG\,([\lambda f.[\lambda x.(h\,(f\,x))]]\,g)$
$\S\backslash\ \ [\lambda f_{uv}.[\lambda x_v.(h_{tu}(f_{uv}x_v))_t]_{(tv)}]g_{uv}$
$\#\qquad = ([\lambda f.[\lambda x.(h\,(f\,x))]]\,g)\,[\lambda x.(h\,(g\,x))]$
$\S s\ \ \%1\ \ 3\ \ \%0$
$\#\qquad = \$HG\,[\lambda x.(h\,(g\,x))]$

$:=\ \ \$HGxF\ \ COMPS_{\backslash 6\backslash 4(\backslash 5\backslash 4)(\backslash 5\backslash 4)\tau\tau\tau}t_\tau v_\tau w_\tau \$HG_{tv}f_{vw}$
$\#$ wff 459 : $COMPS\,t\,v\,w\,\$HG\,f_{tw} \qquad :=\ \ \$HGxF$
$\S{=}\ \ \$HGxF$
$\#\qquad = \$HGxF\,\$HGxF$
$\S s\ \ \%0\ \ 13\ \ \%1$
$\#\qquad = \$HGxF\,(COMPS\,t\,v\,w\,[\lambda x.(h\,(g\,x))]\,f)$
$\S\backslash\ \ COMPS_{\backslash 6\backslash 4(\backslash 5\backslash 4)(\backslash 5\backslash 4)\tau\tau\tau}t_\tau$

\# $= (COMPS\,t)\,[\lambda b.[\lambda c.[\lambda g.[\lambda f.[\lambda x.(g\,(f\,x))]]]]]$
§s %1 48 %0
\# $= \$HGxF\,([\lambda b.[\lambda c.[\lambda g.[\lambda f.[\lambda x.(g\,(f\,x))]]]]]\,v\,w\,[\lambda x.(h\,(g\,x))]\,f)$
§\ $[\lambda b_\tau.[\lambda c_\tau.[\lambda g_{tb}.[\lambda f_{bc}.[\lambda x_c.(g_{tb}(f_{bc}x_c))_t]_{(tc)}]_{(tc(bc))}]_{(tc(bc)(tb))}]_{(t\backslash 4(b\backslash 4)(tb)\tau)}]v_\tau$
\# $= ([\lambda b.[\lambda c.[\lambda g.[\lambda f.[\lambda x.(g\,(f\,x))]]]]]\,v)\,[\lambda c.[\lambda g.[\lambda f.[\lambda x.(g\,(f\,x))]]]]$
§s %1 24 %0
\# $= \$HGxF\,([\lambda c.[\lambda g.[\lambda f.[\lambda x.(g\,(f\,x))]]]]\,w\,[\lambda x.(h\,(g\,x))]\,f)$
§\ $[\lambda c_\tau.[\lambda g_{tv}.[\lambda f_{vc}.[\lambda x_c.(g_{tv}(f_{vc}x_c))_t]_{(tc)}]_{(tc(vc))}]_{(tc(vc)(tv))}]w_\tau$
\# $= ([\lambda c.[\lambda g.[\lambda f.[\lambda x.(g\,(f\,x))]]]]\,w)\,[\lambda g.[\lambda f.[\lambda x.(g\,(f\,x))]]]$
§s %1 12 %0
\# $= \$HGxF\,([\lambda g.[\lambda f.[\lambda x.(g\,(f\,x))]]]\,[\lambda x.(h\,(g\,x))]\,f)$
§\ $[\lambda g_{tv}.[\lambda f_{vw}.[\lambda x_w.(g_{tv}(f_{vw}x_w))_t]_{(tw)}]_{(tw(vw))}][\lambda x_v.(h_{tu}(g_{uv}x_v))_t]$
\# $= ([\lambda g.[\lambda f.[\lambda x.(g\,(f\,x))]]]\,[\lambda x.(h\,(g\,x))])\,[\lambda f.[\lambda x.([\lambda x.(h\,(g\,x))]\,(f\,x))]]$
§s %1 6 %0
\# $= \$HGxF\,([\lambda f.[\lambda x.([\lambda x.(h\,(g\,x))]\,(f\,x))]]\,f)$
§\ $[\lambda f_{vw}.[\lambda x_w.([\lambda x_v.(h_{tu}(g_{uv}x_v))_t](f_{vw}x_w))_t]_{(tw)}]f_{vw}$
\# $= ([\lambda f.[\lambda x.([\lambda x.(h\,(g\,x))]\,(f\,x))]]\,f)\,[\lambda x.([\lambda x.(h\,(g\,x))]\,(f\,x))]$
§s %1 3 %0
\# $= \$HGxF\,[\lambda x.([\lambda x.(h\,(g\,x))]\,(f\,x))]$
§\ $[\lambda x_v.(h_{tu}(g_{uv}x_v))_t](f_{vw}x_w)$
\# $= ([\lambda x.(h\,(g\,x))]\,(f\,x))\,(h\,(g\,(f\,x)))$
§s %1 7 %0
\# $= \$HGxF\,[\lambda x.(h\,(g\,(f\,x)))]$

:= $\$TMP2$ %0
\# wff 505 : $= \$HGxF\,[\lambda x.(h\,(g\,(f\,x)))]_o$:= $\$TMP2$

\#\# .3

%$\$TMP1$
\# $= \$HxGF\,[\lambda x.(h\,(g\,(f\,x)))]$:= $\$TMP1$
\# $=_{o\omega\omega}\$HxGF_\omega[\lambda x_w.(h_{tu}(g_{uv}(f_{vw}x_w)))_t]$:= $\$TMP1$
:= $\$TMP1$
%$\$TMP2$
\# $= \$HGxF\,[\lambda x.(h\,(g\,(f\,x)))]$:= $\$TMP2$
\# $=_{o\omega\omega}\$HGxF_\omega[\lambda x_w.(h_{tu}(g_{uv}(f_{vw}x_w)))_t]$:= $\$TMP2$
:= $\$TMP2$

\#\# use Proof Template A5201b (Swap): A = B → B = A
<< A5201b.r0t.txt
%0
\# $= [\lambda x.(h\,(g\,(f\,x)))]\,\$HGxF$
\# $=_{o\omega\omega}[\lambda x_w.(h_{tu}(g_{uv}(f_{vw}x_w)))_t]\$HGxF_\omega$

§s %4 3 %0
\# $= \$HxGF\,\$HGxF$

:= $\$GF$
:= $\$HG$

:= $\$HxGF$
:= $\$HGxF$

##
Print Result
##

%0
$= (COMPS\,t\,u\,w\,h\,(COMPS\,u\,v\,w\,g\,f))\,(COMPS\,t\,v\,w\,(COMPS\,t\,u\,v\,h\,g)\,f)$
$\ldots$
$\ldots =_{o\omega\omega}(COMPS_{\backslash 6\backslash 4(\backslash 5\backslash 4)(\backslash 5\backslash 4)\tau\tau\tau}t_\tau u_\tau w_\tau h_{tu}(COMPS_{\backslash 6\backslash 4(\backslash 5\backslash 4)(\backslash 5\backslash 4)\tau\tau\tau}u_\tau v_\tau w_\tau g_{uv} f_{vw}))\ldots$
$\ldots(COMPS_{\backslash 6\backslash 4(\backslash 5\backslash 4)(\backslash 5\backslash 4)\tau\tau\tau}t_\tau v_\tau w_\tau(COMPS_{\backslash 6\backslash 4(\backslash 5\backslash 4)(\backslash 5\backslash 4)\tau\tau\tau}t_\tau u_\tau v_\tau h_{tu} g_{uv}) f_{vw})$

2.1.97 Results for File definitions1.r0.txt

##
Basic Definitions
##
##
Source: [Andrews 2002 (ISBN 1-4020-0763-9), p. 212]
##
Copyright (c) 2017 Owl of Minerva Press GmbH. All rights reserved.
Written by Ken Kubota (<mail@kenkubota.de>).
##
This file is part of the publication of the mathematical logic $\mathcal{R}_0$.
For more information, visit: <http://doi.org/10.4444/100.10>
##

Definition of truth
:= $T\ =_{o\omega\omega}=_\omega=_\omega$
wff 12 : $===_o$:= T

Definition of falsehood
:= $F\ =_{o(oo)(oo)}[\lambda x_o.T_o][\lambda x_o.x_o]$
wff 20 : $=[\lambda x.T]\,[\lambda x.x]_o$:= F

Definition of the universal quantifier (with type abstraction)
:= $\forall\ [\lambda t_\tau.[\lambda p_{ot}.(=_{o(ot)(ot)}[\lambda x_t.T_o]p_{ot})_o]_{(o(ot))}]$
wff 29 : $[\lambda t.[\lambda p.(=[\lambda x.T]\,p)]]_{o(o\backslash 3)\tau}$:= $\forall$

Definition of the conjunction
:= $\wedge\ [\lambda x_o.[\lambda y_o.(=_{o\omega\omega}[\lambda g_{ooo}.(g_{ooo}T_oT_o)_o][\lambda g_{ooo}.(g_{ooo}x_oy_o)_o])_o]_{(oo)}]$
wff 47 : $[\lambda x.[\lambda y.(=[\lambda g.(g\,T\,T)]\,[\lambda g.(g\,x\,y)])]]_{ooo}$:= $\wedge$

Definition of the implication
:= $\supset\ [\lambda x_o.[\lambda y_o.(=_{ooo}x_o(\wedge_{ooo}x_oy_o))_o]_{(oo)}]$
wff 53 : $[\lambda x.[\lambda y.(=x\,(\wedge\,x\,y))]]_{ooo}$:= $\supset$

Definition of the negation
$:=\ \sim\ [\lambda a_o.(=_{ooo} F_o a_o)_o]$
wff 57 : $[\lambda a.(= F\, a)]_{oo}$ $:=\ \sim$

Definition of the disjunction
$:=\ \vee\ [\lambda a_o.[\lambda b_o.(\sim_{oo}(\wedge_{ooo}(\sim_{oo}a_o)(\sim_{oo}b_o)))_o]_{(oo)}]$
wff 65 : $[\lambda a.[\lambda b.(\sim(\wedge(\sim a)\,(\sim b)))]]_{ooo}$ $:=\ \vee$

Definition of the existential quantifier (with type abstraction)
$:=\ \exists\ [\lambda t_\tau.[\lambda p_{ot}.(\sim_{oo}(=_{o(ot)(ot)}[\lambda x_t.T_o][\lambda x_t.(\sim_{oo}(p_{ot}x_t))_o]))_o]_{(o(ot))}]$
wff 72 : $[\lambda t.[\lambda p.(\sim(=[\lambda x.T]\,[\lambda x.(\sim(p\,x))]))]]_{o(o\backslash 3)\tau}$ $:=\ \exists$

Definition of inequality
$:=\ \neq\ [\lambda x_\omega.[\lambda y_\omega.(\sim_{oo}(=_{o\omega\omega}x_\omega y_\omega))_o]_{(o\omega)}]$
wff 79 : $[\lambda x.[\lambda y.(\sim(=x\,y))]]_{o\omega\omega}$ $:=\ \neq$

2.1.98 Results for File definitions2.r0.txt

##
Further Definitions
##
##
Source: [Andrews 2002 (ISBN 1-4020-0763-9), pp. 231, 233]
##
Copyright (c) 2017 Owl of Minerva Press GmbH. All rights reserved.
Written by Ken Kubota (<mail@kenkubota.de>).
##
This file is part of the publication of the mathematical logic $\mathcal{R}_0$.
For more information, visit: <http://doi.org/10.4444/100.10>
##

<< definitions1.r0.txt

Definition of the subset
$:=\ \subseteq\ [\lambda t_\tau.[\lambda x_{ot}.[\lambda y_{ot}.(\forall_{o(o\backslash 3)\tau}t_\tau[\lambda z_t.(\supset_{ooo}(x_{ot}z_t)(y_{ot}z_t))_o])_o]_{(o(ot))}]_{(o(ot)(ot))}]$
wff 92 : $[\lambda t.[\lambda x.[\lambda y.(\forall\, t\,[\lambda z.(\supset(x\,z)\,(y\,z))])]]]_{o(o\backslash 4)(o\backslash 3)\tau}$ $:=\ \subseteq$

Definition of the power set
$:=\ \mathcal{P}\ [\lambda t_\tau.[\lambda y_{ot}.[\lambda x_{ot}.(\subseteq_{o(o\backslash 4)(o\backslash 3)\tau}t_\tau x_{ot}y_{ot})_o]_{(o(ot))}]_{(o(ot)(ot))}]$
wff 103 : $[\lambda t.[\lambda y.[\lambda x.(\subseteq t\,x\,y)]]]_{o(o\backslash 4)(o\backslash 3)\tau}$ $:=\ \mathcal{P}$

Definition of the uniqueness quantifier (with type abstraction)
$:=\ \exists_1\ [\lambda t_\tau.[\lambda p_{ot}.(\exists_{o(o\backslash 3)\tau}t_\tau[\lambda y_t.(=_{o(ot)(ot)}p_{ot}(=_{ott}y_t))_o])_o]_{(o(ot))}]$
wff 112 : $[\lambda t.[\lambda p.(\exists\, t\,[\lambda y.(=p\,(=y))])]]_{o(o\backslash 3)\tau}$ $:=\ \exists_1$

2.1.99 Results for File definitions3.r0.txt

##
New Definitions
##
##
Source: [Kubota 2017 (doi: 10.4444/100.10)]
##

Written by Ken Kubota (<mail@kenkubota.de>).
##
This file is part of the publication of the mathematical logic $\mathcal{R}_0$.
For more information, visit: <http://doi.org/10.4444/100.10>
##

<< definitions2.r0.txt

Definition of the universal set
:= $V\ [\lambda x_\omega.T_o]$
wff 113 : $[\lambda x.T]_{o\omega}$:= V

Definition of the empty set
:= $\emptyset\ [\lambda x_\omega.F_o]$
wff 114 : $[\lambda x.F]_{o\omega}$:= $\emptyset$

Definition of the polymorphic identity relation helper function
:= == $[\lambda t_\tau.[\lambda x_t.[\lambda y_t.(=_{ott} x_t y_t)_o]_{(ot)}]_{(ott)}]$
wff 119 : $[\lambda t.[\lambda x.[\lambda y.(= x\, y)]]]_{o\backslash 3\backslash 2\tau}$:= ==

Definition of the polymorphic non-identity relation helper function
:= != = $[\lambda t_\tau.[\lambda x_t.[\lambda y_t.(\sim_{oo}(=_{ott} x_t y_t))_o]_{(ot)}]_{(ott)}]$
wff 126 : $[\lambda t.[\lambda x.[\lambda y.(\sim (= x\, y))]]]_{o\backslash 3\backslash 2\tau}$:= ! ==

Definition of the polymorphic descriptor helper function
:= $I\ [\lambda t_\tau.[\lambda x_{ot}.(\iota_{t(ot)} x_{ot})_t]_{(t(ot))}]$
wff 129 : $[\lambda t.[\lambda x.(\iota\, x)]]_{\backslash 2(o\backslash 3)\tau}$:= I

Definition of exclusive disjunction (logical exclusive "or", XOR)
:= $XOR\ [\lambda x_o.[\lambda y_o.(\sim_{oo}(=_{ooo} x_o y_o))_o]_{(oo)}]$
wff 135 : $[\lambda x.[\lambda y.(\sim (= x\, y))]]_{ooo}$:= XOR

Definition of commutativity
:= $COMMT\ [\lambda t_\tau.[\lambda f_{ttt}.(=_{ott}(f_{ttt} x_t y_t)(f_{ttt} y_t x_t))_o]_{(o(ttt))}]$
wff 147 : $[\lambda t.[\lambda f.(= (f\, x\, y)\, (f\, y\, x))]]_{o(\backslash 4\backslash 4\backslash 3)\tau}$:= $COMMT$

Definition of associativity
:= $ASSOC\ [\lambda t_\tau.[\lambda f_{ttt}.(=_{ott}(f_{ttt}(f_{ttt} x_t y_t) z_t)(f_{ttt} x_t (f_{ttt} y_t z_t)))_o]_{(o(ttt))}]$
wff 159 : $[\lambda t.[\lambda f.(= (f\, (f\, x\, y)\, z)\, (f\, x\, (f\, y\, z)))]]_{o(\backslash 4\backslash 4\backslash 3)\tau}$:= $ASSOC$

2.1.100 Results for File group.r0.txt

##
Groups
##
##
Source: [Kubota 2017 (doi: 10.4444/100.10)]
##

Written by Ken Kubota (<mail@kenkubota.de>).
##
This file is part of the publication of the mathematical logic $\mathcal{R}_0$.
For more information, visit: <http://doi.org/10.4444/100.10>
##

<< basics.r0.txt

.1: Associativity
:= $GrpAsc$ ...
$\ldots\forall_{o(o\backslash 3)\tau} g_\tau [\lambda a_g.(\forall_{o(o\backslash 3)\tau} g_\tau [\lambda b_g.(\forall_{o(o\backslash 3)\tau} g_\tau [\lambda c_g.(=_{ogg}(l_{ggg}(l_{ggg}a_g b_g)c_g)(l_{ggg}a_g(l_{ggg}b_g c_g)))_o])_o])_o]$
wff 233 : $\forall g\,[\lambda a.(\forall g\,[\lambda b.(\forall g\,[\lambda c.(=(l\,(l\,a\,b)\,c)\,(l\,a\,(l\,b\,c)))])])]_o$:= $GrpAsc$

.2: Identity element
:= $GrpIdy\ \forall_{o(o\backslash 3)\tau} g_\tau [\lambda a_g.(\wedge_{ooo}(=_{ogg}(l_{ggg}a_g e_g)a_g)(=_{ogg}(l_{ggg}e_g a_g)a_g))_o]$
wff 245 : $\forall g\,[\lambda a.(\wedge\,(=(l\,a\,e)\,a)\,(=(l\,e\,a)\,a))]_o$:= $GrpIdy$

.3: Inverse element
:= $GrpInv\ \forall_{o(o\backslash 3)\tau} g_\tau [\lambda a_g.(\exists_{o(o\backslash 3)\tau} g_\tau [\lambda b_g.(\wedge_{ooo}(=_{ogg}(l_{ggg}a_g b_g)e_g)(=_{ogg}(l_{ggg}b_g a_g)e_g))_o])_o]$
wff 257 : $\forall g\,[\lambda a.(\exists g\,[\lambda b.(\wedge\,(=(l\,a\,b)\,e)\,(=(l\,b\,a)\,e))])]_o$:= $GrpInv$

##
Definition of group (all three group properties combined)
##

:= $Grp\ [\lambda g_\tau.[\lambda l_{ggg}.(\wedge_{ooo}GrpAsc_o(\exists_{o(o\backslash 3)\tau} g_\tau [\lambda e_g.(\wedge_{ooo}GrpIdy_o GrpInv_o)_o]))_o]_{(o(ggg))}]$
wff 266 : $[\lambda g.[\lambda l.(\wedge\,GrpAsc\,(\exists g\,[\lambda e.(\wedge\,GrpIdy\,GrpInv)]))]]_{o(\backslash 4\backslash 4\backslash 3)\tau}$:= Grp

Group property identity element only (with identity element abstracted)
:= $GrpIdO\ [\lambda g_\tau.[\lambda l_{ggg}.[\lambda e_g.GrpIdy_o]_{(og)}]_{(og(ggg))}]$
wff 270 : $[\lambda g.[\lambda l.[\lambda e.GrpIdy]]]_{o\backslash 3(\backslash 4\backslash 4\backslash 3)\tau}$:= $GrpIdO$

2.1.101 Results for File group_identity_element_unique.r0.txt

##
Uniqueness of the Group Identity Element
##

##
Source: [Kubota 2017 (doi: 10.4444/100.10)]
##

Written by Ken Kubota (<mail@kenkubota.de>).
##
This file is part of the publication of the mathematical logic $\mathcal{R}_0$.
For more information, visit: <http://doi.org/10.4444/100.10>
##

<< basics.r0.txt
<< K8005.r0.txt
<< group.r0.txt

shorthands
:= $\$HYPTH \;\; \wedge_{ooo}(\wedge_{ooo}(Grp_{o(\backslash 4\backslash 4\backslash 3)\tau}g_\tau l_{ggg})(GrpIdO_{o\backslash 3(\backslash 4\backslash 4\backslash 3)\tau}g_\tau l_{ggg}e_g))\ldots$
$\ldots(GrpIdO_{o\backslash 3(\backslash 4\backslash 4\backslash 3)\tau}g_\tau l_{ggg}f_g)$
wff 1446 : $\wedge(\wedge(Grp\,g\,l)\,(GrpIdO\,g\,l\,e))\,(GrpIdO\,g\,l\,f)_o$:= $\$HYPTH$
:= $\$TMPDED \;\; \forall_{o(o\backslash 3)\tau}g_\tau[\lambda a_g.(\wedge_{ooo}(=_{ogg}(l_{ggg}a_g f_g)a_g)(=_{ogg}(l_{ggg}f_g a_g)a_g))_o]$
wff 1457 : $\forall g\,[\lambda a.(\wedge\,(=(l\,a\,f)\,a)\,(=(l\,f\,a)\,a))]_o$:= $\$TMPDED$

.1: Let (g,l) be a group, and e and f identity elements of it

$\%K8005$
$\supset x\,x$:= $K8005$
$\supset_{ooo}x_o x_o$:= $K8005$

use Proof Template A5221 (Sub): B $\rightarrow$ B [x/A]
:= $\$B5221$ %0
wff 1357 : $\supset x\,x_{o,\ldots}$:= $\$B5221 \;\; K8005$
:= $\$T5221 \;\; o$
wff 2 : o_τ := $\$T5221$
:= $\$X5221 \;\; x_o$
wff 16 : x_o := $\$X5221$
:= $\$A5221 \;\; \wedge_{ooo}(\wedge_{ooo}(Grp_{o(\backslash 4\backslash 4\backslash 3)\tau}g_\tau l_{ggg})(GrpIdO_{o\backslash 3(\backslash 4\backslash 4\backslash 3)\tau}g_\tau l_{ggg}e_g))(GrpIdO_{o\backslash 3(\backslash 4\backslash 4\backslash 3)\tau}g_\tau l_{ggg}f_g)$
wff 1446 : $\wedge(\wedge(Grp\,g\,l)\,(GrpIdO\,g\,l\,e))\,(GrpIdO\,g\,l\,f)_o$:= $\$A5221 \;\; \$HYPTH$
<< A5221.r0t.txt
:= $\$B5221$
:= $\$T5221$
:= $\$X5221$
:= $\$A5221$

:= $\$FULLH$ %0
wff 1494 : $\supset \$HYPTH\,\$HYPTH_{o,\ldots}$:= $\$FULLH$

.2: Proof of H $\supset$ e * f = e

$\%\$FULLH$
$\supset \$HYPTH\,\$HYPTH$:= $\$FULLH$

\# $\supset_{ooo}\$HYPTH_o\$HYPTH_o$:= $\$FULLH$

\#\# use Proof Template K8019H: H ⊃ (A ∧ B) → H ⊃ A, H ⊃ B
:= $\$H8019H$ %0
\# wff 1494 : $\supset \$HYPTH\,\$HYPTH_{o,\ldots}$:= $\$FULLH$ $\$H8019H$
<< K8019H.r0t.txt
:= $\$H8019H$
%$\$B8019H$
\# $\supset \$HYPTH\,(GrpIdO\,g\,l\,f)$:= $\$B8019H$
\# $\supset_{ooo}\$HYPTH_o(GrpIdO_{o\backslash 3(\backslash 4\backslash 4\backslash 3)\tau}g_\tau l_{ggg}f_g)$:= $\$B8019H$
:= $\$A8019H$
:= $\$B8019H$
%0
\# $\supset \$HYPTH\,(GrpIdO\,g\,l\,f)$
\# $\supset_{ooo}\$HYPTH_o(GrpIdO_{o\backslash 3(\backslash 4\backslash 4\backslash 3)\tau}g_\tau l_{ggg}f_g)$

§\ $GrpIdO_{o\backslash 3(\backslash 4\backslash 4\backslash 3)\tau}g_\tau$
\# $=(GrpIdO\,g)\,[\lambda l.[\lambda e.GrpIdy]]$
§s %1 12 %0
\# $\supset \$HYPTH\,([\lambda l.[\lambda e.GrpIdy]]\,l\,f)$
§\ $[\lambda l_{ggg}.[\lambda e_g.GrpIdy_o]_{(og)}]l_{ggg}$
\# $=([\lambda l.[\lambda e.GrpIdy]]\,l)\,[\lambda e.GrpIdy]$
§s %1 6 %0
\# $\supset \$HYPTH\,([\lambda e.GrpIdy]\,f)$
§\ $[\lambda e_g.GrpIdy_o]f_g$
\# $=([\lambda e.GrpIdy]\,f)\,\$TMPDED$
§s %1 3 %0
\# $\supset \$HYPTH\,\$TMPDED$

\#\# use Proof Template A5215H (∀ I): H ⊃ ∀ x: B → H ⊃ B [x/a]
:= $\$T5215H$ g_τ
\# wff 1371 : g_τ := $\$T5215H$
:= $\$X5215H$ $a_{\$T5215H}$
\# wff 1375 : $a_{\$T5215H}$:= $\$X5215H$
:= $\$A5215H$ $e_{\$T5215H}$
\# wff 1397 : $e_{\$T5215H}$:= $\$A5215H$
:= $\$H5215H$ %0
\# wff 1872 : $\supset \$HYPTH\,\$TMPDED_o$:= $\$H5215H$
<< A5215H.r0t.txt
:= $\$T5215H$
:= $\$X5215H$
:= $\$A5215H$
:= $\$H5215H$
%0
\# $\supset \$HYPTH\,(\wedge\,(=(l\,e\,f)\,e)\,(=(l\,f\,e)\,e))$
\# $\supset_{ooo}\$HYPTH_o(\wedge_{ooo}(=_{ogg}(l_{ggg}e_gf_g)e_g)(=_{ogg}(l_{ggg}f_ge_g)e_g))$

\#\# use Proof Template K8019H: H ⊃ (A ∧ B) → H ⊃ A, H ⊃ B
:= $\$H8019H$ %0

\# wff 1946 : $\supset \$HYPTH\,(\wedge\,(=(l\,e\,f)\,e)\,(=(l\,f\,e)\,e))_o$:= $\$H8019H$
<< K8019H.r0t.txt
:= $\$H8019H$
%$\$A8019H$
\# $\supset \$HYPTH\,(=(l\,e\,f)\,e)$:= $\$A8019H$
\# $\supset_{ooo}\$HYPTH_o(=_{ogg}(l_{ggg}e_gf_g)e_g)$:= $\$A8019H$
:= $\$A8019H$
:= $\$B8019H$

:= $\$EIDTY$ %0
\# wff 1987 : $\supset \$HYPTH\,(=(l\,e\,f)\,e)_o$:= $\$EIDTY$

\## .3: Proof of H $\supset$ e * f = f

%$\$FULLH$
\# $\supset \$HYPTH\,\$HYPTH$:= $\$FULLH$
\# $\supset_{ooo}\$HYPTH_o\$HYPTH_o$:= $\$FULLH$
:= $\$FULLH$

\## use Proof Template K8019H: H $\supset$ (A $\wedge$ B) $\rightarrow$ H $\supset$ A, H $\supset$ B
:= $\$H8019H$ %0
\# wff 1494 : $\supset \$HYPTH\,\$HYPTH_{o,\ldots}$:= $\$H8019H$
<< K8019H.r0t.txt
:= $\$H8019H$
%$\$A8019H$
\# $\supset \$HYPTH\,(\wedge\,(Grp\,g\,l)\,(GrpIdO\,g\,l\,e))$:= $\$A8019H$
\# $\supset_{ooo}\$HYPTH_o(\wedge_{ooo}(Grp_{o(\backslash 4\backslash 4\backslash 3)\tau}g_\tau l_{ggg})(GrpIdO_{o\backslash 3(\backslash 4\backslash 4\backslash 3)\tau}g_\tau l_{ggg}e_g))$:=
$\$A8019H$
:= $\$A8019H$
:= $\$B8019H$

\## use Proof Template K8019H: H $\supset$ (A $\wedge$ B) $\rightarrow$ H $\supset$ A, H $\supset$ B
:= $\$H8019H$ %0
\# wff 1788 : $\supset \$HYPTH\,(\wedge\,(Grp\,g\,l)\,(GrpIdO\,g\,l\,e))_o$:= $\$H8019H$
<< K8019H.r0t.txt
:= $\$H8019H$
%$\$B8019H$
\# $\supset \$HYPTH\,(GrpIdO\,g\,l\,e)$:= $\$B8019H$
\# $\supset_{ooo}\$HYPTH_o(GrpIdO_{o\backslash 3(\backslash 4\backslash 4\backslash 3)\tau}g_\tau l_{ggg}e_g)$:= $\$B8019H$
:= $\$A8019H$
:= $\$B8019H$

§\ $GrpIdO_{o\backslash 3(\backslash 4\backslash 4\backslash 3)\tau}g_\tau$
\# $=(GrpIdO\,g)\,[\lambda l.[\lambda e.GrpIdy]]$
§s %1 12 %0
\# $\supset \$HYPTH\,([\lambda l.[\lambda e.GrpIdy]]\,l\,e)$
§\ $[\lambda l_{ggg}.[\lambda e_g.GrpIdy_o]_{(og)}]l_{ggg}$
\# $=([\lambda l.[\lambda e.GrpIdy]]\,l)\,[\lambda e.GrpIdy]$
§s %1 6 %0

```
#                ⊃ $HYPTH ([λe.GrpIdy] e)
§\ [λe_g.GrpIdy_o]e_g
#                = ([λe.GrpIdy] e) GrpIdy
§s %1 3 %0
#                ⊃ $HYPTH GrpIdy

## use Proof Template A5215H (∀ I):  H ⊃ ∀ x: B  →  H ⊃ B [x/a]
:= $T5215H g_τ
# wff   1371 :     g_τ     := $T5215H
:= $X5215H a_$T5215H
# wff   1375 :     a_$T5215H     := $X5215H
:= $A5215H f_$T5215H
# wff   1444 :     f_$T5215H     := $A5215H
:= $H5215H %0
# wff   2085 :     ⊃ $HYPTH GrpIdy_o     := $H5215H
<< A5215H.r0t.txt
:= $T5215H
:= $X5215H
:= $A5215H
:= $H5215H
%0
#                ⊃ $HYPTH (∧ (= (l f e) f) (= (l e f) f))
#                ⊃_ooo$HYPTH_o(∧_ooo(=_ogg(l_ggg f_g e_g) f_g)(=_ogg(l_ggg e_g f_g) f_g))

## use Proof Template K8019H:  H ⊃ (A ∧ B)  →  H ⊃ A, H ⊃ B
:= $H8019H %0
# wff   2143 :     ⊃ $HYPTH (∧ (= (l f e) f) (= (l e f) f))_o     := $H8019H
<< K8019H.r0t.txt
:= $H8019H
%$B8019H
#                ⊃ $HYPTH (= (l e f) f)     := $B8019H
#                ⊃_ooo$HYPTH_o(=_ogg(l_ggg e_g f_g) f_g)     := $B8019H
:= $A8019H
:= $B8019H

:= $FIDTY %0
# wff   2209 :     ⊃ $HYPTH (= (l e f) f)_o     := $FIDTY

## .4:  Proof of H ⊃ e = f

%$FIDTY
#                ⊃ $HYPTH (= (l e f) f)     := $FIDTY
#                ⊃_ooo$HYPTH_o(=_ogg(l_ggg e_g f_g) f_g)     := $FIDTY
:= $FIDTY
%$EIDTY
#                ⊃ $HYPTH (= (l e f) e)     := $EIDTY
#                ⊃_ooo$HYPTH_o(=_ogg(l_ggg e_g f_g) e_g)     := $EIDTY
:= $EIDTY
§s' %1 5 %0
```

\# $\supset \$HYPTH\,(= e\,f)$

\## use Proof Template K8025 (Deduction Theorem): (H ∧ I) ⊃ A → H ⊃ (I ⊃ A)
<< K8025.r0t.txt
%0
\# $\supset (\wedge\,(Grp\,g\,l)\,(GrpIdO\,g\,l\,e))\,(\supset (GrpIdO\,g\,l\,f)\,(= e\,f))$
\# $\supset_{ooo}(\wedge_{ooo}(Grp_{o(\backslash4\backslash4\backslash3)\tau}g_{\tau}l_{ggg})(GrpIdO_{o\backslash3(\backslash4\backslash4\backslash3)\tau}g_{\tau}l_{ggg}e_{g}))\ldots$
$\ldots(\supset_{ooo}(GrpIdO_{o\backslash3(\backslash4\backslash4\backslash3)\tau}g_{\tau}l_{ggg}f_{g})(=_{ogg}e_{g}f_{g}))$

\## use Proof Template K8025 (Deduction Theorem): (H ∧ I) ⊃ A → H ⊃ (I ⊃ A)
<< K8025.r0t.txt
%0
\# $\supset (Grp\,g\,l)\,(\supset (GrpIdO\,g\,l\,e)\,(\supset (GrpIdO\,g\,l\,f)\,(= e\,f)))$
\# $\supset_{ooo}(Grp_{o(\backslash4\backslash4\backslash3)\tau}g_{\tau}l_{ggg})\ldots$
$\ldots(\supset_{ooo}(GrpIdO_{o\backslash3(\backslash4\backslash4\backslash3)\tau}g_{\tau}l_{ggg}e_{g})(\supset_{ooo}(GrpIdO_{o\backslash3(\backslash4\backslash4\backslash3)\tau}g_{\tau}l_{ggg}f_{g})(=_{ogg}e_{g}f_{g})))$

:= $GrpIdElUniq$ %0
\# wff 4852 : $\supset (Grp\,g\,l)\,(\supset (GrpIdO\,g\,l\,e)\,(\supset (GrpIdO\,g\,l\,f)\,(= e\,f)))_{o,\ldots}$:= $GrpIdElUniq$

\## undefine local variables
:= $\$HYPTH$
:= $\$TMPDED$

\##
\## Print Result
\##

%0
\# $\supset (Grp\,g\,l)\,(\supset (GrpIdO\,g\,l\,e)\,(\supset (GrpIdO\,g\,l\,f)\,(= e\,f)))$:= $GrpIdElUniq$
\# $\supset_{ooo}(Grp_{o(\backslash4\backslash4\backslash3)\tau}g_{\tau}l_{ggg})\ldots$
$\ldots(\supset_{ooo}(GrpIdO_{o\backslash3(\backslash4\backslash4\backslash3)\tau}g_{\tau}l_{ggg}e_{g})(\supset_{ooo}(GrpIdO_{o\backslash3(\backslash4\backslash4\backslash3)\tau}g_{\tau}l_{ggg}f_{g})(=_{ogg}e_{g}f_{g})))$:= $GrpIdElUniq$

2.1.102 Results for File natural_numbers.r0.txt

\##
\## Peano's Postulates
\##
\##
\## Source: [Andrews 2002 (ISBN 1-4020-0763-9), pp. 258 f.]
\##

\## Written by Ken Kubota (<mail@kenkubota.de>).
\##
\## This file is part of the publication of the mathematical logic $\mathcal{R}_0$.
\## For more information, visit: <http://doi.org/10.4444/100.10>
\##

<< basics.r0.txt

variables used
t: domain (type of the natural numbers)
z: zero
s: successor function
n: set of natural numbers

definition of the lambda abstraction as part of the universal quantifier on natural numbers,
the universal quantifier with dot ([cf. Andrews 2002 (ISBN 1-4020-0763-9), p. 260])
:= $\$DOT\ [\lambda x_t.(\supset_{ooo}(n_{ot}x_t))_{(oo)}]$
wff 211 : $[\lambda x.(\supset (n\, x))]_{oot}$:= $\$DOT$
:= $DOT\ [\lambda t_\tau.[\lambda n_{ot}.\$DOT_{(oot)}]_{(oot(ot))}]$
wff 215 : $[\lambda t.[\lambda n.\$DOT]]_{oo\backslash 3(o\backslash 3)\tau}$:= DOT
§\ $\$DOT_{oot}x_t$
$= (\$DOT\, x)\,(\supset (n\, x))$
:= $\$DOTx$ %0/3
wff 210 : $\supset (n\, x)_{oo,\ldots}$:= $\$DOTx$
:= $DOTx\ [\lambda t_\tau.[\lambda n_{ot}.\$DOTx_{(oo)}]_{(oo(ot))}]$
wff 224 : $[\lambda t.[\lambda n.\$DOTx]]_{oo(o\backslash 3)\tau}$:= $DOTx$

"(P1) There is an entity called 0 which is a natural number."
:= $\$P1\ n_{ot}z_t$
wff 227 : $n\, z_o$:= $\$P1$
:= $P1\ [\lambda t_\tau.[\lambda z_t.[\lambda s_{tt}.[\lambda n_{ot}.\$P1_o]_{(o(ot))}]_{(o(ot)(tt))}]_{(o(ot)(tt)t)}]$
wff 234 : $[\lambda t.[\lambda z.[\lambda s.[\lambda n.\$P1]]]]_{o(o\backslash 5)(\backslash 4\backslash 4)\backslash 2\tau}$:= $P1$

"(P2) Every natural number n has a successor S[_]n which is also a natural number."
:= $\$P2\ \forall_{o(o\backslash 3)\tau}t_\tau[\lambda x_t.(\$DOTx_{oo}(n_{ot}(s_{tt}x_t)))_o]$
wff 245 : $\forall t\,[\lambda x.(\$DOTx\,(n\,(s\, x)))]_o$:= $\$P2$
:= $P2\ [\lambda t_\tau.[\lambda z_t.[\lambda s_{tt}.[\lambda n_{ot}.\$P2_o]_{(o(ot))}]_{(o(ot)(tt))}]_{(o(ot)(tt)t)}]$
wff 249 : $[\lambda t.[\lambda z.[\lambda s.[\lambda n.\$P2]]]]_{o(o\backslash 5)(\backslash 4\backslash 4)\backslash 2\tau}$:= $P2$

"(P3) 0 is not the successor of any natural number."
(formula not verified yet, using a temporary definition)
:= $\$P3\ =_{o\omega\omega}=_\omega=_\omega$
wff 12 : $===_o$:= $\$P3\ \ T$
:= $P3\ [\lambda t_\tau.[\lambda z_t.[\lambda s_{tt}.[\lambda n_{ot}.T_o]_{(o(ot))}]_{(o(ot)(tt))}]_{(o(ot)(tt)t)}]$
wff 253 : $[\lambda t.[\lambda z.[\lambda s.[\lambda n.T]]]]_{o(o\backslash 5)(\backslash 4\backslash 4)\backslash 2\tau}$:= $P3$

"(P4) If n and m are natural numbers with the same successors, then n and m are the same."
(formula not verified yet, using a temporary definition)
:= $\$P4\ =_{o\omega\omega}=_\omega=_\omega$
wff 12 : $===_o$:= $\$P3\ \ \$P4\ \ T$

:= $P4$ $[\lambda t_\tau.[\lambda z_t.[\lambda s_{tt}.[\lambda n_{ot}.T_o]_{(o(ot))}]_{(o(ot)(tt))}]_{(o(ot)(tt)t)}]$
wff 253 : $[\lambda t.[\lambda z.[\lambda s.[\lambda n.T]]]]_{o(o\backslash 5)(\backslash 4\backslash 4)\backslash 2\tau}$:= $P3$ $P4$

"(P5) Principle of Mathematical Induction"
:= $\$P5N$ $\$DOTx_{oo}(\supset_{ooo}(p_{ot}x_t)(p_{ot}(s_{tt}x_t)))$
wff 257 : $\$DOTx\,(\supset (p\,x)\,(p\,(s\,x)))_o$:= $\$P5N$
:= $\$P5T$ $\$DOTx_{oo}(p_{ot}x_t)$
wff 258 : $\$DOTx\,(p\,x)_o$:= $\$P5T$
:= $\$P5$ $\forall_{o(o\backslash 3)\tau}(ot)_\tau[\lambda p_{ot}.(\supset_{ooo}(\wedge_{ooo}(p_{ot}z_t)(\forall_{o(o\backslash 3)\tau}t_\tau[\lambda x_t.\$P5N_o]))(\forall_{o(o\backslash 3)\tau}t_\tau[\lambda x_t.\$P5T_o]))_o]$
wff 271 : $\forall\,(ot)\,[\lambda p.(\supset (\wedge\,(p\,z)\,(\forall\,t\,[\lambda x.\$P5N]))\,(\forall\,t\,[\lambda x.\$P5T]))]_o$:= $\$P5$
:= $P5$ $[\lambda t_\tau.[\lambda z_t.[\lambda s_{tt}.[\lambda n_{ot}.\$P5_o]_{(o(ot))}]_{(o(ot)(tt))}]_{(o(ot)(tt)t)}]$
wff 275 : $[\lambda t.[\lambda z.[\lambda s.[\lambda n.\$P5]]]]_{o(o\backslash 5)(\backslash 4\backslash 4)\backslash 2\tau}$:= $P5$

all of Peano's Postulates combined
:= $\$PEANO$ $\wedge_{ooo}(\wedge_{ooo}(\wedge_{ooo}(\wedge_{ooo}\$P1_o\$P2_o)T_o)T_o)\$P5_o$
wff 283 : $\wedge\,(\wedge\,(\wedge\,(\wedge\,\$P1\,\$P2)\,T)\,T)\,\$P5_o$:= $\$PEANO$
:= $PEANO$ $[\lambda t_\tau.[\lambda z_t.[\lambda s_{tt}.[\lambda n_{ot}.\$PEANO_o]_{(o(ot))}]_{(o(ot)(tt))}]_{(o(ot)(tt)t)}]$
wff 287 : $[\lambda t.[\lambda z.[\lambda s.[\lambda n.\$PEANO]]]]_{o(o\backslash 5)(\backslash 4\backslash 4)\backslash 2\tau}$:= $PEANO$

undefine local variables
:= $\$DOT$
:= $\$DOTx$
:= $\$P1$
:= $\$P2$
:= $\$P3$
:= $\$P4$
:= $\$P5$
:= $\$P5N$
:= $\$P5T$
:= $\$PEANO$

2.1.103 Results for File natural_numbers_andrews.r0.txt

##
Andrews' Definition of Natural Numbers
##
##
Source: [Andrews 2002 (ISBN 1-4020-0763-9), p. 260]
##
Copyright (c) 2017 Owl of Minerva Press GmbH. All rights reserved.
Written by Ken Kubota (<mail@kenkubota.de>).
##
This file is part of the publication of the mathematical logic $\mathcal{R}_0$.
For more information, visit: <http://doi.org/10.4444/100.10>
##

<< natural_numbers.r0.txt

polymorphic sigma
:= $SIGMA\ \ o(ot)$
wff 22 : $o(ot)_\tau$:= $SIGMA$

shorthand for polymorphic sigma
:= $\$S\ \ o(ot)$
wff 22 : $o(ot)_\tau$:= $\$S\ \ SIGMA$

zero
:= $ATZERO\ =_{\$S\,(ot)}[\lambda x_t.F_o]$
wff 289 : $=[\lambda x.F]_{\$S}$:= $ATZERO$
:= $AZERO\ [\lambda t_\tau.ATZERO_{\$S}]$
wff 290 : $[\lambda t.ATZERO]_{o(o\backslash 3)\tau}$:= $AZERO$

successor function
:= $ATSUCC\ [\lambda n_{\$S}.[\lambda p_{ot}.(\exists_{o(o\backslash 3)\tau}t_\tau[\lambda x_t.(\wedge_{ooo}(p_{ot}x_t)(n_{\$S}[\lambda t_t.(\wedge_{ooo}(\sim_{oo}(=_{ott}t_tx_t))(p_{ot}t_t))_o]))_o])_o]_{\$S}]$
wff 306 : $[\lambda n.[\lambda p.(\exists t\,[\lambda x.(\wedge\,(p\,x)\,(n\,[\lambda t.(\wedge\,(\sim(=t\,x))\,(p\,t))]))])]]_{\$S\,\$S}$:= $ATSUCC$
:= $ASUCC\ [\lambda t_\tau.ATSUCC_{(\$S\,\$S)}]$
wff 308 : $[\lambda t.ATSUCC]_{o(o\backslash 4)(o(o\backslash 4))\tau}$:= $ASUCC$

set of natural numbers
:= $\$ANSETZ\ \ p_{o\,\$S}ATZERO_{\$S}$
wff 312 : $p\,ATZERO_o$:= $\$ANSETZ$
:= $\$ANSETS\ \forall_{o(o\backslash 3)\tau}\$S_\tau[\lambda x_{\$S}.(\supset_{ooo}(p_{o\,\$S}x_{\$S})(p_{o\,\$S}(ATSUCC_{\$S\,\$S}x_{\$S})))_o]$
wff 322 : $\forall\,\$S\,[\lambda x.(\supset(p\,x)\,(p\,(ATSUCC\,x)))]_o$:= $\$ANSETS$
:= $ATNSET\ [\lambda n_{\$S}.(\forall_{o(o\backslash 3)\tau}(o\,\$S)_\tau[\lambda p_{o\,\$S}.(\supset_{ooo}(\wedge_{ooo}\$ANSETZ_o\$ANSETS_o)(p_{o\,\$S}n_{\$S}))_o])_o]$
wff 332 : $[\lambda n.(\forall\,(o\,\$S)\,[\lambda p.(\supset(\wedge\,\$ANSETZ\,\$ANSETS)\,(p\,n))])]_{o\,\$S}$:= $ATNSET$
:= $ANSET\ [\lambda t_\tau.ATNSET_{(o\,\$S)}]$
wff 333 : $[\lambda t.ATNSET]_{o(o(o\backslash 4))\tau}$:= $ANSET$

set of finite sets
:= $ATFINI\ [\lambda p_{ot}.(\exists_{o(o\backslash 3)\tau}\$S_\tau[\lambda n_{\$S}.(\wedge_{ooo}(ATNSET_{o\,\$S}n_{\$S})(n_{\$S}p_{ot}))_o])_o]$
wff 343 : $[\lambda p.(\exists\,\$S\,[\lambda n.(\wedge\,(ATNSET\,n)\,(n\,p))])]_{\$S}$:= $ATFINI$
:= $AFINI\ [\lambda t_\tau.ATFINI_{\$S}]$
wff 344 : $[\lambda t.ATFINI]_{o(o\backslash 3)\tau}$:= $AFINI$

definition of the universal quantifier on (Andrews' definition of) natural numbers (with dot)
§= $DOT_{oo\backslash 3(o\backslash 3)\tau}\$S_\tau(ANSET_{o(o(o\backslash 4))\tau}t_\tau)$
$=(DOT\,\$S\,(ANSET\,t))\,(DOT\,\$S\,(ANSET\,t))$
§\ $DOT_{oo\backslash 3(o\backslash 3)\tau}\$S_\tau$
$=(DOT\,\$S)\,[\lambda n.[\lambda x.(\supset(n\,x))]]$
§s %1 6 %0
$=(DOT\,\$S\,(ANSET\,t))\,([\lambda n.[\lambda x.(\supset(n\,x))]]\,(ANSET\,t))$
§\ $[\lambda n_{o\,\$S}.[\lambda x_{\$S}.(\supset_{ooo}(n_{o\,\$S}x_{\$S}))_{(oo)}]_{(oo\,\$S)}](ANSET_{o(o(o\backslash 4))\tau}t_\tau)$
$=([\lambda n.[\lambda x.(\supset(n\,x))]]\,(ANSET\,t))\,[\lambda x.(\supset(ANSET\,t\,x))]$
§s %1 3 %0

\# $= (DOT\,\$S\,(ANSET\,t))\,[\lambda x.(\supset (ANSET\,t\,x))]$
:= $ADOT$ %0/3
\# wff 365 : $[\lambda x.(\supset (ANSET\,t\,x))]_{oo\,\$S,\ldots}$:= $ADOT$
§\ $ADOT_{oo\,\$S}x_{\$S}$
\# $= (ADOT\,x)\,(\supset (ANSET\,t\,x))$
:= $ADOTx$ %0/3
\# wff 364 : $\supset (ANSET\,t\,x)_{oo,\ldots}$:= $ADOTx$

\## undefine local variables
:= $\$S$
:= $\$ANSETZ$
:= $\$ANSETS$

2.1.104 Results for File neumann.r0.txt

\##
\## Definition of natural numbers similar to the idea of John von Neumann
\##
\##
\## Source: [Kubota 2017 (doi: 10.4444/100.10)]
\##
\## Copyright (c) 2017 Owl of Minerva Press GmbH. All rights reserved.
\## Written by Ken Kubota (<mail@kenkubota.de>).
\##
\## This file is part of the publication of the mathematical logic $\mathcal{R}_0$.
\## For more information, visit: <http://doi.org/10.4444/100.10>
\##

<< basics.r0.txt
<< pair0.r0.txt

\## zero (empty set)
:= $NEUMNNO000\ [\lambda x_\omega.F_o]$
\# wff 114 : $[\lambda x.F]_{o\omega}$:= $NEUMNNO000\ \emptyset$

\## successor function
:= $NEUMNSUCCR\ [\lambda x_\omega.(ODPR0_{TYPR0\omega\omega}\emptyset_\omega x_\omega)_{TYPR0}]$
\# wff 293 : $[\lambda x.(ODPR0\,\emptyset\,x)]_{TYPR0\,\omega}$:= $NEUMNSUCCR$

\## predecessor function (= right element function)
:= $NEUMNPREDR\ [\lambda p_{TYPR0}.(p_{TYPR0}[\lambda x_\omega.[\lambda y_\omega.y_\omega]_{(\omega\omega)}])_\omega]$
\# wff 269 : $[\lambda p.(p\,[\lambda x.[\lambda y.y]])]_{\omega\,TYPR0}$:= $NEUMNPREDR\ \ RELE0$

\##
\## Examples: Expand numbers zero, one, two and three
\##

.0
§= $\emptyset$
$=\emptyset\,\emptyset$
:= $NEUMNNO000EXPND$ %0
wff 295 : $=\emptyset\,\emptyset_o$:= $NEUMNNO000EXPND$

.1
:= $NEUMNNO001$ $NEUMNSUCCR_{TYPR0\,\omega}\emptyset_\omega$
wff 296 : $NEUMNSUCCR\,\emptyset_{TYPR0}$:= $NEUMNNO001$
§= $NEUMNNO001$
$=NEUMNNO001\,NEUMNNO001$
§\ $NEUMNNO001$
$=NEUMNNO001\,(ODPR0\,\emptyset\,\emptyset)$
§s %1 3 %0
$=NEUMNNO001\,(ODPR0\,\emptyset\,\emptyset)$
:= $NEUMNNO001EXPND$ %0
wff 300 : $=NEUMNNO001\,(ODPR0\,\emptyset\,\emptyset)_o$:= $NEUMNNO001EXPND$

.2
:= $NEUMNNO002$ $NEUMNSUCCR_{TYPR0\,\omega}NEUMNNO001_\omega$
wff 301 : $NEUMNSUCCR\,NEUMNNO001_{TYPR0}$:= $NEUMNNO002$
§= $NEUMNNO002$
$=NEUMNNO002\,NEUMNNO002$
§\ $NEUMNNO002$
$=NEUMNNO002\,(ODPR0\,\emptyset\,NEUMNNO001)$
§s %1 3 %0
$=NEUMNNO002\,(ODPR0\,\emptyset\,NEUMNNO001)$
§\ $NEUMNNO001$
$=NEUMNNO001\,(ODPR0\,\emptyset\,\emptyset)$:= $NEUMNNO001EXPND$
§s %1 7 %0
$=NEUMNNO002\,(ODPR0\,\emptyset\,(ODPR0\,\emptyset\,\emptyset))$
:= $NEUMNNO002EXPND$ %0
wff 307 : $=NEUMNNO002\,(ODPR0\,\emptyset\,(ODPR0\,\emptyset\,\emptyset))_o$ ...
... := $NEUMNNO002EXPND$

.3
:= $NEUMNNO003$ $NEUMNSUCCR_{TYPR0\,\omega}NEUMNNO002_\omega$
wff 308 : $NEUMNSUCCR\,NEUMNNO002_{TYPR0}$:= $NEUMNNO003$
§= $NEUMNNO003$
$=NEUMNNO003\,NEUMNNO003$
§\ $NEUMNNO003$
$=NEUMNNO003\,(ODPR0\,\emptyset\,NEUMNNO002)$
§s %1 3 %0
$=NEUMNNO003\,(ODPR0\,\emptyset\,NEUMNNO002)$
§\ $NEUMNNO002$
$=NEUMNNO002\,(ODPR0\,\emptyset\,NEUMNNO001)$
§s %1 7 %0
$=NEUMNNO003\,(ODPR0\,\emptyset\,(ODPR0\,\emptyset\,NEUMNNO001))$
§\ $NEUMNNO001$

$= NEUMNNO001\,(ODPR0\,\emptyset\,\emptyset)$:= $NEUMNNO001EXPND$
§s %1 15 %0
$= NEUMNNO003\,(ODPR0\,\emptyset\,(ODPR0\,\emptyset\,(ODPR0\,\emptyset\,\emptyset)))$
:= $NEUMNNO003EXPND$ %0
wff 316 : $= NEUMNNO003\,(ODPR0\,\emptyset\,(ODPR0\,\emptyset\,(ODPR0\,\emptyset\,\emptyset)))_o$:=
$NEUMNNO003EXPND$

##
Expand 3 - 1 = 2 (via predecessor function)
##

define 2 (expanded)
:= $NM002 =_{o\omega\omega} NEUMNNO002_\omega (ODPR0_{TYPR0\,\omega\omega}\emptyset_\omega(ODPR0_{TYPR0\,\omega\omega}\emptyset_\omega\emptyset_\omega))/3$
wff 306 : $ODPR0\,\emptyset\,(ODPR0\,\emptyset\,\emptyset)_{TYPR0,\dots}$:= $NM002$
define 3 (expanded)
:= $NM003 =_{o\omega\omega} NEUMNNO003_\omega (ODPR0_{TYPR0\,\omega\omega}\emptyset_\omega NM002_\omega)/3$
wff 315 : $ODPR0\,\emptyset\,NM002_{TYPR0,\dots}$:= $NM003$

obtain predecessor of three
§= $RELE0_{\omega\,TYPR0} NM003_{TYPR0}$
$= (RELE0\,NM003)\,(RELE0\,NM003)$

expand right element
§\ $RELE0_{\omega\,TYPR0} NM003_{TYPR0}$
$= (RELE0\,NM003)\,(NM003\,[\lambda x.[\lambda y.y]])$
§s %1 3 %0
$= (RELE0\,NM003)\,(NM003\,[\lambda x.[\lambda y.y]])$
§\ $ODPR0_{TYPR0\,\omega\omega}\emptyset_\omega$
$= (ODPR0\,\emptyset)\,[\lambda y.[\lambda g.(g\,\emptyset\,y)]]$
§s %1 12 %0
$= (RELE0\,NM003)\,([\lambda y.[\lambda g.(g\,\emptyset\,y)]]\,NM002\,[\lambda x.[\lambda y.y]])$
§\ $[\lambda y_\omega.[\lambda g_{\omega\omega\omega}.(g_{\omega\omega\omega}\emptyset_\omega y_\omega)_\omega]_{TYPR0}]NM002_\omega$
$= ([\lambda y.[\lambda g.(g\,\emptyset\,y)]]\,NM002)\,[\lambda g.(g\,\emptyset\,NM002)]$
§s %1 6 %0
$= (RELE0\,NM003)\,([\lambda g.(g\,\emptyset\,NM002)]\,[\lambda x.[\lambda y.y]])$
§\ $[\lambda g_{\omega\omega\omega}.(g_{\omega\omega\omega}\emptyset_\omega NM002_\omega)_\omega][\lambda x_\omega.[\lambda y_\omega.y_\omega]_{(\omega\omega)}]$
$= ([\lambda g.(g\,\emptyset\,NM002)]\,[\lambda x.[\lambda y.y]])\,([\lambda x.[\lambda y.y]]\,\emptyset\,NM002)$
§s %1 3 %0
$= (RELE0\,NM003)\,([\lambda x.[\lambda y.y]]\,\emptyset\,NM002)$
§\ $[\lambda x_\omega.[\lambda y_\omega.y_\omega]_{(\omega\omega)}]\emptyset_\omega$
$= ([\lambda x.[\lambda y.y]]\,\emptyset)\,[\lambda y.y]$
§s %1 6 %0
$= (RELE0\,NM003)\,([\lambda y.y]\,NM002)$
§\ $[\lambda y_\omega.y_\omega]NM002_\omega$
$= ([\lambda y.y]\,NM002)\,NM002$
§s %1 3 %0
$= (RELE0\,NM003)\,NM002$

2.1.105 Results for File pair0.r0.txt

##
Ordered Pairs With No Type Variable
##
##
Source: [Andrews 2002 (ISBN 1-4020-0763-9), p. 208]
##

Written by Ken Kubota (<mail@kenkubota.de>).
##
This file is part of the publication of the mathematical logic $\mathcal{R}_0$.
For more information, visit: <http://doi.org/10.4444/100.10>
##

definition of ordered pair (no type variable)
:= $ODPR0\ [\lambda x_\omega.[\lambda y_\omega.[\lambda g_{\omega\omega\omega}.(g_{\omega\omega\omega} x_\omega y_\omega)_\omega]_{(\omega(\omega\omega\omega))}]_{(\omega(\omega\omega\omega)\omega)}]$
wff 22 : $[\lambda x.[\lambda y.[\lambda g.(g\, x\, y)]]]_{\omega(\omega\omega\omega)\omega\omega}$:= $ODPR0$

type of ordered pair (no type variable)
:= $TYPR0\ \omega(\omega\omega\omega)$
wff 19 : $\omega(\omega\omega\omega)_\tau$:= $TYPR0$

example pair and (evaluated) standard pair (no type variable)
:= $XLPR0\ ODPR0_{TYPR0\omega\omega} a_\omega b_\omega$
wff 27 : $ODPR0\, a\, b_{TYPR0}$:= $XLPR0$
§= $XLPR0$
$= XLPR0\, XLPR0$
§\ $ODPR0_{TYPR0\omega\omega} a_\omega$
$= (ODPR0\, a)\, [\lambda y.[\lambda g.(g\, a\, y)]]$
§s %1 6 %0
$= XLPR0\, ([\lambda y.[\lambda g.(g\, a\, y)]]\, b)$
§\ $[\lambda y_\omega.[\lambda g_{\omega\omega\omega}.(g_{\omega\omega\omega} a_\omega y_\omega)_\omega]_{TYPR0}] b_\omega$
$= ([\lambda y.[\lambda g.(g\, a\, y)]]\, b)\, [\lambda g.(g\, a\, b)]$
§s %1 3 %0
$= XLPR0\, [\lambda g.(g\, a\, b)]$
:= $SDPR0$ %0/3
wff 40 : $[\lambda g.(g\, a\, b)]_{TYPR0,\ldots}$:= $SDPR0$
%0
$= XLPR0\, SDPR0$
$=_{o\omega\omega} XLPR0_\omega SDPR0_\omega$

left element function (no type variable)
:= $LELE0\ [\lambda p_{TYPR0}.(p_{TYPR0}[\lambda x_\omega.[\lambda y_\omega.x_\omega]_{(\omega\omega)}])_\omega]$
wff 47 : $[\lambda p.(p\, [\lambda x.[\lambda y.x]])]_{\omega TYPR0}$:= $LELE0$

:= $L $LELE0_{\omega TYPR0} XLPR0_{TYPR0}$
wff 49 : $LELE0\, XLPR0_\omega$:= $L

§= $L
$= \$L\,\$L$
§\ $L
$= \$L\,(XLPR0\,[\lambda x.[\lambda y.x]])$
§s %1 3 %0
$= \$L\,(XLPR0\,[\lambda x.[\lambda y.x]])$
§\ $ODPR0_{TYPR0\,\omega\omega}a_{\omega}$
$= (ODPR0\,a)\,[\lambda y.[\lambda g.(g\,a\,y)]]$
§s %1 12 %0
$= \$L\,([\lambda y.[\lambda g.(g\,a\,y)]]\,b\,[\lambda x.[\lambda y.x]])$
§\ $[\lambda y_{\omega}.[\lambda g_{\omega\omega\omega}.(g_{\omega\omega\omega}a_{\omega}y_{\omega})_{\omega}]_{TYPR0}]b_{\omega}$
$= ([\lambda y.[\lambda g.(g\,a\,y)]]\,b)\,SDPR0$
§s %1 6 %0
$= \$L\,(SDPR0\,[\lambda x.[\lambda y.x]])$
§\ $SDPR0_{TYPR0}[\lambda x_{\omega}.[\lambda y_{\omega}.x_{\omega}]_{(\omega\omega)}]$
$= (SDPR0\,[\lambda x.[\lambda y.x]])\,([\lambda x.[\lambda y.x]]\,a\,b)$
§s %1 3 %0
$= \$L\,([\lambda x.[\lambda y.x]]\,a\,b)$
§\ $[\lambda x_{\omega}.[\lambda y_{\omega}.x_{\omega}]_{(\omega\omega)}]a_{\omega}$
$= ([\lambda x.[\lambda y.x]]\,a)\,[\lambda y.a]$
§s %1 6 %0
$= \$L\,([\lambda y.a]\,b)$
§\ $[\lambda y_{\omega}.a_{\omega}]b_{\omega}$
$= ([\lambda y.a]\,b)\,a$
§s %1 3 %0
$= \$L\,a$

right element function (no type variable)
:= $RELE0\ [\lambda p_{TYPR0}.(p_{TYPR0}[\lambda x_{\omega}.[\lambda y_{\omega}.y_{\omega}]_{(\omega\omega)}])_{\omega}]$
wff 74 : $[\lambda p.(p\,[\lambda x.[\lambda y.y]])]_{\omega\,TYPR0}$:= $RELE0$

:= $R $RELE0_{\omega\,TYPR0}XLPR0_{TYPR0}$
wff 75 : $RELE0\,XLPR0_{\omega}$:= $R
§= $R
$= \$R\,\$R$
§\ $R
$= \$R\,(XLPR0\,[\lambda x.[\lambda y.y]])$
§s %1 3 %0
$= \$R\,(XLPR0\,[\lambda x.[\lambda y.y]])$
§\ $ODPR0_{TYPR0\,\omega\omega}a_{\omega}$
$= (ODPR0\,a)\,[\lambda y.[\lambda g.(g\,a\,y)]]$
§s %1 12 %0
$= \$R\,([\lambda y.[\lambda g.(g\,a\,y)]]\,b\,[\lambda x.[\lambda y.y]])$
§\ $[\lambda y_{\omega}.[\lambda g_{\omega\omega\omega}.(g_{\omega\omega\omega}a_{\omega}y_{\omega})_{\omega}]_{TYPR0}]b_{\omega}$
$= ([\lambda y.[\lambda g.(g\,a\,y)]]\,b)\,SDPR0$
§s %1 6 %0
$= \$R\,(SDPR0\,[\lambda x.[\lambda y.y]])$
§\ $SDPR0_{TYPR0}[\lambda x_{\omega}.[\lambda y_{\omega}.y_{\omega}]_{(\omega\omega)}]$
$= (SDPR0\,[\lambda x.[\lambda y.y]])\,([\lambda x.[\lambda y.y]]\,a\,b)$

§s %1 3 %0
\# $\qquad = \$R\,([\lambda x.[\lambda y.y]]\,a\,b)$
§\ $[\lambda x_\omega.[\lambda y_\omega.y_\omega]_{(\omega\omega)}]a_\omega$
\# $\qquad = ([\lambda x.[\lambda y.y]]\,a)\,[\lambda y.y]$
§s %1 6 %0
\# $\qquad = \$R\,([\lambda y.y]\,b)$
§\ $[\lambda y_\omega.y_\omega]b_\omega$
\# $\qquad = ([\lambda y.y]\,b)\,b$
§s %1 3 %0
\# $\qquad = \$R\,b$

\## undefine local variables
:= $\$L$
:= $\$R$

2.1.106 Results for File pair1.r0.txt

\##
\## Ordered Pairs With One Type Variable
\##
\##
\## Source: [Andrews 2002 (ISBN 1-4020-0763-9), p. 208]
\##

\## Written by Ken Kubota (<mail@kenkubota.de>).
\##
\## This file is part of the publication of the mathematical logic $\mathcal{R}_0$.
\## For more information, visit: <http://doi.org/10.4444/100.10>
\##

\## definition of ordered pair (one type variable)
:= $ODPR1\ \ [\lambda t_\tau.[\lambda x_t.[\lambda y_t.[\lambda g_{ttt}.(g_{ttt}x_t y_t)_t]_{(t(ttt))}]_{(t(ttt)t)}]_{(t(ttt)tt)}]$
\# wff 24 : $\qquad [\lambda t.[\lambda x.[\lambda y.[\lambda g.(g\,x\,y)]]]]_{\backslash 4(\backslash 6\backslash 6\backslash 5)\backslash 3\backslash 2\tau} \qquad := \quad ODPR1$

\## type of ordered pair (one type variable)
:= $TYPR1\ \ [\lambda t_\tau.(t(ttt))_\tau]$
\# wff 36 : $\qquad [\lambda t.(t(ttt))]_{\tau\tau} \qquad := \quad TYPR1$

\## example pair and (evaluated) standard pair (one type variable)
:= $XLPR1\ \ ODPR1_{\backslash 4(\backslash 6\backslash 6\backslash 5)\backslash 3\backslash 2\tau}u_\tau a_u b_u$
\# wff 48 : $\qquad ODPR1\,u\,a\,b_{u(uuu)} \qquad := \quad XLPR1$
§= $XLPR1$
\# $\qquad = XLPR1\,XLPR1$
§\ $ODPR1_{\backslash 4(\backslash 6\backslash 6\backslash 5)\backslash 3\backslash 2\tau}u_\tau$
\# $\qquad = (ODPR1\,u)\,[\lambda x.[\lambda y.[\lambda g.(g\,x\,y)]]]$
§s %1 12 %0
\# $\qquad = XLPR1\,([\lambda x.[\lambda y.[\lambda g.(g\,x\,y)]]]\,a\,b)$
§\ $[\lambda x_u.[\lambda y_u.[\lambda g_{uuu}.(g_{uuu}x_u y_u)_u]_{(u(uuu))}]_{(u(uuu)u)}]a_u$

$= ([\lambda x.[\lambda y.[\lambda g.(g\,x\,y)]]]\,a)\,[\lambda y.[\lambda g.(g\,a\,y)]]$
§s %1 6 %0
$= XLPR1\,([\lambda y.[\lambda g.(g\,a\,y)]]\,b)$
§\ $[\lambda y_u.[\lambda g_{uuu}.(g_{uuu}a_u y_u)_u]_{(u(uuu))}]b_u$
$= ([\lambda y.[\lambda g.(g\,a\,y)]]\,b)\,[\lambda g.(g\,a\,b)]$
§s %1 3 %0
$= XLPR1\,[\lambda g.(g\,a\,b)]$
:= $SDPR1$ %0/3
wff 74 : $[\lambda g.(g\,a\,b)]_{u(uuu),\dots}$:= $SDPR1$
%0
$= XLPR1\,SDPR1$
$=_{o\omega\omega} XLPR1_\omega SDPR1_\omega$

left element function (one type variable)
:= $LELE1$ $[\lambda t_\tau.[\lambda p_{t(ttt)}.(p_{t(ttt)}[\lambda x_t.[\lambda y_t.x_t]_{(tt)}])_t]_{(t(t(ttt)))}]$
wff 83 : $[\lambda t.[\lambda p.(p\,[\lambda x.[\lambda y.x]])]]_{\backslash 2(\backslash 3(\backslash 5\backslash 5\backslash 4))\tau}$:= $LELE1$

:= $\$L$ $LELE1_{\backslash 2(\backslash 3(\backslash 5\backslash 5\backslash 4))\tau}u_\tau SDPR1_{u(uuu)}$
wff 91 : $LELE1\,u\,SDPR1_{u_\tau}$:= $\$L$
§= $\$L$
$= \$L\,\$L$
§\ $LELE1_{\backslash 2(\backslash 3(\backslash 5\backslash 5\backslash 4))\tau}u_\tau$
$= (LELE1\,u)\,[\lambda p.(p\,[\lambda x.[\lambda y.x]])]$
§s %1 6 %0
$= \$L\,([\lambda p.(p\,[\lambda x.[\lambda y.x]])]\,SDPR1)$
§\ $[\lambda p_{u(uuu)}.(p_{u(uuu)}[\lambda x_u.[\lambda y_u.x_u]_{(uu)}])_u]SDPR1_{u(uuu)}$
$= ([\lambda p.(p\,[\lambda x.[\lambda y.x]])]\,SDPR1)\,(SDPR1\,[\lambda x.[\lambda y.x]])$
§s %1 3 %0
$= \$L\,(SDPR1\,[\lambda x.[\lambda y.x]])$
§\ $SDPR1_{u(uuu)}[\lambda x_u.[\lambda y_u.x_u]_{(uu)}]$
$= (SDPR1\,[\lambda x.[\lambda y.x]])\,([\lambda x.[\lambda y.x]]\,a\,b)$
§s %1 3 %0
$= \$L\,([\lambda x.[\lambda y.x]]\,a\,b)$
§\ $[\lambda x_u.[\lambda y_u.x_u]_{(uu)}]a_u$
$= ([\lambda x.[\lambda y.x]]\,a)\,[\lambda y.a]$
§s %1 6 %0
$= \$L\,([\lambda y.a]\,b)$
§\ $[\lambda y_u.a_u]b_u$
$= ([\lambda y.a]\,b)\,a$
§s %1 3 %0
$= \$L\,a$

right element function (one type variable)
:= $RELE1$ $[\lambda t_\tau.[\lambda p_{t(ttt)}.(p_{t(ttt)}[\lambda x_t.[\lambda y_t.y_t]_{(tt)}])_t]_{(t(t(ttt)))}]$
wff 124 : $[\lambda t.[\lambda p.(p\,[\lambda x.[\lambda y.y]])]]_{\backslash 2(\backslash 3(\backslash 5\backslash 5\backslash 4))\tau}$:= $RELE1$

:= $\$R$ $RELE1_{\backslash 2(\backslash 3(\backslash 5\backslash 5\backslash 4))\tau}u_\tau SDPR1_{u(uuu)}$
wff 126 : $RELE1\,u\,SDPR1_{u_\tau}$:= $\$R$
§= $\$R$

\# $=\$R\,\$R$
§\ $RELE1_{\backslash 2(\backslash 3(\backslash 5\backslash 5\backslash 4))_\tau u_\tau}$
\# $=(RELE1\,u)\,[\lambda p.(p\,[\lambda x.[\lambda y.y]])]$
§s %1 6 %0
\# $=\$R\,([\lambda p.(p\,[\lambda x.[\lambda y.y]])]\,SDPR1)$
§\ $[\lambda p_{u(uuu)}.(p_{u(uuu)}\,[\lambda x_u.[\lambda y_u.y_u]_{(uu)}])_u]SDPR1_{u(uuu)}$
\# $=([\lambda p.(p\,[\lambda x.[\lambda y.y]])]\,SDPR1)\,(SDPR1\,[\lambda x.[\lambda y.y]])$
§s %1 3 %0
\# $=\$R\,(SDPR1\,[\lambda x.[\lambda y.y]])$
§\ $SDPR1_{u(uuu)}\,[\lambda x_u.[\lambda y_u.y_u]_{(uu)}]$
\# $=(SDPR1\,[\lambda x.[\lambda y.y]])\,([\lambda x.[\lambda y.y]]\,a\,b)$
§s %1 3 %0
\# $=\$R\,([\lambda x.[\lambda y.y]]\,a\,b)$
§\ $[\lambda x_u.[\lambda y_u.y_u]_{(uu)}]a_u$
\# $=([\lambda x.[\lambda y.y]]\,a)\,[\lambda y.y]$
§s %1 6 %0
\# $=\$R\,([\lambda y.y]\,b)$
§\ $[\lambda y_u.y_u]b_u$
\# $=([\lambda y.y]\,b)\,b$
§s %1 3 %0
\# $=\$R\,b$

\## undefine local variables
:= $\$L$
:= $\$R$

2.1.107 Results for File pair3.r0.txt

\##
\## Ordered Pairs With Three Type Variables
\##
\##
\## Source: [Andrews 2002 (ISBN 1-4020-0763-9), p. 208]
\##
\##
\## Written by Ken Kubota (<mail@kenkubota.de>).
\##
\## This file is part of the publication of the mathematical logic $\mathcal{R}_0$.
\## For more information, visit: <http://doi.org/10.4444/100.10>
\##

\##
\## Comment
\##
\## One might consider placing the two type variables of the pair elements first:
\## PROD := [\t.[\u.[\x:t.[\y:u.[\v.[\g:vut.(gxy)]]]]]],
\## hence
\## PROD a b

would represent the Cartesian product a x b.
##
Source: [cf. https://sourceforge.net/p/hol/mailman/message/35648326/ (Feb. 5, 2017)]
[cf. https://sympa.inria.fr/sympa/arc/coq-club/2017-02/msg00024.html (Feb. 5, 2017)]
##

definition of ordered pair (three type variables)
:= $ODPR3$ ...
... $[\lambda t_\tau.[\lambda x_t.[\lambda u_\tau.[\lambda y_u.[\lambda v_\tau.[\lambda g_{vut}.(g_{vut}x_t y_u)_v]_{(v(vut))}]_{(\backslash 2(\backslash 4ut)\tau)}]_{(\backslash 2(\backslash 4ut)\tau u)}]_{(\backslash 2(\backslash 4\backslash 6t)\tau\backslash 2\tau)}]_{(\backslash 2(\backslash 4\backslash 6t)\tau\backslash 2\tau t)}]$
wff 41 : $[\lambda t.[\lambda x.[\lambda u.[\lambda y.[\lambda v.[\lambda g.(g\,x\,y)]]]]]]_{\backslash 2(\backslash 4\backslash 6\backslash 7)\tau\backslash 2\tau\backslash 2\tau}$:= $ODPR3$

type of ordered pair (three type variables)
:= $TYPR3$ $[\lambda t_\tau.[\lambda u_\tau.[\lambda v_\tau.(v(vut))_\tau]_{(\tau\tau)}]_{(\tau\tau\tau)}]$
wff 54 : $[\lambda t.[\lambda u.[\lambda v.(v(vut))]]]_{\tau\tau\tau\tau}$:= $TYPR3$

example pair and (evaluated) standard pair (three type variables)
:= $XLPR3$ $ODPR3_{\backslash 2(\backslash 4\backslash 6\backslash 7)\tau\backslash 2\tau\backslash 2\tau}t_\tau a_t u_\tau b_u$
wff 61 : $ODPR3\,t\,a\,u\,b_{\backslash 2(\backslash 4ut)\tau}$:= $XLPR3$
§= $XLPR3$
$= XLPR3\,XLPR3$
§\ $ODPR3_{\backslash 2(\backslash 4\backslash 6\backslash 7)\tau\backslash 2\tau\backslash 2\tau}t_\tau$
$= (ODPR3\,t)\,[\lambda x.[\lambda u.[\lambda y.[\lambda v.[\lambda g.(g\,x\,y)]]]]]$
§s %1 24 %0
$= XLPR3\,([\lambda x.[\lambda u.[\lambda y.[\lambda v.[\lambda g.(g\,x\,y)]]]]]\,a\,u\,b)$
§\ $[\lambda x_t.[\lambda u_\tau.[\lambda y_u.[\lambda v_\tau.[\lambda g_{vut}.(g_{vut}x_t y_u)_v]_{(v(vut))}]_{(\backslash 2(\backslash 4ut)\tau)}]_{(\backslash 2(\backslash 4ut)\tau u)}]_{(\backslash 2(\backslash 4\backslash 6t)\tau\backslash 2\tau)}]a_t$
$= ([\lambda x.[\lambda u.[\lambda y.[\lambda v.[\lambda g.(g\,x\,y)]]]]]\,a)\,[\lambda u.[\lambda y.[\lambda v.[\lambda g.(g\,a\,y)]]]]$
§s %1 12 %0
$= XLPR3\,([\lambda u.[\lambda y.[\lambda v.[\lambda g.(g\,a\,y)]]]]\,u\,b)$
§\ $[\lambda u_\tau.[\lambda y_u.[\lambda v_\tau.[\lambda g_{vut}.(g_{vut}a_t y_u)_v]_{(v(vut))}]_{(\backslash 2(\backslash 4ut)\tau)}]_{(\backslash 2(\backslash 4ut)\tau u)}]u_\tau$
$= ([\lambda u.[\lambda y.[\lambda v.[\lambda g.(g\,a\,y)]]]]\,u)\,[\lambda y.[\lambda v.[\lambda g.(g\,a\,y)]]]$
§s %1 6 %0
$= XLPR3\,([\lambda y.[\lambda v.[\lambda g.(g\,a\,y)]]]\,b)$
§\ $[\lambda y_u.[\lambda v_\tau.[\lambda g_{vut}.(g_{vut}a_t y_u)_v]_{(v(vut))}]_{(\backslash 2(\backslash 4ut)\tau)}]b_u$
$= ([\lambda y.[\lambda v.[\lambda g.(g\,a\,y)]]]\,b)\,[\lambda v.[\lambda g.(g\,a\,b)]]$
§s %1 3 %0
$= XLPR3\,[\lambda v.[\lambda g.(g\,a\,b)]]$
:= $SDPR3$ %0/3
wff 88 : $[\lambda v.[\lambda g.(g\,a\,b)]]_{\backslash 2(\backslash 4ut)\tau,\ldots}$:= $SDPR3$
%0
$= XLPR3\,SDPR3$
$=_{o\omega\omega} XLPR3_\omega SDPR3_\omega$

left element function (three type variables)
:= $LELE3$ $[\lambda t_\tau.[\lambda u_\tau.[\lambda p_{\backslash 2(\backslash 4ut)\tau}.(p_{\backslash 2(\backslash 4ut)\tau}t_\tau[\lambda x_t.[\lambda y_u.x_t]_{(tu)}])_t]_{(t(\backslash 2(\backslash 4ut)\tau))}]_{(t(\backslash 2(\backslash 4\backslash 6t)\tau)\tau)}]$
wff 104 : $[\lambda t.[\lambda u.[\lambda p.(p\,t\,[\lambda x.[\lambda y.x]])]]]_{\backslash 3(\backslash 2(\backslash 4\backslash 6\backslash 6)\tau)\tau\tau}$:= $LELE3$

:= \$L $LELE3_{\backslash 3(\backslash 2(\backslash 4\backslash 6\backslash 6)\tau)\tau\tau}t_\tau u_\tau SDPR3_{\backslash 2(\backslash 4ut)\tau}$
wff 114 : $LELE3\,t\,u\,SDPR3_{t_\tau}$:= \$L

§= $\$L$
\# $= \$L\,\$L$
§\ $LELE3_{\backslash 3(\backslash 2(\backslash 4\backslash 6\backslash 6)\tau)\tau\tau}t_\tau$
\# $= (LELE3\,t)\,[\lambda u.[\lambda p.(p\,t\,[\lambda x.[\lambda y.x]])]]$
§s %1 12 %0
\# $= \$L\,([\lambda u.[\lambda p.(p\,t\,[\lambda x.[\lambda y.x]])]]\,u\,SDPR3)$
§\ $[\lambda u_\tau.[\lambda p_{\backslash 2(\backslash 4ut)\tau}.(p_{\backslash 2(\backslash 4ut)\tau}t_\tau[\lambda x_t.[\lambda y_u.x_t]_{(tu)}])_t]_{(t(\backslash 2(\backslash 4ut)\tau))}]u_\tau$
\# $= ([\lambda u.[\lambda p.(p\,t\,[\lambda x.[\lambda y.x]])]]\,u)\,[\lambda p.(p\,t\,[\lambda x.[\lambda y.x]])]$
§s %1 6 %0
\# $= \$L\,([\lambda p.(p\,t\,[\lambda x.[\lambda y.x]])]\,SDPR3)$
§\ $[\lambda p_{\backslash 2(\backslash 4ut)\tau}.(p_{\backslash 2(\backslash 4ut)\tau}t_\tau[\lambda x_t.[\lambda y_u.x_t]_{(tu)}])_t]SDPR3_{\backslash 2(\backslash 4ut)\tau}$
\# $= ([\lambda p.(p\,t\,[\lambda x.[\lambda y.x]])]\,SDPR3)\,(SDPR3\,t\,[\lambda x.[\lambda y.x]])$
§s %1 3 %0
\# $= \$L\,(SDPR3\,t\,[\lambda x.[\lambda y.x]])$
§\ $SDPR3_{\backslash 2(\backslash 4ut)\tau}t_\tau$
\# $= (SDPR3\,t)\,[\lambda g.(g\,a\,b)]$
§s %1 6 %0
\# $= \$L\,([\lambda g.(g\,a\,b)]\,[\lambda x.[\lambda y.x]])$
§\ $[\lambda g_{tut}.(g_{tut}a_tb_u)_t][\lambda x_t.[\lambda y_u.x_t]_{(tu)}]$
\# $= ([\lambda g.(g\,a\,b)]\,[\lambda x.[\lambda y.x]])\,([\lambda x.[\lambda y.x]]\,a\,b)$
§s %1 3 %0
\# $= \$L\,([\lambda x.[\lambda y.x]]\,a\,b)$
§\ $[\lambda x_t.[\lambda y_u.x_t]_{(tu)}]a_t$
\# $= ([\lambda x.[\lambda y.x]]\,a)\,[\lambda y.a]$
§s %1 6 %0
\# $= \$L\,([\lambda y.a]\,b)$
§\ $[\lambda y_u.a_t]b_u$
\# $= ([\lambda y.a]\,b)\,a$
§s %1 3 %0
\# $= \$L\,a$

\#\# right element function (three type variables)
:= $RELE3\ \ [\lambda t_\tau.[\lambda u_\tau.[\lambda p_{\backslash 2(\backslash 4ut)\tau}.(p_{\backslash 2(\backslash 4ut)\tau}u_\tau[\lambda x_t.[\lambda y_u.y_u]_{(uu)}])_u]_{(u(\backslash 2(\backslash 4ut)\tau))}]_{(\backslash 2(\backslash 2(\backslash 4\backslash 6t)\tau)\tau)}]$
\# wff 164 : $[\lambda t.[\lambda u.[\lambda p.(p\,u\,[\lambda x.[\lambda y.y]])]]]_{\backslash 2(\backslash 2(\backslash 4\backslash 6\backslash 6)\tau)\tau\tau}$:= $RELE3$

:= $\$R\ \ RELE3_{\backslash 2(\backslash 2(\backslash 4\backslash 6\backslash 6)\tau)\tau\tau}t_\tau u_\tau SDPR3_{\backslash 2(\backslash 4ut)\tau}$
\# wff 170 : $RELE3\,t\,u\,SDPR3_{u_\tau}$:= $\$R$
§= $\$R$
\# $= \$R\,\$R$
§\ $RELE3_{\backslash 2(\backslash 2(\backslash 4\backslash 6\backslash 6)\tau)\tau\tau}t_\tau$
\# $= (RELE3\,t)\,[\lambda u.[\lambda p.(p\,u\,[\lambda x.[\lambda y.y]])]]$
§s %1 12 %0
\# $= \$R\,([\lambda u.[\lambda p.(p\,u\,[\lambda x.[\lambda y.y]])]]\,u\,SDPR3)$
§\ $[\lambda u_\tau.[\lambda p_{\backslash 2(\backslash 4ut)\tau}.(p_{\backslash 2(\backslash 4ut)\tau}u_\tau[\lambda x_t.[\lambda y_u.y_u]_{(uu)}])_u]_{(u(\backslash 2(\backslash 4ut)\tau))}]u_\tau$
\# $= ([\lambda u.[\lambda p.(p\,u\,[\lambda x.[\lambda y.y]])]]\,u)\,[\lambda p.(p\,u\,[\lambda x.[\lambda y.y]])]$
§s %1 6 %0
\# $= \$R\,([\lambda p.(p\,u\,[\lambda x.[\lambda y.y]])]\,SDPR3)$
§\ $[\lambda p_{\backslash 2(\backslash 4ut)\tau}.(p_{\backslash 2(\backslash 4ut)\tau}u_\tau[\lambda x_t.[\lambda y_u.y_u]_{(uu)}])_u]SDPR3_{\backslash 2(\backslash 4ut)\tau}$
\# $= ([\lambda p.(p\,u\,[\lambda x.[\lambda y.y]])]\,SDPR3)\,(SDPR3\,u\,[\lambda x.[\lambda y.y]])$

§s %1 3 %0
$= \$R\,(SDPR3\,u\,[\lambda x.[\lambda y.y]])$
§\ $SDPR3_{\backslash 2(\backslash 4ut)\tau}u_\tau$
$= (SDPR3\,u)\,[\lambda g.(g\,a\,b)]$
§s %1 6 %0
$= \$R\,([\lambda g.(g\,a\,b)]\,[\lambda x.[\lambda y.y]])$
§\ $[\lambda g_{uut}.(g_{uut}a_t b_u)_u][\lambda x_t.[\lambda y_u.y_u]_{(uu)}]$
$= ([\lambda g.(g\,a\,b)]\,[\lambda x.[\lambda y.y]])\,([\lambda x.[\lambda y.y]]\,a\,b)$
§s %1 3 %0
$= \$R\,([\lambda x.[\lambda y.y]]\,a\,b)$
§\ $[\lambda x_t.[\lambda y_u.y_u]_{(uu)}]a_t$
$= ([\lambda x.[\lambda y.y]]\,a)\,[\lambda y.y]$
§s %1 6 %0
$= \$R\,([\lambda y.y]\,b)$
§\ $[\lambda y_u.y_u]b_u$
$= ([\lambda y.y]\,b)\,b$
§s %1 3 %0
$= \$R\,b$

undefine local variables
:= $L
:= $R

2.1.108 Results for File paradox_cantor.r0e.txt

##
Cantor's paradox
##
##
Source: [Kubota 2017 (doi: 10.4444/100.10)]
##

Written by Ken Kubota (<mail@kenkubota.de>).
##
This file is part of the publication of the mathematical logic $\mathcal{R}_0$.
For more information, visit: <http://doi.org/10.4444/100.10>
##

<< basics.r0.txt
<< A5200t.r0.txt

##
Demonstration of Positive Self-Reference: The universal set contains itself (not a paradox)
##

§= $V_{o\omega}V_\omega$
$= (V\,V)\,(V\,V)$
§\ /3

```
#              =(V V) T
§s %1 5 %0
#              =T (V V)
%T
#              ===       :=  A5200t  T
#              =_{oωω}=_ω=_ω   :=  A5200t  T
§s %0 1 %1
#              V V
%0
#              V V
#              V_{oω} V_ω
```

demonstrate that V now has type V

```
§= _V V
#              =V V
%0
#              =V V
#              =_{oVV} V_V V_V
```

##
Cantor's paradox: The power set of the universal set should be a subset of the universal set
##

obtain power set of universal set (resulting set has type 'o(ow)' – is a set of sets')

:= $PC $\mathcal{P}_{o(o\backslash 4)(o\backslash 3)\tau}\omega_\tau V_{o\omega}$
wff 221 : $\mathcal{P}\,\omega\,V_{o(o\omega)}$:= $PC

power set of the universal set is a subset of ... (resulting function has type 'o(o(ow))')

:= $SPC $\subseteq_{o(o\backslash 4)(o\backslash 3)\tau}(o\omega)_\tau \$PC_{o(o\omega)}$
wff 225 : $\subseteq (o\omega)\,\$PC_{o(o(o\omega))}$:= $SPC

... the universal set (which has type 'ow')

trying to apply the wff (will result in failure)

interactive command for lambda application (with automatic type matching):

___ $SPC V
error 1 [-]: no possible type match for '$SPC' _ 'V'

undefine local variables

:= $PC
:= $SPC

##
Q.E.D.
##

It is not possible to express Cantor's paradox in the formulation $\mathcal{R}_0$ of higher-order logic.
1 error generated

2.1.109 Results for File paradox_russell.r0e.txt

##
Russell's paradox
##
##
Source: [Kubota 2017 (doi: 10.4444/100.10)]
##
Copyright (c) 2017 Owl of Minerva Press GmbH. All rights reserved.
Written by Ken Kubota (<mail@kenkubota.de>).
##
This file is part of the publication of the mathematical logic $\mathcal{R}_0$.
For more information, visit: <http://doi.org/10.4444/100.10>
##

<< basics.r0.txt

##
The set of all sets that are not members of themselves
##

:= $RUSSELL$ $[\lambda x_{o\omega}.(\sim_{oo}(x_{o\omega} x_\omega))_o]$
wff 211 : $[\lambda x.(\sim (x\, x))]_{o(o\omega)}$:= $RUSSELL$

trying to apply the wff onto itself (will result in failure)

interactive command for lambda application (with automatic type matching):
__ $RUSSELL$ $RUSSELL$
error 1 [-]: no possible type match for 'RUSSELL' _ 'RUSSELL'

##
Q.E.D.
##

It is not possible to express Russell's paradox in the formulation $\mathcal{R}_0$ of higher-order logic.
1 error generated

2.1.110 Results for File polymorphism.r0.txt

##
Polymorphism
##
##
Source: [Kubota 2017 (doi: 10.4444/100.10)]
##
Copyright (c) 2017 Owl of Minerva Press GmbH. All rights reserved.
Written by Ken Kubota (<mail@kenkubota.de>).
##
This file is part of the publication of the mathematical logic $\mathcal{R}_0$.
For more information, visit: <http://doi.org/10.4444/100.10>
##

<< basics.r0.txt

testing the polymorphic identity relation (=) and
the polymorphic description operator (i) ...

... with type variable t
§= $t_\tau \;\; \iota_{t(ot)} p_{ot}$
\# $= (\iota\, p)\, (\iota\, p)$
%0
\# $= (\iota\, p)\, (\iota\, p)$
\# $=_{ott} (\iota_{t(ot)} p_{ot}) (\iota_{t(ot)} p_{ot})$

... with type variable a
§= $a_\tau \;\; \iota_{a(oa)} p_{oa}$
\# $= (\iota\, p)\, (\iota\, p)$
%0
\# $= (\iota\, p)\, (\iota\, p)$
\# $=_{oaa} (\iota_{a(oa)} p_{oa}) (\iota_{a(oa)} p_{oa})$

... with type Boole
§= $o \;\; \iota_{o(oo)} p_{oo}$
\# $= (\iota\, p)\, (\iota\, p)$
%0
\# $= (\iota\, p)\, (\iota\, p)$
\# $=_{ooo} (\iota_{o(oo)} p_{oo}) (\iota_{o(oo)} p_{oo})$

2.1.111 Results for File scope_violation_in_lambda_conversion.r0e.txt

##
Scope Violation in Lambda Conversion
##
##
Source: [Kubota 2017 (doi: 10.4444/100.10)]

##
Copyright (c) 2017 Owl of Minerva Press GmbH. All rights reserved.
Written by Ken Kubota (<mail@kenkubota.de>).
##
This file is part of the publication of the mathematical logic $\mathcal{R}_0$.
For more information, visit: <http://doi.org/10.4444/100.10>
##

##
Condition "A is free for x in B"
##
[Andrews 2002 (ISBN 1-4020-0763-9), pp. 218 f. (5207) and p. 213 (definition of term)]
##

§\ $[\lambda x_\omega.[\lambda y_\omega.(=_{o\omega\omega} x_\omega y_\omega)_o]_{(o\omega)}]y_\omega$
error 1 [-]: scope violation in lambda conversion – 'y_ω' is not free for 'x_ω' in '$[\lambda y_\omega.(=_{o\omega\omega} x_\omega y_\omega)_o]$' (wffs 12, 11, 15)
1 error generated

2.1.112 Results for File scope_violation_in_lambda_conversion_type.r0e.txt

##
Scope Violation in Lambda Conversion at Type Level
##
##
Source: [Kubota 2017 (doi: 10.4444/100.10)]
##
Copyright (c) 2017 Owl of Minerva Press GmbH. All rights reserved.
Written by Ken Kubota (<mail@kenkubota.de>).
##
This file is part of the publication of the mathematical logic $\mathcal{R}_0$.
For more information, visit: <http://doi.org/10.4444/100.10>
##

##
Condition "A is free for x in B"
##
[Andrews 2002 (ISBN 1-4020-0763-9), pp. 218 f. (5207) and p. 213 (definition of term)]
##

§\ $[\lambda t_\tau.[\lambda u_\tau.x_t]_{(t\tau)}]u_\tau$
error 1 [-]: scope violation in lambda conversion – 'u_τ' is not free for 't_τ' in '$[\lambda u_\tau.x_t]$' (wffs 11, 4, 13)
1 error generated

2.1.113 Results for File scope_violation_in_substitution.r0e.txt

##
Scope Violation in Substitution
##
##
Source: [Kubota 2017 (doi: 10.4444/100.10)]
##

Written by Ken Kubota (<mail@kenkubota.de>).
##
This file is part of the publication of the mathematical logic $\mathcal{R}_0$.
For more information, visit: <http://doi.org/10.4444/100.10>
##

##
Condition "the occurrence of A in C is not in a wf part [\x.E] of C,
where x is free in a member of H and free in [A = B]"
[Andrews 2002 (ISBN 1-4020-0763-9), p. 214 (Rule R')]
##

<< basics.r0.txt

undefine V to see the formula in detail
:= V

H ⊃ A = B
§! $\supset_{ooo}(p_{o\omega}x_{\omega})(=_{o\omega\omega}T_{\omega}(p_{o\omega}x_{\omega}))$
$\supset (p\,x)\,(= T\,(p\,x))$

H ⊃ C
§! $\supset_{ooo}(p_{o\omega}x_{\omega})(=_{o\omega\omega}[\lambda x_{\omega}.T_o][\lambda x_{\omega}.T_o])$
$\supset (p\,x)\,(= [\lambda x.T]\,[\lambda x.T])$

now try to replace A (first T) in C
§s' %0 7 %1
error 1 [-]: scope violation in substitution – bound variable 'x_{ω}' is free in hypothesis '$p_{o\omega}x_{\omega}$' and free in equation '$=_{o\omega\omega}T_{\omega}(p_{o\omega}x_{\omega})$' (wffs 73, 209, 212)
1 error generated

2.1.114 Results for File scope_violation_in_variable_renaming_conv.r0e.txt

##
Scope Violation in Variable Renaming (Lambda Conversion)
##
##
Source: [Kubota 2017 (doi: 10.4444/100.10)]
##

Copyright (c) 2017 Owl of Minerva Press GmbH. All rights reserved.
Written by Ken Kubota (<mail@kenkubota.de>).
##
This file is part of the publication of the mathematical logic $\mathcal{R}_0$.
For more information, visit: <http://doi.org/10.4444/100.10>
##

##
Condition “z is free for x in A”
##
[Andrews 2002 (ISBN 1-4020-0763-9), pp. 217 f. (5206) and p. 213 (definition of term)]
##

§r $[\lambda x_t.[\lambda z_t.(=_{ott} x_t z_t)_o]_{(ot)}]\ z_t$
error 1 [-]: scope violation in lambda conversion – ‘z_t’ is not free for ‘x_t’ in ‘$[\lambda z_t.(=_{ott} x_t z_t)_o]$’ (wffs 12, 11, 15)
1 error generated

2.1.115 Results for File scope_violation_in_variable_renaming_var.r0e.txt

##
Scope Violation in Variable Renaming (Free Variable)
##
##
Source: [Kubota 2017 (doi: 10.4444/100.10)]
##
Copyright (c) 2017 Owl of Minerva Press GmbH. All rights reserved.
Written by Ken Kubota (<mail@kenkubota.de>).
##
This file is part of the publication of the mathematical logic $\mathcal{R}_0$.
For more information, visit: <http://doi.org/10.4444/100.10>
##

##
Condition “z does not occur free in A”
##
[Andrews 2002 (ISBN 1-4020-0763-9), pp. 217 f. (5206)]
##

§r $[\lambda x_t.(=_{ott} x_t z_t)_o]\ z_t$
error 1 [-]: scope violation in variable renaming – variable ‘z_t’ occurs free in ‘$[\lambda x_t.(=_{ott} x_t z_t)_o]$’ (wffs 13, 15)
1 error generated

2.1.116 Results for File vector.r0.txt

##
Vectors (Dependent Type Theory)
##
##
Source: [Kubota 2017 (doi: 10.4444/100.10)]
##
Copyright (c) 2017 Owl of Minerva Press GmbH. All rights reserved.
Written by Ken Kubota (<mail@kenkubota.de>).
##
This file is part of the publication of the mathematical logic $\mathcal{R}_0$.
For more information, visit: <http://doi.org/10.4444/100.10>
##

<< basics.r0.txt
<< pair3.r0.txt

##
Example: Define Three-Dimensional Vector as Nested Ordered Pair <a,<b,<c,O> > >
##

level 1
:= $TLVL1\ \backslash 2(\backslash 4\omega s)\tau$
\# wff 404 : $\backslash 2(\backslash 4\omega s)\tau_{\tau}$:= $TLVL1$
:= $PLVL1\ ODPR3_{\backslash 2(\backslash 4\backslash 6\backslash 7)\tau\backslash 2\tau\backslash 2\tau}s_{\tau}c_{s}\omega_{\tau}\emptyset_{\omega}$
\# wff 416 : $ODPR3\, s\, c\, \omega\, \emptyset_{TLVL1}$:= $PLVL1$

level 2
:= $TLVL2\ \backslash 2(\backslash 4\, TLVL1 s)\tau$
\# wff 420 : $\backslash 2(\backslash 4\, TLVL1 s)\tau_{\tau}$:= $TLVL2$
:= $PLVL2\ ODPR3_{\backslash 2(\backslash 4\backslash 6\backslash 7)\tau\backslash 2\tau\backslash 2\tau}s_{\tau}b_{s}TLVL1_{\tau}PLVL1_{TLVL1}$
\# wff 425 : $ODPR3\, s\, b\, TLVL1\, PLVL1_{TLVL2}$:= $PLVL2$

level 3
:= $TLVL3\ \backslash 2(\backslash 4\, TLVL2 s)\tau$
\# wff 429 : $\backslash 2(\backslash 4\, TLVL2 s)\tau_{\tau}$:= $TLVL3$
:= $PLVL3\ ODPR3_{\backslash 2(\backslash 4\backslash 6\backslash 7)\tau\backslash 2\tau\backslash 2\tau}s_{\tau}a_{s}TLVL2_{\tau}PLVL2_{TLVL2}$
\# wff 434 : $ODPR3\, s\, a\, TLVL2\, PLVL2_{TLVL3}$:= $PLVL3$

##
Type Depending on Level/Dimension (Dependent Type Theory)
##

type successor function
:= $TZERO\ \omega$
\# wff 1 : ω_{τ} := $TZERO$

:= $TSUCC$ $[\lambda t_{\tau}.[\lambda x_{\tau}.(\backslash 2(\backslash 4xt)\tau)_{\tau}]_{(\tau\tau)}]$
wff 441 : $[\lambda t.[\lambda x.(\backslash 2(\backslash 4xt)\tau)]]_{\tau\tau\tau}$:= $TSUCC$

evaluate type successor function for type s
§\ $TSUCC_{\tau\tau\tau}s_{\tau}$
$=(TSUCC\,s)\,[\lambda x.(\backslash 2(\backslash 4xs)\tau)]$
:= $TSUCCTYPES$ %0/3
wff 447 : $[\lambda x.(\backslash 2(\backslash 4xs)\tau)]_{\tau\tau,\ldots}$:= $TSUCCTYPES$

evaluate types for all three levels
:= $TSUCCNO000$ ω
wff 1 : ω_{τ} := $TSUCCNO000$ $TZERO$
:= $TSUCCNO001$ $TSUCCTYPES_{\tau\tau}\omega_{\tau}$
wff 449 : $TSUCCTYPES\,\omega_{\tau}$:= $TSUCCNO001$
:= $TSUCCNO002$ $TSUCCTYPES_{\tau\tau}TSUCCNO001_{\tau}$
wff 450 : $TSUCCTYPES\,TSUCCNO001_{\tau}$:= $TSUCCNO002$
:= $TSUCCNO003$ $TSUCCTYPES_{\tau\tau}TSUCCNO002_{\tau}$
wff 451 : $TSUCCTYPES\,TSUCCNO002_{\tau}$:= $TSUCCNO003$

level 1
§= $TSUCCNO001$
$=TSUCCNO001\,TSUCCNO001$
§\ $TSUCCNO001$
$=TSUCCNO001\,TLVL1$
§s %1 3 %0
$=TSUCCNO001\,TLVL1$
:= $TSUCCNO001EXPND$ %0
wff 454 : $=TSUCCNO001\,TLVL1_{o}$:= $TSUCCNO001EXPND$

level 2
§= $TSUCCNO002$
$=TSUCCNO002\,TSUCCNO002$
§\ $TSUCCNO002$
$=TSUCCNO002\,(\backslash 2(\backslash 4\,TSUCCNO001s)\tau)$
§s %1 3 %0
$=TSUCCNO002\,(\backslash 2(\backslash 4\,TSUCCNO001s)\tau)$
%$TSUCCNO001EXPND$
$=TSUCCNO001\,TLVL1$:= $TSUCCNO001EXPND$
$=_{o\omega\omega}TSUCCNO001_{\omega}TLVL1_{\omega}$:= $TSUCCNO001EXPND$
§s %1 53 %0
$=TSUCCNO002\,TLVL2$
:= $TSUCCNO002EXPND$ %0
wff 462 : $=TSUCCNO002\,TLVL2_{o}$:= $TSUCCNO002EXPND$

level 3
§= $TSUCCNO003$
$=TSUCCNO003\,TSUCCNO003$
§\ $TSUCCNO003$
$=TSUCCNO003\,(\backslash 2(\backslash 4\,TSUCCNO002s)\tau)$

§s %1 3 %0
$\quad = TSUCCNO003\,(\backslash 2(\backslash 4\, TSUCCNO002 s)\tau)$
%$TSUCCNO002EXPND$
$\quad = TSUCCNO002\, TLVL2 \quad := \quad TSUCCNO002EXPND$
$\quad =_{o\omega\omega} TSUCCNO002_\omega TLVL2_\omega \quad := \quad TSUCCNO002EXPND$
§s %1 53 %0
$\quad = TSUCCNO003\, TLVL3$

##
Obtain Vector Elements
##

first element (left element at top level)
§= $LELE3_{\backslash 3(\backslash 2(\backslash 4\backslash 6\backslash 6)\tau)\tau\tau} s_\tau TLVL2_\tau PLVL3_{TLVL3}$
$\quad = (LELE3\, s\, TLVL2\, PLVL3)\, (LELE3\, s\, TLVL2\, PLVL3)$
§\ $LELE3_{\backslash 3(\backslash 2(\backslash 4\backslash 6\backslash 6)\tau)\tau\tau} s_\tau$
$\quad = (LELE3\, s)\, [\lambda u.[\lambda p.(p\, s\, [\lambda x.[\lambda y.x]])]]$
§s %1 12 %0
$\quad = (LELE3\, s\, TLVL2\, PLVL3)\, ([\lambda u.[\lambda p.(p\, s\, [\lambda x.[\lambda y.x]])]]\, TLVL2\, PLVL3)$
§\ $[\lambda u_\tau.[\lambda p_{\backslash 2(\backslash 4us)\tau}.(p_{\backslash 2(\backslash 4us)\tau} s_\tau [\lambda x_s.[\lambda y_u.x_s]_{(su)}])_s]_{(s(\backslash 2(\backslash 4us)\tau))}] TLVL2_\tau$
$\quad = ([\lambda u.[\lambda p.(p\, s\, [\lambda x.[\lambda y.x]])]]\, TLVL2)\, [\lambda p.(p\, s\, [\lambda x.[\lambda y.x]])]$
§s %1 6 %0
$\quad = (LELE3\, s\, TLVL2\, PLVL3)\, ([\lambda p.(p\, s\, [\lambda x.[\lambda y.x]])]\, PLVL3)$
§\ $[\lambda p_{TLVL3}.(p_{TLVL3} s_\tau [\lambda x_s.[\lambda y_{TLVL2}.x_s]_{(s\,TLVL2)}])_s] PLVL3_{TLVL3}$
$\quad = ([\lambda p.(p\, s\, [\lambda x.[\lambda y.x]])]\, PLVL3)\, (PLVL3\, s\, [\lambda x.[\lambda y.x]])$
§s %1 3 %0
$\quad = (LELE3\, s\, TLVL2\, PLVL3)\, (PLVL3\, s\, [\lambda x.[\lambda y.x]])$
§\ $ODPR3_{\backslash 2(\backslash 4\backslash 6\backslash 7)\tau\backslash 2\tau\backslash 2\tau} s_\tau$
$\quad = (ODPR3\, s)\, [\lambda x.[\lambda u.[\lambda y.[\lambda v.[\lambda g.(g\, x\, y)]]]]]$
§s %1 96 %0
$\quad = (LELE3\, s\, TLVL2\, PLVL3)\, \ldots$
$\ldots ([\lambda x.[\lambda u.[\lambda y.[\lambda v.[\lambda g.(g\, x\, y)]]]]]\, a\, TLVL2\, PLVL2\, s\, [\lambda x.[\lambda y.x]])$
§\ $[\lambda x_s.[\lambda u_\tau.[\lambda y_u.[\lambda v_\tau.[\lambda g_{vus}.(g_{vus} x_s y_u)_v]_{(v(vus))}]_{(\backslash 2(\backslash 4us)\tau)}]_{(\backslash 2(\backslash 4us)\tau u)}]_{(\backslash 2(\backslash 4\backslash 6s)\tau\backslash 2\tau)}] a_s$
$\quad = ([\lambda x.[\lambda u.[\lambda y.[\lambda v.[\lambda g.(g\, x\, y)]]]]]\, a)\, [\lambda u.[\lambda y.[\lambda v.[\lambda g.(g\, a\, y)]]]]$
§s %1 48 %0
$\quad = (LELE3\, s\, TLVL2\, PLVL3)\, \ldots$
$\ldots ([\lambda u.[\lambda y.[\lambda v.[\lambda g.(g\, a\, y)]]]]\, TLVL2\, PLVL2\, s\, [\lambda x.[\lambda y.x]])$
§\ $[\lambda u_\tau.[\lambda y_u.[\lambda v_\tau.[\lambda g_{vus}.(g_{vus} a_s y_u)_v]_{(v(vus))}]_{(\backslash 2(\backslash 4us)\tau)}]_{(\backslash 2(\backslash 4us)\tau u)}] TLVL2_\tau$
$\quad = ([\lambda u.[\lambda y.[\lambda v.[\lambda g.(g\, a\, y)]]]]\, TLVL2)\, [\lambda y.[\lambda v.[\lambda g.(g\, a\, y)]]]$
§s %1 24 %0
$\quad = (LELE3\, s\, TLVL2\, PLVL3)\, ([\lambda y.[\lambda v.[\lambda g.(g\, a\, y)]]]\, PLVL2\, s\, [\lambda x.[\lambda y.x]])$
§\ $[\lambda y_{TLVL2}.[\lambda v_\tau.[\lambda g_{v\,TLVL2s}.(g_{v\,TLVL2s} a_s y_{TLVL2})_v]_{(v(v\,TLVL2s))}]_{TLVL3}] PLVL2_{TLVL2}$
$\quad = ([\lambda y.[\lambda v.[\lambda g.(g\, a\, y)]]]\, PLVL2)\, [\lambda v.[\lambda g.(g\, a\, PLVL2)]]$
§s %1 12 %0
$\quad = (LELE3\, s\, TLVL2\, PLVL3)\, ([\lambda v.[\lambda g.(g\, a\, PLVL2)]]\, s\, [\lambda x.[\lambda y.x]])$
§\ $[\lambda v_\tau.[\lambda g_{v\,TLVL2s}.(g_{v\,TLVL2s} a_s PLVL2_{TLVL2})_v]_{(v(v\,TLVL2s))}] s_\tau$
$\quad = ([\lambda v.[\lambda g.(g\, a\, PLVL2)]]\, s)\, [\lambda g.(g\, a\, PLVL2)]$
§s %1 6 %0

\# $= (LELE3\, s\, TLVL2\, PLVL3)\; ([\lambda g.(g\, a\, PLVL2)]\; [\lambda x.[\lambda y.x]])$
§\ $[\lambda g_{s\,TLVL2s}.(g_{s\,TLVL2s} a_s PLVL2_{TLVL2})_s][\lambda x_s.[\lambda y_{TLVL2}.x_s]_{(s\,TLVL2)}]$
\# $= ([\lambda g.(g\, a\, PLVL2)]\; [\lambda x.[\lambda y.x]])\; ([\lambda x.[\lambda y.x]]\; a\, PLVL2)$
§s %1 3 %0
\# $= (LELE3\, s\, TLVL2\, PLVL3)\; ([\lambda x.[\lambda y.x]]\; a\, PLVL2)$
§\ $[\lambda x_s.[\lambda y_{TLVL2}.x_s]_{(s\,TLVL2)}]a_s$
\# $= ([\lambda x.[\lambda y.x]]\; a)\; [\lambda y.a]$
§s %1 6 %0
\# $= (LELE3\, s\, TLVL2\, PLVL3)\; ([\lambda y.a]\; PLVL2)$
§\ $[\lambda y_{TLVL2}.a_s]PLVL2_{TLVL2}$
\# $= ([\lambda y.a]\; PLVL2)\; a$
§s %1 3 %0
\# $= (LELE3\, s\, TLVL2\, PLVL3)\; a$

\## etc.

\##
\## Finally, one may use the recursion operator R to implement vectors and vector
\## access via an index number, and thus obtain a fully dependent type theory,
\## in which the type depends on an object (the dimension or the index number).
\##
\## For the formal definition of R and some of its applications,
\## see [Andrews 2002 (ISBN 1-4020-0763-9), pp. 281 f., 284].
\##

2.1.117 Results for File xor_associativity.r0.txt

\##
\## Associativity of Exclusive Disjunction (Exclusive OR, XOR)
\##
\##
\## Source: [Kubota 2017 (doi: 10.4444/100.10)]
\##

\## Written by Ken Kubota (<mail@kenkubota.de>).
\##
\## This file is part of the publication of the mathematical logic $\mathcal{R}_0$.
\## For more information, visit: <http://doi.org/10.4444/100.10>
\##

<< basics.r0.txt
<< xor_table.r0.txt

:= \$L $[\lambda a_o.[\lambda b_o.[\lambda c_o.(=_{ooo}(XOR_{ooo}(XOR_{ooo}a_o b_o)c_o)(XOR_{ooo}a_o(XOR_{ooo}b_o c_o)))_o]_{(oo)}]_{(ooo)}]$
\# wff 1652 : $[\lambda a.[\lambda b.[\lambda c.(= (XOR\,(XOR\,a\,b)\,c)\,(XOR\,a\,(XOR\,b\,c)))]]]_{oooo}$:= \$L

\## .1: subcase TT

:= $\$TT$ $\$L_{oooo}T_oT_o$
wff 1655 : $\$LTT_{oo}$:= $\$TT$
§= $\$TT$
$= \$TT\,\$TT$
§\ $\$L_{oooo}T_o$
$= (\$L\,T)\,[\lambda b.[\lambda c.(= (XOR\,(XORT\,b)\,c)\,(XORT\,(XOR\,b\,c)))]]$
§s %1 6 %0
$= \$TT\,([\lambda b.[\lambda c.(= (XOR\,(XORT\,b)\,c)\,(XORT\,(XOR\,b\,c)))]]\,T)$
§\ $[\lambda b_o.[\lambda c_o.(=_{ooo}(XOR_{ooo}(XOR_{ooo}T_ob_o)c_o)(XOR_{ooo}T_o(XOR_{ooo}b_oc_o)))_o]_{(oo)}]T_o$
$= ([\lambda b.[\lambda c.(= (XOR\,(XORT\,b)\,c)\,(XORT\,(XOR\,b\,c)))]]\,T)\ \ldots$
$\ldots [\lambda c.(= (XOR\,(XORT\,T)\,c)\,(XORT\,(XORT\,c)))]$
§s %1 3 %0
$= \$TT\,[\lambda c.(= (XOR\,(XORT\,T)\,c)\,(XORT\,(XORT\,c)))]$

use Proof Template A5222 (Rule of Cases): [\x.A]T, [\x.A]F → A
:= $\$L5222$ %0/3
wff 1677 : $[\lambda c.(= (XOR\,(XORT\,T)\,c)\,(XORT\,(XORT\,c)))]_{oo,\ldots}$:= $\$L5222$
:= $\$X5222$ c_o
wff 1640 : c_o := $\$X5222$
:= $\$T5222$ $\$L5222_{oo}T_o$
wff 1680 : $\$L5222\,T_o$:= $\$T5222$
:= $\$F5222$ $\$L5222_{oo}F_o$
wff 1681 : $\$L5222\,F_o$:= $\$F5222$

case T
§= $_o$ $\$T5222$
$= \$T5222\,\$T5222$
§\ $\$T5222$
$= \$T5222\,(= (XOR\,(XORT\,T)\,T)\,(XORT\,(XORT\,T)))$
§s %1 3 %0
$= \$T5222\,(= (XOR\,(XORT\,T)\,T)\,(XORT\,(XORT\,T)))$
$\%XorTableTTisF$
$= (XORT\,T)\,F$:= $XorTableTTisF$
$=_{ooo}(XOR_{ooo}T_oT_o)F_o$:= $XorTableTTisF$
§s %1 53 %0
$= \$T5222\,(= (XOR\,F\,T)\,(XORT\,(XORT\,T)))$
$\%XorTableFTisT$
$= (XOR\,F\,T)\,T$:= $XorTableFTisT$
$=_{ooo}(XOR_{ooo}F_oT_o)T_o$:= $XorTableFTisT$
§s %1 13 %0
$= \$T5222\,(= T\,(XORT\,(XORT\,T)))$
$\%XorTableTTisF$
$= (XORT\,T)\,F$:= $XorTableTTisF$
$=_{ooo}(XOR_{ooo}T_oT_o)F_o$:= $XorTableTTisF$
§s %1 15 %0
$= \$T5222\,(= T\,(XORT\,F))$
$\%XorTableTFisT$
$= (XORT\,F)\,T$:= $XorTableTFisT$

```
#        =_{ooo}(XOR_{ooo}T_oF_o)T_o      :=  XorTableTFisT
§s %1 7 %0
#        = $T5222 (= T T)
%A5230a
#        = (= T T) T      :=  A5230a
#        =_{ooo}(=_{ooo}T_oT_o)T_o      :=  A5230a
§s %1 3 %0
#        = $T5222 T
## use Proof Template A5201b (Swap):  A = B  →  B = A
<< A5201b.r0t.txt
%0
#        = T $T5222
#        =_{ooo}T_o$T5222_o
%T
#        = = =      :=  A5200t  T
#        =_{oωω}=_ω=_ω      :=  A5200t  T
§s %0 1 %1
#        $L5222 T      :=  $T5222

## case F
§= $F5222
#        = $F5222 $F5222
§\ $F5222
#        = $F5222 (= (XOR (XOR T T) F) (XOR T (XOR T F)))
§s %1 3 %0
#        = $F5222 (= (XOR (XOR T T) F) (XOR T (XOR T F)))
%XorTableTTisF
#        = (XOR T T) F      :=  XorTableTTisF
#        =_{ooo}(XOR_{ooo}T_oT_o)F_o      :=  XorTableTTisF
§s %1 53 %0
#        = $F5222 (= (XOR F F) (XOR T (XOR T F)))
%XorTableFFisF
#        = (XOR F F) F      :=  XorTableFFisF
#        =_{ooo}(XOR_{ooo}F_oF_o)F_o      :=  XorTableFFisF
§s %1 13 %0
#        = $F5222 (= F (XOR T (XOR T F)))
%XorTableTFisT
#        = (XOR T F) T      :=  XorTableTFisT
#        =_{ooo}(XOR_{ooo}T_oF_o)T_o      :=  XorTableTFisT
§s %1 15 %0
#        = $F5222 (= F (XOR T T))
%XorTableTTisF
#        = (XOR T T) F      :=  XorTableTTisF
#        =_{ooo}(XOR_{ooo}T_oT_o)F_o      :=  XorTableTTisF
§s %1 7 %0
#        = $F5222 (= F F)
%A5230d
#        = (= F F) T      :=  A5230d
#        =_{ooo}(=_{ooo}F_oF_o)T_o      :=  A5230d
```

§s %1 3 %0
\# $=\$F5222\,T$
\## use Proof Template A5201b (Swap): A = B → B = A
<< A5201b.r0t.txt
%0
\# $=T\,\$F5222$
\# $=_{o\omega\omega}T_{\omega}\$F5222_{\omega}$
%T
\# $===$:= $A5200t$ T
\# $=_{o\omega\omega}=_{\omega}=_{\omega}$:= $A5200t$ T
§s %0 1 %1
\# $\$L5222\,F$:= $\$F5222$

<< A5222.r0t.txt
:= $\$L5222$
:= $\$X5222$
:= $\$T5222$
:= $\$F5222$

:= $\$TT$
:= $\$TT$ %0
\# wff 1676 : $=(XOR\,(XOR\,T\,T)\,c)\,(XOR\,T\,(XOR\,T\,c))_{o,\ldots}$:= $\$TT$

\## .2: subcase TF

:= $\$TF$ $\$L_{oooo}T_{o}F_{o}$
\# wff 1788 : $\$L\,T\,F_{oo}$:= $\$TF$
§= $\$TF$
\# $=\$TF\,\$TF$
§\ $\$L_{oooo}T_{o}$
\# $=(\$L\,T)\,[\lambda b.[\lambda c.(=(XOR\,(XOR\,T\,b)\,c)\,(XOR\,T\,(XOR\,b\,c)))]]$
§s %1 6 %0
\# $=\$TF\,([\lambda b.[\lambda c.(=(XOR\,(XOR\,T\,b)\,c)\,(XOR\,T\,(XOR\,b\,c)))]]\,F)$
§\ $[\lambda b_{o}.[\lambda c_{o}.(=_{ooo}(XOR_{ooo}(XOR_{ooo}T_{o}b_{o})c_{o})(XOR_{ooo}T_{o}(XOR_{ooo}b_{o}c_{o})))_{o}]_{(oo)}]F_{o}$
\# $=([\lambda b.[\lambda c.(=(XOR\,(XOR\,T\,b)\,c)\,(XOR\,T\,(XOR\,b\,c)))]]\,F)\ldots$
$\ldots[\lambda c.(=(XOR\,(XOR\,T\,F)\,c)\,(XOR\,T\,(XOR\,F\,c)))]$
§s %1 3 %0
\# $=\$TF\,[\lambda c.(=(XOR\,(XOR\,T\,F)\,c)\,(XOR\,T\,(XOR\,F\,c)))]$

\## use Proof Template A5222 (Rule of Cases): [\x.A]T, [\x.A]F → A
:= $\$L5222$ %0/3
\# wff 1800 : $[\lambda c.(=(XOR\,(XOR\,T\,F)\,c)\,(XOR\,T\,(XOR\,F\,c)))]_{oo,\ldots}$:= $\$L5222$
:= $\$X5222$ c_{o}
\# wff 1640 : c_{o} := $\$X5222$
:= $\$T5222$ $\$L5222_{oo}T_{o}$
\# wff 1803 : $\$L5222\,T_{o}$:= $\$T5222$
:= $\$F5222$ $\$L5222_{oo}F_{o}$
\# wff 1804 : $\$L5222\,F_{o}$:= $\$F5222$

case T
§= $\$T5222$
$= \$T5222\, \$T5222$
§\ $\$T5222$
$= \$T5222\, (= (XOR\, (XORT\, F)\, T)\, (XORT\, (XORF\, T)))$
§s %1 3 %0
$= \$T5222\, (= (XOR\, (XORT\, F)\, T)\, (XORT\, (XORF\, T)))$
$\%XorTableTFisT$
$= (XORT\, F)\, T \quad := \quad XorTableTFisT$
$=_{ooo}(XOR_{ooo}T_oF_o)T_o \quad := \quad XorTableTFisT$
§s %1 53 %0
$= \$T5222\, (= (XORT\, T)\, (XORT\, (XORF\, T)))$
$\%XorTableTTisF$
$= (XORT\, T)\, F \quad := \quad XorTableTTisF$
$=_{ooo}(XOR_{ooo}T_oT_o)F_o \quad := \quad XorTableTTisF$
§s %1 13 %0
$= \$T5222\, (= F\, (XORT\, (XORF\, T)))$
$\%XorTableFTisT$
$= (XORF\, T)\, T \quad := \quad XorTableFTisT$
$=_{ooo}(XOR_{ooo}F_oT_o)T_o \quad := \quad XorTableFTisT$
§s %1 15 %0
$= \$T5222\, (= F\, (XORT\, T))$
$\%XorTableTTisF$
$= (XORT\, T)\, F \quad := \quad XorTableTTisF$
$=_{ooo}(XOR_{ooo}T_oT_o)F_o \quad := \quad XorTableTTisF$
§s %1 7 %0
$= \$T5222\, (= F\, F)$
$\%A5230d$
$= (= F\, F)\, T \quad := \quad A5230d$
$=_{ooo}(=_{ooo}F_oF_o)T_o \quad := \quad A5230d$
§s %1 3 %0
$= \$T5222\, T$
use Proof Template A5201b (Swap): A = B → B = A
<< A5201b.r0t.txt
%0
$= T\, \$T5222$
$=_{o\omega\omega}T_\omega \$T5222_\omega$
$\%T$
$= = = \quad := \quad A5200t \quad T$
$=_{o\omega\omega} =_\omega =_\omega \quad := \quad A5200t \quad T$
§s %0 1 %1
$\$L5222\, T \quad := \quad \$T5222$

case F
§= $\$F5222$
$= \$F5222\, \$F5222$
§\ $\$F5222$
$= \$F5222\, (= (XOR\, (XORT\, F)\, F)\, (XORT\, (XORF\, F)))$
§s %1 3 %0

\# $= \$F5222\,(=(XOR\,(XORT\,F)\,F)\,(XORT\,(XORF\,F)))$
%$XorTableTFisT$
\# $=(XORT\,F)\,T$:= $XorTableTFisT$
\# $=_{ooo}(XOR_{ooo}T_oF_o)T_o$:= $XorTableTFisT$
§s %1 53 %0
\# $= \$F5222\,(=(XORT\,F)\,(XORT\,(XORF\,F)))$
%$XorTableTFisT$
\# $=(XORT\,F)\,T$:= $XorTableTFisT$
\# $=_{ooo}(XOR_{ooo}T_oF_o)T_o$:= $XorTableTFisT$
§s %1 13 %0
\# $= \$F5222\,(=T\,(XORT\,(XORF\,F)))$
%$XorTableFFisF$
\# $=(XORF\,F)\,F$:= $XorTableFFisF$
\# $=_{ooo}(XOR_{ooo}F_oF_o)F_o$:= $XorTableFFisF$
§s %1 15 %0
\# $= \$F5222\,(=T\,(XORT\,F))$
%$XorTableTFisT$
\# $=(XORT\,F)\,T$:= $XorTableTFisT$
\# $=_{ooo}(XOR_{ooo}T_oF_o)T_o$:= $XorTableTFisT$
§s %1 7 %0
\# $= \$F5222\,(=T\,T)$
%$A5230a$
\# $=(=T\,T)\,T$:= $A5230a$
\# $=_{ooo}(=_{ooo}T_oT_o)T_o$:= $A5230a$
§s %1 3 %0
\# $= \$F5222\,T$
\## use Proof Template A5201b (Swap): A = B → B = A
<< A5201b.r0t.txt
%0
\# $=T\,\$F5222$
\# $=_{o\omega\omega}T_\omega\$F5222_\omega$
%T
\# $===$:= $A5200t$ T
\# $=_{o\omega\omega}=_\omega=_\omega$:= $A5200t$ T
§s %0 1 %1
\# $\$L5222\,F$:= $\$F5222$

<< A5222.r0t.txt
:= $\$L5222$
:= $\$X5222$
:= $\$T5222$
:= $\$F5222$

:= $\$TF$
:= $\$TF$ %0
\# wff 1799 : $=(XOR\,(XORT\,F)\,c)\,(XORT\,(XORF\,c))_{o,\ldots}$:= $\$TF$

\## .3: subcase FT

:= $\$FT$ $\$L_{oooo}F_oT_o$
wff 1906 : $\$L\,F\,T_{oo}$:= $\$FT$
§= $\$FT$
$= \$FT\,\$FT$
§\ $\$L_{oooo}F_o$
$= (\$L\,F)\,[\lambda b.[\lambda c.(=(XOR\,(XOR\,F\,b)\,c)\,(XOR\,F\,(XOR\,b\,c)))]]$
§s %1 6 %0
$= \$FT\,([\lambda b.[\lambda c.(=(XOR\,(XOR\,F\,b)\,c)\,(XOR\,F\,(XOR\,b\,c)))]]\,T)$
§\ $[\lambda b_o.[\lambda c_o.(=_{ooo}(XOR_{ooo}(XOR_{ooo}F_ob_o)c_o)(XOR_{ooo}F_o(XOR_{ooo}b_oc_o)))_o]_{(oo)}]T_o$
$= ([\lambda b.[\lambda c.(=(XOR\,(XOR\,F\,b)\,c)\,(XOR\,F\,(XOR\,b\,c)))]]\,T)\ldots$
$\ldots[\lambda c.(=(XOR\,(XOR\,F\,T)\,c)\,(XOR\,F\,(XOR\,T\,c)))]$
§s %1 3 %0
$= \$FT\,[\lambda c.(=(XOR\,(XOR\,F\,T)\,c)\,(XOR\,F\,(XOR\,T\,c)))]$

use Proof Template A5222 (Rule of Cases): [\x.A]T, [\x.A]F → A
:= $\$L5222$ %0/3
wff 1927 : $[\lambda c.(=(XOR\,(XOR\,F\,T)\,c)\,(XOR\,F\,(XOR\,T\,c)))]_{oo,\ldots}$:= $\$L5222$
:= $\$X5222$ c_o
wff 1640 : c_o := $\$X5222$
:= $\$T5222$ $\$L5222_{oo}T_o$
wff 1930 : $\$L5222\,T_o$:= $\$T5222$
:= $\$F5222$ $\$L5222_{oo}F_o$
wff 1931 : $\$L5222\,F_o$:= $\$F5222$

case T
§= $\$T5222$
$= \$T5222\,\$T5222$
§\ $\$T5222$
$= \$T5222\,(=(XOR\,(XOR\,F\,T)\,T)\,(XOR\,F\,(XOR\,T\,T)))$
§s %1 3 %0
$= \$T5222\,(=(XOR\,(XOR\,F\,T)\,T)\,(XOR\,F\,(XOR\,T\,T)))$
$\%XorTableFTisT$
$= (XOR\,F\,T)\,T$:= $XorTableFTisT$
$=_{ooo}(XOR_{ooo}F_oT_o)T_o$:= $XorTableFTisT$
§s %1 53 %0
$= \$T5222\,(=(XOR\,T\,T)\,(XOR\,F\,(XOR\,T\,T)))$
$\%XorTableTTisF$
$= (XOR\,T\,T)\,F$:= $XorTableTTisF$
$=_{ooo}(XOR_{ooo}T_oT_o)F_o$:= $XorTableTTisF$
§s %1 13 %0
$= \$T5222\,(=F\,(XOR\,F\,(XOR\,T\,T)))$
$\%XorTableTTisF$
$= (XOR\,T\,T)\,F$:= $XorTableTTisF$
$=_{ooo}(XOR_{ooo}T_oT_o)F_o$:= $XorTableTTisF$
§s %1 15 %0
$= \$T5222\,(=F\,(XOR\,F\,F))$
$\%XorTableFFisF$
$= (XOR\,F\,F)\,F$:= $XorTableFFisF$
$=_{ooo}(XOR_{ooo}F_oF_o)F_o$:= $XorTableFFisF$

§s %1 7 %0
$= \$T5222\,(= F\,F)$
%$A5230d$
$= (= F\,F)\,T \quad := \quad A5230d$
$=_{ooo}(=_{ooo}F_oF_o)T_o \quad := \quad A5230d$
§s %1 3 %0
$= \$T5222\,T$
use Proof Template A5201b (Swap): A = B → B = A
<< A5201b.r0t.txt
%0
$= T\,\$T5222$
$=_{o\omega\omega}T_\omega\$T5222_\omega$
%T
$= = = \quad := \quad A5200t \quad T$
$=_{o\omega\omega}=_\omega=_\omega \quad := \quad A5200t \quad T$
§s %0 1 %1
$\$L5222\,T \quad := \quad \$T5222$

case F
§= $\$F5222$
$= \$F5222\,\$F5222$
§\ $\$F5222$
$= \$F5222\,(= (XOR\,(XOR\,F\,T)\,F)\,(XOR\,F\,(XOR\,T\,F)))$
§s %1 3 %0
$= \$F5222\,(= (XOR\,(XOR\,F\,T)\,F)\,(XOR\,F\,(XOR\,T\,F)))$
%$XorTableFTisT$
$= (XOR\,F\,T)\,T \quad := \quad XorTableFTisT$
$=_{ooo}(XOR_{ooo}F_oT_o)T_o \quad := \quad XorTableFTisT$
§s %1 53 %0
$= \$F5222\,(= (XOR\,T\,F)\,(XOR\,F\,(XOR\,T\,F)))$
%$XorTableTFisT$
$= (XOR\,T\,F)\,T \quad := \quad XorTableTFisT$
$=_{ooo}(XOR_{ooo}T_oF_o)T_o \quad := \quad XorTableTFisT$
§s %1 13 %0
$= \$F5222\,(= T\,(XOR\,F\,(XOR\,T\,F)))$
%$XorTableTFisT$
$= (XOR\,T\,F)\,T \quad := \quad XorTableTFisT$
$=_{ooo}(XOR_{ooo}T_oF_o)T_o \quad := \quad XorTableTFisT$
§s %1 15 %0
$= \$F5222\,(= T\,(XOR\,F\,T))$
%$XorTableFTisT$
$= (XOR\,F\,T)\,T \quad := \quad XorTableFTisT$
$=_{ooo}(XOR_{ooo}F_oT_o)T_o \quad := \quad XorTableFTisT$
§s %1 7 %0
$= \$F5222\,(= T\,T)$
%$A5230a$
$= (= T\,T)\,T \quad := \quad A5230a$
$=_{ooo}(=_{ooo}T_oT_o)T_o \quad := \quad A5230a$
§s %1 3 %0

\# $= \$F5222\,T$
\#\# use Proof Template A5201b (Swap): A = B $\rightarrow$ B = A
<< A5201b.r0t.txt
%0
\# $= T\,\$F5222$
\# $=_{o\omega\omega} T_\omega \$F5222_\omega$
%T
\# $= = =$:= $A5200t$ T
\# $=_{o\omega\omega} =_\omega =_\omega$:= $A5200t$ T
§s %0 1 %1
\# $\$L5222\,F$:= $\$F5222$

<< A5222.r0t.txt
:= $\$L5222$
:= $\$X5222$
:= $\$T5222$
:= $\$F5222$

:= $\$FT$
:= $\$FT$ %0
\# wff 1926 : $= (XOR\,(XOR\,F\,T)\,c)\,(XOR\,F\,(XOR\,T\,c))_{o,\ldots}$:= $\$FT$

\#\# .4: subcase FF

:= $\$FF$ $\$L_{oooo}F_oF_o$
\# wff 2033 : $\$L\,F\,F_{oo}$:= $\$FF$
§= $\$FF$
\# $= \$FF\,\$FF$
§\ $\$L_{oooo}F_o$
\# $= (\$L\,F)\,[\lambda b.[\lambda c.(= (XOR\,(XOR\,F\,b)\,c)\,(XOR\,F\,(XOR\,b\,c)))]]$
§s %1 6 %0
\# $= \$FF\,([\lambda b.[\lambda c.(= (XOR\,(XOR\,F\,b)\,c)\,(XOR\,F\,(XOR\,b\,c)))]]\,F)$
§\ $[\lambda b_o.[\lambda c_o.(=_{ooo}(XOR_{ooo}(XOR_{ooo}F_ob_o)c_o)(XOR_{ooo}F_o(XOR_{ooo}b_oc_o)))_o]_{(oo)}]F_o$
\# $= ([\lambda b.[\lambda c.(= (XOR\,(XOR\,F\,b)\,c)\,(XOR\,F\,(XOR\,b\,c)))]]\,F)\ \ldots$
$\ldots\,[\lambda c.(= (XOR\,(XOR\,F\,F)\,c)\,(XOR\,F\,(XOR\,F\,c)))]$
§s %1 3 %0
\# $= \$FF\,[\lambda c.(= (XOR\,(XOR\,F\,F)\,c)\,(XOR\,F\,(XOR\,F\,c)))]$

\#\# use Proof Template A5222 (Rule of Cases): [\x.A]T, [\x.A]F $\rightarrow$ A
:= $\$L5222$ %0/3
\# wff 2044 : $[\lambda c.(= (XOR\,(XOR\,F\,F)\,c)\,(XOR\,F\,(XOR\,F\,c)))]_{oo,\ldots}$:= $\$L5222$
:= $\$X5222$ c_o
\# wff 1640 : c_o := $\$X5222$
:= $\$T5222$ $\$L5222_{oo}T_o$
\# wff 2047 : $\$L5222\,T_o$:= $\$T5222$
:= $\$F5222$ $\$L5222_{oo}F_o$
\# wff 2048 : $\$L5222\,F_o$:= $\$F5222$

\#\# case T

§= $\$T5222$
\# $= \$T5222\, \$T5222$
§\ $\$T5222$
\# $= \$T5222\, (= (XOR\, (XOR\, F\, F)\, T)\, (XOR\, F\, (XOR\, F\, T)))$
§s %1 3 %0
\# $= \$T5222\, (= (XOR\, (XOR\, F\, F)\, T)\, (XOR\, F\, (XOR\, F\, T)))$
%$XorTableFFisF$
\# $= (XOR\, F\, F)\, F \quad := \quad XorTableFFisF$
\# $=_{ooo}(XOR_{ooo}F_oF_o)F_o \quad := \quad XorTableFFisF$
§s %1 53 %0
\# $= \$T5222\, (= (XOR\, F\, T)\, (XOR\, F\, (XOR\, F\, T)))$
%$XorTableFTisT$
\# $= (XOR\, F\, T)\, T \quad := \quad XorTableFTisT$
\# $=_{ooo}(XOR_{ooo}F_oT_o)T_o \quad := \quad XorTableFTisT$
§s %1 13 %0
\# $= \$T5222\, (= T\, (XOR\, F\, (XOR\, F\, T)))$
%$XorTableFTisT$
\# $= (XOR\, F\, T)\, T \quad := \quad XorTableFTisT$
\# $=_{ooo}(XOR_{ooo}F_oT_o)T_o \quad := \quad XorTableFTisT$
§s %1 15 %0
\# $= \$T5222\, (= T\, (XOR\, F\, T))$
%$XorTableFTisT$
\# $= (XOR\, F\, T)\, T \quad := \quad XorTableFTisT$
\# $=_{ooo}(XOR_{ooo}F_oT_o)T_o \quad := \quad XorTableFTisT$
§s %1 7 %0
\# $= \$T5222\, (= T\, T)$
%$A5230a$
\# $= (= T\, T)\, T \quad := \quad A5230a$
\# $=_{ooo}(=_{ooo}T_oT_o)T_o \quad := \quad A5230a$
§s %1 3 %0
\# $= \$T5222\, T$
\#\# use Proof Template A5201b (Swap): A = B → B = A
<< A5201b.r0t.txt
%0
\# $= T\, \$T5222$
\# $=_{o\omega\omega}T_\omega \$T5222_\omega$
%T
\# $= = = \quad := \quad A5200t \quad T$
\# $=_{o\omega\omega}=_\omega=_\omega \quad := \quad A5200t \quad T$
§s %0 1 %1
\# $\$L5222\, T \quad := \quad \$T5222$

\#\# case F
§= $\$F5222$
\# $= \$F5222\, \$F5222$
§\ $\$F5222$
\# $= \$F5222\, (= (XOR\, (XOR\, F\, F)\, F)\, (XOR\, F\, (XOR\, F\, F)))$
§s %1 3 %0
\# $= \$F5222\, (= (XOR\, (XOR\, F\, F)\, F)\, (XOR\, F\, (XOR\, F\, F)))$

```
%XorTableFFisF
#              = (XOR F F) F        :=   XorTableFFisF
#              =_ooo(XOR_ooo F_o F_o) F_o      :=   XorTableFFisF
§s %1 53 %0
#              = $F5222 (= (XOR F F) (XOR F (XOR F F)))
%XorTableFFisF
#              = (XOR F F) F        :=   XorTableFFisF
#              =_ooo(XOR_ooo F_o F_o) F_o      :=   XorTableFFisF
§s %1 13 %0
#              = $F5222 (= F (XOR F (XOR F F)))
%XorTableFFisF
#              = (XOR F F) F        :=   XorTableFFisF
#              =_ooo(XOR_ooo F_o F_o) F_o      :=   XorTableFFisF
§s %1 15 %0
#              = $F5222 (= F (XOR F F))
%XorTableFFisF
#              = (XOR F F) F        :=   XorTableFFisF
#              =_ooo(XOR_ooo F_o F_o) F_o      :=   XorTableFFisF
§s %1 7 %0
#              = $F5222 (= F F)
%A5230d
#              = (= F F) T       :=   A5230d
#              =_ooo(=_ooo F_o F_o) T_o      :=   A5230d
§s %1 3 %0
#              = $F5222 T
## use Proof Template A5201b (Swap):   A = B   →   B = A
<< A5201b.r0t.txt
%0
#              = T $F5222
#              =_oωω T_ω $F5222_ω
%T
#              = = =       :=   A5200t   T
#              =_oωω =_ω =_ω       :=   A5200t   T
§s %0 1 %1
#              $L5222 F       :=   $F5222

<< A5222.r0t.txt
:=  $L5222
:=  $X5222
:=  $T5222
:=  $F5222

:=  $FF
:=  $FF  %0
# wff    2043  :      = (XOR (XOR F F) c) (XOR F (XOR F c))_o,...       :=   $FF

## .5:   case T

:=  $T  [λb_o.($L_oooo T_o b_o c_o)_o]
```

wff 2150 : $[\lambda b.(\$L\,T\,b\,c)]_{oo}$:= $\$T$
§= $\$T$
$= \$T\,\$T$
§\ $\$L_{oooo}T_o$
$= (\$L\,T)\,[\lambda b.[\lambda c.(= (XOR\,(XORT\,b)\,c)\,(XORT\,(XOR\,b\,c)))]]$
§s %1 28 %0
$= \$T\,[\lambda b.([\lambda b.[\lambda c.(= (XOR\,(XORT\,b)\,c)\,(XORT\,(XOR\,b\,c)))]]\,b\,c)]$
§\ $[\lambda b_o.[\lambda c_o.(=_{ooo}(XOR_{ooo}(XOR_{ooo}T_ob_o)c_o)(XOR_{ooo}T_o(XOR_{ooo}b_oc_o)))_o]_{(oo)}]b_o$
$= ([\lambda b.[\lambda c.(= (XOR\,(XORT\,b)\,c)\,(XORT\,(XOR\,b\,c)))]]\,b)$ ...
... $[\lambda c.(= (XOR\,(XORT\,b)\,c)\,(XORT\,(XOR\,b\,c)))]$
§s %1 14 %0
$= \$T\,[\lambda b.([\lambda c.(= (XOR\,(XORT\,b)\,c)\,(XORT\,(XOR\,b\,c)))]\,c)]$
§\ $[\lambda c_o.(=_{ooo}(XOR_{ooo}(XOR_{ooo}T_ob_o)c_o)(XOR_{ooo}T_o(XOR_{ooo}b_oc_o)))_o]c_o$
$= ([\lambda c.(= (XOR\,(XORT\,b)\,c)\,(XORT\,(XOR\,b\,c)))]\,c)$ ...
... $(= (XOR\,(XORT\,b)\,c)\,(XORT\,(XOR\,b\,c)))$
§s %1 7 %0
$= \$T\,[\lambda b.(= (XOR\,(XORT\,b)\,c)\,(XORT\,(XOR\,b\,c)))]$

use Proof Template A5222 (Rule of Cases): [\x.A]T, [\x.A]F → A
:= $\$L5222$ %0/3
wff 2164 : $[\lambda b.(= (XOR\,(XORT\,b)\,c)\,(XORT\,(XOR\,b\,c)))]_{oo,\ldots}$:= $\$L5222$
:= $\$X5222$ b_o
wff 58 : b_o := $\$X5222$
:= $\$T5222$ $\$L5222_{oo}T_o$
wff 2166 : $\$L5222\,T_o$:= $\$T5222$
:= $\$F5222$ $\$L5222_{oo}F_o$
wff 2167 : $\$L5222\,F_o$:= $\$F5222$

case T
§= $\$T5222$
$= \$T5222\,\$T5222$
§\ $\$T5222$
$= \$T5222\,\$TT$
§s %1 3 %0
$= \$T5222\,\$TT$
use Proof Template A5201b (Swap): A = B → B = A
<< A5201b.r0t.txt
%0
$= \$TT\,\$T5222$
$=_{o\omega\omega}\$TT_\omega\$T5222_\omega$
%$\$TT$
$= (XOR\,(XORT\,T)\,c)\,(XORT\,(XORT\,c))$:= $\$TT$
$=_{ooo}(XOR_{ooo}(XOR_{ooo}T_oT_o)c_o)(XOR_{ooo}T_o(XOR_{ooo}T_oc_o))$:= $\$TT$
:= $\$TT$
§s %0 1 %1
$\$L5222\,T$:= $\$T5222$

case F
§= $\$F5222$

$= \$F5222\,\$F5222$
§\ $\$F5222$
$= \$F5222\,\$TF$
§s %1 3 %0
$= \$F5222\,\$TF$
use Proof Template A5201b (Swap): A = B → B = A
<< A5201b.r0t.txt
%0
$= \$TF\,\$F5222$
$=_{o\omega\omega}\$TF_{\omega}\$F5222_{\omega}$
%$\$TF$
$= (XOR\,(XOR\,T\,F)\,c)\,(XOR\,T\,(XOR\,F\,c))$:= $\$TF$
$=_{ooo}(XOR_{ooo}(XOR_{ooo}T_oF_o)c_o)(XOR_{ooo}T_o(XOR_{ooo}F_oc_o))$:= $\$TF$
:= $\$TF$
§s %0 1 %1
$\$L5222\,F$:= $\$F5222$

<< A5222.r0t.txt
:= $\$L5222$
:= $\$X5222$
:= $\$T5222$
:= $\$F5222$

:= $\$T$
:= $\$T$ %0
wff 1664 : $= (XOR\,(XOR\,T\,b)\,c)\,(XOR\,T\,(XOR\,b\,c))_{o,\ldots}$:= $\$T$

.6: case F

:= $\$F$ $[\lambda b_o.(\$L_{oooo}F_ob_oc_o)_o]$
wff 2256 : $[\lambda b.(\$L\,F\,b\,c)]_{oo}$:= $\$F$
§= $\$F$
$= \$F\,\$F$
§\ $\$L_{oooo}F_o$
$= (\$L\,F)\,[\lambda b.[\lambda c.(= (XOR\,(XOR\,F\,b)\,c)\,(XOR\,F\,(XOR\,b\,c)))]]$
§s %1 28 %0
$= \$F\,[\lambda b.([\lambda b.[\lambda c.(= (XOR\,(XOR\,F\,b)\,c)\,(XOR\,F\,(XOR\,b\,c)))]]\,b\,c)]$
§\ $[\lambda b_o.[\lambda c_o.(=_{ooo}(XOR_{ooo}(XOR_{ooo}F_ob_o)c_o)(XOR_{ooo}F_o(XOR_{ooo}b_oc_o)))_o]_{(oo)}]b_o$
$= ([\lambda b.[\lambda c.(= (XOR\,(XOR\,F\,b)\,c)\,(XOR\,F\,(XOR\,b\,c)))]]\,b)\ \ldots$
$\ldots\,[\lambda c.(= (XOR\,(XOR\,F\,b)\,c)\,(XOR\,F\,(XOR\,b\,c)))]$
§s %1 14 %0
$= \$F\,[\lambda b.([\lambda c.(= (XOR\,(XOR\,F\,b)\,c)\,(XOR\,F\,(XOR\,b\,c)))]\,c)]$
§\ $[\lambda c_o.(=_{ooo}(XOR_{ooo}(XOR_{ooo}F_ob_o)c_o)(XOR_{ooo}F_o(XOR_{ooo}b_oc_o)))_o]c_o$
$= ([\lambda c.(= (XOR\,(XOR\,F\,b)\,c)\,(XOR\,F\,(XOR\,b\,c)))]\,c)\ \ldots$
$\ldots\,(= (XOR\,(XOR\,F\,b)\,c)\,(XOR\,F\,(XOR\,b\,c)))$
§s %1 7 %0
$= \$F\,[\lambda b.(= (XOR\,(XOR\,F\,b)\,c)\,(XOR\,F\,(XOR\,b\,c)))]$

use Proof Template A5222 (Rule of Cases): [\x.A]T, [\x.A]F → A

```
:= $L5222 %0/3
# wff     2270 :      [λb.(= (XOR (XOR F b) c) (XOR F (XOR b c)))]_oo,...      := $L5222
:= $X5222 b_o
# wff     58 :      b_o      := $X5222
:= $T5222 $L5222_oo T_o
# wff     2272 :      $L5222 T_o      := $T5222
:= $F5222 $L5222_oo F_o
# wff     2273 :      $L5222 F_o      := $F5222

## case T
§= $T5222
#                 = $T5222 $T5222
§\ $T5222
#                 = $T5222 $FT
§s %1 3 %0
#                 = $T5222 $FT
## use Proof Template A5201b (Swap):   A = B   →   B = A
<< A5201b.r0t.txt
%0
#                 = $FT $T5222
#                 =_oωω $FT_ω $T5222_ω
%$FT
#                 = (XOR (XOR F T) c) (XOR F (XOR T c))      := $FT
#                 =_ooo(XOR_ooo(XOR_ooo F_o T_o) c_o)(XOR_ooo F_o(XOR_ooo T_o c_o))      := $FT
:= $FT
§s %0 1 %1
#                 $L5222 T      := $T5222

## case F
§= $F5222
#                 = $F5222 $F5222
§\ $F5222
#                 = $F5222 $FF
§s %1 3 %0
#                 = $F5222 $FF
## use Proof Template A5201b (Swap):   A = B   →   B = A
<< A5201b.r0t.txt
%0
#                 = $FF $F5222
#                 =_oωω $FF_ω $F5222_ω
%$FF
#                 = (XOR (XOR F F) c) (XOR F (XOR F c))      := $FF
#                 =_ooo(XOR_ooo(XOR_ooo F_o F_o) c_o)(XOR_ooo F_o(XOR_ooo F_o c_o))      := $FF
:= $FF
§s %0 1 %1
#                 $L5222 F      := $F5222

<< A5222.r0t.txt
:= $L5222
```

$:= \ \$X5222$
$:= \ \$T5222$
$:= \ \$F5222$

$:= \ \$F$
$:= \ \$F$ %0
wff 1915 : $= (XOR\,(XOR\,F\,b)\,c)\,(XOR\,F\,(XOR\,b\,c))_{o,\dots}$ $:= \ \$F$

.7: general case

$:= \ \$R \ [\lambda a_o.(\$L_{oooo} a_o b_o c_o)_o]$
wff 2357 : $[\lambda a.(\$L\,a\,b\,c)]_{oo}$ $:= \ \$R$
§= $\$R$
$= \$R\,\$R$
§\ $\$L_{oooo} a_o$
$= (\$L\,a)\,[\lambda b.[\lambda c.(= (XOR\,(XOR\,a\,b)\,c)\,(XOR\,a\,(XOR\,b\,c)))]]$
§s %1 28 %0
$= \$R\,[\lambda a.([\lambda b.[\lambda c.(= (XOR\,(XOR\,a\,b)\,c)\,(XOR\,a\,(XOR\,b\,c)))]]\,b\,c)]$
§\ $[\lambda b_o.[\lambda c_o.(=_{ooo}(XOR_{ooo}(XOR_{ooo}a_o b_o)c_o)(XOR_{ooo}a_o(XOR_{ooo}b_o c_o)))_o]_{(oo)}]b_o$
$= ([\lambda b.[\lambda c.(= (XOR\,(XOR\,a\,b)\,c)\,(XOR\,a\,(XOR\,b\,c)))]]\,b)\ \dots$
$\dots [\lambda c.(= (XOR\,(XOR\,a\,b)\,c)\,(XOR\,a\,(XOR\,b\,c)))]$
§s %1 14 %0
$= \$R\,[\lambda a.([\lambda c.(= (XOR\,(XOR\,a\,b)\,c)\,(XOR\,a\,(XOR\,b\,c)))]\,c)]$
§\ $[\lambda c_o.(=_{ooo}(XOR_{ooo}(XOR_{ooo}a_o b_o)c_o)(XOR_{ooo}a_o(XOR_{ooo}b_o c_o)))_o]c_o$
$= ([\lambda c.(= (XOR\,(XOR\,a\,b)\,c)\,(XOR\,a\,(XOR\,b\,c)))]\,c)\ \dots$
$\dots (= (XOR\,(XOR\,a\,b)\,c)\,(XOR\,a\,(XOR\,b\,c)))$
§s %1 7 %0
$= \$R\,[\lambda a.(= (XOR\,(XOR\,a\,b)\,c)\,(XOR\,a\,(XOR\,b\,c)))]$

use Proof Template A5222 (Rule of Cases): [\x.A]T, [\x.A]F → A
$:= \ \$L5222$ %0/3
wff 2373 : $[\lambda a.(= (XOR\,(XOR\,a\,b)\,c)\,(XOR\,a\,(XOR\,b\,c)))]_{oo,\dots}$ $:= \ \$L5222$
$:= \ \$X5222 \ a_o$
wff 54 : a_o $:= \ \$X5222$
$:= \ \$T5222 \ \$L5222_{oo} T_o$
wff 2375 : $\$L5222\,T_o$ $:= \ \$T5222$
$:= \ \$F5222 \ \$L5222_{oo} F_o$
wff 2376 : $\$L5222\,F_o$ $:= \ \$F5222$

case T
§= $\$T5222$
$= \$T5222\,\$T5222$
§\ $\$T5222$
$= \$T5222\,\$T$
§s %1 3 %0
$= \$T5222\,\$T$
use Proof Template A5201b (Swap): A = B → B = A
<< A5201b.r0t.txt
%0

\# $= \$T\,\$T5222$
\# $=_{o\omega\omega}\$T_\omega\$T5222_\omega$
%$\$T$
\# $= (XOR\,(XOR\,T\,b)\,c)\,(XOR\,T\,(XOR\,b\,c))$:= $\$T$
\# $=_{ooo}(XOR_{ooo}(XOR_{ooo}T_ob_o)c_o)(XOR_{ooo}T_o(XOR_{ooo}b_oc_o))$:= $\$T$
:= $\$T$
§s %0 1 %1
\# $\$L5222\,T$:= $\$T5222$

\#\# case F
§= $\$F5222$
\# $= \$F5222\,\$F5222$
§\ $\$F5222$
\# $= \$F5222\,\$F$
§s %1 3 %0
\# $= \$F5222\,\$F$
\#\# use Proof Template A5201b (Swap): A = B → B = A
<< A5201b.r0t.txt
%0
\# $= \$F\,\$F5222$
\# $=_{o\omega\omega}\$F_\omega\$F5222_\omega$
%$\$F$
\# $= (XOR\,(XOR\,F\,b)\,c)\,(XOR\,F\,(XOR\,b\,c))$:= $\$F$
\# $=_{ooo}(XOR_{ooo}(XOR_{ooo}F_ob_o)c_o)(XOR_{ooo}F_o(XOR_{ooo}b_oc_o))$:= $\$F$
:= $\$F$
§s %0 1 %1
\# $\$L5222\,F$:= $\$F5222$

<< A5222.r0t.txt
:= $\$L5222$
:= $\$X5222$
:= $\$T5222$
:= $\$F5222$

:= $\$R$
:= $\$L$

\#\# .8: match general definition

\#\# use Proof Template A5220 (Gen): A → ∀ x: A
:= $\$T5220$ o
\# wff 2 : o_τ := $\$T5220$
:= $\$X5220$ c_o
\# wff 1640 : c_o := $\$X5220$
:= $\$A5220$ %0
\# wff 1649 : $= (XOR\,(XOR\,a\,b)\,\$X5220)\,(XOR\,a\,(XOR\,b\,\$X5220))_{o,\ldots}$:= $\$A5220$
<< A5220.r0t.txt
:= $\$T5220$
:= $\$X5220$

:= $\$A5220$

use Proof Template A5220 (Gen): A → ∀ x: A
:= $\$T5220$ o
wff 2 : o_τ := $\$T5220$
:= $\$X5220$ b_o
wff 58 : b_o := $\$X5220$
:= $\$A5220$ %0
wff 2485 : $\forall o\,[\lambda c.(=(XOR\,(XOR\,a\,\$X5220)\,c)\,(XOR\,a\,(XOR\,\$X5220\,c)))]_{o,\ldots}$:=
$\$A5220$
<< A5220.r0t.txt
:= $\$T5220$
:= $\$X5220$
:= $\$A5220$

use Proof Template A5220 (Gen): A → ∀ x: A
:= $\$T5220$ o
wff 2 : o_τ := $\$T5220$
:= $\$X5220$ a_o
wff 54 : a_o := $\$X5220$
:= $\$A5220$ %0
wff 2518 : ...
$\ldots\forall o\,[\lambda b.(\forall o\,[\lambda c.(=(XOR\,(XOR\,\$X5220\,b)\,c)\,(XOR\,\$X5220\,(XOR\,b\,c)))])]_{o,\ldots}$:= $\$A5220$
<< A5220.r0t.txt
:= $\$T5220$
:= $\$X5220$
:= $\$A5220$

:= $XorAssociativity$ %0
wff 2551 : $\forall o\,[\lambda a.(\forall o\,[\lambda b.(\forall o\,[\lambda c.(=(XOR\,(XOR\,a\,b)\,c)\,(XOR\,a\,(XOR\,b\,c)))])])]_{o,\ldots}$
:= $XorAssociativity$

2.1.118 Results for File xor_case_f.r0.txt

##
Proof: (F X A) = A; (A X F) = A; (F X A) = (A X F) ; X = XOR
##
##
Source: [Kubota 2017 (doi: 10.4444/100.10)]
##
Copyright (c) 2017 Owl of Minerva Press GmbH. All rights reserved.
Written by Ken Kubota (<mail@kenkubota.de>).
##
This file is part of the publication of the mathematical logic $\mathcal{R}_0$.
For more information, visit: <http://doi.org/10.4444/100.10>
##

<< basics.r0.txt
<< xor_table.r0.txt

.a (case left): (F X A) = A

use Proof Template A5222 (Rule of Cases): [\x.A]T, [\x.A]F → A
:= $L5222 $[\lambda x_o.(=_{ooo}(XOR_{ooo}F_ox_o)x_o)_o]$
wff 1643 : $[\lambda x.(=(XOR\,F\,x)\,x)]_{oo}$:= $L5222
:= $X5222 x_o
wff 16 : x_o := $X5222
:= $T5222 $L5222$_{oo}T_o$
wff 1644 : $L5222 T_o := $T5222
:= $F5222 $L5222$_{oo}F_o$
wff 1645 : $L5222 F_o := $F5222

subcase T: (F X T) = T
§\ $_o$ $T5222
= $T5222 $XorTableFTisT$
%$XorTableFTisT$
$=(XOR\,F\,T)\,T$:= $XorTableFTisT$
$=_{ooo}(XOR_{ooo}F_oT_o)T_o$:= $XorTableFTisT$
§s %1 13 %0
= $T5222 $(=T\,T)$
%$A5230a$
$=(=T\,T)\,T$:= $A5230a$
$=_{ooo}(=_{ooo}T_oT_o)T_o$:= $A5230a$
§s %1 3 %0
= $T5222 T

use Proof Template A5219d (Rule T): A = T → A
:= $A5219d %0
wff 1649 : = $T5222 T_o := $A5219d
<< A5219d.r0t.txt
:= $A5219d
%0
$L5222 T := $T5222
$L5222$_{oo}T_o$:= $T5222

subcase F: (F X F) = F
§\ $_o$ $F5222
= $F5222 $XorTableFFisF$
%$XorTableFFisF$
$=(XOR\,F\,F)\,F$:= $XorTableFFisF$
$=_{ooo}(XOR_{ooo}F_oF_o)F_o$:= $XorTableFFisF$
§s %1 13 %0
= $F5222 $(=F\,F)$
%$A5230d$
$=(=F\,F)\,T$:= $A5230d$
$=_{ooo}(=_{ooo}F_oF_o)T_o$:= $A5230d$
§s %1 3 %0

\# $=\$F5222\,T$

\## use Proof Template A5219d (Rule T): A = T → A
:= $\$A5219d$ %0
\# wff 1667 : $=\$F5222\,T_o$:= $\$A5219d$
<< A5219d.r0t.txt
:= $\$A5219d$
%0
\# $\$L5222\,F$:= $\$F5222$
\# $\$L5222_{oo}F_o$:= $\$F5222$

<< A5222.r0t.txt
:= $\$L5222$
:= $\$X5222$
:= $\$T5222$
:= $\$F5222$

:= $XorCaseFLeft$ %0
\# wff 1642 : $=(XOR\,F\,x)\,x_{o,\dots}$:= $XorCaseFLeft$

\## .b (case right): (A X F) = A

\## use Proof Template A5222 (Rule of Cases): [\x.A]T, [\x.A]F → A
:= $\$L5222$ $[\lambda x_o.(=_{ooo}(XOR_{ooo}x_oF_o)x_o)_o]$
\# wff 1719 : $[\lambda x.(=(XOR\,x\,F)\,x)]_{oo}$:= $\$L5222$
:= $\$X5222$ x_o
\# wff 16 : x_o := $\$X5222$
:= $\$T5222$ $\$L5222_{oo}T_o$
\# wff 1720 : $\$L5222\,T_o$:= $\$T5222$
:= $\$F5222$ $\$L5222_{oo}F_o$
\# wff 1721 : $\$L5222\,F_o$:= $\$F5222$

\## subcase T: (T X F) = T
§\ $_o$ $\$T5222$
\# $=\$T5222\,XorTableTFisT$
%$XorTableTFisT$
\# $=(XOR\,T\,F)\,T$:= $XorTableTFisT$
\# $=_{ooo}(XOR_{ooo}T_oF_o)T_o$:= $XorTableTFisT$
§s %1 13 %0
\# $=\$T5222\,(=T\,T)$
%$A5230a$
\# $=(=T\,T)\,T$:= $A5230a$
\# $=_{ooo}(=_{ooo}T_oT_o)T_o$:= $A5230a$
§s %1 3 %0
\# $=\$T5222\,T$

\## use Proof Template A5219d (Rule T): A = T → A
:= $\$A5219d$ %0
\# wff 1725 : $=\$T5222\,T_o$:= $\$A5219d$

```
<< A5219d.r0t.txt
:=  $A5219d
%0
#                   $L5222 T        :=  $T5222
#                   $L5222_ooT_o        :=  $T5222

## subcase F:  (F X F) = F
§\ _o $F5222
#                   = $F5222 XorTableFFisF
%XorTableFFisF
#                   = (XOR F F) F       :=  XorTableFFisF
#                   =_ooo(XOR_oooF_oF_o)F_o       :=  XorTableFFisF
§s  %1  13  %0
#                   = $F5222 (= F F)
%A5230d
#                   = (= F F) T      :=  A5230d
#                   =_ooo(=_oooF_oF_o)T_o        :=  A5230d
§s  %1  3  %0
#                   = $F5222 T

## use Proof Template A5219d (Rule T):  A = T  →  A
:=  $A5219d  %0
# wff      1743  :        = $F5222 T_o       :=  $A5219d
<< A5219d.r0t.txt
:=  $A5219d
%0
#                   $L5222 F        :=  $F5222
#                   $L5222_ooF_o        :=  $F5222

<< A5222.r0t.txt
:=  $L5222
:=  $X5222
:=  $T5222
:=  $F5222

:=  XorCaseFRight  %0
# wff      1718  :        = (XOR x F) x_o,...        :=  XorCaseFRight

## .c:  (F X A) = (A X F)

%XorCaseFRight
#                   = (XOR x F) x       :=  XorCaseFRight
#                   =_ooo(XOR_ooox_oF_o)x_o        :=  XorCaseFRight

## use Proof Template A5201b (Swap):  A = B  →  B = A
<< A5201b.r0t.txt
%0
#                   = x (XOR x F)
#                   =_ooox_o(XOR_ooox_oF_o)
```

$\%XorCaseFLeft$
\# $=(XOR\,F\,x)\,x$:= $XorCaseFLeft$
\# $=_{ooo}(XOR_{ooo}F_o x_o)x_o$:= $XorCaseFLeft$
§s %0 3 %1
\# $=(XOR\,F\,x)\,(XOR\,x\,F)$

:= $XorCaseFLeftRight$ %0
\# wff 1793 : $=(XOR\,F\,x)\,(XOR\,x\,F)_o$:= $XorCaseFLeftRight$

2.1.119 Results for File xor_case_t.r0.txt

\#\#
\#\# Proof: (T X A) = ~A; (A X T) = ~A; (T X A) = (A X T) ; X = XOR
\#\#
\#\#
\#\# Source: [Kubota 2017 (doi: 10.4444/100.10)]
\#\#

\#\# Written by Ken Kubota (<mail@kenkubota.de>).
\#\#
\#\# This file is part of the publication of the mathematical logic $\mathcal{R}_0$.
\#\# For more information, visit: <http://doi.org/10.4444/100.10>
\#\#

<< basics.r0.txt
<< xor_table.r0.txt
\#\# Skipping file A5231.r0.txt (already included)

\#\# .a (case left): (T X A) = ~A

\#\# use Proof Template A5222 (Rule of Cases): [\x.A]T, [\x.A]F → A
:= $\$L5222$ $[\lambda x_o.(=_{ooo}(XOR_{ooo}T_o x_o)(\sim_{oo}x_o))_o]$
\# wff 1644 : $[\lambda x.(=(XOR\,T\,x)\,(\sim x))]_{oo}$:= $\$L5222$
:= $\$X5222$ x_o
\# wff 16 : x_o := $\$X5222$
:= $\$T5222$ $\$L5222_{oo}T_o$
\# wff 1645 : $\$L5222\,T_o$:= $\$T5222$
:= $\$F5222$ $\$L5222_{oo}F_o$
\# wff 1646 : $\$L5222\,F_o$:= $\$F5222$

\#\# subcase T: (T X T) = ~T
§\ $_o$ $\$T5222$
\# $=\$T5222\,(=(XOR\,T\,T)\,(\sim T))$
$\%A5231a$
\# $=(\sim T)\,F$:= $A5231a$
\# $=_{ooo}(\sim_{oo}T_o)F_o$:= $A5231a$
§s %1 7 %0

```
#                 = $T5222 XorTableTTisF
%XorTableTTisF
#                 = (XORTT) F       :=  XorTableTTisF
#                 =ooo(XORoooToTo)Fo       :=  XorTableTTisF
§s %1 13 %0
#                 = $T5222 (= F F)
%A5230d
#                 = (= F F) T       :=  A5230d
#                 =ooo(=oooFoFo)To       :=  A5230d
§s %1 3 %0
#                 = $T5222 T

## use Proof Template A5219d (Rule T):  A = T  →  A
:=  $A5219d %0
# wff    1651  :       = $T5222 To      :=  $A5219d
<< A5219d.r0t.txt
:=  $A5219d
%0
#                 $L5222 T      :=  $T5222
#                 $L5222ooTo      :=  $T5222

## subcase F:  (T X F) = ~F
§\ o $F5222
#                 = $F5222 (= (XORT F) (~ F))
%A5231b
#                 = (~ F) T      :=  A5231b
#                 =ooo(~ooFo)To       :=  A5231b
§s %1 7 %0
#                 = $F5222 XorTableTFisT
%XorTableTFisT
#                 = (XORT F) T       :=  XorTableTFisT
#                 =ooo(XORoooToFo)To       :=  XorTableTFisT
§s %1 13 %0
#                 = $F5222 (= T T)
%A5230a
#                 = (= T T) T       :=  A5230a
#                 =ooo(=oooToTo)To       :=  A5230a
§s %1 3 %0
#                 = $F5222 T

## use Proof Template A5219d (Rule T):  A = T  →  A
:=  $A5219d %0
# wff    1670  :       = $F5222 To      :=  $A5219d
<< A5219d.r0t.txt
:=  $A5219d
%0
#                 $L5222 F      :=  $F5222
#                 $L5222ooFo      :=  $F5222
```

<< A5222.r0t.txt
:= $\$L5222$
:= $\$X5222$
:= $\$T5222$
:= $\$F5222$

:= $XorCaseTLeft$ %0
wff 1643 : $=(XORT\,x)\,(\sim x)_{o,\ldots}$:= $XorCaseTLeft$

.b (case right): (A X T) = ~A

use Proof Template A5222 (Rule of Cases): [\x.A]T, [\x.A]F → A
:= $\$L5222$ $[\lambda x_o.(=_{ooo}(XOR_{ooo}x_oT_o)(\sim_{oo}x_o))_o]$
wff 1722 : $[\lambda x.(=(XOR\,x\,T)\,(\sim x))]_{oo}$:= $\$L5222$
:= $\$X5222$ x_o
wff 16 : x_o := $\$X5222$
:= $\$T5222$ $\$L5222_{oo}T_o$
wff 1723 : $\$L5222\,T_o$:= $\$T5222$
:= $\$F5222$ $\$L5222_{oo}F_o$
wff 1724 : $\$L5222\,F_o$:= $\$F5222$

subcase T: (T X T) = ~T
§\ $_o$ $\$T5222$
$=\$T5222\,(=(XORT\,T)\,(\sim T))$
%$A5231a$
$=(\sim T)\,F$:= $A5231a$
$=_{ooo}(\sim_{oo}T_o)F_o$:= $A5231a$
§s %1 7 %0
$=\$T5222\,XorTableTTisF$
%$XorTableTTisF$
$=(XORT\,T)\,F$:= $XorTableTTisF$
$=_{ooo}(XOR_{ooo}T_oT_o)F_o$:= $XorTableTTisF$
§s %1 13 %0
$=\$T5222\,(=F\,F)$
%$A5230d$
$=(=F\,F)\,T$:= $A5230d$
$=_{ooo}(=_{ooo}F_oF_o)T_o$:= $A5230d$
§s %1 3 %0
$=\$T5222\,T$

use Proof Template A5219d (Rule T): A = T → A
:= $\$A5219d$ %0
wff 1729 : $=\$T5222\,T_o$:= $\$A5219d$
<< A5219d.r0t.txt
:= $\$A5219d$
%0
$\$L5222\,T$:= $\$T5222$
$\$L5222_{oo}T_o$:= $\$T5222$

subcase F: (F X T) = ~F
§\ $_o$ $\$F5222$
\# $= \$F5222\,(=(XOR\,F\,T)\,(\sim F))$
%$A5231b$
\# $=(\sim F)\,T$:= $A5231b$
\# $=_{ooo}(\sim_{oo}F_o)T_o$:= $A5231b$
§s %1 7 %0
\# $= \$F5222\,XorTableFTisT$
%$XorTableFTisT$
\# $=(XOR\,F\,T)\,T$:= $XorTableFTisT$
\# $=_{ooo}(XOR_{ooo}F_oT_o)T_o$:= $XorTableFTisT$
§s %1 13 %0
\# $= \$F5222\,(=T\,T)$
%$A5230a$
\# $=(=T\,T)\,T$:= $A5230a$
\# $=_{ooo}(=_{ooo}T_oT_o)T_o$:= $A5230a$
§s %1 3 %0
\# $= \$F5222\,T$

use Proof Template A5219d (Rule T): A = T → A
:= $\$A5219d$ %0
\# wff 1748 : $= \$F5222\,T_o$:= $\$A5219d$
<< A5219d.r0t.txt
:= $\$A5219d$
%0
\# $\$L5222\,F$:= $\$F5222$
\# $\$L5222_{oo}F_o$:= $\$F5222$

<< A5222.r0t.txt
:= $\$L5222$
:= $\$X5222$
:= $\$T5222$
:= $\$F5222$

:= $XorCaseTRight$ %0
\# wff 1721 : $=(XOR\,x\,T)\,(\sim x)_{o,\ldots}$:= $XorCaseTRight$

.c: (T X A) = (A X T)

%$XorCaseTRight$
\# $=(XOR\,x\,T)\,(\sim x)$:= $XorCaseTRight$
\# $=_{ooo}(XOR_{ooo}x_oT_o)(\sim_{oo}x_o)$:= $XorCaseTRight$

use Proof Template A5201b (Swap): A = B → B = A
<< A5201b.r0t.txt
%0
\# $=(\sim x)\,(XOR\,x\,T)$
\# $=_{ooo}(\sim_{oo}x_o)(XOR_{ooo}x_oT_o)$

%$XorCaseTLeft$
$= (XOR\,T\,x)\,(\sim x)$:= $XorCaseTLeft$
$=_{ooo}(XOR_{ooo}T_o x_o)(\sim_{oo} x_o)$:= $XorCaseTLeft$
§s %0 3 %1
$= (XOR\,T\,x)\,(XOR\,x\,T)$

:= $XorCaseTLeftRight$ %0
wff 1799 : $= (XOR\,T\,x)\,(XOR\,x\,T)_o$:= $XorCaseTLeftRight$

2.1.120 Results for File xor_group.r0.txt

##
Group Property of Exclusive Disjunction (Exclusive OR, XOR)
##
##
Source: [Kubota 2017 (doi: 10.4444/100.10)]
##

Written by Ken Kubota (<mail@kenkubota.de>).
##
This file is part of the publication of the mathematical logic $\mathcal{R}_0$.
For more information, visit: <http://doi.org/10.4444/100.10>
##

<< A5229.r0.txt
<< group.r0.txt
<< xor_associativity.r0.txt
<< xor_identity_element.r0.txt
<< xor_inverse_element.r0.txt

shorthands
:= $\$Xab$ $XOR_{ooo}a_o b_o$
wff 1707 : $XOR\,a\,b_o$:= $\$Xab$
:= $\$Xbc$ $XOR_{ooo}b_o c_o$
wff 1712 : $XOR\,b\,c_o$:= $\$Xbc$
:= $\$GrpAsc$ $\forall_{o(o\backslash 3)_\tau o_\tau} \ldots$
$\ldots[\lambda a_o.(\forall_{o(o\backslash 3)_\tau o_\tau}[\lambda b_o.(\forall_{o(o\backslash 3)_\tau o_\tau}[\lambda c_o.(=_{ooo}(l_{ooo}(l_{ooo}a_o b_o)c_o)(l_{ooo}a_o(l_{ooo}b_o c_o)))_o])_o])_o]$
wff 6925 : $\forall o\,[\lambda a.(\forall o\,[\lambda b.(\forall o\,[\lambda c.(= (l\,(l\,a\,b)\,c)\,(l\,a\,(l\,b\,c)))])])]_o$:= $\$GrpAsc$
:= $\$GrpIdy$ $\forall_{o(o\backslash 3)_\tau o_\tau}[\lambda a_o.(\wedge_{ooo}(=_{ooo}(l_{ooo}a_o e_o)a_o)(=_{ooo}(l_{ooo}e_o a_o)a_o))_o]$
wff 6936 : $\forall o\,[\lambda a.(\wedge\,(= (l\,a\,e)\,a)\,(= (l\,e\,a)\,a))]_o$:= $\$GrpIdy$
:= $\$GrpInv$ $\forall_{o(o\backslash 3)_\tau o_\tau}[\lambda a_o.(\exists_{o(o\backslash 3)_\tau o_\tau}[\lambda b_o.(\wedge_{ooo}(=_{ooo}(l_{ooo}a_o b_o)e_o)(=_{ooo}(l_{ooo}b_o a_o)e_o))_o])_o]$
wff 6939 : $\forall o\,[\lambda a.(\exists o\,[\lambda b.(\wedge\,(= (l\,a\,b)\,e)\,(= (l\,b\,a)\,e))])]_o$:= $\$GrpInv$
:= $\$XAsc$ $\forall_{o(o\backslash 3)_\tau o_\tau} \ldots$
$\ldots[\lambda a_o.(\forall_{o(o\backslash 3)_\tau o_\tau}[\lambda b_o.(\forall_{o(o\backslash 3)_\tau o_\tau}[\lambda c_o.(=_{ooo}(XOR_{ooo}\$Xab_o c_o)(XOR_{ooo}a_o\$Xbc_o))_o])_o])_o]$
wff 2616 : $\forall o\,[\lambda a.(\forall o\,[\lambda b.(\forall o\,[\lambda c.(= (XOR\,\$Xab\,c)\,(XOR\,a\,\$Xbc))])])]_{o,\ldots}$:=
$\$XAsc$ $XorAssociativity$
:= $\$XIdy$ $\forall_{o(o\backslash 3)_\tau o_\tau}[\lambda a_o.(\wedge_{ooo}(=_{ooo}(XOR_{ooo}a_o e_o)a_o)(=_{ooo}(XOR_{ooo}e_o a_o)a_o))_o]$
wff 6950 : $\forall o\,[\lambda a.(\wedge\,(= (XOR\,a\,e)\,a)\,(= (XOR\,e\,a)\,a))]_o$:= $\$XIdy$

:= $\$XInv\ \forall_{o(o\backslash 3)\tau}o_\tau[\lambda a_o.(\exists_{o(o\backslash 3)\tau}o_\tau[\lambda b_o.(\wedge_{ooo}(=_{ooo}\$Xab_oe_o)(=_{ooo}(XOR_{ooo}b_oa_o)e_o))_o])_o]$
wff 6953 : $\forall o\,[\lambda a.(\exists o\,[\lambda b.(\wedge\,(=\$Xab\,e)\,(=(XOR\,b\,a)\,e))])]_o$:= $\$XInv$
:= $\$XFIdy\ \forall_{o(o\backslash 3)\tau}o_\tau[\lambda a_o.(\wedge_{ooo}(=_{ooo}(XOR_{ooo}a_oF_o)a_o)(=_{ooo}(XOR_{ooo}F_oa_o)a_o))_o]$
wff 2850 : $\forall o\,[\lambda a.(\wedge\,(=(XOR\,a\,F)\,a)\,(=(XOR\,F\,a)\,a))]_o$:= $\$XFIdy$ $XorIdentityElement$
:= $\$XFInv\ \forall_{o(o\backslash 3)\tau}o_\tau[\lambda a_o.(\exists_{o(o\backslash 3)\tau}o_\tau[\lambda b_o.(\wedge_{ooo}(=_{ooo}\$Xab_oF_o)(=_{ooo}(XOR_{ooo}b_oa_o)F_o))_o])_o]$
wff 6905 : $\forall o\,[\lambda a.(\exists o\,[\lambda b.(\wedge\,(=\$Xab\,F)\,(=(XOR\,b\,a)\,F))])]_{o,\ldots}$:= $\$XFInv$ $XorInverseElement$

.1

§= $Grp_{o(\backslash 4\backslash 4\backslash 3)\tau}o_\tau XOR_{ooo}$
$=(Grp\,o\,XOR)\,(Grp\,o\,XOR)$
§\ $Grp_{o(\backslash 4\backslash 4\backslash 3)\tau}o_\tau$
$=(Grp\,o)\,[\lambda l.(\wedge\,\$GrpAsc\,(\exists o\,[\lambda e.(\wedge\,\$GrpIdy\,\$GrpInv)]))]$
§s %1 6 %0
$=(Grp\,o\,XOR)\,([\lambda l.(\wedge\,\$GrpAsc\,(\exists o\,[\lambda e.(\wedge\,\$GrpIdy\,\$GrpInv)]))]\,XOR)$
§\ $[\lambda l_{ooo}.(\wedge_{ooo}\$GrpAsc_o(\exists_{o(o\backslash 3)\tau}o_\tau[\lambda e_o.(\wedge_{ooo}\$GrpIdy_o\$GrpInv_o)_o]))_o]XOR_{ooo}$
$=([\lambda l.(\wedge\,\$GrpAsc\,(\exists o\,[\lambda e.(\wedge\,\$GrpIdy\,\$GrpInv)]))]\,XOR)\ldots$
$\ldots(\wedge\,\$XAsc\,(\exists o\,[\lambda e.(\wedge\,\$XIdy\,\$XInv)]))$
§s %1 3 %0
$=(Grp\,o\,XOR)\,(\wedge\,\$XAsc\,(\exists o\,[\lambda e.(\wedge\,\$XIdy\,\$XInv)]))$

:= $\$T1$ %0
wff 6977 : $=(Grp\,o\,XOR)\,(\wedge\,\$XAsc\,(\exists o\,[\lambda e.(\wedge\,\$XIdy\,\$XInv)]))_o$:= $\$T1$

.2

§= $[\lambda e_o.(\wedge_{ooo}\$XIdy_o\$XInv_o)_o]F_o$
$=([\lambda e.(\wedge\,\$XIdy\,\$XInv)]\,F)\,([\lambda e.(\wedge\,\$XIdy\,\$XInv)]\,F)$
§\ $[\lambda e_o.(\wedge_{ooo}\$XIdy_o\$XInv_o)_o]F_o$
$=([\lambda e.(\wedge\,\$XIdy\,\$XInv)]\,F)\,(\wedge\,\$XFIdy\,\$XFInv)$
§s %1 3 %0
$=([\lambda e.(\wedge\,\$XIdy\,\$XInv)]\,F)\,(\wedge\,\$XFIdy\,\$XFInv)$

:= $\$T2$ %0
wff 6983 : $=([\lambda e.(\wedge\,\$XIdy\,\$XInv)]\,F)\,(\wedge\,\$XFIdy\,\$XFInv)_o$:= $\$T2$

.3

%$\$XFIdy$
$\forall o\,[\lambda a.(\wedge\,(=(XOR\,a\,F)\,a)\,(=(XOR\,F\,a)\,a))]$:= $\$XFIdy$ $XorIdentityElement$
$\forall_{o(o\backslash 3)\tau}o_\tau[\lambda a_o.(\wedge_{ooo}(=_{ooo}(XOR_{ooo}a_oF_o)a_o)(=_{ooo}(XOR_{ooo}F_oa_o)a_o))_o]$:= $\$XFIdy$ $XorIdentityElement$
use Proof Template A5219b (Rule T): A → A = T
:= $\$A5219b$ %0
wff 2850 : $\forall o\,[\lambda a.(\wedge\,(=(XOR\,a\,F)\,a)\,(=(XOR\,F\,a)\,a))]_o$:= $\$A5219b$ $\$XFIdy$ $XorIdentityElement$

<< A5219b.r0t.txt
:= $\$A5219b$

:= $\$E$ %0
wff 7000 : $=\$XFIdy\,T_o$:= $\$E$

%$\$T2$
$=([\lambda e.(\wedge\,\$XIdy\,\$XInv)]\,F)\,(\wedge\,\$XFIdy\,\$XFInv)$:= $\$T2$
$=_{o\omega\omega}([\lambda e_o.(\wedge_{ooo}\$XIdy_o\$XInv_o)_o]F_o)(\wedge_{ooo}\$XFIdy_o\$XFInv_o)$:= $\$T2$
:= $\$T2$
%$\$E$
$=\$XFIdy\,T$:= $\$E$
$=_{ooo}\$XFIdy_oT_o$:= $\$E$
:= $\$E$
§s %1 13 %0
$=([\lambda e.(\wedge\,\$XIdy\,\$XInv)]\,F)\,(\wedge\,T\,\$XFInv)$

:= $\$T3$ %0
wff 7002 : $=([\lambda e.(\wedge\,\$XIdy\,\$XInv)]\,F)\,(\wedge\,T\,\$XFInv)_o$:= $\$T3$

.4

%$\$XFInv$
$\forall\,o\,[\lambda a.(\exists\,o\,[\lambda b.(\wedge\,(=\$Xab\,F)\,(=(XOR\,b\,a)\,F))])]$:= $\$XFInv$
$XorInverseElement$
$\forall_{o(o\backslash 3)\tau}o_\tau[\lambda a_o.(\exists_{o(o\backslash 3)\tau}o_\tau[\lambda b_o.(\wedge_{ooo}(=_{ooo}\$Xab_oF_o)(=_{ooo}(XOR_{ooo}b_oa_o)F_o))_o])_o]$
:= $\$XFInv$ $XorInverseElement$
use Proof Template A5219b (Rule T): A → A = T
:= $\$A5219b$ %0
wff 6905 : $\forall\,o\,[\lambda a.(\exists\,o\,[\lambda b.(\wedge\,(=\$Xab\,F)\,(=(XOR\,b\,a)\,F))])]_{o,\ldots}$:= $\$A5219b$
$\$XFInv$ $XorInverseElement$
<< A5219b.r0t.txt
:= $\$A5219b$

:= $\$E$ %0
wff 7019 : $=\$XFInv\,T_o$:= $\$E$

%$\$T3$
$=([\lambda e.(\wedge\,\$XIdy\,\$XInv)]\,F)\,(\wedge\,T\,\$XFInv)$:= $\$T3$
$=_{o\omega\omega}([\lambda e_o.(\wedge_{ooo}\$XIdy_o\$XInv_o)_o]F_o)(\wedge_{ooo}T_o\$XFInv_o)$:= $\$T3$
:= $\$T3$
%$\$E$
$=\$XFInv\,T$:= $\$E$
$=_{ooo}\$XFInv_oT_o$:= $\$E$
:= $\$E$
§s %1 7 %0
$=([\lambda e.(\wedge\,\$XIdy\,\$XInv)]\,F)\,A5212$

.5

```
%A5211
#                $= A5212\,T$        :=  $A5211$   $A5229a$
#                $=_{ooo} A5212_o T_o$       :=  $A5211$   $A5229a$
§s  %1  3  %0
#                $= ([\lambda e.(\wedge \$XIdy \$XInv)]\, F)\, T$
## use Proof Template A5201b (Swap):   A = B   →   B = A
<< A5201b.r0t.txt
%0
#                $= T\, ([\lambda e.(\wedge \$XIdy \$XInv)]\, F)$
#                $=_{o\omega\omega} T_\omega ([\lambda e_o.(\wedge_{ooo} \$XIdy_o \$XInv_o)_o] F_o)$
%T
#                $===$        :=  $A5200t$   $T$
#                $=_{o\omega\omega} =_\omega =_\omega$       :=  $A5200t$   $T$
§s  %0  1  %1
#                $[\lambda e.(\wedge \$XIdy \$XInv)]\, F$

## .6

## use Proof Template K8031 (∃ Gen):   ([\x.B]A)   →   ∃ x: B
:=  $T8031  o
# wff      2  :        $o_\tau$        :=  $T8031
:=  $B8031  %0/2
# wff      6973  :        $[\lambda e.(\wedge \$XIdy \$XInv)]_{oo}$        :=  $B8031
:=  $A8031  %0/3
# wff      20  :        $= [\lambda x.T]\, [\lambda x.x]_{o,\ldots}$        :=  $A8031   F
:=  $P8031  $B8031_{oo}F_o$
# wff      6978  :        $\$B8031\, F_{o,\ldots}$        :=  $P8031
<< K8031.r0t.txt
:=  $T8031
:=  $B8031
:=  $A8031

:=  $T6  %0
# wff      6974  :        $\exists\, o\, [\lambda e.(\wedge \$XIdy \$XInv)]_{o,\ldots}$        :=  $T6

## .7

%$T1
#                $= (Grp\, o\, XOR)\, (\wedge \$XAsc \$T6)$        :=  $T1
#                $=_{o\omega\omega} (Grp_{o(\backslash 4\backslash 4\backslash 3)_\tau} o_\tau XOR_{ooo})(\wedge_{ooo} \$XAsc_o \$T6_o)$        :=  $T1
%$T6
#                $\exists\, o\, [\lambda e.(\wedge \$XIdy \$XInv)]$        :=  $T6
#                $\exists_{o(o\backslash 3)_\tau} o_\tau [\lambda e_o.(\wedge_{ooo} \$XIdy_o \$XInv_o)_o]$        :=  $T6
:=  $T6

## use Proof Template A5219b (Rule T):   A   →   A = T
:=  $A5219b  %0
# wff      6974  :        $\exists\, o\, [\lambda e.(\wedge \$XIdy \$XInv)]_{o,\ldots}$        :=  $A5219b
```

<< A5219b.r0t.txt
:= $\$A5219b$

:= $\$TMP$ %0
\# wff 7654 : $=(\exists o\,[\lambda e.(\wedge \$XIdy\,\$XInv)])\,T_o$:= $\$TMP$

%$\$T1$
\# $=(Grp\,o\,XOR)\,(\wedge \$XAsc\,(\exists o\,[\lambda e.(\wedge \$XIdy\,\$XInv)]))$:= $\$T1$
\# $=_{o\omega\omega}(Grp_{o(\backslash 4\backslash 4\backslash 3)\tau}o_\tau XOR_{ooo})\ldots$
$\ldots(\wedge_{ooo}\$XAsc_o(\exists_{o(o\backslash 3)\tau}o_\tau[\lambda e_o.(\wedge_{ooo}\$XIdy_o\$XInv_o)_o]))$:= $\$T1$
:= $\$T1$
%$\$TMP$
\# $=(\exists o\,[\lambda e.(\wedge \$XIdy\,\$XInv)])\,T$:= $\$TMP$
\# $=_{ooo}(\exists_{o(o\backslash 3)\tau}o_\tau[\lambda e_o.(\wedge_{ooo}\$XIdy_o\$XInv_o)_o])T_o$:= $\$TMP$
:= $\$TMP$
§s %1 7 %0
\# $=(Grp\,o\,XOR)\,(\wedge \$XAsc T)$

:= $\$TMP$ %0
\# wff 7656 : $=(Grp\,o\,XOR)\,(\wedge \$XAsc T)_o$:= $\$TMP$

%$\$XAsc$
\# $\forall o\,[\lambda a.(\forall o\,[\lambda b.(\forall o\,[\lambda c.(=(XOR\,\$Xab\,c)\,(XOR\,a\,\$Xbc))])])]$:= $\$XAsc$
$XorAssociativity$
\# $\forall_{o(o\backslash 3)\tau}o_\tau\ldots$
$\ldots[\lambda a_o.(\forall_{o(o\backslash 3)\tau}o_\tau[\lambda b_o.(\forall_{o(o\backslash 3)\tau}o_\tau[\lambda c_o.(=_{ooo}(XOR_{ooo}\$Xab_oc_o)(XOR_{ooo}a_o\$Xbc_o))_o])_o])_o]$:=
$\$XAsc$ $XorAssociativity$

\#\# use Proof Template A5219b (Rule T): A → A = T
:= $\$A5219b$ %0
\# wff 2616 : $\forall o\,[\lambda a.(\forall o\,[\lambda b.(\forall o\,[\lambda c.(=(XOR\,\$Xab\,c)\,(XOR\,a\,\$Xbc))])])]_{o,\ldots}$:=
$\$A5219b$ $\$XAsc$ $XorAssociativity$
<< A5219b.r0t.txt
:= $\$A5219b$

%$\$TMP$
\# $=(Grp\,o\,XOR)\,(\wedge \$XAsc T)$:= $\$TMP$
\# $=_{o\omega\omega}(Grp_{o(\backslash 4\backslash 4\backslash 3)\tau}o_\tau XOR_{ooo})(\wedge_{ooo}\$XAsc_oT_o)$:= $\$TMP$
:= $\$TMP$
%1
\# $=\$XAsc T$
\# $=_{ooo}\$XAsc_oT_o$
§s %1 13 %0
\# $=(Grp\,o\,XOR)\,A5212$

%$A5211$
\# $=A5212\,T$:= $A5211$ $A5229a$
\# $=_{ooo}A5212_oT_o$:= $A5211$ $A5229a$
§s %1 3 %0

\# $=(Grp\,o\,XOR)\,T$
\## use Proof Template A5201b (Swap): A = B → B = A
<< A5201b.r0t.txt
%0
\# $=T\,(Grp\,o\,XOR)$
\# $=_{o\omega\omega}T_{\omega}(Grp_{o(\backslash 4\backslash 4\backslash 3)_{\tau}o_{\tau}}o_{\tau}XOR_{ooo})$
%T
\# $===$:= $A5200t$ T
\# $=_{o\omega\omega}=_{\omega}=_{\omega}$:= $A5200t$ T
§s %0 1 %1
\# $Grp\,o\,XOR$

:= $XorGroup$ %0
\# wff 6955 : $Grp\,o\,XOR_{o,\ldots}$:= $XorGroup$

\## demonstrate that XOR now has type Grp_o
§= $Grp_{o(\backslash 4\backslash 4\backslash 3)_{\tau}o_{\tau}}$ XOR
\# $=XOR\,XOR$
%0
\# $=XOR\,XOR$
\# $=_{o\,(Grp_{o(\backslash 4\backslash 4\backslash 3)_{\tau}o_{\tau}})\,(Grp_{o(\backslash 4\backslash 4\backslash 3)_{\tau}o_{\tau}})}XOR_{Grp_{o(\backslash 4\backslash 4\backslash 3)_{\tau}o_{\tau}}}XOR_{Grp_{o(\backslash 4\backslash 4\backslash 3)_{\tau}o_{\tau}}}$

\## demonstrate that Grp_o now has type tau (type “type”)
§= $_{\tau}$ $Grp_{o(\backslash 4\backslash 4\backslash 3)_{\tau}o_{\tau}}$
\# $=(Grp\,o)\,(Grp\,o)$
%0
\# $=(Grp\,o)\,(Grp\,o)$
\# $=_{o\tau\tau}(Grp_{o(\backslash 4\backslash 4\backslash 3)_{\tau}o_{\tau}})(Grp_{o(\backslash 4\backslash 4\backslash 3)_{\tau}o_{\tau}})$

\## undefine local variables
:= $\$Xab$
:= $\$Xbc$
:= $\$GrpAsc$
:= $\$GrpIdy$
:= $\$GrpInv$
:= $\$XAsc$
:= $\$XIdy$
:= $\$XInv$
:= $\$XFIdy$
:= $\$XFInv$

2.1.121 Results for File xor_group_identity_element_unique.r0.txt

\##
\## Uniqueness of the Group Identity Element of the XOR Group
\##
\##
\## Source: [Kubota 2017 (doi: 10.4444/100.10)]
\##

This file is part of the publication of the mathematical logic $\mathcal{R}_0$.
For more information, visit: <http://doi.org/10.4444/100.10>
##

<< A5223.r0.txt
<< group_identity_element_unique.r0.txt
<< xor_group.r0.txt

shorthands
$:= \ \$GIdOXe\ \ GrpIdO_{o\backslash 3(\backslash 4\backslash 4\backslash 3)\tau} o_\tau XOR_{ooo} e_o$
wff 8467 : $GrpIdO\, o\, XOR\, e_o$ $:= \ \$GIdOXe$
$:= \ \$GIdOXf\ \ GrpIdO_{o\backslash 3(\backslash 4\backslash 4\backslash 3)\tau} o_\tau XOR_{ooo} f_o$
wff 8469 : $GrpIdO\, o\, XOR\, f_o$ $:= \ \$GIdOXf$

.1

$\%GrpIdElUniq$
$\supset (Grp\, g\, l)\, (\supset (GrpIdO\, g\, l\, e)\, (\supset (GrpIdO\, g\, l\, f)\, (= e\, f)))$ $:=$ $GrpIdElUniq$
$\supset_{ooo}(Grp_{o(\backslash 4\backslash 4\backslash 3)\tau} g_\tau l_{ggg}) \ldots$
$\ldots (\supset_{ooo}(GrpIdO_{o\backslash 3(\backslash 4\backslash 4\backslash 3)\tau} g_\tau l_{ggg} e_g)(\supset_{ooo}(GrpIdO_{o\backslash 3(\backslash 4\backslash 4\backslash 3)\tau} g_\tau l_{ggg} f_g)(=_{ogg} e_g f_g)))$ $:=$
$GrpIdElUniq$

use Proof Template A5221 (Sub): B → B [x/A]
$:= \ \$B5221\ \%0$
wff 4852 : $\supset (Grp\, g\, l)\, (\supset (GrpIdO\, g\, l\, e)\, (\supset (GrpIdO\, g\, l\, f)\, (= e\, f)))_{o,\ldots}$ $:= \ \$B5221$
$GrpIdElUniq$
$:= \ \$T5221\ \ \tau$
wff 0 : τ_τ $:= \ \$T5221$
$:= \ \$X5221\ \ g_\tau$
wff 1411 : g_τ $:= \ \$X5221$
$:= \ \$A5221\ \ o$
wff 2 : o_τ $:= \ \$A5221$
<< A5221.r0t.txt
$:= \ \$B5221$
$:= \ \$T5221$
$:= \ \$X5221$
$:= \ \$A5221$
%0
$\supset (Grp\, o\, l)\, (\supset (GrpIdO\, o\, l\, e)\, (\supset (GrpIdO\, o\, l\, f)\, (= e\, f)))$
$\supset_{ooo}(Grp_{o(\backslash 4\backslash 4\backslash 3)\tau} o_\tau l_{ooo}) \ldots$
$\ldots (\supset_{ooo}(GrpIdO_{o\backslash 3(\backslash 4\backslash 4\backslash 3)\tau} o_\tau l_{ooo} e_o)(\supset_{ooo}(GrpIdO_{o\backslash 3(\backslash 4\backslash 4\backslash 3)\tau} o_\tau l_{ooo} f_o)(=_{ooo} e_o f_o)))$

use Proof Template A5221 (Sub): B → B [x/A]
$:= \ \$B5221\ \%0$
wff 8518 : $\supset (Grp\, o\, l)\, (\supset (GrpIdO\, o\, l\, e)\, (\supset (GrpIdO\, o\, l\, f)\, (= e\, f)))_{o,\ldots}$ $:= \ \$B5221$

:= $\$T5221$ ooo
\# wff 35 : ooo_τ := $\$T5221$
:= $\$X5221$ $l_{\$T5221}$
\# wff 6058 : $l_{\$T5221}$:= $\$X5221$
:= $\$A5221$ $[\lambda x_o.[\lambda y_o.(\sim_{oo}(=_{\$T5221}x_o y_o))_o]_{(oo)}]$
\# wff 135 : $[\lambda x.[\lambda y.(\sim(=x\,y))]]_{\$T5221,\ldots}$:= $\$A5221$ XOR
<< A5221.r0t.txt
:= $\$B5221$
:= $\$T5221$
:= $\$X5221$
:= $\$A5221$

:= $\$TMP$ %0
\# wff 8567 : $\supset XorGroup\,(\supset \$GIdOXe\,(\supset \$GIdOXf\,(=e\,f)))_{o,\ldots}$:= $\$TMP$

\#\# .2

$\%XorGroup$
\# $Grp\,o\,XOR$:= $XorGroup$
\# $Grp_{o(\backslash 4\backslash 4\backslash 3)\tau}o_\tau XOR_{ooo}$:= $XorGroup$

\#\# use Proof Template A5219b (Rule T): A → A = T
:= $\$A5219b$ %0
\# wff 7736 : $Grp\,o\,XOR_{o,\ldots}$:= $\$A5219b$ $XorGroup$
<< A5219b.r0t.txt
:= $\$A5219b$
%0
\# $=XorGroup\,T$
\# $=_{ooo}XorGroup_o T_o$

$\%\$TMP$
\# $\supset XorGroup\,(\supset \$GIdOXe\,(\supset \$GIdOXf\,(=e\,f)))$:= $\$TMP$
\# $\supset_{ooo}XorGroup_o(\supset_{ooo}\$GIdOXe_o(\supset_{ooo}\$GIdOXf_o(=_{ooo}e_o f_o)))$:= $\$TMP$
:= $\$TMP$
%1
\# $=XorGroup\,T$
\# $=_{ooo}XorGroup_o T_o$
§s %1 5 %0
\# $\supset T\,(\supset \$GIdOXe\,(\supset \$GIdOXf\,(=e\,f)))$

:= $\$TMP$ %0
\# wff 8587 : $\supset T\,(\supset \$GIdOXe\,(\supset \$GIdOXf\,(=e\,f)))_o$:= $\$TMP$

\#\# use Proof Template A5221 (Sub): B → B [x/A]
:= $\$B5221$ $=_{ooo}(\supset_{ooo}T_o y_o)y_o$
\# wff 823 : $=(\supset T\,y)\,y_{o,\ldots}$:= $\$B5221$ $A5223$
:= $\$T5221$ o
\# wff 2 : o_τ := $\$T5221$
:= $\$X5221$ y_o

\# wff 34 : y_o := $\$X5221$
:= $\$A5221$ %0/3
\# wff 8566 : $\supset \$GIdOXe\,(\supset \$GIdOXf\,(=e\,f))_o$:= $\$A5221$
<< A5221.r0t.txt
:= $\$B5221$
:= $\$T5221$
:= $\$X5221$
:= $\$A5221$
%0
\# $= \$TMP\,(\supset \$GIdOXe\,(\supset \$GIdOXf\,(=e\,f)))$
\# $=_{ooo}\$TMP_o(\supset_{ooo}\$GIdOXe_o(\supset_{ooo}\$GIdOXf_o(=_{ooo}e_of_o)))$

%$\$TMP$
\# $\supset T\,(\supset \$GIdOXe\,(\supset \$GIdOXf\,(=e\,f)))$:= $\$TMP$
\# $\supset_{ooo}T_o(\supset_{ooo}\$GIdOXe_o(\supset_{ooo}\$GIdOXf_o(=_{ooo}e_of_o)))$:= $\$TMP$
:= $\$TMP$
%1
\# $= (\supset T\,(\supset \$GIdOXe\,(\supset \$GIdOXf\,(=e\,f))))\,(\supset \$GIdOXe\,(\supset \$GIdOXf\,(=e\,f)))$
\# $=_{ooo}(\supset_{ooo}T_o(\supset_{ooo}\$GIdOXe_o(\supset_{ooo}\$GIdOXf_o(=_{ooo}e_of_o))))\ldots$
$\ldots(\supset_{ooo}\$GIdOXe_o(\supset_{ooo}\$GIdOXf_o(=_{ooo}e_of_o)))$
§s %1 1 %0
\# $\supset \$GIdOXe\,(\supset \$GIdOXf\,(=e\,f))$

\#\# use Proof Template K8026 (Deduction Theorem Reversed): H ⊃ (I ⊃ A) → (H ∧ I) ⊃ A
<< K8026.r0t.txt
%0
\# $\supset (\wedge \$GIdOXe\,\$GIdOXf)\,(=e\,f)$
\# $\supset_{ooo}(\wedge_{ooo}\$GIdOXe_o\$GIdOXf_o)(=_{ooo}e_of_o)$

:= $XorGrpIdElUniq$ %0
\# wff 8696 : $\supset (\wedge \$GIdOXe\,\$GIdOXf)\,(=e\,f)_{o,\ldots}$:= $XorGrpIdElUniq$

\#\# undefine local variables
:= $\$GIdOXe$
:= $\$GIdOXf$

2.1.122 Results for File xor_identity_element.r0.txt

\#\#
\#\# Neutral Element of Exclusive Disjunction (Exclusive OR, XOR)
\#\#
\#\#
\#\# Source: [Kubota 2017 (doi: 10.4444/100.10)]
\#\#

\#\# Written by Ken Kubota (<mail@kenkubota.de>).
\#\#
\#\# This file is part of the publication of the mathematical logic $\mathcal{R}_0$.

For more information, visit: <http://doi.org/10.4444/100.10>
##

<< basics.r0.txt
<< xor_case_f.r0.txt

%$XorCaseFRight$
$\quad = (XOR\, x\, F)\, x \quad := \quad XorCaseFRight$
$\quad =_{ooo}(XOR_{ooo} x_o F_o) x_o \quad := \quad XorCaseFRight$
%$XorCaseFLeft$
$\quad = (XOR\, F\, x)\, x \quad := \quad XorCaseFLeft$
$\quad =_{ooo}(XOR_{ooo} F_o x_o) x_o \quad := \quad XorCaseFLeft$

use Proof Template K8020: A, B $\rightarrow$ A $\wedge$ B
:= $\$A8020$ %1
wff 1718 : $= (XOR\, x\, F)\, x_{o,\ldots} \quad := \quad \$A8020 \quad XorCaseFRight$
:= $\$B8020$ %0
wff 1642 : $= (XOR\, F\, x)\, x_{o,\ldots} \quad := \quad \$B8020 \quad XorCaseFLeft$
<< K8020.r0t.txt
:= $\$A8020$
:= $\$B8020$
%0
$\quad \wedge\, XorCaseFRight\, XorCaseFLeft$
$\quad \wedge_{ooo} XorCaseFRight_o XorCaseFLeft_o$

use Proof Template A5220 (Gen): A $\rightarrow$ $\forall$ x: A
:= $\$T5220\ o$
wff 2 : $o_\tau \quad := \quad \$T5220$
:= $\$X5220\ x_o$
wff 16 : $x_o \quad := \quad \$X5220$
:= $\$A5220$ %0
wff 1828 : $\wedge\, XorCaseFRight\, XorCaseFLeft_o \quad := \quad \$A5220$
<< A5220.r0t.txt
:= $\$T5220$
:= $\$X5220$
:= $\$A5220$
%0
$\quad \forall\, o\, [\lambda x.(\wedge\, XorCaseFRight\, XorCaseFLeft)]$
$\quad \forall_{o(o\backslash 3)\tau} o_\tau [\lambda x_o.(\wedge_{ooo} XorCaseFRight_o XorCaseFLeft_o)_o]$

§r /3 a_o
$\quad = [\lambda x.(\wedge\, XorCaseFRight\, XorCaseFLeft)] \ldots$
$\ldots [\lambda a.(\wedge\, (= (XOR\, a\, F)\, a)\, (= (XOR\, F\, a)\, a))]$
§s %1 3 %0
$\quad \forall\, o\, [\lambda a.(\wedge\, (= (XOR\, a\, F)\, a)\, (= (XOR\, F\, a)\, a))]$

:= $XorIdentityElement$ %0
wff 1868 : $\forall\, o\, [\lambda a.(\wedge\, (= (XOR\, a\, F)\, a)\, (= (XOR\, F\, a)\, a))]_o \quad := \quad XorIdentityElement$

2.1.123 Results for File xor_inverse_element.r0.txt

##
Inverse Element of Exclusive Disjunction (Exclusive OR, XOR)
##
##
Source: [Kubota 2017 (doi: 10.4444/100.10)]
##

Written by Ken Kubota (<mail@kenkubota.de>).
##
This file is part of the publication of the mathematical logic $\mathcal{R}_0$.
For more information, visit: <http://doi.org/10.4444/100.10>
##

<< basics.r0.txt
<< A5229.r0.txt
<< xor_table.r0.txt
<< group.r0.txt

shorthands
:= $\$T1$ ...
$\ldots [\lambda g_{\tau}.[\lambda l_{ggg}.[\lambda e_g.[\lambda b_g.(\wedge_{ooo}(=_{ogg}(l_{ggg}a_g b_g)e_g)(=_{ogg}(l_{ggg}b_g a_g)e_g))_o]_{(og)}]_{(ogg)}]_{(ogg(ggg))}]o_{\tau}XOR_{ooo}$
wff 1714 : $[\lambda g.[\lambda l.[\lambda e.[\lambda b.(\wedge\,(=(l\,a\,b)\,e)\,(=(l\,b\,a)\,e))]]]]\,o\,XOR_{ooo}$:= $\$T1$
:= $\$T1a$ $[\lambda l_{ooo}.[\lambda e_o.[\lambda b_o.(\wedge_{ooo}(=_{ooo}(l_{ooo}a_o b_o)e_o)(=_{ooo}(l_{ooo}b_o a_o)e_o))_o]_{(oo)}]_{(ooo)}]XOR_{ooo}$
wff 1730 : $[\lambda l.[\lambda e.[\lambda b.(\wedge\,(=(l\,a\,b)\,e)\,(=(l\,b\,a)\,e))]]]\,XOR_{ooo}$:= $\$T1a$
:= $\$T1b$ $[\lambda e_o.[\lambda b_o.(\wedge_{ooo}(=_{ooo}(XOR_{ooo}a_o b_o)e_o)(=_{ooo}(XOR_{ooo}b_o a_o)e_o))_o]_{(oo)}]$
wff 1742 : $[\lambda e.[\lambda b.(\wedge\,(=(XOR\,a\,b)\,e)\,(=(XOR\,b\,a)\,e))]]_{ooo}$:= $\$T1b$

.1

§= $\$T1$
$=\$T1\,\$T1$
§\ $[\lambda g_{\tau}.[\lambda l_{ggg}.[\lambda e_g.[\lambda b_g.(\wedge_{ooo}(=_{ogg}(l_{ggg}a_g b_g)e_g)(=_{ogg}(l_{ggg}b_g a_g)e_g))_o]_{(og)}]_{(ogg)}]_{(ogg(ggg))}]o_{\tau}$
$=([\lambda g.[\lambda l.[\lambda e.[\lambda b.(\wedge\,(=(l\,a\,b)\,e)\,(=(l\,b\,a)\,e))]]]]\,o)\ \ldots$
$\ldots [\lambda l.[\lambda e.[\lambda b.(\wedge\,(=(l\,a\,b)\,e)\,(=(l\,b\,a)\,e))]]]$
§s %1 6 %0
$=\$T1\,\$T1a$
§\ $\$T1a$
$=\$T1a\,\$T1b$
§s %1 3 %0
$=\$T1\,\$T1b$

§= $\$T1b_{ooo}F_o a_o$
$=(\$T1b\,F\,a)\,(\$T1b\,F\,a)$
§\ $\$T1b_{ooo}F_o$
$=(\$T1b\,F)\,[\lambda b.(\wedge\,(=(XOR\,a\,b)\,F)\,(=(XOR\,b\,a)\,F))]$
§s %1 6 %0
$=(\$T1b\,F\,a)\,([\lambda b.(\wedge\,(=(XOR\,a\,b)\,F)\,(=(XOR\,b\,a)\,F))]\,a)$

§\ $[\lambda b_o.(\wedge_{ooo}(=_{ooo}(XOR_{ooo}a_ob_o)F_o)(=_{ooo}(XOR_{ooo}b_oa_o)F_o))_o]a_o$
$= ([\lambda b.(\wedge (= (XOR\,a\,b)\,F)\,(= (XOR\,b\,a)\,F))]\,a)$ ...
... $(\wedge (= (XOR\,a\,a)\,F)\,(= (XOR\,a\,a)\,F))$
§s %1 3 %0
$= (\$T1b\,F\,a)\,(\wedge (= (XOR\,a\,a)\,F)\,(= (XOR\,a\,a)\,F))$

:= $\$ATMP$ %0
wff 1771 : $= (\$T1b\,F\,a)\,(\wedge (= (XOR\,a\,a)\,F)\,(= (XOR\,a\,a)\,F))_o$:= $\$ATMP$

.2

use Proof Template A5222 (Rule of Cases): [\x.A]T, [\x.A]F → A
:= $\$L5222$ $[\lambda a_o.(=_{ooo}(XOR_{ooo}a_oa_o)F_o)_o]$
wff 1772 : $[\lambda a.(= (XOR\,a\,a)\,F)]_{oo}$:= $\$L5222$
:= $\$X5222$ x_o
wff 16 : x_o := $\$X5222$
:= $\$T5222$ $\$L5222_{oo}T_o$
wff 1773 : $\$L5222\,T_o$:= $\$T5222$
:= $\$F5222$ $\$L5222_{oo}F_o$
wff 1774 : $\$L5222\,F_o$:= $\$F5222$

case T
§\ $_o$ $\$T5222$
$= \$T5222\,XorTableTTisF$
%$XorTableTTisF$
$= (XOR\,T\,T)\,F$:= $XorTableTTisF$
$=_{ooo}(XOR_{ooo}T_oT_o)F_o$:= $XorTableTTisF$
§s %1 13 %0
$= \$T5222\,(= F\,F)$
%$A5230d$
$= (= F\,F)\,T$:= $A5230d$
$=_{ooo}(=_{ooo}F_oF_o)T_o$:= $A5230d$
§s %1 3 %0
$= \$T5222\,T$
use Proof Template A5201b (Swap): A = B → B = A
<< A5201b.r0t.txt
%0
$= T\,\$T5222$
$=_{ooo}T_o\$T5222_o$
%T
$= = =$:= $A5200t$ T
$=_{o\omega\omega}=_\omega=_\omega$:= $A5200t$ T
§s %0 1 %1
$\$L5222\,T$:= $\$T5222$

case F
§\ $_o$ $\$F5222$
$= \$F5222\,XorTableFFisF$
%$XorTableFFisF$

\# $=(XOR\,F\,F)\,F$:= $XorTableFFisF$
\# $=_{ooo}(XOR_{ooo}F_oF_o)F_o$:= $XorTableFFisF$
§s %1 13 %0
\# $=\$F5222\,(=F\,F)$
%$A5230d$
\# $=(=F\,F)\,T$:= $A5230d$
\# $=_{ooo}(=_{ooo}F_oF_o)T_o$:= $A5230d$
§s %1 3 %0
\# $=\$F5222\,T$
\## use Proof Template A5201b (Swap): A = B → B = A
<< A5201b.r0t.txt
%0
\# $=T\,\$F5222$
\# $=_{ooo}T_o\$F5222_o$
%T
\# $===$:= $A5200t$ T
\# $=_{o\omega\omega}=_\omega=_\omega$:= $A5200t$ T
§s %0 1 %1
\# $\$L5222\,F$:= $\$F5222$

<< A5222.r0t.txt
:= $\$L5222$
:= $\$X5222$
:= $\$T5222$
:= $\$F5222$
%0
\# $=(XOR\,x\,x)\,F$
\# $=_{ooo}(XOR_{ooo}x_ox_o)F_o$

\## use Proof Template A5221 (Sub): B → B [x/A]
:= $\$B5221$ %0
\# wff 1837 : $=(XOR\,x\,x)\,F_{o,\dots}$:= $\$B5221$
:= $\$T5221$ o
\# wff 2 : o_τ := $\$T5221$
:= $\$X5221$ x_o
\# wff 16 : x_o := $\$X5221$
:= $\$A5221$ a_o
\# wff 54 : a_o := $\$A5221$
<< A5221.r0t.txt
:= $\$B5221$
:= $\$T5221$
:= $\$X5221$
:= $\$A5221$
%0
\# $=(XOR\,a\,a)\,F$
\# $=_{ooo}(XOR_{ooo}a_oa_o)F_o$

\## use Proof Template A5219b (Rule T): A → A = T
:= $\$A5219b$ %0

\# wff 1767 : $=(XOR\, a\, a)\, F_{o,\ldots}$:= $\$A5219b$
<< A5219b.r0t.txt
:= $\$A5219b$

:= $\$ETMP$ %0
\# wff 1899 : $=(=(XOR\, a\, a)\, F)\, T_o$:= $\$ETMP$

%$\$ATMP$
\# $=(\$T1b\, F\, a)\, (\wedge\, (=(XOR\, a\, a)\, F)\, (=(XOR\, a\, a)\, F))$:= $\$ATMP$
\# $=_{o\omega\omega}(\$T1b_{ooo}F_o a_o)(\wedge_{ooo}(=_{ooo}(XOR_{ooo}a_o a_o)F_o)(=_{ooo}(XOR_{ooo}a_o a_o)F_o))$:= $\$ATMP$
:= $\$ATMP$

%$\$ETMP$
\# $=(=(XOR\, a\, a)\, F)\, T$:= $\$ETMP$
\# $=_{ooo}(=_{ooo}(XOR_{ooo}a_o a_o)F_o)T_o$:= $\$ETMP$
§s %1 13 %0
\# $=(\$T1b\, F\, a)\, (\wedge\, T\, (=(XOR\, a\, a)\, F))$

%$\$ETMP$
\# $=(=(XOR\, a\, a)\, F)\, T$:= $\$ETMP$
\# $=_{ooo}(=_{ooo}(XOR_{ooo}a_o a_o)F_o)T_o$:= $\$ETMP$
:= $\$ETMP$
§s %1 7 %0
\# $=(\$T1b\, F\, a)\, A5212$

%$A5211$
\# $=A5212\, T$:= $A5211$ $A5229a$
\# $=_{ooo}A5212_o T_o$:= $A5211$ $A5229a$
§s %1 3 %0
\# $=(\$T1b\, F\, a)\, T$
\#\# use Proof Template A5201b (Swap): A = B → B = A
<< A5201b.r0t.txt
%0
\# $=T\, (\$T1b\, F\, a)$
\# $=_{o\omega\omega}T_\omega(\$T1b_{ooo}F_o a_o)$
%T
\# $===$:= $A5200t$ T
\# $=_{o\omega\omega}=_\omega=_\omega$:= $A5200t$ T
§s %0 1 %1
\# $\$T1b\, F\, a$

§\ $\$T1b_{ooo}F_o$
\# $=(\$T1b\, F)\, [\lambda b.(\wedge\, (=(XOR\, a\, b)\, F)\, (=(XOR\, b\, a)\, F))]$
§s %1 2 %0
\# $[\lambda b.(\wedge\, (=(XOR\, a\, b)\, F)\, (=(XOR\, b\, a)\, F))]\, a$

\#\# .3

use Proof Template K8031 ($\exists$ Gen): ([\x.B]A) $\rightarrow$ $\exists$ x: B
:= $T8031$ o
wff 2 : o_τ := $T8031$
:= $B8031$ %0/2
wff 1760 : $[\lambda b.(\wedge (= (XOR\, a\, b)\, F)\, (= (XOR\, b\, a)\, F))]_{oo,\ldots}$:= $B8031$
:= $A8031$ %0/3
wff 54 : a_o := $A8031$
:= $P8031$ $B8031_{oo}$A8031_o$
wff 1762 : $B8031$A8031_{o,\ldots}$:= $P8031$
<< K8031.r0t.txt
:= $T8031$
:= $B8031$
:= $A8031$
%0
$\exists\, o\, [\lambda b.(\wedge (= (XOR\, a\, b)\, F)\, (= (XOR\, b\, a)\, F))]$
$\exists_{o(o\backslash 3)\tau} o_\tau [\lambda b_o.(\wedge_{ooo}(=_{ooo}(XOR_{ooo} a_o b_o) F_o)(=_{ooo}(XOR_{ooo} b_o a_o) F_o))_o]$

.4

use Proof Template A5220 (Gen): A $\rightarrow$ $\forall$ x: A
:= $T5220$ o
wff 2 : o_τ := $T5220$
:= $X5220$ a_o
wff 54 : a_o := $X5220$
:= $A5220$ %0
wff 3153 : $\exists\, o\, [\lambda b.(\wedge (= (XOR\, \$X5220\, b)\, F)\, (= (XOR\, b\, \$X5220)\, F))]_{o,\ldots}$:=
$A5220$
<< A5220.r0t.txt
:= $T5220$
:= $X5220$
:= $A5220$
%0
$\forall\, o\, [\lambda a.(\exists\, o\, [\lambda b.(\wedge (= (XOR\, a\, b)\, F)\, (= (XOR\, b\, a)\, F))])]$
$\forall_{o(o\backslash 3)\tau} o_\tau \ldots$
$\ldots [\lambda a_o.(\exists_{o(o\backslash 3)\tau} o_\tau [\lambda b_o.(\wedge_{ooo}(=_{ooo}(XOR_{ooo} a_o b_o) F_o)(=_{ooo}(XOR_{ooo} b_o a_o) F_o))_o])_o]$

:= $XorInverseElement$ %0
wff 5778 : $\forall\, o\, [\lambda a.(\exists\, o\, [\lambda b.(\wedge (= (XOR\, a\, b)\, F)\, (= (XOR\, b\, a)\, F))])]_{o,\ldots}$:=
$XorInverseElement$

undefine local variables
:= $T1$
:= $T1a$
:= $T1b$

2.1.124 Results for File xor_table.r0.txt

##
Proof: (T X T) = F; (T X F) = T; (F X T) = T; (F X F) = F ; X = XOR

```
##
##
##  Source: [Kubota 2017 (doi: 10.4444/100.10)]
##
##  Copyright (c) 2017 Owl of Minerva Press GmbH. All rights reserved.
##  Written by Ken Kubota (<mail@kenkubota.de>).
##
##  This file is part of the publication of the mathematical logic R0.
##  For more information, visit: <http://doi.org/10.4444/100.10>
##

<< basics.r0.txt
<< A5230.r0.txt
<< A5231.r0.txt


##
##  Proof
##

## .a:  (T X T) = F
```

$\S{=}\ \ _{o}\ XOR_{ooo}T_{o}T_{o}$
$\#\qquad\qquad = (XOR\,T\,T)\,(XOR\,T\,T)$
$\S\backslash\ \ XOR_{ooo}T_{o}$
$\#\qquad\qquad = (XOR\,T)\,[\lambda y.(\sim(=T\,y))]$
$\S s\ \ \%1\ \ 6\ \ \%0$
$\#\qquad\qquad = (XOR\,T\,T)\,([\lambda y.(\sim(=T\,y))]\,T)$
$\S\backslash\ \ [\lambda y_{o}.(\sim_{oo}(=_{ooo}T_{o}y_{o}))_{o}]T_{o}$
$\#\qquad\qquad = ([\lambda y.(\sim(=T\,y))]\,T)\,(\sim(=T\,T))$
$\S s\ \ \%1\ \ 3\ \ \%0$
$\#\qquad\qquad = (XOR\,T\,T)\,(\sim(=T\,T))$

$\%A5230a$
$\#\qquad\qquad = (=T\,T)\,T \quad := \ A5230a$
$\#\qquad\qquad =_{ooo}(=_{ooo}T_{o}T_{o})T_{o} \quad := \ A5230a$
$\S s\ \ \%1\ \ 7\ \ \%0$
$\#\qquad\qquad = (XOR\,T\,T)\,(\sim T)$

$\%A5231a$
$\#\qquad\qquad = (\sim T)\,F \quad := \ A5231a$
$\#\qquad\qquad =_{ooo}(\sim_{oo}T_{o})F_{o} \quad := \ A5231a$
$\S s\ \ \%1\ \ 3\ \ \%0$
$\#\qquad\qquad = (XOR\,T\,T)\,F$

$:=\ \ XorTableTTisF\ \ \%0$
$\#\ \mathrm{wff}\quad 1601\ :\quad = (XOR\,T\,T)\,F_{o} \quad := \ XorTableTTisF$

```
## .b:  (T X F) = T
```

§= $_o$ $XOR_{ooo}T_oF_o$
$=(XOR\,T\,F)\,(XOR\,T\,F)$
§\ $XOR_{ooo}T_o$
$=(XOR\,T)\,[\lambda y.(\sim(=T\,y))]$
§s %1 6 %0
$=(XOR\,T\,F)\,([\lambda y.(\sim(=T\,y))]\,F)$
§\ $[\lambda y_o.(\sim_{oo}(=_{ooo}T_oy_o))_o]F_o$
$=([\lambda y.(\sim(=T\,y))]\,F)\,(\sim(=T\,F))$
§s %1 3 %0
$=(XOR\,T\,F)\,(\sim(=T\,F))$

%$A5217$
$=(=T\,F)\,F$:= $A5217$ $A5230b$
$=_{ooo}(=_{ooo}T_oF_o)F_o$:= $A5217$ $A5230b$
§s %1 7 %0
$=(XOR\,T\,F)\,(\sim F)$

%$A5231b$
$=(\sim F)\,T$:= $A5231b$
$=_{ooo}(\sim_{oo}F_o)T_o$:= $A5231b$
§s %1 3 %0
$=(XOR\,T\,F)\,T$

:= $XorTableTFisT$ %0
wff 1612 : $=(XOR\,T\,F)\,T_o$:= $XorTableTFisT$

.c: (F X T) = T

§= $_o$ $XOR_{ooo}F_oT_o$
$=(XOR\,F\,T)\,(XOR\,F\,T)$
§\ $XOR_{ooo}F_o$
$=(XOR\,F)\,[\lambda y.(\sim(=F\,y))]$
§s %1 6 %0
$=(XOR\,F\,T)\,([\lambda y.(\sim(=F\,y))]\,T)$
§\ $[\lambda y_o.(\sim_{oo}(=_{ooo}F_oy_o))_o]T_o$
$=([\lambda y.(\sim(=F\,y))]\,T)\,(\sim(=F\,T))$
§s %1 3 %0
$=(XOR\,F\,T)\,(\sim(=F\,T))$

%$A5230c$
$=(=F\,T)\,F$:= $A5230c$
$=_{ooo}(=_{ooo}F_oT_o)F_o$:= $A5230c$
§s %1 7 %0
$=(XOR\,F\,T)\,(\sim F)$

%$A5231b$
$=(\sim F)\,T$:= $A5231b$
$=_{ooo}(\sim_{oo}F_o)T_o$:= $A5231b$

§s %1 3 %0
$=(XOR\,F\,T)\,T$

:= $XorTableFTisT$ %0
wff 1628 : $=(XOR\,F\,T)\,T_o$:= $XorTableFTisT$

.d: (F X F) = F

§= $_o\ XOR_{ooo}F_oF_o$
$=(XOR\,F\,F)\,(XOR\,F\,F)$
§\ $XOR_{ooo}F_o$
$=(XOR\,F)\,[\lambda y.(\sim(=F\,y))]$
§s %1 6 %0
$=(XOR\,F\,F)\,([\lambda y.(\sim(=F\,y))]\,F)$
§\ $[\lambda y_o.(\sim_{oo}(=_{ooo}F_oy_o))_o]F_o$
$=([\lambda y.(\sim(=F\,y))]\,F)\,(\sim(=F\,F))$
§s %1 3 %0
$=(XOR\,F\,F)\,(\sim(=F\,F))$

%$A5230d$
$=(=F\,F)\,T$:= $A5230d$
$=_{ooo}(=_{ooo}F_oF_o)T_o$:= $A5230d$
§s %1 7 %0
$=(XOR\,F\,F)\,(\sim T)$

%$A5231a$
$=(\sim T)\,F$:= $A5231a$
$=_{ooo}(\sim_{oo}T_o)F_o$:= $A5231a$
§s %1 3 %0
$=(XOR\,F\,F)\,F$

:= $XorTableFFisF$ %0
wff 1639 : $=(XOR\,F\,F)\,F_o$:= $XorTableFFisF$

2.2 Source Files

2.2.1 File Makefile

```
##
##  Makefile
##
##
##  The Makefile for this project.
##
##  Copyright (c) 2017 Owl of Minerva Press GmbH. All rights reserved.
##  Written by Ken Kubota (<mail@kenkubota.de>).
##
##  This file is part of the publication of the mathematical logic R0.
```

```
##  For more information, visit: <http://doi.org/10.4444/100.10>
##


# test file (current work)
TEST            = xor_group.r0.txt

# parameters
URLFULL         = "http://doi.org/10.4444/100.10"
URLSHORT        = "doi.org/10.4444/100.10"
WIDTH           = 84
#BLANKLASTPAGE = --variable lastpageblank="on"


#
#  Target and source files
#

# program
PRG        = R0
PRGDEBUG   = --debug --strict

# source
SRCR0      = $(sort $(wildcard *.r0.txt))
SRCR0A     = $(sort $(wildcard *.r0a.txt))
SRCR0T     = $(sort $(wildcard *.r0t.txt))
SRCR0E     = $(sort $(wildcard *.r0e.txt))
RSRC       = $(sort $(SRCR0) $(SRCR0A) $(SRCR0T) $(SRCR0E))
RSRCOUT    = $(sort $(SRCR0) $(SRCR0A) $(SRCR0E))
RSRCCHK    = $(sort $(SRCR0) $(SRCR0A))

# source in other formats
MSOF       = $(RSRC:.txt=.src.md)
HSOF       = $(RSRC:.txt=.src.html)
PSOF       = $(RSRC:.txt=.src.pdf)
SOFFILES   = $(MSOF) $(HSOF) $(PSOF)

# results
SOUT       = $(RSRCOUT:.txt=.out.txt)
MOUT       = $(RSRCOUT:.txt=.out.md)
HOUT       = $(RSRCOUT:.txt=.out.html)
POUT       = $(RSRCOUT:.txt=.out.pdf)
OUTFILES   = $(SOUT) $(MOUT) $(HOUT) $(POUT)

# check
SCHK       = $(RSRCCHK:.txt=.chk.txt)
DOUT       = $(RSRCCHK:.txt=.dbg.txt)
DCHK       = $(RSRCCHK:.txt=.dck.txt)
CHKFILES   = $(sort $(SCHK) $(DOUT) $(DCHK))
```

```
# errors
EOUT       = $(SRCROE:.txt=.out.txt)

# index and summary files
IDXMD        = index.md
IDXHTML      = $(IDXMD:.md=.html)
MATHMD       = math.md
MATHPDF      = $(MATHMD:.md=.pdf)
IDXMATH      = $(IDXMD) $(IDXHTML) $(MATHMD) $(MATHPDF)

# formulae
TEXTMD       = text.md
REFSMD       = references.md
BIBTEX       = literature.bib
FORMULAEMD   = formulae.md
FORMULAETEX  = $(FORMULAEMD:.md=.tex)
FORMULAEPDF  = $(FORMULAEMD:.md=.pdf)
THSFILES     = $(FORMULAEMD) $(FORMULAETEX) $(FORMULAEPDF)

# LaTeX template
LSRC         = mathtemplate.tex

# hyphenation
HYPHEN       = hyphenate

# files for publication (press only)
PRESS        = $(wildcard press.tex lastpage.tex)

ADDFILES     = $(PRG) $(HYPHEN) $(CHKFILES) $(IDXMATH) $(THSFILES)
ALLFILES     = $(SOFFILES) $(OUTFILES) $(ADDFILES)

MATHFILES    = $(IDXMATH) $(SOFFILES) $(OUTFILES)
STRTFILE     = $(IDXHTML)


#
#  Parameters for compilation
#

# text sources
TSOURCE = $(TEXTMD) $(REFSMD) $(BIBTEX)

# Makefile sources
MSOURCE = Makefile

# input parser
PSOURCE = parser.y
PCC = yacc
```

```
PCCFLAGS = -d -t -v
PTARGET = $(PSOURCE:.y=.c)
PINCL = $(PSOURCE:.y=.h)
POUTPUT = $(PSOURCE:.y=.output)
PFILES = $(PTARGET) $(PINCL) $(POUTPUT)

# input scanner
SSOURCE = scanner.yy
SCC = lex
SCCFLAGS =
STMPTARGET = $(SSOURCE:.yy=.c)
STARGET = $(SSOURCE:.yy=.cc)
SFILES = $(STARGET) $(STMPTARGET)

# C++ main program
CSOURCE = R0.hh R0.cc
CC = g++
CCFLAGS = -Wall -ggdb
LIBS =
SUFFIX = cc

# LaTeX sources (export only)
LSOURCE = $(LSRC)

# script sources
KSOURCE = hyphenation.txt

# Markdown interpreter
MCC = pandoc
MCCFLAGS = -s -S -f markdown
MCCHTML = --mathjax="https://cdn.mathjax.org/mathjax/latest/MathJax.js?config=TeX-AM
S-MML_HTMLorMML"
MCCPDF = -N --variable papersize=a4paper --variable geometry="a4paper,includeheadfoo
t,margin=2cm" --variable lang=english --variable fontsize=11pt --latex-engine=xelate
x
MCCPDFFORMULAE = --template=$(LSRC) $(MCCPDF)
MCCPDFMATH     = --template=$(LSRC) $(MCCPDF) --variable mathshort="on"


#
#  All source files for grepping and editing
#

# internal source (not to be exported)
ISRC = $(PRESS) $(TSOURCE)

# export source (with copyright notice)
ESRC = $(MSOURCE) $(PSOURCE) $(SSOURCE) $(CSOURCE) $(LSOURCE) $(KSOURCE) $(RSRC)
```

```
# all source files
SRC = $(ISRC) $(ESRC)


#
#  Main targets
#

.PHONY: view std standard all chk check schk smallcheck error full clean run md edit
 obsolete copyright preliminaries export math formulae browse test


view: formulae

# build standard files
std standard: $(OUTFILES)

# build all
all: $(ALLFILES)

# build check (re-read output)
chk check: clean preliminaries $(CHKFILES)

# build smallcheck
schk smallcheck: clean preliminaries $(SOUT)

# build error (intentional errors: test check for type violations, etc.)
error: $(EOUT)

# full build with previous clean
full: clean all

# clean (including Xcode build)
clean:
        rm -f $(ALLFILES) $(PFILES) $(SFILES)
        rm -f *.aux *.bbl *.bcf *.blg *.fdb_latexmk *.fls *.log *.out *.run.
xml *.toc
        rm -f *.pdf *.html *.tmp *.out.tex

# run program interactively
run: $(PRG)
        ./$(PRG) --allow-additional-axioms basics.r0.txt -

# run program interactively in Markdown mode
md: $(PRG)
        ./$(PRG) --allow-additional-axioms --markdown basics.r0.txt -

edit:
    open -a Xcode Makefile .gitignore $(SRC)
```

```
# stop on use of obsolete (or interactive) commands
CHKT   = $(shell ls -1 *.r0.txt *.r0a.txt *.r0e.txt *.r0t.txt | grep -v "^A5200t.r0.
txt$$")
CHKALL = $(shell ls -1 *.r0.txt *.r0a.txt *.r0e.txt *.r0t.txt)
obsolete:
	@! grep -v "^#" $(CHKT)   | grep "§= =";
	@! grep -v "^#" $(CHKALL) | grep "§sw"; # replace by "<< A5201b.r0t.txt"
	@! grep -v "^#" $(CHKALL) | grep -v "\.txt$$" | grep "<<"

copyright:
	@for i in $(ESRC); do \
		cat $$i | head -n 12 | tail -n 7 | cut -b3- > header.txt.tmp; \
		if ! diff header.txt header.txt.tmp; then echo $$i: copyright notice
 missing; rm -f header.txt.tmp; exit 1; fi; \
	done
	@rm -f header.txt.tmp

preliminaries: obsolete copyright

export: preliminaries

hyphenate: hyphenation.txt
	echo "#!/bin/sh" > $@
	no=`wc -l $< | sed -E 's/ [^ ]+$$//'`; tail -n `expr $$no - 13` $< | perl -n
 -l -e '$$s1 = $$s2 = $$_; $$s1 =~ s/\|//g; $$s2 =~ s/\|/\\ldots\\\\\\\\\ldots{}/g;
 print "export FIND=\x{27}$$s1\x{27}"; print "export REPL=\x{27}$$s2\x{27}"; print "
ruby -i -pe \"gsub(ENV[\x{27}FIND\x{27}], ENV[\x{27}REPL\x{27}])\" \"\$$@\""' >> $@
	chmod a+x $@

chkhyphen: $(TEST:.txt=.out.tex)
	latexmk -pdf $(TEST:.txt=.out.tex)
	-! grep ^Overfull $(TEST:.txt=.out.log)
	open -a Xcode $(TEST:.txt=.out.md)
	open -a Preview $(TEST:.txt=.out.pdf)

math: $(MATHPDF)
	open -a Preview $(MATHPDF)

formulae: $(FORMULAEPDF)
	open -a Preview $(FORMULAEPDF)

browse: preliminaries $(MATHFILES)
	open -a Safari $(STRTFILE)

# test (development)
test:	$(PRG)
	clear
	./$(PRG) --allow-additional-axioms $(TEST) -
```

```
#
#  Standard case (.r0.txt)
#

# redirect stdout to textfile (or delete in case of error)
%.r0.out.txt: %.r0.txt $(PRG) $(RSRC)
	./$(PRG) --strict $< > $@ || (res=$$?; rm -f $@; exit $$res)

# run program a second time with previous output as input and compare result
%.r0.chk.txt: %.r0.out.txt $(PRG) $(RSRC)
	./$(PRG) --strict --allow-additional-axioms $< > $@
	diff $< $@
	rm $@

# textfile output in debug mode
%.r0.dbg.txt: %.r0.txt $(PRG) $(RSRC)
	./$(PRG) --strict $(PRGDEBUG) $< > $@ 2>&1

# run program a second time with previous output as input and compare result
%.r0.dck.txt: %.r0.dbg.txt $(PRG) $(RSRC)
	./$(PRG) --strict $(PRGDEBUG) --allow-additional-axioms $< > $@ 2>&1
	diff $< $@
	rm $< $@

# markdown output (or delete in case of error)
%.r0.out.md: %.r0.txt $(PRG) $(RSRC) hyphenate
	./$(PRG) --strict --markdown $< > $@ || (res=$$?; rm -f $@; exit $$res)
	./hyphenate $@


#
#  Allow for new axioms (.r0a.txt)
#

# redirect stdout to textfile (or delete in case of error)
%.r0a.out.txt: %.r0a.txt $(PRG) $(RSRC)
	./$(PRG) --strict --allow-additional-axioms $< > $@ || (res=$$?; rm -f $@; e
xit $$res)

# run program a second time with previous output as input and compare result
%.r0a.chk.txt: %.r0a.out.txt $(PRG) $(RSRC)
	./$(PRG) --strict --allow-additional-axioms $< > $@
	diff $< $@
	rm $@

# textfile output in debug mode
%.r0a.dbg.txt: %.r0a.txt $(PRG) $(RSRC)
```

```
	./$(PRG) --strict $(PRGDEBUG) --allow-additional-axioms $< > $@ 2>&1

# run program a second time with previous output as input and compare result
%.r0a.dck.txt: %.r0a.dbg.txt $(PRG) $(RSRC)
	./$(PRG) --strict $(PRGDEBUG) --allow-additional-axioms $< > $@ 2>&1
	diff $< $@
	rm $< $@

# markdown output
%.r0a.out.md: %.r0a.txt $(PRG) $(RSRC) hyphenate
	./$(PRG) --strict --allow-additional-axioms --markdown $< > $@
	./hyphenate $@


#
#  Allow for intentional errors (.r0e.txt)
#

# redirect stdout to textfile
%.r0e.out.txt: %.r0e.txt $(PRG) $(RSRC)
	( cat $< | ./$(PRG) --strict --allow-additional-axioms --allow-definition-re
moval > $@ 2>&1 ); ( res=$$?; if [ ! "$$res" == "1" ]; then echo exactly one error e
xpected, not: $$res; rm -f $@; exit 1; else exit 0; fi )
	grep error $@
	@echo "(ignored -- this error is intentional)"

# markdown output
%.r0e.out.md: %.r0e.txt $(PRG) $(RSRC) hyphenate
	( cat $< | ./$(PRG) --strict --allow-additional-axioms --allow-definition-re
moval --markdown > $@ 2>&1 ); ( res=$$?; if [ ! "$$res" == "1" ]; then echo exactly
one error expected, not: $$res; rm -f $@; exit 1; else exit 0; fi )
	grep error $@
	@echo "(ignored -- this error is intentional)"
	./hyphenate $@


#
#  Other cases
#

# run program with input file and display output
%: %.r0.txt $(PRG)
	./$(PRG) $<


.SECONDEXPANSION:


#
```

```
#  Source
#


# create markdown file for source (with workaround for pandoc bug with heading '*')
%.src.md: %.txt
	cat $< | fold -w $(WIDTH) | sed -E 's/^\*/ */g;s/([$$\\^_\*])/\\\1/g;s/^##/\
\\#\\\#/g;s/^#/\\\#/g;s/^(<<<?) ([^ ]+)\.txt$$/\1 [\2.txt](\2.src.md)  /g;s/^(.+)$$/
\1  /g;s/^$$/\\/g' > $@

# create html file for source
%.r0.src.html %.r0a.src.html %.r0e.src.html: $$(subst .src.html,.src.md,$$@)
	echo "File $(@:.src.html=.txt) -- [[Contents]](./$(IDXHTML)) -- Source: [[TX
T]](./$(@:.src.html=.txt)) [[MD]](./$(@:.src.html=.src.md)) [[HTML]](./$(@:.src.html
=.src.html)) [[PDF]](./$(@:.src.html=.src.pdf)) -- Results: [[TXT]](./$(@:.src.html=
.out.txt)) [[MD]](./$(@:.src.html=.out.md)) [[HTML]](./$(@:.src.html=.out.html)) [[P
DF]](./$(@:.src.html=.out.pdf))  \n\\" > $@.1.tmp
	cat $< | sed -E 's/\.md\)/.html)/g' > $@.2.tmp
	cat $@.1.tmp $@.2.tmp | $(MCC) $(MCCFLAGS) $(MCCHTML) -o $@
	rm -f $@.1.tmp $@.2.tmp

%.r0t.src.html: %.r0t.src.md
	echo "File $(@:.src.html=.txt) -- [[Contents]](./$(IDXHTML)) -- Source: [[TX
T]](./$(@:.src.html=.txt)) [[MD]](./$(@:.src.html=.src.md)) [[HTML]](./$(@:.src.html
=.src.html)) [[PDF]](./$(@:.src.html=.src.pdf))  \n\\" > $@.1.tmp
	cat $< | sed -E 's/\.md\)/.html)/g' > $@.2.tmp
	cat $@.1.tmp $@.2.tmp | $(MCC) $(MCCFLAGS) $(MCCHTML) -o $@
	rm -f $@.1.tmp $@.2.tmp

# create pdf file for source
%.r0.src.pdf %.r0a.src.pdf %.r0e.src.pdf: $$(subst .src.pdf,.src.md,$$@) $(LSRC)
	echo "[[Contents]](./$(IDXHTML)) -- Source: [[TXT]](./$(@:.src.pdf=.txt)) [[
MD]](./$(@:.src.pdf=.src.md)) [[HTML]](./$(@:.src.pdf=.src.html)) [[PDF]](./$(@:.src
.pdf=.src.pdf)) -- Results: [[TXT]](./$(@:.src.pdf=.out.txt)) [[MD]](./$(@:.src.pdf=
.out.md)) [[HTML]](./$(@:.src.pdf=.out.html)) [[PDF]](./$(@:.src.pdf=.out.pdf))  \n\
\" > $@.1.tmp
	cat $< | sed -E 's/\.md\)/.pdf)/g' > $@.2.tmp
	cat $@.1.tmp $@.2.tmp | $(MCC) $(MCCFLAGS) $(MCCPDFMATH) --variable lhead="F
ile `echo $(@:.src.pdf=.txt) | sed -E 's/_/\\\\_/g' `" --variable lfoot="\href{$(URL
FULL)}{$(URLSHORT)}" -o $@
	rm -f $@.1.tmp $@.2.tmp

%.r0t.src.pdf: %.r0t.src.md $(LSRC)
	echo "[[Contents]](./$(IDXHTML)) -- Source: [[TXT]](./$(@:.src.pdf=.txt)) [[
MD]](./$(@:.src.pdf=.src.md)) [[HTML]](./$(@:.src.pdf=.src.html)) [[PDF]](./$(@:.src
.pdf=.src.pdf))  \n\\" > $@.1.tmp
	cat $< | sed -E 's/\.md\)/.pdf)/g' > $@.2.tmp
	cat $@.1.tmp $@.2.tmp | $(MCC) $(MCCFLAGS) $(MCCPDFMATH) --variable lhead="F
ile `echo $(@:.src.pdf=.txt) | sed -E 's/_/\\\\_/g' `" --variable lfoot="\href{$(URL
FULL)}{$(URLSHORT)}" -o $@
```

```
    rm -f $@.1.tmp $@.2.tmp


#
#  Results
#

# create html file from markdown output
%.out.html: %.out.md
    echo "Results for File $(@:.out.html=.txt) -- [[Contents]](./$(IDXHTML)) --
Source: [[TXT]](./$(@:.out.html=.txt)) [[MD]](./$(@:.out.html=.src.md)) [[HTML]](./$
(@:.out.html=.src.html)) [[PDF]](./$(@:.out.html=.src.pdf)) -- Results: [[TXT]](./$(
@:.out.html=.out.txt)) [[MD]](./$(@:.out.html=.out.md)) [[HTML]](./$(@:.out.html=.ou
t.html)) [[PDF]](./$(@:.out.html=.out.pdf))  \n\\" > $@.1.tmp
    cat $< | sed -E 's/\.md\)/.html)/g' > $@.2.tmp
    cat $@.1.tmp $@.2.tmp | $(MCC) $(MCCFLAGS) $(MCCHTML) -o $@
    rm -f $@.1.tmp $@.2.tmp

# create pdf (or tex) file from markdown output
%.out.pdf %.out.tex: %.out.md $(LSRC)
    echo "[[Contents]](./$(IDXHTML)) -- Source: [[TXT]](./$(@:.out.pdf=.txt)) [[
MD]](./$(@:.out.pdf=.src.md)) [[HTML]](./$(@:.out.pdf=.src.html)) [[PDF]](./$(@:.out
.pdf=.src.pdf)) -- Results: [[TXT]](./$(@:.out.pdf=.out.txt)) [[MD]](./$(@:.out.pdf=
.out.md)) [[HTML]](./$(@:.out.pdf=.out.html)) [[PDF]](./$(@:.out.pdf=.out.pdf))  \n\
\" > $@.1.tmp
    cat $< | sed -E 's/\.md\)/.pdf)/g' > $@.2.tmp
    cat $@.1.tmp $@.2.tmp | $(MCC) $(MCCFLAGS) $(MCCPDFMATH) --variable lhead="R
esults for File `echo $(@:.out.pdf=.txt) | sed -E 's/_/\\\\_/g' `" --variable lfoot=
"\href{$(URLFULL)}{$(URLSHORT)}" -o $@
    rm -f $@.1.tmp $@.2.tmp


$(IDXHTML): $(IDXMD)
    cat $< | $(MCC) $(MCCFLAGS) $(MCCHTML) -o $@

$(IDXMD): $(SRCRO) $(OUTFILES)
    rm -f $@
    echo "% RO Contents" >> $@
    echo >> $@
    echo "## Summary" >> $@
    echo >> $@
    echo "math -- Results: [[MD]](math.md) [[PDF]](math.pdf)  " >> $@
    echo >> $@
    echo "## Proofs" >> $@
    echo >> $@
    for i in $(RSRCCHK:.txt=); do \
        echo "$$i -- Source: [[TXT]](./$$i.txt) [[MD]](./$$i.src.md) [[HTML]
](./$$i.src.html) [[PDF]](./$$i.src.pdf) -- Results: [[TXT]](./$$i.out.txt) [[MD]](.
/$$i.out.md) [[HTML]](./$$i.out.html) [[PDF]](./$$i.out.pdf)  " >> $@; \
```

```
    done
    echo >> $@
    echo "## Templates" >> $@
    echo >> $@
    for i in $(SRCROT:.txt=); do \
        echo "$$i -- Source: [[TXT]](./$$i.txt) [[MD]](./$$i.src.md) [[HTML]
](./$$i.src.html) [[PDF]](./$$i.src.pdf)  " >> $@; \
    done
    echo >> $@
    echo "## Violations" >> $@
    echo >> $@
    for i in $(SRCROE:.txt=); do \
        echo "$$i -- Source: [[TXT]](./$$i.txt) [[MD]](./$$i.src.md) [[HTML]
](./$$i.src.html) [[PDF]](./$$i.src.pdf) -- Results: [[TXT]](./$$i.out.txt) [[MD]](.
/$$i.out.md) [[HTML]](./$$i.out.html) [[PDF]](./$$i.out.pdf)  " >> $@; \
    done


# print version (remove links)
$(MATHMD): $(MOUT) $(MSOURCE) $(ESRC)
    rm -f $@
    echo "## Results\n\n" >> $@
    for f in $(MOUT:.out.md=); do \
        echo "### Results for File $$f.txt\n" >> $@; \
        cat $$f.out.md | sed -E 's/\[([^\[]+)\]\([^\(]+\)/\1/g;s/\<([^\<]+)\
>/\1/g' >> $@; \
        echo "\n" >> $@; \
    done
    echo "## Source Files\n\n" >> $@
    for f in $(ESRC); do \
        echo "### File $$f\n" >> $@; \
        cat $$f| fold -w $(WIDTH) | sed -E "s/(.*)/    \1/g" >> $@; \
        echo "\n" >> $@; \
    done

$(MATHPDF): $(MATHMD) $(LSRC)
    cat $< | $(MCC) $(MCCFLAGS) $(MCCPDFMATH) --variable title="R0 Summary" --va
riable lhead="R0 Summary" --variable lfoot="\href{$(URLFULL)}{$(URLSHORT)}" --toc -o
 $@

$(FORMULAEMD): $(TEXTMD) $(MATHMD) $(REFSMD) $(MSOURCE)
    cat $(TEXTMD) $(MATHMD) $(REFSMD) > $(FORMULAEMD)

$(FORMULAETEX): $(FORMULAEMD) $(LSRC) $(BIBTEX)
    cat $< | $(MCC) $(MCCFLAGS) $(MCCPDFFORMULAE) $(BLANKLASTPAGE) --variable gr
aphics="on" --variable mathops="on" --variable lhead="Ken Kubota" --variable rhead="
Mathematical Formulae" --variable lfoot="$(URLSHORT)" --variable documentclass="book
" --variable biblio-files="$(BIBTEX)" --toc --biblatex --bibliography=$(BIBTEX) -o $
@
```

```
$(FORMULAEPDF): $(FORMULAETEX) $(BIBTEX) $(PRESS)
    rm -f formulae.bbl
    latexmk -pdf $(FORMULAETEX)
    @! grep ^Overfull $(@:.pdf=.log) || echo "warning: overfull hbox(es) reporte
d in $(@:.pdf=.log)"


#
#  Dependencies
#

# dependencies of main program
R0: $(PTARGET) $(STARGET)

# dependencies of input parser
parser: $(PTARGET) ;
$(PTARGET): $(PSOURCE) R0.hh
    $(PCC) $(PCCFLAGS) -o $(PTARGET) $(PSOURCE)

# dependencies of input scanner
scanner: $(STARGET) ;
$(STARGET): $(SSOURCE) parser.h
    $(SCC) $(SCCFLAGS) -o $(STMPTARGET) $(SSOURCE)
    cat $(STMPTARGET) | sed 's/static int yyinput/static inline int yyinput/g' >
 $(STARGET)

# dependencies of C++ source files
%: %.cc %.hh
    $(CC) $(CCFLAGS) $(LIBS) $< -o $@
```

2.2.2 File parser.y

```
//
//  parser.y
//
//
//  The grammar definition for the parser generator.
//
//  Copyright (c) 2017 Owl of Minerva Press GmbH. All rights reserved.
//  Written by Ken Kubota (<mail@kenkubota.de>).
//
//  This file is part of the publication of the mathematical logic R0.
//  For more information, visit: <http://doi.org/10.4444/100.10>
//


%{
    #include "R0.hh"
```

```
    int yylex (void);
    void yyerror (char const *);

    const char* interactive_only_err_msg = "use of implicit expressions and syst
em commands allowed in interactive mode only";
    inline bool is_interactive() { return (interactive && display); }
    void exception_handler(nonfatal_exception &e);
    void error_handler(const string& _errmsg);

    wff_reference last_wff;
%}

%union {
    string* str_ptr;
    reference_data_type ref;
}


/* Bison declarations. */

// commands
%token CMD_IDENTIFY CMD_IDENTIFY_PRIME CMD_LMBD_CONV CMD_SUBST CMD_SUBST_PRIME CMD_R
NVAR CMD_AXIOM CMD_LMBD_CONV_AND_SUBST CMD_LMBD_CONV_AND_SUBST_PRIME CMD_RNVAR_AND_S
UBST CMD_SWAP CMD_DEFINE CMD_HELP CMD_PRINT CMD_ALL CMD_ID CMD_WFFS CMD_DEFS CMD_THM
S CMD_STACK CMD_QUIT

// token types
%token <str_ptr> TOK_IDENTIFIER TOK_VARIABLE TOK_STRING TOK_COMMENT_EXTERNAL TOK_COM
MENT_INTERNAL
%token <ref> TOK_INDEX

// improper symbols
%token TOK_SEMICOLON
%left TOK_LMBD_APPL_IMPL TOK_LMBD_APPL_EXPL TOK_LMBD_ABST_IMPL TOK_LMBD_ABST_EXPL
%left TOK_DOT TOK_COMMA TOK_OPEN_CURLY TOK_CLOSE_CURLY TOK_OPEN_BRCKT TOK_CLOSE_BRCK
T
%token TOK_TAU TOK_SLASH TOK_OPEN_PAREN TOK_CLOSE_PAREN TOK_ENDLINE
%token TOK_PERCENT TOK_DOUBLE_STAR TOK_STAR

// undefined symbol or scanner error messages
%token <str_ptr> TOK_UNDEF TOK_LEX_ERROR

// references
%type <ref> primary lambda wff type theorem


%% /* The grammar follows. */
```

```
input:
          /* empty */
        | input line
;

line:
          end                         {   if (display)
                                          cout <<
MIF1("\\") << endl; }
        | cmd end
        | TOK_COMMENT_INTERNAL end            // do not print
        | TOK_COMMENT_EXTERNAL end            {   string comme
nt(*$1); delete $1;
                                      if (display
|| debug)
                                          cout <<
MIF2("\\#\\#","##") << markdown_escape(comment) << MENDL << endl; }
        | error end                   {   yyerrok; }
;

end:
          TOK_ENDLINE
        | TOK_SEMICOLON
        | err                         {   YYERROR; }
;

cmd:    // commands that are allowed in both interactive and batch mode
        // identification
          CMD_IDENTIFY wff {
            try {
            if (display || debug) cout << "§" << MDOL << "=" << MSPC <<
(*wff_reference($2)).get_name() << MDOL << MENDL << endl;
            const thm_reference& r = th.rule_identification(wff_referenc
e($2));
            th.input_stack.push(r);
            if (display || debug) { r.print(cout, false); cout << MENDL
<< endl; }
            }
            catch(nonfatal_exception &e) { exception_handler(e); YYERROR; }
}
        // identification with polymorphic identity relation
        | CMD_IDENTIFY type wff {
            try {
            if (display || debug) cout << "§" << MDOL << "=" << MSPC <<
MIF2("{}_{","{") << (*wff_reference($2)).get_name() << "}" << MSPC << (*wff_referenc
e($3)).get_name() << MDOL << MENDL << endl;
            const thm_reference& r = th.rule_identification(wff_referenc
e($3), wff_reference($2));
            th.input_stack.push(r);
```

```
            if (display || debug) { r.print(cout, false); cout << MENDL
<< endl; }
            }
            catch(nonfatal_exception &e) { exception_handler(e); YYERROR; }
}
        // identification with polymorphic identity relation
        | CMD_IDENTIFY type TOK_SLASH TOK_INDEX {
            try {
            if (display || debug) cout << "§" << MDOL << "=" << MSPC <<
MIF2("{}_{","{") << (*wff_reference($2)).get_name() << "}" << MSPC << MSLASH << $4 <
< MDOL << MENDL << endl;
            const thm_reference& r = th.rule_identification(thm_referenc
e(th.input_stack[0].get_no()).get_wff_reference().get_wff_part($4),  wff_reference($
2));
            th.input_stack.push(r);
            if (display || debug) { r.print(cout, false); cout << MENDL
<< endl; }
            }
            catch(nonfatal_exception &e) { exception_handler(e); YYERROR; }
}
        // identification with hypothesis
        | CMD_IDENTIFY_PRIME TOK_SLASH TOK_INDEX {
            try {
            if (display || debug) cout << "§" << MDOL << "='" << MSPC <<
 MSLASH << $3 << MDOL << MENDL << endl;
            // recalculate node reference (starting point is right hand
side of formula)
            node_reference no($3); no.push_front();
            wff_reference wff = thm_reference(th.input_stack[0].get_no()
).get_wff_reference().get_wff_part(no.to_ulong());
            const thm_reference& r = th.rule_identification(wff);
            th.input_stack.push(r);
            if (display || debug) { r.print(cout, false); cout << MENDL
<< endl; }
            }
            catch(nonfatal_exception &e) { exception_handler(e); YYERROR; }
}
        // identification with hypothesis and polymorphic identity relation
        | CMD_IDENTIFY_PRIME type TOK_SLASH TOK_INDEX {
            try {
            if (display || debug) cout << "§" << MDOL << "='" << MSPC <<
 "{" << (*wff_reference($2)).get_name() << "}" << MSPC << MSLASH << $4 << MDOL << ME
NDL << endl;
            // recalculate node reference (starting point is right hand
side of formula)
            node_reference no($4); no.push_front();
            wff_reference wff = thm_reference(th.input_stack[0].get_no()
).get_wff_reference().get_wff_part(no.to_ulong());
            const thm_reference& r = th.rule_identification(wff, wff_ref
```

```
erence($2));
            th.input_stack.push(r);
            if (display || debug) { r.print(cout, false); cout << MENDL
<< endl; }
            }
            catch(nonfatal_exception &e) { exception_handler(e); YYERROR; }
}
        // lambda conversion
        | CMD_LMBD_CONV wff {
            try {
            if (display || debug) cout << "§" << MDOL << MBSLASH << MSPC
 << (*wff_reference($2)).get_name() << MDOL << MENDL << endl;
            const thm_reference& r = th.rule_lambda_conversion(wff_refer
ence($2));
            th.input_stack.push(r);
            if (display || debug) { r.print(cout, false); cout << MENDL
<< endl; }
            }
            catch(nonfatal_exception &e) { exception_handler(e); YYERROR; }
}
        | CMD_LMBD_CONV TOK_SLASH TOK_INDEX {
            try {
            if (display || debug) cout << "§" << MDOL << MBSLASH << MSPC
 << MSLASH << $3 << MDOL << MENDL << endl;
            const thm_reference& r = th.rule_lambda_conversion(thm_refer
ence(th.input_stack[0].get_no()).get_wff_reference().get_wff_part($3));
            th.input_stack.push(r);
            if (display || debug) { r.print(cout, false); cout << MENDL
<< endl; }
            }
            catch(nonfatal_exception &e) { exception_handler(e); YYERROR; }
}
        | CMD_LMBD_CONV TOK_PERCENT TOK_INDEX TOK_SLASH TOK_INDEX {
            try {
            if (display || debug) cout << "§" << MDOL << MBSLASH << MSPC
 << MPERC << $3 << MSLASH << $5 << MDOL << MENDL << endl;
            const thm_reference& r = th.rule_lambda_conversion(thm_refer
ence(th.input_stack[$3].get_no()).get_wff_reference().get_wff_part($5));
            th.input_stack.push(r);
            if (display || debug) { r.print(cout, false); cout << MENDL
<< endl; }
            }
            catch(nonfatal_exception &e) { exception_handler(e); YYERROR; }
}
        // lambda conversion with polymorphic identity relation
        | CMD_LMBD_CONV type wff {
            try {
            if (display || debug) cout << "§" << MDOL << MBSLASH << MSPC
 << MIF2("{}_{","{") << (*wff_reference($2)).get_name() << "}" << MSPC << (*(wff_ref
```

```
erence($3))).get_name() << MDOL << MENDL << endl;
            const thm_reference& r = th.rule_lambda_conversion(wff_refer
ence($3), wff_reference($2));
            th.input_stack.push(r);
            if (display || debug) { r.print(cout, false); cout << MENDL
<< endl; }
            }
            catch(nonfatal_exception &e) { exception_handler(e); YYERROR; }
}
        // substitution using part number index
        | CMD_SUBST theorem TOK_INDEX theorem {
            try {
            if (display || debug) cout << "§" << MDOL << "s" << MSPC <<
MPERC << th.input_stack.get_index($2) << MSPC << $3 << MSPC << MPERC << th.input_sta
ck.get_index($4) << MDOL << MENDL << endl;
            const thm_reference& r = th.rule_substitution_r(thm_referenc
e($2), $3, thm_reference($4));
            th.input_stack.push(r);
            if (display || debug) { r.print(cout, false); cout << MENDL
<< endl; }
            }
            catch(nonfatal_exception &e) { exception_handler(e); YYERROR; }
}
        // substitution using part number index
        | CMD_SUBST_PRIME theorem TOK_INDEX theorem {
            try {
            if (display || debug) cout << "§" << MDOL << "s'" << MSPC <<
 MPERC << th.input_stack.get_index($2) << MSPC << $3 << MSPC << MPERC << th.input_st
ack.get_index($4) << MDOL << MENDL << endl;
            // recalculate node reference (starting point is right hand
side of formula)
            node_reference no($3); no.push_front();
            const thm_reference& r = th.rule_substitution_r_prime(thm_re
ference($2), no.to_ulong(), thm_reference($4));
            th.input_stack.push(r);
            if (display || debug) { r.print(cout, false); cout << MENDL
<< endl; }
            }
            catch(nonfatal_exception &e) { exception_handler(e); YYERROR; }
}
        // substitution using occurence number index
        | CMD_SUBST theorem theorem TOK_INDEX {
            try {
            if (!is_interactive()) throw syntax_exception(interactive_on
ly_err_msg); // depends on context
            reference_data_type no = 0;
            if (!th.get_part_no_for_substitution(no, thm_reference($2),
thm_reference($3), $4)) throw syntax_exception("wff not found");
            if (display || debug) cout << "§" << MDOL << "s" << MSPC <<
```

```
MPERC << th.input_stack.get_index($2) << MSPC << no << MSPC << MPERC << th.input_sta
ck.get_index($3) << MDOL << MENDL << endl;
            const thm_reference& r = th.rule_substitution_r(thm_referenc
e($2), no, thm_reference($3));
            th.input_stack.push(r);
            if (display || debug) { r.print(cout, false); cout << MENDL
<< endl; }
            }
            catch(nonfatal_exception &e) { exception_handler(e); YYERROR; }
}
        // substitution using standard values: §s %1 %0 0 (first occurence)
        | CMD_SUBST {
            try {
            if (!is_interactive()) throw syntax_exception(interactive_on
ly_err_msg); // depends on context
            reference_data_type no = 0;
            if (!th.get_part_no_for_substitution(no, th.input_stack[1].g
et_no(), th.input_stack[0].get_no(), 0)) throw syntax_exception("wff not found");
            if (display || debug) cout << "§" << MDOL << "s" << MSPC <<
MPERC << "1" << MSPC << no << MSPC << MPERC << "0" << MDOL << MENDL << endl;
            const thm_reference& r = th.rule_substitution_r(th.input_sta
ck[1].get_no(), no, th.input_stack[0].get_no());
            th.input_stack.push(r);
            if (display || debug) { r.print(cout, false); cout << MENDL
<< endl; }
            }
            catch(nonfatal_exception &e) { exception_handler(e); YYERROR; }
}
        // rename variable
        | CMD_RNVAR wff wff {
            try {
            if (display || debug)
            cout << "§" << MDOL << "r" << MSPC << (*(wff_reference($2)))
.get_name() << MSPC << (*(wff_reference($3))).get_name() << MDOL << MENDL << endl;
            const thm_reference& r = th.rule_rename_bound_variable(wff_r
eference($2), wff_reference($3));
            th.input_stack.push(r);
            if (display || debug) { r.print(cout, false); cout << MENDL
<< endl; }
            }
            catch(nonfatal_exception &e) { exception_handler(e); YYERROR; }
}
        | CMD_RNVAR TOK_SLASH TOK_INDEX wff {
            try {
            if (display || debug) cout << "§" << MDOL << "r" << MSPC <<
MSLASH << $3 << MSPC << (*(wff_reference($4))).get_name() << MDOL << MENDL << endl;
            wff_reference wff = thm_reference(th.input_stack[0].get_no()
).get_wff_reference().get_wff_part($3);
            const thm_reference& r = th.rule_rename_bound_variable(wff,
```

```
wff_reference($4));
            th.input_stack.push(r);
            if (display || debug) { r.print(cout, false); cout << MENDL
<< endl; }
            }
            catch(nonfatal_exception &e) { exception_handler(e); YYERROR; }
}
        // lambda conversion and substitution
        | CMD_LMBD_CONV_AND_SUBST TOK_SLASH TOK_INDEX {
            try {
            wff_reference wff = thm_reference(th.input_stack[0].get_no()
).get_wff_reference().get_wff_part($3);
            if (display || debug) cout << "§" << MDOL << MBSLASH << MSPC
 << (*wff).get_name() << MDOL << MENDL << endl;
            const thm_reference& r1 = th.rule_lambda_conversion(wff);
            th.input_stack.push(r1);
            if (display || debug) { r1.print(cout, false); cout << MENDL
 << endl; }
            if (display || debug) cout << "§" << MDOL << "s" << MSPC <<
MPERC << th.input_stack.get_index(th.input_stack[1]) << MSPC << $3 << MSPC << MPERC
<< th.input_stack.get_index(th.input_stack[0]) << MDOL << MENDL << endl;
            const thm_reference& r2 = th.rule_substitution_r(th.input_st
ack[1].get_no(), $3, th.input_stack[0].get_no());
            th.input_stack.push(r2);
            if (display || debug) { r2.print(cout, false); cout << MENDL
 << endl; }
            }
            catch(nonfatal_exception &e) { exception_handler(e); YYERROR; }
}
        // lambda conversion and substitution
        | CMD_LMBD_CONV_AND_SUBST_PRIME TOK_SLASH TOK_INDEX {
            try {
            // recalculate node reference (starting point is right hand
side of formula)
            node_reference no($3); no.push_front();
            wff_reference wff = thm_reference(th.input_stack[0].get_no()
).get_wff_reference().get_wff_part(no.to_ulong());
            if (display || debug) cout << "§" << MDOL << MBSLASH << MSPC
 << (*wff).get_name() << MDOL << MENDL << endl;
            const thm_reference& r1 = th.rule_lambda_conversion(wff);
            th.input_stack.push(r1);
            if (display || debug) { r1.print(cout, false); cout << MENDL
 << endl; }
            if (display || debug) cout << "§" << MDOL << "s" << MSPC <<
MPERC << th.input_stack.get_index(th.input_stack[1]) << MSPC << no.to_ulong() << MSP
C << MPERC << th.input_stack.get_index(th.input_stack[0]) << MDOL << MENDL << endl;
            const thm_reference& r2 = th.rule_substitution_r(th.input_st
ack[1].get_no(), no.to_ulong(), th.input_stack[0].get_no());
            th.input_stack.push(r2);
```

```
            if (display || debug) { r2.print(cout, false); cout << MENDL
 << endl; }
            }
            catch(nonfatal_exception &e) { exception_handler(e); YYERROR; }
}
        // rename variable and substitution
        | CMD_RNVAR_AND_SUBST TOK_SLASH TOK_INDEX wff {
            try {
            wff_reference wff = thm_reference(th.input_stack[0].get_no()
).get_wff_reference().get_wff_part($3);
            if (display || debug) cout << "§" << MDOL << "r" << MSPC <<
MSLASH << $3 << MSPC << (*(wff_reference($4))).get_name() << MDOL << MENDL << endl;
            const thm_reference& r1 = th.rule_rename_bound_variable(wff,
 wff_reference($4));
            th.input_stack.push(r1);
            if (display || debug) { r1.print(cout, false); cout << MENDL
 << endl; }
            if (display || debug) cout << "§" << MDOL << "s" << MSPC <<
MPERC << th.input_stack.get_index(th.input_stack[1]) << MSPC << $3 << MSPC << MPERC
<< th.input_stack.get_index(th.input_stack[0]) << MDOL << MENDL << endl;
            const thm_reference& r2 = th.rule_substitution_r(th.input_st
ack[1].get_no(), $3, th.input_stack[0].get_no());
            th.input_stack.push(r2);
            if (display || debug) { r2.print(cout, false); cout << MENDL
 << endl; }
            }
            catch(nonfatal_exception &e) { exception_handler(e); YYERROR; }
}
        | CMD_AXIOM wff {
            try {
            if (display || debug) cout << "§" << MDOL << "!" << MSPC <<
(*wff_reference($2)).get_name() << MDOL << MENDL << endl;
            const thm_reference& r = th.add_axiom(wff_reference($2));
            th.input_stack.push(r);
            if (display || debug) { r.print(cout, false); cout << MENDL
<< endl; } }
            catch(nonfatal_exception &e) { exception_handler(e); YYERROR; }
}
        | CMD_SWAP {
            try {
            wff_reference wff = thm_reference(th.input_stack[0].get_no()
).get_wff_reference().get_wff_part((reference_data_type)5);
            if (display || debug) cout << "§" << MDOL << "=" << MSPC <<
(*wff).get_name() << MDOL << MENDL << endl;
            const thm_reference& r1 = th.rule_identification(wff);
            th.input_stack.push(r1);
            if (display || debug) { r1.print(cout, false); cout << MENDL
 << endl; }
            if (display || debug) cout << "§" << MDOL << "s" << MSPC <<
```

```
MPERC << "0" << MSPC << "5" << MSPC << MPERC << "1" << MDOL << MENDL << endl;
            const thm_reference& r2 = th.rule_substitution_r(th.input_st
ack[0].get_no(), 5, th.input_stack[1].get_no());
            th.input_stack.push(r2);
            if (display || debug) { r2.print(cout, false); cout << MENDL
 << endl; }
            }
            catch(nonfatal_exception &e) { exception_handler(e); YYERROR; }
}
        | CMD_DEFINE TOK_IDENTIFIER wff {
            try { // use get_sub_name() here avoiding tautological output
            if (display || debug) cout << ":" << MDOL<< "="  << MSPC <<
MESC((*$2)) << MSPC << (*wff_reference($3)).get_sub_name() << MDOL << MENDL << endl;
            (*(wff_reference($3))).add_definiton(*$2);
            last_wff = wff_reference($3);
            delete $2;
            if (display || debug) { wff_reference($3).print(cout, false)
; cout << MENDL << endl; }
            }
            catch(nonfatal_exception &e) { exception_handler(e); YYERROR; }
}
        | CMD_DEFINE TOK_IDENTIFIER TOK_PERCENT TOK_INDEX {
            try {
            if (display || debug) cout << ":" << MDOL<< "="  << MSPC <<
MESC((*$2)) << MSPC << MPERC << $4 << MDOL << MENDL << endl;
            wff_reference ref(thm_reference(th.input_stack[$4].get_no())
.get_wff_reference());
            (*ref).add_definiton(*$2);
            last_wff = ref;
            delete $2;
            if (display || debug) { ref.print(cout, false); cout << MEND
L << endl; }
            }
            catch(nonfatal_exception &e) { exception_handler(e); YYERROR; }
}
        | CMD_DEFINE TOK_IDENTIFIER primary TOK_SLASH TOK_INDEX {
            try { // use get_sub_name() here avoiding tautological output
            if (display || debug) cout << ":" << MDOL<< "="  << MSPC <<
MESC((*$2)) << MSPC << (*wff_reference($3)).get_sub_name() << MSLASH << $5 << MDOL <
< MENDL << endl;
            wff_reference ref(wff_reference($3).get_wff_part($5));
            (*ref).add_definiton(*$2);
            last_wff = ref;
            delete $2;
            if (display || debug) { ref.print(cout, false); cout << MEND
L << endl; }
            }
            catch(nonfatal_exception &e) { exception_handler(e); YYERROR; }
}
```

```
        | CMD_DEFINE TOK_IDENTIFIER wff TOK_SLASH TOK_INDEX {
            try { // use get_sub_name() here avoiding tautological output
            if (display || debug) cout << ":" << MDOL<< "="  << MSPC <<
MESC((*$2)) << MSPC << (*wff_reference($3)).get_sub_name() << MSLASH << $5 << MDOL <
< MENDL << endl;
            wff_reference ref(wff_reference($3).get_wff_part($5));
            (*ref).add_definiton(*$2);
            last_wff = ref;
            delete $2;
            if (display || debug) { ref.print(cout, false); cout << MEND
L << endl; }
            }
            catch(nonfatal_exception &e) { exception_handler(e); YYERROR; }
}
        | CMD_DEFINE TOK_IDENTIFIER TOK_PERCENT TOK_INDEX TOK_SLASH TOK_INDE
X {
            try {
            if (display || debug) cout << ":" << MDOL<< "="  << MSPC <<
MESC((*$2)) << MSPC << MPERC << $4 << MSLASH << $6 << MDOL << MENDL << endl;
            wff_reference ref(thm_reference(th.input_stack[$4].get_no())
.get_wff_reference().get_wff_part($6));
            (*ref).add_definiton(*$2);
            last_wff = ref;
            delete $2;
            if (display || debug) { ref.print(cout, false); cout << MEND
L << endl; }
            }
            catch(nonfatal_exception &e) { exception_handler(e); YYERROR; }
}
        | CMD_DEFINE TOK_IDENTIFIER {
            try {
            if (display || debug) cout << ":" << MDOL<< "="  << MSPC <<
MESC((*$2)) << MDOL << MENDL << endl;
            if (!th.rm_def(*$2)) {
                string msg;
                msg += "no such definition symbol '" + *$2 + "'";
                throw syntax_exception(msg);
            }
            delete $2;
            }
            catch(nonfatal_exception &e) { exception_handler(e); YYERROR; }
}
        // commands that are allowed in interactive mode only since results
may depend on context
        | CMD_HELP {
            try {
            cout << "help" << MENDL << endl;
            if (!is_interactive()) throw syntax_exception(interactive_on
ly_err_msg);
```

```
            cout << "## math commands:    §= §\\ §s !!" << MENDL << endl
            << "## syntax commands: := <<" << MENDL << endl
            << "## system commands: ? :h[elp] :[a]ll :w[ffs] :d[efinitio
ns] :t[heorems]" << MENDL << endl
            << "##                   :p[rint] :s[tack] :i[d] :q[uit]" <
< MENDL << endl;
            }
            catch(nonfatal_exception &e) { exception_handler(e); YYERROR; }
}
        | wff {
            try {
            if( !is_interactive()) throw syntax_exception(interactive_on
ly_err_msg);
            last_wff = $1;
            wff_reference($1).print(cout, false); cout << MENDL << endl;
            wff_reference($1).print(cout, true); cout << MENDL << endl;
            }
            catch(nonfatal_exception &e) { exception_handler(e); YYERROR; }
}
        | CMD_PRINT wff {
            try {
            if (!is_interactive()) throw syntax_exception(interactive_on
ly_err_msg);
            last_wff = $2;
            wff_reference($2).print(cout, false); cout << MENDL << endl;
            wff_reference($2).print(cout, true); cout << MENDL << endl;
            wff_reference($2).print_types(cout);
            wff_reference($2).print_wff_parts(cout); cout << MENDL << en
dl;
            }
            catch(nonfatal_exception &e) { exception_handler(e); YYERROR; }
}
        | TOK_PERCENT TOK_INDEX {
            try {
            thm_reference t_ref(th.input_stack[$2].get_no());
            th.input_stack.push(t_ref);
            if (display || debug) {
                cout << MPERC << MDOL << $2 << MDOL << MENDL << endl;
                t_ref.print(cout, false); cout << MENDL << endl;
                t_ref.print(cout, true); cout << MENDL << endl;
            }
            }
            catch(nonfatal_exception &e) { exception_handler(e); YYERROR; }
}
        | theorem {
            try {
            thm_reference t_ref($1);
            th.input_stack.push(t_ref);
            if (display || debug) {
```

```
                if ((*t_ref).has_definition()) cout << MPERC << MDOL <<
(*t_ref).get_name() << MDOL << MENDL << endl;
                else cout << MPERC << MDOL << th.input_stack.get_index($
1) << MDOL << MENDL << endl;
                t_ref.print(cout, false); cout << MENDL << endl;
                t_ref.print(cout, true); cout << MENDL << endl;
            }
            }
            catch(nonfatal_exception &e) { exception_handler(e); YYERROR; }
}
        | CMD_PRINT theorem {
            try {
            if (!is_interactive()) throw syntax_exception(interactive_on
ly_err_msg);
            thm_reference($2).print(cout, false); cout << MENDL << endl;
            thm_reference($2).print(cout, true); cout << MENDL << endl;
            thm_reference($2).get_wff_reference().print_wff_parts(cout);
 cout << MENDL << endl;
            }
            catch(nonfatal_exception &e) { exception_handler(e); YYERROR; }
}
        | CMD_PRINT {
            try {
            if (!is_interactive()) throw syntax_exception(interactive_on
ly_err_msg);
            thm_reference(th.input_stack[0].get_no()).print(cout, false)
; cout << MENDL << endl;
            thm_reference(th.input_stack[0].get_no()).print(cout, true);
 cout << MENDL << endl;
            thm_reference(th.input_stack[0].get_no()).get_wff_reference(
).print_wff_parts(cout); cout << MENDL << endl; }
            catch(nonfatal_exception &e) { exception_handler(e); YYERROR; }
}
        | CMD_ALL                    {   try {   if (
!is_interactive()) throw syntax_exception(interactive_only_err_msg);

    th.print(cout, false); cout << MENDL << endl; }
                                     catch(nonfat
al_exception &e) { exception_handler(e); YYERROR; } }
        | CMD_ID wff                     {   try {   if (
!is_interactive()) throw syntax_exception(interactive_only_err_msg);

    cout << (*wff_reference($2)).id<< MENDL << endl; }
                                     catch(nonfat
al_exception &e) { exception_handler(e); YYERROR; } }
        | CMD_WFFS                   {   try {   if (
!is_interactive()) throw syntax_exception(interactive_only_err_msg);

    th.print_wffs(cout, false); }
```

```
                                        catch(nonfat
al_exception &e) { exception_handler(e); YYERROR; } }
        | CMD_WFFS wff                    {   try {   if (
!is_interactive()) throw syntax_exception(interactive_only_err_msg);

    wff_reference($2).print(cout, false); cout << MENDL << endl; }
                                        catch(nonfat
al_exception &e) { exception_handler(e); YYERROR; } }
        | CMD_DEFS                    {   try {   if (
!is_interactive()) throw syntax_exception(interactive_only_err_msg);

    th.print_defs(cout, false); }
                                        catch(nonfat
al_exception &e) { exception_handler(e); YYERROR; } }
        | CMD_THMS                    {   try {   if (
!is_interactive()) throw syntax_exception(interactive_only_err_msg);

    th.print_thms(cout, false); }
                                        catch(nonfat
al_exception &e) { exception_handler(e); YYERROR; } }
        | CMD_STACK                   {   try {   if (
!is_interactive()) throw syntax_exception(interactive_only_err_msg);

    th.input_stack.print(cout); }
                                        catch(nonfat
al_exception &e) { exception_handler(e); YYERROR; } }
        | CMD_STACK TOK_INDEX                 {   try {   if (
!is_interactive()) throw syntax_exception(interactive_only_err_msg);

    th.input_stack.print(cout, $2); }
                                        catch(nonfat
al_exception &e) { exception_handler(e); YYERROR; } }
        | CMD_QUIT                    {   try {   if (
!is_interactive()) throw syntax_exception(interactive_only_err_msg);

    print_status(cerr); exit(errors); }
                                        catch(nonfat
al_exception &e) { exception_handler(e); YYERROR; } }
;

lambda:
          TOK_LMBD_ABST_EXPL wff type wff type {
            try {
            if (!is_interactive()) throw syntax_exception(interactive_on
ly_err_msg);
            if (!(*wff_reference($2)).is_variable()) {
                string msg;
                msg += "first argument not a variable in lambda abstract
ion: " + (*wff_reference($2)).get_name();
```

```
                throw syntax_exception(msg);
            }
            wff_reference type_left($3);
            wff_reference left = theory::variable::obtain_variable((*wff
_reference($2)).get_variable().id_short, type_left).get_no();
            wff_reference type_right($5);
            wff_reference right = theory::wff::obtain_wff(wff_reference(
$4), type_right).get_no();
            $$ = theory::lambda_abstraction::obtain_lambda_abstraction(l
eft, type_left, right, type_right).get_no(); }
            catch(nonfatal_exception &e) { exception_handler(e); YYERROR; }
}
        | TOK_LMBD_ABST_IMPL wff wff {
            try {
            if (!is_interactive()) throw syntax_exception(interactive_on
ly_err_msg);
            wff_reference left($2); wff_reference type_left((*left).get_
single_type());
            wff_reference right($3); wff_reference type_right((*right).g
et_single_type());
            if (display || debug) cout << "\\\\" << MDOL << MSPC << (*le
ft).get_name() << MSPC << (*right).get_name() << MDOL << MENDL << endl;
            $$ = theory::lambda_abstraction::obtain_lambda_abstraction(l
eft, type_left, right, type_right).get_no(); }
            catch(nonfatal_exception &e) { exception_handler(e); YYERROR; }
}
        | TOK_OPEN_BRCKT TOK_LMBD_ABST_EXPL wff type TOK_DOT wff type TOK_CL
OSE_BRCKT {
            try {
            if (!(*wff_reference($3)).is_variable()) {
                string msg;
                msg += "first argument not a variable in lambda abstract
ion: " + (*wff_reference($3)).get_name();
                throw syntax_exception(msg);
            }
            wff_reference type_left($4);
            wff_reference left = theory::variable::obtain_variable((*wff
_reference($3)).get_variable().id_short, type_left).get_no();
            wff_reference type_right($7);
            wff_reference right = theory::wff::obtain_wff(wff_reference(
$6), type_right).get_no();
            $$ = theory::lambda_abstraction::obtain_lambda_abstraction(l
eft, type_left, right, type_right).get_no(); }
            catch(nonfatal_exception &e) { exception_handler(e); YYERROR; }
}
        | TOK_LMBD_APPL_EXPL wff type wff type  {
            try {
            wff_reference type_left($3);
            wff_reference left = theory::wff::obtain_wff(wff_reference($
```

```
2), type_left).get_no();
            wff_reference type_right($5);
            wff_reference right = theory::wff::obtain_wff(wff_reference(
$4), type_right).get_no();
            $$ = theory::lambda_application::obtain_lambda_application(l
eft, type_left, right, type_right).get_no(); }
            catch(nonfatal_exception &e) { exception_handler(e); YYERROR; }
}
        | TOK_LMBD_APPL_IMPL wff wff {
            try {
            if (!is_interactive()) throw syntax_exception(interactive_on
ly_err_msg);
            wff_reference left($2); wff_reference right($3);
            if (display || debug) cout << "__" << MDOL << MSPC << (*left
).get_name() << MSPC << (*right).get_name() << MDOL << MENDL << endl;
            pair<wff_reference, wff_reference> p = theory::lambda_applic
ation::match_lambda_application(left, right);
            $$ = theory::lambda_application::obtain_lambda_application(l
eft, p.first, right, p.second).get_no();
            }
            catch(nonfatal_exception &e) { exception_handler(e); YYERROR; }
}
        | TOK_OPEN_PAREN wff type TOK_LMBD_APPL_EXPL wff type TOK_CLOSE_PARE
N {
            try {
            wff_reference type_left($3);
            wff_reference left = theory::wff::obtain_wff(wff_reference($
2), type_left).get_no();
            wff_reference type_right($6);
            wff_reference right = theory::wff::obtain_wff(wff_reference(
$5), type_right).get_no();
            $$ = theory::lambda_application::obtain_lambda_application(l
eft, type_left, right, type_right).get_no(); }
            catch(nonfatal_exception &e) { exception_handler(e); YYERROR; }
}
;

wff:
          primary
        | lambda
        | type
        | TOK_VARIABLE TOK_OPEN_CURLY type TOK_CLOSE_CURLY {
            try { $$ = theory::variable::obtain_variable(*$1, wff_reference(
$3)).get_no(); delete $1; }
            catch(nonfatal_exception &e) { exception_handler(e); YYERROR; }
}
        | primary TOK_SLASH TOK_INDEX {
            try { $$ = wff_reference($1).get_wff_part($3).get_no(); }
            catch(nonfatal_exception &e) { exception_handler(e); YYERROR; }
```

```
}
        | TOK_SLASH TOK_INDEX {
            try { $$ = thm_reference(th.input_stack[0].get_no()).get_wff_ref
erence().get_wff_part($2).get_no(); }
            catch(nonfatal_exception &e) { exception_handler(e); YYERROR; }
}
        | TOK_OPEN_PAREN wff TOK_CLOSE_PAREN { $$ = $2; }
;

type:
          primary {
            try {
            wff_reference ref($1);
            if (!(*ref).is_type()) throw syntax_exception(string("not a
type: '") + (*ref).get_name() + "'");
            else $$ = $1;
            }
            catch(nonfatal_exception &e) { exception_handler(e); YYERROR; }
}
        | lambda {
            try {
            wff_reference ref($1);
            if (!(*ref).is_type()) throw syntax_exception(string("not a
type: '") + (*ref).get_name() + "'");
            else $$ = $1;
            }
            catch(nonfatal_exception &e) { exception_handler(e); YYERROR; }
}
        | TOK_VARIABLE TOK_OPEN_CURLY TOK_TAU TOK_CLOSE_CURLY {
            try {
            $$ = theory::variable::obtain_variable(*$1, th.tau.ref.get_n
o()).get_no(); delete $1;
            }
            catch(nonfatal_exception &e) { exception_handler(e); YYERROR; }
}
        | TOK_OPEN_CURLY type TOK_CLOSE_CURLY { $$ = $2; }
        | TOK_OPEN_CURLY type TOK_COMMA type TOK_CLOSE_CURLY {
            try {
            $$ = theory::composed_type::obtain_composed_type(wff_referen
ce($2), wff_reference($4)).get_no(); }
            catch(nonfatal_exception &e) { exception_handler(e); YYERROR; }
}
        | TOK_DOUBLE_STAR wff {
            try {
            wff_reference ref($2);
            if (!(*ref).is_lambda_application()) throw syntax_exception(
string("not a lambda application: '") + (*ref).get_name() + "'");
            else $$ = (*ref).get_lambda_application().type_right.get_no(
);
```

```
            }
            catch(nonfatal_exception &e) { exception_handler(e); YYERROR; }
}
        | TOK_STAR wff {
            try {
            wff_reference ref($2);
            if (!(*ref).is_variable()) throw syntax_exception(string("no
t a variable: '") + (*ref).get_name() + "'");
            else $$ = (*ref).get_variable().type.get_no();
            }
            catch(nonfatal_exception &e) { exception_handler(e); YYERROR; }
}
;

primary:
          TOK_TAU                        {   $$ = th.tau.
ref.get_no(); }
        | TOK_IDENTIFIER {
            try {
            if (string(*$1)=="$_") {
                $$ = last_wff.get_no();
            }
            else if (!th.wff_exists(*$1)) {
                string msg;
                msg += "undefined token '" + *$1 + "'";
                delete $1;
                throw syntax_exception(msg);
            }
            else {
                $$ = th.find_wff(*$1).get_no(); delete $1;
            }
            }
            catch(nonfatal_exception &e) { exception_handler(e); YYERROR; }
}
        | TOK_INDEX                      {   try {   if (
!is_interactive()) throw syntax_exception(interactive_only_err_msg);
                                             $$ =
 $1; }
                                         catch(nonfat
al_exception &e) { exception_handler(e); YYERROR; } }
        | theorem                        {   try {   $$ =
 thm_reference($1).get_wff_reference().get_no(); }
                                         catch(nonfat
al_exception &e) { exception_handler(e); YYERROR; } }
        | err                            {   YYERROR; }
;

theorem:
          TOK_PERCENT TOK_INDEX {
```

```
            try { $$ = th.input_stack[$2].get_no(); }
            catch(nonfatal_exception &e) { exception_handler(e); YYERROR; }
}
        | TOK_PERCENT TOK_IDENTIFIER {
            try {
            if (!th.thm_exists(*$2)) {
                string msg;
                msg += string("no proven theorem '") + MPERC + *$2 + "'"
;
                delete $2;
                throw syntax_exception(msg);
            }
            else {
                $$ = th.find_thm(*$2).get_no(); delete $2;
            }
            }
            catch(nonfatal_exception &e) { exception_handler(e); YYERROR; }
}
;

err:
          TOK_UNDEF {
            try {
            string msg;
            msg += "undefined character or token '" + *$1 + "'";
            delete $1;
            throw syntax_exception(msg);
            }
            catch(nonfatal_exception &e) { exception_handler(e); YYERROR; }
}
        | TOK_LEX_ERROR                         { error_handler(stri
ng(*$1)); delete $1; YYERROR; }


%%


// called by yyparse on error
void yyerror(char const *s) {
    error_handler(string(s));
}

// called manually on catch
void exception_handler(nonfatal_exception &e) {
    error_handler(e.what());
}

void error_handler(const string& _errmsg) {
    string err_no_str;
```

```
    if (errors != EXIT_CODE_MAX_NUMBER_OF_ERRORS) {
    errors++;
    err_no_str=to_string(errors);
    }
    else {
    err_no_str=to_string(errors)+"+";
    }
    cerr << MIF2("\\# ","# ") << "error " << err_no_str << " [" << current_filename
<< "]" << ": " << _errmsg << MENDL << endl;
    if ( (!interactive) && errors>=max_errors ) {
    cerr << MIF2("\\# ","# ") << "file stack: "<< MENDL << endl; for (int i=incl
ude_level; i>= 0; i--) cerr << MIF2("\\# ","# ") << "[" << i << "] " << level_filena
me(include_level-i)<< MENDL << endl;
    throw nonfatal_exception("too many errors, stopping file parsing ...");
    }
}
```

2.2.3 File scanner.yy

```
/*
//  scanner.yy
//
//
//  The token definitions for the lexical analyser generator.
//
//  Copyright (c) 2017 Owl of Minerva Press GmbH. All rights reserved.
//  Written by Ken Kubota (<mail@kenkubota.de>).
//
//  This file is part of the publication of the mathematical logic R0.
//  For more information, visit: <http://doi.org/10.4444/100.10>
*/


%option noyywrap
%{
    #include "R0.hh"
    #include "parser.h"
    bool display_include;
%}

%x incl
%%


"##"[^\n]*        yylval.str_ptr = new string(yytext+2); return TOK_COMMENT_EX
TERNAL;
"#"[^\n]*         yylval.str_ptr = new string(yytext+1); return TOK_COMMENT_IN
TERNAL;
```

```
"§="            return CMD_IDENTIFY;
"§='"           return CMD_IDENTIFY_PRIME;
"§\\"           return CMD_LMBD_CONV;
"§s"            return CMD_SUBST;
"§s'"           return CMD_SUBST_PRIME;
"§r"            return CMD_RNVAR;

"§!"            return CMD_AXIOM;

"§\\s"          return CMD_LMBD_CONV_AND_SUBST;
"§\\s'"         return CMD_LMBD_CONV_AND_SUBST_PRIME;
"§rs"           return CMD_RNVAR_AND_SUBST;
"§sw"           return CMD_SWAP;

":="            return CMD_DEFINE;

"<<"            display_include = false; BEGIN(incl);
"<<<"           display_include = true;  BEGIN(incl);
">>>"[\n]       {
                if (display || debug) {
                cout << ">>>" << MENDL << endl;
                }
            }

"?"         return CMD_HELP;
":h"            return CMD_HELP;
":help"         return CMD_HELP;

":p"            return CMD_PRINT;
":print"        return CMD_PRINT;

":a"            return CMD_ALL;
":all"          return CMD_ALL;

":i"            return CMD_ID;
":id"           return CMD_ID;

":w"            return CMD_WFFS;
":wffs"         return CMD_WFFS;

":d"            return CMD_DEFS;
":definitions"      return CMD_DEFS;

":t"            return CMD_THMS;
":theorems"     return CMD_THMS;

":s"            return CMD_STACK;
":stack"        return CMD_STACK;
```

```
":q"            return CMD_QUIT;
":quit"         return CMD_QUIT;

"\^"            return TOK_TAU;

[$]?[A-Z][A-Za-z0-9']*  yylval.str_ptr = new string(yytext); return TOK_IDENTIFIER;
"$_"            yylval.str_ptr = new string(yytext); return TOK_IDENTIFIER;
"o"         yylval.str_ptr = new string(yytext); return TOK_IDENTIFIER;
"i"         yylval.str_ptr = new string(yytext); return TOK_IDENTIFIER;
[@=&!|]         yylval.str_ptr = new string(yytext); return TOK_IDENTIFIER;
"=>"            yylval.str_ptr = new string(yytext); return TOK_IDENTIFIER;
"!="            yylval.str_ptr = new string(yytext); return TOK_IDENTIFIER;
"=="            yylval.str_ptr = new string(yytext); return TOK_IDENTIFIER;
"!=="           yylval.str_ptr = new string(yytext); return TOK_IDENTIFIER;

[$]?[a-z][a-z0-9']* yylval.str_ptr = new string(yytext); return TOK_VARIABLE;
[\\][0-9]+      yylval.str_ptr = new string(yytext); return TOK_VARIABLE;

\"[^\n]+\"      yylval.str_ptr = new string(yytext+1); yylval.str_ptr->resiz
e(yylval.str_ptr->length()-1); return TOK_STRING;

[0-9]+          yylval.ref=atoi(yytext); return TOK_INDEX;

"%"         return TOK_PERCENT;
"**"            return TOK_DOUBLE_STAR;
"*"         return TOK_STAR;

"_"         return TOK_LMBD_APPL_EXPL;
"__"            return TOK_LMBD_APPL_IMPL;
"\\"            return TOK_LMBD_ABST_EXPL;
"\\\\"          return TOK_LMBD_ABST_IMPL;
"\."            return TOK_DOT;
","         return TOK_COMMA;
"{"         return TOK_OPEN_CURLY;
"}"         return TOK_CLOSE_CURLY;
"["         return TOK_OPEN_BRCKT;
"]"         return TOK_CLOSE_BRCKT;

"/"         return TOK_SLASH;
"("         return TOK_OPEN_PAREN;
")"         return TOK_CLOSE_PAREN;

";"         return TOK_SEMICOLON;

[\n]            return TOK_ENDLINE;

[ \t]           // ignore whitespaces

<incl>[ \t]*        /* eat the whitespace */
```

```
<incl>[^ \t\n]+[\n] { /* got the include file name */
                string filename(yytext);
                filename.resize(filename.length()-1);
                // check if already included
                if (!register_included_file(filename)) {
                if (display)
                    cout << COMMENT_HEADER_2 << "Skipping file " <<
filename << " (already included)" << MENDL << endl;
                if (debug)
                    cerr << COMMENT_HEADER_2 << "Skipping file " <<
filename << " (already included)" << MENDL << endl;
                }
                else {
                FILE *yyin_bak = yyin;
                yyin = fopen( filename.c_str(), "r" );
                if ( ! yyin ) {
                    yyin = yyin_bak;
                    BEGIN(INITIAL);
                    string msg;
                    msg += "unable to open file '"+filename+"' while
 parsing '"+current_filename+"'";
                    throw performance_exception(msg);
                }
                else {
                    if (display || debug) {
                    cout << (display_include || debug ? MIF2("\\
#\\# <","## <") : "" ) <<  "<< " << MIF1("[") << filename << MIF1("](") << MIF1(md_s
uffix(filename)) << MIF1(")") << MENDL << endl;
                    }
                    if ((display && display_include) || debug) {
                    cout << COMMENT_HEADER_2 << "Include begin (
" << filename << ") [oldfile=(" << current_filename  << ")]" << MENDL << endl;
                    }
                    include_level++;
                    level_display.push_back(display_include);
                    level_filename.push_back(filename);
                    recalc_display();
                    if (debug) {
                    cerr << COMMENT_HEADER_2 << "Entering includ
e level " << include_level << " (" << filename << ") [oldlevel=" << include_level-1
<<  " (" << previous_filename  << "), display=" << (display ? "true" : "false" ) <<
"]" << MENDL << endl;
                    }
                    yypush_buffer_state(yy_create_buffer( yyin, YY_B
UF_SIZE ));
                }
                }
                BEGIN(INITIAL);
            }
```

```
<<EOF>>          {   if (include_level>0) {
                 if (debug) {
                     cerr << COMMENT_HEADER_2 << "Leaving include lev
el " << include_level << " (" << level_filename[include_level-1] << ") [newlevel=" <
< include_level-1 << " (" << previous_filename  << "), display=" << (display ? "true
" : "false" ) << "]" << MENDL << endl;
                 }
                 include_level--;
                 level_display.pop_back();
                 level_filename.pop_back();
                 recalc_display();
                 if ((display && level_display[include_level]) || deb
ug) {
                     cout << COMMENT_HEADER_2 << "Include end (" << l
evel_filename[include_level] << ") [newfile=(" << current_filename << ")]" << MENDL
<< endl;
                 }
                 if (display || debug) {
                     if (level_display[include_level] || debug)
                     cout << ">>>" << MENDL << endl;
                 }
                 }
                 fclose(yyin);
                 yypop_buffer_state();
                 if ( !YY_CURRENT_BUFFER ) yyterminate();
             }

.            yylval.str_ptr = new string(yytext); return TOK_UNDEF;


%%
```

2.2.4 File R0.hh

```
//
//  R0.hh
//
//
//  The C++ header file for R0.
//
//  Copyright (c) 2017 Owl of Minerva Press GmbH. All rights reserved.
//  Written by Ken Kubota (<mail@kenkubota.de>).
//
//  This file is part of the publication of the mathematical logic R0.
//  For more information, visit: <http://doi.org/10.4444/100.10>
//
```

```
#ifndef _RO_HH_
#define _RO_HH_


#include <limits>
#include <vector>
#include <set>
#include <string>


using namespace std;


extern bool debug;

extern unsigned long errors;

extern string main_file;
extern set<string> included_files;

extern unsigned int include_level;
extern vector<bool> level_display;
extern vector<string> level_filename;
extern bool display;

bool recalc_display();
bool register_included_file(const string& _filename);
string to_markdown(const string& _str);
string markdown_escape(const string& _str);
string md_suffix(const string& _filename);


// number of statements in the stack to be printed if not specified otherwise
#define std_stack_print 8

// number of errors until non-interactive mode stops
#define max_errors 3


// data type for numbering mathematical objects
#define reference_data_type unsigned long int

// maximum number stands for undefined
#define undefined (reference_data_type) ULONG_MAX


// maximum length of node references in bits
// (may not exceed 32 bits, the smallest unsigned long used for integer representati
on)
```

```
#define node_reference_bits 8


// general macros
#define MAX(x,y)          (x>y ? x : y)

// macros for Markdown output
#define MIF1(x)           (file_format==FORMAT_MARKDOWN ? x : "")
#define MIF2(x,y)         (file_format==FORMAT_MARKDOWN ? x : y)
#define MIFP(x)           (file_format==FORMAT_PLAIN ? x : "")
#define MESC(x)           MIF2(to_markdown(x),x)
#define COMMENT_HEADER        MIF2("\\# ","# ")
#define COMMENT_HEADER_2      MIF2("\\#\\# ","## ")
#define MENDL             MIF2("  ","")
#define MTAB              MIF2("\\qquad "," ")
#define MSPC              MIF2("\\;\\;"," ")
#define MTYPESEP          MIF2(",\\,"," ")
#define MPERC             MIF2("\\%","%")
#define MSLASH            MIF2("/","/")
#define MBSLASH           MIF2("\\backslash ","\\")
#define MDOL              MIF1("$")

#define level_filename(x)   ( x<include_level ? level_filename[include_level-(x+1)]
: main_file )
#define current_filename    level_filename(0)
#define previous_filename   level_filename(1)


#endif /* _R0_HH_ */
```

2.2.5 File R0.cc

```
//
//  R0.cc
//
//
//  The C++ source file for R0.
//
//  Copyright (c) 2017 Owl of Minerva Press GmbH. All rights reserved.
//  Written by Ken Kubota (<mail@kenkubota.de>).
//
//  This file is part of the publication of the mathematical logic R0.
//  For more information, visit: <http://doi.org/10.4444/100.10>
//



//
//  Bibliography
//
```

```
//  Peter B. Andrews
//  An Introduction to Mathematical Logic and Type Theory: To Truth Through Proof
//  2nd ed., Dordrecht/Boston/London 2002
//  ISBN 1-4020-0763-9
//  doi: 10.1007/978-94-015-9934-4 (<http://doi.org/10.1007/978-94-015-9934-4>)
//
//  Ken Kubota
//  Mathematical Formulae
//  Berlin 2017
//  ISBN 978-3-943334-07-4
//  doi: 10.4444/100.3 (<http://doi.org/10.4444/100.3>)
//


#include "R0.hh"

#include <map>
#include <deque>
#include <bitset>

#include <fstream>
#include <sstream>
#include <iostream>


using namespace std;


bool parser_debug = false; // change this for parser debugging information
bool scope_violation_debug = true; // change this to hide types in error messages
bool print_debug = false; // change this to show display calculation


enum FORMATS
{
    FORMAT_PLAIN, FORMAT_MARKDOWN
};

bool debug = false;
bool strict = false; // produce error after warnings
unsigned int file_format = FORMAT_PLAIN;
bool allow_additional_axioms = false;
bool allow_definition_removal = false;

bool interactive = true;

unsigned long errors = 0;

string main_file;
```

```
set<string> included_files;

unsigned int include_level = 0;
vector<bool> level_display;
vector<string> level_filename;
bool display = true;

const char *STRING_STDIN_FILE  = "-";
const char *STRING_AXIOMS_FILE = "axioms.r0.txt";

#define EXIT_CODE_MAX                  255 // Unix standard
#define EXIT_CODE_INTERNAL_ERROR           EXIT_CODE_MAX
#define EXIT_CODE_TEMPORARY_DEFINITIONS_LEFT    EXIT_CODE_MAX-1
#define EXIT_CODE_MAX_NUMBER_OF_ERRORS      EXIT_CODE_MAX-2

#define INPUT_STACK_SIZE                8

#define SCREEN_WIDTH                   80
#define SPACING_INDENT_WIDTH                8
#define SPACING_INDENT_WIDTH_DIST           1
#define SPACING_NUMBER_WIDTH                4
#define SPACING_NUMBER_WIDTH_DIST           4
#define SPACING_NAME_WIDTH             30
#define SPACING_NAME_WIDTH_DIST             2
#define SPACING_TYPES_WIDTH             9
#define SPACING_TYPES_WIDTH_DIST            2
#define SPACING_DEFS_WIDTH             18

#define SPACING_REGULAR_INDENT (SPACING_INDENT_WIDTH+SPACING_INDENT_WIDTH_DIST+SPACI
NG_NUMBER_WIDTH+SPACING_NUMBER_WIDTH_DIST)


// basic mathematical entities
static const char *ID_TAU                  = "TAU";
static const char *ID_OMEGA                = "OMEGA";
static const char *ID_BOOLE                = "BOOLE";
static const char *ID_IDENTITY             = "IDENTITY";
static const char *ID_DESCRIPTOR           = "DESCRIPTOR";

static const char *DEF_TAU                 = "^";
static const char *DEF_OMEGA               = "@";
static const char *DEF_BOOLE               = "o";
static const char *DEF_IDENTITY            = "=";
static const char *DEF_DESCRIPTOR          = "i";

static const char *MARKDOWN_DEF_TAU        = "{\\tau}";
static const char *MARKDOWN_DEF_OMEGA      = "{\\omega}";
static const char *MARKDOWN_DEF_BOOLE      = "o";
// no Markdown definition for "=" (identical)
```

```
static const char *MARKDOWN_DEF_DESCRIPTOR  = "{\\iota}";

// Markdown output conversion for comments
#define DEFS_NUMBER                13
static const char* DEFS_PLAIN_TO_MARKDOWN[DEFS_NUMBER][2] = {
    {"  "    , "$\\quad$"},
    {"--\\>", "$\\to$"},
    {"=\\>" , "$\\supset$"},
    {"~"    , "$\\sim$"},
    {"&"    , "$\\land$"},
    {"|"    , "$\\lor$"},
    {"OP"   , "$\\oplus$"},
    {"ALL"  , "$\\forall$"},
    {"EXI1" , "$\\exists_1$"},
    {"EXI"  , "$\\exists$"},
    {"IOTA" , "$\\iota$"},
    {"R0"   , "$\\mathcal{R}_{0}$"},
    {"[...]", "$[\\ldots]$"}};

// check if file already included (or parsed)
bool register_included_file(const string& _filename) {
    bool skip = false;
    string suffix(_filename);
    suffix.erase(0,suffix.length()-7);
    if (suffix == ".r0.txt") {
    if (included_files.find(_filename) != included_files.end())
        skip = true;
    else
        included_files.insert(_filename);
    }
    return (!skip);
}

string to_string(unsigned int u) {
    stringstream s;
    s << u;
    return s.str();
}

bool to_int(const string& _str, unsigned int& _result) {
    string::const_iterator i = _str.begin();
    if (i == _str.end()) return false;
    for (_result = 0; i != _str.end(); ++i) {
        if (*i < '0' || *i > '9') return false;
        _result *= 10;
        _result += *i - '0';
    }
    return true;
}
```

```
string to_markdown(const string& _str) {
    if      (_str==DEF_TAU       ) return(MARKDOWN_DEF_TAU                    );
    else if (_str==DEF_OMEGA     ) return(MARKDOWN_DEF_OMEGA                  );
    else if (_str==DEF_DESCRIPTOR) return(MARKDOWN_DEF_DESCRIPTOR             );
    else if (_str=="A"           ) return("{\\forall}"                        );
    else if (_str=="&"           ) return("{\\land}"                          );
    else if (_str=="=>"          ) return("{\\supset}"                        );
    else if (_str=="!"           ) return("{\\sim}"                           );
    else if (_str=="|"           ) return("{\\lor}"                           );
    else if (_str=="!="          ) return("{\\neq}"                           );
    else if (_str=="E"           ) return("{\\exists}"                        );
    else if (_str=="E1"          ) return("{\\exists_1}"                      );
    else if (_str=="0"           ) return("{\\emptyset}"                      );
    else if (_str=="SBSET"       ) return("{\\subseteq}"                      );
    else if (_str=="PWSET"       ) return("{\\mathcal{P}}"                    );
    else if (_str=="{"           ) return("\\{"                               );
    else if (_str=="}"           ) return("\\}"                               );
    else if (_str[0]=='$'        ) return(string("\\")+_str                   );
    else if (_str[0]=='\\'       ) return(string("\\backslash")+_str.substr(1));
    else return _str;
}

string markdown_escape(const string& _str) {
    if (file_format!=FORMAT_MARKDOWN) return _str;
    string str(_str);
    for(int i=0; i<str.length(); i++)
    if (str[i]=='$' || str[i]=='\\' || str[i]=='^' || str[i]=='_' || str[i]=='*'
 || str[i]=='<' || str[i]=='>') str.insert(i++,1,'\\');
    std::size_t found;
    string str2;
    for(int i=0; i<DEFS_NUMBER; i++)
    while ((found=str.find(str2 = DEFS_PLAIN_TO_MARKDOWN[i][0]))!=std::string::n
pos) str.replace(str.find(str2), str2.length(), DEFS_PLAIN_TO_MARKDOWN[i][1]);
    return str;
}

bool recalc_display() { // test for command display
    if (level_display.size()>0) for (size_t i=0; i<level_display.size(); i++) if (!l
evel_display[i]) return display = false;
    return display = true;
}

void force_display_position(ostream& _out, unsigned int& _position, unsigned int _li
mit, unsigned int _additional_indent, bool _thm=false) {
    if (print_debug) {
    cout << "[force_display_position(" << _position << "," << _limit << "," << _
additional_indent << "): " << (_position>_limit ? "true" : "false]" ) << endl;
    _out.width(_position); _out << right << ">";
```

```
    }
    if (file_format==FORMAT_PLAIN) {
    unsigned int indent = _limit+_additional_indent;  // indent subsequent line
    if (_position>_limit) { _out << endl; _out.width(indent); _out << left << CO
MMENT_HEADER; } // force linebreak
    else { _out.width(indent-_position); _out << left << ""; }
    _position = indent;
    }
}

// error exceptions
class my_exception: public exception {
public:
    my_exception(const string& _msg) : msg(_msg) {}
    virtual ~my_exception() throw() {}
    const string msg;
    virtual const char* what() const throw() { return msg.c_str(); }
};

class fatal_exception: public my_exception { // stop program execution
public:
    fatal_exception(const string& _msg) : my_exception(_msg) {}
};

class internal_exception: public fatal_exception { // logical consistency checks
public:
    internal_exception(const string& _msg) : fatal_exception(_msg) {}
};

class nonfatal_exception: public my_exception { // continue program execution
public:
    nonfatal_exception(const string& _msg) : my_exception(_msg) {}
};

class syntax_exception: public nonfatal_exception {
public:
    syntax_exception(const string& _msg) : nonfatal_exception(_msg) {}
};

class performance_exception: public nonfatal_exception { // e.g., i/o problems
public:
    performance_exception(const string& _msg) : nonfatal_exception(_msg) {}
};


extern class theory th;

class theory {
```

```
public:

    class wff;
    class node_reference;

    // class for referencing wffs/theorems (use of numbers instead of pointers)
    class reference {

    protected:

    reference_data_type no;

    virtual void check_for_limit(reference_data_type _no) const = 0;

    public:

    reference() : no(undefined) {}

    reference_data_type get_no() const { check_for_limit(no); return no; }
    bool is_defined() const { return no!=undefined; }
    bool is_undefined() const { return no==undefined; }

    reference& operator= (reference_data_type _no) { check_for_limit(_no); no=_n
o; return *this; }
    bool operator== (const reference& _ref) const { return no==_ref.no; }
    bool operator!= (const reference& _ref) const { return no!=_ref.no; }
    bool operator< (const reference& _ref) const { return no<_ref.no; }

    virtual ostream& print(ostream& _out, bool _typeflag=true, bool _indexflag=f
alse, unsigned int _level=0, bool _left=true, int _index=0) const = 0;
    operator string() const { ostringstream out; print(out, false); return out.s
tr(); }

    };

    // class for referencing wffs (use of numbers instead of pointers)
    class wff_reference : public reference {

    protected:

    virtual void check_for_limit(reference_data_type _no) const {
        if (_no>=th.wff_ptrs.size() || _no>=undefined) throw syntax_exception("i
llegal wff reference");
    }

    public:

    wff_reference() : reference() {}
    wff_reference(reference_data_type _no) { check_for_limit(_no); no=_no; }
```

```
    theory::wff& operator* () const { check_for_limit(no); return *(th.wff_ptrs[
no]); }

    wff_reference& operator++ (int) { no++; if (no==th.wff_ptrs.size()) no=undef
ined; return *this; }
    static wff_reference begin() { if (th.wff_ptrs.size()>0) return wff_referenc
e(0); else return wff_reference(); }
    static wff_reference end() { return wff_reference(); }

    virtual ostream& print(ostream& _out, bool _typeflag=true, bool _indexflag=f
alse, unsigned int _level=0, bool _left=true, int _index=0) const {
        string header(COMMENT_HEADER); header+=(_typeflag ? "w typd" : "wff");
        if (file_format==FORMAT_PLAIN) _out.width(SPACING_INDENT_WIDTH);
        _out << left << header << MDOL << MTAB;
        if (file_format==FORMAT_PLAIN) _out.width(SPACING_NUMBER_WIDTH);
        _out << right << no << MSPC << ":" << MSPC << MTAB;
        (**this).print(_out, _typeflag, _level, _left, false, SPACING_INDENT_WID
TH+SPACING_NUMBER_WIDTH+5, 0);
        _out << MDOL;
        return _out;
    }

    virtual wff_reference get_wff_part(const reference_data_type& _no) const {
        return (**this).get_wff_part(node_reference(), _no);
    }

    virtual bool get_wff_part_no(node_reference& _node, const wff_reference& _pa
rt, reference_data_type& _occurence_no) const {
        _node = node_reference();
        return (**this).get_wff_part_no(_node, _part, _occurence_no);
    }

    virtual wff_reference substitute_wff_part_no(signed int _no, const wff_refer
ence& _wff_part, const wff_reference& _wff_subst, const wff_reference& _h=wff_refere
nce(), const wff_reference& _a_b=wff_reference(), const set<wff_reference>& _bound_v
ars=set<wff_reference>()) const {
        wff_reference ref = (**this).substitute_wff_part_no(node_reference(), _n
o, _wff_part, _wff_subst, _h, _a_b, _bound_vars);
        if(_no>=0) throw syntax_exception("illegal wff part number");
        return ref;
    }

    virtual ostream& print_types(ostream& _out) const { (**this).print_types(_ou
t); return _out; }
    virtual ostream& print_wff_parts(ostream& _out) const { (**this).print_wff_p
arts(_out, node_reference()); return _out; }

    };
```

```
    class no_match_for_substitution_exception: public syntax_exception {
    protected:
    const wff_reference x, y;
    static string create_msg(const wff_reference& _x, const wff_reference& _y) {
 return string("wffs do not match for substitution:'")+(*_x).get_name(scope_violatio
n_debug)+"' != '"+(*_y).get_name(scope_violation_debug)+"'"+" (wffs "+to_string(_x.g
et_no())+", "+to_string(_y.get_no())+")"; }
    public:
    no_match_for_substitution_exception(const wff_reference& _x, const wff_refer
ence& _y) : syntax_exception(create_msg(_x, _y)), x(_x), y(_y) {}
    };

    class substitution_result_missing_expected_type_exception: public internal_excep
tion {
    protected:
    const wff_reference x, y;
    static string create_msg(const wff_reference& _x, const wff_reference& _y) {
 return string("substitution result '")+(*_x).get_name(scope_violation_debug)+"' doe
s not have type '"+(*_y).get_name(scope_violation_debug)+"'"+" (wffs "+to_string(_x.
get_no())+", "+to_string(_y.get_no())+")"; }
    public:
    substitution_result_missing_expected_type_exception(const wff_reference& _x,
 const wff_reference& _y) : internal_exception(create_msg(_x, _y)), x(_x), y(_y) {}
    };

    class scope_violation_exception: public syntax_exception {
    public:
    scope_violation_exception(const string& _msg) : syntax_exception(_msg) {}
    };

    class scope_violation_in_substitution_exception: public scope_violation_exceptio
n {
    protected:
    const wff_reference x, H, AB;
    static string create_msg(const wff_reference& _x, const wff_reference& _H, c
onst wff_reference& _AB) { return string("scope violation in substitution -- bound v
ariable '")+MDOL+(*_x).get_name(scope_violation_debug)+MDOL+"' is free in hypothesis
 '"+MDOL+(*_H).get_name(scope_violation_debug)+MDOL+"' and free in equation '"+MDOL+
(*_AB).get_name(scope_violation_debug)+MDOL+"'"+" (wffs "+to_string(_x.get_no())+",
"+to_string(_H.get_no())+", "+to_string(_AB.get_no())+")"; }
    public:
    scope_violation_in_substitution_exception(const wff_reference& _x, const wff
_reference& _H, const wff_reference& _AB) : scope_violation_exception(create_msg(_x,
 _H, _AB)), x(_x), H(_H), AB(_AB) {}
    };

    class scope_violation_in_lambda_conversion_exception: public scope_violation_exc
eption {
```

```
    protected:
    const wff_reference A, x, B;
    static string create_msg(const wff_reference& _A, const wff_reference& _x, c
onst wff_reference& _B = wff_reference()) { return string("scope violation in lambda
 conversion -- '")+MDOL+(*_A).get_name(scope_violation_debug)+MDOL+"' is not free fo
r '"+MDOL+(*_x).get_name(scope_violation_debug)+MDOL+(_B.is_defined() ? string("' in
 '")+MDOL+(*_B).get_name(scope_violation_debug)+MDOL+"'" : "'")+" (wffs "+to_string(
_A.get_no())+", "+to_string(_x.get_no())+(_B.is_defined() ? ", "+to_string(_B.get_no
()) : "")+")"; }
    static inline wff_reference choose_b(const scope_violation_in_lambda_convers
ion_exception& _e, const wff_reference& _B) { return ( _e.B.is_defined() ? _e.B : _B
); }
    public:
    scope_violation_in_lambda_conversion_exception(const wff_reference& _A, cons
t wff_reference& _x) : scope_violation_exception(create_msg(_A, _x)), A(_A), x(_x) {
}
    scope_violation_in_lambda_conversion_exception(const scope_violation_in_lamb
da_conversion_exception& _e, const wff_reference& _B) : scope_violation_exception(cr
eate_msg(_e.A, _e.x, choose_b(_e, _B))), A(_e.A), x(_e.x), B(choose_b(_e, _B)) {}
    };

    class scope_violation_in_variable_renaming_exception: public scope_violation_exc
eption {
    protected:
    const wff_reference v, A;
    static string create_msg(const wff_reference& _v, const wff_reference& _A) {
 return string("scope violation in variable renaming -- variable '")+MDOL+(*_v).get_
name(scope_violation_debug)+MDOL+"' occurs free in '"+MDOL+(*_A).get_name(scope_viol
ation_debug)+MDOL+"'"+" (wffs "+to_string(_v.get_no())+", "+to_string(_A.get_no())+"
)"; }
    public:
    scope_violation_in_variable_renaming_exception(const wff_reference& _v, cons
t wff_reference& _A) : scope_violation_exception(create_msg(_v, _A)), v(_v), A(_A) {
}
    };


    // class for referencing thms (use of numbers instead of pointers)
    class thm_reference : public reference {

    protected:

    virtual void check_for_limit(reference_data_type _no) const {
        if (_no>=th.thm_ptrs.size() || _no>=undefined) throw syntax_exception("i
llegal wff reference");
    }

    public:
```

```
    thm_reference() : reference() {}
    thm_reference(reference_data_type _no) { check_for_limit(_no); no=_no; }

    theory::wff& operator* () const { check_for_limit(no); return *(th.thm_ptrs[
no]); }

    thm_reference& operator++ (int) { no++; if (no==th.thm_ptrs.size()) no=undef
ined; return *this; }
    static thm_reference begin() { if (th.thm_ptrs.size()>0) return thm_referenc
e(0); else return thm_reference(); }
    static thm_reference end() { return thm_reference(); }
    const wff_reference& get_wff_reference () const { return (*(th.thm_ptrs[no])
).ref; }

    virtual ostream& print(ostream& _out, bool _typeflag=true, bool _indexflag=f
alse, unsigned int _level=0, bool _left=true, int _index=0) const {
        string header(COMMENT_HEADER); if (_indexflag) header+="thm";
        if (file_format==FORMAT_PLAIN) _out.width(SPACING_INDENT_WIDTH);
        _out << left << header << MDOL << MTAB;
        ostringstream os;
        if (_indexflag) os << MIF2("\\%","%") << _index;
        else os << MSPC << MSPC;
        if (file_format==FORMAT_PLAIN) _out.width(SPACING_NUMBER_WIDTH);
        _out << right << os.str() << MSPC << (_indexflag ? ":" : MSPC) << MSPC <
< MTAB;
        (**this).print(_out, _typeflag, 0, false, true, SPACING_REGULAR_INDENT,
0);
        _out << MDOL;
        return _out;
    }

    };

    // class for referencing wff nodes (use of binary trees respectively the integer
 representation of the binary number)
    class node_reference : public bitset<node_reference_bits> {

    public:

    node_reference() : bitset<node_reference_bits>(1) {}
    node_reference(unsigned long _u) : bitset<node_reference_bits>(_u) {}
    node_reference(string _str) : bitset<node_reference_bits>(_str) {}

    size_t length() const {
        unsigned int u = node_reference_bits;
        do {
        u--;
        if (test(u)) return u+1;
        } while (u>0);
```

```
        return 0;
    }

    bool maximum_reached() const {
        return (length()==node_reference_bits);
    }

    node_reference& push_back(bool _b) {
        if (maximum_reached()) throw syntax_exception("maximum node size exceede
d");
        operator<<=(1);
        operator|=(_b);
        return *this;
    }

    node_reference& push_front() {
        if (maximum_reached()) throw syntax_exception("maximum node size exceede
d");
        set(length()); // add a heading bit
        return *this;
    }

    ostream& print(ostream& _out) const {
        _out << to_string() << " [";
        _out.width(2);
        _out.fill(' ');
        _out << length() << " bits]: " << to_ulong();
        return _out;
    }

    ostream& print_no(ostream& _out) const {
        _out << to_string() << " (";
        _out.width(3);
        _out.fill(' ');
        _out << right << to_ulong() << ")";
        return _out;
    }

    operator string() const { ostringstream out; print(out); return out.str(); }

    };

    // stack for backward thm_reference in input
    class stack {

    protected:

    deque<thm_reference> deq;
```

```
    public:

    thm_reference operator[] (unsigned int _u) const { if (_u>=deq.size()) throw
  syntax_exception("illegal stack reference"); else return deq[_u]; }
    thm_reference push(thm_reference _ref) { deq.push_front(_ref); if (deq.size(
)>INPUT_STACK_SIZE) deq.pop_back(); return _ref; }
    reference_data_type size() const { return deq.size(); }

    reference_data_type get_index(const thm_reference& _ref) {
        reference_data_type sz = size();
        reference_data_type u = 0;
        do {
        if (deq[u]==_ref)
            return u;
        } while (++u<sz);
        throw syntax_exception("theorem not in stack (recall with: %THEOREM)");
    }

    ostream& print(ostream& _out, unsigned int _max=std_stack_print) const {
        unsigned int sz = size();
        if (sz==0) {
        _out << COMMENT_HEADER << "stack empty" << endl;
        }
        else {
        unsigned int u=sz-1;
        if (_max!=0) u=(u>_max-1 ? _max-1 : u);
        do {
            deq[u].print(_out, false, true, 0, true, u);
            _out << endl;
        } while (u--);
        }
        return _out;
    }

    operator string() const { ostringstream out; print(out); return out.str(); }

    } input_stack;


protected:

    //
    // initialize these members first so following members can register here
    //

    // wffs of this theory (for avoiding duplicate creation)
    vector<wff*> wff_ptrs; // all well-formed formulas
    vector<wff*> thm_ptrs; // proven theorems
```

```
    // map of created composed objects (for avoiding duplicate creation)
    map<pair<wff_reference, wff_reference>, wff_reference> composed_types;
    map<pair<pair<wff_reference, wff_reference>, pair<wff_reference, wff_reference>
>, wff_reference> lambda_abstractions;
    map<pair<pair<wff_reference, wff_reference>, pair<wff_reference, wff_reference>
>, wff_reference> lambda_applications;

    // add a theorem
    thm_reference add_theorem(const wff_reference& _ref) {
    if (thm_exists((*_ref).id))
        return(find_thm((*_ref).id));
    else {
        thm_ptrs.push_back(&(*_ref));
        // check for new types
        thm_reference ref = thm_reference(thm_ptrs.size()-1);
        check_for_new_type(ref);
        return ref;
    }
    }

    // check for a proposition that introduces a new type
    void check_for_new_type(const thm_reference& _ref) {
    if ((*_ref).is_lambda_application()) { // check for lambda application
        lambda_application& appl = (*_ref).get_lambda_application();
        const wff_reference& type_left = appl.type_left;
        if ((*type_left).is_composed_type() && // check left argument for set (a
nything to boole)
        (*type_left).get_composed_type().left==th.boole.ref) {
        wff_reference newtype = appl.left;
        if (!(*newtype).is_variable()) (*newtype).add_type(th.tau.ref); // m
ake newtype a new type: add tau (unless it is a variable -- variables only have one
type)
        if (!(*appl.right).is_variable()) (*appl.right).add_type(newtype); /
/ object is of the new type (unless it is a variable -- variables only have one type
)
        }
    }
    }


public:

    // forward declaration for typecasts in entity::get_type() etc.
    class variable;
    class lambda_abstraction;
    class lambda_application;
    class composed_type;

    //
```

```
    // All well-formed formulas
    //
    // All names for entitites at the object level: both variables and constants, in
cluding types
    // in dependent type theory, types are objects, so the object level is complete
and universal.
    // An object may have several wffs that denote the same object (e.g., '2' may be
 written as '1+1').
    // A wff may have several types (e.g., '2' is a natural number and belongs to th
e universal type omega).
    //
    class wff {

    public:

    // the unique ID string within the theory
    const string id;

    // the reference to this wff
    const wff_reference ref;

    protected:

    // the definitions (names) this wff has
    set<string> defs;
    // the shortest definition that was declared (to be used for printing the wf
f)
    string shortest_def;
    // the markdown definition that was declared (to be used for printing the wf
f)
    string markdown_def;

    // the types of the object denoted by this wff has (excluding trivial omega)
    set<wff_reference> types;


    // register wff and return reference
    static wff_reference register_wff(wff* _wff) {
        // here only do an additional check for formal security against programm
ing mistakes at the drawback of performance
        // (this check may be omitted after intensive testing of the program)
        if (debug)
        if (th.wff_exists(_wff->id))
            throw internal_exception("wff to be created already exists");
        th.wff_ptrs.push_back(_wff);
        return wff_reference(th.wff_ptrs.size()-1);
    }

    void internal_add_type(const wff_reference& _ref, const wff_reference& _ref2
```

```
=wff_reference()) {
        if(types.find(_ref)==types.end()) types.insert(_ref);
        if (_ref2.is_defined()) if(types.find(_ref2)==types.end()) types.insert(
_ref2);
    }


    public:

    wff(const string& _id) : id(_id), ref(register_wff(this)) {
        // main check for existing wff implemented within the specific derived c
lasses (e.g., class composed_type)
        // an additional check is done in: static wff_reference wff::register_wf
f(wff* _wff) - see above
    }

    // method for constructing the volatile name for printing (depending on defi
nitions of subentities only)
    virtual string get_sub_name(bool _typeflag=true, unsigned int _level=0, bool
 _left=true) const { return MESC(id); }

    // volatile name for printing (depending on definitions of entity and subent
ities)
    string get_name(bool _typeflag=true, unsigned int _level=0, bool _left=true)
 const { return(!shortest_def.empty() ? ((file_format==FORMAT_MARKDOWN && has_markdo
wn_definition()) ? markdown_def : MESC(shortest_def)) : get_sub_name(_typeflag, _lev
el, _left)); }

    // determine whether this wff needs space for printing
    bool needs_space() {
        return !(has_markdown_definition() || (!has_definition() && (is_composed
_type() || (is_type_variable() && get_variable().id_short.length()==1) || is_depende
nt_type_variable())));
    }

    // check for existing definition
    bool has_definition() const { return !defs.empty(); }

    // add a definition token
    void add_definiton(const string& _str) {
        // allow existing identical definition for this symbol
        if (*this==_str || !th.wff_exists(_str)) {
        // add definition to wff
        defs.insert(_str);
        // find shortest definition for printing
        if (shortest_def.empty() || _str.length() < shortest_def.length()) s
hortest_def = _str;
        }
        else {
```

```
        if (th.wff_exists(_str)) throw syntax_exception(string("definition s
ymbol '")+_str+"' exists already");
        }
    }

    // remove a definition token
    reference_data_type remove_definiton(const string& _str) {
        if (_str[0]!='$' && !allow_definition_removal) throw syntax_exception("d
efinition removal not allowed without flag");
        reference_data_type no = defs.erase(_str);
        if (no!=0) {
        // find shortest definition for printing
        string shortest;
        for (set<string>::iterator iter=defs.begin(); iter!=defs.end(); iter
++) {
            if (shortest.empty()) shortest=(*iter);
            else if ((*iter).length() < shortest.length()) shortest=(*iter);
        }
        shortest_def = shortest;
        }
        return no;
    }

    // check for existing definition
    bool has_markdown_definition() const { return !markdown_def.empty(); }

    // add a definition token
    void add_markdown_definiton(const string& _str) {
        // allow existing identical definition for this symbol
        if (markdown_def != _str) {
        if(markdown_def.empty())
            markdown_def = _str;
        else
            throw syntax_exception(string("markdown definition symbol exists
 already for '")+get_name(false)+"'");
        }
    }

    bool has_temporary_definition() const {
        for (set<string>::iterator iter=defs.begin(); iter!=defs.end(); iter++)
        if ((*iter)[0]=='$') return true;
        return false;
    }

    bool operator== (const wff& _wff) const { return id==_wff.id; }
    bool operator== (const string& _id) const {
        if (id==_id) return true;
        else {
        set<string>::iterator iter;
```

```
        for (iter=defs.begin(); iter!=defs.end(); iter++)
            if (*iter==_id) return true;
        }
        return false;

    }
    bool operator!= (const wff& _wff) const { return !(*this==_wff); }
    bool operator!= (const string& _id) const { return !(*this==_id); }

    bool is_type() { return( has_type(th.tau.ref) ); }

    virtual bool is_variable() const = 0;
    virtual bool is_type_variable() const = 0;
    virtual bool is_dependent_type_variable() const = 0;
    virtual bool is_predefined_non_variable() const = 0;
    virtual bool is_composed() const = 0;
    virtual bool is_lambda_abstraction() const = 0;
    virtual bool is_lambda_application() const = 0;
    virtual bool is_composed_type() const = 0;

    virtual bool has_free_variable(const wff_reference& _ref) const = 0;

    virtual wff_reference matches_for_lambda_application(const wff_reference& _r
ight, unsigned int _mode=0) const {
        // exact match only
        if (_mode==0) {
        if (ref==_right)
            return _right;
        }
        // allow for omega
        else {
        if (ref==th.omega.ref)
            return th.omega.ref;
        }
        return wff_reference();
    }

    variable& get_variable() { return *(dynamic_cast<variable*>(this)); }
    lambda_abstraction& get_lambda_abstraction() { return *(dynamic_cast<lambda_
abstraction*>(this)); }
    lambda_application& get_lambda_application() { return *(dynamic_cast<lambda_
application*>(this)); }
    composed_type& get_composed_type() { return *(dynamic_cast<composed_type*>(t
his)); }

    const variable& get_const_variable() const { return *(dynamic_cast<const var
iable*>(this)); }

    virtual bool has_type(const wff_reference& _ref) {
```

```
        if (_ref==th.omega.ref) return true; // omega is not included in list
        if (types.find(_ref)!=types.end()) return true; // check list
        for (set<wff_reference>::iterator it = types.begin(); it != types.end();
 it++) {
        wff_reference ref = (*_ref).matches_for_lambda_application(*it, 1);
// check for matching omegas
        if (ref.is_defined()) { add_type(ref); return true; }
        }
        return false;
    }

    void add_type(const wff_reference& _ref) { internal_add_type(_ref); }
    void add_types(const wff_reference& _ref) { for (set<wff_reference>::iterato
r it = (*_ref).get_types().begin(); it != (*_ref).get_types().end(); it++) internal_
add_type(*it); }

    // get types (possibly without trivial omega, which may not be included)
    const set<wff_reference>& get_types() const { return types; }

    // get single type (or throw exception if more than one type exists)
    wff_reference get_single_type() const {
        if (types.size()==1)
        return *types.begin();
        else if (types.size()==0)
        return th.omega.ref;
        else
        throw syntax_exception("wff has no or more than one type, explicit t
ype name required");
    }

    // get wff by base and type
    static wff_reference obtain_wff(const wff_reference& _base, const wff_refere
nce& _type) {
        // for variables ignore type info, since it is part of the name (and sec
ond appearance of a type belongs to a lambda abstraction or application)
        if ((*_base).is_variable())
        return _base;
        else {
        // check for necessary conditions first
        if (!(*_base).has_type(_type)) {
            string msg;
            msg += "wff '"+(*_base).get_name()+"' does not have type '"+(*_t
ype).get_name()+"'";
            throw syntax_exception(msg);
        }
        return _base;
        }
    }
```

```
    virtual wff_reference get_wff_part(const node_reference& _node, const refere
nce_data_type& _no) const {
        if ((unsigned)_no==_node.to_ulong())
        return ref;
        else
        return wff_reference(); // return undefined reference
    }

    virtual bool get_wff_part_no(node_reference& _node, const wff_reference& _pa
rt, reference_data_type& _occurence_no) const {
        if (_part==ref)
        if (_occurence_no--==0)
            return true;
        return false;
    }

    virtual wff_reference substitute_wff_part_no_standard_check(const node_refer
ence& _node, signed int& _no, const wff_reference& _wff_part, const wff_reference& _
wff_subst, const wff_reference& _h, const wff_reference& _a_b, const set<wff_referen
ce>& _bound_vars) const {
        if ((unsigned)_no==_node.to_ulong()) {
        if (_wff_part==ref) {
            // check for scope violation in Rule R' [cf. Andrews 2002 (ISBN
1-4020-0763-9), p. 214 (Rule R')]
            if (_h.is_defined())
            for (set<wff_reference>::iterator iter=_bound_vars.begin();
iter!=_bound_vars.end(); iter++)
                if ((*_h).has_free_variable(*iter) && (*_a_b).has_free_v
ariable(*iter))
                throw scope_violation_in_substitution_exception(*ite
r, _h, _a_b);
            _no=-1;
            return _wff_subst;
        }
        else throw no_match_for_substitution_exception(ref, _wff_part);
        }
        return wff_reference();
    }

    virtual wff_reference substitute_wff_part_no(const node_reference& _node, si
gned int& _no, const wff_reference& _wff_part, const wff_reference& _wff_subst, cons
t wff_reference& _h, const wff_reference& _a_b, const set<wff_reference>& _bound_var
s) const {
        wff_reference chkref = substitute_wff_part_no_standard_check(_node, _no,
 _wff_part, _wff_subst, _h, _a_b, _bound_vars);
        if (chkref.is_defined()) return chkref;
        return ref;
    }
```

```
    virtual ostream& print_types(ostream& _out) const {
        set<wff_reference>::iterator type_iter;
        unsigned int u = 0;
        for (type_iter=types.begin(); type_iter!=types.end(); type_iter++) {
        unsigned int position = 0;
        ostringstream out;
        out << "type     # ";
        out.width(3); out << right << u << ":  ";
        out << ((**type_iter).is_lambda_application() && (**type_iter).get_l
ambda_application().left==th.identity.ref ? "[trivial] " : "" ) << (**type_iter).get
_name();
        _out << out.str(); position += out.str().length();
        force_display_position(_out, position, SPACING_REGULAR_INDENT+SPACIN
G_NAME_WIDTH+SPACING_NAME_WIDTH_DIST+SPACING_TYPES_WIDTH, SPACING_TYPES_WIDTH_DIST);
        cout << " =  " << (*type_iter).get_no() << endl;
        u++;
        }
        return _out;
    }

    virtual ostream& print_wff_parts(ostream& _out, const node_reference& _node)
 const {
        print_wff_parts_this(_out, _node);
        return _out;
    }

    virtual ostream& print_wff_parts_this(ostream& _out, const node_reference& _
node) const {
        _out << "/";
        _node.print_no(_out);
        print(_out, false, 0, true, false, SPACING_REGULAR_INDENT, 2);
        _out << endl;
        return _out;
    }

    virtual wff_reference recursively_substitute_variable(const wff_reference& _
var_ref, const wff_reference& _body_ref, const set<wff_reference>& _bound_vars) = 0;

    virtual wff_reference recursively_substitute_type_variable_at_base_level(con
st wff_reference& _var_ref, const wff_reference& _body_ref, const set<wff_reference>
& _bound_vars) = 0;

    virtual wff_reference recursively_substitute_type_variable_at_type_level(con
st wff_reference& _var_ref, const wff_reference& _body_ref, const set<wff_reference>
& _bound_vars) = 0;

    virtual wff_reference recursively_substitute_by_dependent_type(const wff_ref
erence& _var_ref, unsigned int _level=1) { return ref; }
```

```
    virtual wff_reference recursively_substitute_dependent_type_variable(const w
ff_reference& _var_ref, unsigned int _level=1) { return ref; }

    ostream& print(ostream& _out, bool _typeflag=true, unsigned int _level=0, bo
ol _left=true, bool _thm=false,
             unsigned int _indent=SPACING_REGULAR_INDENT,
             unsigned int _indent_first=SPACING_REGULAR_INDENT) const {

        bool show_types = (MIF2((_thm && (_level==0) ? false : _typeflag != (_le
vel==0)),_typeflag));
        bool show_defs = !defs.empty();
        unsigned int position = _indent;

        if (file_format==FORMAT_PLAIN) { _out.width(_indent_first); _out << righ
t << ""; } // indent first line
        else if (file_format==FORMAT_MARKDOWN && show_types) _out << "{";

        if (print_debug) {
        cout << "[print(" << _indent << "," << _indent_first << ")]" << endl
;
        if (file_format==FORMAT_PLAIN) { _out.width(_indent); _out << right
<< ">"; }
        }

        string sub_name = get_sub_name(_typeflag, _level);
        _out << sub_name; position += sub_name.length();

        if (file_format==FORMAT_MARKDOWN && show_types) _out << "}";

        if (show_types) {
        force_display_position(_out, position, _indent+SPACING_NAME_WIDTH, S
PACING_NAME_WIDTH_DIST, _thm);
        string types_str;
        types_str += MIF2("_{","{");
        if (types.size()==0) {
            types_str += MIF2(""," ")+th.omega.get_name();
        }
        else if (types.size()==1) {
            types_str += (file_format==FORMAT_MARKDOWN ? "" : MTYPESEP) + (*
get_single_type()).get_name();
        }
        else {
            bool trivial_identity_occured = false; // skip omega-bool trivia
ls in output (from simple identifications)
            wff_reference md_type; // type chosen for markdown print (only o
ne)
            set<wff_reference>::iterator type_iter;
            for (type_iter=types.begin(); type_iter!=types.end(); type_iter+
+) {
```

```
            if ((**type_iter).is_lambda_application() && (**type_iter).g
et_lambda_application().left==th.identity.ref) // trivial type (skip)
                trivial_identity_occured = true;
            else { // non-trivial type
                if (file_format==FORMAT_MARKDOWN) { // markdown format:
print first type only (but prefer other type than tau)
                if (!md_type.is_defined() || md_type==th.tau.ref) md
_type = *type_iter;
                }
                else { // plain format: print all types (except trivial
types)
                types_str += MTYPESEP + (**type_iter).get_name();
                }
                // types_str += (file_format==FORMAT_MARKDOWN && !real_t
ype_occured ? "" : MTYPESEP) + (**type_iter).get_name();
                // real_type_occured = true;
            }
            }
            if (file_format==FORMAT_MARKDOWN) { types_str += (*md_type).get_
name(); }
            if (trivial_identity_occured) types_str += MTYPESEP + string("..
.");
        }
        types_str += MIF2("}"," }");
        _out << types_str; position += types_str.length();
        }

        if (show_defs) {
        force_display_position(_out, position, _indent+SPACING_NAME_WIDTH+SP
ACING_NAME_WIDTH_DIST+SPACING_TYPES_WIDTH, SPACING_TYPES_WIDTH_DIST, _thm);
        string d(MIF2("\\qquad$ $\\mathrel{\\mathop:}= ",":="));
        set<string>::iterator iter;
        for (iter=defs.begin(); iter!=defs.end(); iter++)
            if (!(is_predefined_non_variable() && *iter==shortest_def )) d +
= MIF2("\\;\\;$ $","  ") + MESC((*iter));
        _out << d;
        }
        return _out;
    }

    operator string() const { ostringstream out; print(out, false); return out.s
tr(); }

    };

    //
    // simple objects (both variables and non-variables)
    //
    class simple : public wff {
```

```
    public:

    simple(const string& _id) : wff(_id) {}

    virtual bool is_composed() const { return false; }

    virtual wff_reference substitute_wff_subpart_no(signed int& _no, const wff_r
eference& _wff_part, const wff_reference& _wff_subst) const {
        return ref; // simple wff has no subparts: nothing to do
    }

    };

    //
    // composed objects (non-variables only)
    //
    class composed : public wff {

    public:

    const wff_reference left, right;

    composed(const string& _id, const wff_reference& _left, const wff_reference&
 _right) : wff(_id), left(_left), right(_right) {}

    virtual bool is_variable() const { return false; }
    virtual bool is_type_variable() const { return false; }
    virtual bool is_dependent_type_variable() const { return false; }
    virtual bool is_predefined_non_variable() const { return false; }
    virtual bool is_composed() const { return true; }

    bool needs_inner_space(bool _typeflag) const { // decide whether spacing is
needed between printing of two objects
        return (!_typeflag && ((*left).needs_space() || (*right).needs_space()))
;
    }

    virtual wff_reference get_wff_part(const node_reference& _node, const refere
nce_data_type& _no) const {
        if ((unsigned)_no==_node.to_ulong())
        return ref;
        if (!_node.maximum_reached()) {
        wff_reference ref;
        // left node
        node_reference left_node(_node);
        left_node.push_back(false);
        ref = (*left).get_wff_part(left_node, _no); if (ref.is_defined()) {
return ref; }
```

```
        // right node
        node_reference right_node(_node);
        right_node.push_back(true);
        ref = (*right).get_wff_part(right_node, _no); if (ref.is_defined())
{ return ref; }
        }
        else {
        // cout << "# maximum reached -- skipping nodes ..." << endl;
        }
        return wff_reference(); // return undefined reference
    }

    virtual bool get_wff_part_no(node_reference& _node, const wff_reference& _pa
rt, reference_data_type& _occurence_no) const {
        if (_part==ref)
        if(_occurence_no--==0)
            return true;
        if (!_node.maximum_reached()) {
        // left node
        node_reference left_node(_node);
        left_node.push_back(false);
        if ((*left).get_wff_part_no(left_node, _part, _occurence_no)) {
            _node = left_node;
            return true;
        }
        // right node
        node_reference right_node(_node);
        right_node.push_back(true);
        if ((*right).get_wff_part_no(right_node, _part, _occurence_no)) {
            _node = right_node;
            return true;
        }
        }
        else {
        // cout << "# maximum reached -- skipping nodes ..." << endl;
        }
        return false;
    }

    virtual ostream& print_wff_parts(ostream& _out, const node_reference& _node)
 const {
        // first print this node itself
        print_wff_parts_this(_out, _node);
        if (!_node.maximum_reached()) {
        // print left node
        node_reference left_node(_node);
        left_node.push_back(false);
        (*left).print_wff_parts(_out, left_node);
        // print right node
```

```
        node_reference right_node(_node);
        right_node.push_back(true);
        (*right).print_wff_parts(_out, right_node);
        }
        else {
        // _out << "# maximum reached -- skipping nodes ..." << endl;
        }
        return _out;
    }

    };

    //
    // variables
    //
    class variable : public simple {

    protected:

    variable(const string& _id, const string& _id_short, const wff_reference& _t
ype, unsigned int _type_variable_dependency) : simple(_id), id_short(_id_short), typ
e(_type), type_variable_dependency(_type_variable_dependency) {
        add_type(_type);
    }

    // method for constructing the static id for identification
    static string create_id(const string& _name, const wff_reference& _type) {
        string str;
        str += _name+"{"+(*_type).id+"}";
        return str;
    }

    static wff_reference find_var(const string& _str) {
        return th.find_wff(_str);
    }

    static bool var_exists(const string& _str) {
        return th.wff_exists(_str);
    }


    public:

    // the name without the type
    const string id_short;

    // the reference to the type
    const wff_reference type;
```

```
    // dependency level, if type variable, otherwise zero
    unsigned int type_variable_dependency;

    // method for constructing the volatile name for printing (depending on defi
nitions of entity and subentities)
    virtual string get_sub_name(bool _typeflag=true, unsigned int _level=0, bool
 _left=true) const {
        string str=MESC(id_short); // in Markdown skip at higher levels avoiding
 double subscript
        if (MIF2(_typeflag && _level==0,_typeflag)) str += MIF2("_{","{")+(*type
).get_name(_typeflag, _level+1)+"}";
        return str;
    }

    virtual bool is_variable() const { return true; }
    virtual bool is_type_variable() const { return type==th.tau.ref; }
    virtual bool is_dependent_type_variable() const { return id_short[0]=='\\';
}
    virtual bool is_predefined_non_variable() const { return false; }
    virtual bool is_lambda_abstraction() const { return false; }
    virtual bool is_lambda_application() const { return false; }
    virtual bool is_composed_type() const { return false; }

    virtual bool has_free_variable(const wff_reference& _ref) const { return (_r
ef == ref); }

    static wff_reference obtain_variable(const string& _name, const wff_referenc
e& _type, unsigned int _level=0) {
        string name = create_id(_name, _type);
        if (var_exists(name)) { // first check for existing variable
        wff_reference var = find_var(name);
        if (!(*var).is_variable()) throw internal_exception(string("variable
 argument is not a variable in declaration of variable '")+_name+"{"+(*_type).get_na
me()+"}' ");
        if (!(*var).has_type(_type)) throw internal_exception(string("variab
le does not have declared type in declaration of variable '")+_name+"{"+(*_type).get
_name()+"}' ");
        return var;
        }
        else { // create new variable
        if (!(*_type).is_type()) throw syntax_exception(string("type argumen
t is not a type in declaration of variable '")+_name+"{"+(*_type).get_name()+"}' ");
        unsigned int level(_level), parsed_number; // parse dependent type n
umber
        if (_name[0]=='\\' && to_int(_name.substr(1), parsed_number)) level
= parsed_number;
        return (*(new variable(name, _name, _type, level))).ref;
        }
    }
```

```
    virtual wff_reference recursively_substitute_variable(const wff_reference& _
var_ref, const wff_reference& _body_ref, const set<wff_reference>& _bound_vars) {
        // substitute base first
        wff_reference new_base = ref;
        if (ref==_var_ref) {
        check_for_scope_violation_in_lambda_conversion(_var_ref, _body_ref,
_bound_vars);
        new_base = _body_ref;
        }
        // now substitute type
        wff_reference new_type = (*new_base).recursively_substitute_type_variabl
e_at_base_level(_var_ref, _body_ref, _bound_vars);
        return new_type;
    }

    virtual wff_reference recursively_substitute_type_variable_at_base_level(con
st wff_reference& _var_ref, const wff_reference& _body_ref, const set<wff_reference>
& _bound_vars) {
        wff_reference new_type = (*type).recursively_substitute_type_variable_at
_type_level(_var_ref, _body_ref, _bound_vars);
        if (type!=new_type) return obtain_variable(id_short, new_type);
        else return ref;
    }

    virtual wff_reference recursively_substitute_type_variable_at_type_level(con
st wff_reference& _var_ref, const wff_reference& _body_ref, const set<wff_reference>
& _bound_vars) {
        if (ref==_var_ref) {
        check_for_scope_violation_in_lambda_conversion(_var_ref, _body_ref,
_bound_vars);
        return _body_ref;
        }
        else return ref;
    }

    virtual wff_reference recursively_substitute_by_dependent_type(const wff_ref
erence& _var_ref, unsigned int _level=1) {
        if(ref==_var_ref) return obtain_variable(string("\\")+::to_string(_level
), th.tau.ref, _level);
        else return ref;
    }

    virtual wff_reference recursively_substitute_dependent_type_variable(const w
ff_reference& _var_ref, unsigned int _level=1) {
        return (type_variable_dependency == _level ? _var_ref : ref);
    }

    // check for scope violation in lambda conversion
```

```
    void check_for_scope_violation_in_lambda_conversion(const wff_reference& _va
r_ref, const wff_reference& _body_ref, const set<wff_reference>& _bound_vars) {
        // provide that "A is free for x in B" [Andrews 2002 (ISBN 1-4020-0763-9
), pp. 218 f. (5207) and p. 213 (definition of term)]
        set<wff_reference>::iterator iter;
        for (iter=_bound_vars.begin(); iter!=_bound_vars.end(); iter++)
        if ((*_body_ref).has_free_variable(*iter))
            throw scope_violation_in_lambda_conversion_exception(_body_ref,
_var_ref);
    }
    };

    //
    // all non-variables (including functions and types that are not variables, and
composed wffs)
    //
    class simple_non_variable : public simple {

    public:

    simple_non_variable(const string& _id) : simple(_id) {}

    virtual bool is_variable() const { return false; }
    virtual bool is_type_variable() const { return false; }
    virtual bool is_dependent_type_variable() const { return false; }

    virtual bool has_free_variable(const wff_reference& _ref) const { return fal
se; }
    };

    class predefined_non_variable : public simple_non_variable {

    public:

    predefined_non_variable(const string& _id, const char *def=NULL, const char
*md_def=NULL) : simple_non_variable(_id) {
        if (def!=NULL) add_definiton(def);
        if (md_def!=NULL) add_markdown_definiton(md_def);
    }

    virtual string get_sub_name(bool _typeflag=true, unsigned int _level=0, bool
 _left=true) const { return get_name(_typeflag, _level, _left); }

    virtual bool is_predefined_non_variable() const { return true; }
    virtual bool is_lambda_abstraction() const { return false; }
    virtual bool is_lambda_application() const { return false; }
    virtual bool is_composed_type() const { return false; }

    virtual wff_reference recursively_substitute_variable(const wff_reference& _
```

```
var_ref, const wff_reference& _body_ref, const set<wff_reference>& _bound_vars) {
        return ref;
    }

    virtual wff_reference recursively_substitute_type_variable_at_base_level(con
st wff_reference& _var_ref, const wff_reference& _body_ref, const set<wff_reference>
& _bound_vars) {
        return ref;
    }

    virtual wff_reference recursively_substitute_type_variable_at_type_level(con
st wff_reference& _var_ref, const wff_reference& _body_ref, const set<wff_reference>
& _bound_vars) {
        return ref;
    }

    };

    class tau_class : public predefined_non_variable {
    public:
    tau_class() : predefined_non_variable(ID_TAU, DEF_TAU, MARKDOWN_DEF_TAU) { i
nternal_add_type(th.tau.ref); }
    } tau;

    class omega_class : public predefined_non_variable {
    public:
    omega_class() : predefined_non_variable(ID_OMEGA, DEF_OMEGA, MARKDOWN_DEF_OM
EGA) { internal_add_type(th.tau.ref); }
    } omega;

    class boole_class : public predefined_non_variable {
    public:
    boole_class() : predefined_non_variable(ID_BOOLE, DEF_BOOLE, MARKDOWN_DEF_BO
OLE) { internal_add_type(th.tau.ref); }
    } boole;

    // types composed from two other types (types for lambda application)
    class composed_type : public composed {

    protected:

    composed_type(const string& _id, const wff_reference& _left, const wff_refer
ence& _right) :
    composed(_id, _left, _right) {
        register_composed_type(_left, _right, ref);
    }

    // method for constructing the static id for identification
    static string create_id(const wff_reference& _left, const wff_reference& _ri
```

```
ght) {
        string str;
        str += "{"+(*_left).id+","+(*_right).id+"}";
        return str;
    }

    // method for constructing the volatile name for printing (depending on defi
nitions of entity and subentities)
    virtual string get_sub_name(bool _typeflag=true, unsigned int _level=0, bool
 _left=true) const {
        string str;
        if (file_format==FORMAT_PLAIN) {
        str += "{"+(*left).get_name(_typeflag, _level+1)+","+(*right).get_na
me(_typeflag, _level+1, false)+"}";
        }
        else if (file_format==FORMAT_MARKDOWN) {
        str += string(_left ? "{" : "{(")+(*left).get_name(_typeflag, _level
+1)+(needs_inner_space(false) ? "\\," : "")+(*right).get_name(_typeflag, _level+1, f
alse)+string(_left ? "}" : ")}");
        }
        return str;
    }

    static bool composed_type_exists(const wff_reference& _left, const wff_refer
ence& _right) {
        pair<wff_reference, wff_reference> p(_left, _right);
        return (th.composed_types.find(p)!=th.composed_types.end());
    }

    static const wff_reference& find_composed_type(const wff_reference& _left, c
onst wff_reference& _right) {
        pair<wff_reference, wff_reference> p(_left, _right);
        map<pair<wff_reference, wff_reference>, wff_reference>::iterator iter =
th.composed_types.find(p);
        if (iter!=th.composed_types.end()) return (*iter).second;
        else throw internal_exception("illegal access to wff reference");
    }

    static void register_composed_type(const wff_reference& _left, const wff_ref
erence& _right, const wff_reference& _new) {
        pair<wff_reference, wff_reference> p(_left, _right);
        th.composed_types.insert(pair<pair<wff_reference, wff_reference>, wff_re
ference>(p, _new));
    }


    public:

    virtual bool is_lambda_abstraction() const { return false; }
```

```
    virtual bool is_lambda_application() const { return false; }
    virtual bool is_composed_type() const { return true; }

    virtual bool has_free_variable(const wff_reference& _ref) const { return (*l
eft).has_free_variable(_ref) || (*right).has_free_variable(_ref); }

    virtual wff_reference matches_for_lambda_application(const wff_reference& _r
ight, unsigned int _mode=0) const {
        // exact match only
        if (_mode==0) {
        if (ref==_right)
            return _right;
        }
        // allow for omega
        else {
        if ((*_right).is_composed_type()) {
            wff_reference l = (*left).matches_for_lambda_application((*_righ
t).get_composed_type().left, _mode);
            wff_reference r = (*right).matches_for_lambda_application((*_rig
ht).get_composed_type().right, _mode);
            if (l.is_defined() && r.is_defined()) {
            return composed_type::obtain_composed_type(l, r);
            }
        }
        }
        return wff_reference();
    }

    static wff_reference obtain_composed_type(const wff_reference& _left, const
wff_reference& _right) {
        // create new variable
        if (!(*_left).is_type()) {
        string msg;
        msg += "left argument is not a type in declaration of composed type
{"+(*_left).get_name()+","+(*_right).get_name()+"}' ";
        throw syntax_exception(msg);
        }
        if (!(*_right).is_type()) {
        string msg;
        msg += "right argument is not a type in declaration of composed type
 {"+(*_left).get_name()+","+(*_right).get_name()+"}' ";
        throw syntax_exception(msg);
        }
        // obtain entity
        wff_reference r = ( composed_type_exists(_left, _right) ? // check for e
xistance
                    find_composed_type(_left, _right) : // look up existi
ng entity
                    (*(new composed_type(create_id(_left, _right), _left,
```

```
  _right))).ref // or create new
                    );
        // specific for class composed_type: add type tau
        (*r).get_composed_type().add_type(th.tau.ref);
        return r;
    }

    virtual wff_reference recursively_substitute_variable(const wff_reference& _
var_ref, const wff_reference& _body_ref, const set<wff_reference>& _bound_vars) {
        return obtain_composed_type((*left).recursively_substitute_variable(_var
_ref, _body_ref, _bound_vars),
                    (*right).recursively_substitute_variable(_va
r_ref, _body_ref, _bound_vars));
    }

    virtual wff_reference recursively_substitute_type_variable_at_base_level(con
st wff_reference& _var_ref, const wff_reference& _body_ref, const set<wff_reference>
& _bound_vars) {
        return ref; // type is tau anyway
    }

    virtual wff_reference recursively_substitute_type_variable_at_type_level(con
st wff_reference& _var_ref, const wff_reference& _body_ref, const set<wff_reference>
& _bound_vars) {
        return obtain_composed_type((*left).recursively_substitute_type_variable
_at_type_level(_var_ref, _body_ref, _bound_vars),
                    (*right).recursively_substitute_type_variabl
e_at_type_level(_var_ref, _body_ref, _bound_vars));
    }

    virtual wff_reference recursively_substitute_by_dependent_type(const wff_ref
erence& _var_ref, unsigned int _level=1) {
        return obtain_composed_type((*left).recursively_substitute_by_dependent_
type(_var_ref, _level+1),
                    (*right).recursively_substitute_by_dependent
_type(_var_ref, _level+1));
    }

    virtual wff_reference recursively_substitute_dependent_type_variable(const w
ff_reference& _var_ref, unsigned int _level=1) {
        return obtain_composed_type((*left).recursively_substitute_dependent_typ
e_variable(_var_ref, _level+1),
                    (*right).recursively_substitute_dependent_ty
pe_variable(_var_ref, _level+1));
    }

    virtual wff_reference substitute_wff_subpart_no(signed int& _no, const wff_r
eference& _wff_part, const wff_reference& _wff_subst) const {
        return ref; // composed type has no base subparts: nothing to do
```

```
    }

    virtual wff_reference substitute_wff_part_no(const node_reference& _node, si
gned int& _no, const wff_reference& _wff_part, const wff_reference& _wff_subst, cons
t wff_reference& _h, const wff_reference& _a_b, const set<wff_reference>& _bound_var
s) const {
        wff_reference chkref = substitute_wff_part_no_standard_check(_node, _no,
 _wff_part, _wff_subst, _h, _a_b, _bound_vars);
        if (chkref.is_defined()) return chkref;
        if (!_node.maximum_reached()) {
        // left node
        node_reference left_node(_node);
        left_node.push_back(false);
        wff_reference new_left = (*left).substitute_wff_part_no(left_node, _
no, _wff_part, _wff_subst, _h, _a_b, _bound_vars);
        // right node
        node_reference right_node(_node);
        right_node.push_back(true);
        wff_reference new_right = (*right).substitute_wff_part_no(right_node
, _no, _wff_part, _wff_subst, _h, _a_b, _bound_vars);
        if (new_left!=left || new_right!=right) {
            return obtain_composed_type(new_left, new_right);
        }
        }
        else {
        // cout << "# maximum reached -- skipping nodes ..." << endl;
        }
        return ref;
    }

    };

    class identity_subtype1_class : public composed_type { // {o,@}
    public:
    identity_subtype1_class() :
    composed_type(create_id(th.boole.ref, th.omega.ref), th.boole.ref, th.omega.
ref) { internal_add_type(th.tau.ref); }
    } identity_subtype1;

    class identity_subtype2_class : public composed_type { // {o,t{^}}
    public:
    identity_subtype2_class() :
    composed_type(create_id(th.boole.ref, variable::obtain_variable("t", th.tau.
ref)), th.boole.ref, variable::obtain_variable("t", th.tau.ref)) { internal_add_type
(th.tau.ref); }
    } identity_subtype2;

    class identity_type1_class : public composed_type { // {{o,@},@}
    public:
```

```
    identity_type1_class() :
    composed_type(create_id(th.identity_subtype1.ref, th.omega.ref), th.identity
_subtype1.ref, th.omega.ref) { internal_add_type(th.tau.ref); }
    } identity_type1;

    class identity_type2_class : public composed_type { // {{o,t{^}},t{^}}
    public:
    identity_type2_class() :
    composed_type(create_id(th.identity_subtype2.ref, variable::obtain_variable(
"t", th.tau.ref)), th.identity_subtype2.ref, variable::obtain_variable("t", th.tau.r
ef)) { internal_add_type(th.tau.ref); }
    } identity_type2;

    class identity_class : public predefined_non_variable {

    public:

    const wff_reference type, type2;

    identity_class() : predefined_non_variable(ID_IDENTITY, DEF_IDENTITY), type(
th.identity_type1.ref), type2(th.identity_type2.ref) { internal_add_type(type, type2
); }

    // check for type
    virtual bool has_type(const wff_reference& _ref) {
        if (_ref==th.omega.ref) return true; // omega is not included in list
        if (types.find(_ref)!=types.end()) return true; // check list
        if (is_polymorphic_identity_type(_ref)) { add_type(_ref); return true; }
        return false;
    }

    // obtain polymorphic identity type corresponding to given type
    static wff_reference obtain_type(const wff_reference& _type_ref) {
        return composed_type::obtain_composed_type(composed_type::obtain_compose
d_type(th.boole.ref, _type_ref), _type_ref);
    }

    virtual wff_reference recursively_substitute_variable(const wff_reference& _
var_ref, const wff_reference& _body_ref, const set<wff_reference>& _bound_vars) {
        wff_reference new_type = (*type2).recursively_substitute_type_variable_a
t_type_level(_var_ref, _body_ref, _bound_vars);
        add_type(new_type);
        return ref; // substitute type (add new type) only
    }

    protected:

    static inline bool is_polymorphic_identity_type(const wff_reference& _ref) {
        return (*_ref).is_composed_type() && (*(*_ref).get_composed_type().left)
```

```
.is_composed_type() && (*(*_ref).get_composed_type().left).get_composed_type().left=
=th.boole.ref && (*(*_ref).get_composed_type().left).get_composed_type().right==(*_r
ef).get_composed_type().right;
    }

    } identity;

    class descriptor_type_class : public composed_type { // {t{^},{o,t{^}}}

    public:

    descriptor_type_class() :
    composed_type(create_id(variable::obtain_variable("t", th.tau.ref), th.ident
ity_subtype2.ref), variable::obtain_variable("t", th.tau.ref), th.identity_subtype2.
ref) { internal_add_type(th.tau.ref); }

    } descriptor_type;

    class descriptor_class : public predefined_non_variable {

    public:

    const wff_reference type; // main type

    descriptor_class() : predefined_non_variable(ID_DESCRIPTOR, DEF_DESCRIPTOR,
MARKDOWN_DEF_DESCRIPTOR), type(th.descriptor_type.ref) { internal_add_type(type); }

    // check for type
    virtual bool has_type(const wff_reference& _ref) {
        if (_ref==th.omega.ref) return true; // omega is not included in list
        if (types.find(_ref)!=types.end()) return true; // check list
        if (is_polymorphic_descriptor_type(_ref)) { add_type(_ref); return true;
 }
        return false;
    }

    virtual wff_reference recursively_substitute_variable(const wff_reference& _
var_ref, const wff_reference& _body_ref, const set<wff_reference>& _bound_vars) {
        wff_reference new_type = (*type).recursively_substitute_type_variable_at
_type_level(_var_ref, _body_ref, _bound_vars);
        add_type(new_type);
        return ref; // substitute type (add new type) only
    }

    protected:

    static inline bool is_polymorphic_descriptor_type(const wff_reference& _ref)
 {
        return (*_ref).is_composed_type() && (*(*_ref).get_composed_type().right
```

```
).is_composed_type() && (*(*_ref).get_composed_type().right).get_composed_type().lef
t==th.boole.ref && (*(*_ref).get_composed_type().right).get_composed_type().right==(
*_ref).get_composed_type().left;
    }

    } descriptor;

    class lambda_abstraction : public composed {

    protected:

    lambda_abstraction(const string& _id,
               const wff_reference& _left, const wff_reference& _type_le
ft,
               const wff_reference& _right, const wff_reference& _type_r
ight) :
    composed(_id, _left, _right), type_left(_type_left), type_right(_type_right)
 {
        register_lambda_abstraction(_left, _type_left, _right, _type_right, ref)
;
    }

    // method for constructing the static id for identification
    static string create_id(const wff_reference& _left, const wff_reference& _ty
pe_left,
                const wff_reference& _right, const wff_reference& _t
ype_right) {
        string str;
        str += "[\\"+(*_left).id+"{"+(*_type_left).id+"."+(*_right).id+"{"+(*_ty
pe_right).id+"}"+"]";
        return str;
    }

    // method for constructing the volatile name for printing (depending on defi
nitions of subentities only)
    virtual string get_sub_name(bool _typeflag=true, unsigned int _level=0, bool
 _left=true) const {
        string str;
        str += MIF2("[{\\lambda}","[\\")+(*left).get_name(_typeflag, _level+1, t
rue);
        if (_typeflag)
        str += MIF2("_{","{")+(*type_left).get_name(_typeflag, _level+1, tru
e)+"}";
        str += "."+(*right).get_name(_typeflag, _level+1, false);
        if (_typeflag)
        str += MIF2("_{","{")+(*type_right).get_name(_typeflag, _level+1, fa
lse)+"}";
        str += "]";
        return str;
```

```
    }

    static bool lambda_abstraction_exists(const wff_reference& _left, const wff_
reference& _type_left,
                              const wff_reference& _right, const wff
_reference& _type_right) {
        pair<wff_reference, wff_reference> pl(_left, _type_left);
        pair<wff_reference, wff_reference> pr(_right, _type_right);
        pair<pair<wff_reference, wff_reference>, pair<wff_reference, wff_referen
ce> > p(pl, pr);
        return (th.lambda_abstractions.find(p)!=th.lambda_abstractions.end());
    }


    static const wff_reference& find_lambda_abstraction(const wff_reference& _le
ft, const wff_reference& _type_left,
                                    const wff_reference& _ri
ght, const wff_reference& _type_right) {
        pair<wff_reference, wff_reference> pl(_left, _type_left);
        pair<wff_reference, wff_reference> pr(_right, _type_right);
        pair<pair<wff_reference, wff_reference>, pair<wff_reference, wff_referen
ce> > p(pl, pr);
        map<pair<pair<wff_reference, wff_reference>, pair<wff_reference, wff_ref
erence> >, wff_reference>::iterator iter = th.lambda_abstractions.find(p);
        if (iter!=th.lambda_abstractions.end())
        return (*iter).second;
        else {
        throw internal_exception("illegal access to wff reference");
        }
    }

    static void register_lambda_abstraction(const wff_reference& _left, const wf
f_reference& _type_left,
                            const wff_reference& _right, const w
ff_reference& _type_right,
                            const wff_reference& _new) {
        pair<wff_reference, wff_reference> pl(_left, _type_left);
        pair<wff_reference, wff_reference> pr(_right, _type_right);
        pair<pair<wff_reference, wff_reference>, pair<wff_reference, wff_referen
ce> > p(pl, pr);
        th.lambda_abstractions.insert(pair<pair<pair<wff_reference, wff_referenc
e>, pair<wff_reference, wff_reference> >, wff_reference>(p, _new));
    }


    public:

    const wff_reference type_left, type_right;
```

```
    virtual bool is_lambda_abstraction() const { return true; }
    virtual bool is_lambda_application() const { return false; }
    virtual bool is_composed_type() const { return false; }

    virtual bool has_free_variable(const wff_reference& _ref) const { return _re
f!=left && (*right).has_free_variable(_ref); }

    static string obtain_lambda_abstraction_msg(const wff_reference& _left, cons
t wff_reference& _type_left, const wff_reference& _right, const wff_reference& _type
_right) {
        string msg;
        msg += "lambda abstraction ( \\ "+(*_left).get_name()+" { "+(*_type_left
).get_name()+" } . "+(*_right).get_name()+" { "+(*_type_right).get_name()+" } ) (wff
s "+to_string(_left.get_no())+", "+to_string(_type_left.get_no())+", "+to_string(_ri
ght.get_no())+", "+to_string(_type_right.get_no())+")";
        return msg;
    }

    static wff_reference obtain_lambda_abstraction(const wff_reference& _left, c
onst wff_reference& _type_left, const wff_reference& _right, const wff_reference& _t
ype_right) {
        // check for necessary conditions first
        if (!(*_left).has_type(_type_left)) {
        string msg;
        msg += "first operand does not have specified type needed for " + ob
tain_lambda_abstraction_msg(_left, _type_left, _right, _type_right);
        throw syntax_exception(msg);
        }
        if (!(*_right).has_type(_type_right)) {
        string msg;
        msg += "second operand does not have specified type needed for " + o
btain_lambda_abstraction_msg(_left, _type_left, _right, _type_right);
        throw syntax_exception(msg);
        }
        if (!(*_left).is_variable()) {
        string msg;
        msg += "first operand is not a variable needed for " + obtain_lambda
_abstraction_msg(_left, _type_left, _right, _type_right);
        }
        // look up existing entity
        if (lambda_abstraction_exists(_left, _type_left, _right, _type_right)) {
        return find_lambda_abstraction(_left, _type_left, _right, _type_righ
t);
        }
        // or otherwise create new
        else {
        wff_reference r = (*(new lambda_abstraction(create_id(_left, _type_l
eft, _right, _type_right),
                                    _left, _type_left, _righ
```

```
t, _type_right))).ref;
        // specific for class lambda_abstraction: add resulting type
        wff_reference new_type_right = (_type_left == th.tau.ref ? (*_type_r
ight).recursively_substitute_by_dependent_type(_left) : _type_right);
        (*r).get_lambda_abstraction().add_type(composed_type::obtain_compose
d_type(new_type_right, _type_left));
        return r;
        }
    }

    virtual wff_reference recursively_substitute_variable(const wff_reference& _
var_ref, const wff_reference& _body_ref, const set<wff_reference>& _bound_vars) {
        if (_var_ref==left) return ref; // ignore case for bound var is the var
to replace
        wff_reference new_left = (*left).recursively_substitute_variable(_var_re
f, _body_ref, _bound_vars);
        wff_reference new_type_left = (*type_left).recursively_substitute_type_v
ariable_at_type_level(_var_ref, _body_ref, _bound_vars);
        if (type_right!=th.tau.ref || (*right).is_variable() || (*right).is_comp
osed_type()) { // non-variable types in lambda abstractions are not affected by subs
titution
        set<wff_reference> bound_vars = _bound_vars; bound_vars.insert(left)
; // insert abstracted variable to bound variables
        wff_reference new_right = (*right).recursively_substitute_variable(_
var_ref, _body_ref, bound_vars);
        wff_reference new_type_right = (*type_right).recursively_substitute_
type_variable_at_type_level(_var_ref, _body_ref, bound_vars);
        return obtain_lambda_abstraction(new_left, new_type_left, new_right,
 new_type_right);
        }
        else return obtain_lambda_abstraction(new_left, new_type_left, right, ty
pe_right);
    }

    virtual wff_reference recursively_substitute_type_variable_at_base_level(con
st wff_reference& _var_ref, const wff_reference& _body_ref, const set<wff_reference>
& _bound_vars) {
        return ref; // lambda abstraction has automatic type determination depen
ding on arguments (already done in recursively_substitute_variable)
    }

    virtual wff_reference recursively_substitute_type_variable_at_type_level(con
st wff_reference& _var_ref, const wff_reference& _body_ref, const set<wff_reference>
& _bound_vars) {
        return ref; // lambda abstractions as types are not affected by substitu
tion
    }

    virtual wff_reference substitute_wff_part_no(const node_reference& _node, si
```

```
gned int& _no, const wff_reference& _wff_part, const wff_reference& _wff_subst, cons
t wff_reference& _h, const wff_reference& _a_b, const set<wff_reference>& _bound_var
s) const {
        wff_reference chkref = substitute_wff_part_no_standard_check(_node, _no,
 _wff_part, _wff_subst, _h, _a_b, _bound_vars);
        if (chkref.is_defined()) return chkref;
        if (!_node.maximum_reached()) {
        // left node
        node_reference left_node(_node);
        left_node.push_back(false);
        if ((unsigned)_no==left_node.to_ulong()) {
            if (_wff_part==(*left).ref) throw syntax_exception("substitution
 of variable immediately preceded by lambda");
            else throw no_match_for_substitution_exception(_wff_part, left);
        }
        // right node
        node_reference right_node(_node);
        right_node.push_back(true);
        // insert abstracted variable to bound variables
        set<wff_reference> bound_vars = _bound_vars;
        bound_vars.insert(left);
        wff_reference new_right = (*right).substitute_wff_part_no(right_node
, _no, _wff_part, _wff_subst, _h, _a_b, bound_vars);
        if (new_right!=right) {
            if (!(*new_right).has_type(type_right)) throw substitution_resul
t_missing_expected_type_exception(new_right, type_right);
            return obtain_lambda_abstraction(left, type_left,
                                 new_right, type_right);
        }
        }
        else {
        // cout << "# maximum reached -- skipping nodes ..." << endl;
        }
        return ref;
    }

    // rename bound variable [cf. Andrews 2002 (ISBN 1-4020-0763-9), pp. 217 f.
(5206)]
    wff_reference rename_bound_variable(const wff_reference& _new_var) const {
        if (!(*_new_var).is_variable()) {
        throw syntax_exception("variable renaming requires a variable as sec
ond argument");
        }
        wff_reference type_new_var = (*_new_var).get_variable().type;
        if (type_left != type_new_var) {
        throw syntax_exception("variable renaming with variable type mismatc
h: '"+(*type_left).get_name()+"' != '"+(*type_new_var).get_name()+"'");
        }
        if (has_free_variable(_new_var)) {
```

```
        throw scope_violation_in_variable_renaming_exception(_new_var, ref);
        }
        try {
        set<wff_reference> bound_vars;
        wff_reference new_right = (*right).recursively_substitute_variable(l
eft, _new_var, bound_vars);
        wff_reference new_type_right = (*type_right).recursively_substitute_
variable(left, _new_var, bound_vars);
        return obtain_lambda_abstraction(_new_var, type_new_var,
                      new_right, new_type_right);
        } catch (scope_violation_in_lambda_conversion_exception& e) {
        throw scope_violation_in_lambda_conversion_exception(e, right);
        }
    }
    };

    class lambda_application : public composed {

    protected:

    lambda_application(const string& _id,
             const wff_reference& _left, const wff_reference& _type_le
ft,
             const wff_reference& _right, const wff_reference& _type_r
ight) :
    composed(_id, _left, _right), type_left(_type_left), type_right(_type_right)
 {
        register_lambda_application(_left, _type_left, _right, _type_right, ref)
;
    }

    // method for constructing the static id for identification
    static string create_id(const wff_reference& _left, const wff_reference& _ty
pe_left,
              const wff_reference& _right, const wff_reference& _t
ype_right) {
        string str;
        str += "("+(*_left).id+"{"+(*_type_left).id+"}"+"_"+(*_right).id+"{"+(*_
type_right).id+"}"+")";
        return str;
    }

    // method for constructing the volatile name for printing (depending on defi
nitions of subentities only)
    virtual string get_sub_name(bool _typeflag=true, unsigned int _level=0, bool
 _left=true) const {
        string str;
        if (file_format==FORMAT_PLAIN) {
        str += "("+(*left).get_name(_typeflag, _level+1);
```

```
        if (_typeflag)
            str += "{"+(*type_left).get_name(_typeflag, _level+1)+"}";
        str += "_"+(*right).get_name(_typeflag, _level+1, false);
        if (_typeflag)
            str += "{"+(*type_right).get_name(_typeflag, _level+1)+"}";
        str += ")";
        }
        else if (file_format==FORMAT_MARKDOWN) {
        if (!_left) str += "(";
        str += "{" + (*left).get_name(_typeflag, _level+1) + "}";
        if (_typeflag && ((*left).has_definition() || !((*left).is_lambda_ab
straction() || (*left).is_lambda_application()))) str += "_{"+(*type_left).get_name(
_typeflag, _level+1)+"}"; // do not print type of intermediate wffs
        str += string(needs_inner_space(_typeflag) ? "\\," : "") + "{"+ (*ri
ght).get_name(_typeflag, _level+1, false) + "}";
        if (_typeflag && ((*right).has_definition() || !((*right).is_lambda_
abstraction() || (*right).is_lambda_application()))) str += "_{"+(*type_right).get_n
ame(_typeflag, _level+1)+"}"; // do not print type of intermediate wffs
        if (!_left) str += ")";
        }
        return str;
    }

    static void register_lambda_application(const wff_reference& _left, const wf
f_reference& _type_left,
                            const wff_reference& _right, const w
ff_reference& _type_right,
                            const wff_reference& _new) {
        pair<wff_reference, wff_reference> pl(_left, _type_left);
        pair<wff_reference, wff_reference> pr(_right, _type_right);
        pair<pair<wff_reference, wff_reference>, pair<wff_reference, wff_referen
ce> > p(pl, pr);
        th.lambda_applications.insert(pair<pair<pair<wff_reference, wff_referenc
e>, pair<wff_reference, wff_reference> >, wff_reference>(p, _new));
    }


    public:

    const wff_reference type_left, type_right;

    virtual bool is_lambda_abstraction() const { return false; }
    virtual bool is_lambda_application() const { return true; }
    virtual bool is_composed_type() const { return false; }

    virtual bool has_free_variable(const wff_reference& _ref) const { return (*l
eft).has_free_variable(_ref) || (*right).has_free_variable(_ref); }

    static bool lambda_application_exists(const wff_reference& _left, const wff_
```

```
reference& _type_left,
                            const wff_reference& _right, const wff
_reference& _type_right) {
        pair<wff_reference, wff_reference> pl(_left, _type_left);
        pair<wff_reference, wff_reference> pr(_right, _type_right);
        pair<pair<wff_reference, wff_reference>, pair<wff_reference, wff_referen
ce> > p(pl, pr);
        return (th.lambda_applications.find(p)!=th.lambda_applications.end());
    }

    static const wff_reference& find_lambda_application(const wff_reference& _le
ft, const wff_reference& _type_left,
                                  const wff_reference& _ri
ght, const wff_reference& _type_right) {
        pair<wff_reference, wff_reference> pl(_left, _type_left);
        pair<wff_reference, wff_reference> pr(_right, _type_right);
        pair<pair<wff_reference, wff_reference>, pair<wff_reference, wff_referen
ce> > p(pl, pr);
        map<pair<pair<wff_reference, wff_reference>, pair<wff_reference, wff_ref
erence> >, wff_reference>::iterator iter = th.lambda_applications.find(p);
        if (iter!=th.lambda_applications.end())
        return (*iter).second;
        else {
        throw internal_exception("illegal access to wff reference");
        }
    }

    static pair<wff_reference, wff_reference> match_lambda_application(const wff
_reference& _left, const wff_reference& _right) {
        bool matched = false;
        pair<wff_reference, wff_reference> ret;
        set<wff_reference> types_left = (*_left).get_types();
        set<wff_reference> types_right = (*_right).get_types();
        // check for matching types
        for (unsigned int mode=0; !matched && mode<=1; mode++) {
        for (set<wff_reference>::iterator it1 = types_left.begin(); it1 != t
ypes_left.end(); it1++) {
            if ((**it1).is_composed_type()) {
            for (set<wff_reference>::iterator it2 = types_right.begin();
 it2 != types_right.end(); it2++) {
                wff_reference type_in_ref = (**it1).get_composed_type().
right;
                wff_reference r = (*type_in_ref).matches_for_lambda_appl
ication(*it2, mode);
                if (r.is_defined()) {
                (*_right).add_type(r);
                if (matched) throw syntax_exception(string("more tha
n one possible match for '"+(*_left).get_name()+"' _ '"+(*_right).get_name()+"'"));
                else {
```

```
                    matched = true;
                    ret.first = *it1;
                    ret.second = r;
                }
                }
            }
            }
        }
        }
        if (!matched) throw syntax_exception(string("no possible type match for
'"+(*_left).get_name()+"' _ '"+(*_right).get_name()+"'"));
        return ret;
    }

    static string obtain_lambda_application_msg(const wff_reference& _left, cons
t wff_reference& _type_left, const wff_reference& _right, const wff_reference& _type
_right) {
        string msg;
        msg += "lambda application ( "+(*_left).get_name(false)+" { "+(*_type_le
ft).get_name(false)+" } _ "+(*_right).get_name(false)+" { "+(*_type_right).get_name(
false)+" } ) (wffs "+to_string(_left.get_no())+", "+to_string(_type_left.get_no())+"
, "+to_string(_right.get_no())+", "+to_string(_type_right.get_no())+")";
        return msg;
    }

    static wff_reference obtain_lambda_application(const wff_reference& _left, c
onst wff_reference& _type_left, const wff_reference& _right, const wff_reference& _t
ype_right) {
        // check for necessary conditions first
        if (!(*_left).has_type(_type_left)) {
        string msg;
        msg += "first operand does not have specified type needed for " + ob
tain_lambda_application_msg(_left, _type_left, _right, _type_right);
        throw syntax_exception(msg);
        }
        if (!(*_right).has_type(_type_right)) {
        string msg;
        msg += "second operand does not have specified type needed for " + o
btain_lambda_application_msg(_left, _type_left, _right, _type_right);
        throw syntax_exception(msg);
        }
        if (!(*_type_left).is_composed_type()) {
        string msg;
        msg += "type of first operand is not a composed type needed for " +
obtain_lambda_application_msg(_left, _type_left, _right, _type_right);
        throw syntax_exception(msg);
        }
        // check for matching types
        wff_reference type_out_ref = (*_type_left).get_composed_type().left;
```

```
        wff_reference type_in_ref = (*_type_left).get_composed_type().right;
        if (!(type_in_ref==_type_right)) {
        string msg;
        msg += "type mismatch in " + obtain_lambda_application_msg(_left, _t
ype_left, _right, _type_right);
        throw syntax_exception(msg);
        }
        // obtain entity
        bool exists = lambda_application_exists(_left, _type_left, _right, _type
_right);
        wff_reference result = ( exists ? // check for existance
                        find_lambda_application(_left, _type_left, _righ
t, _type_right) : // look up existing entity
                        (*(new lambda_application(create_id(_left, _type
_left, _right, _type_right),
                                     _left, _type_left, _ri
ght, _type_right))).ref ); // otherwise create new
        wff_reference type = type_out_ref;
        if (_type_right == th.tau.ref) type = (*type).recursively_substitute_dep
endent_type_variable(_right);
        (*result).add_type(type);
        return result;
    }

    virtual wff_reference recursively_substitute_variable(const wff_reference& _
var_ref, const wff_reference& _body_ref, const set<wff_reference>& _bound_vars) {
        wff_reference new_left = (*left).recursively_substitute_variable(_var_re
f, _body_ref, _bound_vars);
        wff_reference new_type_left = (*type_left).recursively_substitute_type_v
ariable_at_type_level(_var_ref, _body_ref, _bound_vars);
        if (type_right!=th.tau.ref || (*right).is_variable() || (*right).is_comp
osed_type()) { // non-variable types in lambda applications are not affected by subs
titution
        wff_reference new_right = (*right).recursively_substitute_variable(_
var_ref, _body_ref, _bound_vars);
        wff_reference new_type_right = (*type_right).recursively_substitute_
type_variable_at_type_level(_var_ref, _body_ref, _bound_vars);
        return obtain_lambda_application(new_left, new_type_left, new_right,
 new_type_right);
        }
        else return obtain_lambda_application(new_left, new_type_left, right, ty
pe_right);
    }

    virtual wff_reference recursively_substitute_type_variable_at_base_level(con
st wff_reference& _var_ref, const wff_reference& _body_ref, const set<wff_reference>
& _bound_vars) {
        return ref; // lambda application has automatic type determination depen
ding on arguments (already done in recursively_substitute_variable)
```

```
    }

    virtual wff_reference recursively_substitute_type_variable_at_type_level(con
st wff_reference& _var_ref, const wff_reference& _body_ref, const set<wff_reference>
& _bound_vars) {
        return ref; // lambda applications as types are not affected by substitu
tion
    }

    virtual wff_reference substitute_wff_part_no(const node_reference& _node, si
gned int& _no, const wff_reference& _wff_part, const wff_reference& _wff_subst, cons
t wff_reference& _h, const wff_reference& _a_b, const set<wff_reference>& _bound_var
s) const {
        wff_reference chkref = substitute_wff_part_no_standard_check(_node, _no,
 _wff_part, _wff_subst, _h, _a_b, _bound_vars);
        if (chkref.is_defined()) return chkref;
        if (!_node.maximum_reached()) {
        // left node
        node_reference left_node(_node);
        left_node.push_back(false);
        wff_reference new_left = (*left).substitute_wff_part_no(left_node, _
no, _wff_part, _wff_subst, _h, _a_b, _bound_vars);
        // right node
        node_reference right_node(_node);
        right_node.push_back(true);
        wff_reference new_right = (*right).substitute_wff_part_no(right_node
, _no, _wff_part, _wff_subst, _h, _a_b, _bound_vars);
        if (new_left!=left || new_right!=right) {
            if (!(*new_left).has_type(type_left)) throw substitution_result_
missing_expected_type_exception(new_left, type_left);
            if (!(*new_right).has_type(type_right)) throw substitution_resul
t_missing_expected_type_exception(new_right, type_right);
            return obtain_lambda_application(new_left, type_left,
                              new_right, type_right);
        }
        }
        else {
        // cout << "# maximum reached -- skipping nodes ..." << endl;
        }
        return ref;
    }

    };

    // Functions for class th (theory)

    wff_reference find_wff(const string& _str) {
    for (wff_reference ref=wff_reference::begin(); ref!=wff_reference::end(); re
f++)
```

```
        if (*ref==_str)
        return ref;
    return wff_reference();
    }

    bool wff_exists(const string& _str) {
    return (find_wff(_str).is_defined());
    }

    reference_data_type rm_def(const string& _str) {
    reference_data_type r;
    for (wff_reference ref=wff_reference::begin(); ref!=wff_reference::end(); re
f++) {
        if ((r = (*ref).remove_definiton(_str)) > 0)
        return r;
    }
    return 0;
    }

    thm_reference find_thm(const string& _str) {
    for (thm_reference ref=thm_reference::begin(); ref!=thm_reference::end(); re
f++)
        if (*ref==_str)
        return ref;
    return thm_reference();
    }

    bool thm_exists(const string& _str) {
    return (find_thm(_str).is_defined());
    }

    // Tool for Rule 2 (Lambda Conversion): beta-reduction
    static wff_reference do_lambda_conversion(const wff_reference& _left, const wff_
reference& _right) {
    if (!(*_left).is_lambda_abstraction()) {
        throw syntax_exception("lambda conversion requires a lambda abstraction
as left part of wff");
    }
    const lambda_abstraction& abst = (*(_left)).get_lambda_abstraction();
    wff_reference var_ref = abst.left;
    wff_reference body_ref = abst.right;
    // set up empty list of bound variables for recursion through formula
    set<wff_reference> bound_vars;
    // do the beta-reduction
    wff_reference ref;
    try {
        ref = (*body_ref).recursively_substitute_variable(var_ref, _right, bound
_vars);
    } catch (scope_violation_in_lambda_conversion_exception& e) {
```

```
        throw scope_violation_in_lambda_conversion_exception(e, body_ref);
    }
    return ref;
    }

    // Tool for Rule 3 (Substitution by Identity): find wff part number by occurence
 number
    static bool get_part_no_for_substitution(reference_data_type& _no, const thm_ref
erence& _ref, const thm_reference& _idt, reference_data_type _occurence_no) {
    if ((*_idt).is_lambda_application()) {
        lambda_application& appl1 = (*_idt).get_lambda_application();
        if ((*(appl1.left)).is_lambda_application()) {
        lambda_application& appl2 = (*(appl1.left)).get_lambda_application()
;
        if ((*(appl1.left)).is_lambda_application()) {
            if (appl2.left==th.identity.ref) {
            const wff_reference& wff_part = appl2.right;
            node_reference node;
            bool r = _ref.get_wff_reference().get_wff_part_no(node, wff_
part, _occurence_no);
            _no = node.to_ulong();
            return r;
            }
        }
        }
    }
    throw syntax_exception("identification theorem (equation) required");
    }

    // Rule 1 (Identification): e.g.,  A = A
    // [cf. Andrews 2002 (ISBN 1-4020-0763-9), p. 215 (5200)]
    thm_reference rule_identification(const wff_reference& _ref, const wff_reference
& _type_ref=th.omega.ref) {
    return add_theorem(
       theory::lambda_application::obtain_lambda_application(
        theory::lambda_application::obtain_lambda_application(
            th.identity.ref, identity_class::obtain_type(_type_ref),
            _ref, _type_ref), composed_type::obtain_composed_type(th.boole.r
ef, _type_ref),
        _ref, _type_ref));
    }

    // Rule 2 (Lambda Conversion): e.g.,  [\x.x+1]y = y+1
    // [cf. Andrews 2002 (ISBN 1-4020-0763-9), pp. 218 f. (5207)] (first step of bet
a-reduction [cf. p. 219])
    thm_reference rule_lambda_conversion(const wff_reference& _ref, const wff_refere
nce& _type_ref=th.omega.ref) {
    // check for necessary conditions
    if (!(*_ref).is_lambda_application()) {
```

```
        throw syntax_exception("lambda conversion requires a lambda application"
);
    }
    const lambda_application& appl = (*_ref).get_lambda_application();
    return add_theorem(
        theory::lambda_application::obtain_lambda_application(
        theory::lambda_application::obtain_lambda_application(
            th.identity.ref, identity_class::obtain_type(_type_ref),
            _ref, _type_ref), composed_type::obtain_composed_type(th.boole.r
ef, _type_ref),
        do_lambda_conversion(appl.left, appl.right),
        _type_ref)
    );
    }

    // Rule 3a (Substitution by Identity): e.g.,  C, (A = B)  ->  C<A/B> (one occure
nce)
    // [cf. Andrews 2002 (ISBN 1-4020-0763-9), p. 213 (Rule R)]
    thm_reference rule_substitution_r(const thm_reference& _ref, signed int _no, con
st thm_reference& _idt) {
    if ((*_idt).is_lambda_application()) {
        lambda_application& appl1 = (*_idt).get_lambda_application();
        if ((*(appl1.left)).is_lambda_application()) {
        lambda_application& appl2 = (*(appl1.left)).get_lambda_application()
;
        if (appl2.left==th.identity.ref) {
            const wff_reference& wff_part = appl2.right;
            const wff_reference& wff_subst = appl1.right;
            return add_theorem(_ref.get_wff_reference().substitute_wff_part_
no(_no, wff_part, wff_subst));
        }
        }
    }
    throw syntax_exception("identification theorem (equation) required");
    }

    // Rule 3b (Substitution by Identity with Hypotheses): e.g.,  (H => C), (H => (A
 = B))  ->  H => C<A/B> (one occurence)
    // [cf. Andrews 2002 (ISBN 1-4020-0763-9), p. 214 (Rule R')]
    thm_reference rule_substitution_r_prime(const thm_reference& _ref, signed int _n
o, const thm_reference& _idt) {
    wff_reference implication = find_wff("=>");
    if (implication.is_undefined())
        throw internal_exception("implication not defined");
    wff_reference c_hypo_ref, c_ref;
    if ((*_ref).is_lambda_application()) {
        lambda_application& appl_c_1 = (*_ref).get_lambda_application();
        if ((*(appl_c_1.left)).is_lambda_application()) {
        lambda_application& appl_c_2 = (*(appl_c_1.left)).get_lambda_applica
```

```
tion();
        if (appl_c_2.left==implication) {
            c_hypo_ref = appl_c_2.right;
            c_ref = appl_c_1.right;
        }
        }
    }
    if (c_ref.is_undefined())
        throw syntax_exception("implication required as first argument");
    wff_reference a_b_hypo_ref, a_b_ref;
    if ((*_idt).is_lambda_application()) {
        lambda_application& appl_a_b_1 = (*_idt).get_lambda_application();
        if ((*(appl_a_b_1.left)).is_lambda_application()) {
        lambda_application& appl_a_b_2 = (*(appl_a_b_1.left)).get_lambda_app
lication();
        if (appl_a_b_2.left==implication) {
            a_b_hypo_ref = appl_a_b_2.right;
            a_b_ref = appl_a_b_1.right;
        }
        }
    }
    if (a_b_ref.is_undefined())
        throw syntax_exception("implication required as second argument");
    if (c_hypo_ref != a_b_hypo_ref)
        throw syntax_exception("identical hypothesis required");
    if ((*a_b_ref).is_lambda_application()) {
        lambda_application& appl1 = (*a_b_ref).get_lambda_application();
        if ((*(appl1.left)).is_lambda_application()) {
        lambda_application& appl2 = (*(appl1.left)).get_lambda_application()
;
        if (appl2.left==th.identity.ref) {
            const wff_reference& wff_part = appl2.right;
            const wff_reference& wff_subst = appl1.right;
            // set up empty list of bound variables for recursion through fo
rmula
            set<wff_reference> bound_vars;
            // call substitute_wff_part_no with six arguments: check for res
trictions (scope violation)
            return add_theorem(_ref.get_wff_reference().substitute_wff_part_
no(_no, wff_part, wff_subst, a_b_hypo_ref, a_b_ref, bound_vars));
        }
        }
    }
    throw syntax_exception("identification theorem (equation) required");
    }

    // Rule 4 (Alphabetic Change of Bound Variables): e.g.,  [\x.A] = [\z.A<x/z>]
    // [cf. Andrews 2002 (ISBN 1-4020-0763-9), pp. 217 f. (5206)] (first step of alp
ha-conversion [cf. p. 219])
```

```
    thm_reference rule_rename_bound_variable(const wff_reference& _ref, const wff_re
ference& _new_var) {
    if (!(*_ref).is_lambda_abstraction()) {
        throw syntax_exception("variable renaming requires a lambda abstraction
as first argument");
    }
    return add_theorem(
        theory::lambda_application::obtain_lambda_application(
            theory::lambda_application::obtain_lambda_application(
                th.identity.ref, th.identity.type,
                _ref, th.omega.ref), th.identity_subtype1.ref,
            (*_ref).get_lambda_abstraction().rename_bound_variable(_new_var)
, th.omega.ref));
    }

    // Add Axiom
    thm_reference add_axiom(const wff_reference& _ref) {
    if (current_filename != STRING_AXIOMS_FILE && !allow_additional_axioms) thro
w syntax_exception("new axioms not allowed without flag");
    else if (!(*_ref).has_type(th.boole.ref)) throw syntax_exception("axiom must
 have type BOOLE");
    else return add_theorem(_ref);
    }

    bool has_temporary_definition() const {
    for (wff_reference ref=wff_reference::begin(); ref!=wff_reference::end(); re
f++)
        if ((*ref).has_temporary_definition()) return true;
    return false;
    }

    ostream& print(ostream& _out, bool _typeflag=true, unsigned int _level=0, bool _
left=true) const {
    print_header(_out, _typeflag);
    print_wffs(_out, _typeflag);
    print_defs(_out, _typeflag);
    print_thms(_out, _typeflag);
    return _out;
    }
    operator string() const { ostringstream out; print(out, false); return out.str()
; }

    // print wffs
    virtual ostream& print_wffs(ostream& _out, bool _typeflag=true, unsigned int _le
vel=0, bool _left=true) const {

    _out << COMMENT_HEADER << "Wffs:" << endl;
    int ctr=0;
    for (wff_reference ref=wff_reference::begin(); ref!=wff_reference::end(); re
```

```
f++) {
        ref.print(_out, _typeflag) << endl;
        ctr++;
    }
    if (ctr==0) {
        _out.width(SPACING_REGULAR_INDENT);
        _out << left << COMMENT_HEADER;
        _out << "(none)" << endl;
    }
    _out << COMMENT_HEADER << endl;

    return _out;
    }
    string wffs() const { ostringstream out; print_wffs(out); return out.str(); }

    // print definitions
    ostream& print_defs(ostream& _out, bool _typeflag=true, unsigned int _level=0, b
ool _left=true) const {

    _out << COMMENT_HEADER << "Definitions:" << endl;
    int ctr=0;
    for (wff_reference ref=wff_reference::begin(); ref!=wff_reference::end(); re
f++) {
        if ((*ref).has_definition()) {
        ref.print(_out, _typeflag) << endl;
        ctr++;
        }
    }
    if (ctr==0) {
        _out.width(SPACING_REGULAR_INDENT);
        _out << left << COMMENT_HEADER;
        _out << "(none)" << endl;
    }
    _out << COMMENT_HEADER << endl;

    return _out;
    }
    string defs() const { ostringstream out; print_defs(out); return out.str(); }

    // print theorems
    ostream& print_thms(ostream& _out, bool _typeflag=true) const {

    _out << COMMENT_HEADER << "Theorems:" << endl;
    int ctr=0;
    for (thm_reference ref=thm_reference::begin(); ref!=thm_reference::end(); re
f++) {
        ref.print(_out, _typeflag) << endl;
        ctr++;
    }
```

```
    if (ctr==0) {
        _out.width(SPACING_REGULAR_INDENT);
        _out << left << COMMENT_HEADER;
        _out << "(none)" << endl;
    }
    _out << COMMENT_HEADER << endl;

    return _out;
    }
    string thms() const { ostringstream out; print_thms(out); return out.str(); }

    // print header
    ostream& print_header(ostream& _out,
              unsigned int _indent=SPACING_REGULAR_INDENT,
              unsigned int _indent_first=SPACING_REGULAR_INDENT) const {
    _out << COMMENT_HEADER << endl;
    _out << COMMENT_HEADER << "Theory 'R0'" << endl;
    string line;
    for (int i=0; i<SCREEN_WIDTH-2; i++)
        line += "_";
    _out << COMMENT_HEADER << line << endl;
    _out << COMMENT_HEADER << endl;

    return _out;
    }

} th;

typedef theory::wff_reference wff_reference;
typedef theory::thm_reference thm_reference;
typedef theory::node_reference node_reference;
typedef theory::stack stack;


// stream operators
ostream& operator <<(ostream& _out, const wff_reference& _ref) { return _ref.print(_
out); }
ostream& operator <<(ostream& _out, const thm_reference& _ref) { return _ref.print(_
out); }
ostream& operator <<(ostream& _out, const node_reference& _ref) { return _ref.print(
_out); }
ostream& operator <<(ostream& _out, const stack& _stack) { return _stack.print(_out)
; }
ostream& operator <<(ostream& _out, const theory& _th) { return _th.print(_out); }


ostream& print_status(ostream& _out, bool _printtmpdefs=false) {
    if (errors) _out << MIF2("\\# ","# ") << errors << (errors==EXIT_CODE_MAX_NUMBER
_OF_ERRORS ? " or more" : "") << " error" << (errors==1 ? "" : "s") << " generated"
```

```
<< MENDL << endl;
    if (th.has_temporary_definition()) {
    _out << MIF2("\\# ","# ") << "warning: temporary definitions left" << MENDL
<< endl;
    for (wff_reference ref=wff_reference::begin(); ref!=wff_reference::end(); re
f++)
        if ((*ref).has_temporary_definition()) ref.print(_out, false) << MENDL <
< endl;
    }
    return _out;
}


#include "scanner.cc"
#include "parser.c"


// replace filename suffix
string md_suffix(const string& _filename) {
    string filename(_filename);
    filename.erase(filename.end()-4,filename.end());
    filename += ".src.md";
    return filename;
}

// we have to use the classic C buffer since lex (the scanner) requires it
void parse_buffer(FILE* _outfile, FILE* _infile, const string& _filename) {
    main_file = _filename;
    yyout = _outfile;
    // directly manipulating yyin is allowed only in special cases, we use the safe
way here
    yyrestart(_infile);
    yyparse();
    // pacify compilers like gcc when the user code never invokes yyunput.
    if (/*CONSTCOND*/ 0) yyunput(0, NULL);
}

// we have to use the classic C buffer since lex (the scanner) requires it
void parse_file(FILE* _outfile, const string& _filename) {
    bool interactive_bak = interactive;
    bool skip = false;
    FILE* _infile=NULL;
    if (_filename==STRING_STDIN_FILE) {
    interactive = true;
    _infile = stdin;
    }
    else {
    interactive = false;
    if (register_included_file(_filename)) {
```

```
        _infile = fopen(_filename.c_str(), "r");
        if (_infile==NULL) {
        string os;
        os+="unable to open file '"+_filename+"'";
        throw performance_exception(os);
        }
    }
    else {
        if (display) cout << "## Skipping file " << _filename << " (already incl
uded)" << MENDL << endl;
        if (debug)   cerr << "## Skipping file " << _filename << " (already incl
uded)" << MENDL << endl;
    }
    }
    try {
    if (!skip) parse_buffer(_outfile, _infile, _filename);
    } catch(internal_exception &e) {
    while (include_level>0) { fclose(yyin); yypop_buffer_state(); include_level-
-; }
    if (_infile!=NULL) fclose(_infile);
    throw e;
    } catch(fatal_exception &e) { // close files
    while (include_level>0) { fclose(yyin); yypop_buffer_state(); include_level-
-; }
    if (_infile!=NULL) fclose(_infile);
    throw e;
    }
    interactive = interactive_bak;
}

// main routine
int main(int argc, char **argv) {
    try {
    if (parser_debug) { yydebug=1; }
    ++argv, --argc;  // skip over program name
    while (argc>0) {
        string arg = argv[0]; // check for options and files
        if (arg=="--help" || arg=="-h") {
        cout << "usage: R0 [-h|--help] [-d|--debug] [-s|--strict] [-m|--mark
down] [--allow-additional-axioms]" << endl;
        return 0;
        }
        else if (arg=="--debug" || arg=="-d") debug = true;
        else if (arg=="--strict" || arg=="-s") strict = true;
        else if (arg=="--markdown" || arg=="-m") file_format = FORMAT_MARKDOWN;
        else if (arg=="--allow-additional-axioms") allow_additional_axioms = tru
e;
        else if (arg=="--allow-definition-removal") allow_definition_removal = t
rue;
```

```
        else { // file (not an option)
        interactive = false;
        parse_file(stdout, argv[0]);
        }
        ++argv, --argc;  // next argument
    }
    if (interactive) parse_buffer(stdout, stdin, STRING_STDIN_FILE); // parse st
andard input
    if (strict && th.has_temporary_definition()) {
        print_status(cerr, true);
        cerr << MIF2("\\# ","# ") << "exiting with code " << EXIT_CODE_TEMPORARY
_DEFINITIONS_LEFT << " (temporary definitions left)" << MENDL << endl;
        exit(EXIT_CODE_TEMPORARY_DEFINITIONS_LEFT);
    }
    } catch(nonfatal_exception &e) { // maximum number of errors reached
    cerr << MIF2("\\# ","# ") << "error: " << e.msg << MENDL << endl;
    } catch(internal_exception &e) { // internal error
    cerr << MIF2("\\# ","# ") << "internal error: " << e.msg << MENDL << endl;
    cerr << MIF2("\\# ","# ") << "exiting with code " << EXIT_CODE_INTERNAL_ERRO
R << " (internal error)" << MENDL << endl;
    exit(EXIT_CODE_INTERNAL_ERROR);
    }
    print_status(cout);
    return errors;
}
```

2.2.6 File mathtemplate.tex

```
%%
%%  LaTeX template
%%
%%
%%  The LaTeX template for this project.
%%
%%  Copyright (c) 2017 Owl of Minerva Press GmbH. All rights reserved.
%%  Written by Ken Kubota (<mail@kenkubota.de>).
%%
%%  This file is part of the publication of the mathematical logic R0.
%%  For more information, visit: <http://doi.org/10.4444/100.10>
%%


\documentclass[$if(fontsize)$$fontsize$,$endif$$if(lang)$$lang$,$endif$]{$documentcl
ass$}
\usepackage[T1]{fontenc}
\usepackage{lmodern}
\usepackage{amssymb,amsmath}
\usepackage{ifxetex,ifluatex}
\usepackage{fixltx2e} % provides \textsubscript
```

```
% use upquote if available, for straight quotes in verbatim environments
\IfFileExists{upquote.sty}{\usepackage{upquote}}{}
\ifnum 0\ifxetex 1\fi\ifluatex 1\fi=0 % if pdftex
  \usepackage[utf8]{inputenc}
$if(euro)$
  \usepackage{eurosym}
$endif$
\else % if luatex or xelatex
  \ifxetex
    \usepackage{mathspec}
    \usepackage{xltxtra,xunicode}
  \else
    \usepackage{fontspec}
  \fi
  \defaultfontfeatures{Mapping=tex-text,Scale=MatchLowercase}
  \newcommand{\euro}{€}
$if(mainfont)$
    \setmainfont{$mainfont$}
$endif$
$if(sansfont)$
    \setsansfont{$sansfont$}
$endif$
$if(monofont)$
    \setmonofont[Mapping=tex-ansi]{$monofont$}
$endif$
$if(mathfont)$
    \setmathfont(Digits,Latin,Greek){$mathfont$}
$endif$
\fi
% use microtype if available
\IfFileExists{microtype.sty}{\usepackage{microtype}}{}

% Forbid hyphenation
\exhyphenpenalty=10000
\hyphenpenalty=10000
% Allow hyphenation
%\input{$if(makepath)$$makepath$/$endif$hyphenation}

\usepackage{lastpage}

$if(mathops)$
\DeclareMathOperator{\fuv}{fuv}
\DeclareMathOperator{\fuvbase}{fuvbase}
\DeclareMathOperator{\vectype}{vectype}
$endif$

% Include package geometry first
$if(geometry)$
\usepackage[$for(geometry)$$geometry$$sep$,$endfor$]{geometry}
```

```
$endif$

% Then include package fancyhdr (after package geometry)
\usepackage{fancyhdr}
% fancy
\pagestyle{fancy}
\pagenumbering{arabic}
$if(mathshort)$
% general page layout
\lhead{$lhead$}
\chead{}
\rhead{$rhead$}
\lfoot{$lfoot$}
\cfoot{$cfoot$}
\rfoot{\thepage/\pageref*{LastPage}} % suppress hyperlink by using \pageref* instead
 of \pageref
\renewcommand{\footrulewidth}{0.4pt}
\setlength{\headheight}{13.6pt}
$else$
% general page layout
\fancyhead{} % clear all header fields
\fancyhead[LE]{$lhead$}
\fancyhead[RO]{$rhead$}
\fancyfoot{} % clear all footer fields
\fancyfoot[LE,RO]{\thepage}
\setlength{\headheight}{13.6pt}
% plain
\fancypagestyle{plain}{%
\renewcommand{\headrulewidth}{0.0pt}
\fancyhead{} % clear all header fields
\fancyfoot{} % clear all footer fields
\fancyfoot[LE,RO]{\thepage}
}
$endif$

% Adjust indentation for section
\usepackage{tocloft}
\cftsetindents{subsection}{0.5in}{$if(mathshort)$0.45in$else$0.55in$endif$}

% Prevent splitting of tables (extra command required before each table)
\usepackage{needspace}

% For definition symbol \colonequals (:=)
\usepackage{colonequals}

$if(natbib)$
\usepackage{natbib}
\bibliographystyle{$if(biblio-style)$$biblio-style$$else$plainnat$endif$}
$endif$
```

```
$if(biblatex)$
\usepackage[backend=biber,style=authoryear,natbib=true]{biblatex}
$if(biblio-files)$
\bibliography{$biblio-files$}
$endif$
% square brackets for cite references
\renewcommand*{\mkbibparens}[1]{{\ifcitation{\bibleftbracket#1\bibrightbracket}%
{\bibleftparen#1\bibrightparen}}}
\renewcommand*{\bibopenparen}[1]{{\ifcitation{\bibleftbracket#1}{\bibleftparen#1}}}
\renewcommand*{\bibcloseparen}{{\ifcitation{\bibrightbracket}{\bibrightparen}}}
$endif$
$if(listings)$
\usepackage{listings}
$endif$
$if(lhs)$
\lstnewenvironment{code}{\lstset{language=Haskell,basicstyle=\small\ttfamily}}{}
$endif$
$if(highlighting-macros)$
$highlighting-macros$
$endif$
$if(verbatim-in-note)$
\usepackage{fancyvrb}
$endif$
$if(tables)$
\usepackage{longtable,booktabs}
$endif$
$if(graphics)$
\usepackage{graphicx}
% Redefine \includegraphics so that, unless explicit options are
% given, the image width will not exceed the width of the page.
% Images get their normal width if they fit onto the page, but
% are scaled down if they would overflow the margins.
\makeatletter
\def\ScaleIfNeeded{%
  \ifdim\Gin@nat@width>\linewidth
    \linewidth
  \else
    \Gin@nat@width
  \fi
}
\makeatother
\let\Oldincludegraphics\includegraphics
{%
 \catcode`\@=11\relax%
 \gdef\includegraphics{\@ifnextchar[{\Oldincludegraphics}{\Oldincludegraphics[width=
\ScaleIfNeeded]}}%
}%
$endif$
\ifxetex
```

```
  \usepackage[setpagesize=false, % page size defined by xetex
              unicode=false, % unicode breaks when used with xetex
              xetex]{hyperref}
\else
  \usepackage[unicode=true]{hyperref}
\fi
\hypersetup{breaklinks=true,
            bookmarks=true,
            pdfauthor={$author-meta$},
            pdftitle={$title-meta$},
            colorlinks=$if(mathshort)$true$else$false$endif$,
            citecolor=$if(citecolor)$$citecolor$$else$green$endif$,
            urlcolor=$if(urlcolor)$$urlcolor$$else$blue$endif$,
            linkcolor=$if(linkcolor)$$linkcolor$$else$$if(mathshort)$black$else$mage
nta$endif$$endif$,
            pdfborder={0 0 0}}
\newcommand{\myhref}[2][blue]{\href{#2}{\color{#1}{#2}}} % individual color
\urlstyle{same}  % don't use monospace font for urls
$if(links-as-notes)$
% Make links footnotes instead of hotlinks:
\renewcommand{\href}[2]{#2\footnote{\url{#1}}}
$endif$
$if(strikeout)$
\usepackage[normalem]{ulem}
% avoid problems with \sout in headers with hyperref:
\pdfstringdefDisableCommands{\renewcommand{\sout}{}}
$endif$
\setlength{\parindent}{0pt}
\setlength{\parskip}{6pt plus 2pt minus 1pt}
\setlength{\emergencystretch}{3em}  % prevent overfull lines
$if(numbersections)$
\setcounter{secnumdepth}{5}
$else$
\setcounter{secnumdepth}{0}
$endif$
$if(verbatim-in-note)$
\VerbatimFootnotes % allows verbatim text in footnotes
$endif$
$if(lang)$
\ifxetex
  \usepackage{polyglossia}
  \setmainlanguage{$mainlang$}
\else
  \usepackage[$lang$]{babel}
\fi
$endif$
$for(header-includes)$
$header-includes$
$endfor$
```

```
$if(mathshort)$
% remove section numbering
\renewcommand{\thesection}{}
\renewcommand{\thesubsection}{\arabic{subsection}}
\makeatletter
\def\@seccntformat#1{\csname #1ignore\expandafter\endcsname\csname the#1\endcsname\q
uad}
\let\sectionignore\@gobbletwo
\let\latex@numberline\numberline
\def\numberline#1{\if\relax#1\relax\else\latex@numberline{#1}\fi}
\makeatother
$endif$

$if(title)$
\title{$title$}
$endif$
$if(subtitle)$
\subtitle{$subtitle$}
$endif$
\author{$for(author)$$author$$sep$ \and $endfor$}
\date{$date$}

\begin{document}

$if(mathshort)$
$else$
%\null\thispagestyle{empty}\newpage
%\null\thispagestyle{empty}\newpage
$endif$

$if(title)$
\maketitle
\thispagestyle{empty}
$endif$

$if(mathshort)$
$else$
\IfFileExists{press}{\input{press}}{\null\thispagestyle{empty}}\newpage
$endif$

$for(include-before)$
$include-before$

$endfor$
$if(toc)$
{
\hypersetup{linkcolor=black}
\setcounter{tocdepth}{$toc-depth$}
```

```
$if(mathshort)$
\begingroup
\let\clearpage\relax
\tableofcontents\thispagestyle{fancy}
\endgroup
$else$
\tableofcontents
$endif$
}
$endif$
$body$

$if(natbib)$
$if(biblio-files)$
$if(biblio-title)$
$if(book-class)$
\renewcommand\bibname{$biblio-title$}
$else$
\renewcommand\refname{$biblio-title$}
$endif$
$endif$
\bibliography{$biblio-files$}

$endif$
$endif$
$if(biblatex)$
\printbibliography[heading=bibintoc$if(biblio-title)$,title=$biblio-title$$endif$]

$endif$
$for(include-after)$
$include-after$

$endfor$

$if(mathshort)$
$else$
\newpage
$if(lastpageblank)$
\null\thispagestyle{empty}\newpage
$else$
\IfFileExists{lastpage}{\input{lastpage}}{\null}\thispagestyle{empty}\newpage
$endif$
$endif$

\end{document}
```

2.2.7 File hyphenation.txt

```
##
```

```
##  Hyphenation of formulas
##
##
##  The hyphenation file for this project.
##
##  Copyright (c) 2017 Owl of Minerva Press GmbH. All rights reserved.
##  Written by Ken Kubota (<mail@kenkubota.de>).
##
##  This file is part of the publication of the mathematical logic R0.
##  For more information, visit: <http://doi.org/10.4444/100.10>
##

{{=}_{{{oo}o}}{({{=}_{{{o{(\$T5210\,\$T5210)}}{(\$T5210\,\$T5210)}}}{[{\lambda}y_{\$
T5210}.y_{\$T5210}]}}{[{\lambda}y_{\$T5210}.y_{\$T5210}]}})}}|{({{{\forall}}_{{{o{(o\
backslash3)}}{\tau}}}{\$T5210}_{{\tau}}}{[{\lambda}x_{\$T5210}.({{=}_{{{o\,\$T5210}\
,\$T5210}}{({[{\lambda}y_{\$T5210}.y_{\$T5210}]}{x}_{\$T5210})}}{({[{\lambda}y_{\$T5
210}.y_{\$T5210}]}{x}_{\$T5210})})_{o}]})}
{[{\lambda}p_{{o\,\$T5215}}.({{=}_{{{o{(o\,\$T5215)}}{(o\,\$T5215)}}}{[{\lambda}\$X5
215_{\$T5215}.T_{o}]}}{p}_{{o\,\$T5215}})_{o}]}|{[{\lambda}\$X5215_{\$T5215}.({{=}_{
{{o{\omega}}{\omega}}}{\$X5215}_{{\omega}}}{\$X5215}_{{\omega}})_{o}]}
{{=}\,{({[{\lambda}p.({{=}\,{[{\lambda}\$X5215.T]}}\,{p})]}\,{[{\lambda}\$X5215.({{=
}\,{\$X5215}}\,{\$X5215})]})}}\,|{({{=}\,{[{\lambda}\$X5215.T]}}\,{[{\lambda}\$X5215
.({{=}\,{\$X5215}}\,{\$X5215})]})}
{{=}_{{{oo}o}}{({{{\land}}_{{{oo}o}}{({[{\lambda}x_{o}.({{=}_{{{oo}o}}{({{{\land}}_{
{{oo}o}}{T}_{o}}{x}_{o})}}{x}_{o})_{o}]}{T}_{o})}}{({[{\lambda}x_{o}.({{=}_{{{oo}o}}
{({{{\land}}_{{{oo}o}}{T}_{o}}{x}_{o})}}{x}_{o})_{o}]}{F}_{o})})}}|{({{{\forall}}_{{
{o{(o\backslash3)}}{\tau}}}{o}_{{\tau}}}{[{\lambda}x_{o}.({[{\lambda}x_{o}.({{=}_{{{
oo}o}}{({{{\land}}_{{{oo}o}}{T}_{o}}{x}_{o})}}{x}_{o})_{o}]}{x}_{o})_{o}]})}
{{=}_{{{oo}o}}{({{{\land}}_{{{oo}o}}{({[{\lambda}x_{o}.({{=}_{{{oo}o}}{T}_{o}}{x}_{o
})_{o}]}{T}_{o})}}{({[{\lambda}x_{o}.({{=}_{{{oo}o}}{T}_{o}}{x}_{o})_{o}]}{F}_{o})})
}}|{({{{\forall}}_{{{o{(o\backslash3)}}{\tau}}}{o}_{{\tau}}}{[{\lambda}x_{o}.({[{\la
mbda}x_{o}.({{=}_{{{oo}o}}{T}_{o}}{x}_{o})_{o}]}{x}_{o})_{o}]})}
{{=}_{{{oo}o}}{({{{\land}}_{{{oo}o}}{({{=}_{{{oo}o}}{T}_{o}}{T}_{o})}}{({{=}_{{{oo}o
}}{T}_{o}}{F}_{o})})}}|{({{{\forall}}_{{{o{(o\backslash3)}}{\tau}}}{o}_{{\tau}}}{[{\
lambda}x_{o}.({{=}_{{{oo}o}}{T}_{o}}{x}_{o})_{o}]})}
{{=}_{{{oo}o}}{({{{\land}}_{{{oo}o}}{({[{\lambda}x_{o}.({{=}_{{{oo}o}}{({{=}_{{{oo}o
}}{T}_{o}}{x}_{o})}}{x}_{o})_{o}]}{T}_{o})}}{({[{\lambda}x_{o}.({{=}_{{{oo}o}}{({{=}
_{{{oo}o}}{T}_{o}}{x}_{o})}}{x}_{o})_{o}]}{F}_{o})})}}|{({{{\forall}}_{{{o{(o\backsl
ash3)}}{\tau}}}{o}_{{\tau}}}{[{\lambda}x_{o}.({[{\lambda}x_{o}.({{=}_{{{oo}o}}{({{=}
_{{{oo}o}}{T}_{o}}{x}_{o})}}{x}_{o})_{o}]}{x}_{o})_{o}]})}
{{{\supset}}_{{{oo}o}}{({{=}_{{{o\,\$AA2}\,\$AA2}}{\$XA2}_{\$AA2}}{\$YA2}_{\$AA2})}}
|{({{=}_{{{oo}o}}{({\$HA2}_{{o\,\$AA2}}{\$XA2}_{\$AA2})}}{({\$HA2}_{{o\,\$AA2}}{\$YA
2}_{\$AA2})})}
{{=}_{{{oo}o}}{({{=}_{{{o{(\$AA3\,b)}}{(\$AA3\,b)}}}{f}_{{\$AA3\,b}}}{g}_{{\$AA3\,b}
})}}|{({{{\forall}}_{{{o{(o\backslash3)}}{\tau}}}{b}_{{\tau}}}{[{\lambda}x_{b}.({{=}
_{{{o\,\$AA3}\,\$AA3}}{({f}_{{\$AA3\,b}}{x}_{b})}}{({g}_{{\$AA3\,b}}{x}_{b})})_{o}]}
)}
{{=}_{{{oo}o}}{({{=}_{{{o{(\$AA3\,\$BA3)}}{(\$AA3\,\$BA3)}}}{f}_{{\$AA3\,\$BA3}}}{g}
_{{\$AA3\,\$BA3}})}}|{({{{\forall}}_{{{o{(o\backslash3)}}{\tau}}}{\$BA3}_{{\tau}}}{[
```

```
{\lambda}x_{\$BA3}.({{=}_{{{o\,\$AA3}\,\$AA3}}{({f}_{{\$AA3\,\$BA3}}{x}_{\$BA3})}}{(
{g}_{{\$AA3\,\$BA3}}{x}_{\$BA3})})_{o}]})}
{{=}_{{{oo}o}}{({{=}_{{{o{(\$AA3\,\$BA3)}}{(\$AA3\,\$BA3)}}}{\$FA3}_{{\$AA3\,\$BA3}}
}{g}_{{\$AA3\,\$BA3}})}}|{({{{\forall}}_{{{o{(o\backslash3)}}{\tau}}}{\$BA3}_{{\tau}
}}{[{\lambda}x_{\$BA3}.({{=}_{{{o\,\$AA3}\,\$AA3}}{({\$FA3}_{{\$AA3\,\$BA3}}{x}_{\$B
A3})}}{({g}_{{\$AA3\,\$BA3}}{x}_{\$BA3})})_{o}]})}
{{=}_{{{oo}o}}{({{=}_{{{o{(\$AA3\,\$BA3)}}{(\$AA3\,\$BA3)}}}{\$FA3}_{{\$AA3\,\$BA3}}
}{\$GA3}_{{\$AA3\,\$BA3}})}}|{({{{\forall}}_{{{o{(o\backslash3)}}{\tau}}}{\$BA3}_{{\
tau}}}{[{\lambda}x_{\$BA3}.({{=}_{{{o\,\$AA3}\,\$AA3}}{({\$FA3}_{{\$AA3\,\$BA3}}{x}_
{\$BA3})}}{({\$GA3}_{{\$AA3\,\$BA3}}{x}_{\$BA3})})_{o}]})}
{{{\supset}}_{{{oo}o}}{({{=}_{{{o\,\$AA2}\,\$AA2}}{\$XA2}_{\$AA2}}{y}_{\$AA2})}}|{({
{=}_{{{oo}o}}{({\$HA2}_{{o\,\$AA2}}{\$XA2}_{\$AA2})}}{({\$HA2}_{{o\,\$AA2}}{y}_{\$AA
2})})}
{{=}_{{{oo}o}}{({{=}_{{{o{(ab)}}{(ab)}}}{f}_{{ab}}}{[{\lambda}\$Y5205_{b}.({f}_{{ab}
}{\$Y5205}_{b})_{a}]})}}|{({{{\forall}}_{{{o{(o\backslash3)}}{\tau}}}{b}_{{\tau}}}{[
{\lambda}x_{b}.({{=}_{{{oa}a}}{({f}_{{ab}}{x}_{b})}}{({[{\lambda}\$Y5205_{b}.({f}_{{
ab}}{\$Y5205}_{b})_{a}]}{x}_{b})})_{o}]})}
{{=}_{{{oo}o}}{({{=}_{{{o{(ab)}}{(ab)}}}{f}_{{ab}}}{[{\lambda}\$Y5205_{b}.({f}_{{ab}
}{\$Y5205}_{b})_{a}]})}}|{({{{\forall}}_{{{o{(o\backslash3)}}{\tau}}}{b}_{{\tau}}}{[
{\lambda}x_{b}.({{=}_{{{oa}a}}{({f}_{{ab}}{x}_{b})}}{({f}_{{ab}}{x}_{b})})_{o}]})}
{[{\lambda}p_{{o\,\$T5215H}}.({{=}_{{{o{(o\,\$T5215H)}}{(o\,\$T5215H)}}}{[{\lambda}\
$X5215H_{\$T5215H}.T_{o}]}}{p}_{{o\,\$T5215H}})_{o}]}|{[{\lambda}\$X5215H_{\$T5215H}
.({{=}_{{{o{\omega}}{\omega}}}{\$X5215H}_{{\omega}}}{\$X5215H}_{{\omega}})_{o}]}
{{=}\,{({[{\lambda}p.({{=}\,{[{\lambda}\$X5215H.T]}}\,{p})]}\,{[{\lambda}\$X5215H.({
{=}\,{\$X5215H}}\,{\$X5215H})]})}}\,|{({{=}\,{[{\lambda}\$X5215H.T]}}\,{[{\lambda}\$
X5215H.({{=}\,{\$X5215H}}\,{\$X5215H})]})}
{{{\supset}}_{{{oo}o}}{h}_{o}}|{({{=}_{{{o{\omega}}{\omega}}}{({[{\lambda}\$X5215H_{
\$T5215H}.T_{o}]}{\$A5215H}_{\$T5215H})}}{({[{\lambda}\$X5215H_{\$T5215H}.T_{o}]}{\$
A5215H}_{\$T5215H})})}
{{{\supset}}_{{{oo}o}}{h}_{o}}|{({{=}_{{{o{(o\,\$T5215H)}}{(o\,\$T5215H)}}}{[{\lambd
a}\$X5215H_{\$T5215H}.T_{o}]}}{[{\lambda}\$X5215H_{\$T5215H}.({{=}_{{{o{\omega}}{\om
ega}}}{\$X5215H}_{{\omega}}}{\$X5215H}_{{\omega}})_{o}]})}
{{{\supset}}_{{{oo}o}}{h}_{o}}|{({{=}_{{{o{\omega}}{\omega}}}{({{{\forall}}_{{{o{(o\
backslash3)}}{\tau}}}{\$T5220H}_{{\tau}}}{[{\lambda}\$X5220H_{\$T5220H}.a_{o}]})}}{(
{{{\forall}}_{{{o{(o\backslash3)}}{\tau}}}{\$T5220H}_{{\tau}}}{[{\lambda}\$X5220H_{\
$T5220H}.a_{o}]})})}
{{=}_{{{o{\omega}}{\omega}}}{({{{\exists_1}}_{{{o{(o\backslash3)}}{\tau}}}{t}_{{\tau
}}}{[{\lambda}y_{t}.({p}_{{ot}}{y}_{t})_{o}]})}}|{({{{\exists}}_{{{o{(o\backslash3)}
}{\tau}}}{t}_{{\tau}}}{[{\lambda}y_{t}.({{{\forall}}_{{{o{(o\backslash3)}}{\tau}}}{t
}_{{\tau}}}{[{\lambda}z_{t}.({{=}_{{{oo}o}}{({p}_{{ot}}{z}_{t})}}{({{=}_{{{ot}t}}{y}
_{t}}{z}_{t})})_{o}]})_{o}]})}
{{=}\,{({[{\lambda}f.({{=}\,{({{f}\,{\$A8013}}\,{\$B8013})}}\,{({{f}\,{\$B8013}}\,{\
$A8013})})]}\,{{\land}})}}\,|{({{=}\,{({{{\land}}\,{\$A8013}}\,{\$B8013})}}\,{({{{\l
and}}\,{\$B8013}}\,{\$A8013})})}
{{=}\,{({[{\lambda}\$A8013.[{\lambda}ytmp.({{=}\,{\$A8013}}\,{({{{\land}}\,{\$A8013}
}\,{ytmp})})]]}\,{\$B8013})}}\,|{[{\lambda}ytmp.({{=}\,{\$B8013}}\,{({{{\land}}\,{\$
B8013}}\,{ytmp})})]}
{{=}\,{({[{\lambda}\$A8013.[{\lambda}ytmp.({{=}\,{\$A8013}}\,{({{{\land}}\,{\$A8013}
}\,{ytmp})})]]}\,{\$A8013})}}\,|{[{\lambda}ytmp.({{=}\,{\$A8013}}\,{({{{\land}}\,{\$
```

```
A8013}}\,{ytmp})})]}
{{=}\,{[{\lambda}t.[{\lambda}p.({{\sim}}\,{({{{\exists}}\,{t}}\,{[{\lambda}x.({{\sim
}}\,{({p}\,{x})})]})})]]}}\,|{[{\lambda}t.[{\lambda}p.({{\sim}}\,{({[{\lambda}p.({{\
sim}}\,{({{=}\,{[{\lambda}x.T]}}\,{[{\lambda}x.({{\sim}}\,{({p}\,{x})})]})})]}\,{[{\
lambda}x.({{\sim}}\,{({p}\,{x})})]})})]]}
{{=}\,{({[{\lambda}p.({{\sim}}\,{({{=}\,{[{\lambda}x.T]}}\,{[{\lambda}x.({{\sim}}\,{
({p}\,{x})})]})})]}\,{[{\lambda}x.({{\sim}}\,{({p}\,{x})})]})}}\,|{({{\sim}}\,{({{=}
\,{[{\lambda}x.T]}}\,{[{\lambda}x.({{\sim}}\,{({[{\lambda}x.({{\sim}}\,{({p}\,{x})})
]}\,{x})})]})})}
{{=}\,{[{\lambda}t.[{\lambda}p.({{\sim}}\,{({{{\exists}}\,{t}}\,{[{\lambda}x.({{\sim
}}\,{({p}\,{x})})]})})]]}}\,|{[{\lambda}t.[{\lambda}p.({{\sim}}\,{({{\sim}}\,{({{=}\
,{[{\lambda}x.T]}}\,{[{\lambda}x.({{\sim}}\,{({[{\lambda}x.({{\sim}}\,{({p}\,{x})})]
}\,{x})})]})})})]]}$
\# wff$\qquad 1779\;\;:\;\;\qquad |{{{=}\,{[{\lambda}t.[{\lambda}p.({{\sim}}\,{({{{\
exists}}\,{t}}\,{[{\lambda}x.({{\sim}}\,{({p}\,{x})})]})})]]}}\,{[{\lambda}t.[{\lamb
da}p.({{\sim}}\,{({{\sim}}\,{({{=}\,{[{\lambda}x.T]}}\,{[{\lambda}x.({{\sim}}\,{({[{
\lambda}x.({{\sim}}\,{({p}\,{x})})]}\,{x})})]})})})]]}}_{o}
{{=}_{{{oo}o}}{({{\sim}}_{{oo}}{({{\sim}}_{{oo}}{({{=}_{{{o{(ot)}}{(ot)}}}{[{\lambda
}x_{t}.T_{o}]}}{[{\lambda}x_{t}.({{\sim}}_{{oo}}{({[{\lambda}x_{t}.({{\sim}}_{{oo}}{
({p}_{{ot}}{x}_{t})})_{o}]}{x}_{t})})_{o}]})})})}}|{({{=}_{{{o{(ot)}}{(ot)}}}{[{\lam
bda}x_{t}.T_{o}]}}{[{\lambda}x_{t}.({{\sim}}_{{oo}}{({[{\lambda}x_{t}.({{\sim}}_{{oo
}}{({p}_{{ot}}{x}_{t})})_{o}]}{x}_{t})})_{o}]})}
\# $\qquad \;\;\;\;\;\;\;\;\;\;\qquad {{=}\,{[{\lambda}t.[{\lambda}p.({{\sim}}\,{({{
{\exists}}\,{t}}\,{[{\lambda}x.({{\sim}}\,{({p}\,{x})})]})})]]}}\,|{[{\lambda}t.[{\l
ambda}p.({{\sim}}\,{({{\sim}}\,{({{=}\,{[{\lambda}x.T]}}\,{[{\lambda}x.({{\sim}}\,{(
{[{\lambda}x.({{\sim}}\,{({p}\,{x})})]}\,{x})})]})})})]]}
{{=}_{{{o{({o{(o\backslash3)}}{\tau})}}{({o{(o\backslash3)}}{\tau})}}}{[{\lambda}t_{
{\tau}}.[{\lambda}p_{{ot}}.({{\sim}}_{{oo}}{({{{\exists}}_{{{o{(o\backslash3)}}{\tau
}}}{t}_{{\tau}}}{[{\lambda}x_{t}.({{\sim}}_{{oo}}{({p}_{{ot}}{x}_{t})})_{o}]})})_{o}
]_{{(o{(ot)})}}]}}|{[{\lambda}t_{{\tau}}.[{\lambda}p_{{ot}}.({{\sim}}_{{oo}}{({{\sim
}}_{{oo}}{({{=}_{{{o{(ot)}}{(ot)}}}{[{\lambda}x_{t}.T_{o}]}}{[{\lambda}x_{t}.({{\sim
}}_{{oo}}{({[{\lambda}x_{t}.({{\sim}}_{{oo}}{({p}_{{ot}}{x}_{t})})_{o}]}{x}_{t})})_{
o}]})})})_{o}]_{{(o{(ot)})}}]}
{{=}_{{{o{({o{(o\backslash3)}}{\tau})}}{({o{(o\backslash3)}}{\tau})}}}{[{\lambda}t_{
{\tau}}.[{\lambda}p_{{ot}}.({{\sim}}_{{oo}}{({{{\exists}}_{{{o{(o\backslash3)}}{\tau
}}}{t}_{{\tau}}}{[{\lambda}x_{t}.({{\sim}}_{{oo}}{({p}_{{ot}}{x}_{t})})_{o}]})})_{o}
]_{{(o{(ot)})}}]}}|{[{\lambda}t_{{\tau}}.[{\lambda}p_{{ot}}.({{=}_{{{o{(ot)}}{(ot)}}
}{[{\lambda}x_{t}.T_{o}]}}{[{\lambda}x_{t}.({{\sim}}_{{oo}}{({{\sim}}_{{oo}}{({p}_{{
ot}}{x}_{t})})})_{o}]})_{o}]_{{(o{(ot)})}}]}
{{=}_{{{o{({o{(o\backslash3)}}{\tau})}}{({o{(o\backslash3)}}{\tau})}}}{[{\lambda}t_{
{\tau}}.[{\lambda}p_{{ot}}.({{\sim}}_{{oo}}{({{{\exists}}_{{{o{(o\backslash3)}}{\tau
}}}{t}_{{\tau}}}{[{\lambda}x_{t}.({{\sim}}_{{oo}}{({p}_{{ot}}{x}_{t})})_{o}]})})_{o}
]_{{(o{(ot)})}}]}}|{[{\lambda}t_{{\tau}}.[{\lambda}p_{{ot}}.({{=}_{{{o{(ot)}}{(ot)}}
}{[{\lambda}x_{t}.T_{o}]}}{[{\lambda}x_{t}.({p}_{{ot}}{x}_{t})_{o}]})_{o}]_{{(o{(ot)
})}}]}
{{=}_{{{o{({oo}o)}}{({oo}o)}}}{[{\lambda}a_{o}.[{\lambda}b_{o}.({{\sim}}_{{oo}}{({{{
\lor}}_{{{oo}o}}{({{\sim}}_{{oo}}{a}_{o})}}{({{\sim}}_{{oo}}{b}_{o})})})_{o}]_{{(oo)
}}]}}|{[{\lambda}a_{o}.[{\lambda}b_{o}.({{\sim}}_{{oo}}{({{\sim}}_{{oo}}{({{{\land}}
_{{{oo}o}}{({{\sim}}_{{oo}}{({{\sim}}_{{oo}}{a}_{o})})}}{({{\sim}}_{{oo}}{({{\sim}}_
```

```
{{oo}}{b}_{o})})})})})_{o}]_{{(oo)}}]}
{{=}_{{{o{({oo}o)}}{({oo}o)}}}{[{\lambda}a_{o}.[{\lambda}b_{o}.({{\sim}}_{{oo}}{({{{
\lor}}_{{{oo}o}}{({{\sim}}_{{oo}}{a}_{o})}}{({{\sim}}_{{oo}}{b}_{o})})})_{o}]_{{(oo)
}}]}}|{[{\lambda}a_{o}.[{\lambda}b_{o}.({{\sim}}_{{oo}}{({{\sim}}_{{oo}}{({{{\land}}
_{{{oo}o}}{a}_{o}}{({{\sim}}_{{oo}}{({{\sim}}_{{oo}}{b}_{o})})})})})_{o}]_{{(oo)}}]}
{{=}_{{{o{({oo}o)}}{({oo}o)}}}{[{\lambda}a_{o}.[{\lambda}b_{o}.({{\sim}}_{{oo}}{({{{
\lor}}_{{{oo}o}}{({{\sim}}_{{oo}}{a}_{o})}}{({{\sim}}_{{oo}}{b}_{o})})})_{o}]_{{(oo)
}}]}}|{[{\lambda}a_{o}.[{\lambda}b_{o}.({{\sim}}_{{oo}}{({{\sim}}_{{oo}}{({{{\land}}
_{{{oo}o}}{a}_{o}}{b}_{o})})})_{o}]_{{(oo)}}]}
{{=}_{{{o{({oo}o)}}{({oo}o)}}}{[{\lambda}a_{o}.[{\lambda}b_{o}.({{\sim}}_{{oo}}{({{{
\lor}}_{{{oo}o}}{({{\sim}}_{{oo}}{a}_{o})}}{({{\sim}}_{{oo}}{b}_{o})})})_{o}]_{{(oo)
}}]}}|{[{\lambda}a_{o}.[{\lambda}b_{o}.({{{\land}}_{{{oo}o}}{a}_{o}}{b}_{o})_{o}]_{{
(oo)}}]}
{{{\supset}}_{{{oo}o}}{h}_{o}}|{({{=}_{{{o{\omega}}{\omega}}}{({[{\lambda}g_{{{oo}o}
}.({{g}_{{{oo}o}}{T}_{o}}{T}_{o})_{o}]}{[{\lambda}x_{o}.[{\lambda}y_{o}.x_{o}]_{{(oo
)}}]})}}{({[{\lambda}g_{{{oo}o}}.({{g}_{{{oo}o}}{T}_{o}}{T}_{o})_{o}]}{[{\lambda}x_{
o}.[{\lambda}y_{o}.x_{o}]_{{(oo)}}]})})}
{{{\supset}}_{{{oo}o}}{h}_{o}}|{({{=}_{{{o{\omega}}{\omega}}}{({[{\lambda}g_{{{oo}o}
}.({{g}_{{{oo}o}}{T}_{o}}{T}_{o})_{o}]}{[{\lambda}x_{o}.[{\lambda}y_{o}.y_{o}]_{{(oo
)}}]})}}{({[{\lambda}g_{{{oo}o}}.({{g}_{{{oo}o}}{T}_{o}}{T}_{o})_{o}]}{[{\lambda}x_{
o}.[{\lambda}y_{o}.y_{o}]_{{(oo)}}]})})}
{{=}_{{{o{({o{(o\backslash3)}}{\tau})}}{({o{(o\backslash3)}}{\tau})}}}{{\exists}}_{{
{o{(o\backslash3)}}{\tau}}}}|{[{\lambda}\$T8028_{{\tau}}.[{\lambda}p_{{o\,\$T8028}}.
({{\sim}}_{{oo}}{({{{\forall}}_{{{o{(o\backslash3)}}{\tau}}}{\$T8028}_{{\tau}}}{[{\l
ambda}x_{\$T8028}.({{\sim}}_{{oo}}{({p}_{{o\,\$T8028}}{x}_{\$T8028})})_{o}]})})_{o}]
_{{(o{(o\,\$T8028)})}}]}
{{{\supset}}\,{({{\sim}}\,{({{{\exists}}\,{\$T8028}}\,{\$B8028})})}}\,|{({{\sim}}\,{
({{[{\lambda}\$T8028.[{\lambda}p.({{\sim}}\,{({{{\forall}}\,{\$T8028}}\,{[{\lambda}x
.({{\sim}}\,{({p}\,{x})})]})})]]}\,{\$T8028}}\,{\$B8028})})}
{[{\lambda}\$T8028_{{\tau}}.[{\lambda}p_{{o\,\$T8028}}.({{\sim}}_{{oo}}{({{{\forall}
}_{{{o{(o\backslash3)}}{\tau}}}{\$T8028}_{{\tau}}}{[{\lambda}x_{\$T8028}.({{\sim}}_{
{oo}}{({p}_{{o\,\$T8028}}{x}_{\$T8028})})_{o}]})})_{o}]_{{(o{(o\,\$T8028)})}}]}|{\$T
8028}_{{\tau}}}
{{=}\,{({[{\lambda}\$T8028.[{\lambda}p.({{\sim}}\,{({{{\forall}}\,{\$T8028}}\,{[{\la
mbda}x.({{\sim}}\,{({p}\,{x})})]})})]]}\,{\$T8028})}}\,|{[{\lambda}p.({{\sim}}\,{({{
{\forall}}\,{\$T8028}}\,{[{\lambda}x.({{\sim}}\,{({p}\,{x})})]})})]}
{{=}_{{{oo}o}}{({{\sim}}_{{oo}}{({{\sim}}_{{oo}}{({{{\forall}}_{{{o{(o\backslash3)}}
{\tau}}}{\$T8028}_{{\tau}}}{[{\lambda}x_{\$T8028}.({{\sim}}_{{oo}}{({\$B8028}_{{o\,\
$T8028}}{x}_{\$T8028})})_{o}]})})})}}|{({{{\forall}}_{{{o{(o\backslash3)}}{\tau}}}{\
$T8028}_{{\tau}}}{[{\lambda}x_{\$T8028}.({{\sim}}_{{oo}}{({\$B8028}_{{o\,\$T8028}}{x
}_{\$T8028})})_{o}]})}
{{{\supset}}_{{{oo}o}}{({{\sim}}_{{oo}}{({{{\exists}}_{{{o{(o\backslash3)}}{\tau}}}{
\$T8028}_{{\tau}}}{\$B8028}_{{o\,\$T8028}})})}}|{({{\sim}}_{{oo}}{({{\sim}}_{{oo}}{(
{{{\forall}}_{{{o{(o\backslash3)}}{\tau}}}{\$T8028}_{{\tau}}}{[{\lambda}x_{\$T8028}.
({{\sim}}_{{oo}}{({\$B8028}_{{o\,\$T8028}}{x}_{\$T8028})})_{o}]})})})}
{{{\supset}}_{{{oo}o}}{({{\sim}}_{{oo}}{({{{\exists}}_{{{o{(o\backslash3)}}{\tau}}}{
\$T8028}_{{\tau}}}{\$B8028}_{{o\,\$T8028}})})}}|{({{=}_{{{o{(o\,\$T8028)}}{(o\,\$T80
28)}}}{[{\lambda}x_{\$T8028}.T_{o}]}}{[{\lambda}x_{\$T8028}.({{\sim}}_{{oo}}{({\$B80
28}_{{o\,\$T8028}}{x}_{\$T8028})})_{o}]})}
```

```
{{{\supset}}_{{{oo}o}}{\$H8028}_{o}}|{({{{\supset}}_{{{oo}o}}{({{\sim}}_{{oo}}{({{{\
exists}}_{{{o{(o\backslash3)}}{\tau}}}{\$T8028}_{{\tau}}}{\$B8028}_{{o\,\$T8028}})})
}}{({{\sim}}_{{oo}}{({\$B8028}_{{o\,\$T8028}}{\$A8028}_{\$T8028})})})}
{{{\supset}}_{{{oo}o}}{({{\sim}}_{{oo}}{({{{\exists}}_{{{o{(o\backslash3)}}{\tau}}}{
\$T8028}_{{\tau}}}{\$B8028}_{{o\,\$T8028}})})}}|{({{=}_{{{oo}o}}{({[{\lambda}x_{\$T8
028}.T_{o}]}{\$A8028}_{\$T8028})}}{({[{\lambda}x_{\$T8028}.T_{o}]}{\$A8028}_{\$T8028
})})}
{{{\supset}}_{{{oo}o}}{({{{\land}}_{{{oo}o}}{({{\sim}}_{{oo}}{({{{\exists}}_{{{o{(o\
backslash3)}}{\tau}}}{\$T8028}_{{\tau}}}{\$B8028}_{{o\,\$T8028}})})}}{\$H8028}_{o})}
}|{({\$B8028}_{{o\,\$T8028}}{\$A8028}_{\$T8028})}
{{{\supset}}_{{{oo}o}}{({{{\land}}_{{{oo}o}}{\$H8028}_{o}}{({{\sim}}_{{oo}}{({{{\exi
sts}}_{{{o{(o\backslash3)}}{\tau}}}{\$T8028}_{{\tau}}}{\$B8028}_{{o\,\$T8028}})})})}
}|{({\$B8028}_{{o\,\$T8028}}{\$A8028}_{\$T8028})}
{{{\supset}}_{{{oo}o}}{({{{\land}}_{{{oo}o}}{\$H8028}_{o}}{({{\sim}}_{{oo}}{({{{\exi
sts}}_{{{o{(o\backslash3)}}{\tau}}}{\$T8028}_{{\tau}}}{\$B8028}_{{o\,\$T8028}})})})}
}|{({{=}_{{{oo}o}}{({\$B8028}_{{o\,\$T8028}}{\$A8028}_{\$T8028})}}{T}_{o})}
{{{\supset}}_{{{oo}o}}{({{{\land}}_{{{oo}o}}{\$H8028}_{o}}{({{\sim}}_{{oo}}{({{{\exi
sts}}_{{{o{(o\backslash3)}}{\tau}}}{\$T8028}_{{\tau}}}{\$B8028}_{{o\,\$T8028}})})})}
}|{({{\sim}}_{{oo}}{({\$B8028}_{{o\,\$T8028}}{\$A8028}_{\$T8028})})}
{{=}_{{{oo}o}}{({{{\supset}}_{{{oo}o}}{({{\sim}}_{{oo}}{({{{\exists}}_{{{o{(o\backsl
ash3)}}{\tau}}}{\$T8028}_{{\tau}}}{\$B8028}_{{o\,\$T8028}})})}}{F}_{o})}}|{({{\sim}}
_{{oo}}{({{\sim}}_{{oo}}{({{{\exists}}_{{{o{(o\backslash3)}}{\tau}}}{\$T8028}_{{\tau
}}}{\$B8028}_{{o\,\$T8028}})})})}
{{=}_{{{oo}o}}{({{{\supset}}_{{{oo}o}}{({{\sim}}_{{oo}}{({{{\exists}}_{{{o{(o\backsl
ash3)}}{\tau}}}{\$T8028}_{{\tau}}}{\$B8028}_{{o\,\$T8028}})})}}{F}_{o})}}|{({{\sim}}
_{{oo}}{({{\sim}}_{{oo}}{({{{\exists}}_{{{o{(o\backslash3)}}{\tau}}}{\$T8028}_{{\tau
}}}{\$B8028}_{{o\,\$T8028}})})})}
{{{\supset}}_{{{oo}o}}{\$H8028}_{o}}|{({{=}_{{{oo}o}}{({{{\supset}}_{{{oo}o}}{({{\si
m}}_{{oo}}{({{{\exists}}_{{{o{(o\backslash3)}}{\tau}}}{\$T8028}_{{\tau}}}{\$B8028}_{
{o\,\$T8028}})})}}{F}_{o})}}{({{{\exists}}_{{{o{(o\backslash3)}}{\tau}}}{\$T8028}_{{
\tau}}}{\$B8028}_{{o\,\$T8028}})})}
\# $\qquad \;\;\;\;\;\;\;\;\;\;\qquad |{{=}_{{{o{({o{(o\backslash3)}}{\tau})}}{({o{(
o\backslash3)}}{\tau})}}}{[{\lambda}t_{{\tau}}.[{\lambda}p_{{ot}}.({{\sim}}_{{oo}}{(
{{\sim}}_{{oo}}{({{=}_{{{o{(ot)}}{(ot)}}}{[{\lambda}x_{t}.T_{o}]}}{[{\lambda}x_{t}.(
{{\sim}}_{{oo}}{({[{\lambda}x_{t}.({{\sim}}_{{oo}}{({p}_{{ot}}{x}_{t})})_{o}]}{x}_{t
})})_{o}]})})})_{o}]_{{(o{(ot)})}}]}}|{[{\lambda}t_{{\tau}}.[{\lambda}p_{{ot}}.({{\s
im}}_{{oo}}{({{{\exists}}_{{{o{(o\backslash3)}}{\tau}}}{t}_{{\tau}}}{[{\lambda}x_{t}
.({{\sim}}_{{oo}}{({p}_{{ot}}{x}_{t})})_{o}]})})_{o}]_{{(o{(ot)})}}]}
{{=}_{{{o{({o{(o\backslash3)}}{\tau})}}{({o{(o\backslash3)}}{\tau})}}}{[{\lambda}t_{
{\tau}}.[{\lambda}p_{{ot}}.({{=}_{{{o{(ot)}}{(ot)}}}{[{\lambda}x_{t}.T_{o}]}}{[{\lam
bda}x_{t}.({{\sim}}_{{oo}}{({{\sim}}_{{oo}}{({p}_{{ot}}{x}_{t})})})_{o}]})_{o}]_{{(o
{(ot)})}}]}}|{[{\lambda}t_{{\tau}}.[{\lambda}p_{{ot}}.({{\sim}}_{{oo}}{({{{\exists}}
_{{{o{(o\backslash3)}}{\tau}}}{t}_{{\tau}}}{[{\lambda}x_{t}.({{\sim}}_{{oo}}{({p}_{{
ot}}{x}_{t})})_{o}]})})_{o}]_{{(o{(ot)})}}]}
{{=}_{{{o{({o{(o\backslash3)}}{\tau})}}{({o{(o\backslash3)}}{\tau})}}}{[{\lambda}t_{
{\tau}}.[{\lambda}p_{{ot}}.({{=}_{{{o{(ot)}}{(ot)}}}{[{\lambda}x_{t}.T_{o}]}}{[{\lam
bda}x_{t}.({p}_{{ot}}{x}_{t})_{o}]})_{o}]_{{(o{(ot)})}}]}}|{[{\lambda}t_{{\tau}}.[{\
lambda}p_{{ot}}.({{\sim}}_{{oo}}{({{{\exists}}_{{{o{(o\backslash3)}}{\tau}}}{t}_{{\t
au}}}{[{\lambda}x_{t}.({{\sim}}_{{oo}}{({p}_{{ot}}{x}_{t})})_{o}]})})_{o}]_{{(o{(ot)
```

```
})}}]}
{{=}_{{{oo}o}}{({{{\exists_1}}_{{{o{(o\backslash3)}}{\tau}}}{t}_{{\tau}}}{[{\lambda}
y_{t}.({p}_{{ot}}{y}_{t})_{o}]})}}|{({{{\exists}}_{{{o{(o\backslash3)}}{\tau}}}{t}_{
{\tau}}}{[{\lambda}y_{t}.({{{\forall}}_{{{o{(o\backslash3)}}{\tau}}}{t}_{{\tau}}}{[{
\lambda}z_{t}.({{=}_{{{oo}o}}{({p}_{{ot}}{z}_{t})}}{({{=}_{{{ot}t}}{y}_{t}}{z}_{t})}
)_{o}]})_{o}]})}
:$=\;\;COND\;\;|[{\lambda}t_{{\tau}}.[{\lambda}x_{t}.[{\lambda}y_{t}.[{\lambda}p_{o}
.({{\iota}}_{{t{(ot)}}}{[{\lambda}q_{t}.({{{\lor}}_{{{oo}o}}{({{{\land}}_{{{oo}o}}{p
}_{o}}{({{=}_{{{ot}t}}{x}_{t}}{q}_{t})})}}{({{{\land}}_{{{oo}o}}{({{\sim}}_{{oo}}{p}
_{o})}}{({{=}_{{{ot}t}}{y}_{t}}{q}_{t})})})_{o}]})_{t}]_{{(to)}}]_{{({to}t)}}]_{{({
to}t}t)}}]$
{{=}\,{({[{\lambda}x.[{\lambda}y.[{\lambda}p.({{\iota}}\,{[{\lambda}q.({{{\lor}}\,{(
{{{\land}}\,{p}}\,{({{=}\,{x}}\,{q})})}}\,{({{{\land}}\,{({{\sim}}\,{p})}}\,{({{=}\,
{y}}\,{q})})})]})]]]}\,{x})}}\,|{[{\lambda}y.[{\lambda}p.({{\iota}}\,{[{\lambda}q.({
{{\lor}}\,{({{{\land}}\,{p}}\,{({{=}\,{x}}\,{q})})}}\,{({{{\land}}\,{({{\sim}}\,{p})
}}\,{({{=}\,{y}}\,{q})})})]})]]}
{{=}\,{({[{\lambda}y.[{\lambda}p.({{\iota}}\,{[{\lambda}q.({{{\lor}}\,{({{{\land}}\,
{p}}\,{({{=}\,{x}}\,{q})})}}\,{({{{\land}}\,{({{\sim}}\,{p})}}\,{({{=}\,{y}}\,{q})})
})]})]]}\,{y})}}\,|{[{\lambda}p.({{\iota}}\,{[{\lambda}q.({{{\lor}}\,{({{{\land}}\,{
p}}\,{({{=}\,{x}}\,{q})})}}\,{({{{\land}}\,{({{\sim}}\,{p})}}\,{({{=}\,{y}}\,{q})})}
)]})]}
{{=}\,{({[{\lambda}p.({{\iota}}\,{[{\lambda}q.({{{\lor}}\,{({{{\land}}\,{p}}\,{({{=}
\,{x}}\,{q})})}}\,{({{{\land}}\,{({{\sim}}\,{p})}}\,{({{=}\,{y}}\,{q})})})]})]}\,{T}
)}}\,|{({{\iota}}\,{[{\lambda}q.({{{\lor}}\,{({{{\land}}\,{T}}\,{({{=}\,{x}}\,{q})})
}}\,{({{{\land}}\,{({{\sim}}\,{T})}}\,{({{=}\,{y}}\,{q})})})]})}
{{=}_{{{oo}o}}{({{{\lor}}_{{{oo}o}}{({{{\land}}_{{{oo}o}}{T}_{o}}{({{=}_{{{ot}t}}{x}
_{t}}{q}_{t})})}}{({{{\land}}_{{{oo}o}}{({{\sim}}_{{oo}}{T}_{o})}}{({{=}_{{{ot}t}}{y
}_{t}}{q}_{t})})})}}|{({{{\lor}}_{{{oo}o}}{({{{\land}}_{{{oo}o}}{T}_{o}}{({{=}_{{{ot
}t}}{x}_{t}}{q}_{t})})}}{({{{\land}}_{{{oo}o}}{F}_{o}}{({{=}_{{{ot}t}}{y}_{t}}{q}_{t
})})})}
{{=}_{{{ot}t}}{({{{{COND}_{{{{{\backslash4o}\backslash3}\backslash2}{\tau}}}{t}_{{\t
au}}}{x}_{t}}{y}_{t}}{T}_{o})}}|{({{\iota}}_{{t{(ot)}}}{[{\lambda}q_{t}.({{{\lor}}_{
{{oo}o}}{({{{\land}}_{{{oo}o}}{T}_{o}}{({{=}_{{{ot}t}}{x}_{t}}{q}_{t})})}}{({{{\land
}}_{{{oo}o}}{({{\sim}}_{{oo}}{T}_{o})}}{({{=}_{{{ot}t}}{y}_{t}}{q}_{t})})})_{o}]})}
{{=}\,{({[{\lambda}p.({{\iota}}\,{[{\lambda}q.({{{\lor}}\,{({{{\land}}\,{p}}\,{({{=}
\,{x}}\,{q})})}}\,{({{{\land}}\,{({{\sim}}\,{p})}}\,{({{=}\,{y}}\,{q})})})]})]}\,{F}
)}}\,|{({{\iota}}\,{[{\lambda}q.({{{\lor}}\,{({{{\land}}\,{F}}\,{({{=}\,{x}}\,{q})})
}}\,{({{{\land}}\,{({{\sim}}\,{F})}}\,{({{=}\,{y}}\,{q})})})]})}
{{=}_{{{ot}t}}{({{{{COND}_{{{{{\backslash4o}\backslash3}\backslash2}{\tau}}}{t}_{{\t
au}}}{x}_{t}}{y}_{t}}{F}_{o})}}|{({{\iota}}_{{t{(ot)}}}{[{\lambda}q_{t}.({{{\lor}}_{
{{oo}o}}{({{{\land}}_{{{oo}o}}{F}_{o}}{({{=}_{{{ot}t}}{x}_{t}}{q}_{t})})}}{({{{\land
}}_{{{oo}o}}{T}_{o}}{({{=}_{{{ot}t}}{y}_{t}}{q}_{t})})})_{o}]})}
{{=}_{{{oo}o}}{({{{\forall}}_{{{o{(o\backslash3)}}{\tau}}}{{(o\,\$S)}}_{{\tau}}}{[{\
lambda}p_{{o\,\$S}}.({{{\supset}}_{{{oo}o}}{\$ANBOTH}_{o}}{\$ANSETZ}_{o})_{o}]})}}|{
({{ANSET}_{{{o{(o{(o\backslash4)})}}{\tau}}}{t}_{{\tau}}}{({AZERO}_{{{o{(o\backslash
3)}}{\tau}}}{t}_{{\tau}})})}
{{{\supset}}_{{{oo}o}}{({{{\land}}_{{{oo}o}}{({{ANSET}_{{{o{(o{(o\backslash4)})}}{\t
au}}}{t}_{{\tau}}}{x}_{\$S})}}{\$ANBOTH}_{o})}}|{({{{\land}}_{{{oo}o}}{({{ANSET}_{{{
o{(o{(o\backslash4)})}}{\tau}}}{t}_{{\tau}}}{x}_{\$S})}}{\$ANBOTH}_{o})}
```

```
{{{\supset}}_{{{oo}o}}{({{{\land}}_{{{oo}o}}{({{ANSET}_{{{o{(o{(o\backslash4)})}}{\t
au}}}{t}_{{\tau}}}{x}_{\$S})}}{\$ANBOTH}_{o})}}|{({{{\supset}}_{{{oo}o}}{({p}_{{o\,\
$S}}{x}_{\$S})}}{({p}_{{o\,\$S}}{({ATSUCC}_{{\$S\,\$S}}{x}_{\$S})})})}
:$=\;\;COMPS\;\;|[{\lambda}a_{{\tau}}.[{\lambda}b_{{\tau}}.[{\lambda}c_{{\tau}}.[{\l
ambda}g_{{ab}}.[{\lambda}f_{{bc}}.[{\lambda}x_{c}.({g}_{{ab}}{({f}_{{bc}}{x}_{c})})_
{a}]_{{(ac)}}]_{{({ac}{(bc)})}}]_{{({{ac}{(bc)}}{(ab)})}}]_{{({{{a\backslash4}{(b\ba
ckslash4)}}{(ab)}}{\tau})}}]_{{({{{{a\backslash4}{(\backslash5\backslash4)}}{(a\back
slash4)}}{\tau}}{\tau})}}]$
\# $\qquad \;\;\;\;\;\;\;\;\;\;\qquad |{{=}_{{{o{\omega}}{\omega}}}{({{{{{COMPS}_{{{
{{{{\backslash6\backslash4}{(\backslash5\backslash4)}}{(\backslash5\backslash4)}}{\t
au}}{\tau}}{\tau}}}{t}_{{\tau}}}{u}_{{\tau}}}{w}_{{\tau}}}{h}_{{tu}}}{({{{{{COMPS}_{
{{{{{{\backslash6\backslash4}{(\backslash5\backslash4)}}{(\backslash5\backslash4)}}{
\tau}}{\tau}}{\tau}}}{u}_{{\tau}}}{v}_{{\tau}}}{w}_{{\tau}}}{g}_{{uv}}}{f}_{{vw}})})
}}|{({{{{{COMPS}_{{{{{{{\backslash6\backslash4}{(\backslash5\backslash4)}}{(\backsla
sh5\backslash4)}}{\tau}}{\tau}}{\tau}}}{t}_{{\tau}}}{v}_{{\tau}}}{w}_{{\tau}}}{({{{{
{COMPS}_{{{{{{{\backslash6\backslash4}{(\backslash5\backslash4)}}{(\backslash5\backs
lash4)}}{\tau}}{\tau}}{\tau}}}{t}_{{\tau}}}{u}_{{\tau}}}{v}_{{\tau}}}{h}_{{tu}}}{g}_
{{uv}})}}{f}_{{vw}})}$
:$=\;\;ODPR3\;\;|[{\lambda}t_{{\tau}}.[{\lambda}x_{t}.[{\lambda}u_{{\tau}}.[{\lambda
}y_{u}.[{\lambda}v_{{\tau}}.[{\lambda}g_{{{vu}t}}.({{g}_{{{vu}t}}{x}_{t}}{y}_{u})_{v
}]_{{(v{({vu}t)})}}]_{{({\backslash2{({\backslash4u}t)}}{\tau})}}]_{{({{\backslash2{
({\backslash4u}t)}}{\tau}}u)}}]_{{({{{\backslash2{({\backslash4\backslash6}t)}}{\tau
}}\backslash2}{\tau})}}]_{{({{{{\backslash2{({\backslash4\backslash6}t)}}{\tau}}\bac
kslash2}{\tau}}t)}}]$
{{=}\,{({{{LELE3}\,{s}}\,{\$TLVL2}}\,{\$PLVL3})}}\,|{({{{{{[{\lambda}x.[{\lambda}u.[
{\lambda}y.[{\lambda}v.[{\lambda}g.({{g}\,{x}}\,{y})]]]]]}\,{a}}\,{\$TLVL2}}\,{\$PLV
L2}}\,{s}}\,{[{\lambda}x.[{\lambda}y.x]]})}
{{=}\,{({{{LELE3}\,{s}}\,{\$TLVL2}}\,{\$PLVL3})}}\,|{({{{{[{\lambda}u.[{\lambda}y.[{
\lambda}v.[{\lambda}g.({{g}\,{a}}\,{y})]]]]}\,{\$TLVL2}}\,{\$PLVL2}}\,{s}}\,{[{\lamb
da}x.[{\lambda}y.x]]})}
:$=\;\;GrpAsc\;\;|{{{\forall}}_{{{o{(o\backslash3)}}{\tau}}}{g}_{{\tau}}}{[{\lambda}
a_{g}.({{{\forall}}_{{{o{(o\backslash3)}}{\tau}}}{g}_{{\tau}}}{[{\lambda}b_{g}.({{{\
forall}}_{{{o{(o\backslash3)}}{\tau}}}{g}_{{\tau}}}{[{\lambda}c_{g}.({{=}_{{{og}g}}{
({{l}_{{{gg}g}}{({{l}_{{{gg}g}}{a}_{g}}{b}_{g})}}{c}_{g})}}{({{l}_{{{gg}g}}{a}_{g}}{
({{l}_{{{gg}g}}{b}_{g}}{c}_{g})})})_{o}]})_{o}]})_{o}]}$
:$=\;\;\$HYPTH\;\;{{{\land}}_{{{oo}o}}{({{{\land}}_{{{oo}o}}{({{Grp}_{{{o{({\backsla
sh4\backslash4}\backslash3)}}{\tau}}}{g}_{{\tau}}}{l}_{{{gg}g}})}}{({{{GrpId0}_{{{{o
\backslash3}{({\backslash4\backslash4}\backslash3)}}{\tau}}}{g}_{{\tau}}}{l}_{{{gg}g
}}}{e}_{g})})}}|{({{{GrpId0}_{{{{o\backslash3}{({\backslash4\backslash4}\backslash3)
}}{\tau}}}{g}_{{\tau}}}{l}_{{{gg}g}}}{f}_{g})}$
\# $\qquad \;\;\;\;\;\;\;\;\;\;\qquad {{{\supset}}_{{{oo}o}}{({{{\land}}_{{{oo}o}}{(
{{Grp}_{{{o{({\backslash4\backslash4}\backslash3)}}{\tau}}}{g}_{{\tau}}}{l}_{{{gg}g}
})}}{({{{GrpId0}_{{{{o\backslash3}{({\backslash4\backslash4}\backslash3)}}{\tau}}}{g
}_{{\tau}}}{l}_{{{gg}g}}}{e}_{g})})}}|{({{{\supset}}_{{{oo}o}}{({{{GrpId0}_{{{{o\bac
kslash3}{({\backslash4\backslash4}\backslash3)}}{\tau}}}{g}_{{\tau}}}{l}_{{{gg}g}}}{
f}_{g})}}{({{=}_{{{og}g}}{e}_{g}}{f}_{g})})}$
\# $\qquad \;\;\;\;\;\;\;\;\;\;\qquad {{{\supset}}_{{{oo}o}}{({{Grp}_{{{o{({\backsla
sh4\backslash4}\backslash3)}}{\tau}}}{g}_{{\tau}}}{l}_{{{gg}g}})}}|{({{{\supset}}_{{
{oo}o}}{({{{GrpId0}_{{{{o\backslash3}{({\backslash4\backslash4}\backslash3)}}{\tau}}
```

```
}{g}_{{\tau}}}{l}_{{{gg}g}}}{e}_{g}))}}{({{{\supset}}_{{{oo}o}}{({{{GrpIdO}_{{{{o\bac
kslash3}{({\backslash4\backslash4}\backslash3))}}{\tau}}}{g}_{{\tau}}}{l}_{{{gg}g}}}{
f}_{g})}}{({{=}_{{{og}g}}{e}_{g}}{f}_{g})})})}$
\# $\qquad \;\;\;\;\;\;\;\;\;\;\qquad {{{\supset}}_{{{oo}o}}{({{Grp}_{{{o{({\backsla
sh4\backslash4}\backslash3))}}{\tau}}}{g}_{{\tau}}}{l}_{{{gg}g}})}}|{({{{\supset}}_{{
{oo}o}}{({{{GrpIdO}_{{{{o\backslash3}{({\backslash4\backslash4}\backslash3))}}{\tau}}
}{g}_{{\tau}}}{l}_{{{gg}g}}}{e}_{g}))}}{({{{\supset}}_{{{oo}o}}{({{{GrpIdO}_{{{{o\bac
kslash3}{({\backslash4\backslash4}\backslash3))}}{\tau}}}{g}_{{\tau}}}{l}_{{{gg}g}}}{
f}_{g})}}{({{=}_{{{og}g}}{e}_{g}}{f}_{g})})})}\qquad$ $\mathrel{\mathop:}= \;\;$ $Gr
pIdElUniq$
\# $\qquad \;\;\;\;\;\;\;\;\;\;\qquad {{=}\,{({[{\lambda}b.[{\lambda}c.({{=}\,{({{XO
R}\,{({{XOR}\,{T}}\,{b})}}\,{c})}}\,{({{XOR}\,{T}}\,{({{XOR}\,{b}}\,{c})})})]]}\,{T}
)}}\,|{[{\lambda}c.({{=}\,{({{XOR}\,{({{XOR}\,{T}}\,{T})}}\,{c})}}\,{({{XOR}\,{T}}\,
{({{XOR}\,{T}}\,{c})})})]}$
\# $\qquad \;\;\;\;\;\;\;\;\;\;\qquad {{=}\,{({[{\lambda}b.[{\lambda}c.({{=}\,{({{XO
R}\,{({{XOR}\,{T}}\,{b})}}\,{c})}}\,{({{XOR}\,{T}}\,{({{XOR}\,{b}}\,{c})})})]]}\,{F}
)}}\,|{[{\lambda}c.({{=}\,{({{XOR}\,{({{XOR}\,{T}}\,{F})}}\,{c})}}\,{({{XOR}\,{T}}\,
{({{XOR}\,{F}}\,{c})})})]}$
\# $\qquad \;\;\;\;\;\;\;\;\;\;\qquad {{=}\,{({[{\lambda}b.[{\lambda}c.({{=}\,{({{XO
R}\,{({{XOR}\,{F}}\,{b})}}\,{c})}}\,{({{XOR}\,{F}}\,{({{XOR}\,{b}}\,{c})})})]]}\,{T}
)}}\,|{[{\lambda}c.({{=}\,{({{XOR}\,{({{XOR}\,{F}}\,{T})}}\,{c})}}\,{({{XOR}\,{F}}\,
{({{XOR}\,{T}}\,{c})})})]}$
\# $\qquad \;\;\;\;\;\;\;\;\;\;\qquad {{=}\,{({[{\lambda}b.[{\lambda}c.({{=}\,{({{XO
R}\,{({{XOR}\,{F}}\,{b})}}\,{c})}}\,{({{XOR}\,{F}}\,{({{XOR}\,{b}}\,{c})})})]]}\,{F}
)}}\,|{[{\lambda}c.({{=}\,{({{XOR}\,{({{XOR}\,{F}}\,{F})}}\,{c})}}\,{({{XOR}\,{F}}\,
{({{XOR}\,{F}}\,{c})})})]}$
\# $\qquad \;\;\;\;\;\;\;\;\;\;\qquad {{=}\,{({[{\lambda}c.({{=}\,{({{XOR}\,{({{XOR}
\,{T}}\,{b})}}\,{c})}}\,{({{XOR}\,{T}}\,{({{XOR}\,{b}}\,{c})})})]}\,{c})}}\,|{({{=}\
,{({{XOR}\,{({{XOR}\,{T}}\,{b})}}\,{c})}}\,{({{XOR}\,{T}}\,{({{XOR}\,{b}}\,{c})})})}
$
\# $\qquad \;\;\;\;\;\;\;\;\;\;\qquad {{=}\,{({[{\lambda}b.[{\lambda}c.({{=}\,{({{XO
R}\,{({{XOR}\,{F}}\,{b})}}\,{c})}}\,{({{XOR}\,{F}}\,{({{XOR}\,{b}}\,{c})})})]]}\,{b}
)}}\,|{[{\lambda}c.({{=}\,{({{XOR}\,{({{XOR}\,{F}}\,{b})}}\,{c})}}\,{({{XOR}\,{F}}\,
{({{XOR}\,{b}}\,{c})})})]}$
\# $\qquad \;\;\;\;\;\;\;\;\;\;\qquad {{=}\,{({[{\lambda}c.({{=}\,{({{XOR}\,{({{XOR}
\,{F}}\,{b})}}\,{c})}}\,{({{XOR}\,{F}}\,{({{XOR}\,{b}}\,{c})})})]}\,{c})}}\,|{({{=}\
,{({{XOR}\,{({{XOR}\,{F}}\,{b})}}\,{c})}}\,{({{XOR}\,{F}}\,{({{XOR}\,{b}}\,{c})})})}
$
\# $\qquad \;\;\;\;\;\;\;\;\;\;\qquad {{=}\,{({[{\lambda}b.[{\lambda}c.({{=}\,{({{XO
R}\,{({{XOR}\,{a}}\,{b})}}\,{c})}}\,{({{XOR}\,{a}}\,{({{XOR}\,{b}}\,{c})})})]]}\,{b}
)}}\,|{[{\lambda}c.({{=}\,{({{XOR}\,{({{XOR}\,{a}}\,{b})}}\,{c})}}\,{({{XOR}\,{a}}\,
{({{XOR}\,{b}}\,{c})})})]}$
\# $\qquad \;\;\;\;\;\;\;\;\;\;\qquad {{=}\,{({[{\lambda}c.({{=}\,{({{XOR}\,{({{XOR}
\,{a}}\,{b})}}\,{c})}}\,{({{XOR}\,{a}}\,{({{XOR}\,{b}}\,{c})})})]}\,{c})}}\,|{({{=}\
,{({{XOR}\,{({{XOR}\,{a}}\,{b})}}\,{c})}}\,{({{XOR}\,{a}}\,{({{XOR}\,{b}}\,{c})})})}
$
\# $\qquad \;\;\;\;\;\;\;\;\;\;\qquad {{=}\,{({[{\lambda}b.[{\lambda}c.({{=}\,{({{XO
R}\,{({{XOR}\,{T}}\,{b})}}\,{c})}}\,{({{XOR}\,{T}}\,{({{XOR}\,{b}}\,{c})})})]]}\,{b}
)}}\,|{[{\lambda}c.({{=}\,{({{XOR}\,{({{XOR}\,{T}}\,{b})}}\,{c})}}\,{({{XOR}\,{T}}\,
```

```
{({{XOR}\,{b}}\,{c})})})]}$
\# $\qquad \;\;\;\;\;\;\;\;\;\;\qquad {{=}\,{[{\lambda}x.({{{\land}}\,{XorCaseFRight
}}\,{XorCaseFLeft})]}}\,|{[{\lambda}a.({{{\land}}\,{({{=}\,{({{XOR}\,{a}}\,{F})}}\,{
a})}}\,{({{=}\,{({{XOR}\,{F}}\,{a})}}\,{a})})]}$
\# $\qquad \;\;\;\;\;\;\;\;\;\;\qquad {{=}\,{({[{\lambda}g.[{\lambda}l.[{\lambda}e.[
{\lambda}b.({{{\land}}\,{({{=}\,{({{l}\,{a}}\,{b})}}\,{e})}}\,{({{=}\,{({{l}\,{b}}\,
{a})}}\,{e})})]]]]}\,{o})}}\,|{[{\lambda}l.[{\lambda}e.[{\lambda}b.({{{\land}}\,{({{
=}\,{({{l}\,{a}}\,{b})}}\,{e})}}\,{({{=}\,{({{l}\,{b}}\,{a})}}\,{e})})]]]}$
\# $\qquad \;\;\;\;\;\;\;\;\;\;\qquad {{=}\,{({[{\lambda}b.({{{\land}}\,{({{=}\,{({{
XOR}\,{a}}\,{b})}}\,{F})}}\,{({{=}\,{({{XOR}\,{b}}\,{a})}}\,{F})})]}\,{a})}}\,|{({{{
\land}}\,{({{=}\,{({{XOR}\,{a}}\,{a})}}\,{F})}}\,{({{=}\,{({{XOR}\,{a}}\,{a})}}\,{F}
)})}$
\# $\qquad \;\;\;\;\;\;\;\;\;\;\qquad {{{\forall}}_{{{o{(o\backslash3)}}{\tau}}}{o}_
{{\tau}}}|{[{\lambda}a_{o}.({{{\exists}}_{{{o{(o\backslash3)}}{\tau}}}{o}_{{\tau}}}{
[{\lambda}b_{o}.({{{\land}}_{{{oo}o}}{({{=}_{{{oo}o}}{({{XOR}_{{{oo}o}}{a}_{o}}{b}_{
o})}}{F}_{o})}}{({{=}_{{{oo}o}}{({{XOR}_{{{oo}o}}{b}_{o}}{a}_{o})}}{F}_{o})})_{o}]})
_{o}]}$
:$=\;\;\$GrpAsc\;\;{{{\forall}}_{{{o{(o\backslash3)}}{\tau}}}{o}_{{\tau}}}|{[{\lambd
a}a_{o}.({{{\forall}}_{{{o{(o\backslash3)}}{\tau}}}{o}_{{\tau}}}{[{\lambda}b_{o}.({{
{\forall}}_{{{o{(o\backslash3)}}{\tau}}}{o}_{{\tau}}}{[{\lambda}c_{o}.({{=}_{{{oo}o}
}{({{l}_{{{oo}o}}{({{l}_{{{oo}o}}{a}_{o}}{b}_{o})}}{c}_{o})}}{({{l}_{{{oo}o}}{a}_{o}
}{({{l}_{{{oo}o}}{b}_{o}}{c}_{o})})})_{o}]})_{o}]})_{o}]}$
:$=\;\;\$XAsc\;\;{{{\forall}}_{{{o{(o\backslash3)}}{\tau}}}{o}_{{\tau}}}|{[{\lambda}
a_{o}.({{{\forall}}_{{{o{(o\backslash3)}}{\tau}}}{o}_{{\tau}}}{[{\lambda}b_{o}.({{{\
forall}}_{{{o{(o\backslash3)}}{\tau}}}{o}_{{\tau}}}{[{\lambda}c_{o}.({{=}_{{{oo}o}}{
({{XOR}_{{{oo}o}}{\$Xab}_{o}}{c}_{o})}}{({{XOR}_{{{oo}o}}{a}_{o}}{\$Xbc}_{o})})_{o}]
})_{o}]})_{o}]}$
\# $\qquad \;\;\;\;\;\;\;\;\;\;\qquad {{=}\,{({[{\lambda}l.({{{\land}}\,{\$GrpAsc}}\
,{({{{\exists}}\,{o}}\,{[{\lambda}e.({{{\land}}\,{\$GrpIdy}}\,{\$GrpInv})]})})]}\,{X
OR})}}\,|{({{{\land}}\,{\$XAsc}}\,{({{{\exists}}\,{o}}\,{[{\lambda}e.({{{\land}}\,{\
$XIdy}}\,{\$XInv})]})})}$
\# $\qquad \;\;\;\;\;\;\;\;\;\;\qquad {{{\forall}}_{{{o{(o\backslash3)}}{\tau}}}{o}_
{{\tau}}}|{[{\lambda}a_{o}.({{{\forall}}_{{{o{(o\backslash3)}}{\tau}}}{o}_{{\tau}}}{
[{\lambda}b_{o}.({{{\forall}}_{{{o{(o\backslash3)}}{\tau}}}{o}_{{\tau}}}{[{\lambda}c
_{o}.({{=}_{{{oo}o}}{({{XOR}_{{{oo}o}}{\$Xab}_{o}}{c}_{o})}}{({{XOR}_{{{oo}o}}{a}_{o
}}{\$Xbc}_{o})})_{o}]})_{o}]})_{o}]}\qquad$ $\mathrel{\mathop:}= \;\;$ $\$XAsc\;\;$
$XorAssociativity$
\# $\qquad \;\;\;\;\;\;\;\;\;\;\qquad {{{\supset}}_{{{oo}o}}{({{Grp}_{{{o{({\backsla
sh4\backslash4}\backslash3)}}{\tau}}}{o}_{{\tau}}}{l}_{{{oo}o}})}}|{({{{\supset}}_{{
{oo}o}}{({{{GrpIdO}_{{{{o\backslash3}{({\backslash4\backslash4}\backslash3)}}{\tau}}
}{o}_{{\tau}}}{l}_{{{oo}o}}}{e}_{o})}}{({{{\supset}}_{{{oo}o}}{({{{GrpIdO}_{{{{o\bac
kslash3}{({\backslash4\backslash4}\backslash3)}}{\tau}}}{o}_{{\tau}}}{l}_{{{oo}o}}}{
f}_{o})}}{({{=}_{{{oo}o}}{e}_{o}}{f}_{o})})})}$
\# $\qquad \;\;\;\;\;\;\;\;\;\;\qquad {{=}_{{{oo}o}}{({{{\supset}}_{{{oo}o}}{T}_{o}}
{({{{\supset}}_{{{oo}o}}{\$GIdOXe}_{o}}{({{{\supset}}_{{{oo}o}}{\$GIdOXf}_{o}}{({{=}
_{{{oo}o}}{e}_{o}}{f}_{o})})})})}}|{({{{\supset}}_{{{oo}o}}{\$GIdOXe}_{o}}{({{{\sups
et}}_{{{oo}o}}{\$GIdOXf}_{o}}{({{=}_{{{oo}o}}{e}_{o}}{f}_{o})})})}$
\# wff$\qquad 307\;\;:\;\;\qquad {{{=}\,{NEUMNN0002}}\,{({{ODPRO}\,{{\emptyset}}}\,{
({{ODPRO}\,{{\emptyset}}}\,{{\emptyset}})})}}_{o}\qquad|$ $\mathrel{\mathop:}= \;\;$
```

```
$NEUMNNO002EXPND$
\# $\qquad \;\;\;\;\;\;\;\;\;\;\qquad {{=}\,{({[{\lambda}xtmp.[{\lambda}ytmp.({{=}\,
{xtmp}}\,{({{{\land}}\,{xtmp}}\,{ytmp})})]]}\,{\$H8003})}}\,|{[{\lambda}ytmp.({{=}\,
{\$H8003}}\,{({{{\land}}\,{\$H8003}}\,{ytmp})})]}$
\# $\qquad \;\;\;\;\;\;\;\;\;\;\qquad {{{\supset}}_{{{oo}o}}{({{{\forall}}_{{{o{(o\b
ackslash3)}}{\tau}}}{\$T5226}_{{\tau}}}{[{\lambda}\$X5226_{\$T5226}.\$B5226_{o}]})}}
|{({[{\lambda}\$X5226_{\$T5226}.\$B5226_{o}]}{\$A5226}_{\$T5226})}$
\# $\qquad \;\;\;\;\;\;\;\;\;\;\qquad {{=}_{{{oo}o}}{({{{\land}}_{{{oo}o}}{T}_{o}}{(
{{{\exists}}_{{{o{(o\backslash3)}}{\tau}}}{t}_{{\tau}}}{[{\lambda}y_{t}.({{=}_{{{o{(
ot)}}{(ot)}}}{p}_{{ot}}}{({=}_{{{ot}t}}{y}_{t})})_{o}]})})}}|{({{{\exists}}_{{{o{(o\
backslash3)}}{\tau}}}{t}_{{\tau}}}{[{\lambda}y_{t}.({{=}_{{{o{(ot)}}{(ot)}}}{p}_{{ot
}}}{({=}_{{{ot}t}}{y}_{t})})_{o}]})}$
\# $\qquad \;\;\;\;\;\;\;\;\;\;\qquad {{=}\,{({{{LELE3}\,{s}}\,{TLVL2}}\,{PLVL3})}}\
,|{({{{{{[{\lambda}x.[{\lambda}u.[{\lambda}y.[{\lambda}v.[{\lambda}g.({{g}\,{x}}\,{y
})]]]]]}\,{a}}\,{TLVL2}}\,{PLVL2}}\,{s}}\,{[{\lambda}x.[{\lambda}y.x]]})}$
\# $\qquad \;\;\;\;\;\;\;\;\;\;\qquad {{=}\,{({{{LELE3}\,{s}}\,{TLVL2}}\,{PLVL3})}}\
,|{({{{{[{\lambda}u.[{\lambda}y.[{\lambda}v.[{\lambda}g.({{g}\,{a}}\,{y})]]]]}\,{TLV
L2}}\,{PLVL2}}\,{s}}\,{[{\lambda}x.[{\lambda}y.x]]})}$
\# $\qquad \;\;\;\;\;\;\;\;\;\;\qquad {{{\supset}}_{{{oo}o}}{({{{\forall}}_{{{o{(o\b
ackslash3)}}{\tau}}}{t}_{{\tau}}}{[{\lambda}x_{t}.({{{\exists_1}}_{{{o{(o\backslash3
)}}{\tau}}}{u}_{{\tau}}}{[{\lambda}y_{u}.({{p}_{{{ou}t}}{x}_{t}}{y}_{u})_{o}]})_{o}]
})}}|{({{{\exists}}_{{{o{(o\backslash3)}}{\tau}}}{{(ut)}}_{{\tau}}}{[{\lambda}f_{{ut
}}.({{{\forall}}_{{{o{(o\backslash3)}}{\tau}}}{t}_{{\tau}}}{[{\lambda}x_{t}.({{p}_{{
{ou}t}}{x}_{t}}{({f}_{{ut}}{x}_{t})})_{o}]})_{o}]})}\qquad$ $\mathrel{\mathop:}= \;\
;$ $K8033$
\# $\qquad \;\;\;\;\;\;\;\;\;\;\qquad {{{\exists}}_{{{o{(o\backslash3)}}{\tau}}}{{(t
{(ot)})}}_{{\tau}}}|{[{\lambda}j_{{t{(ot)}}}.({{{\forall}}_{{{o{(o\backslash3)}}{\ta
u}}}{{(ot)}}_{{\tau}}}{[{\lambda}p_{{ot}}.({{{\supset}}_{{{oo}o}}{({{{\exists}}_{{{o
{(o\backslash3)}}{\tau}}}{t}_{{\tau}}}{[{\lambda}x_{t}.({p}_{{ot}}{x}_{t})_{o}]})}}{
({p}_{{ot}}{({j}_{{t{(ot)}}}{p}_{{ot}})})})_{o}]})_{o}]}\qquad$ $\mathrel{\mathop:}=
 \;\;$ $AC$
\# $\qquad \;\;\;\;\;\;\;\;\;\;\qquad {{{\exists}}_{{{o{(o\backslash3)}}{\tau}}}{{(u
{(ou)})}}_{{\tau}}}|{[{\lambda}j_{{u{(ou)}}}.({{{\forall}}_{{{o{(o\backslash3)}}{\ta
u}}}{{(ou)}}_{{\tau}}}{[{\lambda}p_{{ou}}.({{{\supset}}_{{{oo}o}}{({{{\exists}}_{{{o
{(o\backslash3)}}{\tau}}}{u}_{{\tau}}}{[{\lambda}x_{u}.({p}_{{ou}}{x}_{u})_{o}]})}}{
({p}_{{ou}}{({j}_{{u{(ou)}}}{p}_{{ou}})})})_{o}]})_{o}]}$
:$=\;\;\$H8013H\;\;{{{\exists}}_{{{o{(o\backslash3)}}{\tau}}}{{(u{(ou)})}}_{{\tau}}}
|{[{\lambda}j_{{u{(ou)}}}.({{{\forall}}_{{{o{(o\backslash3)}}{\tau}}}{{(ou)}}_{{\tau
}}}{[{\lambda}p_{{ou}}.({{{\supset}}_{{{oo}o}}{({{{\exists}}_{{{o{(o\backslash3)}}{\
tau}}}{u}_{{\tau}}}{[{\lambda}x_{u}.({p}_{{ou}}{x}_{u})_{o}]})}}{({p}_{{ou}}{({j}_{{
u{(ou)}}}{p}_{{ou}})})})_{o}]})_{o}]}$
\# $\qquad \;\;\;\;\;\;\;\;\;\;\qquad {{=}\,{({[{\lambda}f.({{=}\,{({{f}\,{\$A8013H}
}\,{\$B8013H})}}\,{({{f}\,{\$B8013H}}\,{\$A8013H})})]}\,{{\land}})}}|\,{({{=}\,{({{{
\land}}\,{\$A8013H}}\,{\$B8013H})}}\,{({{{\land}}\,{\$B8013H}}\,{\$A8013H})})}$
\# $\qquad \;\;\;\;\;\;\;\;\;\;\qquad {{=}\,{[{\lambda}\$B8013H.({{=}\,{\$A8013H}}\,
{({{{\land}}\,{\$A8013H}}\,{\$B8013H})})]}}|\,{[{\lambda}ytmp.({{=}\,{\$A8013H}}\,{(
{{{\land}}\,{\$A8013H}}\,{ytmp})})]}$
\# $\qquad \;\;\;\;\;\;\;\;\;\;\qquad {{{\supset}}\,{\$H8013H}}|\,{({{[{\lambda}\$A8
013H.[{\lambda}ytmp.({{=}\,{\$A8013H}}\,{({{{\land}}\,{\$A8013H}}\,{ytmp})})]]}\,{\$
```

```
B8013H}}\,{\$A8013H})}$
\# $\qquad \;\;\;\;\;\;\;\;\;\;\qquad {{=}\,{({[{\lambda}\$A8013H.[{\lambda}ytmp.({{
=}\,{\$A8013H}}\,{({{{\land}}\,{\$A8013H}}\,{ytmp})})]]}\,{\$B8013H})}}|\,{[{\lambda
}ytmp.({{=}\,{\$B8013H}}\,{({{{\land}}\,{\$B8013H}}\,{ytmp})})]}$
\# $\qquad \;\;\;\;\;\;\;\;\;\;\qquad {{=}\,{({[{\lambda}ytmp.({{=}\,{\$B8013H}}\,{(
{{{\land}}\,{\$B8013H}}\,{ytmp})})]}\,{\$A8013H})}}|\,{({{=}\,{\$B8013H}}\,{({{{\lan
d}}\,{\$B8013H}}\,{\$A8013H})})}$
\# $\qquad \;\;\;\;\;\;\;\;\;\;\qquad {{{\supset}}\,{\$H8013H}}|\,{({{[{\lambda}\$A8
013H.[{\lambda}ytmp.({{=}\,{\$A8013H}}\,{({{{\land}}\,{\$A8013H}}\,{ytmp})})]]}\,{\$
A8013H}}\,{\$B8013H})}$
\# $\qquad \;\;\;\;\;\;\;\;\;\;\qquad {{=}\,{({[{\lambda}\$A8013H.[{\lambda}ytmp.({{
=}\,{\$A8013H}}\,{({{{\land}}\,{\$A8013H}}\,{ytmp})})]]}\,{\$A8013H})}}|\,{[{\lambda
}ytmp.({{=}\,{\$A8013H}}\,{({{{\land}}\,{\$A8013H}}\,{ytmp})})]}$
\# $\qquad \;\;\;\;\;\;\;\;\;\;\qquad {{=}\,{({[{\lambda}ytmp.({{=}\,{\$A8013H}}\,{(
{{{\land}}\,{\$A8013H}}\,{ytmp})})]}\,{\$B8013H})}}|\,{({{=}\,{\$A8013H}}\,{({{{\lan
d}}\,{\$A8013H}}\,{\$B8013H})})}$
\# $\qquad \;\;\;\;\;\;\;\;\;\;\qquad {{{\supset}}_{{oo}o}}{\$HYP5312}_{o}}|{({{=}_
{{oo}o}}{({{=}_{{o{(ot)}}{(ot)}}}{p}_{{ot}}}{({=}_{{ot}t}}{({{\iota}}_{{t{(ot)}}}
{p}_{{ot}})})})}}{({{{\forall}}_{{o{(o\backslash3)}}{\tau}}}{t}_{{\tau}}}{[{\lambda
}x_{t}.({{=}_{{oo}o}}{({p}_{{ot}}{x}_{t})}}{({{=}_{{ot}t}}{({{\iota}}_{{t{(ot)}}}{
p}_{{ot}})}}{x}_{t})})_{o}]})})}$
\# $\qquad \;\;\;\;\;\;\;\;\;\;\qquad {{{\supset}}_{{oo}o}}{({{{\land}}_{{oo}o}}{T
}_{o}}{({{{\exists}}_{{o{(o\backslash3)}}{\tau}}}{t}_{{\tau}}}{[{\lambda}y_{t}.\$HY
P5312_{o}]})})}}|{({{{\forall}}_{{o{(o\backslash3)}}{\tau}}}{t}_{{\tau}}}{[{\lambda
}z_{t}.({{=}_{{oo}o}}{({p}_{{ot}}{z}_{t})}}{({{=}_{{ot}t}}{({{\iota}}_{{t{(ot)}}}{
p}_{{ot}})}}{z}_{t})})_{o}]})}$
\# $\qquad \;\;\;\;\;\;\;\;\;\;\qquad {{{\supset}}_{{oo}o}}{({{{\land}}_{{oo}o}}{T
}_{o}}{({{{\exists}}_{{o{(o\backslash3)}}{\tau}}}{t}_{{\tau}}}{[{\lambda}y_{t}.\$HY
P5312_{o}]})})}}|{({{{\forall}}_{{o{(o\backslash3)}}{\tau}}}{t}_{{\tau}}}{[{\lambda
}z_{t}.({{=}_{{oo}o}}{({p}_{{ot}}{z}_{t})}}{({{=}_{{ot}t}}{({{\iota}}_{{t{(ot)}}}{
p}_{{ot}})}}{z}_{t})})_{o}]})}\qquad$ $\mathrel{\mathop:}= \;\;$ $\$LTMP5312$
\# $\qquad \;\;\;\;\;\;\;\;\;\;\qquad {{{\supset}}_{{oo}o}}{({{\sim}}_{{oo}}{({{{\e
xists}}_{{o{(o\backslash3)}}{\tau}}}{\$T8028}_{{\tau}}}{\$B8028}_{{o\,\$T8028}})})}
}|{({{{\supset}}_{{oo}o}}{\$H8028}_{o}}{({\$B8028}_{{o\,\$T8028}}{\$A8028}_{\$T8028
})})}$
\# wff$\qquad 5331\;\;:\;\;\qquad {{{{\supset}}\,{\$H8028}}\,{({{{\supset}}\,{({{\si
m}}\,{({{{\exists}}\,{\$T8028}}\,{\$B8028})})}}\,{F})}}_{o,\,...}\qquad|$ $\mathrel{
\mathop:}= \;\;$ $\$DTMP8028\;\;$ $\$HA8004$
\# $\qquad \;\;\;\;\;\;\;\;\;\;\qquad| {{{\supset}}_{{oo}o}}{({{{\land}}_{{oo}o}}{
\$HTMP}_{o}}{({{{\land}}_{{oo}o}}{({{ANSET}_{\$T4}{t}_{{\tau}}}{x}_{\$S})}}{({p}_{{
o\,\$S}}{x}_{\$S})})})}}{({\$P}_{{o\,\$S}}{({{ASUCC}_{\$T44}{t}_{{\tau}}}{x}_{\$S})}
)}\qquad$ $\mathrel{\mathop:}= \;\;$ $\$TMP$
\# $\qquad \;\;\;\;\;\;\;\;\;\;\qquad {{{\supset}}_{{oo}o}}{h}_{o}}|{({{=}_{{o{(o\
,\$T5215H)}}{(o\,\$T5215H)}}}{[{\lambda}\$X5215H_{\$T5215H}.T_{o}]}}{\$A2TMP5215H}_{
{o\,\$T5215H}})}\qquad$ $\mathrel{\mathop:}= \;\;$ $\$ATMP5215H$
\# $\qquad \;\;\;\;\;\;\;\;\;\;\qquad {{{\supset}}_{{oo}o}}{({{{\forall}}_{{o{(o\b
ackslash3)}}{\tau}}}{t}_{{\tau}}}{[{\lambda}z_{t}.({{=}_{{oo}o}}{({p}_{{ot}}{z}_{t}
)}}{({{=}_{{ot}t}}{y}_{t}}{z}_{t})})_{o}]})}}|{({{{\forall}}_{{o{(o\backslash3)}}{
\tau}}}{t}_{{\tau}}}{[{\lambda}z_{t}.({{=}_{{oo}o}}{({p}_{{ot}}{z}_{t})}}{({{=}_{{{
```

```
ot}t}}{y}_{t}}{z}_{t})})_{o}]})}\qquad$ $\mathrel{\mathop:}= \;\;$ $\$TMP5310$
\# $\qquad \;\;\;\;\;\;\;\;\;\;\qquad {{{{P1}_{{{{{o{(o\backslash5)}}{(\backslash4\b
ackslash4)}}\backslash2}{\tau}}}{\$S}_{{\tau}}}{({AZERO}_{{{o{(o\backslash3)}}{\tau}
}}{t}_{{\tau}})}}{({ASUCC}_{{{{o{(o\backslash4)}}{(o{(o\backslash4)})}}{\tau}}}{t}_{
{\tau}})}}|{({ANSET}_{{{o{(o{(o\backslash4)})}}{\tau}}}{t}_{{\tau}})}\qquad$ $\mathr
el{\mathop:}= \;\;$ $\$P1APP\;\;$ $A6100$
\# $\qquad \;\;\;\;\;\;\;\;\;\;\qquad {{{{P2}_{{{{{o{(o\backslash5)}}{(\backslash4\b
ackslash4)}}\backslash2}{\tau}}}{\$S}_{{\tau}}}{({AZERO}_{{{o{(o\backslash3)}}{\tau}
}}{t}_{{\tau}})}}{({ASUCC}_{{{{o{(o\backslash4)}}{(o{(o\backslash4)})}}{\tau}}}{t}_{
{\tau}})}}|{({ANSET}_{{{o{(o{(o\backslash4)})}}{\tau}}}{t}_{{\tau}})}\qquad$ $\mathr
el{\mathop:}= \;\;$ $\$P2APP\;\;$ $A6101$
\# $\qquad \;\;\;\;\;\;\;\;\;\;\qquad {ADOTx}_{{oo}}|{({{{\forall}}_{{{o{(o\backslas
h3)}}{\tau}}}{{(o\,\$S)}}_{{\tau}}}{[{\lambda}p_{{o\,\$S}}.({{{\supset}}_{{{oo}o}}{\
$ANBOTH}_{o}}{({p}_{{o\,\$S}}{({ATSUCC}_{{\$S\,\$S}}{x}_{\$S})})})_{o}]})}\qquad$ $\
mathrel{\mathop:}= \;\;$ $\$TMP6101$
\# $\qquad \;\;\;\;\;\;\;\;\;\;\qquad {{{\forall}}_{{{o{(o\backslash3)}}{\tau}}}{\$S
}_{{\tau}}}|{[{\lambda}x_{\$S}.({ADOTx}_{{oo}}{({{ANSET}_{{{o{(o{(o\backslash4)})}}{
\tau}}}{t}_{{\tau}}}{({{ASUCC}_{{{{o{(o\backslash4)}}{(o{(o\backslash4)})}}{\tau}}}{
t}_{{\tau}}}{x}_{\$S})})})_{o}]}\qquad$ $\mathrel{\mathop:}= \;\;$ $\$TMP6101$
\# $\qquad \;\;\;\;\;\;\;\;\;\;\qquad {{=}_{{{oo}o}}{({{{\supset}}_{{{oo}o}}{({{\sim
}}_{{oo}}{({{{\exists}}_{{{o{(o\backslash3)}}{\tau}}}{\$T8028}_{{\tau}}}{\$B8028}_{{
o\,\$T8028}})})}}{F}_{o})}}|{({{{\exists}}_{{{o{(o\backslash3)}}{\tau}}}{\$T8028}_{{
\tau}}}{\$B8028}_{{o\,\$T8028}})}\qquad$ $\mathrel{\mathop:}= \;\;$ $\$TTMP8028$
\# wff$\qquad 2518\;\;:\;\;\qquad| {{{{\forall}}\,{o}}\,{[{\lambda}b.({{{\forall}}\,
{o}}\,{[{\lambda}c.({{=}\,{({{XOR}\,{({{XOR}\,{\$X5220}}\,{b})}}\,{c})}}\,{({{XOR}\,
{\$X5220}}\,{({{XOR}\,{b}}\,{c})})})]})]}}_{o,\,...}\qquad$ $\mathrel{\mathop:}= \;\
;$ $\$A5220$
:$=\;\;\$T1\;\;|{{[{\lambda}g_{{\tau}}.[{\lambda}l_{{{gg}g}}.[{\lambda}e_{g}.[{\lamb
da}b_{g}.({{{\land}}_{{{oo}o}}{({{=}_{{{og}g}}{({{l}_{{{gg}g}}{a}_{g}}{b}_{g})}}{e}_
{g})}}{({{=}_{{{og}g}}{({{l}_{{{gg}g}}{b}_{g}}{a}_{g})}}{e}_{g})})_{o}]_{{(og)}}]_{{
({og}g)}}]_{{({{og}g}{({gg}g)})}}]}{o}_{{\tau}}}{XOR}_{{{oo}o}}$
\# $\qquad \;\;\;\;\;\;\;\;\;\;\qquad {{=}_{{{o{\omega}}{\omega}}}{({{Grp}_{{{o{({\b
ackslash4\backslash4}\backslash3)}}{\tau}}}{o}_{{\tau}}}{XOR}_{{{oo}o}})}}|{({{{\lan
d}}_{{{oo}o}}{\$XAsc}_{o}}{({{{\exists}}_{{{o{(o\backslash3)}}{\tau}}}{o}_{{\tau}}}{
[{\lambda}e_{o}.({{{\land}}_{{{oo}o}}{\$XIdy}_{o}}{\$XInv}_{o})_{o}]})})}\qquad$ $\m
athrel{\mathop:}= \;\;$ $\$T1$
```

2.2.8 File A5200t.r0.txt

```
##
##  Proof A5200t:  T  (special case of A = A)
##
##
##  Source: [Andrews 2002 (ISBN 1-4020-0763-9), p. 215]
##
##  Copyright (c) 2017 Owl of Minerva Press GmbH. All rights reserved.
##  Written by Ken Kubota (<mail@kenkubota.de>).
##
##  This file is part of the publication of the mathematical logic R0.
```

```
##  For more information, visit: <http://doi.org/10.4444/100.10>
##

<< definitions1.r0.txt


##
##  Proof
##

§= =

:= A5200t %0


##
##  Q.E.D.
##

%0
```

2.2.9 File A5201b.r0a.txt

```
##
##  Proof Template A5201b (Swap):  A = B  -->  B = A
##      for any A, B of any type T
##
##  Source: [Andrews 2002 (ISBN 1-4020-0763-9), p. 215]
##
##  Copyright (c) 2017 Owl of Minerva Press GmbH. All rights reserved.
##  Written by Ken Kubota (<mail@kenkubota.de>).
##
##  This file is part of the publication of the mathematical logic R0.
##  For more information, visit: <http://doi.org/10.4444/100.10>
##


##
##  Assumptions and Resulting Syntactical Variables
##

## the assumption as last theorem on stack (%0)
§! ((={{{o,o},o}}_a{o}{o}){{o,o}}_b{o}{o})


##
##  Include Proof Template
##
```

```
<<< A5201b.r0t.txt


##
##  Q.E.D.
##

%0
```

2.2.10 File A5201b.r0t.txt

```
##
##  Proof Template A5201b (Swap):  A = B  -->  B = A
##      for any A, B of any type T
##
##  Source: [Andrews 2002 (ISBN 1-4020-0763-9), p. 215]
##
##  Copyright (c) 2017 Owl of Minerva Press GmbH. All rights reserved.
##  Written by Ken Kubota (<mail@kenkubota.de>).
##
##  This file is part of the publication of the mathematical logic R0.
##  For more information, visit: <http://doi.org/10.4444/100.10>
##



##
##  Proof Template
##

## use polymorphic identity relation with type of right side of given equation ({**%
0})
§= {**%0} /5

## now replace left hand side of new equation
§s %0 5 %1
```

2.2.11 File A5201bH.r0a.txt

```
##
##  Proof Template A5201bH (SwapH):  H => (A = B)  -->  H => (B = A)
##      for any A, B of any type T
##
##  Source: [Andrews 2002 (ISBN 1-4020-0763-9), p. 215]
##
##  Copyright (c) 2017 Owl of Minerva Press GmbH. All rights reserved.
##  Written by Ken Kubota (<mail@kenkubota.de>).
##
##  This file is part of the publication of the mathematical logic R0.
```

```
##  For more information, visit: <http://doi.org/10.4444/100.10>
##

<< basics.r0.txt


##
##  Assumptions and Resulting Syntactical Variables
##

## the assumption as last theorem on stack (%0)
§! ((=>{{{o,o},o}}_h{o}{o}){{o,o}}_((={{{o,o},o}}_a{o}{o}){{o,o}}_b{o}{o}){o})


##
##  Include Proof Template
##

<<< A5201bH.r0t.txt


##
##  Q.E.D.
##

%0
```

2.2.12 File A5201bH.r0t.txt

```
##
##  Proof Template A5201bH (SwapH):  H => (A = B)  -->  H => (B = A)
##      for any A, B of any type T
##
##  Source: [Andrews 2002 (ISBN 1-4020-0763-9), p. 215]
##
##  Copyright (c) 2017 Owl of Minerva Press GmbH. All rights reserved.
##  Written by Ken Kubota (<mail@kenkubota.de>).
##
##  This file is part of the publication of the mathematical logic R0.
##  For more information, visit: <http://doi.org/10.4444/100.10>
##


##
##  Exception: Forward Reference
##
##  Because of the different rules of inference, unlike in Q0,
##  this theorem (with hypothesis) cannot be inferred from
##  previous theorems only, but depends on new theorems.
```

```
##
##  Dependencies (selection):
##
##            K8003 << K8000a, A5219b, A5221
##                     K8000a << A5222, A5229a, A5229c
##                                        A5221 << A5220
##                                                  A5229c << A5227 << A5226 << A5225
##
```

```
##
##  Proof Template
##

:= $TMPswapH %0

## use polymorphic identity relation with type of right side of given equation ({**%
0})
§=' {**%0/3} /5

## use Proof Template K8003 (Intro):  A  -->  H => A
:= $A8003 %0
:= $H8003 %1/5
<< K8003.r0t.txt
:= $A8003; := $H8003
%0

%$TMPswapH; := $TMPswapH

## now replace left hand side of new equation
§s' %1 5 %0
```

2.2.13 File A5205.r0.txt

```
##
##  Proof Template A5205:  f = [\y.fy]
##      for any y of type b and f of type ab
##
##  Source: [Andrews 2002 (ISBN 1-4020-0763-9), p. 217]
##
##  Copyright (c) 2017 Owl of Minerva Press GmbH. All rights reserved.
##  Written by Ken Kubota (<mail@kenkubota.de>).
##
##  This file is part of the publication of the mathematical logic R0.
##  For more information, visit: <http://doi.org/10.4444/100.10>
##


##
```

```
##  Define Syntactical Variables
##

## the variable of type b{^} to be used
:= $Y5205 y{b{^}}


##
##  Include Proof Template
##


<<< A5205.r0t.txt
:= A5205 %0


##
##  Undefine Syntactical Variables
##

:= $Y5205


##
##  Q.E.D.
##

%0
```

2.2.14 File A5205.r0t.txt

```
##
##  Proof Template A5205:  f = [\y.fy]
##      for any y of type b and f of type ab
##
##  Source: [Andrews 2002 (ISBN 1-4020-0763-9), p. 217]
##
##  Copyright (c) 2017 Owl of Minerva Press GmbH. All rights reserved.
##  Written by Ken Kubota (<mail@kenkubota.de>).
##
##  This file is part of the publication of the mathematical logic R0.
##  For more information, visit: <http://doi.org/10.4444/100.10>
##

<< axioms.r0.txt


##
##  Exception: Forward Reference
```

```
##
##  The original proof 5205 uses some of the Axiom Schemata 4_1 - 4_5
##  (indirectly via 5203 and 5204), which are not available in R0,
##  but replaced by Rule 2 (Lambda Conversion) [5207].
##  Therefore the use of the Rule of Substitution (A5221) is required here.
##
##  For historical purposes, and since the proof did not change otherwise,
##  the proof number 5205 was not altered.
##



##
##  Proof Template
##

## .1

%A3

## use Proof Template A5221 (Sub):  B  -->  B [x/A]
:= $B5221 %0
:= $T5221 {a{^},b{^}}
:= $X5221 g{$T5221}
:= $A5221 [\$Y5205{b{^}}.(f{{a{^},b{^}}}{{a{^},b{^}}}_$Y5205{b{^}}){a{^}}]
<< A5221.r0t.txt
:= $B5221; := $T5221; := $X5221; := $A5221
%0

## .2

§\ ([\$Y5205{b{^}}.(f{{a{^},b{^}}}{{a{^},b{^}}}_$Y5205{b{^}}){a{^}}]{{a{^},b{^}}}_x{
b{^}}{b{^}})

## .3

§s %1 31 %0
:= $TMP5205 %0

## .4

%A3

## use Proof Template A5221 (Sub):  B  -->  B [x/A]
:= $B5221 %0
:= $T5221 {a{^},b{^}}
:= $X5221 g{$T5221}
:= $A5221 f{$T5221}
<< A5221.r0t.txt
:= $B5221; := $T5221; := $X5221; := $A5221
```

```
%0

## .5

%$TMP5205; := $TMP5205

## use Proof Template A5201b (Swap):  A = B  -->  B = A
<< A5201b.r0t.txt
%0

§s %4 3 %0

§= {a{^},b{^}} f{{a{^},b{^}}}
§s %0 1 %1
```

2.2.15 File A5209.r0a.txt

```
##
##  Proof Template A5209 (incl. A5204):  B = C  -->  (B = C) [x/A]
##      (Substitution of a Free Variable on Both Sides of an Equation)
##
##  Source: [Andrews 2002 (ISBN 1-4020-0763-9), p. 220 (217)]
##
##  Copyright (c) 2017 Owl of Minerva Press GmbH. All rights reserved.
##  Written by Ken Kubota (<mail@kenkubota.de>).
##
##  This file is part of the publication of the mathematical logic R0.
##  For more information, visit: <http://doi.org/10.4444/100.10>
##



##
##  Define Syntactical Variables
##

## type of both sides of the equation (of b and c)
:= $M5209 m{^}

## type of the variable and the substitution term
:= $T5209 t{^}

## the variable to be replaced
:= $X5209 x{$T5209}

## substitution term
:= $A5209 a{$T5209}

## assumption (equation b=c)
:= $E5209 ((={{{o,$M5209},$M5209}}_(bs{{$M5209,$T5209}}{{$M5209,$T5209}}_$X5209{$T52
```

```
09}){$M5209}){{o,$M5209}}_(cs{{$M5209,$T5209}}{{$M5209,$T5209}}_$X5209{$T5209}){$M52
09})


##
##  Assumptions and Resulting Syntactical Variables
##

§! $E5209


##
##  Include Proof Template
##

<<< A5209.r0t.txt


##
##  Undefine Syntactical Variables
##

:= $M5209; := $E5209; := $T5209; := $X5209; := $A5209


##
##  Q.E.D.
##

%0
```

2.2.16 File A5209.r0t.txt

```
##
##  Proof Template A5209 (incl. A5204):  B = C  -->  (B = C) [x/A]
##      (Substitution of a Free Variable on Both Sides of an Equation)
##
##  Source: [Andrews 2002 (ISBN 1-4020-0763-9), p. 220 (217)]
##
##  Copyright (c) 2017 Owl of Minerva Press GmbH. All rights reserved.
##  Written by Ken Kubota (<mail@kenkubota.de>).
##
##  This file is part of the publication of the mathematical logic R0.
##  For more information, visit: <http://doi.org/10.4444/100.10>
##


##
##  Proof Template
```

```
##

## extract b and c
:= $B5209 $E5209/5
:= $C5209 $E5209/3

## .1

§= {$M5209} ([\$X5209{$T5209}.$B5209{$M5209}]{{$M5209,$T5209}}_$A5209{$T5209})

## .2

%$E5209
§s %1 13 %0

## .3

§\ /5

## .4

§\ %1/3

## .5

§s %2 5 %1
§s %0 3 %1

## undefine local variables
:= $B5209; := $C5209
```

2.2.17 File A5210.r0.txt

```
##
##  Proof Template A5210:  T = (B = B)
##      for any B of any type
##
##  Source: [Andrews 2002 (ISBN 1-4020-0763-9), p. 220]
##
##  Copyright (c) 2017 Owl of Minerva Press GmbH. All rights reserved.
##  Written by Ken Kubota (<mail@kenkubota.de>).
##
##  This file is part of the publication of the mathematical logic R0.
##  For more information, visit: <http://doi.org/10.4444/100.10>
##



##
##  Define Syntactical Variables
```

```
##

## type of the wff
:= $T5210 t{^}

## the wff
:= $B5210 b{$T5210}


##
##  Include Proof Template
##

<<< A5210.r0t.txt


##
##  Undefine Syntactical Variables
##

:= $T5210; := $B5210


##
##  Q.E.D.
##

%0
```

2.2.18 File A5210.r0t.txt

```
##
##  Proof Template A5210:  T = (B = B)
##      for any B of any type
##
##  Source: [Andrews 2002 (ISBN 1-4020-0763-9), p. 220]
##
##  Copyright (c) 2017 Owl of Minerva Press GmbH. All rights reserved.
##  Written by Ken Kubota (<mail@kenkubota.de>).
##
##  This file is part of the publication of the mathematical logic R0.
##  For more information, visit: <http://doi.org/10.4444/100.10>
##


##
##  Proof Template
##
```

```
## .1

## use Proof Template:  Axiom 3 Substitutions
:= $AA3 $T5210
:= $BA3 $T5210
:= $FA3 [\y{$T5210}{$T5210}.y{$T5210}{$T5210}]
:= $GA3 [\y{$T5210}{$T5210}.y{$T5210}{$T5210}]
<< axiom3_substitutions.r0t.txt
:= $AA3; := $BA3; := $FA3; := $GA3
%0

## .2

§= {$T5210,$T5210} [\y{$T5210}{$T5210}.y{$T5210}{$T5210}]
§s %0 1 %1
§\ ([\y{$T5210}{$T5210}.y{$T5210}{$T5210}]{{$T5210,$T5210}}_x{$T5210}{$T5210})
§s %1 29 %0
§s %0 15 %1
§\ (A{{{o,{o,\3{^}}},^}}_$T5210{^})
§s %1 2 %0
§\ %0
§s %1 1 %0

## .3

:= $LxT5210 [\x{$T5210}{$T5210}.T{o}]
§= {o} ($LxT5210{{o,$T5210}}_$B5210{$T5210})
§s %0 6 %1

## .4

§\ /5
§\ %1/3
§s %2 5 %1
§s %0 3 %1

## undefine local variables
:= $LxT5210
```

2.2.19 File A5211.r0.txt

```
##
##  Proof A5211:  (T & T) = T
##
##
##  Source: [Andrews 2002 (ISBN 1-4020-0763-9), p. 220]
##
##  Copyright (c) 2017 Owl of Minerva Press GmbH. All rights reserved.
##  Written by Ken Kubota (<mail@kenkubota.de>).
```

```
##
##  This file is part of the publication of the mathematical logic R0.
##  For more information, visit: <http://doi.org/10.4444/100.10>
##

<< axioms.r0.txt


##
##  Proof
##

## .1

%A1

## use Proof Template A5209 (incl. A5204):  B = C  -->  (B = C) [x/A]
:= $M5209 o
:= $E5209 %0
:= $T5209 {o,o}
:= $X5209 g{{o,o}}
:= $A5209 [\y{o}{o}.T{o}]
<< A5209.r0t.txt
:= $M5209; := $E5209; := $T5209; := $X5209; := $A5209
%0

## .2

§\s /21
§\s /11
§\s /15
:= $ATMP5211 %0

## .3

§= ((A{{{o,{o,\3{^}}},^}}_o{^}){{o,{o,o}}}_[\x{o}{o}.T{o}]{{o,o}})
§\s /10
§\ {o} /5
§s %1 5 %0
:= $BTMP5211 %0

## use Proof Template A5210:  T = (B = B)
:= $T5210 {o,o}
:= $B5210 /21
<< A5210.r0t.txt
:= $T5210; := $B5210
%0

%$BTMP5211
```

```
§s %1 3 %0

## .4

%$ATMP5211
§= T
§s %0 5 %2
§s %2 3 %0

:= A5211 %0

## undefine local variables
:= $ATMP5211; := $BTMP5211


##
##  Q.E.D.
##

%0
```

2.2.20 File A5212.r0.txt

```
##
##  Proof A5212:  T & T
##
##
##  Source: [Andrews 2002 (ISBN 1-4020-0763-9), p. 220]
##
##  Copyright (c) 2017 Owl of Minerva Press GmbH. All rights reserved.
##  Written by Ken Kubota (<mail@kenkubota.de>).
##
##  This file is part of the publication of the mathematical logic R0.
##  For more information, visit: <http://doi.org/10.4444/100.10>
##

<< A5200t.r0.txt
<< A5211.r0.txt


##
##  Proof
##

%A5211
## use Proof Template A5201b (Swap):  A = B  -->  B = A
<< A5201b.r0t.txt
%0
%A5200t
```

```
§s %0 1 %1

:= A5212 %0


##
##  Q.E.D.
##

%0
```

2.2.21 File A5213.r0a.txt

```
##
##  Proof Template A5213:  A = B and C = D  -->  (A = B) & (C = D)
##      for any A, B of type T and any C, D of type U
##
##  Source: [Andrews 2002 (ISBN 1-4020-0763-9), pp. 220 f.]
##
##  Copyright (c) 2017 Owl of Minerva Press GmbH. All rights reserved.
##  Written by Ken Kubota (<mail@kenkubota.de>).
##
##  This file is part of the publication of the mathematical logic R0.
##  For more information, visit: <http://doi.org/10.4444/100.10>
##



##
##  Define Syntactical Variables
##

## type of A, B
:= $T5213 t{^}

## A = B
:= $AB5213 ((={{{o,@},@}}_a{$T5213}{@}){{o,@}}_b{$T5213}{@})

## type of C, D
:= $U5213 u{^}

## C = D
:= $CD5213 ((={{{o,@},@}}_c{$U5213}{@}){{o,@}}_d{$U5213}{@})


##
##  Assumptions and Resulting Syntactical Variables
##

§! $AB5213
```

```
§! $CD5213


##
##  Include Proof Template
##


<<< A5213.r0t.txt


##
##  Undefine Syntactical Variables
##

:= $T5213; := $AB5213; := $U5213; := $CD5213


##
##  Q.E.D.
##

%0
```

2.2.22 File A5213.r0t.txt

```
##
##  Proof Template A5213:  A = B and C = D  -->  (A = B) & (C = D)
##      for any A, B of type T and any C, D of type U
##
##  Source: [Andrews 2002 (ISBN 1-4020-0763-9), pp. 220 f.]
##
##  Copyright (c) 2017 Owl of Minerva Press GmbH. All rights reserved.
##  Written by Ken Kubota (<mail@kenkubota.de>).
##
##  This file is part of the publication of the mathematical logic R0.
##  For more information, visit: <http://doi.org/10.4444/100.10>
##

<< A5212.r0.txt


##
##  Proof Template
##

## .1

%$AB5213
```

```
## .2

## use Proof Template A5210:  T = (B = B)
:= $T5210 $T5213
:= $B5210 $AB5213/5
<< A5210.r0t.txt
:= $T5210; := $B5210
%0

%$AB5213
§s %1 7 %0
:= $TMP5213 %0

## .3

%$CD5213

## .4

## use Proof Template A5210:  T = (B = B)
:= $T5210 $U5213
:= $B5210 $CD5213/5
<< A5210.r0t.txt
:= $T5210; := $B5210
%0

%$CD5213
§s %1 7 %0

## .5

%A5212
%$TMP5213
§s %1 5 %0
§s %0 3 %3

## undefine local variables
:= $TMP5213
```

2.2.23 File A5214.r0.txt

```
##
##  Proof A5214:  (T & F) = F
##
##
##  Source: [Andrews 2002 (ISBN 1-4020-0763-9), p. 221]
##
##  Copyright (c) 2017 Owl of Minerva Press GmbH. All rights reserved.
```

```
##  Written by Ken Kubota (<mail@kenkubota.de>).
##
##  This file is part of the publication of the mathematical logic R0.
##  For more information, visit: <http://doi.org/10.4444/100.10>
##

<< axioms.r0.txt


##
##  Proof
##

## .1

%A1

## use Proof Template A5209 (incl. A5204):  B = C  -->  (B = C) [x/A]
:= $M5209 o
:= $E5209 %0
:= $T5209 {o,o}
:= $X5209 g{{o,o}}
:= $A5209 [\x{o}{o}.x{o}{o}]
<< A5209.r0t.txt
:= $M5209; := $E5209; := $T5209; := $X5209; := $A5209
%0

## .2

§\s /21
§\s /11
§\s /15

§= /3
§\s /6
§\s /3
§s %5 3 %0

:= A5214 %0


##
##  Q.E.D.
##

%0
```

2.2.24 File A5215.r0a.txt

```
##
##  Proof Template A5215 (ALL I):  ALL x: B  -->  B [x/a]
##      (Universal Instantiation)
##
##  Source: [Andrews 2002 (ISBN 1-4020-0763-9), p. 221]
##
##  Copyright (c) 2017 Owl of Minerva Press GmbH. All rights reserved.
##  Written by Ken Kubota (<mail@kenkubota.de>).
##
##  This file is part of the publication of the mathematical logic R0.
##  For more information, visit: <http://doi.org/10.4444/100.10>
##



##
##  Define Syntactical Variables
##

<< definitions1.r0.txt

## type of the variable and the substitution term
:= $T5215 t{^}

## the variable to be replaced
:= $X5215 x{$T5215}

## substitution term
:= $A5215 a{$T5215}

## hypothesis:  ALL x of type t: B  (in this example, B is defined as x=x)
:= $H5215 ((A{{{o,{o,\3{^}}},^}}_$T5215{^}){{o,{o,$T5215}}}_[\$X5215{$T5215}.((={{{o
,@},@}}_$X5215{@}){{o,@}}_$X5215{@}){o}]{{o,$T5215}})


##
##  Assumptions and Resulting Syntactical Variables
##

§! $H5215


##
##  Include Proof Template
##

<<< A5215.r0t.txt
```

```
##
##  Undefine Syntactical Variables
##

:= $T5215; := $X5215; := $A5215; := $H5215


##
##  Q.E.D.
##

%0
```

2.2.25 File A5215.r0t.txt

```
##
##  Proof Template A5215 (ALL I):  ALL x: B  -->  B [x/a]
##      (Universal Instantiation)
##
##  Source: [Andrews 2002 (ISBN 1-4020-0763-9), p. 221]
##
##  Copyright (c) 2017 Owl of Minerva Press GmbH. All rights reserved.
##  Written by Ken Kubota (<mail@kenkubota.de>).
##
##  This file is part of the publication of the mathematical logic R0.
##  For more information, visit: <http://doi.org/10.4444/100.10>
##

<< A5200t.r0.txt


##
##  Proof Template
##

## .1

%$H5215
§\s /2
§\s /1

## .2

§= ([\x{$T5215}{$T5215}.T{o}]{{o,$T5215}}_$A5215{$T5215})
§s %0 6 %1

## .3
```

```
§\s /5
§\s /3

## .4

%A5200t
§s %0 1 %1
```

2.2.26 File A5215H.r0a.txt

```
##
## Proof Template A5215H (ALL I):  H => ALL x: B  -->  H => B [x/a]
##     (Universal Instantiation)
##
## Source: [Andrews 2002 (ISBN 1-4020-0763-9), p. 221]
##
## Copyright (c) 2017 Owl of Minerva Press GmbH. All rights reserved.
## Written by Ken Kubota (<mail@kenkubota.de>).
##
## This file is part of the publication of the mathematical logic R0.
## For more information, visit: <http://doi.org/10.4444/100.10>
##


##
## Define Syntactical Variables
##

<< definitions1.r0.txt

## type of the variable and the substitution term
:= $T5215H t{^}

## the variable to be replaced
:= $X5215H x{$T5215H}

## substitution term
:= $A5215H a{$T5215H}

## hypothesis:  H => ALL x of type t: B  (in this example, B is defined as x=x)
:= $H5215H ((=>{{{o,o},o}}_h{o}{o}){{o,o}}_((A{{{o,{o,\3{^}}},^}}_$T5215H{^}){{o,{o,
$T5215H}}}_[\$X5215H{$T5215H}.((={{{o,@},@}}_$X5215H{@}){{o,@}}_$X5215H{@}){o}]{{o,$
T5215H}}){o})


##
## Assumptions and Resulting Syntactical Variables
##
```

```
§! $H5215H


##
## Include Proof Template
##

<<< A5215H.r0t.txt


##
## Undefine Syntactical Variables
##

:= $T5215H; := $X5215H; := $A5215H; := $H5215H


##
## Q.E.D.
##

%0
```

2.2.27 File A5215H.r0t.txt

```
##
## Proof Template A5215H (ALL I):  H => ALL x: B  -->  H => B [x/a]
##     (Universal Instantiation)
##
## Source: [Andrews 2002 (ISBN 1-4020-0763-9), p. 221]
##
## Copyright (c) 2017 Owl of Minerva Press GmbH. All rights reserved.
## Written by Ken Kubota (<mail@kenkubota.de>).
##
## This file is part of the publication of the mathematical logic R0.
## For more information, visit: <http://doi.org/10.4444/100.10>
##

<< A5200t.r0.txt


##
## Exception: Forward Reference
##
## Because of the different rules of inference, unlike in Q0,
## this theorem (with hypothesis) cannot be inferred from
## previous theorems only, but depends on new theorems.
##
## Dependencies (selection):
```

```
##
##  K8004 << K8003
##           K8003 << K8000a, A5219b, A5221
##                    K8000a << A5222, A5229a, A5229c
##                                    A5221 << A5220
##                                              A5229c << A5227 << A5226 << A5225
##


##
##  Proof Template
##

## .1

%$H5215H
§\s' /2
§\s' /1
:= $ATMP5215H %0

## .2

§= ([\x{$T5215H}{$T5215H}.T{o}]{{o,$T5215H}}_$A5215H{$T5215H})

## use Proof Template K8004 (Trans):  (H OP A), B  -->  H => B
:= $HA8004 $H5215H
:= $B8004 %0
<< K8004.r0t.txt
:= $HA8004; := $B8004
%0

:= $BTMP5215H %0

## shorthand to avoid overlong line
:= $A2TMP5215H [\$X5215H{$T5215H}.((={{o,@},@}}_$X5215H{@}){{o,@}}_$X5215H{@}){o}]

%$ATMP5215H; := $ATMP5215H
%$BTMP5215H; := $BTMP5215H
§s' %0 6 %1

## undefine shorthand
:= $A2TMP5215H

## .3

§\s' /5
§\s' /3
:= $ATMP5215H %0
```

```
## .4

## use Proof Template K8004 (Trans):  (H OP A), B  -->  H => B
:= $HA8004 $H5215H
:= $B8004 %A5200t
<< K8004.r0t.txt
:= $HA8004; := $B8004
%0

:= $BTMP5215H %0
%$ATMP5215H; := $ATMP5215H
%$BTMP5215H; := $BTMP5215H
§s' %0 1 %1
```

2.2.28 File A5216.r0.txt

```
##
##  Proof Template A5216:  (T & A) = A
##
##
##  Source: [Andrews 2002 (ISBN 1-4020-0763-9), p. 221]
##
##  Copyright (c) 2017 Owl of Minerva Press GmbH. All rights reserved.
##  Written by Ken Kubota (<mail@kenkubota.de>).
##
##  This file is part of the publication of the mathematical logic R0.
##  For more information, visit: <http://doi.org/10.4444/100.10>
##



##
##  Define Syntactical Variables
##

## the proposition
:= $A5216 a{o}



##
##  Include Proof Template
##

<<< A5216.r0t.txt


##
##  Undefine Syntactical Variables
##
```

```
:= $A5216


##
##  Q.E.D.
##

%0
```

2.2.29 File A5216.r0t.txt

```
##
##  Proof Template A5216:  (T & A) = A
##
##
##  Source: [Andrews 2002 (ISBN 1-4020-0763-9), p. 221]
##
##  Copyright (c) 2017 Owl of Minerva Press GmbH. All rights reserved.
##  Written by Ken Kubota (<mail@kenkubota.de>).
##
##  This file is part of the publication of the mathematical logic R0.
##  For more information, visit: <http://doi.org/10.4444/100.10>
##

<< A5211.r0.txt
<< A5214.r0.txt


##
##  Proof Template
##

## .1

%A1

## use Proof Template A5209 (incl. A5204):  B = C  -->  (B = C) [x/A]
:= $M5209 o
:= $E5209 %0
:= $T5209 {o,o}
:= $X5209 g{{o,o}}
:= $A5209 [\x{o}{o}.((={{{o,o},o}}_((&{{{o,o},o}}_T{o}){{o,o}}_x{o}{o}){o}){{o,o}}_x
{o}{o}){o}]
<< A5209.r0t.txt
:= $M5209; := $E5209; := $T5209; := $X5209; := $A5209
%0

## .2
```

```
§\s /21
§\s /11
§\s /15
:= $TMP5216 %0

## .3

## use Proof Template A5213:  A = B and C = D  -->  (A = B) & (C = D)
:= $T5213 o
:= $AB5213 A5211
:= $U5213 o
:= $CD5213 A5214
<< A5213.r0t.txt
:= $T5213; := $AB5213; := $U5213; := $CD5213
%0

## .4

%$TMP5216
§s %1 1 %0

## .5

## use Proof Template A5215 (ALL I):  ALL x: B  -->  B [x/a]
:= $T5215 o
:= $X5215 x{o}
:= $A5215 $A5216
:= $H5215 %0
<< A5215.r0t.txt
:= $T5215; := $X5215; := $A5215; := $H5215
%0

## undefine local variables
:= $TMP5216
```

2.2.30 File A5217.r0.txt

```
##
##  Proof A5217:  (T = F) = F
##
##
##  Source: [Andrews 2002 (ISBN 1-4020-0763-9), pp. 221 f.]
##
##  Copyright (c) 2017 Owl of Minerva Press GmbH. All rights reserved.
##  Written by Ken Kubota (<mail@kenkubota.de>).
##
##  This file is part of the publication of the mathematical logic R0.
##  For more information, visit: <http://doi.org/10.4444/100.10>
##
```

```
<< axioms.r0.txt


##
## Proof
##

## .1

%A1

## use Proof Template A5209 (incl. A5204):  B = C  -->  (B = C) [x/A]
:= $M5209 o
:= $E5209 %0
:= $T5209 {o,o}
:= $X5209 g{{o,o}}
:= $A5209 [\x{o}{o}.((={{{o,o},o}}_T{o}){{o,o}}_x{o}{o}){o}]
<< A5209.r0t.txt
:= $M5209; := $E5209; := $T5209; := $X5209; := $A5209
%0

## .2

§\s /21
§\s /11
§\s /15
:= $ATMP5217 %0

## .3

## use Proof Template A5210:  T = (B = B)
:= $T5210 o
:= $B5210 T
<< A5210.r0t.txt
:= $T5210; := $B5210
%0

%$ATMP5217
§= T
§s %0 5 %2
§s %2 21 %0

:= $BTMP5217 %0

## .4

## use Proof Template A5216:  (T & A) = A
:= $A5216 ((={{{o,o},o}}_T{o}){{o,o}}_F{o})
```

```
<< A5216.r0t.txt
:= $A5216
%0

%$BTMP5217
§s %0 5 %1

:= $CTMP5217 %0

## .5

## use Proof Template:  Axiom 3 Substitutions
:= $AA3 o
:= $BA3 o
:= $FA3 [\x{o}{o}.T{o}]
:= $GA3 [\x{o}{o}.x{o}{o}]
<< axiom3_substitutions.r0t.txt
:= $AA3; := $BA3; := $FA3; := $GA3
%0

## .6

§\s /61
§\s /31

## .7

## use Proof Template A5201b (Swap):  A = B  -->  B = A
<< A5201b.r0t.txt
%0
%$CTMP5217
§s %0 3 %1

:= A5217 %0

## undefine local variables
:= $ATMP5217; := $BTMP5217; := $CTMP5217


##
##  Q.E.D.
##

%0
```

2.2.31 File A5218.r0.txt

```
##
##  Proof Template A5218:  (T = A) = A
```

```
##
##
##  Source: [Andrews 2002 (ISBN 1-4020-0763-9), p. 222]
##
##  Copyright (c) 2017 Owl of Minerva Press GmbH. All rights reserved.
##  Written by Ken Kubota (<mail@kenkubota.de>).
##
##  This file is part of the publication of the mathematical logic R0.
##  For more information, visit: <http://doi.org/10.4444/100.10>
##


##
##  Define Syntactical Variables
##

## the bool wff
:= $A5218 a{o}


##
##  Include Proof Template
##

<<< A5218.r0t.txt


##
##  Undefine Syntactical Variables
##

:= $A5218


##
##  Q.E.D.
##

%0
```

2.2.32 File A5218.r0t.txt

```
##
##  Proof Template A5218:  (T = A) = A
##
##
##  Source: [Andrews 2002 (ISBN 1-4020-0763-9), p. 222]
##
##  Copyright (c) 2017 Owl of Minerva Press GmbH. All rights reserved.
```

```
##  Written by Ken Kubota (<mail@kenkubota.de>).
##
##  This file is part of the publication of the mathematical logic R0.
##  For more information, visit: <http://doi.org/10.4444/100.10>
##

<< A5217.r0.txt


##
##  Proof Template
##

## .1

%A1

## use Proof Template A5209 (incl. A5204):  B = C  -->  (B = C) [x/A]
:= $M5209 o
:= $E5209 %0
:= $T5209 {o,o}
:= $X5209 g{{o,o}}
:= $A5209 [\x{o}{o}.((={{{o,o},o}}_((={{{o,o},o}}_T{o}){{o,o}}_x{o}{o}){o}){{o,o}}_x
{o}{o}){o}]
<< A5209.r0t.txt
:= $E5209; := $T5209; := $X5209; := $A5209
%0

§\s /21
§\s /11
§\s /15
:= $TMP5218 %0

## .2

## use Proof Template A5210:  T = (B = B)
:= $T5210 o
:= $B5210 T
<< A5210.r0t.txt
:= $T5210; := $B5210
%0

## .3

## use Proof Template A5201b (Swap):  A = B  -->  B = A
<< A5201b.r0t.txt
%0

## use Proof Template A5213:  A = B and C = D  -->  (A = B) & (C = D)
```

```
:= $T5213 o
:= $AB5213 %0
:= $U5213 o
:= $CD5213 A5217
<< A5213.r0t.txt
:= $T5213; := $AB5213; := $U5213; := $CD5213
%0

## .4

%$TMP5218
§s %1 1 %0

## .5

## use Proof Template A5215 (ALL I):  ALL x: B  -->  B [x/a]
:= $T5215 o
:= $X5215 x{o}
:= $A5215 $A5218
:= $H5215 %0
<< A5215.r0t.txt
:= $T5215; := $X5215; := $A5215; := $H5215
%0

## undefine local variables
:= $TMP5218
```

2.2.33 File A5219a.r0a.txt

```
##
##  Proof Template A5219a (Rule T):  A  -->  T = A
##
##
##  Source: [Andrews 2002 (ISBN 1-4020-0763-9), p. 222]
##
##  Copyright (c) 2017 Owl of Minerva Press GmbH. All rights reserved.
##  Written by Ken Kubota (<mail@kenkubota.de>).
##
##  This file is part of the publication of the mathematical logic R0.
##  For more information, visit: <http://doi.org/10.4444/100.10>
##


##
##  Define Syntactical Variables
##

<< basics.r0.txt
```

```
## the assumption
:= $A5219a a{o}


##
##  Assumptions and Resulting Syntactical Variables
##

§! $A5219a


##
##  Include Proof Template
##

<<< A5219a.r0t.txt


##
##  Undefine Syntactical Variables
##

:= $A5219a


##
##  Q.E.D.
##

%0
```

2.2.34 File A5219a.r0t.txt

```
##
##  Proof Template A5219a (Rule T):  A  -->  T = A
##
##
##  Source: [Andrews 2002 (ISBN 1-4020-0763-9), p. 222]
##
##  Copyright (c) 2017 Owl of Minerva Press GmbH. All rights reserved.
##  Written by Ken Kubota (<mail@kenkubota.de>).
##
##  This file is part of the publication of the mathematical logic R0.
##  For more information, visit: <http://doi.org/10.4444/100.10>
##
```

```
## Empty lines are needed for comparison between A5219a-A5219d and A5219aH-A5219dH.


##
## Proof Template
##

## use Proof Template A5218:  (T = A) = A
:= $A5218 $A5219a
<< A5218.r0t.txt
:= $A5218
%0

## use Proof Template A5201b (Swap):  A = B  -->  B = A
<< A5201b.r0t.txt
%0









%$A5219a





§s %0 1 %1
```

2.2.35 File A5219aH.r0a.txt

```
##
## Proof Template A5219aH (Rule T):  H => A  -->  H => (T = A)
##
##
## Source: [Andrews 2002 (ISBN 1-4020-0763-9), p. 222]
##
## Copyright (c) 2017 Owl of Minerva Press GmbH. All rights reserved.
## Written by Ken Kubota (<mail@kenkubota.de>).
##
## This file is part of the publication of the mathematical logic R0.
```

```
## For more information, visit: <http://doi.org/10.4444/100.10>
##


##
## Define Syntactical Variables
##

<< basics.r0.txt

## the assumption
:= $A5219aH ((=>{{{o,o},o}}_h{o}{o}){{o,o}}_a{o}{o})


##
## Assumptions and Resulting Syntactical Variables
##

§! $A5219aH


##
## Include Proof Template
##

<<< A5219aH.r0t.txt


##
## Undefine Syntactical Variables
##

:= $A5219aH


##
## Q.E.D.
##

%0
```

2.2.36 File A5219aH.r0t.txt

```
##
## Proof Template A5219aH (Rule T): H => A --> H => (T = A)
##
##
## Source: [Andrews 2002 (ISBN 1-4020-0763-9), p. 222]
##
```

```
##  Copyright (c) 2017 Owl of Minerva Press GmbH. All rights reserved.
##  Written by Ken Kubota (<mail@kenkubota.de>).
##
##  This file is part of the publication of the mathematical logic R0.
##  For more information, visit: <http://doi.org/10.4444/100.10>
##



##
##  Exception: Forward Reference
##
##  (See comment in Proof Template A5215H.)
##

##  Empty lines are needed for comparison between A5219a-A5219d and A5219aH-A5219dH.


##
##  Proof Template
##

## use Proof Template A5218:  (T = A) = A
:= $A5218 $A5219aH/3
<< A5218.r0t.txt
:= $A5218
%0

## use Proof Template A5201b (Swap):  A = B  -->  B = A
<< A5201b.r0t.txt
%0

## use Proof Template K8004 (Trans):  (H OP A), B  -->  H => B
:= $HA8004 $A5219aH
:= $B8004 %0
<< K8004.r0t.txt
:= $HA8004; := $B8004
%0


%$A5219aH






§s' %0 1 %1
```

2.2.37 File A5219b.r0a.txt

```
##
##  Proof Template A5219b (Rule T):  A  -->  A = T
##
##
##  Source: [Andrews 2002 (ISBN 1-4020-0763-9), p. 222]
##
##  Copyright (c) 2017 Owl of Minerva Press GmbH. All rights reserved.
##  Written by Ken Kubota (<mail@kenkubota.de>).
##
##  This file is part of the publication of the mathematical logic R0.
##  For more information, visit: <http://doi.org/10.4444/100.10>
##



##
##  Define Syntactical Variables
##

<< basics.r0.txt

## the assumption
:= $A5219b a{o}


##
##  Assumptions and Resulting Syntactical Variables
##

§! $A5219b


##
##  Include Proof Template
##

<<< A5219b.r0t.txt


##
##  Undefine Syntactical Variables
##

:= $A5219b


##
##  Q.E.D.
```

```
##

%0
```

2.2.38 File A5219b.r0t.txt

```
##
##  Proof Template A5219b (Rule T):  A  -->  A = T
##
##
##  Source: [Andrews 2002 (ISBN 1-4020-0763-9), p. 222]
##
##  Copyright (c) 2017 Owl of Minerva Press GmbH. All rights reserved.
##  Written by Ken Kubota (<mail@kenkubota.de>).
##
##  This file is part of the publication of the mathematical logic R0.
##  For more information, visit: <http://doi.org/10.4444/100.10>
##









##  Empty lines are needed for comparison between A5219a-A5219d and A5219aH-A5219dH.


##
##  Proof Template
##

## use Proof Template A5218:  (T = A) = A
:= $A5218 $A5219b
<< A5218.r0t.txt
:= $A5218
%0

## use Proof Template A5201b (Swap):  A = B  -->  B = A
<< A5201b.r0t.txt
%0
```

```
%$A5219b




§s %0 1 %1

## use Proof Template A5201b (Swap):  A = B  -->  B = A
<< A5201b.r0t.txt
%0
```

2.2.39 File A5219bH.r0a.txt

```
##
##  Proof Template A5219bH (Rule T):  H => A  -->  H => (A = T)
##
##
##  Source: [Andrews 2002 (ISBN 1-4020-0763-9), p. 222]
##
##  Copyright (c) 2017 Owl of Minerva Press GmbH. All rights reserved.
##  Written by Ken Kubota (<mail@kenkubota.de>).
##
##  This file is part of the publication of the mathematical logic R0.
##  For more information, visit: <http://doi.org/10.4444/100.10>
##



##
##  Define Syntactical Variables
##

<< basics.r0.txt

## the assumption
:= $A5219bH ((=>{{{o,o},o}}_h{o}{o}){{o,o}}_a{o}{o})



##
##  Assumptions and Resulting Syntactical Variables
##

§! $A5219bH



##
##  Include Proof Template
```

```
##

<<< A5219bH.r0t.txt


##
##  Undefine Syntactical Variables
##

:= $A5219bH


##
##  Q.E.D.
##

%0
```

2.2.40 File A5219bH.r0t.txt

```
##
##  Proof Template A5219bH (Rule T):  H => A  -->  H => (A = T)
##
##
##  Source: [Andrews 2002 (ISBN 1-4020-0763-9), p. 222]
##
##  Copyright (c) 2017 Owl of Minerva Press GmbH. All rights reserved.
##  Written by Ken Kubota (<mail@kenkubota.de>).
##
##  This file is part of the publication of the mathematical logic R0.
##  For more information, visit: <http://doi.org/10.4444/100.10>
##


##
##  Exception: Forward Reference
##
##  (See comment in Proof Template A5215H.)
##

##  Empty lines are needed for comparison between A5219a-A5219d and A5219aH-A5219dH.


##
##  Proof Template
##

## use Proof Template A5218:  (T = A) = A
:= $A5218 $A5219bH/3
```

```
<< A5218.r0t.txt
:= $A5218
%0

## use Proof Template A5201b (Swap):  A = B  -->  B = A
<< A5201b.r0t.txt
%0

## use Proof Template K8004 (Trans):  (H OP A), B  -->  H => B
:= $HA8004 $A5219bH
:= $B8004 %0
<< K8004.r0t.txt
:= $HA8004; := $B8004
%0


%$A5219bH




§s' %0 1 %1

## use Proof Template A5201bH (SwapH):  H => (A = B)  -->  H => (B = A)
<< A5201bH.r0t.txt
%0
```

2.2.41 File A5219c.r0a.txt

```
##
##  Proof Template A5219c (Rule T):  T = A  -->  A
##
##
##  Source: [Andrews 2002 (ISBN 1-4020-0763-9), p. 222]
##
##  Copyright (c) 2017 Owl of Minerva Press GmbH. All rights reserved.
##  Written by Ken Kubota (<mail@kenkubota.de>).
##
##  This file is part of the publication of the mathematical logic R0.
##  For more information, visit: <http://doi.org/10.4444/100.10>
##


##
##  Define Syntactical Variables
##

<< basics.r0.txt
```

```
## the assumption
:= $A5219c ((={{{o,o},o}}_T{o}){{o,o}}_a{o}{o})


##
##  Assumptions and Resulting Syntactical Variables
##

§! $A5219c


##
##  Include Proof Template
##

<<< A5219c.r0t.txt


##
##  Undefine Syntactical Variables
##

:= $A5219c


##
##  Q.E.D.
##

%0
```

2.2.42 File A5219c.r0t.txt

```
##
##  Proof Template A5219c (Rule T):  T = A  -->  A
##
##
##  Source: [Andrews 2002 (ISBN 1-4020-0763-9), p. 222]
##
##  Copyright (c) 2017 Owl of Minerva Press GmbH. All rights reserved.
##  Written by Ken Kubota (<mail@kenkubota.de>).
##
##  This file is part of the publication of the mathematical logic R0.
##  For more information, visit: <http://doi.org/10.4444/100.10>
##
```

```
## Empty lines are needed for comparison between A5219a-A5219d and A5219aH-A5219dH.


##
## Proof Template
##

## use Proof Template A5218:  (T = A) = A
:= $A5218 $A5219c/3
<< A5218.r0t.txt
:= $A5218
%0













%$A5219c






§s %0 1 %1
```

2.2.43 File A5219cH.r0a.txt

```
##
## Proof Template A5219cH (Rule T):  H => (T = A)  -->  H => A
##
##
## Source: [Andrews 2002 (ISBN 1-4020-0763-9), p. 222]
##
## Copyright (c) 2017 Owl of Minerva Press GmbH. All rights reserved.
## Written by Ken Kubota (<mail@kenkubota.de>).
##
```

```
##  This file is part of the publication of the mathematical logic R0.
##  For more information, visit: <http://doi.org/10.4444/100.10>
##



##
##  Define Syntactical Variables
##

<< basics.r0.txt

## the assumption
:= $A5219cH ((=>{{{o,o},o}}_h{o}{o}){{o,o}}_((={{{o,o},o}}_T{o}){{o,o}}_a{o}{o}){o})


##
##  Assumptions and Resulting Syntactical Variables
##

§! $A5219cH


##
##  Include Proof Template
##

<<< A5219cH.r0t.txt


##
##  Undefine Syntactical Variables
##

:= $A5219cH


##
##  Q.E.D.
##

%0
```

2.2.44 File A5219cH.r0t.txt

```
##
##  Proof Template A5219cH (Rule T):  H => (T = A)  -->  H => A
##
##
##  Source: [Andrews 2002 (ISBN 1-4020-0763-9), p. 222]
```

```
##
## Copyright (c) 2017 Owl of Minerva Press GmbH. All rights reserved.
## Written by Ken Kubota (<mail@kenkubota.de>).
##
## This file is part of the publication of the mathematical logic R0.
## For more information, visit: <http://doi.org/10.4444/100.10>
##



##
## Exception: Forward Reference
##
## (See comment in Proof Template A5215H.)
##

## Empty lines are needed for comparison between A5219a-A5219d and A5219aH-A5219dH.


##
## Proof Template
##

## use Proof Template A5218:  (T = A) = A
:= $A5218 $A5219cH/7
<< A5218.r0t.txt
:= $A5218
%0





## use Proof Template K8004 (Trans):  (H OP A), B  -->  H => B
:= $HA8004 $A5219cH
:= $B8004 %0
<< K8004.r0t.txt
:= $HA8004; := $B8004
%0


%$A5219cH




§s' %0 1 %1
```

2.2.45 File A5219d.r0a.txt

```
##
##  Proof Template A5219d (Rule T):  A = T  -->  A
##
##
##  Source: [Andrews 2002 (ISBN 1-4020-0763-9), p. 222]
##
##  Copyright (c) 2017 Owl of Minerva Press GmbH. All rights reserved.
##  Written by Ken Kubota (<mail@kenkubota.de>).
##
##  This file is part of the publication of the mathematical logic R0.
##  For more information, visit: <http://doi.org/10.4444/100.10>
##



##
##  Define Syntactical Variables
##

<< basics.r0.txt

## the assumption
:= $A5219d ((={{{o,o},o}}_a{o}{o}){{o,o}}_T{o})


##
##  Assumptions and Resulting Syntactical Variables
##

§! $A5219d


##
##  Include Proof Template
##

<<< A5219d.r0t.txt


##
##  Undefine Syntactical Variables
##

:= $A5219d


##
##  Q.E.D.
```

```
##

%0
```

2.2.46 File A5219d.r0t.txt

```
##
##  Proof Template A5219d (Rule T):  A = T  -->  A
##
##
##  Source: [Andrews 2002 (ISBN 1-4020-0763-9), p. 222]
##
##  Copyright (c) 2017 Owl of Minerva Press GmbH. All rights reserved.
##  Written by Ken Kubota (<mail@kenkubota.de>).
##
##  This file is part of the publication of the mathematical logic R0.
##  For more information, visit: <http://doi.org/10.4444/100.10>
##







##  Empty lines are needed for comparison between A5219a-A5219d and A5219aH-A5219dH.


##
##  Proof Template
##

## use Proof Template A5218:  (T = A) = A
:= $A5218 $A5219d/5
<< A5218.r0t.txt
:= $A5218
%0
```

```
:= $TMP5219d %0
%$A5219d
## use Proof Template A5201b (Swap):  A = B  -->  B = A
<< A5201b.r0t.txt
%0
%$TMP5219d; := $TMP5219d

§s %1 1 %0
```

2.2.47 File A5219dH.r0a.txt

```
##
##  Proof Template A5219dH (Rule T):  H => (A = T)  -->  H => A
##
##
##  Source: [Andrews 2002 (ISBN 1-4020-0763-9), p. 222]
##
##  Copyright (c) 2017 Owl of Minerva Press GmbH. All rights reserved.
##  Written by Ken Kubota (<mail@kenkubota.de>).
##
##  This file is part of the publication of the mathematical logic R0.
##  For more information, visit: <http://doi.org/10.4444/100.10>
##



##
##  Define Syntactical Variables
##

<< basics.r0.txt

## the assumption
:= $A5219dH ((=>{{{o,o},o}}_h{o}{o}){{o,o}}_((={{{o,o},o}}_a{o}{o}){{o,o}}_T{o}){o})


##
##  Assumptions and Resulting Syntactical Variables
##

§! $A5219dH


##
##  Include Proof Template
##

<<< A5219dH.r0t.txt
```

```
##
##  Undefine Syntactical Variables
##

:= $A5219dH


##
##  Q.E.D.
##

%0
```

2.2.48 File A5219dH.r0t.txt

```
##
##  Proof Template A5219dH (Rule T):  H => (A = T)  -->  H => A
##
##
##  Source: [Andrews 2002 (ISBN 1-4020-0763-9), p. 222]
##
##  Copyright (c) 2017 Owl of Minerva Press GmbH. All rights reserved.
##  Written by Ken Kubota (<mail@kenkubota.de>).
##
##  This file is part of the publication of the mathematical logic R0.
##  For more information, visit: <http://doi.org/10.4444/100.10>
##



##
##  Exception: Forward Reference
##
##  (See comment in Proof Template A5215H.)
##

##  Empty lines are needed for comparison between A5219a-A5219d and A5219aH-A5219dH.



##
##  Proof Template
##

## use Proof Template A5218:  (T = A) = A
:= $A5218 $A5219dH/13
<< A5218.r0t.txt
:= $A5218
%0
```

```
## use Proof Template K8004 (Trans):  (H OP A), B  -->  H => B
:= $HA8004 $A5219dH
:= $B8004 %0
<< K8004.r0t.txt
:= $HA8004; := $B8004
%0

:= $TMP5219dH %0
%$A5219dH
## use Proof Template A5201bH (SwapH):  H => (A = B)  -->  H => (B = A)
<< A5201bH.r0t.txt
%0
%$TMP5219dH; := $TMP5219dH

§s' %1 1 %0
```

2.2.49 File A5220.r0a.txt

```
##
##  Proof Template A5220 (Gen):  A  -->  ALL x: A
##      for any x of any type (Rule of Universal Generalization)
##
##  Source: [Andrews 2002 (ISBN 1-4020-0763-9), p. 222]
##
##  Copyright (c) 2017 Owl of Minerva Press GmbH. All rights reserved.
##  Written by Ken Kubota (<mail@kenkubota.de>).
##
##  This file is part of the publication of the mathematical logic R0.
##  For more information, visit: <http://doi.org/10.4444/100.10>
##


##
##  Define Syntactical Variables
##

## type of variable
:= $T5220 t{^}

## the variable
:= $X5220 x{$T5220}

## the proposition
:= $A5220 a{o}
```

```
##
##  Assumptions and Resulting Syntactical Variables
##

§! $A5220


##
##  Include Proof Template
##

<<< A5220.r0t.txt


##
##  Undefine Syntactical Variables
##

:= $T5220; := $X5220; := $A5220


##
##  Q.E.D.
##

%0
```

2.2.50 File A5220.r0t.txt

```
##
##  Proof Template A5220 (Gen):  A  -->  ALL x: A
##      for any x of any type (Rule of Universal Generalization)
##
##  Source: [Andrews 2002 (ISBN 1-4020-0763-9), p. 222]
##
##  Copyright (c) 2017 Owl of Minerva Press GmbH. All rights reserved.
##  Written by Ken Kubota (<mail@kenkubota.de>).
##
##  This file is part of the publication of the mathematical logic R0.
##  For more information, visit: <http://doi.org/10.4444/100.10>
##


##
##  Proof Template
##

## .1
```

```
%$A5220

## .2

## use Proof Template A5219a (Rule T):  A  -->  T = A
:= $A5219a %0
<< A5219a.r0t.txt
:= $A5219a
%0

## .3

§= {o,$T5220} [\$X5220{$T5220}.T{o}]
§rs /5 x{$T5220}

## .4

§s %0 7 %3
§= ((A{{{o,{o,\3{^}}},^}}_$T5220{^}){{o,{o,$T5220}}}_[\$X5220{$T5220}.$A5220{o}]{{o,
$T5220}})
§\s /10
§\s /5
§s %5 1 %0
```

2.2.51 File A5220H.r0a.txt

```
##
##  Proof Template A5220H (Gen):  (H => A)  -->  (H => ALL x: A)
##      for any x of any type (Rule of Universal Generalization), provided x is not
free in H
##
##  Source: [Andrews 2002 (ISBN 1-4020-0763-9), p. 222]
##
##  Copyright (c) 2017 Owl of Minerva Press GmbH. All rights reserved.
##  Written by Ken Kubota (<mail@kenkubota.de>).
##
##  This file is part of the publication of the mathematical logic R0.
##  For more information, visit: <http://doi.org/10.4444/100.10>
##


##
##  Define Syntactical Variables
##

<< basics.r0.txt

## type of variable
```

```
:= $T5220H t{^}

## the variable
:= $X5220H x{$T5220H}

## the proposition
:= $A5220H ((=>{{{o,o},o}}_h{o}{o}){{o,o}}_a{o}{o})


##
##  Assumptions and Resulting Syntactical Variables
##

§! $A5220H


##
##  Include Proof Template
##

<<< A5220H.r0t.txt


##
##  Undefine Syntactical Variables
##

:= $T5220H; := $X5220H; := $A5220H


##
##  Q.E.D.
##

%0
```

2.2.52 File A5220H.r0t.txt

```
##
##  Proof Template A5220H (Gen):  (H => A)  -->  (H => ALL x: A)
##      for any x of any type (Rule of Universal Generalization), provided x is not
free in H
##
##  Source: [Andrews 2002 (ISBN 1-4020-0763-9), p. 222]
##
##  Copyright (c) 2017 Owl of Minerva Press GmbH. All rights reserved.
##  Written by Ken Kubota (<mail@kenkubota.de>).
##
##  This file is part of the publication of the mathematical logic R0.
```

```
##  For more information, visit: <http://doi.org/10.4444/100.10>
##



##
##  Exception: Forward Reference
##
##  (See comment in Proof Template A5215H.)
##



##
##  Proof Template
##

## .1

%$A5220H

## .2

## use Proof Template A5219aH (Rule T):  H => A  -->  H => (T = A)
:= $A5219aH %0
<< A5219aH.r0t.txt
:= $A5219aH
%0

:= $HTMP5220H %0

## .3

§= {o,$T5220H} [\$X5220H{$T5220H}.T{o}]
§rs /5 x{$T5220H}
:= $TTMP5220H %0

## use Proof Template K8004 (Trans):  (H OP A), B  -->  H => B
:= $HA8004 $HTMP5220H
:= $B8004 $TTMP5220H; := $TTMP5220H
<< K8004.r0t.txt
:= $HA8004; := $B8004
%0

## .4

%$HTMP5220H; := $HTMP5220H
§s' %1 7 %0
:= $HTMP5220H %0
§= ((A{{{o,{o,\3{^}}},^}}_$T5220H{^}){{o,{o,$T5220H}}}_[\$X5220H{$T5220H}.$A5220H/3{
o}]{{o,$T5220H}})
```

```
## use Proof Template K8004 (Trans):  (H OP A), B  -->  H => B
:= $HA8004 $HTMP5220H
:= $B8004 %0
<< K8004.r0t.txt
:= $HA8004; := $B8004
%0

§\s' /10
§\s' /5
%$HTMP5220H; := $HTMP5220H
§s' %0 1 %1
```

2.2.53 File A5221.r0a.txt

```
##
##  Proof Template A5221 (Sub):  B  -->  B [x/A]
##      (Rule of Substitution)
##
##  Source: [Andrews 2002 (ISBN 1-4020-0763-9), pp. 222 f.]
##
##  Copyright (c) 2017 Owl of Minerva Press GmbH. All rights reserved.
##  Written by Ken Kubota (<mail@kenkubota.de>).
##
##  This file is part of the publication of the mathematical logic R0.
##  For more information, visit: <http://doi.org/10.4444/100.10>
##



##
##  Define Syntactical Variables
##

<< basics.r0.txt

## assumption
:= $B5221 (g{{o,o}}{{o,o}}_x{o}{o})

## type of the variable and the substitution term
:= $T5221 o

## the variable to be replaced
:= $X5221 x{$T5221}

## substitution term
:= $A5221 F


##
```

```
##  Assumptions and Resulting Syntactical Variables
##

§! $B5221


##
##  Include Proof Template
##

<<< A5221.r0t.txt


##
##  Undefine Syntactical Variables
##

:= $B5221; := $T5221; := $X5221; := $A5221


##
##  Q.E.D.
##

%0
```

2.2.54 File A5221.r0t.txt

```
##
##  Proof Template A5221 (Sub):  B  -->  B [x/A]
##      (Rule of Substitution)
##
##  Source: [Andrews 2002 (ISBN 1-4020-0763-9), pp. 222 f.]
##
##  Copyright (c) 2017 Owl of Minerva Press GmbH. All rights reserved.
##  Written by Ken Kubota (<mail@kenkubota.de>).
##
##  This file is part of the publication of the mathematical logic R0.
##  For more information, visit: <http://doi.org/10.4444/100.10>
##


##
##  Proof Template
##

## .1

%$B5221
```

```
## .2

## use Proof Template A5220 (Gen):  A  -->  ALL x: A
:= $T5220 $T5221
:= $X5220 $X5221
:= $A5220 %0
<< A5220.r0t.txt
:= $T5220; := $X5220; := $A5220
%0

## .3

## use Proof Template A5215 (ALL I):  ALL x: B  -->  B [x/a]
:= $T5215 $T5221
:= $X5215 $X5221
:= $A5215 $A5221
:= $H5215 %0
<< A5215.r0t.txt
:= $T5215; := $X5215; := $A5215; := $H5215
%0
```

2.2.55 File A5221H.r0a.txt

```
##
##  Proof Template A5221H (Sub):  H => B  -->  H => B [x/A]
##      (Rule of Substitution)
##
##  Source: [Andrews 2002 (ISBN 1-4020-0763-9), pp. 222 f.]
##
##  Copyright (c) 2017 Owl of Minerva Press GmbH. All rights reserved.
##  Written by Ken Kubota (<mail@kenkubota.de>).
##
##  This file is part of the publication of the mathematical logic R0.
##  For more information, visit: <http://doi.org/10.4444/100.10>
##


##
##  Define Syntactical Variables
##

<< basics.r0.txt

## assumption
:= $B5221H ((=>{{{o,o},o}}_h{o}{o}){{o,o}}_(g{{o,o}}{{o,o}}_x{o}{o}){o})

## type of the variable and the substitution term
:= $T5221H o
```

```
## the variable to be replaced
:= $X5221H x{$T5221H}

## substitution term
:= $A5221H F


##
##  Assumptions and Resulting Syntactical Variables
##

§! $B5221H


##
##  Include Proof Template
##

<<< A5221H.r0t.txt


##
##  Undefine Syntactical Variables
##

:= $B5221H; := $T5221H; := $X5221H; := $A5221H


##
##  Q.E.D.
##

%0
```

2.2.56 File A5221H.r0t.txt

```
##
##  Proof Template A5221H (Sub):  H => B  -->  H => B [x/A]
##      (Rule of Substitution)
##
##  Source: [Andrews 2002 (ISBN 1-4020-0763-9), pp. 222 f.]
##
##  Copyright (c) 2017 Owl of Minerva Press GmbH. All rights reserved.
##  Written by Ken Kubota (<mail@kenkubota.de>).
##
##  This file is part of the publication of the mathematical logic R0.
##  For more information, visit: <http://doi.org/10.4444/100.10>
##
```

```
##
##  Proof Template
##

## .1

%$B5221H

## .2

## use Proof Template A5220H (Gen):  (H => A)  -->  (H => ALL x: A)
:= $T5220H $T5221H
:= $X5220H $X5221H
:= $A5220H %0
<< A5220H.r0t.txt
:= $T5220H; := $X5220H; := $A5220H
%0

## .3

## use Proof Template A5215H (ALL I):  H => ALL x: B  -->  H => B [x/a]
:= $T5215H $T5221H
:= $X5215H $X5221H
:= $A5215H $A5221H
:= $H5215H %0
<< A5215H.r0t.txt
:= $T5215H; := $X5215H; := $A5215H; := $H5215H
%0
```

2.2.57 File A5222.r0a.txt

```
##
##  Proof Template A5222 (Rule of Cases):  [\x.A]T, [\x.A]F  -->  A
##      for any x of type bool
##
##  Source: [Andrews 2002 (ISBN 1-4020-0763-9), p. 223]
##
##  Copyright (c) 2017 Owl of Minerva Press GmbH. All rights reserved.
##  Written by Ken Kubota (<mail@kenkubota.de>).
##
##  This file is part of the publication of the mathematical logic R0.
##  For more information, visit: <http://doi.org/10.4444/100.10>
##

<< basics.r0.txt
```

```
##
##  Define Syntactical Variables
##

## the lambda abstraction
:= $L5222 [\x{o}{o}.a{o}{o}]

## the variable to be used in place of the one abstracted
:= $X5222 x{o}

## assumption 1
:= $T5222 ($L5222{{o,o}}_T{o})

## assumption 2
:= $F5222 ($L5222{{o,o}}_F{o})


##
##  Assumptions and Resulting Syntactical Variables
##

§! $T5222
§! $F5222


##
##  Include Proof Template
##

<<< A5222.r0t.txt


##
##  Undefine Syntactical Variables
##

:= $L5222; := $X5222; := $T5222; := $F5222


##
##  Q.E.D.
##

%0
```

2.2.58 File A5222.r0t.txt

```
##
##  Proof Template A5222 (Rule of Cases):   [\x.A]T, [\x.A]F  -->  A
```

```
##      for any x of type bool
##
##  Source: [Andrews 2002 (ISBN 1-4020-0763-9), p. 223]
##
##  Copyright (c) 2017 Owl of Minerva Press GmbH. All rights reserved.
##  Written by Ken Kubota (<mail@kenkubota.de>).
##
##  This file is part of the publication of the mathematical logic R0.
##  For more information, visit: <http://doi.org/10.4444/100.10>
##

<< A5212.r0.txt


##
##  Proof Template
##

## .1

%$T5222

## use Proof Template A5219a (Rule T):  A  -->  T = A
:= $A5219a %0
<< A5219a.r0t.txt
:= $A5219a
%0

:= $ATMP5222 %0

## .2

%$F5222

## use Proof Template A5219a (Rule T):  A  -->  T = A
:= $A5219a %0
<< A5219a.r0t.txt
:= $A5219a
%0

## .3

%A5212

## .4

§s %0 3 %1
%$ATMP5222
§s %1 5 %0
```

```
:= $BTMP5222 %0

## .5

%A1

## use Proof Template A5221 (Sub):  B  -->  B [x/A]
:= $B5221 %0
:= $T5221 {o,o}
:= $X5221 g{$T5221}
:= $A5221 $L5222
<< A5221.r0t.txt
:= $B5221; := $T5221; := $X5221; := $A5221
%0

§\s /15

## .6

%$BTMP5222
§s %0 1 %1

## .7

## use Proof Template A5215 (ALL I):  ALL x: B  -->  B [x/a]
:= $T5215 o
:= $X5215 x{$T5215}
:= $A5215 $X5222
:= $H5215 %0
<< A5215.r0t.txt
:= $T5215; := $X5215; := $A5215; := $H5215
%0

## undefine local variables
:= $ATMP5222; := $BTMP5222
```

2.2.59 File A5223.r0.txt

```
##
##  Proof A5223:  (T => y) = y
##      with y of type o
##
##  Source: [Andrews 2002 (ISBN 1-4020-0763-9), pp. 223 f.]
##
##  Copyright (c) 2017 Owl of Minerva Press GmbH. All rights reserved.
##  Written by Ken Kubota (<mail@kenkubota.de>).
##
##  This file is part of the publication of the mathematical logic R0.
```

```
##  For more information, visit: <http://doi.org/10.4444/100.10>
##

<< basics.r0.txt


##
## Proof
##

## .1

§= {o} ((=>{{{o,o},o}}_T{o}){{o,o}}_y{o}{o})
§\s /6
§\s /3
:= $ATMP5223 %0

## .2

## use Proof Template A5218:  (T = A) = A
:= $A5218 /7
<< A5218.r0t.txt
:= $A5218
%0

%$ATMP5223
§s %0 3 %1
:= $BTMP5223 %0

## .3

## use Proof Template A5216:  (T & A) = A
:= $A5216 y{o}
<< A5216.r0t.txt
:= $A5216
%0

%$BTMP5223
§s %0 3 %1

:= A5223 %0

## undefine local variables
:= $ATMP5223; := $BTMP5223


##
##  Q.E.D.
##
```

%0

2.2.60 File A5224.r0a.txt

```
##
##  Proof A5224 (MP):  A, (A => B)  -->  B
##      (Modus Ponens)
##
##  Source: [Andrews 2002 (ISBN 1-4020-0763-9), p. 224]
##
##  Copyright (c) 2017 Owl of Minerva Press GmbH. All rights reserved.
##  Written by Ken Kubota (<mail@kenkubota.de>).
##
##  This file is part of the publication of the mathematical logic R0.
##  For more information, visit: <http://doi.org/10.4444/100.10>
##



##
##  Define Syntactical Variables
##

<< basics.r0.txt

## the proposition A
:= $A5224 a{o}

## the proposition A => B
:= $AB5224 ((=>{{{o,o},o}}_a{o}{o}){{o,o}}_b{o}{o})


##
##  Assumptions and Resulting Syntactical Variables
##

§! $A5224
§! $AB5224


##
##  Include Proof Template
##

<<< A5224.r0t.txt


##
##  Undefine Syntactical Variables
```

```
##

:= $AB5224; := $A5224


##
##  Q.E.D.
##

%0
```

2.2.61 File A5224.r0t.txt

```
##
##  Proof A5224 (MP):  A, (A => B)  -->  B
##      (Modus Ponens)
##
##  Source: [Andrews 2002 (ISBN 1-4020-0763-9), p. 224]
##
##  Copyright (c) 2017 Owl of Minerva Press GmbH. All rights reserved.
##  Written by Ken Kubota (<mail@kenkubota.de>).
##
##  This file is part of the publication of the mathematical logic R0.
##  For more information, visit: <http://doi.org/10.4444/100.10>
##

<< A5223.r0.txt


##
##  Proof Template
##

## .1

%$AB5224

## .2

## use Proof Template A5219b (Rule T):  A  -->  A = T
:= $A5219b %$A5224
<< A5219b.r0t.txt
:= $A5219b
%0

## .3

%$AB5224
§s %0 5 %1
```

```
:= $TMP5224 %0

## .4

## use Proof Template A5221 (Sub):  B  -->  B [x/A]
:= $B5221 %A5223
:= $T5221 o
:= $X5221 y{o}
:= $A5221 %0/3
<< A5221.r0t.txt
:= $B5221; := $T5221; := $X5221; := $A5221
%0

%$TMP5224; := $TMP5224
§s %0 1 %1
```

2.2.62 File A5224H.r0a.txt

```
##
##  Proof A5224H (MP):  H => A, H => (A => B)  -->  H => B
##      (Modus Ponens)
##
##  Source: [Andrews 2002 (ISBN 1-4020-0763-9), p. 224]
##
##  Copyright (c) 2017 Owl of Minerva Press GmbH. All rights reserved.
##  Written by Ken Kubota (<mail@kenkubota.de>).
##
##  This file is part of the publication of the mathematical logic R0.
##  For more information, visit: <http://doi.org/10.4444/100.10>
##



##
##  Define Syntactical Variables
##

<< basics.r0.txt

## the proposition H => A
:= $A5224H ((=>{{{o,o},o}}_h{o}{o}){{o,o}}_a{o}{o})

## the proposition H => (A => B)
:= $AB5224H ((=>{{{o,o},o}}_h{o}{o}){{o,o}}_((=>{{{o,o},o}}_a{o}{o}){{o,o}}_b{o}{o})
{o})


##
##  Assumptions and Resulting Syntactical Variables
```

```
##

§! $A5224H
§! $AB5224H


##
##  Include Proof Template
##

<<< A5224H.r0t.txt


##
##  Undefine Syntactical Variables
##

:= $AB5224H; := $A5224H


##
##  Q.E.D.
##

%0
```

2.2.63 File A5224H.r0t.txt

```
##
##  Proof A5224H (MP):  H => A, H => (A => B)  -->  H => B
##      (Modus Ponens)
##
##  Source: [Andrews 2002 (ISBN 1-4020-0763-9), p. 224]
##
##  Copyright (c) 2017 Owl of Minerva Press GmbH. All rights reserved.
##  Written by Ken Kubota (<mail@kenkubota.de>).
##
##  This file is part of the publication of the mathematical logic R0.
##  For more information, visit: <http://doi.org/10.4444/100.10>
##

<< A5223.r0.txt


##
##  Proof Template
##

## .1
```

```
%$AB5224H

## .2

## use Proof Template A5219bH (Rule T):  H => A  -->  H => (A = T)
:= $A5219bH %$A5224H
<< A5219bH.r0t.txt
:= $A5219bH
%0

## .3

%$AB5224H
§s' %0 5 %1

:= $TMP5224H %0

## .4

## use Proof Template A5221 (Sub):  B  -->  B [x/A]
:= $B5221 %A5223
:= $T5221 o
:= $X5221 y{o}
:= $A5221 %0/7
<< A5221.r0t.txt
:= $B5221; := $T5221; := $X5221; := $A5221
%0

## use Proof Template K8004 (Trans):  (H OP A), B  -->  H => B
:= $HA8004 $TMP5224H
:= $B8004 %0
<< K8004.r0t.txt
:= $HA8004; := $B8004
%0

%$TMP5224H; := $TMP5224H
§s' %0 1 %1
```

2.2.64 File A5225.r0.txt

```
##
##  Proof A5225:  ALL x: f  =>  f x
##      for any x of any type a and any f of any type oa
##
##  Source: [Andrews 2002 (ISBN 1-4020-0763-9), p. 224]
##
##  Copyright (c) 2017 Owl of Minerva Press GmbH. All rights reserved.
##  Written by Ken Kubota (<mail@kenkubota.de>).
```

```
##
##  This file is part of the publication of the mathematical logic R0.
##  For more information, visit: <http://doi.org/10.4444/100.10>
##

<< axioms.r0.txt



##
##  Proof
##

## .1

## use Proof Template:  Axiom 2 Substitutions
:= $AA2 {o,a{^}}
:= $HA2 [\f{{o,a{^}}}{{o,a{^}}}.(f{{o,a{^}}}{{o,a{^}}}_x{a{^}}{a{^}}){o}]
:= $XA2 [\x{a{^}}{a{^}}.T{o}]
:= $YA2 f{{o,a{^}}}
<< axiom2_substitutions.r0t.txt
:= $AA2; := $HA2; := $XA2; := $YA2
%0

§= ((A{{{o,{o,\3{^}}},^}}_a{^}{^}){{o,{o,a{^}}}}_f{{o,a{^}}}{{o,a{^}}})
§\s /6
§\s /3
§= /5
§s %0 5 %1
§s %7 5 %0

## .2

§\s /13
§\s /13
§\s /7
:= $TMP5225 %0

## .3

## use Proof Template A5218:  (T = A) = A
:= $A5218 /7
<< A5218.r0t.txt
:= $A5218
%0

%$TMP5225
§s %0 3 %1

:= A5225 %0
```

```
## undefine local variables
:= $TMP5225


##
##  Q.E.D.
##

%0
```

2.2.65 File A5226.r0a.txt

```
##
##  Proof Template A5226:  ALL x: B  =>  B [x/a]
##      for any x of any type a and any A, B of type oa
##
##  Source: [Andrews 2002 (ISBN 1-4020-0763-9), p. 224]
##
##  Copyright (c) 2017 Owl of Minerva Press GmbH. All rights reserved.
##  Written by Ken Kubota (<mail@kenkubota.de>).
##
##  This file is part of the publication of the mathematical logic R0.
##  For more information, visit: <http://doi.org/10.4444/100.10>
##


##
##  Define Syntactical Variables
##

## type of the variable
:= $T5226 t{^}

## the variable to be replaced
:= $X5226 x{$T5226}

## substitution term
:= $A5226 a{$T5226}

## the proposition (in this example, B is defined as x=x)
:= $B5226 ((={{{o,@},@}}_$X5226{@}){{o,@}}_$X5226{@})


##
##  Assumptions and Resulting Syntactical Variables
##

§! $B5226
```

```
##
##  Include Proof Template
##

<<< A5226.r0t.txt


##
##  Undefine Syntactical Variables
##

:= $T5226; := $X5226; := $A5226; := $B5226


##
##  Q.E.D.
##

%0
```

2.2.66 File A5226.r0t.txt

```
##
##  Proof Template A5226:  ALL x: B  =>  B [x/a]
##      for any x of any type a and any A, B of type oa
##
##  Source: [Andrews 2002 (ISBN 1-4020-0763-9), p. 224]
##
##  Copyright (c) 2017 Owl of Minerva Press GmbH. All rights reserved.
##  Written by Ken Kubota (<mail@kenkubota.de>).
##
##  This file is part of the publication of the mathematical logic R0.
##  For more information, visit: <http://doi.org/10.4444/100.10>
##

<< A5225.r0.txt


##
##  Proof Template
##

%A5225

## .1a Replace type a in A5225

## use Proof Template A5221 (Sub):  B  -->  B [x/A]
```

```
:= $B5221 %0
:= $T5221 ^
:= $X5221 a{^}
:= $A5221 $T5226
<< A5221.r0t.txt
:= $B5221; := $T5221; := $X5221; := $A5221
%0

## .1b Replace variable x in A5225

## use Proof Template A5221 (Sub):  B  -->  B [x/A]
:= $B5221 %0
:= $T5221 $T5226
:= $X5221 x{$T5226}
:= $A5221 $A5226
<< A5221.r0t.txt
:= $B5221; := $T5221; := $X5221; := $A5221
%0

## .1c Replace variable f in A5225

## use Proof Template A5221 (Sub):  B  -->  B [x/A]
:= $B5221 %0
:= $T5221 {{o,$T5226}}
:= $X5221 f{{o,$T5226}}
:= $A5221 [\$X5226{$T5226}.$B5226{o}]
<< A5221.r0t.txt
:= $B5221; := $T5221; := $X5221; := $A5221
%0

## .2

§\s /3
```

2.2.67 File A5227.r0.txt

```
##
##  Proof A5227:  F => x
##      with x of type o
##
##  Source: [Andrews 2002 (ISBN 1-4020-0763-9), p. 224]
##
##  Copyright (c) 2017 Owl of Minerva Press GmbH. All rights reserved.
##  Written by Ken Kubota (<mail@kenkubota.de>).
##
##  This file is part of the publication of the mathematical logic R0.
##  For more information, visit: <http://doi.org/10.4444/100.10>
##
```

```
##
## Proof
##

## use Proof Template A5226:  ALL x: B  =>  B [x/a]
:= $T5226 o
:= $X5226 x{$T5226}
:= $A5226 x{$T5226}
:= $B5226 x{o}
<< A5226.r0t.txt
:= $T5226; := $X5226; := $A5226; := $B5226
%0

§\s /10
§\s /5

:= A5227 %0


##
##  Q.E.D.
##

%0
```

2.2.68 File A5228.r0.txt

```
##
##  Proof A5228:  (T => T) = T;  (T => F) = F;  (F => T) = T;  (F => F) = T
##
##
##  Source: [Andrews 2002 (ISBN 1-4020-0763-9), p. 224]
##
##  Copyright (c) 2017 Owl of Minerva Press GmbH. All rights reserved.
##  Written by Ken Kubota (<mail@kenkubota.de>).
##
##  This file is part of the publication of the mathematical logic R0.
##  For more information, visit: <http://doi.org/10.4444/100.10>
##

<< A5223.r0.txt
<< A5227.r0.txt


##
## Proof
##
```

```
## .a:  (T => T) = T

## use Proof Template A5221 (Sub):  B  -->  B [x/A]
:= $B5221 %A5223
:= $T5221 o
:= $X5221 y{$T5221}
:= $A5221 T
<< A5221.r0t.txt
:= $B5221; := $T5221; := $X5221; := $A5221
%0

:= A5228a %0

## .b:  (T => F) = F

## use Proof Template A5221 (Sub):  B  -->  B [x/A]
:= $B5221 %A5223
:= $T5221 o
:= $X5221 y{o}
:= $A5221 F
<< A5221.r0t.txt
:= $B5221; := $T5221; := $X5221; := $A5221
%0

:= A5228b %0

## .c:  (F => T) = T

## use Proof Template A5221 (Sub):  B  -->  B [x/A]
:= $B5221 %A5227
:= $T5221 o
:= $X5221 x{o}
:= $A5221 T
<< A5221.r0t.txt
:= $B5221; := $T5221; := $X5221; := $A5221
%0

## use Proof Template A5219b (Rule T):  A  -->  A = T
:= $A5219b %0
<< A5219b.r0t.txt
:= $A5219b
%0

:= A5228c %0

## .d:  (F => F) = T

## use Proof Template A5221 (Sub):  B  -->  B [x/A]
:= $B5221 %A5227
```

```
:= $T5221 o
:= $X5221 x{o}
:= $A5221 F
<< A5221.r0t.txt
:= $B5221; := $T5221; := $X5221; := $A5221
%0

## use Proof Template A5219b (Rule T):  A  -->  A = T
:= $A5219b %0
<< A5219b.r0t.txt
:= $A5219b
%0

:= A5228d %0


##
##  Q.E.D.
##

## %A5228a
%A5228a

## %A5228b
%A5228b

## %A5228c
%A5228c

## %A5228d
%A5228d
```

2.2.69 File A5229.r0.txt

```
##
##  Proof A5229:  (T & T) = T;  (T & F) = F;  (F & T) = F;  (F & F) = F
##
##
##  Source: [Andrews 2002 (ISBN 1-4020-0763-9), p. 225]
##
##  Copyright (c) 2017 Owl of Minerva Press GmbH. All rights reserved.
##  Written by Ken Kubota (<mail@kenkubota.de>).
##
##  This file is part of the publication of the mathematical logic R0.
##  For more information, visit: <http://doi.org/10.4444/100.10>
##

<< A5227.r0.txt
```

```
##
##  Proof
##

## .a:  (T & T) = T

## use Proof Template A5216:  (T & A) = A
:= $A5216 T
<< A5216.r0t.txt
:= $A5216
:= A5229a %0

## .b:  (T & F) = F

## use Proof Template A5216:  (T & A) = A
:= $A5216 F
<< A5216.r0t.txt
:= $A5216

:= A5229b %0

## .c:  (F & T) = F

%A5227

## use Proof Template A5221 (Sub):  B  -->  B [x/A]
:= $B5221 %0
:= $T5221 o
:= $X5221 x{o}
:= $A5221 T
<< A5221.r0t.txt
:= $B5221; := $T5221; := $X5221; := $A5221
%0

§\s /2
§\s /1
§= {o} F
§s %0 5 %1

:= A5229c %0

## .d:  (F & F) = F

%A5227

## use Proof Template A5221 (Sub):  B  -->  B [x/A]
:= $B5221 %0
:= $T5221 o
```

```
:= $X5221 x{o}
:= $A5221 F
<< A5221.r0t.txt
:= $B5221; := $T5221; := $X5221; := $A5221
%0

§\s /2
§\s /1
§= {o} F
§s %0 5 %1

:= A5229d %0


##
##  Q.E.D.
##

## %A5229a
%A5229a

## %A5229b
%A5229b

## %A5229c
%A5229c

## %A5229d
%A5229d
```

2.2.70 File A5230.r0.txt

```
##
##  Proof A5230:  (T = T) = T;  (T = F) = F;  (F = T) = F;  (F = F) = T
##
##
##  Source: [Andrews 2002 (ISBN 1-4020-0763-9), p. 225]
##
##  Copyright (c) 2017 Owl of Minerva Press GmbH. All rights reserved.
##  Written by Ken Kubota (<mail@kenkubota.de>).
##
##  This file is part of the publication of the mathematical logic R0.
##  For more information, visit: <http://doi.org/10.4444/100.10>
##

<< basics.r0.txt
<< A5229.r0.txt
```

```
##
## Proof
##

## .a:  (T = T) = T

## use Proof Template A5218:  (T = A) = A
:= $A5218 T
<< A5218.r0t.txt
:= $A5218
%0

:= A5230a %0

## .b:  (T = F) = F

## use Proof Template A5218:  (T = A) = A
:= $A5218 F
<< A5218.r0t.txt
:= $A5218
%0

:= A5230b %0

## .c:  (F = T) = F

## .1

## use Proof Template:  Axiom 2 Substitutions
:= $AA2 o
:= $HA2 [\x{o}{o}.((={{{o,o},o}}_x{o}{o}){{o,o}}_F{o}){o}]
:= $XA2 F
:= $YA2 T
<< axiom2_substitutions.r0t.txt
:= $AA2; := $HA2; := $XA2; := $YA2
%0

## .2

§\s /13
§\s /7

:= $ATMP5230 %0

## .3a

## use Proof Template A5210:  T = (B = B)
:= $T5210 o
:= $B5210 F
```

```
<< A5210.r0t.txt
:= $T5210; := $B5210
%0

## use Proof Template A5201b (Swap):  A = B  -->  B = A
<< A5201b.r0t.txt
%0

:= $BTMP5230 %0

## .3b

## use Proof Template A5218:  (T = A) = A
:= $A5218 F
<< A5218.r0t.txt
:= $A5218
%0

:= $CTMP5230 %0

## .3c

%$ATMP5230
%$BTMP5230
§s %1 13 %0
%$CTMP5230
§s %1 7 %0
§s %0 3 %1

## .4

§\s /2
§\s /1

:= $DTMP5230 %0

## .5

## use Proof Template A5222 (Rule of Cases):  [\x.A]T, [\x.A]F  -->  A
:= $L5222 [\x{o}{o}.((={{{o,o},o}}_((&{{{o,o},o}}_x{o}{o}){{o,o}}_F{o}){o}){{o,o}}_F
{o}){o}]
:= $X5222 x{o}
:= $T5222 ($L5222{{o,o}}_T{o})
:= $F5222 ($L5222{{o,o}}_F{o})

## Case T
§\ $T5222
## use Proof Template A5201b (Swap):  A = B  -->  B = A
<< A5201b.r0t.txt
```

```
%0
%A5229b
§s %0 1 %1

## Case F
§\ $F5222
## use Proof Template A5201b (Swap):  A = B  -->  B = A
<< A5201b.r0t.txt
%0
%A5229d
§s %0 1 %1

<< A5222.r0t.txt
:= $L5222; := $X5222; := $T5222; := $F5222
%0

## .6

## use Proof Template A5221 (Sub):  B  -->  B [x/A]
:= $B5221 %0
:= $T5221 {o}
:= $X5221 x{$T5221}
:= $A5221 ((={{{o,o},o}}_F{o}){{o,o}}_T{o})
<< A5221.r0t.txt
:= $B5221; := $T5221; := $X5221; := $A5221
%0

## .7

%$DTMP5230
## use Proof Template A5201b (Swap):  A = B  -->  B = A
<< A5201b.r0t.txt
%0
§s %4 5 %0

:= A5230c %0

## .d:  (F = F) = T

## use Proof Template A5210:  T = (B = B)
:= $T5210 o
:= $B5210 F
<< A5210.r0t.txt
:= $T5210; := $B5210
%0

## use Proof Template A5201b (Swap):  A = B  -->  B = A
<< A5201b.r0t.txt
%0
```

```
:= A5230d %0

## undefine local variables
:= $ATMP5230; := $BTMP5230; := $CTMP5230; := $DTMP5230


##
##  Q.E.D.
##

## %A5230a
%A5230a

## %A5230b
%A5230b

## %A5230c
%A5230c

## %A5230d
%A5230d
```

2.2.71 File A5231.r0.txt

```
##
##  Proof A5231:  ~ T = F;  ~ F = T
##
##
##  Source: [Andrews 2002 (ISBN 1-4020-0763-9), p. 225]
##
##  Copyright (c) 2017 Owl of Minerva Press GmbH. All rights reserved.
##  Written by Ken Kubota (<mail@kenkubota.de>).
##
##  This file is part of the publication of the mathematical logic R0.
##  For more information, visit: <http://doi.org/10.4444/100.10>
##

<< A5230.r0.txt


##
## Proof
##


## .a:  ~ T = F

§\ {o} (!{{o,o}}_T{o})
```

```
%A5230c
§s %1 3 %0

:= A5231a %0


## .b:  ~ F = T

§\ {o} (!{{o,o}}_F{o})
%A5230d
§s %1 3 %0

:= A5231b %0


##
##  Q.E.D.
##

## %A5231a
%A5231a

## %A5231b
%A5231b
```

2.2.72 File A5232.r0.txt

```
##
##  Proof A5232:  T | T = T;  T | F = T;  F | T = T;  F | F = F
##
##
##  Source: [Andrews 2002 (ISBN 1-4020-0763-9), p. 225]
##
##  Copyright (c) 2017 Owl of Minerva Press GmbH. All rights reserved.
##  Written by Ken Kubota (<mail@kenkubota.de>).
##
##  This file is part of the publication of the mathematical logic R0.
##  For more information, visit: <http://doi.org/10.4444/100.10>
##

<< A5231.r0.txt


##
## Proof
##

## .a:  T | T = T
```

```
§= {o} ((|{{{o,o},o}}_T{o}){{o,o}}_T{o})

§\s /6
§\s /3

%A5231a
§s %1 29 %0

%A5231a
§s %1 15 %0

%A5229d
§s %1 7 %0

%A5231b
§s %1 3 %0

:= A5232a %0

## .b:  T | F = T

§= {o} ((|{{{o,o},o}}_T{o}){{o,o}}_F{o})

§\s /6
§\s /3

%A5231a
§s %1 29 %0

%A5231b
§s %1 15 %0

%A5229c
§s %1 7 %0

%A5231b
§s %1 3 %0

:= A5232b %0

## .c:  F | T = T

§= {o} ((|{{{o,o},o}}_F{o}){{o,o}}_T{o})

§\s /6
§\s /3

%A5231b
§s %1 29 %0
```

```
%A5231a
§s %1 15 %0

%A5229b
§s %1 7 %0

%A5231b
§s %1 3 %0

:= A5232c %0

## .d:  F | F = F

§= {o} ((|{{{o,o},o}}_F{o}){{o,o}}_F{o})

§\s /6
§\s /3

%A5231b
§s %1 29 %0

%A5231b
§s %1 15 %0

%A5229a
§s %1 7 %0

%A5231a
§s %1 3 %0

:= A5232d %0


##
##  Q.E.D.
##

## %A5232a
%A5232a

## %A5232b
%A5232b

## %A5232c
%A5232c

## %A5232d
%A5232d
```

2.2.73 File A5245.r0a.txt

```
##
##  Proof Template A5245 (Rule C):  H => EXI x: B, (H & (B [x/y])) => A  -->  H => A
##      for any x, y of any type, provided y is not free in H, EXI x: B or A
##
##  Source: [Andrews 2002 (ISBN 1-4020-0763-9), p. 230 (5245)]
##
##  Copyright (c) 2017 Owl of Minerva Press GmbH. All rights reserved.
##  Written by Ken Kubota (<mail@kenkubota.de>).
##
##  This file is part of the publication of the mathematical logic R0.
##  For more information, visit: <http://doi.org/10.4444/100.10>
##

<< basics.r0.txt



##
##  Define Syntactical Variables
##

## type of variable
:= $T5245 t{^}

## name of variable in assumption 1
:= $X5245 x{$T5245}

## name of variable in assumption 2
:= $Y5245 y{$T5245}

## assumption 1:  H => EXI x: B
:= $B5245 ((=>{{{o,o},o}}_h{o}{o}){{o,o}}_((E{{{o,{o,\3{^}}},^}}_$T5245{^}){{o,{o,$T
5245}}}_[\$X5245{$T5245}.(b{{o,$T5245}}{{o,$T5245}}_$X5245{$T5245}){o}]{{o,$T5245}})
{o})

## assumption 2:  (H & (B [x/y])) => A
:= $A5245 ((=>{{{o,o},o}}_((&{{{o,o},o}}_h{o}{o}){{o,o}}_(b{{o,$T5245}}{{o,$T5245}}_
$Y5245{$T5245}){o}){o}){{o,o}}_a{o}{o})



##
##  Assumptions and Resulting Syntactical Variables
##

§! $B5245
§! $A5245
```

```
##
##  Include Proof Template
##

<<< A5245.r0t.txt


##
##  Undefine Syntactical Variables
##

:= $T5245; := $X5245; := $Y5245; := $B5245; := $A5245


##
##  Q.E.D.
##

%0
```

2.2.74 File A5245.r0t.txt

```
##
##  Proof Template A5245 (Rule C):  H => EXI x: B, (H & (B [x/y])) => A  -->  H => A
##      for any x, y of any type, provided y is not free in H, EXI x: B or A
##
##  Source: [Andrews 2002 (ISBN 1-4020-0763-9), p. 230 (5245)]
##
##  Copyright (c) 2017 Owl of Minerva Press GmbH. All rights reserved.
##  Written by Ken Kubota (<mail@kenkubota.de>).
##
##  This file is part of the publication of the mathematical logic R0.
##  For more information, visit: <http://doi.org/10.4444/100.10>
##


##
##  Proof Template
##

## .1

%$A5245

## .2

## use Proof Template K8030 (EXI Rule):  (H & B) => A  -->  (H & EXI x: B) => A
:= $T8030 $T5245
:= $X8030 $Y5245
```

```
:= $A8030 %0
<< K8030.r0t.txt
:= $T8030; := $X8030; := $A8030;
%0

## .3

## use Proof Template K8025 (Deduction Theorem):  (H & I) => A  -->  H => (I => A)
<< K8025.r0t.txt
%0

## .4

§rs /27 $X5245

## .5

%$B5245

## .6

## use Proof Template A5224H (MP):  H => A, H => (A => B)  -->  H => B
:= $A5224H %0
:= $AB5224H %1
<< A5224H.r0t.txt
:= $AB5224H; := $A5224H
%0
```

2.2.75 File A5304.r0.txt

```
##
##  Proof A5304:  EXI1 y: P y  =  EXI y: P = ( = y )
##
##
##  Source: [Andrews 2002 (ISBN 1-4020-0763-9), p. 233]
##
##  Copyright (c) 2017 Owl of Minerva Press GmbH. All rights reserved.
##  Written by Ken Kubota (<mail@kenkubota.de>).
##
##  This file is part of the publication of the mathematical logic R0.
##  For more information, visit: <http://doi.org/10.4444/100.10>
##

<< basics.r0.txt
<< A5205.r0.txt


##
## Proof
```

```
##

## .1

§= {o} ((E1{{{o,{o,\3{^}}},^}}_t{^}{^}){{o,{o,t{^}}}}_[\y{t{^}}{t{^}}.(p{{o,t{^}}}{{
o,t{^}}}_y{t{^}}{t{^}}){o}]{{o,t{^}}})
§\s /6
§\s /3
:= $TMP5304 %0

## .2

## use Proof Template:  A5205 Substitutions
:= $AA5205 o
:= $BA5205 t{^}
:= $FA5205 p{{$AA5205,$BA5205}}
<< a5205_substitutions.r0t.txt
:= $AA5205; := $BA5205; := $FA5205
%0

## use Proof Template A5201b (Swap):  A = B  -->  B = A
<< A5201b.r0t.txt
%0

%$TMP5304; := $TMP5304
§s %0 61 %1

:= A5304 %0


##
##  Q.E.D.
##

%0
```

2.2.76 File A5305.r0.txt

```
##
##  Proof A5305:  EXI1 y: P y  =  EXI y: ALL z: P z = (y = z)
##
##
##  Source: [Andrews 2002 (ISBN 1-4020-0763-9), p. 233]
##
##  Copyright (c) 2017 Owl of Minerva Press GmbH. All rights reserved.
##  Written by Ken Kubota (<mail@kenkubota.de>).
##
##  This file is part of the publication of the mathematical logic R0.
##  For more information, visit: <http://doi.org/10.4444/100.10>
```

```
##

<< basics.r0.txt
<< A5304.r0.txt


##
## Proof
##

## .1

## use Proof Template:  Axiom 3 Substitutions
:= $AA3 o
:= $BA3 t{^}
:= $FA3 p{{o,t{^}}}
:= $GA3 (={{{o,t{^}},t{^}}}_y{t{^}}{t{^}})
<< axiom3_substitutions.r0t.txt
:= $AA3; := $BA3; := $FA3; := $GA3
%0

## .2

%A5304
§s %0 15 %1
§rs /31 z{t{^}}

:= A5305 %0


##
##  Q.E.D.
##

%0
```

2.2.77 File A5310.r0.txt

```
##
##  Proof A5310:  ( ALL z: P z = (y = z) ) => ( IOTA P = y )
##
##
##  Source: [Andrews 2002 (ISBN 1-4020-0763-9), p. 235]
##
##  Copyright (c) 2017 Owl of Minerva Press GmbH. All rights reserved.
##  Written by Ken Kubota (<mail@kenkubota.de>).
##
##  This file is part of the publication of the mathematical logic R0.
##  For more information, visit: <http://doi.org/10.4444/100.10>
```

```
##

<< basics.r0.txt
<< K8005.r0.txt


##
## Proof
##

## .1

%K8005

## use Proof Template A5221 (Sub):  B  -->  B [x/A]
:= $B5221 %0
:= $T5221 o
:= $X5221 x{$T5221}
:= $A5221 ((A{{{o,{o,\3{^}}},^}}_t{^}{^}){{o,{o,t{^}}}}_[\z{t{^}}{t{^}}.((={{{o,o},o
}}_(p{{o,t{^}}}{{o,t{^}}}_z{t{^}}{t{^}}){o}){{o,o}}_((={{{o,t{^}},t{^}}}_y{t{^}}{t{^
}}){{o,t{^}}}_z{t{^}}{t{^}}){o}){o}]{{o,t{^}}})
<< A5221.r0t.txt
:= $B5221; := $T5221; := $X5221; := $A5221

:= $TMP5310 %0

## .2

## use Proof Template:  Axiom 3 Substitutions
:= $AA3 o
:= $BA3 t{^}
:= $FA3 p{{o,t{^}}}
:= $GA3 (={{{o,t{^}},t{^}}}_y{t{^}}{t{^}})
<< axiom3_substitutions.r0t.txt
:= $AA3; := $BA3; := $FA3; := $GA3
%0

§rs /7 z{t{^}}

## use Proof Template A5201b (Swap):  A = B  -->  B = A
<< A5201b.r0t.txt
%0

%$TMP5310; := $TMP5310
§s %0 3 %1
:= $TMP5310 %0

## .3
```

```
§= {t{^}} (i{{t{^},{o,t{^}}}}_p{{o,t{^}}}{{o,t{^}}})

## use Proof Template K8004 (Trans):  (H OP A), B  -->  H => B
:= $HA8004 %1
:= $B8004 %0
<< K8004.r0t.txt
:= $HA8004; := $B8004
%0

%$TMP5310; := $TMP5310
§s' %1 7 %0
%A5
§s %1 7 %0


##
##  Q.E.D.
##

%0
```

2.2.78 File A5311.r0.txt

```
##
##  Proof A5311:  ( EXI1 y: P y ) => ( P (IOTA P) )
##
##
##  Source: [Andrews 2002 (ISBN 1-4020-0763-9), p. 235]
##
##  Copyright (c) 2017 Owl of Minerva Press GmbH. All rights reserved.
##  Written by Ken Kubota (<mail@kenkubota.de>).
##
##  This file is part of the publication of the mathematical logic R0.
##  For more information, visit: <http://doi.org/10.4444/100.10>
##


<< basics.r0.txt
<< A5304.r0.txt
<< K8000.r0.txt
<< K8005.r0.txt


##
## Proof
##

## .1
```

```
%K8005

## use Proof Template A5221 (Sub):  B  -->  B [x/A]
:= $B5221 %0
:= $T5221 o
:= $X5221 x{$T5221}
:= $A5221 ((={{{o,{o,t{^}}},{o,t{^}}}}_p{{o,t{^}}}{{o,t{^}}})({{o,{o,t{^}}}}_(={{{o,t
{^}},t{^}}}_y{t{^}}{t{^}}){{o,t{^}}})
<< A5221.r0t.txt
:= $B5221; := $T5221; := $X5221; := $A5221

:= $TMP5311 %0
:= $LTMP5311 %0

## .2

§= {o} (p{{o,t{^}}}{{o,t{^}}}_y{t{^}}{t{^}})

## use Proof Template K8004 (Trans):  (H OP A), B  -->  H => B
:= $HA8004 %1
:= $B8004 %0
<< K8004.r0t.txt
:= $HA8004; := $B8004
%0

%$TMP5311; := $TMP5311
§s' %1 6 %0

## .3

## use Proof Template A5201bH (SwapH):  H => (A = B)  -->  H => (B = A)
<< A5201bH.r0t.txt
%0

:= $TMP5311 %0
§= {t{^}} y{t{^}}

## use Proof Template K8004 (Trans):  (H OP A), B  -->  H => B
:= $HA8004 %1
:= $B8004 %0
<< K8004.r0t.txt
:= $HA8004; := $B8004
%0

%$TMP5311; := $TMP5311
§s' %1 1 %0
:= $TMP5311 %0

## .4
```

```
%A5

## use Proof Template A5201b (Swap):  A = B  -->  B = A
<< A5201b.r0t.txt
%0

%$TMP5311; := $TMP5311
§s %0 7 %1
:= $TMP5311 %0

## .5

%$LTMP5311; := $LTMP5311

## use Proof Template A5201bH (SwapH):  H => (A = B)  -->  H => (B = A)
<< A5201bH.r0t.txt
%0

%$TMP5311; := $TMP5311
§s' %0 7 %1
:= $TMP5311 %0

## .6

%K8000b

## use Proof Template A5221 (Sub):  B  -->  B [x/A]
:= $B5221 %0
:= $T5221 o
:= $X5221 x{$T5221}
:= $A5221 %1/5
<< A5221.r0t.txt
:= $B5221; := $T5221; := $X5221; := $A5221
%0

## use Proof Template A5201b (Swap):  A = B  -->  B = A
<< A5201b.r0t.txt
%0

%$TMP5311; := $TMP5311
§s %0 5 %1

## use Proof Template K8030 (EXI Rule):  (H & B) => A  -->  (H & EXI x: B) => A
:= $T8030 t{^}
:= $X8030 y{$T8030}
:= $A8030 %0
<< K8030.r0t.txt
:= $T8030; := $X8030; := $A8030;
```

```
:= $TMP5311 %0

%K8000b

## use Proof Template A5221 (Sub):  B  -->  B [x/A]
:= $B5221 %0
:= $T5221 o
:= $X5221 x{$T5221}
:= $A5221 %1/11
<< A5221.r0t.txt
:= $B5221; := $T5221; := $X5221; := $A5221
%0

%$TMP5311; := $TMP5311
§s %0 5 %1
:= $LTMP5311 %0

## .7

%A5304

## use Proof Template A5201b (Swap):  A = B  -->  B = A
<< A5201b.r0t.txt
%0

%$LTMP5311; := $LTMP5311
§s %0 5 %1

:= A5311 %0


##
##  Q.E.D.
##

%0
```

2.2.79 File A5312.r0.txt

```
##
##  Proof A5312:  EXI1 y: P y  =>  ALL z: P z = ( IOTA P = z )
##
##
##  Source: [Andrews 2002 (ISBN 1-4020-0763-9), p. 235]
##
##  Copyright (c) 2017 Owl of Minerva Press GmbH. All rights reserved.
##  Written by Ken Kubota (<mail@kenkubota.de>).
##
```

```
##  This file is part of the publication of the mathematical logic R0.
##  For more information, visit: <http://doi.org/10.4444/100.10>
##

<< basics.r0.txt
<< K8005.r0.txt
<< A5304.r0.txt


##
## Proof
##

## .1

:= $HYP5312 ((={{{o,{o,t{^}}},{o,t{^}}}}_p{{o,t{^}}}{{o,t{^}}}){{o,{o,t{^}}}}_(={{{o
,t{^}},t{^}}}_y{t{^}}{t{^}}){{o,t{^}}})

%K8005

## use Proof Template A5221 (Sub):  B  -->  B [x/A]
:= $B5221 %0
:= $T5221 o
:= $X5221 x{$T5221}
:= $A5221 $HYP5312
<< A5221.r0t.txt
:= $B5221; := $T5221; := $X5221; := $A5221
%0

:= $ATMP5312 %0

## .2

%A5

## use Proof Template K8003 (Intro):  A  -->  H => A
:= $A8003 %0
:= $H8003 $ATMP5312/5
<< K8003.r0t.txt
:= $A8003; := $H8003
%0

:= $LTMP5312 %0

%$ATMP5312

## use Proof Template A5201bH (SwapH):  H => (A = B)  -->  H => (B = A)
<< A5201bH.r0t.txt
%0
```

```
%$LTMP5312; := $LTMP5312
§s' %0 11 %1

:= $BTMP5312 %0

## .3

## use Proof Template A5201bH (SwapH):  H => (A = B)  -->  H => (B = A)
<< A5201bH.r0t.txt
%0

%$ATMP5312
§s' %0 7 %1

:= $CTMP5312 %0

## .4

## use Proof Template:  Axiom 3 Substitutions
:= $AA3 o
:= $BA3 t{^}
:= $FA3 $CTMP5312/13
:= $GA3 $CTMP5312/7
<< axiom3_substitutions.r0t.txt
:= $AA3; := $BA3; := $FA3; := $GA3
%0

## use Proof Template K8003 (Intro):  A  -->  H => A
:= $A8003 %0
:= $H8003 $ATMP5312/5
<< K8003.r0t.txt
:= $A8003; := $H8003
%0

%$CTMP5312
§s' %0 1 %1

§rs /7 z{t{^}}

:= $DTMP5312 %0

## .5

## use Proof Template A5216:  (T & A) = A
:= $A5216 $HYP5312
<< A5216.r0t.txt
:= $A5216
%0
```

```
§= /5
§s %0 5 %1

%$DTMP5312
§s %0 5 %1

## use Proof Template K8030 (EXI Rule):  (H & B) => A  -->  (H & EXI x: B) => A
:= $T8030 t{^}
:= $X8030 y{$T8030}
:= $A8030 %0
<< K8030.r0t.txt
:= $T8030; := $X8030; := $A8030;
%0

:= $LTMP5312 %0

## use Proof Template A5216:  (T & A) = A
:= $A5216 %0/11
<< A5216.r0t.txt
:= $A5216
%0

%$LTMP5312; := $LTMP5312
§s %0 5 %1

## .6

%A5304

§= /5
§s %0 5 %1

§s %3 5 %0

:= A5312 %0

## undefine local variables
:= $HYP5312; := $ATMP5312; := $BTMP5312; := $CTMP5312; := $DTMP5312


##
##  Q.E.D.
##

%0
```

2.2.80 File A5313.r0.txt

```
##
##  Proof A5313:  (C_t_x_y_T = x) & (C_t_x_y_F = y)
##
##
##  Source: [Andrews 2002 (ISBN 1-4020-0763-9), pp. 235 f.]
##
##  Copyright (c) 2017 Owl of Minerva Press GmbH. All rights reserved.
##  Written by Ken Kubota (<mail@kenkubota.de>).
##
##  This file is part of the publication of the mathematical logic R0.
##  For more information, visit: <http://doi.org/10.4444/100.10>
##



##
##  "C[t]xyp can be read 'if p then x, else y'." [Andrews 2002, p. 235]
##


<< basics.r0.txt
<< A5205.r0.txt
<< A5231.r0.txt
<< K8000.r0.txt
<< K8001.r0.txt
<< K8010.r0.txt


:= COND [\t{^}{^}.[\x{t{^}}{t{^}}.[\y{t{^}}{t{^}}.[\p{o}{o}.(i{{t{^},{o,t{^}}}}_[\q{
t{^}}{t{^}}.((|{{{o,o},o}}_((&{{{o,o},o}}_p{o}{o}){{o,o}}_((={{{o,t{^}},t{^}}}_x{t{^
}}{t{^}}){{o,t{^}}}_q{t{^}}{t{^}}){o}){o}){{o,o}}_((&{{{o,o},o}}_(!{{o,o}}_p{o}{o}){
o}){{o,o}}_((={{{o,t{^}},t{^}}}_y{t{^}}{t{^}}){{o,t{^}}}_q{t{^}}{t{^}}){o}){o}){o}]{
{o,t{^}}}){t{^}}]{{t{^},o}}]{{{t{^},o},t{^}}}]{{{{t{^},o},t{^}},t{^}}}]


##
## Proof
##

## .1

§= {t{^}} ((((COND{{{{{\4{^},o},\3{^}},\2{^}},^}}_t{^}{^}){{{{t{^},o},t{^}},t{^}}}_x
{t{^}}{t{^}}){{{t{^},o},t{^}}}_y{t{^}}{t{^}}){{t{^},o}}_T{o})
§\s /24
§\s /12
§\s /6
§\s /3
:= $LTMP5313 %0
```

```
## .2

§= {o} /15

%A5231a
§s %1 29 %0
:= $TMP5313 %0

%K8001b

## use Proof Template A5221 (Sub):  B  -->  B [x/A]
:= $B5221 %0
:= $T5221 o
:= $X5221 x{$T5221}
:= $A5221 %1/15
<< A5221.r0t.txt
:= $B5221; := $T5221; := $X5221; := $A5221
%0

%$TMP5313; := $TMP5313
§s %0 7 %1
:= $TMP5313 %0

%K8000b

## use Proof Template A5221 (Sub):  B  -->  B [x/A]
:= $B5221 %0
:= $T5221 o
:= $X5221 x{$T5221}
:= $A5221 %1/43
<< A5221.r0t.txt
:= $B5221; := $T5221; := $X5221; := $A5221
%0

%$TMP5313; := $TMP5313
§s %0 13 %1
:= $TMP5313 %0

%K8010a

## use Proof Template A5221 (Sub):  B  -->  B [x/A]
:= $B5221 %0
:= $T5221 o
:= $X5221 x{$T5221}
:= $A5221 %1/13
<< A5221.r0t.txt
:= $B5221; := $T5221; := $X5221; := $A5221
%0
```

```
%$TMP5313; := $TMP5313
§s %0 3 %1

## .3

%$LTMP5313; := $LTMP5313
§s %0 15 %1
:= $TMP5313 %0

## .4

## use Proof Template:  A5205 Substitutions
:= $AA5205 o
:= $BA5205 t{^}
:= $FA5205 (={{{o,t{^}},t{^}}}_x{t{^}}{t{^}})
<< a5205_substitutions.r0t.txt
:= $AA5205; := $BA5205; := $FA5205
%0

§rs /3 q{t{^}}

## use Proof Template A5201b (Swap):  A = B  -->  B = A
<< A5201b.r0t.txt
%0

%$TMP5313; := $TMP5313
§s %0 7 %1
:= $TMP5313 %0

## .5

%A5

## use Proof Template A5221 (Sub):  B  -->  B [x/A]
:= $B5221 %0
:= $T5221 t{^}
:= $X5221 y{$T5221}
:= $A5221 x{$T5221}
<< A5221.r0t.txt
:= $B5221; := $T5221; := $X5221; := $A5221
%0

%$TMP5313; := $TMP5313
§s %0 3 %1
:= $LTMP5313 %0

## .6
```

```
§= {t{^}} ((((COND{{{{{\4{^},o},\3{^}},\2{^}},^}}_t{^}{^}){{{{t{^},o},t{^}},t{^}}}_x
{t{^}}{t{^}}){{{t{^},o},t{^}}}_y{t{^}}{t{^}}){{t{^},o}}_F{o})
§\s /24
§\s /12
§\s /6
§\s /3

%A5231b
§s %1 125 %0
:= $TMP5313 %0

%K8001b

## use Proof Template A5221 (Sub):  B  -->  B [x/A]
:= $B5221 %0
:= $T5221 o
:= $X5221 x{$T5221}
:= $A5221 %1/123
<< A5221.r0t.txt
:= $B5221; := $T5221; := $X5221; := $A5221
%0

%$TMP5313; := $TMP5313
§s %0 61 %1
:= $TMP5313 %0

%K8000b

## use Proof Template A5221 (Sub):  B  -->  B [x/A]
:= $B5221 %0
:= $T5221 o
:= $X5221 x{$T5221}
:= $A5221 %1/63
<< A5221.r0t.txt
:= $B5221; := $T5221; := $X5221; := $A5221
%0

%$TMP5313; := $TMP5313
§s %0 31 %1
:= $TMP5313 %0

%K8010b

## use Proof Template A5221 (Sub):  B  -->  B [x/A]
:= $B5221 %0
:= $T5221 o
:= $X5221 x{$T5221}
:= $A5221 %1/31
<< A5221.r0t.txt
```

```
:= $B5221; := $T5221; := $X5221; := $A5221
%0

%$TMP5313; := $TMP5313
§s %0 15 %1
:= $TMP5313 %0

## .7

## use Proof Template:  A5205 Substitutions
:= $AA5205 o
:= $BA5205 t{^}
:= $FA5205 (={{{o,t{^}},t{^}}}_z{t{^}}{t{^}})
<< a5205_substitutions.r0t.txt
:= $AA5205; := $BA5205; := $FA5205
%0

§rs /3 q{t{^}}

## use Proof Template A5221 (Sub):  B  -->  B [x/A]
:= $B5221 %0
:= $T5221 t{^}
:= $X5221 z{$T5221}
:= $A5221 y{$T5221}
<< A5221.r0t.txt
:= $B5221; := $T5221; := $X5221; := $A5221
%0

## use Proof Template A5201b (Swap):  A = B  -->  B = A
<< A5201b.r0t.txt
%0

%$TMP5313; := $TMP5313
§s %0 7 %1

%A5
§s %1 3 %0

## .8

## use Proof Template K8020:  A, B  -->  A & B
:= $A8020 %$LTMP5313; := $LTMP5313
:= $B8020 %0
<< K8020.r0t.txt
:= $A8020; := $B8020

:= A5313 %0
```

```
##
## Q.E.D.
##

%0
```

2.2.81 File A53X08.r0a.txt

```
##
## Proof A53X08: AC [t/b] => ALL x: EXI y: p_x_y = EXI f: ALL x: p_x_(f_x)
##
##
## Source: [cf. Andrews 2002 (ISBN 1-4020-0763-9), p. 237 (X5308)]
##
## Copyright (c) 2017 Owl of Minerva Press GmbH. All rights reserved.
## Written by Ken Kubota (<mail@kenkubota.de>).
##
## This file is part of the publication of the mathematical logic R0.
## For more information, visit: <http://doi.org/10.4444/100.10>
##


##
## Axioms
##

<< axiom_of_choice.r0a.txt


##
## Proof
##

## .1

%AC

## .2

## use Proof Template A5221 (Sub): B --> B [x/A]
:= $B5221 AC
:= $T5221 ^
:= $X5221 t{$T5221}
:= $A5221 u{$T5221}
<< A5221.r0t.txt
:= $B5221; := $T5221; := $X5221; := $A5221
%0

:= $AC53X08 %0
```

```
## left-hand side of the equation (equivalence)
:= $A53X08 ((A{{{o,{o,\3{^}}},^}}_t{^}{^}){{o,{o,t{^}}}}_[\x{t{^}}{t{^}}.((E{{{o,{o,
\3{^}}},^}}_u{^}{^}){{o,{o,u{^}}}}_[\y{u{^}}{u{^}}.((p{{{o,u{^}},t{^}}}{{{o,u{^}},t{
^}}}_x{t{^}}{t{^}}){{o,u{^}}}_y{u{^}}{u{^}}){o}]{{o,u{^}}}){o}]{{o,t{^}}})

## right-hand side of the equation (equivalence)
:= $B53X08 ((E{{{o,{o,\3{^}}},^}}_{u{^},t{^}}{^}){{o,{o,{u{^},t{^}}}}}_[\f{{u{^},t{^
}}}{{u{^},t{^}}}.((A{{{o,{o,\3{^}}},^}}_t{^}{^}){{o,{o,t{^}}}}_[\x{t{^}}{t{^}}.((p{{
{o,u{^}},t{^}}}{{{o,u{^}},t{^}}}_x{t{^}}{t{^}}){{o,u{^}}}_(f{{u{^},t{^}}}{{u{^},t{^}
}}_x{t{^}}{t{^}}){u{^}}){o}]{{o,t{^}}}){o}]{{o,{u{^},t{^}}}})

## .3

<< A53X08a.r0a.txt
:= $ATMP53X08 %0

## .4

<< A53X08b.r0a.txt
:= $BTMP53X08 %0

## .5

%$ATMP53X08; := $ATMP53X08
%$BTMP53X08; := $BTMP53X08

## use Proof Template K8013H:  H => (A => B), H => (B => A)  -->  H => (A = B)
:= $H8013H /5
:= $A8013H /7
:= $B8013H /13
<< K8013H.r0t.txt
:= $H8013H; := $A8013H; := $B8013H
%0


##
##  Q.E.D.
##

%0


##
##  Undefine Syntactical Variables
##

:= $AC53X08; := $A53X08; := $B53X08
```

2.2.82 File A53X08a.r0a.txt

```
##
##  Proof A53X08a (Part A => B):  AC [t/b]  =>  ALL x: EXI y: p_x_y = EXI f: ALL x:
p_x_(f_x)
##
##
##  Source: [cf. Andrews 2002 (ISBN 1-4020-0763-9), p. 237 (X5308)]
##
##  Copyright (c) 2017 Owl of Minerva Press GmbH. All rights reserved.
##  Written by Ken Kubota (<mail@kenkubota.de>).
##
##  This file is part of the publication of the mathematical logic R0.
##  For more information, visit: <http://doi.org/10.4444/100.10>
##

<< K8005.r0.txt


##
##  Axioms
##

<< axiom_of_choice.r0a.txt


##
##  Define Syntactical Variables
##

## left-hand side of the equation (equivalence)
:= $A53X08A ((A{{{o,{o,\3{^}}},^}}_t{^}{^}){{o,{o,t{^}}}}_[\x{t{^}}{t{^}}.((E{{{o,{o
,\3{^}}},^}}_u{^}{^}){{o,{o,u{^}}}}_[\y{u{^}}{u{^}}.((p{{{o,u{^}},t{^}}}{{{o,u{^}},t
{^}}}_x{t{^}}{t{^}}){{o,u{^}}}_y{u{^}}{u{^}}){o}]{{o,u{^}}}){o}]{{o,t{^}}})

## right-hand side of the equation (equivalence)
:= $B53X08A ((E{{{o,{o,\3{^}}},^}}_{u{^},t{^}}{^}){{o,{o,{u{^},t{^}}}}}_[\f{{u{^},t{
^}}}{{u{^},t{^}}}.((A{{{o,{o,\3{^}}},^}}_t{^}{^}){{o,{o,t{^}}}}_[\x{t{^}}{t{^}}.((p{
{{o,u{^}},t{^}}}{{{o,u{^}},t{^}}}_x{t{^}}{t{^}}){{o,u{^}}}_(f{{u{^},t{^}}}{{u{^},t{^
}}}_x{t{^}}{t{^}}){u{^}}){o}]{{o,t{^}}}){o}]{{o,{u{^},t{^}}}})


##
## Proof
##

## .1

## use Proof Template A5221 (Sub):  B  -->  B [x/A]
```

```
:= $B5221 AC
:= $T5221 ^
:= $X5221 t{$T5221}
:= $A5221 u{$T5221}
<< A5221.r0t.txt
:= $B5221; := $T5221; := $X5221; := $A5221
:= $AC53X08A %0

## part of the Axiom of Choice (after the existential quantifier)
:= $C53X08A %0/7

## hypotheses
:= $HYP1 ((&{{{o,o},o}}_((&{{{o,o},o}}_$AC53X08A{o}){{o,o}}_$A53X08A{o}){o}){{o,o}}_
$C53X08A{o})
:= $HYP2 ((&{{{o,o},o}}_((&{{{o,o},o}}_$AC53X08A{o}){{o,o}}_$B53X08A{o}){o}){{o,o}}_
$C53X08A{o})

## .2

%K8005

## use Proof Template A5221 (Sub):  B  -->  B [x/A]
:= $B5221 %0
:= $T5221 o
:= $X5221 x{$T5221}
:= $A5221 $HYP1
<< A5221.r0t.txt
:= $B5221; := $T5221; := $X5221; := $A5221
%0

## use Proof Template K8019H:  H => (A & B)  -->  H => A, H => B
:= $H8019H %0
<< K8019H.r0t.txt
:= $H8019H
:= $ATMP53X08A %$B8019H
%$A8019H
:= $A8019H; := $B8019H

## .3

## use Proof Template K8019H:  H => (A & B)  -->  H => A, H => B
:= $H8019H %0
<< K8019H.r0t.txt
:= $H8019H
%$B8019H
:= $A8019H; := $B8019H

## use Proof Template A5215H (ALL I):  H => ALL x: B  -->  H => B [x/a]
:= $T5215H t{^}
```

```
:= $X5215H x{$T5215H}
:= $A5215H x{$T5215H}
:= $H5215H %0
<< A5215H.r0t.txt
:= $T5215H; := $X5215H; := $A5215H; := $H5215H
%0

:= $BTMP53X08A %0

%$ATMP53X08A; := $ATMP53X08A

## .4

## use Proof Template A5215H (ALL I):  H => ALL x: B  -->  H => B [x/a]
:= $T5215H {o,u{^}}
:= $X5215H p{$T5215H}
:= $A5215H (p{{{o,u{^}},t{^}}}{{{o,u{^}},t{^}}}_x{t{^}}{t{^}})
:= $H5215H %0
<< A5215H.r0t.txt
:= $T5215H; := $X5215H; := $A5215H; := $H5215H
%0

## .5

§rs /27 y{u{^}}

## .6

%$BTMP53X08A; := $BTMP53X08A

## use Proof Template A5224H (MP):  H => A, H => (A => B)  -->  H => B
:= $A5224H %0
:= $AB5224H %1
<< A5224H.r0t.txt
:= $AB5224H; := $A5224H
%0

## use Proof Template A5220H (Gen):  (H => A)  -->  (H => ALL x: A)
:= $T5220H t{^}
:= $X5220H x{$T5220H}
:= $A5220H %0
<< A5220H.r0t.txt
:= $T5220H; := $X5220H; := $A5220H
%0

## reduce [\x.(j_(p_x))]_x
§\ ([\x{t{^}}{t{^}}.(j{{u{^},{o,u{^}}}}{{u{^},{o,u{^}}}}_(p{{{o,u{^}},t{^}}}{{{o,u{^
}},t{^}}}_x{t{^}}{t{^}}){{o,u{^}}})}{u{^}}]{{u{^},t{^}}}_x{t{^}}{t{^}})
§= /5
```

```
§s %0 5 %1
§s %3 31 %0

§\ ([\f{{u{^},t{^}}}{{u{^},t{^}}}.((A{{{o,{o,\3{^}}},^}}_t{^}{^}){{o,{o,t{^}}}}_[\x{
t{^}}{t{^}}.((p{{{o,u{^}},t{^}}}{{{o,u{^}},t{^}}}_x{t{^}}{t{^}}){{o,u{^}}}_(f{{u{^},
t{^}}}{{u{^},t{^}}}_x{t{^}}{t{^}}){u{^}}){o}]{{o,t{^}}}){o}]{{o,{u{^},t{^}}}}_[\x{t{
^}}{t{^}}.(j{{u{^},{o,u{^}}}}{{u{^},{o,u{^}}}}_(p{{{o,u{^}},t{^}}}{{{o,u{^}},t{^}}}_
x{t{^}}{t{^}}){{o,u{^}}}){u{^}}]{{u{^},t{^}}})
§= /5
§s %0 5 %1
§s %3 3 %0

## use Proof Template K8028 (EXI GenH):  H => ([\x.B]A)  -->  H => EXI x: B
:= $H8028 $HYP1
:= $T8028 {u{^},t{^}}
:= $B8028 %0/6
:= $A8028 %0/7
<< K8028.r0t.txt
:= $H8028; := $T8028; := $B8028; := $A8028
%0

:= $ATMP53X08A %0

## .7

%K8005

## use Proof Template A5221 (Sub):  B  -->  B [x/A]
:= $B5221 %0
:= $T5221 o
:= $X5221 x{$T5221}
:= $A5221 $HYP1/5
<< A5221.r0t.txt
:= $B5221; := $T5221; := $X5221; := $A5221
%0

## use Proof Template K8019H:  H => (A & B)  -->  H => A, H => B
:= $H8019H %0
<< K8019H.r0t.txt
:= $H8019H
%$A8019H
:= $A8019H; := $B8019H

%$ATMP53X08A; := $ATMP53X08A

## use Proof Template A5245 (Rule C):  H => EXI x: B, (H & (B [x/y])) => A  -->  H =
> A
:= $T5245 {u{^},{o,u{^}}}
:= $X5245 j{$T5245}
```

```
:= $Y5245 j{$T5245}
:= $B5245 %1
:= $A5245 %0
<< A5245.r0t.txt
:= $T5245; := $X5245; := $Y5245; := $B5245; := $A5245
%0

## use Proof Template K8025 (Deduction Theorem):  (H & I) => A  -->  H => (I => A)
<< K8025.r0t.txt
%0


##
##  Q.E.D.
##

%0


##
##  Undefine Syntactical Variables
##

:= $A53X08A; := $B53X08A; := $AC53X08A; := $C53X08A; := $HYP1; := $HYP2
```

2.2.83 File A53X08b.r0a.txt

```
##
##  Proof A53X08b (Part B => A):  AC [t/b]  =>  ALL x: EXI y: p_x_y = EXI f: ALL x:
p_x_(f_x)
##
##
##  Source: [cf. Andrews 2002 (ISBN 1-4020-0763-9), p. 237 (X5308)]
##
##  Copyright (c) 2017 Owl of Minerva Press GmbH. All rights reserved.
##  Written by Ken Kubota (<mail@kenkubota.de>).
##
##  This file is part of the publication of the mathematical logic R0.
##  For more information, visit: <http://doi.org/10.4444/100.10>
##

<< K8005.r0.txt


##
##  Axioms
##

<< axiom_of_choice.r0a.txt
```

```
##
##  Define Syntactical Variables
##

## left-hand side of the equation (equivalence)
:= $A53X08B ((A{{{o,{o,\3{^}}},^}}_t{^}{^}){{o,{o,t{^}}}}_[\x{t{^}}{t{^}}.((E{{{o,{o
,\3{^}}},^}}_u{^}{^}){{o,{o,u{^}}}}_[\y{u{^}}{u{^}}.((p{{{o,u{^}},t{^}}}{{{o,u{^}},t
{^}}}_x{t{^}}{t{^}}){{o,u{^}}}_y{u{^}}{u{^}}){o}]{{o,u{^}}}){o}]{{o,t{^}}})

## right-hand side of the equation (equivalence)
:= $B53X08B ((E{{{o,{o,\3{^}}},^}}_{u{^},t{^}}{^}){{o,{o,{u{^},t{^}}}}}_[\f{{u{^},t{
^}}}{{u{^},t{^}}}.((A{{{o,{o,\3{^}}},^}}_t{^}{^}){{o,{o,t{^}}}}_[\x{t{^}}{t{^}}.((p{
{{o,u{^}},t{^}}}{{{o,u{^}},t{^}}}_x{t{^}}{t{^}}){{o,u{^}}}_(f{{u{^},t{^}}}{{u{^},t{^
}}}_x{t{^}}{t{^}}){u{^}}){o}]{{o,t{^}}}){o}]{{o,{u{^},t{^}}}})


##
## Proof
##

## .1

## use Proof Template A5221 (Sub):  B  -->  B [x/A]
:= $B5221 AC
:= $T5221 ^
:= $X5221 t{$T5221}
:= $A5221 u{$T5221}
<< A5221.r0t.txt
:= $B5221; := $T5221; := $X5221; := $A5221
:= $AC53X08B %0

## part of the Axiom of Choice (after the existential quantifier)
:= $C53X08B %0/7

## hypotheses
:= $HYP1 ((&{{{o,o},o}}_((&{{{o,o},o}}_$AC53X08B{o}){{o,o}}_$A53X08B{o}){o}){{o,o}}_
$C53X08B{o})
:= $HYP2 ((&{{{o,o},o}}_((&{{{o,o},o}}_$AC53X08B{o}){{o,o}}_$B53X08B{o}){o}){{o,o}}_
$C53X08B{o})

## .2

%K8005

## use Proof Template A5221 (Sub):  B  -->  B [x/A]
:= $B5221 %0
:= $T5221 o
```

```
:= $X5221 x{$T5221}
:= $A5221 $HYP2
<< A5221.r0t.txt
:= $B5221; := $T5221; := $X5221; := $A5221
%0

## use Proof Template K8019H:  H => (A & B)  -->  H => A, H => B
:= $H8019H %0
<< K8019H.r0t.txt
:= $H8019H
%$A8019H
:= $A8019H; := $B8019H

## use Proof Template K8019H:  H => (A & B)  -->  H => A, H => B
:= $H8019H %0
<< K8019H.r0t.txt
:= $H8019H
%$B8019H
:= $A8019H; := $B8019H

:= $BTMP53X08B %0
:= $D53X08B /15

## .3

%K8005

## use Proof Template A5221 (Sub):  B  -->  B [x/A]
:= $B5221 %0
:= $T5221 o
:= $X5221 x{$T5221}
:= $A5221 ((&{{{o,o},o}}_$HYP2{o}){{o,o}}_$D53X08B{o})
<< A5221.r0t.txt
:= $B5221; := $T5221; := $X5221; := $A5221
%0

## use Proof Template K8019H:  H => (A & B)  -->  H => A, H => B
:= $H8019H %0
<< K8019H.r0t.txt
:= $H8019H
%$B8019H
:= $A8019H; := $B8019H
%0

## .4

## use Proof Template A5215H (ALL I):  H => ALL x: B  -->  H => B [x/a]
:= $T5215H t{^}
:= $X5215H x{$T5215H}
```

```
:= $A5215H x{$T5215H}
:= $H5215H %0
<< A5215H.r0t.txt
:= $T5215H; := $X5215H; := $A5215H; := $H5215H
%0

## .5

§\ ([\y{u{^}}{u{^}}.((p{{{o,u{^}},t{^}}}{{{o,u{^}},t{^}}}_x{t{^}}{t{^}}){{o,u{^}}}_y
{u{^}}{u{^}}){o}]{{o,u{^}}}_(f{{u{^},t{^}}}{{u{^},t{^}}}_x{t{^}}{t{^}}){u{^}})
§= /5
§s %0 5 %1

§s %3 3 %0

## use Proof Template K8028 (EXI GenH):  H => ([\x.B]A)  -->  H => EXI x: B
:= $H8028 ((&{{{o,o},o}}_$HYP2{o}){{o,o}}_$D53X08B{o})
:= $T8028 u{^}
:= $B8028 %0/6
:= $A8028 %0/7
<< K8028.r0t.txt
:= $H8028; := $T8028; := $B8028; := $A8028
%0

## .6

%$BTMP53X08B; := $BTMP53X08B

## use Proof Template A5245 (Rule C):  H => EXI x: B, (H & (B [x/y])) => A  -->  H =
> A
:= $T5245 {u{^},t{^}}
:= $X5245 f{$T5245}
:= $Y5245 f{$T5245}
:= $B5245 %0
:= $A5245 %1
<< A5245.r0t.txt
:= $T5245; := $X5245; := $Y5245; := $B5245; := $A5245
%0

## use Proof Template A5220H (Gen):  (H => A)  -->  (H => ALL x: A)
:= $T5220H t{^}
:= $X5220H x{$T5220H}
:= $A5220H %0
<< A5220H.r0t.txt
:= $T5220H; := $X5220H; := $A5220H
%0

:= $ATMP53X08B %0
```

```
## .7

%K8005

## use Proof Template A5221 (Sub):  B  -->  B [x/A]
:= $B5221 %0
:= $T5221 o
:= $X5221 x{$T5221}
:= $A5221 $HYP2/5
<< A5221.r0t.txt
:= $B5221; := $T5221; := $X5221; := $A5221
%0

## use Proof Template K8019H:  H => (A & B)  -->  H => A, H => B
:= $H8019H %0
<< K8019H.r0t.txt
:= $H8019H
%$A8019H
:= $A8019H; := $B8019H

%$ATMP53X08B; := $ATMP53X08B

## use Proof Template A5245 (Rule C):  H => EXI x: B, (H & (B [x/y])) => A  -->  H =
> A
:= $T5245 {u{^},{o,u{^}}}
:= $X5245 j{$T5245}
:= $Y5245 j{$T5245}
:= $B5245 %1
:= $A5245 %0
<< A5245.r0t.txt
:= $T5245; := $X5245; := $Y5245; := $B5245; := $A5245
%0

## use Proof Template K8025 (Deduction Theorem):  (H & I) => A  -->  H => (I => A)
<< K8025.r0t.txt
%0


##
##  Q.E.D.
##

%0


##
##  Undefine Syntactical Variables
##
```

```
:= $A53X08B; := $B53X08B; := $AC53X08B; := $C53X08B; := $HYP1; := $HYP2; := $D53X08B
```

2.2.84 File A6100.r0.txt

```
##
##  Proof A6100:  Peano's Postulate No. 1 for Andrews' Definition of Natural Numbers
##
##
##  Source: [Andrews 2002 (ISBN 1-4020-0763-9), p. 261]
##
##  Copyright (c) 2017 Owl of Minerva Press GmbH. All rights reserved.
##  Written by Ken Kubota (<mail@kenkubota.de>).
##
##  This file is part of the publication of the mathematical logic R0.
##  For more information, visit: <http://doi.org/10.4444/100.10>
##

<< natural_numbers_andrews.r0.txt
<< K8005.r0.txt


## shorthands
:= $S SIGMA
:= $ANSETZ (p{{o,$S}}{{o,$S}}_ATZERO{$S})
:= $ANSETS ((A{{{o,{o,\3{^}}},^}}_$S{^}){{o,{o,$S}}}_[\x{$S}{$S}.((=>{{{o,o},o}}_(p{
{o,$S}}{{o,$S}}_x{$S}{$S}){o}){{o,o}}_(p{{o,$S}}{{o,$S}}_(ATSUCC{{$S,$S}}_x{$S}{$S})
{$S}){o}){o}]{{o,$S}})
:= $ANBOTH ((&{{{o,o},o}}_$ANSETZ{o}){{o,o}}_$ANSETS{o})
:= $P1APP ((((P1{{{{{o,{o,\5{^}}},{\4{^},\4{^}}},\2{^}},^}}_$S{^}){{{{o,{o,$S}},{$S,
$S}},$S}}_(AZERO{{{o,{o,\3{^}}},^}}_t{^}{^}){$S}){{{o,{o,$S}},{$S,$S}}}_(ASUCC{{{{o,
{o,\4{^}}},{o,{o,\4{^}}}},^}}_t{^}{^}){{$S,$S}}){{o,{o,$S}}}_(ANSET{{{o,{o,{o,\4{^}}
}},^}}_t{^}{^}){{o,$S}})

## .0: expand Peano's postulate

§= {o} ((((P1{{{{{o,{o,\5{^}}},{\4{^},\4{^}}},\2{^}},^}}_$S{^}){{{{o,{o,$S}},{$S,$S}
},$S}}_(AZERO{{{o,{o,\3{^}}},^}}_t{^}{^}){$S}){{{o,{o,$S}},{$S,$S}}}_(ASUCC{{{{o,{o,
\4{^}}},{o,{o,\4{^}}}},^}}_t{^}{^}){{$S,$S}}){{o,{o,$S}}}_(ANSET{{{o,{o,{o,\4{^}}}},
^}}_t{^}{^}){{o,$S}})
§\s /24
§\s /12
§\s /6
§\s /3

:= $DTMP6100 %0

## .1

§= {o} /3
```

```
§\s /6
§\s /3

§\s /63

## use Proof Template A5201b (Swap):  A = B  -->  B = A
<< A5201b.r0t.txt
%0

:= $TMP6100 %0

%K8005

## use Proof Template A5221 (Sub):  B  -->  B [x/A]
:= $B5221 %0
:= $T5221 o
:= $X5221 x{$T5221}
:= $A5221 %1/93
<< A5221.r0t.txt
:= $B5221; := $T5221; := $X5221; := $A5221
%0

## use Proof Template K8019H:  H => (A & B)  -->  H => A, H => B
:= $H8019H %0
<< K8019H.r0t.txt
:= $H8019H
%$A8019H
:= $A8019H; := $B8019H
%0

## use Proof Template A5220 (Gen):  A  -->  ALL x: A
:= $T5220 {{o,$S}}
:= $X5220 p{$T5220}
:= $A5220 %0
<< A5220.r0t.txt
:= $T5220; := $X5220; := $A5220
%0

%$TMP6100; := $TMP6100
§s %1 1 %0

## .2: match general definition

:= $TMP6100 %0
%$DTMP6100; := $DTMP6100

## use Proof Template A5201b (Swap):  A = B  -->  B = A
<< A5201b.r0t.txt
```

```
%0

%$TMP6100; := $TMP6100
§s %0 1 %1

:= A6100 %0


##
##  Q.E.D.
##

%0

## undefine local variables
:= $S; := $ANSETZ; := $ANSETS; := $ANBOTH; := $P1APP
```

2.2.85 File A6101.r0.txt

```
##
##  Proof A6101:  Peano's Postulate No. 2 for Andrews' Definition of Natural Numbers
##
##
##  Source: [Andrews 2002 (ISBN 1-4020-0763-9), p. 261]
##
##  Copyright (c) 2017 Owl of Minerva Press GmbH. All rights reserved.
##  Written by Ken Kubota (<mail@kenkubota.de>).
##
##  This file is part of the publication of the mathematical logic R0.
##  For more information, visit: <http://doi.org/10.4444/100.10>
##

<< natural_numbers_andrews.r0.txt
<< K8005.r0.txt


## shorthands
:= $S SIGMA
:= $ANSETZ (p{{o,$S}}{{o,$S}}_ATZERO{$S})
:= $ANSETS ((A{{{o,{o,\3{^}}},^}}_$S{^}){{o,{o,$S}}}_[\x{$S}{$S}.((=>{{{o,o},o}}_(p{
{o,$S}}{{o,$S}}_x{$S}{$S}){o}){{o,o}}_(p{{o,$S}}{{o,$S}}_(ATSUCC{{$S,$S}}_x{$S}{$S})
{$S}){o}){o}]{{o,$S}})
:= $ANBOTH ((&{{{o,o},o}}_$ANSETZ{o}){{o,o}}_$ANSETS{o})
:= $P2APP ((((P2{{{{{o,{o,\5{^}}},{\4{^},\4{^}}},\2{^}},^}}_$S{^}){{{{o,{o,$S}},{$S,
$S}},$S}}_(AZERO{{{o,{o,\3{^}}},^}}_t{^}{^}){$S}){{{o,{o,$S}},{$S,$S}}}_(ASUCC{{{{o,
{o,\4{^}}},{o,{o,\4{^}}}},^}}_t{^}{^}){{$S,$S}}){{o,{o,$S}}}_(ANSET{{{o,{o,{o,\4{^}}
}},^}}_t{^}{^}){{o,$S}})
:= $ANSETS2 ((A{{{o,{o,\3{^}}},^}}_$S{^}){{o,{o,$S}}}_[\x{$S}{$S}.((=>{{{o,o},o}}_(n
{{o,$S}}{{o,$S}}_x{$S}{$S}){o}){{o,o}}_(n{{o,$S}}{{o,$S}}_(s{{$S,$S}}{{$S,$S}}_x{$S}
```

```
{$S}){$S}){o}){o}]{{o,$S}})
:= $ANSETS3 ((A{{{o,{o,\3{^}}},^}}_$S{^}){{o,{o,$S}}}_[\x{$S}{$S}.((=>{{{o,o},o}}_(n
{{o,$S}}{{o,$S}}_x{$S}{$S}){o}){{o,o}}_(n{{o,$S}}{{o,$S}}_((ASUCC{{{{o,{o,\4{^}}},{o
,{o,\4{^}}}},^}}_t{^}{^}){{$S,$S}}_x{$S}{$S}){$S}){o}){o}]{{o,$S}})
:= $ANSETx ((A{{{o,{o,\3{^}}},^}}_{o,$S}{^}){{o,{o,{o,$S}}}}_[\p{{o,$S}}{{o,$S}}.((=
>{{{o,o},o}}_$ANBOTH{o}){{o,o}}_(p{{o,$S}}{{o,$S}}_x{$S}{$S}){o}){o}]{{o,{o,$S}}})

## .0: expand Peano's Postulate

§= ((((P2{{{{{o,{o,\5{^}}},{\4{^},\4{^}}},\2{^}},^}}_$S{^}){{{{o,{o,$S}},{$S,$S}},$S
}}_(AZERO{{{o,{o,\3{^}}},^}}_t{^}{^}){$S}){{{o,{o,$S}},{$S,$S}}}_(ASUCC{{{{o,{o,\4{^
}}},{o,{o,\4{^}}}},^}}_t{^}{^}){{$S,$S}}){{o,{o,$S}}}_(ANSET{{{o,{o,{o,\4{^}}}},^}}_
t{^}{^}){{o,$S}})
§\s /24
§\s /12
§\s /6
§\s /3

:= $DTMP6101 %0

## .1

%K8005

## use Proof Template A5221 (Sub):  B  -->  B [x/A]
:= $B5221 %0
:= $T5221 o
:= $X5221 x{$T5221}
:= $A5221 ATNSET/61
<< A5221.r0t.txt
:= $B5221; := $T5221; := $X5221; := $A5221

:= $TMP6101 %0

%K8005

## use Proof Template A5221 (Sub):  B  -->  B [x/A]
:= $B5221 %0
:= $T5221 o
:= $X5221 x{$T5221}
:= $A5221 ((ANSET{{{o,{o,{o,\4{^}}}},^}}_t{^}{^}){{o,$S}}_x{$S}{$S})
<< A5221.r0t.txt
:= $B5221; := $T5221; := $X5221; := $A5221
%0

%$TMP6101; := $TMP6101

## use Proof Template K8004 (Trans):  (H OP A), B  -->  H => B
:= $HA8004 %1
```

```
:= $B8004 %0
<< K8004.r0t.txt
:= $HA8004; := $B8004
%0

## use Proof Template K8026 (Deduction Theorem Reversed):  H => (I => A)  -->  (H &
I) => A
<< K8026.r0t.txt
%0

:= $LTMP6101 %0

## .2

%K8005

## use Proof Template A5221 (Sub):  B  -->  B [x/A]
:= $B5221 %0
:= $T5221 o
:= $X5221 x{$T5221}
:= $A5221 %1/5
<< A5221.r0t.txt
:= $B5221; := $T5221; := $X5221; := $A5221
%0

## use Proof Template K8019H:  H => (A & B)  -->  H => A, H => B
:= $H8019H %0
<< K8019H.r0t.txt
:= $H8019H
%$A8019H
:= $A8019H; := $B8019H
%0

§\s /6
§\s /3

## use Proof Template A5215H (ALL I):  H => ALL x: B  -->  H => B [x/a]
:= $T5215H {{o,$S}}
:= $X5215H p{$T5215H}
:= $A5215H p{$T5215H}
:= $H5215H %0
<< A5215H.r0t.txt
:= $T5215H; := $X5215H; := $A5215H; := $H5215H
%0

## use Proof Template A5224H (MP):  H => A, H => (A => B)  -->  H => B
:= $AB5224H %0
:= $A5224H %$LTMP6101
<< A5224H.r0t.txt
```

```
:= $AB5224H; := $A5224H

:= $TMP6101 %0

## .3

%$LTMP6101; := $LTMP6101

## use Proof Template K8019H:  H => (A & B)  -->  H => A, H => B
:= $H8019H %0
<< K8019H.r0t.txt
:= $H8019H
%$B8019H
:= $A8019H; := $B8019H
%0

## use Proof Template A5215H (ALL I):  H => ALL x: B  -->  H => B [x/a]
:= $T5215H {$S}
:= $X5215H x{$T5215H}
:= $A5215H x{$T5215H}
:= $H5215H %0
<< A5215H.r0t.txt
:= $T5215H; := $X5215H; := $A5215H; := $H5215H
%0

## use Proof Template A5224H (MP):  H => A, H => (A => B)  -->  H => B
:= $AB5224H %0
:= $A5224H %$TMP6101; := $TMP6101
<< A5224H.r0t.txt
:= $AB5224H; := $A5224H
%0

## .4

## use Proof Template K8025 (Deduction Theorem):  (H & I) => A  -->  H => (I => A)
<< K8025.r0t.txt
%0

## use Proof Template A5220H (Gen):  (H => A)  -->  (H => ALL x: A)
:= $T5220H {{o,$S}}
:= $X5220H p{$T5220H}
:= $A5220H %0
<< A5220H.r0t.txt
:= $T5220H; := $X5220H; := $A5220H

:= $TMP6101 %0

§= ((ANSET{{{o,{o,{o,\4{^}}}},^}}_t{^}{^}){{o,$S}}_((ASUCC{{{{o,{o,\4{^}}},{o,{o,\4{
^}}}},^}}_t{^}{^}){{$S,$S}}_x{$S}{$S}){$S})
```

```
§\s /10
§\s /5
§\s /190

%$TMP6101; := $TMP6101
§s %0 3 %1

## .5: match general definition

## use Proof Template A5220H (Gen):  (H => A)  -->  (H => ALL x: A)
:= $T5220 $S
:= $X5220 x{$T5220}
:= $A5220 %0
<< A5220.r0t.txt
:= $T5220; := $X5220; := $A5220
%0

:= $TMP6101 %0
%$DTMP6101; := $DTMP6101

## use Proof Template A5201b (Swap):  A = B  -->  B = A
<< A5201b.r0t.txt
%0

%$TMP6101; := $TMP6101
§s %0 1 %1

:= A6101 %0


##
##  Q.E.D.
##

%0

## undefine local variables
:= $S; := $ANSETZ; := $ANSETS; := $ANBOTH; := $P2APP; := $ANSETS2; := $ANSETS3; := $
ANSETx
```

2.2.86 File A6102.r0.txt

```
##
##  Proof A6102:  Peano's Postulate No. 5 for Andrews' Definition of Natural Numbers
##
##
##  Source: [Andrews 2002 (ISBN 1-4020-0763-9), p. 262]
##
##  Copyright (c) 2017 Owl of Minerva Press GmbH. All rights reserved.
```

```
##  Written by Ken Kubota (<mail@kenkubota.de>).
##
##  This file is part of the publication of the mathematical logic R0.
##  For more information, visit: <http://doi.org/10.4444/100.10>
##

<< natural_numbers_andrews.r0.txt
<< K8021.r0.txt
<< A6100.r0.txt
<< A6101.r0.txt


## definition of P
:= $S SIGMA
:= $P [\t{$S}{$S}.((&{{{o,o},o}}_((ANSET{{{o,{o,{o,\4{^}}}}},^}}_t{^}{^}){{o,$S}}_t{$
S}{$S}){o}){{o,o}}_(p{{o,$S}}{{o,$S}}_t{$S}{$S}){o}){o}]

## shorthands
:= $T3 {{o,{o,\3{^}}},^}
:= $T4 {{o,{o,{o,\4{^}}}},^}
:= $T44 {{{o,{o,\4{^}}},{o,{o,\4{^}}}},^}
:= $To2S {o,{o,$S}}
:= $To2S3 {$To2S,{$S,$S}}
:= $P5APP ((((P5{{{{{o,{o,\5{^}}},{\4{^},\4{^}}},\2{^}},^}}_$S{^}){{$To2S3,$S}}_(AZE
RO{$T3}_t{^}{^}){$S}){$To2S3}_(ASUCC{$T44}_t{^}{^}){{$S,$S}}){$To2S}_(ANSET{$T4}_t{^
}{^}){{o,$S}})
:= $ANSETZ (p{{o,$S}}{{o,$S}}_ATZERO{$S})
:= $ANSETS ((A{$T3}_$S{^}){$To2S}_[\x{$S}{$S}.((=>{{{o,o},o}}_(p{{o,$S}}{{o,$S}}_x{$
S}{$S}){o}){{o,o}}_(p{{o,$S}}{{o,$S}}_(ATSUCC{{$S,$S}}_x{$S}{$S}){$S}){o}){o}]{{o,$S
}})
:= $ANBOTH ((&{{{o,o},o}}_$ANSETZ{o}){{o,o}}_$ANSETS{o})
:= $ANSETS2 ((A{$T3}_$S{^}){$To2S}_[\x{$S}{$S}.((=>{{{o,o},o}}_(n{{o,$S}}{{o,$S}}_x{
$S}{$S}){o}){{o,o}}_(n{{o,$S}}{{o,$S}}_(s{{$S,$S}}{{$S,$S}}_x{$S}{$S}){$S}){o}){o}]{
{o,$S}})
:= $ANSETS3 ((A{$T3}_$S{^}){$To2S}_[\x{$S}{$S}.((=>{{{o,o},o}}_(n{{o,$S}}{{o,$S}}_x{
$S}{$S}){o}){{o,o}}_(n{{o,$S}}{{o,$S}}_((ASUCC{$T44}_t{^}{^}){{$S,$S}}_x{$S}{$S}){$S
}){o}){o}]{{o,$S}})
:= $ANSETx ((A{$T3}_{o,$S}{^}){{o,$To2S}}_[\p{{o,$S}}{{o,$S}}.((=>{{{o,o},o}}_$ANBOT
H{o}){{o,o}}_(p{{o,$S}}{{o,$S}}_x{$S}{$S}){o}){o}]{$To2S})
:= $ZRO (p{{o,$S}}{{o,$S}}_z{$S}{$S})
:= $SCC ((A{$T3}_$S{^}){$To2S}_[\x{$S}{$S}.((=>{{{o,o},o}}_(n{{o,$S}}{{o,$S}}_x{$S}{
$S}){o}){{o,o}}_((=>{{{o,o},o}}_(p{{o,$S}}{{o,$S}}_x{$S}{$S}){o}){{o,o}}_(p{{o,$S}}{
{o,$S}}_(s{{$S,$S}}{{$S,$S}}_x{$S}{$S}){$S}){o}){o}){o}]{{o,$S}})
:= $ALL ((A{$T3}_$S{^}){$To2S}_[\x{$S}{$S}.((=>{{{o,o},o}}_(n{{o,$S}}{{o,$S}}_x{$S}{
$S}){o}){{o,o}}_(p{{o,$S}}{{o,$S}}_x{$S}{$S}){o}){o}]{{o,$S}})
:= $IDC ((A{$T3}_{o,$S}{^}){{o,$To2S}}_[\p{{o,$S}}{{o,$S}}.((=>{{{o,o},o}}_((&{{{o,o
},o}}_$ZRO{o}){{o,o}}_$SCC{o}){o}){{o,o}}_$ALL{o}){o}]{$To2S})
:= $P5S ((([\z{$S}{$S}.[\s{{$S,$S}}{{$S,$S}}.[\n{{o,$S}}{{o,$S}}.$IDC{o}]{$To2S}]{$T
o2S3}]{{$To2S3,$S}}_(AZERO{$T3}_t{^}{^}){$S}){$To2S3}_(ASUCC{$T44}_t{^}{^}){{$S,$S}}
```

```
){$To2S}_(ANSET{$T4}_t{^}{^}){{o,$S}})
:= $IDCO ((A{$T3}_{o,$S}{^}){{o,$To2S}}_[\p{{o,$S}}{{o,$S}}.((=>{{{o,o},o}}_((&{{{o,
o},o}}_(p{{o,$S}}{{o,$S}}_(AZERO{$T3}_t{^}{^}){$S}){o}){{o,o}}_$SCC{o}){o}){{o,o}}_$
ALL{o}){o}]{$To2S})
:= $P5SO [\s{{$S,$S}}{{$S,$S}}.[\n{{o,$S}}{{o,$S}}.((A{$T3}_{o,$S}{^}){{o,$To2S}}_[\
p{{o,$S}}{{o,$S}}.((=>{{{o,o},o}}_((&{{{o,o},o}}_(p{{o,$S}}{{o,$S}}_(AZERO{$T3}_t{^}
{^}){$S}){o}){{o,o}}_$SCC{o}){o}){{o,o}}_$ALL{o}){o}]{$To2S}){o}]{$To2S}]
:= $ZRO2 (p{{o,$S}}{{o,$S}}_(AZERO{$T3}_t{^}{^}){$S})
:= $SCC2 ((A{$T3}_$S{^}){$To2S}_[\x{$S}{$S}.((=>{{{o,o},o}}_(n{{o,$S}}{{o,$S}}_x{$S}
{$S}){o}){{o,o}}_((=>{{{o,o},o}}_(p{{o,$S}}{{o,$S}}_x{$S}{$S}){o}){{o,o}}_(p{{o,$S}}
{{o,$S}}_((ASUCC{$T44}_t{^}{^}){{$S,$S}}_x{$S}{$S}){$S}){o}){o}){o}]{{o,$S}})
:= $P5SOSC [\n{{o,$S}}{{o,$S}}.((A{$T3}_{o,$S}{^}){{o,$To2S}}_[\p{{o,$S}}{{o,$S}}.((
=>{{{o,o},o}}_((&{{{o,o},o}}_(p{{o,$S}}{{o,$S}}_(AZERO{$T3}_t{^}{^}){$S}){o}){{o,o}}
_$SCC2{o}){o}){{o,o}}_$ALL{o}){o}]{$To2S}){o}]
:= $SCC3 ((A{$T3}_$S{^}){$To2S}_[\x{$S}{$S}.(ADOTx{{o,o}}_((=>{{{o,o},o}}_(p{{o,$S}}
{{o,$S}}_x{$S}{$S}){o}){{o,o}}_(p{{o,$S}}{{o,$S}}_((ASUCC{$T44}_t{^}{^}){{$S,$S}}_x{
$S}{$S}){$S}){o}){o}){o}]{{o,$S}})
:= $ALL3 ((A{$T3}_$S{^}){$To2S}_[\x{$S}{$S}.(ADOTx{{o,o}}_(p{{o,$S}}{{o,$S}}_x{$S}{$
S}){o}){o}]{{o,$S}})
:= $P5SOSCST ((A{$T3}_{o,$S}{^}){{o,$To2S}}_[\p{{o,$S}}{{o,$S}}.((=>{{{o,o},o}}_((&{
{{o,o},o}}_$ZRO2{o}){{o,o}}_$SCC3{o}){o}){{o,o}}_$ALL3{o}){o}]{$To2S})
:= $STSC ((ANSET{$T4}_t{^}{^}){{o,$S}}_((ASUCC{$T44}_t{^}{^}){{$S,$S}}_x{$S}{$S}){$S
})
:= $HPTMP ((=>{{{o,o},o}}_((&{{{o,o},o}}_((&{{{o,o},o}}_$ZRO2{o}){{o,o}}_$SCC3{o}){o
}){{o,o}}_((ANSET{$T4}_t{^}{^}){{o,$S}}_x{$S}{$S}){o}){o}){{o,o}}_$STSC{o})
:= $SCCP ((A{$T3}_$S{^}){$To2S}_[\x{$S}{$S}.((=>{{{o,o},o}}_($P{{o,$S}}_x{$S}{$S}){o
}){{o,o}}_($P{{o,$S}}_((ASUCC{$T44}_t{^}{^}){{$S,$S}}_x{$S}{$S}){$S}){o}){o}]{{o,$S}
})
:= $SCCPT ((A{$T3}_$S{^}){$To2S}_[\x{$S}{$S}.((=>{{{o,o},o}}_($P{{o,$S}}_x{$S}{$S}){
o}){{o,o}}_($P{{o,$S}}_(ATSUCC{{$S,$S}}_x{$S}{$S}){$S}){o}){o}]{{o,$S}})
:= $ZROSCCT ((&{{{o,o},o}}_($P{{o,$S}}_ATZERO{$S}){o}){{o,o}}_$SCCPT{o})
:= $HTMP2 (&{{{o,o},o}}_((&{{{o,o},o}}_((&{{{o,o},o}}_$ZRO2{o}){{o,o}}_$SCC3{o}){o})
{{o,o}}_((ANSET{$T4}_t{^}{^}){{o,$S}}_x{$S}{$S}){o}){o})

## .0: expand Peano's postulate

§= ((((P5{{{{{o,{o,\5{^}}},{\4{^},\4{^}}},\2{^}},^}}_$S{^}){{$To2S3,$S}}_(AZERO{{{o,
{o,\3{^}}},^}}_t{^}{^}){$S}){$To2S3}_(ASUCC{{{{o,{o,\4{^}}},{o,{o,\4{^}}}},^}}_t{^}{
^}){{$S,$S}}){$To2S}_(ANSET{{{o,{o,{o,\4{^}}}},^}}_t{^}{^}){{o,$S}})
§\s /24
§\s /12
§\s /6
§\s /3

:= $DOTMP %0

## .1

%K8005
```

```
## use Proof Template A5221 (Sub):  B  -->  B [x/A]
:= $B5221 %0
:= $T5221 o
:= $X5221 x{$T5221}
:= $A5221 ((&{{{o,o},o}}_(p{{o,$S}}{{o,$S}}_(AZERO{{{o,{o,\3{^}}},^}}_t{^}{^}){$S}){
o}){{o,o}}_((A{{{o,{o,\3{^}}},^}}_$S{^}){$To2S}_[\x{$S}{$S}.((=>{{{o,o},o}}_((ANSET{
{{o,{o,{o,\4{^}}}},^}}_t{^}{^}){{o,$S}}_x{$S}{$S}){o}){{o,o}}_((=>{{{o,o},o}}_(p{{o,
$S}}{{o,$S}}_x{$S}{$S}){o}){{o,o}}_(p{{o,$S}}{{o,$S}}_((ASUCC{{{{o,{o,\4{^}}},{o,{o,
\4{^}}}},^}}_t{^}{^}){{$S,$S}}_x{$S}{$S}){$S}){o}){o}){o}]{{o,$S}}){o})
<< A5221.r0t.txt
:= $B5221; := $T5221; := $X5221; := $A5221

:= $HTMP %0/5
:= $D1TMP %0

## .2

%K8005

## use Proof Template A5221 (Sub):  B  -->  B [x/A]
:= $B5221 %0
:= $T5221 o
:= $X5221 x{$T5221}
:= $A5221 ((ANSET{{{o,{o,{o,\4{^}}}},^}}_t{^}{^}){{o,$S}}_y{$S}{$S})
<< A5221.r0t.txt
:= $B5221; := $T5221; := $X5221; := $A5221
%0

§\s /6
§\s /3

## use Proof Template A5215H (ALL I):  H => ALL x: B  -->  H => B [x/a]
:= $T5215H {{o,$S}}
:= $X5215H p{$T5215H}
:= $A5215H $P
:= $H5215H %0
<< A5215H.r0t.txt
:= $T5215H; := $X5215H; := $A5215H; := $H5215H

:= $D2TMP %0

## .3

%$D1TMP; := $D1TMP

## use Proof Template K8019H:  H => (A & B)  -->  H => A, H => B
:= $H8019H %0
<< K8019H.r0t.txt
```

```
:= $H8019H
:= $ATMP %$A8019H
:= $BTMP %$B8019H
:= $A8019H; := $B8019H

%$ATMP
%A6100

## use Proof Template K8004 (Trans):  (H OP A), B  -->  H => B
:= $HA8004 %1
:= $B8004 %0
<< K8004.r0t.txt
:= $HA8004; := $B8004
%0

§\s /24
§\s /12
§\s /6
§\s /3

%$ATMP; := $ATMP

## use Proof Template K8020H:  H => A, H => B  -->  H => (A & B)
:= $A8020H %1
:= $B8020H %0
<< K8020H.r0t.txt
:= $A8020H; := $B8020H
%0

§\s /27
§\s /15

:= $TMP %0
§\ ($P{{o,$S}}_ATZERO{$S})

## use Proof Template A5201b (Swap):  A = B  -->  B = A
<< A5201b.r0t.txt
%0

%$TMP; := $TMP

§s %0 3 %1

:= $D3TMP %0

## .4

%$BTMP
```

```
%A6101

## use Proof Template K8004 (Trans):  (H OP A), B  -->  H => B
:= $HA8004 %1
:= $B8004 %0
<< K8004.r0t.txt
:= $HA8004; := $B8004
%0

§\s /24
§\s /12
§\s /6
§\s /3

## use Proof Template A5215H (ALL I):  H => ALL x: B  -->  H => B [x/a]
:= $T5215H {$S}
:= $X5215H x{$T5215H}
:= $A5215H x{$T5215H}
:= $H5215H %0
<< A5215H.r0t.txt
:= $T5215H; := $X5215H; := $A5215H; := $H5215H
%0

## use Proof Template K8026 (Deduction Theorem Reversed):  H => (I => A)  -->  (H &
I) => A
<< K8026.r0t.txt
%0

:= $TMP %0

%K8005

## use Proof Template A5221 (Sub):  B  -->  B [x/A]
:= $B5221 %0
:= $T5221 o
:= $X5221 x{$T5221}
:= $A5221 (p{{o,$S}}{{o,$S}}_x{$S}{$S})
<< A5221.r0t.txt
:= $B5221; := $T5221; := $X5221; := $A5221
%0

%$TMP; := $TMP

## use Proof Template K8004 (Trans):  (H OP A), B  -->  H => B
:= $HA8004 %1
:= $B8004 %0
<< K8004.r0t.txt
:= $HA8004; := $B8004
%0
```

```
## use Proof Template K8026 (Deduction Theorem Reversed):  H => (I => A)  -->  (H &
I) => A
<< K8026.r0t.txt
%0

## use Proof Template K8027:  (A & B) => C  -->  (B & A) => C
<< K8027.r0t.txt
%0

:= $ATMP %0

%$BTMP; := $BTMP

## use Proof Template A5215H (ALL I):  H => ALL x: B  -->  H => B [x/a]
:= $T5215H {$S}
:= $X5215H x{$T5215H}
:= $A5215H x{$T5215H}
:= $H5215H %0
<< A5215H.r0t.txt
:= $T5215H; := $X5215H; := $A5215H; := $H5215H
%0

## use Proof Template K8026 (Deduction Theorem Reversed):  H => (I => A)  -->  (H &
I) => A
<< K8026.r0t.txt
%0

## use Proof Template K8026 (Deduction Theorem Reversed):  H => (I => A)  -->  (H &
I) => A
<< K8026.r0t.txt
%0

:= $BTMP %0

%$ATMP; := $ATMP
%$BTMP; := $BTMP

## use Proof Template K8020H:  H => A, H => B  -->  H => (A & B)
:= $A8020H %1
:= $B8020H %0
<< K8020H.r0t.txt
:= $A8020H; := $B8020H

:= $TMP %0
§\ ($P{{o,$S}}_((ASUCC{{{{o,{o,\4{^}}},{o,{o,\4{^}}}},^}}_t{^}{^}){{$S,$S}}_x{$S}{$S
}){$S})

## use Proof Template A5201b (Swap):  A = B  -->  B = A
```

```
<< A5201b.r0t.txt
%0

%$TMP; := $TMP

§s %0 3 %1

:= $HSWHYTMP %0

%K8021
§\s /2
§\s /1

:= $SWHYTMP %0
%$HSWHYTMP
%$SWHYTMP; := $SWHYTMP

## use Proof Template A5221 (Sub):  B  -->  B [x/A]
:= $B5221 %0
:= $T5221 o
:= $X5221 x{$T5221}
:= $A5221 %1/85
<< A5221.r0t.txt
:= $B5221; := $T5221; := $X5221; := $A5221

:= $SWHYTMP %0
%$HSWHYTMP
%$SWHYTMP; := $SWHYTMP

## use Proof Template A5221 (Sub):  B  -->  B [x/A]
:= $B5221 %0
:= $T5221 o
:= $X5221 y{$T5221}
:= $A5221 %1/43
<< A5221.r0t.txt
:= $B5221; := $T5221; := $X5221; := $A5221

:= $SWHYTMP %0
%$HSWHYTMP
%$SWHYTMP; := $SWHYTMP

## use Proof Template A5221 (Sub):  B  -->  B [x/A]
:= $B5221 %0
:= $T5221 o
:= $X5221 z{$T5221}
:= $A5221 %1/11
<< A5221.r0t.txt
:= $B5221; := $T5221; := $X5221; := $A5221
```

```
%$HSWHYTMP; := $HSWHYTMP
§s %0 5 %1

:= $TMP %0
§\ ($P{{o,$S}}_x{$S}{$S})

## use Proof Template A5201b (Swap):  A = B  -->  B = A
<< A5201b.r0t.txt
%0

%$TMP; := $TMP

§s %0 11 %1

## use Proof Template K8025 (Deduction Theorem):  (H & I) => A  -->  H => (I => A)
<< K8025.r0t.txt
%0

## use Proof Template A5220H (Gen):  (H => A)  -->  (H => ALL x: A)
:= $T5220H {$S}
:= $X5220H x{$T5220H}
:= $A5220H %0
<< A5220H.r0t.txt
:= $T5220H; := $X5220H; := $A5220H

:= $D4TMP %0

## .5

%$D3TMP; := $D3TMP
%$D4TMP; := $D4TMP

## use Proof Template K8020H:  H => A, H => B  -->  H => (A & B)
:= $A8020H %1
:= $B8020H %0
<< K8020H.r0t.txt
:= $A8020H; := $B8020H

:= $TMP %0
%$D2TMP
%$TMP

## use Proof Template K8004 (Trans):  (H OP A), B  -->  H => B
:= $HA8004 %1
:= $B8004 %0
<< K8004.r0t.txt
:= $HA8004; := $B8004
%0
```

```
## use Proof Template K8026 (Deduction Theorem Reversed):  H => (I => A)  -->  (H &
I) => A
<< K8026.r0t.txt
%0

## use Proof Template K8027:  (A & B) => C  -->  (B & A) => C
<< K8027.r0t.txt
%0

§\s /254
:= $ATMP %0

%$TMP; := $TMP
%$D2TMP; := $D2TMP

## use Proof Template K8004 (Trans):  (H OP A), B  -->  H => B
:= $HA8004 %1
:= $B8004 %0
<< K8004.r0t.txt
:= $HA8004; := $B8004
%0

## use Proof Template K8026 (Deduction Theorem Reversed):  H => (I => A)  -->  (H &
I) => A
<< K8026.r0t.txt
%0

:= $ABTMP %0

%$ABTMP; := $ABTMP
%$ATMP; := $ATMP

## use Proof Template A5224H (MP):  H => A, H => (A => B)  -->  H => B
:= $AB5224H %1
:= $A5224H %0
<< A5224H.r0t.txt
:= $AB5224H; := $A5224H
%0

## .6

§\s /3

## use Proof Template K8019H:  H => (A & B)  -->  H => A, H => B
:= $H8019H %0
<< K8019H.r0t.txt
:= $H8019H
%$B8019H
:= $A8019H; := $B8019H
```

```
## use Proof Template K8025 (Deduction Theorem):  (H & I) => A  -->  H => (I => A)
<< K8025.r0t.txt
%0

## use Proof Template A5220H (Gen):  (H => A)  -->  (H => ALL x: A)
:= $T5220H {$S}
:= $X5220H y{$T5220H}
:= $A5220H %0
<< A5220H.r0t.txt
:= $T5220H; := $X5220H; := $A5220H
%0

§rs /7 x{$S}

## use Proof Template A5220 (Gen):  A  -->  ALL x: A
:= $T5220 {{o,$S}}
:= $X5220 p{$T5220}
:= $A5220 %0
<< A5220.r0t.txt
:= $T5220; := $X5220; := $A5220
%0

## .7: Match general definition

:= $TMP %0
%$D0TMP; := $D0TMP

## use Proof Template A5201b (Swap):  A = B  -->  B = A
<< A5201b.r0t.txt
%0

%$TMP; := $TMP
§s %0 1 %1

:= A6102 %0


##
##  Q.E.D.
##

%0

## undefine local variables
:= $S; := $T3; := $T4; := $T44; := $To2S; := $To2S3
:= $P5APP; := $ANSETZ; := $ANSETS; := $ANBOTH; := $ANSETS2; := $ANSETS3; := $ANSETx
:= $ZRO; := $SCC; := $ALL; := $IDC; := $P5S; := $IDC0; := $P5S0; := $ZRO2; := $SCC2
:= $P5S0SC; := $SCC3; := $ALL3; := $P5S0SCST; := $STSC; := $HPTMP
```

```
:= $SCCP; := $SCCPT; := $ZROSCCT; := $HTMP2
:= $P; := $HTMP
```

2.2.87 File K8000.r0.txt

```
##
##  Proof K8000:  (A & T) = A;  (T & A) = A;  (A & T) = (T & A)
##
##
##  Source: [Kubota 2017 (doi: 10.4444/100.10)]
##
##  Copyright (c) 2017 Owl of Minerva Press GmbH. All rights reserved.
##  Written by Ken Kubota (<mail@kenkubota.de>).
##
##  This file is part of the publication of the mathematical logic R0.
##  For more information, visit: <http://doi.org/10.4444/100.10>
##

<< basics.r0.txt
<< A5229.r0.txt


##
##  Proof
##

## .a:  (A & T) = A

## use Proof Template A5222 (Rule of Cases):  [\x.A]T, [\x.A]F  -->  A
:= $L5222 [\x{o}{o}.((={{{o,o},o}}_((&{{{o,o},o}}_x{o}{o}){{o,o}}_T{o}){o}){{o,o}}_x
{o}{o}){o}]
:= $X5222 x{o}
:= $T5222 ($L5222{{o,o}}_T{o})
:= $F5222 ($L5222{{o,o}}_F{o})

## case T:  (T & T) = T
§\ $T5222
## use Proof Template A5201b (Swap):  A = B  -->  B = A
<< A5201b.r0t.txt
%0
%A5229a
§s %0 1 %1

## case F:  (F & T) = F
§\ $F5222
## use Proof Template A5201b (Swap):  A = B  -->  B = A
<< A5201b.r0t.txt
%0
%A5229c
```

```
§s %0 1 %1

<< A5222.r0t.txt
:= $L5222; := $X5222; := $T5222; := $F5222
%0

:= K8000a %0

## .b:  (T & A) = A  [= A5216]

## use Proof Template A5222 (Rule of Cases):  [\x.A]T, [\x.A]F  -->  A
:= $L5222 [\x{o}{o}.((={{{o,o},o}}_((&{{{o,o},o}}_T{o}){{o,o}}_x{o}{o}){o}){{o,o}}_x
{o}{o}){o}]
:= $X5222 x{o}
:= $T5222 ($L5222{{o,o}}_T{o})
:= $F5222 ($L5222{{o,o}}_F{o})

## case T:  (T & T) = T
§\ $T5222
## use Proof Template A5201b (Swap):  A = B  -->  B = A
<< A5201b.r0t.txt
%0
%A5229a
§s %0 1 %1

## case F:  (T & F) = F
§\ $F5222
## use Proof Template A5201b (Swap):  A = B  -->  B = A
<< A5201b.r0t.txt
%0
%A5229b
§s %0 1 %1

<< A5222.r0t.txt
:= $L5222; := $X5222; := $T5222; := $F5222
%0

:= K8000b %0

## .c:  (A & T) = (T & A)

%K8000b
## use Proof Template A5201b (Swap):  A = B  -->  B = A
<< A5201b.r0t.txt
%0
%K8000a
§s %0 3 %1
%0
```

```
:= K8000c %0


##
##  Q.E.D.
##

## %K8000a
%K8000a

## %K8000b
%K8000b

## %K8000c
%K8000c
```

2.2.88 File K8001.r0.txt

```
##
##  Proof K8001:  (A & F) = F;  (F & A) = F;  (A & F) = (F & A)
##
##
##  Source: [Kubota 2017 (doi: 10.4444/100.10)]
##
##  Copyright (c) 2017 Owl of Minerva Press GmbH. All rights reserved.
##  Written by Ken Kubota (<mail@kenkubota.de>).
##
##  This file is part of the publication of the mathematical logic R0.
##  For more information, visit: <http://doi.org/10.4444/100.10>
##

<< basics.r0.txt
<< A5229.r0.txt


##
##  Proof
##

## .a:  (A & F) = F

## use Proof Template A5222 (Rule of Cases):  [\x.A]T, [\x.A]F  -->  A
:= $L5222 [\x{o}{o}.((={{{o,o},o}}_((&{{{o,o},o}}_x{o}{o}){{o,o}}_F{o}){o}){{o,o}}_F
{o}){o}]
:= $X5222 x{o}
:= $T5222 ($L5222{{o,o}}_T{o})
:= $F5222 ($L5222{{o,o}}_F{o})

## Case T:  (T & F) = F
```

```
§\ $T5222
## use Proof Template A5201b (Swap):  A = B  -->  B = A
<< A5201b.r0t.txt
%0
%A5229b
§s %0 1 %1

## Case F:  (F & F) = F
§\ $F5222
## use Proof Template A5201b (Swap):  A = B  -->  B = A
<< A5201b.r0t.txt
%0
%A5229d
§s %0 1 %1

<< A5222.r0t.txt
:= $L5222; := $X5222; := $T5222; := $F5222
%0

:= K8001a %0

## .b:  (F & A) = F

## use Proof Template A5222 (Rule of Cases):  [\x.A]T, [\x.A]F  -->  A
:= $L5222 [\x{o}{o}.((={{{o,o},o}}_((&{{{o,o},o}}_F{o}){{o,o}}_x{o}{o}){o}){{o,o}}_F
{o}){o}]
:= $X5222 x{o}
:= $T5222 ($L5222{{o,o}}_T{o})
:= $F5222 ($L5222{{o,o}}_F{o})

## Case T:  (F & T) = F
§\ $T5222
## use Proof Template A5201b (Swap):  A = B  -->  B = A
<< A5201b.r0t.txt
%0
%A5229c
§s %0 1 %1

## Case F:  (F & F) = F
§\ $F5222
## use Proof Template A5201b (Swap):  A = B  -->  B = A
<< A5201b.r0t.txt
%0
%A5229d
§s %0 1 %1

<< A5222.r0t.txt
:= $L5222; := $X5222; := $T5222; := $F5222
%0
```

```
:= K8001b %0

## .c:   (A & F) = (F & A)

%K8001b
## use Proof Template A5201b (Swap):  A = B  -->  B = A
<< A5201b.r0t.txt
%0
%K8001a
§s %0 3 %1
%0

:= K8001c %0


##
##  Q.E.D.
##

## %K8001a
%K8001a

## %K8001b
%K8001b

## %K8001c
%K8001c
```

2.2.89 File K8002.r0.txt

```
##
##  Proof K8002:  (A & A) = A
##
##
##  Source: [Kubota 2017 (doi: 10.4444/100.10)]
##
##  Copyright (c) 2017 Owl of Minerva Press GmbH. All rights reserved.
##  Written by Ken Kubota (<mail@kenkubota.de>).
##
##  This file is part of the publication of the mathematical logic R0.
##  For more information, visit: <http://doi.org/10.4444/100.10>
##

<< basics.r0.txt
<< A5229.r0.txt


##
```

```
## Proof
##

## use Proof Template A5222 (Rule of Cases):  [\x.A]T, [\x.A]F  -->  A
:= $L5222 [\x{o}{o}.((={{{o,o},o}}_((&{{{o,o},o}}_x{o}{o}){{o,o}}_x{o}{o}){o}){{o,o}
}_x{o}{o}){o}]
:= $X5222 x{o}
:= $T5222 ($L5222{{o,o}}_T{o})
:= $F5222 ($L5222{{o,o}}_F{o})

## Case T:  (T & T) = T
§\ $T5222
## use Proof Template A5201b (Swap):  A = B  -->  B = A
<< A5201b.r0t.txt
%0
%A5211
§s %0 1 %1

## Case F:  (F & F) = F
§\ $F5222
## use Proof Template A5201b (Swap):  A = B  -->  B = A
<< A5201b.r0t.txt
%0
%A5229d
§s %0 1 %1

<< A5222.r0t.txt
:= $L5222; := $X5222; := $T5222; := $F5222
%0

:= K8002 %0


##
##  Q.E.D.
##

%0
```

2.2.90 File K8003.r0a.txt

```
##
##  Proof Template K8003 (Intro):  A  -->  H => A
##      (Hypothesis Introduction)
##
##  Source: [Kubota 2017 (doi: 10.4444/100.10)]
##
##  Copyright (c) 2017 Owl of Minerva Press GmbH. All rights reserved.
##  Written by Ken Kubota (<mail@kenkubota.de>).
```

```
##
##  This file is part of the publication of the mathematical logic R0.
##  For more information, visit: <http://doi.org/10.4444/100.10>
##


##
##  Define Syntactical Variables
##

## the theorem A
:= $A8003 a{o}

## the hypotheses H
:= $H8003 h{o}


##
##  Assumptions and Resulting Syntactical Variables
##

§! $A8003


##
##  Include Proof Template
##

<<< K8003.r0t.txt


##
##  Undefine Syntactical Variables
##

:= $A8003; := $H8003


##
##  Q.E.D.
##

%0
```

2.2.91 File K8003.r0t.txt

```
##
##  Proof Template K8003 (Intro):  A  -->  H => A
##      (Hypothesis Introduction)
```

```
##
##  Source: [Kubota 2017 (doi: 10.4444/100.10)]
##
##  Copyright (c) 2017 Owl of Minerva Press GmbH. All rights reserved.
##  Written by Ken Kubota (<mail@kenkubota.de>).
##
##  This file is part of the publication of the mathematical logic R0.
##  For more information, visit: <http://doi.org/10.4444/100.10>
##

<< basics.r0.txt
<< K8000.r0.txt


##
##  Proof Template
##

## .1

§= ((=>{{{o,o},o}}_$H8003{o}){{o,o}}_$A8003{o})
§rs /12 xtmp{o}
§rs /25 ytmp{o}
§\s /6
§\s /3
:= $TMP8003 %0

## .2

## use Proof Template A5219b (Rule T):  A  -->  A = T
:= $A5219b $A8003
<< A5219b.r0t.txt
:= $A5219b
%0

## .3

%$TMP8003; := $TMP8003
§s %0 15 %1
:= $TMP8003 %0

## .4

## use Proof Template A5221 (Sub):  B  -->  B [x/A]
:= $B5221 K8000a
:= $T5221 o
:= $X5221 x{$T5221}
:= $A5221 $H8003
<< A5221.r0t.txt
```

```
:= $B5221; := $T5221; := $X5221; := $A5221
%0

## .5

%$TMP8003; := $TMP8003
§s %0 7 %1
## use Proof Template A5201b (Swap):  A = B  -->  B = A
<< A5201b.r0t.txt
%0
§= {o} $H8003
§s %0 1 %1
```

2.2.92 File K8004.r0a.txt

```
##
##  Proof Template K8004 (Trans):  (H OP A), B  -->  H => B
##      for any operator OP, including "=>" and "=" (Hypothesis Transfer)
##
##  Source: [Kubota 2017 (doi: 10.4444/100.10)]
##
##  Copyright (c) 2017 Owl of Minerva Press GmbH. All rights reserved.
##  Written by Ken Kubota (<mail@kenkubota.de>).
##
##  This file is part of the publication of the mathematical logic R0.
##  For more information, visit: <http://doi.org/10.4444/100.10>
##

<< basics.r0.txt


##
##  Define Syntactical Variables
##

## hypothesis in theorem
:= $HA8004 ((=>{{{o,o},o}}_h{o}{o}){{o,o}}_a{o}{o})

## proposition
:= $B8004 b{o}


##
##  Assumptions and Resulting Syntactical Variables
##

§! $HA8004
§! $B8004
```

```
##
##  Include Proof Template
##

<<< K8004.r0t.txt


##
##  Undefine Syntactical Variables
##

:= $HA8004; := $B8004


##
##  Q.E.D.
##

%0
```

2.2.93 File K8004.r0t.txt

```
##
##  Proof Template K8004 (Trans):  (H OP A), B  -->  H => B
##      for any operator OP, including "=>" and "=" (Hypothesis Transfer)
##
##  Source: [Kubota 2017 (doi: 10.4444/100.10)]
##
##  Copyright (c) 2017 Owl of Minerva Press GmbH. All rights reserved.
##  Written by Ken Kubota (<mail@kenkubota.de>).
##
##  This file is part of the publication of the mathematical logic R0.
##  For more information, visit: <http://doi.org/10.4444/100.10>
##


##
##  Proof Template
##

## use Proof Template K8003 (Intro):  A  -->  H => A
:= $A8003 $B8004
:= $H8003 $HA8004/5
<< K8003.r0t.txt
:= $A8003; := $H8003
%0
```

2.2.94 File K8005.r0.txt

```
##
##  Proof K8005:  H => H
##
##
##  Source: [Kubota 2017 (doi: 10.4444/100.10)]
##
##  Copyright (c) 2017 Owl of Minerva Press GmbH. All rights reserved.
##  Written by Ken Kubota (<mail@kenkubota.de>).
##
##  This file is part of the publication of the mathematical logic R0.
##  For more information, visit: <http://doi.org/10.4444/100.10>
##

<< basics.r0.txt
<< K8002.r0.txt


##
##  Proof
##

## .1

§= ((=>{{{o,o},o}}_x{o}{o}){{o,o}}_x{o}{o})
§\s /6
§\s /3

## .2

%K8002
§s %1 7 %0

## .3

## use Proof Template A5201b (Swap):  A = B  -->  B = A
<< A5201b.r0t.txt
%0

§= {o} x{o}
§s %0 1 %1

:= K8005 %0


##
##  Q.E.D.
##
```

```
%0
```

2.2.95 File K8006.r0.txt

```
##
##  Proof Template K8006:  (A * T) = (T * A), (A * F) = (F * A)  -->  (A * B) = (B *
 A)
##      for any Boolean relation *
##
##  Source: [Kubota 2017 (doi: 10.4444/100.10)]
##
##  Copyright (c) 2017 Owl of Minerva Press GmbH. All rights reserved.
##  Written by Ken Kubota (<mail@kenkubota.de>).
##
##  This file is part of the publication of the mathematical logic R0.
##  For more information, visit: <http://doi.org/10.4444/100.10>
##



##
##  Define Syntactical Variables
##

<< K8000.r0.txt
<< K8001.r0.txt

## the Boolean relation
:= $R8006 &

## the theorem for case T (using variables x and y)
:= $T8006 K8000c

## the theorem for case F (using variables x and y)
:= $F8006 K8001c



##
##  Proof Template
##

<<< K8006.r0t.txt



##
##  Undefine Syntactical Variables
##

:= $R8006; := $T8006; := $F8006
```

```
##
##  Q.E.D.
##

%0
```

2.2.96 File K8006.r0t.txt

```
##
##  Proof Template K8006:  (A * T) = (T * A), (A * F) = (F * A)  -->  (A * B) = (B *
 A)
##      for any Boolean relation *
##
##  Source: [Kubota 2017 (doi: 10.4444/100.10)]
##
##  Copyright (c) 2017 Owl of Minerva Press GmbH. All rights reserved.
##  Written by Ken Kubota (<mail@kenkubota.de>).
##
##  This file is part of the publication of the mathematical logic R0.
##  For more information, visit: <http://doi.org/10.4444/100.10>
##

<< basics.r0.txt


##
##  Proof Template
##

## use Proof Template A5222 (Rule of Cases):  [\x.A]T, [\x.A]F  -->  A
:= $L5222TMP [\y{o}{o}.((={{{o,o},o}}_(($R8006{{{o,o},o}}_x{o}{o}){{o,o}}_y{o}{o}){o
}){{o,o}}_(($R8006{{{o,o},o}}_y{o}{o}){{o,o}}_x{o}{o}){o}){o}]
:= $X5222TMP y{o}
:= $T5222TMP ($L5222TMP{{o,o}}_T{o})
:= $F5222TMP ($L5222TMP{{o,o}}_F{o})

## case T:  (A & T) = (T & A)
§\ $T5222TMP
## use Proof Template A5201b (Swap):  A = B  -->  B = A
<< A5201b.r0t.txt
%0
%$T8006
§s %0 1 %1

## case F:  (A & F) = (F & A)
§\ $F5222TMP
## use Proof Template A5201b (Swap):  A = B  -->  B = A
```

```
<< A5201b.r0t.txt
%0
%$F8006
§s %0 1 %1

## replace free variable x by variable a avoiding a name collision

:= $L5222 [\y{o}{o}.((={{{o,o},o}}_(($R8006{{{o,o},o}}_a{o}{o}){{o,o}}_y{o}{o}){o}){
{o,o}}_(($R8006{{{o,o},o}}_y{o}{o}){{o,o}}_a{o}{o}){o}){o}]

:= $X5222 y{o}

## use Proof Template A5221 (Sub):  B  -->  B [x/A]
:= $B5221 %$T5222TMP
:= $T5221 {o}
:= $X5221 x{o}
:= $A5221 a{o}
<< A5221.r0t.txt
:= $B5221; := $T5221; := $X5221; := $A5221
:= $T5222 %0
%0

## use Proof Template A5221 (Sub):  B  -->  B [x/A]
:= $B5221 %$F5222TMP
:= $T5221 {o}
:= $X5221 x{o}
:= $A5221 a{o}
<< A5221.r0t.txt
:= $B5221; := $T5221; := $X5221; := $A5221
:= $F5222 %0
%0

:= $L5222TMP; := $X5222TMP; := $T5222TMP; := $F5222TMP

## now actually use Proof Template A5222 (Rule of Cases):  [\x.A]T, [\x.A]F  -->  A
<< A5222.r0t.txt
:= $L5222; := $X5222; := $T5222; := $F5222
%0

## replace back

## use Proof Template A5221 (Sub):  B  -->  B [x/A]
:= $B5221 %0
:= $T5221 {o}
:= $X5221 a{o}
:= $A5221 x{o}
<< A5221.r0t.txt
:= $B5221; := $T5221; := $X5221; := $A5221
%0
```

```
## match general definition
§= ((COMMT{{{o,{{\4{^},\4{^}},\3{^}}},^}}_o{^}){{o,{{o,o},o}}}_$R8006{{{o,o},o}})
§\s /10
§\s /5
§s %5 1 %0
```

2.2.97 File K8007.r0.txt

```
##
##  Proof K8007:  (A & B) = (B & A)
##
##
##  Source: [Kubota 2017 (doi: 10.4444/100.10)]
##
##  Copyright (c) 2017 Owl of Minerva Press GmbH. All rights reserved.
##  Written by Ken Kubota (<mail@kenkubota.de>).
##
##  This file is part of the publication of the mathematical logic R0.
##  For more information, visit: <http://doi.org/10.4444/100.10>
##

<< K8000.r0.txt
<< K8001.r0.txt


##
##  Proof
##

## use Proof Template K8006:  (A * T) = (T * A), (A * F) = (F * A)  -->  (A * B) = (
B * A)
:= $R8006 &
:= $T8006 K8000c
:= $F8006 K8001c
<< K8006.r0t.txt
:= $R8006; := $T8006; := $F8006

:= K8007 %0


##
##  Q.E.D.
##

%0
```

2.2.98 File K8008.r0.txt

```
##
## Proof K8008:  ~ ~ A = A
##
##
## Source: [Kubota 2017 (doi: 10.4444/100.10)]
##
## Copyright (c) 2017 Owl of Minerva Press GmbH. All rights reserved.
## Written by Ken Kubota (<mail@kenkubota.de>).
##
## This file is part of the publication of the mathematical logic R0.
## For more information, visit: <http://doi.org/10.4444/100.10>
##

<< basics.r0.txt
<< A5231.r0.txt


##
## Proof
##

## use Proof Template A5222 (Rule of Cases):  [\x.A]T, [\x.A]F  -->  A
:= $L5222 [\a{o}{o}.((={{{o,o},o}}_(!{{o,o}}_(!{{o,o}}_a{o}{o}){o}){o}){{o,o}}_a{o}{
o}){o}]
:= $X5222 x{o}
:= $T5222 ($L5222{{o,o}}_T{o})
:= $F5222 ($L5222{{o,o}}_F{o})

## case T: ~ ~ T = T
§= $T5222
§\s /3
%A5231a
§s %1 27 %0

## use Proof Template A5201b (Swap):  A = B  -->  B = A
<< A5201b.r0t.txt
%0

%A5231b
§s %0 1 %1

## case F: ~ ~ F = F
§= $F5222
§\s /3
%A5231b
§s %1 27 %0
```

```
## use Proof Template A5201b (Swap):  A = B  -->  B = A
<< A5201b.r0t.txt
%0

%A5231a
§s %0 1 %1

<< A5222.r0t.txt
:= $L5222; := $X5222; := $T5222; := $F5222

:= K8008 %0


##
##  Q.E.D.
##

%0
```

2.2.99 File K8009.r0.txt

```
##
##  Proof K8009:  (A | T) = T;  (T | A) = T;  (A | T) = (T | A)
##
##
##  Source: [Kubota 2017 (doi: 10.4444/100.10)]
##
##  Copyright (c) 2017 Owl of Minerva Press GmbH. All rights reserved.
##  Written by Ken Kubota (<mail@kenkubota.de>).
##
##  This file is part of the publication of the mathematical logic R0.
##  For more information, visit: <http://doi.org/10.4444/100.10>
##

<< basics.r0.txt
<< A5231.r0.txt
<< K8001.r0.txt


##
##  Proof
##

## .a:  (A | T) = T

§= {o} ((|{{{o,o},o}}_x{o}{o}){{o,o}}_T{o})
§\s /6
§\s /3
%A5231a
```

```
§s %1 15 %0
:= $TMP8006 %0

## use Proof Template A5221 (Sub):  B  -->  B [x/A]
:= $B5221 %K8001a
:= $T5221 o
:= $X5221 x{$T5221}
:= $A5221 (!{{o,o}}_x{o}{o})
<< A5221.r0t.txt
:= $B5221; := $T5221; := $X5221; := $A5221
%0

%$TMP8006; := $TMP8006
§s %0 7 %1

%A5231b
§s %1 3 %0

:= K8009a %0

## .b:  (T | A) = T

§= {o} ((|{{{o,o},o}}_T{o}){{o,o}}_x{o}{o})
§\s /6
§\s /3
%A5231a
§s %1 29 %0
:= $TMP8006 %0

## use Proof Template A5221 (Sub):  B  -->  B [x/A]
:= $B5221 %K8001b
:= $T5221 o
:= $X5221 x{$T5221}
:= $A5221 (!{{o,o}}_x{o}{o})
<< A5221.r0t.txt
:= $B5221; := $T5221; := $X5221; := $A5221
%0

%$TMP8006; := $TMP8006
§s %0 7 %1

%A5231b
§s %1 3 %0

:= K8009b %0

## .c:  (A | T) = (T | A)

%K8009b
```

```
## use Proof Template A5201b (Swap):  A = B  -->  B = A
<< A5201b.r0t.txt
%0

%K8009a
§s %0 3 %1

:= K8009c %0


##
##  Q.E.D.
##

## %K8009a
%K8009a

## %K8009b
%K8009b

## %K8009c
%K8009c
```

2.2.100 File K8010.r0.txt

```
##
##  Proof K8010:  (A | F) = A;  (F | A) = A;  (A | F) = (F | A)
##
##
##  Source: [Kubota 2017 (doi: 10.4444/100.10)]
##
##  Copyright (c) 2017 Owl of Minerva Press GmbH. All rights reserved.
##  Written by Ken Kubota (<mail@kenkubota.de>).
##
##  This file is part of the publication of the mathematical logic R0.
##  For more information, visit: <http://doi.org/10.4444/100.10>
##

<< A5231.r0.txt
<< K8000.r0.txt
<< K8008.r0.txt


##
##  Proof
##

## .a:  (A | F) = A
```

```
§= {o} ((|{{{o,o},o}}_x{o}{o}){{o,o}}_F{o})
§\s /6
§\s /3
%A5231b
§s %1 15 %0
:= $TMP8010 %0

## use Proof Template A5221 (Sub):  B  -->  B [x/A]
:= $B5221 %K8000a
:= $T5221 o
:= $X5221 x{$T5221}
:= $A5221 (!{{o,o}}_x{o}{o})
<< A5221.r0t.txt
:= $B5221; := $T5221; := $X5221; := $A5221
%0

%$TMP8010; := $TMP8010
§s %0 7 %1

%K8008
§s %1 3 %0

:= K8010a %0

## .b:  (F | A) = A

§= {o} ((|{{{o,o},o}}_F{o}){{o,o}}_x{o}{o})
§\s /6
§\s /3
%A5231b
§s %1 29 %0
:= $TMP8010 %0

## use Proof Template A5216:  (T & A) = A
:= $A5216 (!{{o,o}}_x{o}{o})
<< A5216.r0t.txt
:= $A5216
%0

%$TMP8010; := $TMP8010
§s %0 7 %1

%K8008
§s %1 3 %0

:= K8010b %0

## .c:  (A | F) = (F | A)
```

```
%K8010b

## use Proof Template A5201b (Swap):  A = B  -->  B = A
<< A5201b.r0t.txt
%0

%K8010a
§s %0 3 %1

:= K8010c %0


##
##  Q.E.D.
##

## %K8010a
%K8010a

## %K8010b
%K8010b

## %K8010c
%K8010c
```

2.2.101 File K8011.r0.txt

```
##
##  Proof K8011:  (A | A) = A
##
##
##  Source: [Kubota 2017 (doi: 10.4444/100.10)]
##
##  Copyright (c) 2017 Owl of Minerva Press GmbH. All rights reserved.
##  Written by Ken Kubota (<mail@kenkubota.de>).
##
##  This file is part of the publication of the mathematical logic R0.
##  For more information, visit: <http://doi.org/10.4444/100.10>
##

<< A5232.r0.txt


##
##  Proof
##

## use Proof Template A5222 (Rule of Cases):  [\x.A]T, [\x.A]F  -->  A
```

```
:= $L5222 [\x{o}{o}.((={{{o,o},o}}_((|{{{o,o},o}}_x{o}{o}){{o,o}}_x{o}{o}){o}){{o,o}
}_x{o}{o}){o}]
:= $X5222 x{o}
:= $T5222 ($L5222{{o,o}}_T{o})
:= $F5222 ($L5222{{o,o}}_F{o})

## case T:  (T | T) = T
§\ $T5222

## use Proof Template A5201b (Swap):  A = B  -->  B = A
<< A5201b.r0t.txt
%0

%A5232a
§s %0 1 %1

## case F:  (F | F) = F
§\ $F5222

## use Proof Template A5201b (Swap):  A = B  -->  B = A
<< A5201b.r0t.txt
%0

%A5232d
§s %0 1 %1

<< A5222.r0t.txt
:= $L5222; := $X5222; := $T5222; := $F5222

:= K8011 %0


##
##  Q.E.D.
##

%0
```

2.2.102 File K8012.r0.txt

```
##
##  Proof K8012:  (A | B) = (B | A)
##
##
##  Source: [Kubota 2017 (doi: 10.4444/100.10)]
##
##  Copyright (c) 2017 Owl of Minerva Press GmbH. All rights reserved.
##  Written by Ken Kubota (<mail@kenkubota.de>).
##
```

```
##  This file is part of the publication of the mathematical logic R0.
##  For more information, visit: <http://doi.org/10.4444/100.10>
##

<< K8009.r0.txt
<< K8010.r0.txt


##
##  Proof
##

## use Proof Template K8006:  (A * T) = (T * A), (A * F) = (F * A)  -->  (A * B) = (
B * A)
:= $R8006 |
:= $T8006 K8009c
:= $F8006 K8010c
<< K8006.r0t.txt
:= $R8006; := $T8006; := $F8006

:= K8012 %0


##
##  Q.E.D.
##

%0
```

2.2.103 File K8013.r0a.txt

```
##
##  Proof Template K8013:  A => B, B => A  -->  A = B
##
##
##  Source: [Kubota 2017 (doi: 10.4444/100.10)]
##
##  Copyright (c) 2017 Owl of Minerva Press GmbH. All rights reserved.
##  Written by Ken Kubota (<mail@kenkubota.de>).
##
##  This file is part of the publication of the mathematical logic R0.
##  For more information, visit: <http://doi.org/10.4444/100.10>
##


##
##  Define Syntactical Variables
##
```

```
## proposition A
:= $A8013 x{o}

## proposition B
:= $B8013 y{o}


##
##  Assumptions and Resulting Syntactical Variables
##

<< basics.r0.txt

§! ((=>{{{o,o},o}}_$A8013{o}){{o,o}}_$B8013{o})
§! ((=>{{{o,o},o}}_$B8013{o}){{o,o}}_$A8013{o})


##
##  Proof Template
##

<<< K8013.r0t.txt


##
##  Undefine Syntactical Variables
##

:= $A8013; := $B8013


##
##  Q.E.D.
##

%0
```

2.2.104 File K8013.r0t.txt

```
##
##  Proof Template K8013:  A => B, B => A  -->  A = B
##
##
##  Source: [Kubota 2017 (doi: 10.4444/100.10)]
##
##  Copyright (c) 2017 Owl of Minerva Press GmbH. All rights reserved.
##  Written by Ken Kubota (<mail@kenkubota.de>).
##
##  This file is part of the publication of the mathematical logic R0.
```

```
##  For more information, visit: <http://doi.org/10.4444/100.10>
##

<< K8007.r0.txt


##
##  Proof Template
##

## .1

## assumption 1
:= $HTMP8013 ((=>{{{o,o},o}}_$A8013{o}){{o,o}}_$B8013{o})

## assumption 2
:= $ITMP8013 ((=>{{{o,o},o}}_$B8013{o}){{o,o}}_$A8013{o})

%K8007
§\s /2
§\s /1

## .2

## use Proof Template A5221 (Sub):  B  -->  B [x/A]
:= $B5221 %0
:= $T5221 o
:= $X5221 y{o}
:= $A5221 ytmp{o}
<< A5221.r0t.txt
:= $B5221; := $T5221; := $X5221; := $A5221
%0

## use Proof Template A5221 (Sub):  B  -->  B [x/A]
:= $B5221 %0
:= $T5221 o
:= $X5221 x{o}
:= $A5221 $B8013
<< A5221.r0t.txt
:= $B5221; := $T5221; := $X5221; := $A5221
%0

## use Proof Template A5221 (Sub):  B  -->  B [x/A]
:= $B5221 %0
:= $T5221 o
:= $X5221 ytmp{o}
:= $A5221 $A8013
<< A5221.r0t.txt
:= $B5221; := $T5221; := $X5221; := $A5221
```

```
%0

## .3

%$ITMP8013
§rs /9 ytmp{o}
§\s /2
§\s /1
§s %0 3 %7

## use Proof Template A5201b (Swap):  A = B  -->  B = A
<< A5201b.r0t.txt
%0

## .4

%$HTMP8013
§rs /9 ytmp{o}
§\s /2
§\s /1
§s %0 3 %7

## undefine local variables
:= $HTMP8013; := $ITMP8013
```

2.2.105 File K8013H.r0a.txt

```
##
##  Proof Template K8013H:  H => (A => B), H => (B => A)  -->  H => (A = B)
##
##
##  Source: [Kubota 2017 (doi: 10.4444/100.10)]
##
##  Copyright (c) 2017 Owl of Minerva Press GmbH. All rights reserved.
##  Written by Ken Kubota (<mail@kenkubota.de>).
##
##  This file is part of the publication of the mathematical logic R0.
##  For more information, visit: <http://doi.org/10.4444/100.10>
##



##
##  Define Syntactical Variables
##

## hypotheses H
:= $H8013H h{o}

## proposition A
```

```
:= $A8013H x{o}

## proposition B
:= $B8013H y{o}


##
##  Assumptions and Resulting Syntactical Variables
##

<< basics.r0.txt

§! ((=>{{{o,o},o}}_$H8013H{o}){{o,o}}_((=>{{{o,o},o}}_$A8013H{o}){{o,o}}_$B8013H{o})
{o})
§! ((=>{{{o,o},o}}_$H8013H{o}){{o,o}}_((=>{{{o,o},o}}_$B8013H{o}){{o,o}}_$A8013H{o})
{o})


##
##  Proof Template
##

<<< K8013H.r0t.txt


##
##  Undefine Syntactical Variables
##

:= $H8013H; := $A8013H; := $B8013H


##
##  Q.E.D.
##

%0
```

2.2.106 File K8013H.r0t.txt

```
##
##  Proof Template K8013H:  H => (A => B), H => (B => A)  -->  H => (A = B)
##
##
##  Source: [Kubota 2017 (doi: 10.4444/100.10)]
##
##  Copyright (c) 2017 Owl of Minerva Press GmbH. All rights reserved.
##  Written by Ken Kubota (<mail@kenkubota.de>).
##
```

```
##  This file is part of the publication of the mathematical logic R0.
##  For more information, visit: <http://doi.org/10.4444/100.10>
##

<< K8007.r0.txt


##
##  Proof Template
##

## .1

## assumption 1
:= $HTMP8013H ((=>{{{o,o},o}}_$H8013H{o}){{o,o}}_((=>{{{o,o},o}}_$A8013H{o}){{o,o}}_
$B8013H{o}){o})

## assumption 2
:= $ITMP8013H ((=>{{{o,o},o}}_$H8013H{o}){{o,o}}_((=>{{{o,o},o}}_$B8013H{o}){{o,o}}_
$A8013H{o}){o})

%K8007
§\s /2
§\s /1

## .2

## use Proof Template A5221 (Sub):  B  -->  B [x/A]
:= $B5221 %0
:= $T5221 o
:= $X5221 y{o}
:= $A5221 ytmp{o}
<< A5221.r0t.txt
:= $B5221; := $T5221; := $X5221; := $A5221
%0

## use Proof Template A5221 (Sub):  B  -->  B [x/A]
:= $B5221 %0
:= $T5221 o
:= $X5221 x{o}
:= $A5221 $B8013H
<< A5221.r0t.txt
:= $B5221; := $T5221; := $X5221; := $A5221
%0

## use Proof Template A5221 (Sub):  B  -->  B [x/A]
:= $B5221 %0
:= $T5221 o
:= $X5221 ytmp{o}
```

```
:= $A5221 $A8013H
<< A5221.r0t.txt
:= $B5221; := $T5221; := $X5221; := $A5221
%0

## .3

%$ITMP8013H
§rs /25 ytmp{o}
§\s /6
§\s /3
§s %0 7 %7

## use Proof Template A5201bH (SwapH):  H => (A = B)  -->  H => (B = A)
<< A5201bH.r0t.txt
%0

## .4

%$HTMP8013H
§rs /25 ytmp{o}
§\s /6
§\s /3
§s' %0 3 %7

## undefine local variables
:= $HTMP8013H; := $ITMP8013H
```

2.2.107 File K8014.r0.txt

```
##
##  Proof K8014:  (x = y)  =  (y = x)
##      for any x, y of any type
##
##  Source: [cf. Andrews 2002 (ISBN 1-4020-0763-9), pp. 232 f. (5302)]
##
##  Copyright (c) 2017 Owl of Minerva Press GmbH. All rights reserved.
##  Written by Ken Kubota (<mail@kenkubota.de>).
##
##  This file is part of the publication of the mathematical logic R0.
##  For more information, visit: <http://doi.org/10.4444/100.10>
##

<< basics.r0.txt
<< K8005.r0.txt


##
##  Proof
```

```
##

## .1

%K8005

## use Proof Template A5221 (Sub):  B  -->  B [x/A]
:= $B5221 %0
:= $T5221 o
:= $X5221 x{$T5221}
:= $A5221 ((={{{o,t{^}},t{^}}}_x{t{^}}{t{^}}){{o,t{^}}}_y{t{^}}{t{^}})
<< A5221.r0t.txt
:= $B5221; := $T5221; := $X5221; := $A5221
%0

## use Proof Template A5201bH (SwapH):  H => (A = B)  -->  H => (B = A)
<< A5201bH.r0t.txt
%0

## .2

%K8005

## use Proof Template A5221 (Sub):  B  -->  B [x/A]
:= $B5221 %0
:= $T5221 o
:= $X5221 x{o}
:= $A5221 ((={{{o,t{^}},t{^}}}_y{t{^}}{t{^}}){{o,t{^}}}_x{t{^}}{t{^}})
<< A5221.r0t.txt
:= $B5221; := $T5221; := $X5221; := $A5221
%0

## use Proof Template A5201bH (SwapH):  H => (A = B)  -->  H => (B = A)
<< A5201bH.r0t.txt
%0

## .3

## use Proof Template K8013:  A => B, B => A  -->  A = B
:= $A8013 ((={{{o,t{^}},t{^}}}_x{t{^}}{t{^}}){{o,t{^}}}_y{t{^}}{t{^}})
:= $B8013 ((={{{o,t{^}},t{^}}}_y{t{^}}{t{^}}){{o,t{^}}}_x{t{^}}{t{^}})
<< K8013.r0t.txt
:= $A8013; := $B8013
%0

:= K8014 %0


##
```

```
## Q.E.D.
##

%0
```

2.2.108 File K8015.r0.txt

```
##
## Proof K8015:  (A => F)  =  (~ A)
##     (Proof by Contradiction)
##
## Source: [Kubota 2017 (doi: 10.4444/100.10)]
##
## Copyright (c) 2017 Owl of Minerva Press GmbH. All rights reserved.
## Written by Ken Kubota (<mail@kenkubota.de>).
##
## This file is part of the publication of the mathematical logic R0.
## For more information, visit: <http://doi.org/10.4444/100.10>
##

<< K8014.r0.txt


##
## Proof
##

§= {o} ((=>{{{o,o},o}}_x{o}{o}){{o,o}}_F{o})
§\s /6
§\s /3

%K8001a
§s %1 7 %0
:= $TMP8015 %0

%K8014

## use Proof Template A5221 (Sub):  B  -->  B [x/A]
:= $B5221 %0
:= $T5221 ^
:= $X5221 t{$T5221}
:= $A5221 o
<< A5221.r0t.txt
:= $B5221; := $T5221; := $X5221; := $A5221
%0

## use Proof Template A5221 (Sub):  B  -->  B [x/A]
:= $B5221 %0
:= $T5221 o
```

```
:= $X5221 y{$T5221}
:= $A5221 F
<< A5221.r0t.txt
:= $B5221; := $T5221; := $X5221; := $A5221
%0

%$TMP8015; := $TMP8015
§s %0 3 %1

§\ (!{{o,o}}_x{o}{o})

## use Proof Template A5201b (Swap):  A = B  -->  B = A
<< A5201b.r0t.txt
%0

§s %4 3 %0

:= K8015 %0


##
##  Q.E.D.
##

%0
```

2.2.109 File K8016.r0.txt

```
##
##  Proof K8016:  ALL x: Px  =  ~ EXI x: ~ Px
##
##
##  Source: [Kubota 2017 (doi: 10.4444/100.10)]
##
##  Copyright (c) 2017 Owl of Minerva Press GmbH. All rights reserved.
##  Written by Ken Kubota (<mail@kenkubota.de>).
##
##  This file is part of the publication of the mathematical logic R0.
##  For more information, visit: <http://doi.org/10.4444/100.10>
##

<< basics.r0.txt
<< A5205.r0.txt
<< K8008.r0.txt


##
##  Proof
##
```

```
## shorthands
:= $NE [\t{^}{^}.[\p{{o,t{^}}}{{o,t{^}}}.(!{{o,o}}_((E{{{o,{o,\3{^}}},^}}_t{^}{^}){{
o,{o,t{^}}}}_[\x{t{^}}{t{^}}.(!{{o,o}}_(p{{o,t{^}}}{{o,t{^}}}_x{t{^}}{t{^}}){o}){o}]
{{o,t{^}}}){o}){o}]{{o,{o,t{^}}}}]
:= $DN (!{{o,o}}_(!{{o,o}}_((={{{o,{o,t{^}}},{o,t{^}}}}_[\x{t{^}}{t{^}}.T{o}]{{o,t{^
}}}){{o,{o,t{^}}}}_[\x{t{^}}{t{^}}.(!{{o,o}}_([\x{t{^}}{t{^}}.(!{{o,o}}_(p{{o,t{^}}}
{{o,t{^}}}_x{t{^}}{t{^}}){o}){o}]{{o,t{^}}}_x{t{^}}{t{^}}){o}){o}]{{o,t{^}}}){o}){o}
)
:= $LT [\t{^}{^}.[\p{{o,t{^}}}{{o,t{^}}}.((={{{o,{o,t{^}}},{o,t{^}}}}_[\x{t{^}}{t{^}
}.T{o}]{{o,t{^}}}){{o,{o,t{^}}}}_[\x{t{^}}{t{^}}.(!{{o,o}}_(!{{o,o}}_(p{{o,t{^}}}{{o
,t{^}}}_x{t{^}}{t{^}}){o}){o}){o}]{{o,t{^}}}){o}]{{o,{o,t{^}}}}]

## .1

§= {{o,{o,\3{^}}},^} [\t{^}{^}.[\p{{o,t{^}}}{{o,t{^}}}.(!{{o,o}}_((E{{{o,{o,\3{^}}},
^}}_t{^}{^}){{o,{o,t{^}}}}_[\x{t{^}}{t{^}}.(!{{o,o}}_(p{{o,t{^}}}{{o,t{^}}}_x{t{^}}{
t{^}}){o}){o}]{{o,t{^}}}){o}){o}]{{o,{o,t{^}}}}]
§\s /94
§\s /47
:= $TMP8016 %0

## .2

%K8008

## use Proof Template A5221 (Sub):  B  -->  B [x/A]
:= $B5221 %0
:= $T5221 o
:= $X5221 x{$T5221}
:= $A5221 ((={{{o,{o,t{^}}},{o,t{^}}}}_[\x{t{^}}{t{^}}.T{o}]{{o,t{^}}}){{o,{o,t{^}}}
}_[\x{t{^}}{t{^}}.(!{{o,o}}_([\x{t{^}}{t{^}}.(!{{o,o}}_(p{{o,t{^}}}{{o,t{^}}}_x{t{^}
}{t{^}}){o}){o}]{{o,t{^}}}_x{t{^}}{t{^}}){o}){o}]{{o,t{^}}})
<< A5221.r0t.txt
:= $B5221; := $T5221; := $X5221; := $A5221
%0

%$TMP8016; := $TMP8016
§s %0 23 %1
§\s /191
:= $TMP8016 %0

## .3

%K8008

## use Proof Template A5221 (Sub):  B  -->  B [x/A]
:= $B5221 %0
:= $T5221 o
```

```
:= $X5221 x{$T5221}
:= $A5221 (p{{o,t{^}}}{{o,t{^}}}_x{t{^}}{t{^}})
<< A5221.r0t.txt
:= $B5221; := $T5221; := $X5221; := $A5221
%0

%$TMP8016; := $TMP8016
§s %0 95 %1
:= $TMP8016 %0

## .4

## use Proof Template:  A5205 Substitutions
:= $AA5205 o
:= $BA5205 t{^}
:= $FA5205 p{{$AA5205,$BA5205}}
<< a5205_substitutions.r0t.txt
:= $AA5205; := $BA5205; := $FA5205
%0

## .5

## use Proof Template A5201b (Swap):  A = B  -->  B = A
<< A5201b.r0t.txt
%0

§rs /5 x{t{^}}

%$TMP8016; := $TMP8016
§s %0 47 %1

:= K8016 %0

## undefine local variables
:= $NE; := $DN; := $LT


##
##  Q.E.D.
##

%0
```

2.2.110 File K8017.r0.txt

```
##
##  Proof K8017:  EXI x: Px  =  ~ ALL x: ~ Px
##
##
```

```
##  Source: [Kubota 2017 (doi: 10.4444/100.10)]
##
##  Copyright (c) 2017 Owl of Minerva Press GmbH. All rights reserved.
##  Written by Ken Kubota (<mail@kenkubota.de>).
##
##  This file is part of the publication of the mathematical logic R0.
##  For more information, visit: <http://doi.org/10.4444/100.10>
##

<< basics.r0.txt


##
##  Proof
##

§= {{o,{o,\3{^}}},^} [\t{^}{^}.[\p{{o,t{^}}}{{o,t{^}}}.(!{{o,o}}_((A{{{o,{o,\3{^}}},
^}}_t{^}{^}){{o,{o,t{^}}}}_[\x{t{^}}{t{^}}.(!{{o,o}}_(p{{o,t{^}}}{{o,t{^}}}_x{t{^}}{
t{^}}){o}){o}]{{o,t{^}}}){o}){o}]{{o,{o,t{^}}}}]
§\s /94
§\s /47

:= K8017 %0


##
##  Q.E.D.
##

%0
```

2.2.111 File K8018.r0.txt

```
##
##  Proof K8018:  (A & B)  =  ~ ( (~ A) | (~ B) )
##
##
##  Source: [Kubota 2017 (doi: 10.4444/100.10)]
##
##  Copyright (c) 2017 Owl of Minerva Press GmbH. All rights reserved.
##  Written by Ken Kubota (<mail@kenkubota.de>).
##
##  This file is part of the publication of the mathematical logic R0.
##  For more information, visit: <http://doi.org/10.4444/100.10>
##

<< basics.r0.txt
<< A5205.r0.txt
<< K8008.r0.txt
```

```
##
## Proof
##

## .1

§= {{o,o},o} [\a{o}{o}.[\b{o}{o}.(!{{o,o}}_((|{{{o,o},o}}_(!{{o,o}}_a{o}{o}){o}){{o,
o}}_(!{{o,o}}_b{o}{o}){o}){o}){o}]{{o,o}}]
§\s /62
§\s /31
:= $TMP8018 %0

## .2

%K8008

## use Proof Template A5221 (Sub):  B  -->  B [x/A]
:= $B5221 %0
:= $T5221 o
:= $X5221 x{$T5221}
:= $A5221 a{$T5221}
<< A5221.r0t.txt
:= $B5221; := $T5221; := $X5221; := $A5221
%0

%$TMP8018; := $TMP8018
§s %0 253 %1
:= $TMP8018 %0

## .3

%K8008

## use Proof Template A5221 (Sub):  B  -->  B [x/A]
:= $B5221 %0
:= $T5221 o
:= $X5221 x{$T5221}
:= $A5221 b{$T5221}
<< A5221.r0t.txt
:= $B5221; := $T5221; := $X5221; := $A5221
%0

%$TMP8018; := $TMP8018
§s %0 127 %1
:= $TMP8018 %0

## .4
```

```
%K8008

## use Proof Template A5221 (Sub):  B  -->  B [x/A]
:= $B5221 %0
:= $T5221 o
:= $X5221 x{$T5221}
:= $A5221 ((&{{{o,o},o}}_a{o}{o}){{o,o}}_b{o}{o})
<< A5221.r0t.txt
:= $B5221; := $T5221; := $X5221; := $A5221
%0

%$TMP8018; := $TMP8018
§s %0 15 %1
:= $TMP8018 %0

## .5

## use Proof Template:  A5205 Substitutions
:= $AA5205 o
:= $BA5205 o
:= $FA5205 (&{{{o,o},o}}_a{o}{o})
<< a5205_substitutions.r0t.txt
:= $AA5205; := $BA5205; := $FA5205
%0

§rs /3 b{o}

## use Proof Template A5201b (Swap):  A = B  -->  B = A
<< A5201b.r0t.txt
%0

%$TMP8018; := $TMP8018
§s %0 7 %1
:= $TMP8018 %0

## .6

## use Proof Template:  A5205 Substitutions
:= $AA5205 {o,o}
:= $BA5205 o
:= $FA5205 &
<< a5205_substitutions.r0t.txt
:= $AA5205; := $BA5205; := $FA5205
%0

§rs /3 a{o}

## use Proof Template A5201b (Swap):  A = B  -->  B = A
```

```
<< A5201b.r0t.txt
%0

%$TMP8018; := $TMP8018
§s %0 3 %1

## use Proof Template A5201b (Swap):  A = B  -->  B = A
<< A5201b.r0t.txt
%0

:= K8018 %0


##
##  Q.E.D.
##

%0
```

2.2.112 File K8019.r0a.txt

```
##
##  Proof Template K8019:  A & B  -->  A, B
##
##
##  Source: [Kubota 2017 (doi: 10.4444/100.10)]
##
##  Copyright (c) 2017 Owl of Minerva Press GmbH. All rights reserved.
##  Written by Ken Kubota (<mail@kenkubota.de>).
##
##  This file is part of the publication of the mathematical logic R0.
##  For more information, visit: <http://doi.org/10.4444/100.10>
##



##
##  Define Syntactical Variables
##

<< basics.r0.txt

## the assumption:  (A & B)
:= $H8019 ((&{{o,o},o}}_x{o}{o}){o,o}}_y{o}{o})


##
##  Assumptions and Resulting Syntactical Variables
##
```

```
§! $H8019


##
##  Include Proof Template
##

<<< K8019.r0t.txt


##
##  Undefine Syntactical Variables
##

:= $H8019


##
##  Q.E.D.
##

%$A8019
%$B8019


##
##  Undefine Results
##

:= $A8019; := $B8019
```

2.2.113 File K8019.r0t.txt

```
##
##  Proof Template K8019:  A & B  -->  A, B
##
##
##  Source: [Kubota 2017 (doi: 10.4444/100.10)]
##
##  Copyright (c) 2017 Owl of Minerva Press GmbH. All rights reserved.
##  Written by Ken Kubota (<mail@kenkubota.de>).
##
##  This file is part of the publication of the mathematical logic R0.
##  For more information, visit: <http://doi.org/10.4444/100.10>
##

<< A5200t.r0.txt
```

```
##
##  Proof Template
##

## .1

%$H8019
§\s /2
§\s /1
§= ([\g{{{o,o},o}}{{{o,o},o}}.((g{{{o,o},o}}{{{o,o},o}}_T{o}){{o,o}}_T{o}){o}]{{o,{{
o,o},o}}}_[\x{o}{o}.[\y{o}{o}.x{o}{o}]{{o,o}}]{{{o,o},o}})
§s %0 6 %1
§\s /5
§\s /3
§\s /10
§\s /6
§\s /5
§\s /3
%A5200t
§s %0 1 %1
:= $A8019 %0

## .2

%$H8019
§\s /2
§\s /1
§= ([\g{{{o,o},o}}{{{o,o},o}}.((g{{{o,o},o}}{{{o,o},o}}_T{o}){{o,o}}_T{o}){o}]{{o,{{
o,o},o}}}_[\x{o}{o}.[\y{o}{o}.y{o}{o}]{{o,o}}]{{{o,o},o}})
§s %0 6 %1
§rs /15 z{o}
§\s /5
§\s /3
§\s /10
§\s /6
§\s /5
§\s /3
%A5200t
§s %0 1 %1
:= $B8019 %0
```

2.2.114 File K8019H.r0a.txt

```
##
##  Proof Template K8019H:  H => (A & B)  -->  H => A, H => B
##
##
##  Source: [Kubota 2017 (doi: 10.4444/100.10)]
##
```

```
##  Copyright (c) 2017 Owl of Minerva Press GmbH. All rights reserved.
##  Written by Ken Kubota (<mail@kenkubota.de>).
##
##  This file is part of the publication of the mathematical logic R0.
##  For more information, visit: <http://doi.org/10.4444/100.10>
##



##
##  Define Syntactical Variables
##

<< basics.r0.txt

## the assumption:  H => (A & B)
:= $H8019H ((=>{{{o,o},o}}_h{o}{o}){{o,o}}_((&{{{o,o},o}}_x{o}{o}){{o,o}}_y{o}{o}){o
})



##
##  Assumptions and Resulting Syntactical Variables
##

§! $H8019H



##
##  Include Proof Template
##

<<< K8019H.r0t.txt



##
##  Undefine Syntactical Variables
##

:= $H8019H



##
##  Q.E.D.
##

%$A8019H
%$B8019H



##
```

```
##  Undefine Results
##

:= $A8019H; := $B8019H
```

2.2.115 File K8019H.r0t.txt

```
##
##  Proof Template K8019H:  H => (A & B)  -->  H => A, H => B
##
##
##  Source: [Kubota 2017 (doi: 10.4444/100.10)]
##
##  Copyright (c) 2017 Owl of Minerva Press GmbH. All rights reserved.
##  Written by Ken Kubota (<mail@kenkubota.de>).
##
##  This file is part of the publication of the mathematical logic R0.
##  For more information, visit: <http://doi.org/10.4444/100.10>
##

<< A5200t.r0.txt


##
##  Proof Template
##

## .1:  H => T

%A5200t

## use Proof Template K8003 (Intro):  A  -->  H => A
:= $A8003 %0
:= $H8003 $H8019H/5
<< K8003.r0t.txt
:= $A8003; := $H8003

:= $TTMP8019H %0

## .2:  H => A

%$H8019H
§\s' /2
§\s' /1
:= $TMP8019H %0

§= ([\g{{{o,o},o}}{{{o,o},o}}.((g{{{o,o},o}}{{{o,o},o}}_T{o}){{o,o}}_T{o}){o}]{{o,{{
o,o},o}}}_[\x{o}{o}.[\y{o}{o}.x{o}{o}]{{o,o}}]{{{o,o},o}})
```

```
## use Proof Template K8003 (Intro):  A  -->  H => A
:= $A8003 %0
:= $H8003 $H8019H/5
<< K8003.r0t.txt
:= $A8003; := $H8003
%0

%$TMP8019H; := $TMP8019H
§s' %1 6 %0
§\s' /5
§\s' /3
§\s' /10
§\s' /6
§\s' /5
§\s' /3

%$TTMP8019H
§s' %0 1 %1

:= $A8019H %0

## .3:  H => B

%$H8019H
§\s' /2
§\s' /1
:= $TMP8019H %0

§= ([\g{{{o,o},o}}{{{o,o},o}}.((g{{{o,o},o}}{{{o,o},o}}_T{o}){{o,o}}_T{o}){o}]{{o,{{
o,o},o}}}_[\x{o}{o}.[\y{o}{o}.y{o}{o}]{{o,o}}]{{{o,o},o}})

## use Proof Template K8003 (Intro):  A  -->  H => A
:= $A8003 %0
:= $H8003 $H8019H/5
<< K8003.r0t.txt
:= $A8003; := $H8003
%0

%$TMP8019H; := $TMP8019H
§s' %1 6 %0
§\s' /5
§\s' /3
§\s' /10
§\s' /6
§\s' /5
§\s' /3

%$TTMP8019H
§s' %0 1 %1
```

```
:= $B8019H %0

## undefine local variables
:= $TTMP8019H
```

2.2.116 File K8020.r0a.txt

```
##
##  Proof Template K8020:  A, B  -->  A & B
##
##
##  Source: [Kubota 2017 (doi: 10.4444/100.10)]
##
##  Copyright (c) 2017 Owl of Minerva Press GmbH. All rights reserved.
##  Written by Ken Kubota (<mail@kenkubota.de>).
##
##  This file is part of the publication of the mathematical logic R0.
##  For more information, visit: <http://doi.org/10.4444/100.10>
##


##
##  Define Syntactical Variables
##

## assumption 1
:= $A8020 x{o}

## assumption 2
:= $B8020 y{o}


##
##  Assumptions and Resulting Syntactical Variables
##

§! $A8020
§! $B8020


##
##  Include Proof Template
##


<<< K8020.r0t.txt
```

```
##
##  Undefine Syntactical Variables
##

:= $A8020; := $B8020


##
##  Q.E.D.
##

%0
```

2.2.117 File K8020.r0t.txt

```
##
##  Proof Template K8020:  A, B  -->  A & B
##
##
##  Source: [Kubota 2017 (doi: 10.4444/100.10)]
##
##  Copyright (c) 2017 Owl of Minerva Press GmbH. All rights reserved.
##  Written by Ken Kubota (<mail@kenkubota.de>).
##
##  This file is part of the publication of the mathematical logic R0.
##  For more information, visit: <http://doi.org/10.4444/100.10>
##

<< A5212.r0.txt


##
##  Proof Template
##

## .1

%A5212

:= $TTMP8020 %0

## .2

%$A8020

## use Proof Template A5219a (Rule T):  A  -->  T = A
:= $A5219a %0
<< A5219a.r0t.txt
:= $A5219a
```

```
:= $ATMP8020 %0

## .3

%$B8020

## use Proof Template A5219a (Rule T):  A  -->  T = A
:= $A5219a %0
<< A5219a.r0t.txt
:= $A5219a

:= $BTMP8020 %0

## .4

%$TTMP8020; := $TTMP8020
%$ATMP8020; := $ATMP8020
§s %1 5 %0
%$BTMP8020; := $BTMP8020
§s %1 3 %0
```

2.2.118 File K8020H.r0a.txt

```
##
##  Proof Template K8020H:  H => A, H => B  -->  H => (A & B)
##
##
##  Source: [Kubota 2017 (doi: 10.4444/100.10)]
##
##  Copyright (c) 2017 Owl of Minerva Press GmbH. All rights reserved.
##  Written by Ken Kubota (<mail@kenkubota.de>).
##
##  This file is part of the publication of the mathematical logic R0.
##  For more information, visit: <http://doi.org/10.4444/100.10>
##



##
##  Define Syntactical Variables
##

<< basics.r0.txt

## assumption 1
:= $A8020H ((=>{{{o,o},o}}_h{o}{o}){{o,o}}_x{o}{o})

## assumption 2
:= $B8020H ((=>{{{o,o},o}}_h{o}{o}){{o,o}}_y{o}{o})
```

```
##
##  Assumptions and Resulting Syntactical Variables
##

§! $A8020H
§! $B8020H


##
##  Include Proof Template
##

<<< K8020H.r0t.txt


##
##  Undefine Syntactical Variables
##

:= $A8020H; := $B8020H


##
##  Q.E.D.
##

%0
```

2.2.119 File K8020H.r0t.txt

```
##
##  Proof Template K8020H:  H => A, H => B  -->  H => (A & B)
##
##
##  Source: [Kubota 2017 (doi: 10.4444/100.10)]
##
##  Copyright (c) 2017 Owl of Minerva Press GmbH. All rights reserved.
##  Written by Ken Kubota (<mail@kenkubota.de>).
##
##  This file is part of the publication of the mathematical logic R0.
##  For more information, visit: <http://doi.org/10.4444/100.10>
##

<< A5212.r0.txt


##
```

```
##  Proof Template
##

## .1

%A5212

## use Proof Template K8003 (Intro):  A  -->  H => A
:= $A8003 %0
:= $H8003 $A8020H/5
<< K8003.r0t.txt
:= $A8003; := $H8003

:= $TTMP8020H %0

## .2

%$A8020H

## use Proof Template A5219aH (Rule T):  H => A  -->  H => (T = A)
:= $A5219aH %0
<< A5219aH.r0t.txt
:= $A5219aH

:= $ATMP8020H %0

## .3

%$B8020H

## use Proof Template A5219aH (Rule T):  H => A  -->  H => (T = A)
:= $A5219aH %0
<< A5219aH.r0t.txt
:= $A5219aH

:= $BTMP8020H %0

## .4

%$TTMP8020H; := $TTMP8020H
%$ATMP8020H; := $ATMP8020H
§s' %1 5 %0
%$BTMP8020H; := $BTMP8020H
§s' %1 3 %0
```

2.2.120 File K8021.r0.txt

```
##
##  Proof K8021:  (A & B) & C  =  A & (B & C)
```

```
##
##
##  Source: [Kubota 2017 (doi: 10.4444/100.10)]
##
##  Copyright (c) 2017 Owl of Minerva Press GmbH. All rights reserved.
##  Written by Ken Kubota (<mail@kenkubota.de>).
##
##  This file is part of the publication of the mathematical logic R0.
##  For more information, visit: <http://doi.org/10.4444/100.10>
##

<< basics.r0.txt
<< K8005.r0.txt


##
##  Proof
##

## .1a

%K8005

## use Proof Template A5221 (Sub):  B  -->  B [x/A]
:= $B5221 %0
:= $T5221 o
:= $X5221 x{$T5221}
:= $A5221 ((&{{{o,o},o}}_((&{{{o,o},o}}_a{o}{o}){{o,o}}_b{o}{o}){o}){{o,o}}_c{o}{o})
<< A5221.r0t.txt
:= $B5221; := $T5221; := $X5221; := $A5221
%0

## .1b

## use Proof Template K8019H:  H => (A & B)  -->  H => A, H => B
:= $H8019H %0
<< K8019H.r0t.txt
:= $H8019H
:= $ABTMP8021 $A8019H
:= $CTMP8021 $B8019H
:= $A8019H; := $B8019H
%0

## .1c

%$ABTMP8021

## use Proof Template K8019H:  H => (A & B)  -->  H => A, H => B
:= $H8019H %0
```

```
<< K8019H.r0t.txt
:= $H8019H
:= $ATMP8021 $A8019H
:= $BTMP8021 $B8019H
:= $A8019H; := $B8019H
%0

:= $ABTMP8021

## .1d

%$BTMP8021; := $BTMP8021
%$CTMP8021; := $CTMP8021

## use Proof Template K8020H:  H => A, H => B  -->  H => (A & B)
:= $A8020H %1
:= $B8020H %0
<< K8020H.r0t.txt
:= $A8020H; := $B8020H

:= $BCTMP8020 %0

%$ATMP8021; := $ATMP8021
%$BCTMP8020; := $BCTMP8020

## use Proof Template K8020H:  H => A, H => B  -->  H => (A & B)
:= $A8020H %1
:= $B8020H %0
<< K8020H.r0t.txt
:= $A8020H; := $B8020H

:= $ABC1TMP8020 %0

## .2a

%K8005

## use Proof Template A5221 (Sub):  B  -->  B [x/A]
:= $B5221 %0
:= $T5221 o
:= $X5221 x{$T5221}
:= $A5221 ((&{{{o,o},o}}_a{o}{o}){{o,o}}_((&{{{o,o},o}}_b{o}{o}){{o,o}}_c{o}{o}){o})
<< A5221.r0t.txt
:= $B5221; := $T5221; := $X5221; := $A5221
%0

## .2b

## use Proof Template K8019H:  H => (A & B)  -->  H => A, H => B
```

```
:= $H8019H %0
<< K8019H.r0t.txt
:= $H8019H
:= $ATMP8021 $A8019H
:= $BCTMP8020 $B8019H
:= $A8019H; := $B8019H
%0

## .2c

%$BCTMP8020

## use Proof Template K8019H:  H => (A & B)  -->  H => A, H => B
:= $H8019H %0
<< K8019H.r0t.txt
:= $H8019H
:= $BTMP8021 $A8019H
:= $CTMP8021 $B8019H
:= $A8019H; := $B8019H
%0

:= $BCTMP8020

## .2d

%$ATMP8021; := $ATMP8021
%$BTMP8021; := $BTMP8021

## use Proof Template K8020H:  H => A, H => B  -->  H => (A & B)
:= $A8020H %1
:= $B8020H %0
<< K8020H.r0t.txt
:= $A8020H; := $B8020H

:= $ABTMP8021 %0

%$ABTMP8021; := $ABTMP8021
%$CTMP8021; := $CTMP8021

## use Proof Template K8020H:  H => A, H => B  -->  H => (A & B)
:= $A8020H %1
:= $B8020H %0
<< K8020H.r0t.txt
:= $A8020H; := $B8020H

:= $ABC2TMP8020 %0

## .3
```

```
## use Proof Template K8013:  A => B, B => A  -->  A = B
:= $A8013 $ABC1TMP8020/5
:= $B8013 $ABC2TMP8020/5
<< K8013.r0t.txt
:= $A8013; := $B8013
%0

:= $ABC1TMP8020
:= $ABC2TMP8020

## .4: Rename variables

## use Proof Template A5221 (Sub):  B  -->  B [x/A]
:= $B5221 %0
:= $T5221 {o}
:= $X5221 a{$T5221}
:= $A5221 x{$T5221}
<< A5221.r0t.txt
:= $B5221; := $T5221; := $X5221; := $A5221
%0

## use Proof Template A5221 (Sub):  B  -->  B [x/A]
:= $B5221 %0
:= $T5221 {o}
:= $X5221 b{$T5221}
:= $A5221 y{$T5221}
<< A5221.r0t.txt
:= $B5221; := $T5221; := $X5221; := $A5221
%0

## use Proof Template A5221 (Sub):  B  -->  B [x/A]
:= $B5221 %0
:= $T5221 {o}
:= $X5221 c{$T5221}
:= $A5221 z{$T5221}
<< A5221.r0t.txt
:= $B5221; := $T5221; := $X5221; := $A5221
%0

:= $TMP8020 %0

## .5: Match general definition

§= ((ASSOC{{{o,{{\4{^},\4{^}},\3{^}}},^}}_o{^}){{o,{{o,o},o}}}_&{{{o,o},o}})
§\s /6
§\s /3

## use Proof Template A5201b (Swap):  A = B  -->  B = A
<< A5201b.r0t.txt
```

```
%0

%$TMP8020; := $TMP8020
§s %0 1 %1

:= K8021 %0


##
##  Q.E.D.
##

%0
```

2.2.121 File K8022.r0.txt

```
##
##  Proof K8022:  A => B  =  (~ A) | B
##
##
##  Source: [Kubota 2017 (doi: 10.4444/100.10)]
##
##  Copyright (c) 2017 Owl of Minerva Press GmbH. All rights reserved.
##  Written by Ken Kubota (<mail@kenkubota.de>).
##
##  This file is part of the publication of the mathematical logic R0.
##  For more information, visit: <http://doi.org/10.4444/100.10>
##

<< basics.r0.txt
<< A5200t.r0.txt
<< A5205.r0.txt
<< A5228.r0.txt
<< A5230.r0.txt
<< A5231.r0.txt
<< A5232.r0.txt



##
##  Proof
##

## .1: main case T

## use Proof Template A5222 (Rule of Cases):  [\x.A]T, [\x.A]F  -->  A
§\ ([\x{o}{o}.[\y{o}{o}.((={{{o,o},o}}_((=>{{{o,o},o}}_x{o}{o}){{o,o}}_y{o}{o}){o}){
{o,o}}_((|{{{o,o},o}}_(!{{o,o}}_x{o}{o}){o}){{o,o}}_y{o}{o}){o}){o}]{{o,o}}]{{{o,o},
o}}_T{o})
 := $L5222 %0/3
```

```
:= $X5222 y{o}
:= $T5222 ($L5222{{o,o}}_T{o})
:= $F5222 ($L5222{{o,o}}_F{o})

## case T
§= $T5222
§\s /3
%A5231a
§s %1 29 %0
%A5228a
§s %1 13 %0
%A5232c
§s %1 7 %0
%A5230a
§s %1 3 %0
## use Proof Template A5201b (Swap):  A = B  -->  B = A
<< A5201b.r0t.txt
%0
%A5200t
§s %0 1 %1

## case F
§= $F5222
§\s /3
%A5231a
§s %1 29 %0
%A5228b
§s %1 13 %0
%A5232d
§s %1 7 %0
%A5230d
§s %1 3 %0
## use Proof Template A5201b (Swap):  A = B  -->  B = A
<< A5201b.r0t.txt
%0
%A5200t
§s %0 1 %1

<< A5222.r0t.txt
:= $L5222; := $X5222; := $T5222; := $F5222

:= $TTMP8022 %0

## .2: main case F

## use Proof Template A5222 (Rule of Cases):  [\x.A]T, [\x.A]F  -->  A
§\ ([\x{o}{o}.[\y{o}{o}.((={{{o,o},o}}_((=>{{{o,o},o}}_x{o}{o}){{o,o}}_y{o}{o}){o}){
{o,o}}_((|{{{o,o},o}}_(!{{o,o}}_x{o}{o}){o}){{o,o}}_y{o}{o}){o}){o}]{{o,o}}]{{{o,o},
o}}_F{o})
```

```
:= $L5222 %0/3
:= $X5222 y{o}
:= $T5222 ($L5222{{o,o}}_T{o})
:= $F5222 ($L5222{{o,o}}_F{o})

## case T
§= $T5222
§\s /3
%A5231b
§s %1 29 %0
%A5228c
§s %1 13 %0
%A5232a
§s %1 7 %0
%A5230a
§s %1 3 %0
## use Proof Template A5201b (Swap):  A = B  -->  B = A
<< A5201b.r0t.txt
%0
%A5200t
§s %0 1 %1

## case F
§= $F5222
§\s /3
%A5231b
§s %1 29 %0
%A5228d
§s %1 13 %0
%A5232b
§s %1 7 %0
%A5230a
§s %1 3 %0
## use Proof Template A5201b (Swap):  A = B  -->  B = A
<< A5201b.r0t.txt
%0
%A5200t
§s %0 1 %1

<< A5222.r0t.txt
:= $L5222; := $X5222; := $T5222; := $F5222

:= $FTMP8022 %0

## .3

## use Proof Template A5222 (Rule of Cases):  [\x.A]T, [\x.A]F  -->  A
:= $L5222 [\x{o}{o}.((={{{o,o},o}}_((=>{{{o,o},o}}_x{o}{o}){{o,o}}_y{o}{o}){o}){{o,o
}}_((|{{{o,o},o}}_(!{{o,o}}_x{o}{o}){o}){{o,o}}_y{o}{o}){o}){o}]
```

```
:= $X5222 x{o}
:= $T5222 ($L5222{{o,o}}_T{o})
:= $F5222 ($L5222{{o,o}}_F{o})

## case T
§= $T5222
§\s /3
## use Proof Template A5201b (Swap):  A = B  -->  B = A
<< A5201b.r0t.txt
%0
%$TTMP8022; := $TTMP8022
§s %0 1 %1

## case F
§= $F5222
§\s /3
## use Proof Template A5201b (Swap):  A = B  -->  B = A
<< A5201b.r0t.txt
%0
%$FTMP8022; := $FTMP8022
§s %0 1 %1

<< A5222.r0t.txt
:= $L5222; := $X5222; := $T5222; := $F5222
%0

## .4

§= {{o,o},o} [\x{o}{o}.[\y{o}{o}.((=>{{{o,o},o}}_x{o}{o}){{o,o}}_y{o}{o}){o}]{{o,o}}
]
§s %0 15 %1
:= $TMP8022 %0

## use Proof Template:  A5205 Substitutions
:= $AA5205 o
:= $BA5205 o
:= $FA5205 (=>{{{o,o},o}}_x{o}{o})
<< a5205_substitutions.r0t.txt
:= $AA5205; := $BA5205; := $FA5205
%0

## use Proof Template A5201b (Swap):  A = B  -->  B = A
<< A5201b.r0t.txt
%0
%$TMP8022; := $TMP8022
§s %0 11 %1
:= $TMP8022 %0

## use Proof Template:  A5205 Substitutions
```

```
:= $AA5205 {o,o}
:= $BA5205 o
:= $FA5205 =>
<< a5205_substitutions.r0t.txt
:= $AA5205; := $BA5205; := $FA5205
%0

§rs /3 x{o}
## use Proof Template A5201b (Swap):  A = B  -->  B = A
<< A5201b.r0t.txt
%0
%$TMP8022; := $TMP8022
§s %0 5 %1

:= K8022 %0


##
##  Q.E.D.
##

%0
```

2.2.122 File K8023.r0.txt

```
##
##  Proof K8023:  (A | B) | C  =  A | (B | C)
##
##
##  Source: [Kubota 2017 (doi: 10.4444/100.10)]
##
##  Copyright (c) 2017 Owl of Minerva Press GmbH. All rights reserved.
##  Written by Ken Kubota (<mail@kenkubota.de>).
##
##  This file is part of the publication of the mathematical logic R0.
##  For more information, visit: <http://doi.org/10.4444/100.10>
##

<< basics.r0.txt
<< A5200t.r0.txt
<< A5230.r0.txt
<< A5232.r0.txt


##
##  Proof
##

:= $LTMP8023 [\x{o}{o}.[\y{o}{o}.[\z{o}{o}.((={{{o,o},o}}_((|{{{o,o},o}}_((|{{{o,o},
```

```
o}}_x{o}{o}){{o,o}}_y{o}{o}){o}){{o,o}}_z{o}{o}){o}){{o,o}}_((|{{{o,o},o}}_x{o}{o}){
{o,o}}_((|{{{o,o},o}}_y{o}{o}){{o,o}}_z{o}{o}){o}){o}){o}]{{o,o}}]{{{o,o},o}}]

## .1:  Subcase TT

:= $TTTMP8023 (($LTMP8023{{{{o,o},o},o}}_T{o}){{{o,o},o}}_T{o})
§= $TTTMP8023
§\s /6
§\s /3

## use Proof Template A5222 (Rule of Cases):  [\x.A]T, [\x.A]F  -->  A
:= $L5222 %0/3
:= $X5222 z{o}
:= $T5222 ($L5222{{o,o}}_T{o})
:= $F5222 ($L5222{{o,o}}_F{o})

## case T
§= $T5222
§\s /3
%A5232a
§s %1 53 %0
%A5232a
§s %1 13 %0
%A5232a
§s %1 15 %0
%A5232a
§s %1 7 %0
%A5230a
§s %1 3 %0
## use Proof Template A5201b (Swap):  A = B  -->  B = A
<< A5201b.r0t.txt
%0
%A5200t
§s %0 1 %1

## case F
§= $F5222
§\s /3
%A5232a
§s %1 53 %0
%A5232b
§s %1 13 %0
%A5232b
§s %1 15 %0
%A5232a
§s %1 7 %0
%A5230a
§s %1 3 %0
## use Proof Template A5201b (Swap):  A = B  -->  B = A
```

```
<< A5201b.r0t.txt
%0
%A5200t
§s %0 1 %1

<< A5222.r0t.txt
:= $L5222; := $X5222; := $T5222; := $F5222

:= $TTTMP8023
:= $TTTMP8023 %0

## .2:  Subcase TF

:= $TFTMP8023 (($LTMP8023{{{{o,o},o},o}}_T{o}){{{o,o},o}}_F{o})
§= $TFTMP8023
§\s /6
§\s /3

## use Proof Template A5222 (Rule of Cases):  [\x.A]T, [\x.A]F  -->  A
:= $L5222 %0/3
:= $X5222 z{o}
:= $T5222 ($L5222{{o,o}}_T{o})
:= $F5222 ($L5222{{o,o}}_F{o})

## case T
§= $T5222
§\s /3
%A5232b
§s %1 53 %0
%A5232a
§s %1 13 %0
%A5232c
§s %1 15 %0
%A5232a
§s %1 7 %0
%A5230a
§s %1 3 %0
## use Proof Template A5201b (Swap):  A = B  -->  B = A
<< A5201b.r0t.txt
%0
%A5200t
§s %0 1 %1

## case F
§= $F5222
§\s /3
%A5232b
§s %1 53 %0
%A5232b
```

```
§s %1 13 %0
%A5232d
§s %1 15 %0
%A5232b
§s %1 7 %0
%A5230a
§s %1 3 %0
## use Proof Template A5201b (Swap):  A = B  -->  B = A
<< A5201b.r0t.txt
%0
%A5200t
§s %0 1 %1

<< A5222.r0t.txt
:= $L5222; := $X5222; := $T5222; := $F5222

:= $TFTMP8023
:= $TFTMP8023 %0

## .3:  Subcase FT

:= $FTTMP8023 (($LTMP8023{{{o,o},o},o}}_F{o}){{{o,o},o}}_T{o})
§= $FTTMP8023
§\s /6
§\s /3

## use Proof Template A5222 (Rule of Cases):  [\x.A]T, [\x.A]F  -->  A
:= $L5222 %0/3
:= $X5222 z{o}
:= $T5222 ($L5222{{o,o}}_T{o})
:= $F5222 ($L5222{{o,o}}_F{o})

## case T
§= $T5222
§\s /3
%A5232c
§s %1 53 %0
%A5232a
§s %1 13 %0
%A5232a
§s %1 15 %0
%A5232c
§s %1 7 %0
%A5230a
§s %1 3 %0
## use Proof Template A5201b (Swap):  A = B  -->  B = A
<< A5201b.r0t.txt
%0
%A5200t
```

```
§s %0 1 %1

## case F
§= $F5222
§\s /3
%A5232c
§s %1 53 %0
%A5232b
§s %1 13 %0
%A5232b
§s %1 15 %0
%A5232c
§s %1 7 %0
%A5230a
§s %1 3 %0
## use Proof Template A5201b (Swap):  A = B  -->  B = A
<< A5201b.r0t.txt
%0
%A5200t
§s %0 1 %1

<< A5222.r0t.txt
:= $L5222; := $X5222; := $T5222; := $F5222

:= $FTTMP8023
:= $FTTMP8023 %0

## .4:  Subcase FF

:= $FFTMP8023 (($LTMP8023{{{{o,o},o},o}}_F{o}){{{o,o},o}}_F{o})
§= $FFTMP8023
§\s /6
§\s /3

## use Proof Template A5222 (Rule of Cases):  [\x.A]T, [\x.A]F  -->  A
:= $L5222 %0/3
:= $X5222 z{o}
:= $T5222 ($L5222{{o,o}}_T{o})
:= $F5222 ($L5222{{o,o}}_F{o})

## case T
§= $T5222
§\s /3
%A5232d
§s %1 53 %0
%A5232c
§s %1 13 %0
%A5232c
§s %1 15 %0
```

```
%A5232c
§s %1 7 %0
%A5230a
§s %1 3 %0
## use Proof Template A5201b (Swap):  A = B  -->  B = A
<< A5201b.r0t.txt
%0
%A5200t
§s %0 1 %1

## case F
§= $F5222
§\s /3
%A5232d
§s %1 53 %0
%A5232d
§s %1 13 %0
%A5232d
§s %1 15 %0
%A5232d
§s %1 7 %0
%A5230d
§s %1 3 %0
## use Proof Template A5201b (Swap):  A = B  -->  B = A
<< A5201b.r0t.txt
%0
%A5200t
§s %0 1 %1

<< A5222.r0t.txt
:= $L5222; := $X5222; := $T5222; := $F5222

:= $FFTMP8023
:= $FFTMP8023 %0

## .5:  Case T

:= $TTMP8023 [\y{o}{o}.((($LTMP8023{{{o,o},o},o}}_T{o}){{{o,o},o}}_y{o}{o}){{o,o}}_
z{o}{o}){o}]
§= $TTMP8023
§\s /28
§\s /14
§\s /7

## use Proof Template A5222 (Rule of Cases):  [\x.A]T, [\x.A]F  -->  A
:= $L5222 %0/3
:= $X5222 y{o}
:= $T5222 ($L5222{{o,o}}_T{o})
:= $F5222 ($L5222{{o,o}}_F{o})
```

```
## case T
§= $T5222
§\s /3
## use Proof Template A5201b (Swap):  A = B  -->  B = A
<< A5201b.r0t.txt
%0
%$TTTMP8023; := $TTTMP8023
§s %0 1 %1

## case F
§= $F5222
§\s /3
## use Proof Template A5201b (Swap):  A = B  -->  B = A
<< A5201b.r0t.txt
%0
%$TFTMP8023; := $TFTMP8023
§s %0 1 %1

<< A5222.r0t.txt
:= $L5222; := $X5222; := $T5222; := $F5222

:= $TTMP8023
:= $TTMP8023 %0

## .6:  Case F

:= $FTMP8023 [\y{o}{o}.((($LTMP8023{{{o,o},o},o}}_F{o}){{{o,o},o}}_y{o}{o}){{o,o}}_
z{o}{o}){o}]
§= $FTMP8023
§\s /28
§\s /14
§\s /7

## use Proof Template A5222 (Rule of Cases):  [\x.A]T, [\x.A]F  -->  A
:= $L5222 %0/3
:= $X5222 y{o}
:= $T5222 ($L5222{{o,o}}_T{o})
:= $F5222 ($L5222{{o,o}}_F{o})

## case T
§= $T5222
§\s /3
## use Proof Template A5201b (Swap):  A = B  -->  B = A
<< A5201b.r0t.txt
%0
%$FTTMP8023; := $FTTMP8023
§s %0 1 %1
```

```
## case F
§= $F5222
§\s /3
## use Proof Template A5201b (Swap):  A = B  -->  B = A
<< A5201b.r0t.txt
%0
%$FFTMP8023; := $FFTMP8023
§s %0 1 %1

<< A5222.r0t.txt
:= $L5222; := $X5222; := $T5222; := $F5222

:= $FTMP8023
:= $FTMP8023 %0

## .7:  General case

:= $TMP8023 [\x{o}{o}.((($LTMP8023{{{{o,o},o},o}}_x{o}{o}){{{o,o},o}}_y{o}{o}){{o,o}
}_z{o}{o}){o}]
§= $TMP8023
§\s /28
§\s /14
§\s /7

## use Proof Template A5222 (Rule of Cases):  [\x.A]T, [\x.A]F  -->  A
:= $L5222 %0/3
:= $X5222 x{o}
:= $T5222 ($L5222{{o,o}}_T{o})
:= $F5222 ($L5222{{o,o}}_F{o})

## case T
§= $T5222
§\s /3
## use Proof Template A5201b (Swap):  A = B  -->  B = A
<< A5201b.r0t.txt
%0
%$TTMP8023; := $TTMP8023
§s %0 1 %1

## case F
§= $F5222
§\s /3
## use Proof Template A5201b (Swap):  A = B  -->  B = A
<< A5201b.r0t.txt
%0
%$FTMP8023; := $FTMP8023
§s %0 1 %1

<< A5222.r0t.txt
```

```
:= $L5222; := $X5222; := $T5222; := $F5222

:= $TMP8023
:= $TMP8023 %0

## .8:  Match general definition

§= ((ASSOC{{{o,{{\4{^},\4{^}},\3{^}}},^}}_o{^}){{o,{{o,o},o}}}_|{{{o,o},o}})
§\s /6
§\s /3
## use Proof Template A5201b (Swap):  A = B  -->  B = A
<< A5201b.r0t.txt
%0
%$TMP8023; := $TMP8023
§s %0 1 %1

:= K8023 %0


:= $LTMP8023


##
##  Q.E.D.
##

%0
```

2.2.123 File K8024.r0.txt

```
##
##  Proof K8024 (Generalized Deduction Theorem):  (H & I) => A  =  H => (I => A)
##
##
##  Source: [Kubota 2017 (doi: 10.4444/100.10)]
##
##  Copyright (c) 2017 Owl of Minerva Press GmbH. All rights reserved.
##  Written by Ken Kubota (<mail@kenkubota.de>).
##
##  This file is part of the publication of the mathematical logic R0.
##  For more information, visit: <http://doi.org/10.4444/100.10>
##

<< basics.r0.txt
<< K8008.r0.txt
<< K8018.r0.txt
<< K8022.r0.txt
<< K8023.r0.txt
```

```
##
## Proof
##

## .1

%K8022
§= {o} ((=>{{{o,o},o}}_((&{{{o,o},o}}_h{o}{o}){{o,o}}_j{o}{o}){o}){{o,o}}_x{o}{o})
§s %0 12 %1
§\s /6
§\s /3

## .2

%K8018
§s %1 108 %0
§\s /54
§\s /27

:= $TMP8025 %0

## .3

%K8008

## use Proof Template A5221 (Sub): B --> B [x/A]
:= $B5221 %0
:= $T5221 o
:= $X5221 x{$T5221}
:= $A5221 ((|{{{o,o},o}}_(!{{o,o}}_h{o}{o}){o}){{o,o}}_(!{{o,o}}_j{o}{o}){o})
<< A5221.r0t.txt
:= $B5221; := $T5221; := $X5221; := $A5221
%0

%$TMP8025; := $TMP8025
§s %0 13 %1
:= $TMP8025 %0

## .4

%K8023
§\s /2
§\s /1

## use Proof Template A5221 (Sub): B --> B [x/A]
:= $B5221 %0
:= $T5221 o
:= $X5221 x{$T5221}
```

```
:= $A5221 (!{{o,o}}_h{o}{o})
<< A5221.r0t.txt
:= $B5221; := $T5221; := $X5221; := $A5221
%0

## use Proof Template A5221 (Sub):  B  -->  B [x/A]
:= $B5221 %0
:= $T5221 o
:= $X5221 y{$T5221}
:= $A5221 (!{{o,o}}_j{o}{o})
<< A5221.r0t.txt
:= $B5221; := $T5221; := $X5221; := $A5221
%0

## use Proof Template A5221 (Sub):  B  -->  B [x/A]
:= $B5221 %0
:= $T5221 o
:= $X5221 z{$T5221}
:= $A5221 x{o}
<< A5221.r0t.txt
:= $B5221; := $T5221; := $X5221; := $A5221
%0

%$TMP8025; := $TMP8025
§s %0 3 %1
:= $TMP8025 %0

## .5

%K8022
§= ((=>{{{o,o},o}}_j{o}{o}){{o,o}}_x{o}{o})
§s %0 12 %1
§\s /6
§\s /3

## use Proof Template A5201b (Swap):  A = B  -->  B = A
<< A5201b.r0t.txt
%0

%$TMP8025; := $TMP8025
§s %0 7 %1
:= $TMP8025 %0

## .6

%K8022
§= ((=>{{{o,o},o}}_h{o}{o}){{o,o}}_((=>{{{o,o},o}}_j{o}{o}){{o,o}}_x{o}{o}){o})
§s %0 12 %1
§\s /6
```

```
§\s /3

## use Proof Template A5201b (Swap):  A = B  -->  B = A
<< A5201b.r0t.txt
%0

%$TMP8025; := $TMP8025
§s %0 3 %1

:= K8024 %0


##
##  Q.E.D.
##

%0
```

2.2.124 File K8025.r0a.txt

```
##
##  Proof Template K8025 (Deduction Theorem):  (H & I) => A  -->  H => (I => A)
##
##
##  Source: [cf. Andrews 2002 (ISBN 1-4020-0763-9), pp. 228 f. (5240)]
##
##  Copyright (c) 2017 Owl of Minerva Press GmbH. All rights reserved.
##  Written by Ken Kubota (<mail@kenkubota.de>).
##
##  This file is part of the publication of the mathematical logic R0.
##  For more information, visit: <http://doi.org/10.4444/100.10>
##



##
##  Assumptions and Resulting Syntactical Variables
##

<< basics.r0.txt

## the assumption as last theorem on stack (%0)
§!  ((=>{{{o,o},o}}_((&{{{o,o},o}}_h{o}{o}){{o,o}}_j{o}{o}){o}){{o,o}}_x{o}{o})


##
##  Include Proof Template
##

<<< K8025.r0t.txt
```

```
##
##  Q.E.D.
##

%0
```

2.2.125 File K8025.r0t.txt

```
##
##  Proof Template K8025 (Deduction Theorem):  (H & I) => A  -->  H => (I => A)
##
##
##  Source: [cf. Andrews 2002 (ISBN 1-4020-0763-9), pp. 228 f. (5240)]
##
##  Copyright (c) 2017 Owl of Minerva Press GmbH. All rights reserved.
##  Written by Ken Kubota (<mail@kenkubota.de>).
##
##  This file is part of the publication of the mathematical logic R0.
##  For more information, visit: <http://doi.org/10.4444/100.10>
##


## define variable first (before inclusion of file)
:= $STMPDED8025 %0



##
##  Proof Template
##

<< K8024.r0.txt
%K8024

:= $TMPDED8025 %0
%$STMPDED8025
%$TMPDED8025; := $TMPDED8025

## use Proof Template A5221 (Sub):  B  -->  B [x/A]
:= $B5221 %0
:= $T5221 o
:= $X5221 h{$T5221}
:= $A5221 %1/21
<< A5221.r0t.txt
:= $B5221; := $T5221; := $X5221; := $A5221

:= $TMPDED8025 %0
%$STMPDED8025
```

```
%$TMPDED8025; := $TMPDED8025

## use Proof Template A5221 (Sub):  B  -->  B [x/A]
:= $B5221 %0
:= $T5221 o
:= $X5221 j{$T5221}
:= $A5221 %1/11
<< A5221.r0t.txt
:= $B5221; := $T5221; := $X5221; := $A5221

:= $TMPDED8025 %0
%$STMPDED8025
%$TMPDED8025; := $TMPDED8025

## use Proof Template A5221 (Sub):  B  -->  B [x/A]
:= $B5221 %0
:= $T5221 o
:= $X5221 x{$T5221}
:= $A5221 %1/3
<< A5221.r0t.txt
:= $B5221; := $T5221; := $X5221; := $A5221
%0

%$STMPDED8025; := $STMPDED8025
§s %0 1 %1
```

2.2.126 File K8026.r0a.txt

```
##
##  Proof Template K8026 (Deduction Theorem Reversed):  H => (I => A)  -->  (H & I)
=> A
##
##
##  Source: [Kubota 2017 (doi: 10.4444/100.10)]
##
##  Copyright (c) 2017 Owl of Minerva Press GmbH. All rights reserved.
##  Written by Ken Kubota (<mail@kenkubota.de>).
##
##  This file is part of the publication of the mathematical logic R0.
##  For more information, visit: <http://doi.org/10.4444/100.10>
##


##
##  Assumptions and Resulting Syntactical Variables
##

<< basics.r0.txt
```

```
## the assumption as last theorem on stack (%0)
§!  ((=>{{{o,o},o}}_h{o}{o}){{o,o}}_((=>{{{o,o},o}}_j{o}{o}){{o,o}}_x{o}{o}){o})


##
##  Include Proof Template
##

<<< K8026.r0t.txt


##
##  Q.E.D.
##

%0
```

2.2.127 File K8026.r0t.txt

```
##
##  Proof Template K8026 (Deduction Theorem Reversed):  H => (I => A)  -->  (H & I)
=> A
##
##
##  Source: [Kubota 2017 (doi: 10.4444/100.10)]
##
##  Copyright (c) 2017 Owl of Minerva Press GmbH. All rights reserved.
##  Written by Ken Kubota (<mail@kenkubota.de>).
##
##  This file is part of the publication of the mathematical logic R0.
##  For more information, visit: <http://doi.org/10.4444/100.10>
##


## define variable first (before inclusion of file)
:= $STMPDED %0


##
##  Proof Template
##

<< K8024.r0.txt
%K8024

:= $TMPDED %0
%$STMPDED
%$TMPDED; := $TMPDED
```

```
## use Proof Template A5221 (Sub):  B  -->  B [x/A]
:= $B5221 %0
:= $T5221 o
:= $X5221 h{$T5221}
:= $A5221 %1/5
<< A5221.r0t.txt
:= $B5221; := $T5221; := $X5221; := $A5221

:= $TMPDED %0
%$STMPDED
%$TMPDED; := $TMPDED

## use Proof Template A5221 (Sub):  B  -->  B [x/A]
:= $B5221 %0
:= $T5221 o
:= $X5221 j{$T5221}
:= $A5221 %1/13
<< A5221.r0t.txt
:= $B5221; := $T5221; := $X5221; := $A5221

:= $TMPDED %0
%$STMPDED
%$TMPDED; := $TMPDED

## use Proof Template A5221 (Sub):  B  -->  B [x/A]
:= $B5221 %0
:= $T5221 o
:= $X5221 x{$T5221}
:= $A5221 %1/7
<< A5221.r0t.txt
:= $B5221; := $T5221; := $X5221; := $A5221
%0

## use Proof Template A5201b (Swap):  A = B  -->  B = A
<< A5201b.r0t.txt
%0

%$STMPDED; := $STMPDED
§s %0 1 %1
```

2.2.128 File K8027.r0a.txt

```
##
##  Proof Template K8027:  (A & B) => C  -->  (B & A) => C
##      (Hypotheses Swap)
##
##  Source: [Kubota 2017 (doi: 10.4444/100.10)]
##
##  Copyright (c) 2017 Owl of Minerva Press GmbH. All rights reserved.
```

```
##  Written by Ken Kubota (<mail@kenkubota.de>).
##
##  This file is part of the publication of the mathematical logic R0.
##  For more information, visit: <http://doi.org/10.4444/100.10>
##


##
##  Assumptions and Resulting Syntactical Variables
##

<< basics.r0.txt

## the assumption as last theorem on stack (%0)
§! ((=>{{{o,o},o}}_((&{{{o,o},o}}_a{o}{o}){{o,o}}_b{o}{o}){o}){{o,o}}_c{o}{o})


##
##  Include Proof Template
##

<<< K8027.r0t.txt


##
##  Q.E.D.
##

%0
```

2.2.129 File K8027.r0t.txt

```
##
##  Proof Template K8027:  (A & B) => C  -->  (B & A) => C
##      (Hypotheses Swap)
##
##  Source: [Kubota 2017 (doi: 10.4444/100.10)]
##
##  Copyright (c) 2017 Owl of Minerva Press GmbH. All rights reserved.
##  Written by Ken Kubota (<mail@kenkubota.de>).
##
##  This file is part of the publication of the mathematical logic R0.
##  For more information, visit: <http://doi.org/10.4444/100.10>
##


## define variable first (before inclusion of file)
:= $HTMPSWPHYP %0
```

```
##
##  Proof Template
##

<< K8007.r0.txt
%K8007
§\s /2
§\s /1

:= $TMPSWPHYP %0
%$HTMPSWPHYP
%$TMPSWPHYP; := $TMPSWPHYP

## use Proof Template A5221 (Sub):  B  -->  B [x/A]
:= $B5221 %0
:= $T5221 o
:= $X5221 x{$T5221}
:= $A5221 %1/21
<< A5221.r0t.txt
:= $B5221; := $T5221; := $X5221; := $A5221

:= $TMPSWPHYP %0
%$HTMPSWPHYP
%$TMPSWPHYP; := $TMPSWPHYP

## use Proof Template A5221 (Sub):  B  -->  B [x/A]
:= $B5221 %0
:= $T5221 o
:= $X5221 y{$T5221}
:= $A5221 %1/11
<< A5221.r0t.txt
:= $B5221; := $T5221; := $X5221; := $A5221
%0

%$HTMPSWPHYP; := $HTMPSWPHYP
§s %0 5 %1
```

2.2.130 File K8028.r0a.txt

```
##
##  Proof Template K8028 (EXI GenH):  H => ([\x.B]A)  -->  H => EXI x: B
##      for any x of any type (Rule of Existential Generalization -- with hypothesis
)
##
##  Source: [cf. Andrews 2002 (ISBN 1-4020-0763-9), p. 229 (5242)]
##
##  Copyright (c) 2017 Owl of Minerva Press GmbH. All rights reserved.
##  Written by Ken Kubota (<mail@kenkubota.de>).
```

```
##
##  This file is part of the publication of the mathematical logic R0.
##  For more information, visit: <http://doi.org/10.4444/100.10>
##


##
##  Define Syntactical Variables
##

## hypothesis:  H
:= $H8028 h{o}

## type of substitute
:= $T8028 t{^}

## proposition:  [\x.B]
:= $B8028 b{{o,$T8028}}

## substitute:  A
:= $A8028 a{$T8028}


##
##  Assumptions and Resulting Syntactical Variables
##

<< basics.r0.txt

## given proposition
§! ((=>{{{o,o},o}}_$H8028{o}){{o,o}}_($B8028{{o,$T8028}}_$A8028{$T8028}){o})


##
##  Proof Template
##

<<< K8028.r0t.txt


##
##  Undefine Syntactical Variables
##

:= $H8028; := $T8028; := $B8028; := $A8028


##
##  Q.E.D.
```

```
##

%0
```

2.2.131 File K8028.r0t.txt

```
##
## Proof Template K8028 (EXI GenH): H => ([\x.B]A) --> H => EXI x: B
##     for any x of any type (Rule of Existential Generalization -- with hypothesis
)
##
## Source: [cf. Andrews 2002 (ISBN 1-4020-0763-9), p. 229 (5242)]
##
## Copyright (c) 2017 Owl of Minerva Press GmbH. All rights reserved.
## Written by Ken Kubota (<mail@kenkubota.de>).
##
## This file is part of the publication of the mathematical logic R0.
## For more information, visit: <http://doi.org/10.4444/100.10>
##

## given proposition
:= $PTMP8028 ((=>{{{o,o},o}}_$H8028{o}){{o,o}}_($B8028{{o,$T8028}}_$A8028{$T8028}){o
})

<< basics.r0.txt
<< A5205.r0.txt
<< A5231.r0.txt
<< K8005.r0.txt
<< K8008.r0.txt
<< K8015.r0.txt
<< K8017.r0.txt



##
## Proof Template
##

## .1

%K8005

## use Proof Template A5221 (Sub): B --> B [x/A]
:= $B5221 %0
:= $T5221 o
:= $X5221 x{$T5221}
:= $A5221 (!{{o,o}}_((E{{{o,{o,\3{^}}},^}}_$T8028{^}){{o,{o,$T8028}}}_$B8028{{o,$T80
28}}){o})
<< A5221.r0t.txt
:= $B5221; := $T5221; := $X5221; := $A5221
```

```
%0

## .2

%K8017
§s %1 28 %0
§\s /14
§\s /7
:= $TTMP8028 %0

## .3

%K8008

## use Proof Template A5221 (Sub):  B  -->  B [x/A]
:= $B5221 %0
:= $T5221 o
:= $X5221 x{$T5221}
:= $A5221 %1/15
<< A5221.r0t.txt
:= $B5221; := $T5221; := $X5221; := $A5221
%0

%$TTMP8028; := $TTMP8028
§s %0 3 %1
§\s /6
§\s /3
:= $TTMP8028 %0

## .4

§= {o} ([\x{$T8028}{$T8028}.T{o}]{{o,$T8028}}_$A8028{$T8028})

## use Proof Template K8003 (Intro):  A  -->  H => A
:= $A8003 %0
:= $H8003 (!{{o,o}}_((E{{{o,{o,\3{^}}},^}}_$T8028{^}){{o,{o,$T8028}}}_$B8028{{o,$T80
28}}){o})
<< K8003.r0t.txt
:= $A8003; := $H8003

:= $HTMP8028 %0

%$TTMP8028; := $TTMP8028
§s' %1 6 %0
§\s /13
§\s /7

## use Proof Template A5219cH (Rule T):  H => (T = A)  -->  H => A
:= $A5219cH %0
```

```
<< A5219cH.r0t.txt
 := $A5219cH

 := $TTMP8028 %0
%$PTMP8028
%$TTMP8028; := $TTMP8028

## use Proof Template K8004 (Trans):  (H OP A), B  -->  H => B
:= $HA8004 %1
:= $B8004 %0
<< K8004.r0t.txt
:= $HA8004; := $B8004
%0

## use Proof Template K8026 (Deduction Theorem Reversed):  H => (I => A)  -->  (H &
I) => A
<< K8026.r0t.txt
:= $NTMP8028 %0

## .5

%$HTMP8028; := $HTMP8028
%$PTMP8028; := $PTMP8028

## use Proof Template K8004 (Trans):  (H OP A), B  -->  H => B
:= $HA8004 %1
:= $B8004 %0
<< K8004.r0t.txt
:= $HA8004; := $B8004
%0

## use Proof Template K8026 (Deduction Theorem Reversed):  H => (I => A)  -->  (H &
I) => A
<< K8026.r0t.txt
%0

## .6

## use Proof Template K8027:  (A & B) => C  -->  (B & A) => C
<< K8027.r0t.txt
%0

## .7

## use Proof Template A5219bH (Rule T):  H => A  -->  H => (A = T)
:= $A5219bH %0
<< A5219bH.r0t.txt
:= $A5219bH
%0
```

```
%$NTMP8028; := $NTMP8028
§s' %0 3 %1
:= $NTMP8028 %0

%A5231a

## use Proof Template K8004 (Trans):  (H OP A), B  -->  H => B
:= $HA8004 %1
:= $B8004 %0
<< K8004.r0t.txt
:= $HA8004; := $B8004
%0

%$NTMP8028; := $NTMP8028
§s' %0 1 %1

## .8

<< K8025.r0t.txt
:= $DTMP8028 %0

%K8015

## use Proof Template A5221 (Sub):  B  -->  B [x/A]
:= $B5221 %0
:= $T5221 o
:= $X5221 x{$T5221}
:= $A5221 %1/13
<< A5221.r0t.txt
:= $B5221; := $T5221; := $X5221; := $A5221
%0

:= $TTMP8028 %0

## .9

%K8008

## use Proof Template A5221 (Sub):  B  -->  B [x/A]
:= $B5221 %0
:= $T5221 o
:= $X5221 x{$T5221}
:= $A5221 %1/15
<< A5221.r0t.txt
:= $B5221; := $T5221; := $X5221; := $A5221
%0

%$TTMP8028; := $TTMP8028
```

```
§s %0 3 %1
:= $TTMP8028 %0

%$DTMP8028
%$TTMP8028; := $TTMP8028

## use Proof Template K8004 (Trans):  (H OP A), B  -->  H => B
:= $HA8004 %1
:= $B8004 %0
<< K8004.r0t.txt
:= $HA8004; := $B8004
%0

%$DTMP8028; := $DTMP8028
§s' %0 1 %1
```

2.2.132 File K8029.r0.txt

```
##
##  Proof K8029:  A => B  =  (~ B) => (~ A)
##
##
##  Source: [Kubota 2017 (doi: 10.4444/100.10)]
##
##  Copyright (c) 2017 Owl of Minerva Press GmbH. All rights reserved.
##  Written by Ken Kubota (<mail@kenkubota.de>).
##
##  This file is part of the publication of the mathematical logic R0.
##  For more information, visit: <http://doi.org/10.4444/100.10>
##

<< basics.r0.txt
<< K8008.r0.txt
<< K8012.r0.txt
<< K8022.r0.txt


##
##  Proof
##

## .1

§= {o} ((=>{{{o,o},o}}_x{o}{o}){{o,o}}_y{o}{o})
%K8022
§s %1 12 %0
§\s /6
§\s /3
:= $TMP8029 %0
```

```
## .2

%K8012
§\s /2
§\s /1

## use Proof Template A5221 (Sub):  B  -->  B [x/A]
:= $B5221 %0
:= $T5221 o
:= $X5221 x{$T5221}
:= $A5221 (!{{$T5221,$T5221}}_x{$T5221}{$T5221})
<< A5221.r0t.txt
:= $B5221; := $T5221; := $X5221; := $A5221
%0

%$TMP8029; := $TMP8029
§s %0 3 %1
:= $LTMP8029 %0

## .3

§= {o} ((=>{{{o,o},o}}_(!{{o,o}}_y{o}{o}){o}){{o,o}}_(!{{o,o}}_x{o}{o}){o})
%K8022
§s %1 12 %0
§rs /25 z{o}
§\s /6
§\s /3
:= $TMP8029 %0

## .4

%K8008

## use Proof Template A5221 (Sub):  B  -->  B [x/A]
:= $B5221 %0
:= $T5221 o
:= $X5221 x{$T5221}
:= $A5221 y{$T5221}
<< A5221.r0t.txt
:= $B5221; := $T5221; := $X5221; := $A5221
%0

%$TMP8029; := $TMP8029
§s %0 13 %1

## use Proof Template A5201b (Swap):  A = B  -->  B = A
<< A5201b.r0t.txt
%0
```

```
%$LTMP8029; := $LTMP8029
§s %0 3 %1

:= K8029 %0


##
##  Q.E.D.
##

%0
```

2.2.133 File K8030.r0a.txt

```
##
##  Proof Template K8030 (EXI Rule):  (H & B) => A  -->  (H & EXI x: B) => A
##      for any x of any type, provided x is not free in H or in A (Existential Rule
)
##
##  Source: [cf. Andrews 2002 (ISBN 1-4020-0763-9), p. 230 (5244)]
##
##  Copyright (c) 2017 Owl of Minerva Press GmbH. All rights reserved.
##  Written by Ken Kubota (<mail@kenkubota.de>).
##
##  This file is part of the publication of the mathematical logic R0.
##  For more information, visit: <http://doi.org/10.4444/100.10>
##


##
##  Define Syntactical Variables
##

<< basics.r0.txt

## type of variable
:= $T8030 u{^}

## the variable
:= $X8030 x{$T8030}

## the proposition
:= $A8030 ((=>{{{o,o},o}}_((&{{{o,o},o}}_h{o}{o}){{o,o}}_b{o}{o}){o}){{o,o}}_a{o}{o}
)


##
##  Assumptions and Resulting Syntactical Variables
```

```
##

§! $A8030


##
##  Proof Template
##

<<< K8030.r0t.txt


##
##  Undefine Syntactical Variables
##

:= $T8030; := $X8030; := $A8030;


##
##  Q.E.D.
##

%0
```

2.2.134 File K8030.r0t.txt

```
##
##  Proof Template K8030 (EXI Rule):  (H & B) => A  -->  (H & EXI x: B) => A
##      for any x of any type, provided x is not free in H or in A (Existential Rule
)
##
##  Source: [cf. Andrews 2002 (ISBN 1-4020-0763-9), p. 230 (5244)]
##
##  Copyright (c) 2017 Owl of Minerva Press GmbH. All rights reserved.
##  Written by Ken Kubota (<mail@kenkubota.de>).
##
##  This file is part of the publication of the mathematical logic R0.
##  For more information, visit: <http://doi.org/10.4444/100.10>
##


## define variable first (save before inclusion of files)
:= $TMP8030 %0

<< K8008.r0.txt
<< K8016.r0.txt
<< K8017.r0.txt
<< K8029.r0.txt
```

```
## shorthands
:= $PTMP8030 [\p{{o,$T8030}}{{o,$T8030}}.(!{{o,o}}_((E{{{o,{o,\3{^}}},^}}_$T8030{^})
{{o,{o,$T8030}}}_[\$X8030{$T8030}.(!{{o,o}}_(p{{o,$T8030}}{{o,$T8030}}_$X8030{$T8030
}){o}){o}]{{o,$T8030}}){o}){o}]
:= $PBTMP8030 ([\$X8030{$T8030}.(!{{o,o}}_b{o}{o}){o}]{{o,$T8030}}_$X8030{$T8030})
:= $ETMP8030 ((E{{{o,{o,\3{^}}},^}}_$T8030{^}){{o,{o,$T8030}}}_[\$X8030{$T8030}.b{o}
{o}]{{o,$T8030}})

%$TMP8030; := $TMP8030


##
##  Proof Template
##

## .1

## use Proof Template K8025 (Deduction Theorem):  (H & I) => A  -->  H => (I => A)
<< K8025.r0t.txt
:= $TMP8030 %0

## .2

%K8029

## use Proof Template A5221 (Sub):  B  -->  B [x/A]
:= $B5221 %0
:= $T5221 o
:= $X5221 x{$T5221}
:= $A5221 $TMP8030/13
<< A5221.r0t.txt
:= $B5221; := $T5221; := $X5221; := $A5221
%0

## use Proof Template A5221 (Sub):  B  -->  B [x/A]
:= $B5221 %0
:= $T5221 o
:= $X5221 y{$T5221}
:= $A5221 $TMP8030/7
<< A5221.r0t.txt
:= $B5221; := $T5221; := $X5221; := $A5221
%0

%$TMP8030; := $TMP8030
§s %0 3 %1

## use Proof Template K8026 (Deduction Theorem Reversed):  H => (I => A)  -->  (H &
I) => A
```

```
<< K8026.r0t.txt
%0

## .3

## use Proof Template A5220H (Gen):  (H => A)  -->  (H => ALL x: A)
:= $T5220H $T8030
:= $X5220H $X8030
:= $A5220H %0
<< A5220H.r0t.txt
:= $T5220H; := $X5220H; := $A5220H
%0

## .4

%K8016
§= (A{{{o,{o,\3{^}}},^}}_$T8030{^})
§s %0 6 %1
§\s /3
§s %5 6 %0
§\s /3
§\s /63
§rs /15 $X8030
:= $TMP8030 %0

## .5

%K8008

## use Proof Template A5221 (Sub):  B  -->  B [x/A]
:= $B5221 %0
:= $T5221 o
:= $X5221 x{$T5221}
:= $A5221 %1/127
<< A5221.r0t.txt
:= $B5221; := $T5221; := $X5221; := $A5221
%0

%$TMP8030; := $TMP8030
§s %0 31 %1
<< K8025.r0t.txt
:= $TMP8030 %0

## .6

%K8029

## use Proof Template A5221 (Sub):  B  -->  B [x/A]
:= $B5221 %0
```

```
:= $T5221 o
:= $X5221 y{$T5221}
:= $A5221 $TMP8030/27
<< A5221.r0t.txt
:= $B5221; := $T5221; := $X5221; := $A5221
%0

## use Proof Template A5221 (Sub):  B  -->  B [x/A]
:= $B5221 %0
:= $T5221 o
:= $X5221 x{$T5221}
:= $A5221 $TMP8030/15
<< A5221.r0t.txt
:= $B5221; := $T5221; := $X5221; := $A5221
%0

## use Proof Template A5201b (Swap):  A = B  -->  B = A
<< A5201b.r0t.txt
%0

%$TMP8030; := $TMP8030

§s %0 3 %1

## use Proof Template K8026 (Deduction Theorem Reversed):  H => (I => A)  -->  (H &
I) => A
<< K8026.r0t.txt
%0

## undefine local variables
:= $PTMP8030; := $PBTMP8030; := $ETMP8030
```

2.2.135 File K8031.r0a.txt

```
##
##  Proof Template K8031 (EXI Gen):  ([\x.B]A)  -->  EXI x: B
##      for any x of any type (Rule of Existential Generalization -- without hypothe
sis)
##
##  Source: [cf. Andrews 2002 (ISBN 1-4020-0763-9), p. 229 (5242)]
##
##  Copyright (c) 2017 Owl of Minerva Press GmbH. All rights reserved.
##  Written by Ken Kubota (<mail@kenkubota.de>).
##
##  This file is part of the publication of the mathematical logic R0.
##  For more information, visit: <http://doi.org/10.4444/100.10>
##
```

```
##
## Define Syntactical Variables
##

## type of the variable and the substitute
:= $T8031 t{^}

## proposition: [\x.B]
:= $B8031 b{{o,$T8031}}

## substitute: A
:= $A8031 a{$T8031}


##
## Assumptions and Resulting Syntactical Variables
##

<< basics.r0.txt

## given proposition
§! ($B8031{{o,$T8031}}_$A8031{$T8031})


##
## Proof Template
##

<<< K8031.r0t.txt


##
## Undefine Syntactical Variables
##

:= $T8031; := $B8031; := $A8031


##
## Q.E.D.
##

%0
```

2.2.136 File K8031.r0t.txt

```
##
## Proof Template K8031 (EXI Gen): ([\x.B]A) --> EXI x: B
##     for any x of any type (Rule of Existential Generalization -- without hypothe
```

```
sis)
##
##  Source: [cf. Andrews 2002 (ISBN 1-4020-0763-9), p. 229 (5242)]
##
##  Copyright (c) 2017 Owl of Minerva Press GmbH. All rights reserved.
##  Written by Ken Kubota (<mail@kenkubota.de>).
##
##  This file is part of the publication of the mathematical logic R0.
##  For more information, visit: <http://doi.org/10.4444/100.10>
##

<< A5223.r0.txt


##
##  Proof Template
##

:= $P8031 ($B8031{{o,$T8031}}_$A8031{$T8031})

## .1

## use Proof Template K8003 (Intro):  A  -->  H => A
:= $A8003 $P8031
:= $H8003 T
<< K8003.r0t.txt
:= $A8003; := $H8003
%0

## .2

## use Proof Template K8028 (EXI GenH):  H => ([\x.B]A)  -->  H => EXI x: B
:= $H8028 T
:= $T8028 $T8031
:= $B8028 $B8031
:= $A8028 $A8031
<< K8028.r0t.txt
:= $H8028; := $T8028; := $B8028; := $A8028

:= $TTMP8031 %0

## .3

## use Proof Template A5221 (Sub):  B  -->  B [x/A]
:= $B5221 %A5223
:= $T5221 o
:= $X5221 y{o}
:= $A5221 %0/3
<< A5221.r0t.txt
```

```
:= $B5221; := $T5221; := $X5221; := $A5221
%0

%$TTMP8031; := $TTMP8031
§s %0 1 %1

## undefine local variables
:= $P8031
```

2.2.137 File K8032.r0a.txt

```
##
## Proof Template K8032 (=> ALL Rule): H => (A => B)  --> H => (A => ALL x: B)
##
##
## Source: [cf. Andrews 2002 (ISBN 1-4020-0763-9), p. 227 (5237)]
##
## Copyright (c) 2017 Owl of Minerva Press GmbH. All rights reserved.
## Written by Ken Kubota (<mail@kenkubota.de>).
##
## This file is part of the publication of the mathematical logic R0.
## For more information, visit: <http://doi.org/10.4444/100.10>
##



##
## Define Syntactical Variables
##

<< basics.r0.txt

## proposition: H => (A => B)
:= $P8032 ((=>{{{o,o},o}}_h{o}{o}){{o,o}}_((=>{{{o,o},o}}_a{o}{o}){{o,o}}_b{o}{o}){o
})

## type of variable
:= $T8032 o

## the variable
:= $X8032 x{$T8032}


##
## Assumptions and Resulting Syntactical Variables
##

## given proposition
§! $P8032
```

```
##
##  Proof Template
##

<<< K8032.r0t.txt


##
##  Undefine Syntactical Variables
##

:= $P8032; := $T8032; := $X8032


##
##  Q.E.D.
##

%0
```

2.2.138 File K8032.r0t.txt

```
##
##  Proof Template K8032 (=> ALL Rule):  H => (A => B)  -->  H => (A => ALL x: B)
##
##
##  Source: [cf. Andrews 2002 (ISBN 1-4020-0763-9), p. 227 (5237)]
##
##  Copyright (c) 2017 Owl of Minerva Press GmbH. All rights reserved.
##  Written by Ken Kubota (<mail@kenkubota.de>).
##
##  This file is part of the publication of the mathematical logic R0.
##  For more information, visit: <http://doi.org/10.4444/100.10>
##



##
##  Proof Template
##

## the assumption as last theorem on stack (%0)
%$P8032

## use Proof Template K8026 (Deduction Theorem Reversed):  H => (I => A)  -->  (H &
I) => A
<< K8026.r0t.txt
%0
```

```
## use Proof Template A5220H (Gen):  (H => A)  -->  (H => ALL x: A)
:= $T5220H $T8032
:= $X5220H $X8032
:= $A5220H %0
<< A5220H.r0t.txt
:= $T5220H; := $X5220H; := $A5220H
%0

## use Proof Template K8025 (Deduction Theorem):  (H & I) => A  -->  H => (I => A)
<< K8025.r0t.txt
%0
```

2.2.139 File K8033.r0.txt

```
##
##  Proof K8033:  ALL x: EXI1 y: P x y  =>  EXI f: ALL x: P x (f x)
##
##
##  Source: [cf. https://sourceforge.net/p/hol/mailman/message/35361865/ (Sep. 11, 2
016)]
##
##  Copyright (c) 2017 Owl of Minerva Press GmbH. All rights reserved.
##  Written by Ken Kubota (<mail@kenkubota.de>).
##
##  This file is part of the publication of the mathematical logic R0.
##  For more information, visit: <http://doi.org/10.4444/100.10>
##

<< basics.r0.txt
<< K8005.r0.txt
<< A5311.r0.txt



##
##  Proof
##

## .1

:= $HYP8033 ((A{{{o,{o,\3{^}}},^}}_t{^}{^}){{o,{o,t{^}}}}_[\x{t{^}}{t{^}}.((E1{{{o,{
o,\3{^}}},^}}_u{^}{^}){{o,{o,u{^}}}}_[\y{u{^}}{u{^}}.((p{{{o,u{^}},t{^}}}{{{o,u{^}},
t{^}}}_x{t{^}}{t{^}}){{o,u{^}}}_y{u{^}}{u{^}}){o}]{{o,u{^}}}){o}]{{o,t{^}}})

%K8005

## use Proof Template A5221 (Sub):  B  -->  B [x/A]
:= $B5221 %0
:= $T5221 o
:= $X5221 x{$T5221}
```

```
:= $A5221 $HYP8033
<< A5221.r0t.txt
:= $B5221; := $T5221; := $X5221; := $A5221
%0

## .2

## use Proof Template A5215H (ALL I):  H => ALL x: B  -->  H => B [x/a]
:= $T5215H t{^}
:= $X5215H x{$T5215H}
:= $A5215H x{$T5215H}
:= $H5215H %0
<< A5215H.r0t.txt
:= $T5215H; := $X5215H; := $A5215H; := $H5215H
%0

:= $LTMP8033 %0

## .3

%A5311

## use Proof Template A5221 (Sub):  B  -->  B [x/A]
:= $B5221 %0
:= $T5221 ^
:= $X5221 t{$T5221}
:= $A5221 u{$T5221}
<< A5221.r0t.txt
:= $B5221; := $T5221; := $X5221; := $A5221
%0

## use Proof Template A5221 (Sub):  B  -->  B [x/A]
:= $B5221 %0
:= $T5221 {o,u{^}}
:= $X5221 p{$T5221}
:= $A5221 (p{{{o,u{^}},t{^}}}{{{o,u{^}},t{^}}}_x{t{^}}{t{^}})
<< A5221.r0t.txt
:= $B5221; := $T5221; := $X5221; := $A5221
%0

## use Proof Template K8003 (Intro):  A  -->  H => A
:= $A8003 %0
:= $H8003 $LTMP8033/5
<< K8003.r0t.txt
:= $A8003; := $H8003
%0

## .4
```

```
%$LTMP8033; := $LTMP8033

## use Proof Template A5224H (MP):  H => A, H => (A => B)  -->  H => B
:= $A5224H %0
:= $AB5224H %1
<< A5224H.r0t.txt
:= $AB5224H; := $A5224H
%0

## .5

§\ ([\x{t{^}}{t{^}}.(i{{u{^},{o,u{^}}}}_(p{{{o,u{^}},t{^}}}{{{o,u{^}},t{^}}}_x{t{^}}
{t{^}}){{o,u{^}}}}){u{^}}]{{u{^},t{^}}}_x{t{^}}{t{^}})

§= /5
§s %0 5 %1

§s %3 7 %0

## .6

## use Proof Template A5220H (Gen):  (H => A)  -->  (H => ALL x: A)
:= $T5220H t{^}
:= $X5220H x{$T5220H}
:= $A5220H %0
<< A5220H.r0t.txt
:= $T5220H; := $X5220H; := $A5220H
%0

## .7

§\ ([\f{{u{^},t{^}}}{{u{^},t{^}}}.((A{{{o,{o,\3{^}}},^}}_t{^}{^}){{o,{o,t{^}}}}_[\x{
t{^}}{t{^}}.((p{{{o,u{^}},t{^}}}{{{o,u{^}},t{^}}}_x{t{^}}{t{^}}){{o,u{^}}}}_(f{{u{^},
t{^}}}{{u{^},t{^}}}_x{t{^}}{t{^}}){u{^}}){o}]{{o,t{^}}}){o}]{{o,{u{^},t{^}}}}_[\x{t{
^}}{t{^}}.(i{{u{^},{o,u{^}}}}_(p{{{o,u{^}},t{^}}}{{{o,u{^}},t{^}}}_x{t{^}}{t{^}}){{o
,u{^}}}}){u{^}}]{{u{^},t{^}}})

§= /5
§s %0 5 %1

§s %3 3 %0

## .8

## use Proof Template K8028 (EXI GenH):  H => ([\x.B]A)  -->  H => EXI x: B
:= $H8028 $HYP8033
:= $T8028 {u{^},t{^}}
:= $B8028 %0/6
:= $A8028 %0/7
```

```
<< K8028.r0t.txt
:= $H8028; := $T8028; := $B8028; := $A8028
%0

:= K8033 %0

## undefine local variables
:= $HYP8033


##
##  Q.E.D.
##

%0
```

2.2.140 File a5205_substitutions.r0.txt

```
##
##  Proof Template:  A5205 Substitutions
##
##
##  Source: [Kubota 2017 (doi: 10.4444/100.10)]
##
##  Copyright (c) 2017 Owl of Minerva Press GmbH. All rights reserved.
##  Written by Ken Kubota (<mail@kenkubota.de>).
##
##  This file is part of the publication of the mathematical logic R0.
##  For more information, visit: <http://doi.org/10.4444/100.10>
##



##
##  Define Syntactical Variables
##

## replacement for type a (alpha) in A5205
:= $AA5205 o

## replacement for type b (beta) in A5205
:= $BA5205 t{^}

## replacement for f() in A5205
:= $FA5205 p{{$AA5205,$BA5205}}


##
##  Include Proof Template
##
```

```
<<< a5205_substitutions.r0t.txt


##
##  Undefine Syntactical Variables
##

:= $AA5205; := $BA5205; := $FA5205


##
##  Q.E.D.
##

%0
```

2.2.141 File a5205_substitutions.r0t.txt

```
##
##  Proof Template:  Axiom 2 Substitutions
##
##
##  Source: [Kubota 2017 (doi: 10.4444/100.10)]
##
##  Copyright (c) 2017 Owl of Minerva Press GmbH. All rights reserved.
##  Written by Ken Kubota (<mail@kenkubota.de>).
##
##  This file is part of the publication of the mathematical logic R0.
##  For more information, visit: <http://doi.org/10.4444/100.10>
##

<< A5205.r0.txt


##
##  Proof Template
##

## .1

%A5205

## .1a Replace type a (alpha) in A5205

## use Proof Template A5221 (Sub):  B  -->  B [x/A]
:= $B5221 %0
:= $T5221 ^
:= $X5221 a{$T5221}
```

```
:= $A5221 $AA5205
<< A5221.r0t.txt
:= $B5221; := $T5221; := $X5221; := $A5221
%0

## .1b Replace type b (beta) in A5205

## use Proof Template A5221 (Sub):  B  -->  B [x/A]
:= $B5221 %0
:= $T5221 ^
:= $X5221 b{$T5221}
:= $A5221 $BA5205
<< A5221.r0t.txt
:= $B5221; := $T5221; := $X5221; := $A5221
%0

## .1c Replace f() in A5205

## use Proof Template A5221 (Sub):  B  -->  B [x/A]
:= $B5221 %0
:= $T5221 {$AA5205,$BA5205}
:= $X5221 f{$T5221}
:= $A5221 $FA5205
<< A5221.r0t.txt
:= $B5221; := $T5221; := $X5221; := $A5221
%0
```

2.2.142 File axiom2_substitutions.r0.txt

```
##
##  Proof Template:  Axiom 2 Substitutions
##
##
##  Source: [Kubota 2017 (doi: 10.4444/100.10)]
##
##  Copyright (c) 2017 Owl of Minerva Press GmbH. All rights reserved.
##  Written by Ken Kubota (<mail@kenkubota.de>).
##
##  This file is part of the publication of the mathematical logic R0.
##  For more information, visit: <http://doi.org/10.4444/100.10>
##


##
##  Define Syntactical Variables
##

<< basics.r0.txt
```

```
## replacement for type a (alpha) in Axiom 2
:= $AA2 {o,a{^}}

## replacement for h() in Axiom 2
:= $HA2 [\f{{o,a{^}}}{{o,a{^}}}.(f{{o,a{^}}}{{o,a{^}}}_x{a{^}}{a{^}}){o}]

## replacement for x in Axiom 2
:= $XA2 [\x{a{^}}{a{^}}.T{o}]

## replacement for y in Axiom 2
:= $YA2 f{{o,a{^}}}


##
##  Include Proof Template
##

<<< axiom2_substitutions.r0t.txt


##
##  Undefine Syntactical Variables
##

:= $AA2; := $HA2; := $XA2; := $YA2


##
##  Q.E.D.
##

%0
```

2.2.143 File axiom2_substitutions.r0t.txt

```
##
##  Proof Template:  Axiom 2 Substitutions
##
##
##  Source: [Kubota 2017 (doi: 10.4444/100.10)]
##
##  Copyright (c) 2017 Owl of Minerva Press GmbH. All rights reserved.
##  Written by Ken Kubota (<mail@kenkubota.de>).
##
##  This file is part of the publication of the mathematical logic R0.
##  For more information, visit: <http://doi.org/10.4444/100.10>
##

<< axioms.r0.txt
```

```
##
##  Proof Template
##

## .1

## Axiom 2: One of the Basic Properties of Equality
%A2

## .1a Replace type a (alpha) in Axiom 2

## use Proof Template A5221 (Sub):  B  -->  B [x/A]
:= $B5221 %0
:= $T5221 ^
:= $X5221 a{^}
:= $A5221 $AA2
<< A5221.r0t.txt
:= $B5221; := $T5221; := $X5221; := $A5221
%0

## .1b Replace h() in Axiom 2

## use Proof Template A5221 (Sub):  B  -->  B [x/A]
:= $B5221 %0
:= $T5221 {{o,$AA2}}
:= $X5221 h{{o,$AA2}}
:= $A5221 $HA2
<< A5221.r0t.txt
:= $B5221; := $T5221; := $X5221; := $A5221
%0

## .1c Replace x in Axiom 2

## use Proof Template A5221 (Sub):  B  -->  B [x/A]
:= $B5221 %0
:= $T5221 {$AA2}
:= $X5221 x{$AA2}
:= $A5221 $XA2
<< A5221.r0t.txt
:= $B5221; := $T5221; := $X5221; := $A5221
%0

## .1d Replace y in Axiom 2

## use Proof Template A5221 (Sub):  B  -->  B [x/A]
:= $B5221 %0
:= $T5221 {$AA2}
```

```
:= $X5221 y{$AA2}
:= $A5221 $YA2
<< A5221.r0t.txt
:= $B5221; := $T5221; := $X5221; := $A5221
%0
```

2.2.144 File axiom3_substitutions.r0.txt

```
##
##  Proof Template:  Axiom 3 Substitutions
##
##
##  Source: [Kubota 2017 (doi: 10.4444/100.10)]
##
##  Copyright (c) 2017 Owl of Minerva Press GmbH. All rights reserved.
##  Written by Ken Kubota (<mail@kenkubota.de>).
##
##  This file is part of the publication of the mathematical logic R0.
##  For more information, visit: <http://doi.org/10.4444/100.10>
##



##
##  Define Syntactical Variables
##

## replacement for type a (alpha) in Axiom 3
:= $AA3 t{^}

## replacement for type b (beta) in Axiom 3
:= $BA3 u{^}

## replacement for f() in Axiom 3
:= $FA3 y{{$AA3,$BA3}}

## replacement for g() in Axiom 3
:= $GA3 z{{$AA3,$BA3}}



##
##  Include Proof Template
##

<<< axiom3_substitutions.r0t.txt


##
##  Undefine Syntactical Variables
##
```

```
:= $AA3; := $BA3; := $FA3; := $GA3


##
## Q.E.D.
##

%0
```

2.2.145 File axiom3_substitutions.r0t.txt

```
##
## Proof Template: Axiom 3 Substitutions
##
##
## Source: [Kubota 2017 (doi: 10.4444/100.10)]
##
## Copyright (c) 2017 Owl of Minerva Press GmbH. All rights reserved.
## Written by Ken Kubota (<mail@kenkubota.de>).
##
## This file is part of the publication of the mathematical logic R0.
## For more information, visit: <http://doi.org/10.4444/100.10>
##

<< axioms.r0.txt


##
## Proof Template
##

## .1

## Axiom 3: Axiom of Extensionality
%A3

## .1a Replace type a (alpha) in Axiom 3

## use Proof Template A5209 (incl. A5204): B = C --> (B = C) [x/A]
:= $M5209 o
:= $E5209 %0
:= $T5209 ^
:= $X5209 a{^}
:= $A5209 $AA3
<< A5209.r0t.txt
:= $M5209; := $E5209; := $T5209; := $X5209; := $A5209
%0
```

```
## .1b Replace type b (beta) in Axiom 3

## use Proof Template A5209 (incl. A5204):  B = C  -->  (B = C) [x/A]
:= $M5209 o
:= $E5209 %0
:= $T5209 ^
:= $X5209 b{^}
:= $A5209 $BA3
<< A5209.r0t.txt
:= $M5209; := $E5209; := $T5209; := $X5209; := $A5209
%0

## .1c Replace f() in Axiom 3

## use Proof Template A5209 (incl. A5204):  B = C  -->  (B = C) [x/A]
:= $M5209 o
:= $E5209 %0
:= $T5209 {{$AA3,$BA3}}
:= $X5209 f{$T5209}
:= $A5209 $FA3
<< A5209.r0t.txt
:= $M5209; := $E5209; := $T5209; := $X5209; := $A5209
%0

## .1d Replace g() in Axiom 3

## use Proof Template A5209 (incl. A5204):  B = C  -->  (B = C) [x/A]
:= $M5209 o
:= $E5209 %0
:= $T5209 {{$AA3,$BA3}}
:= $X5209 g{$T5209}
:= $A5209 $GA3
<< A5209.r0t.txt
:= $M5209; := $E5209; := $T5209; := $X5209; := $A5209
%0
```

2.2.146 File axiom_of_choice.r0a.txt

```
##
##  Axiom of Choice
##
##
##  Source: [Andrews 2002 (ISBN 1-4020-0763-9), p. 236]
##
##  Copyright (c) 2017 Owl of Minerva Press GmbH. All rights reserved.
##  Written by Ken Kubota (<mail@kenkubota.de>).
##
##  This file is part of the publication of the mathematical logic R0.
##  For more information, visit: <http://doi.org/10.4444/100.10>
```

```
##

<< definitions1.r0.txt


##
##  Axiom of Choice
##

:= AC ((E{{{o,{o,\3{^}}},^}}_{t{^},{o,t{^}}}{^}){{o,{o,{t{^},{o,t{^}}}}}}_[\j{{t{^},
{o,t{^}}}}{{t{^},{o,t{^}}}}.((A{{{o,{o,\3{^}}},^}}_{o,t{^}}{^}){{o,{o,{o,t{^}}}}}_[\
p{{o,t{^}}}{{o,t{^}}}.((=>{{{o,o},o}}_((E{{{o,{o,\3{^}}},^}}_t{^}{^}){{o,{o,t{^}}}}_
[\x{t{^}}{t{^}}.(p{{o,t{^}}}{{o,t{^}}}_x{t{^}}{t{^}}){o}]{{o,t{^}}}){o}){{o,o}}_(p{{
o,t{^}}}{{o,t{^}}}_(j{{t{^},{o,t{^}}}}{{t{^},{o,t{^}}}}_p{{o,t{^}}}{{o,t{^}}}){t{^}}
){o}){o}]{{o,{o,t{^}}}}){o}]{{o,{t{^},{o,t{^}}}}})
§! AC
```

2.2.147 File axioms.r0.txt

```
##
##  Axioms
##
##
##  Source: [Andrews 2002 (ISBN 1-4020-0763-9), p. 213]
##
##  Copyright (c) 2017 Owl of Minerva Press GmbH. All rights reserved.
##  Written by Ken Kubota (<mail@kenkubota.de>).
##
##  This file is part of the publication of the mathematical logic R0.
##  For more information, visit: <http://doi.org/10.4444/100.10>
##

<< definitions1.r0.txt


##
##  Axiom 1: Truth and Falsehood Are the Only Truth Values
##

:= A1 ((={{{o,o},o}}_((&{{{o,o},o}}_(g{{o,o}}{{o,o}}_T{o}){o}){{o,o}}_(g{{o,o}}{{o,o
}}_F{o}){o}){o}){{o,o}}_((A{{{o,{o,\3{^}}},^}}_o{^}){{o,{o,o}}}_[\x{o}{o}.(g{{o,o}}{
{o,o}}_x{o}{o}){o}]{{o,o}}){o})
§! A1


##
##  Axiom 2: One of the Basic Properties of Equality
##
```

```
:= A2 ((=>{{{o,o},o}}_((={{{o,a{^}},a{^}}}_x{a{^}}{a{^}}){{o,a{^}}}_y{a{^}}{a{^}}){o
}){{o,o}}_((={{{o,o},o}}_(h{{o,a{^}}}{{o,a{^}}}_x{a{^}}{a{^}}){o}){{o,o}}_(h{{o,a{^}
}}{{o,a{^}}}_y{a{^}}{a{^}}){o}){o})
§! A2


##
##  Axiom 3: Axiom of Extensionality
##

:= A3 ((={{{o,o},o}}_((={{{o,{a{^},b{^}}},{a{^},b{^}}}}_f{{a{^},b{^}}}{{a{^},b{^}}})
{{o,{a{^},b{^}}}}_g{{a{^},b{^}}}{{a{^},b{^}}}){o}){{o,o}}_((A{{{o,{o,\3{^}}},^}}_b{^
}{^}){{o,{o,b{^}}}}_[\x{b{^}}{b{^}}.((={{{o,a{^}},a{^}}}_(f{{a{^},b{^}}}{{a{^},b{^}}
}_x{b{^}}{b{^}}){a{^}}){{o,a{^}}}_(g{{a{^},b{^}}}{{a{^},b{^}}}_x{b{^}}{b{^}}){a{^}})
{o}]{{o,b{^}}}){o})
§! A3


##
##  Axiom 4: Axiom of Lambda Conversion
##

## Replaced by Rule 2 (Lambda Conversion)
## [cf. Andrews 2002 (ISBN 1-4020-0763-9), pp. 218 f. (5207)]
##
## "5207 could be taken as an axiom schema in place of 4_1 - 4_5,
##  and for some purposes this would be desirable,
##  since 5207 has a conceptual simplicity and unity
##  which is not apparent in 4_1 - 4_5." [Andrews 2002, p. 214]


##
##  Axiom 5: Axiom of Descriptions
##

:= A5 ((={{{o,t{^}},t{^}}}_(i{{t{^},{o,t{^}}}}_(={{{o,t{^}},t{^}}}_y{t{^}}{t{^}}){{o
,t{^}}}){t{^}}){{o,t{^}}}_y{t{^}}{t{^}})
§! A5
```

2.2.148 File basics.r0.txt

```
##
##  Basics
##
##
##  Source: [Kubota 2017 (doi: 10.4444/100.10)]
##
##  Copyright (c) 2017 Owl of Minerva Press GmbH. All rights reserved.
##  Written by Ken Kubota (<mail@kenkubota.de>).
```

```
##
##  This file is part of the publication of the mathematical logic R0.
##  For more information, visit: <http://doi.org/10.4444/100.10>
##

<< definitions1.r0.txt
<< definitions2.r0.txt
<< definitions3.r0.txt
<< axioms.r0.txt
```

2.2.149 File composition.r0.txt

```
##
##  Associativity of the Composition of Functions
##
##
##  Source: [Kubota 2017 (doi: 10.4444/100.10)]
##
##  Copyright (c) 2017 Owl of Minerva Press GmbH. All rights reserved.
##  Written by Ken Kubota (<mail@kenkubota.de>).
##
##  This file is part of the publication of the mathematical logic R0.
##  For more information, visit: <http://doi.org/10.4444/100.10>
##

<< basics.r0.txt


:= COMPS [\a{^}{^}.[\b{^}{^}.[\c{^}{^}.[\g{{a{^},b{^}}}{{a{^},b{^}}}.[\f{{b{^},c{^}}
}{{b{^},c{^}}}.[\x{c{^}}{c{^}}.(g{{a{^},b{^}}}{{a{^},b{^}}}_(f{{b{^},c{^}}}{{b{^},c{
^}}}_x{c{^}}{c{^}})(b{^}})(a{^}}]{{a{^},c{^}}}]{{{a{^},c{^}},{b{^},c{^}}}}]{{{{a{^},
c{^}},{b{^},c{^}}},{a{^},b{^}}}}]{{{{{a{^},\4{^}},{b{^},\4{^}}},{a{^},b{^}}},^}}]{{{
{{{a{^},\4{^}},{\5{^},\4{^}}},{a{^},\4{^}}},^},^}}]

## .1

:= $GF (((((COMPS{{{{{{{\6{^},\4{^}},{\5{^},\4{^}}},{\5{^},\4{^}}},^},^},^}}_u{^}{^}
){{{{{{u{^},\4{^}},{\5{^},\4{^}}},{u{^},\4{^}}},^},^}}_v{^}{^}){{{{{u{^},\4{^}},{v{^
},\4{^}}},{u{^},v{^}}},^}}_w{^}{^}){{{{u{^},w{^}},{v{^},w{^}}},{u{^},v{^}}}}_g{{u{^}
,v{^}}}{{u{^},v{^}}}){{{u{^},w{^}},{v{^},w{^}}}}_f{{v{^},w{^}}}{{v{^},w{^}}})
§= $GF
§\s /48
§\s /24
§\s /12
§\s /6
§\s /3

:= $HxGF (((((COMPS{{{{{{{\6{^},\4{^}},{\5{^},\4{^}}},{\5{^},\4{^}}},^},^},^}}_t{^}{
^}){{{{{{t{^},\4{^}},{\5{^},\4{^}}},{t{^},\4{^}}},^},^}}_u{^}{^}){{{{{t{^},\4{^}},{u
```

```
{^},\4{^}}},{t{^},u{^}}},^}}_w{^}{^}){{{{t{^},w{^}},{u{^},w{^}}},{t{^},u{^}}}}_h{{t{
^},u{^}}}{{t{^},u{^}}}){{{t{^},w{^}},{u{^},w{^}}}}_$GF{{u{^},w{^}}})
§= $HxGF
§s %0 7 %1
§\s /48
§\s /24
§\s /12
§\s /6
§\s /3
§\s /15

:= $TMP1 %0

## .2

:= $HG (((((COMPS{{{{{{{\6{^},\4{^}},{\5{^},\4{^}}},{\5{^},\4{^}}},^},^},^}}_t{^}{^}
){{{{{{t{^},\4{^}},{\5{^},\4{^}}},{t{^},\4{^}}},^},^}}_u{^}{^}){{{{{t{^},\4{^}},{u{^
},\4{^}}},{t{^},u{^}}},^}}_v{^}{^}){{{{t{^},v{^}},{u{^},v{^}}},{t{^},u{^}}}}_h{{t{^}
,u{^}}}{{t{^},u{^}}}){{{t{^},v{^}},{u{^},v{^}}}}_g{{u{^},v{^}}}{{u{^},v{^}}})
§= $HG
§\s /48
§\s /24
§\s /12
§\s /6
§\s /3

:= $HGxF (((((COMPS{{{{{{{\6{^},\4{^}},{\5{^},\4{^}}},{\5{^},\4{^}}},^},^},^}}_t{^}{
^}){{{{{{t{^},\4{^}},{\5{^},\4{^}}},{t{^},\4{^}}},^},^}}_v{^}{^}){{{{{t{^},\4{^}},{v
{^},\4{^}}},{t{^},v{^}}},^}}_w{^}{^}){{{{t{^},w{^}},{v{^},w{^}}},{t{^},v{^}}}}_$HG{{
t{^},v{^}}}){{{t{^},w{^}},{v{^},w{^}}}}_f{{v{^},w{^}}}{{v{^},w{^}}})
§= $HGxF
§s %0 13 %1
§\s /48
§\s /24
§\s /12
§\s /6
§\s /3
§\s /7

:= $TMP2 %0

## .3

%$TMP1; := $TMP1
%$TMP2; := $TMP2

## use Proof Template A5201b (Swap):  A = B  -->  B = A
<< A5201b.r0t.txt
%0
```

```
§s %4 3 %0

:= $GF; := $HG; := $HxGF; := $HGxF


##
##  Print Result
##

%0
```

2.2.150 File definitions1.r0.txt

```
##
##  Basic Definitions
##
##
##  Source: [Andrews 2002 (ISBN 1-4020-0763-9), p. 212]
##
##  Copyright (c) 2017 Owl of Minerva Press GmbH. All rights reserved.
##  Written by Ken Kubota (<mail@kenkubota.de>).
##
##  This file is part of the publication of the mathematical logic R0.
##  For more information, visit: <http://doi.org/10.4444/100.10>
##

## Definition of truth
:= T ((={{{o,@},@}}_={@}){{o,@}}_={@})

## Definition of falsehood
:= F ((={{o,{o,o}},{o,o}}_[\x{o}{o}.T{o}]{o,o}){{o,{o,o}}}_[\x{o}{o}.x{o}{o}]{o,o})

## Definition of the universal quantifier (with type abstraction)
:= A [\t{^}{^}.[\p{{o,t{^}}}{{o,t{^}}}.((={{{o,{o,t{^}}},{o,t{^}}}}_[\x{t{^}}{t{^}}.
T{o}]{{o,t{^}}}){{o,{o,t{^}}}}_p{{o,t{^}}}{{o,t{^}}}){o}]{{o,{o,t{^}}}}]

## Definition of the conjunction
:= & [\x{o}{o}.[\y{o}{o}.((={{{o,@},@}}_[\g{{{o,o},o}}{{{o,o},o}}.((g{{{o,o},o}}{{{o
,o},o}}_T{o}){{o,o}}_T{o}){o}]{@}){{o,@}}_[\g{{{o,o},o}}{{{o,o},o}}.((g{{{o,o},o}}{{
{o,o},o}}_x{o}{o}){{o,o}}_y{o}{o}){o}]{@}){o}]{{o,o}}]

## Definition of the implication
:= => [\x{o}{o}.[\y{o}{o}.((={{{o,o},o}}_x{o}{o}){{o,o}}_((&{{{o,o},o}}_x{o}{o}){{o,
o}}_y{o}{o}){o}){o}]{{o,o}}]

## Definition of the negation
:= ! [\a{o}{o}.((={{{o,o},o}}_F{o}){{o,o}}_a{o}{o}){o}]
```

```
## Definition of the disjunction
:= | [\a{o}{o}.[\b{o}{o}.(!{{o,o}}_((&{{{o,o},o}}_(!{{o,o}}_a{o}{o}){o}){{o,o}}_(!{{
o,o}}_b{o}{o}){o}){o}){o}]{{o,o}}]

## Definition of the existential quantifier (with type abstraction)
:= E [\t{^}{^}.[\p{{o,t{^}}}{{o,t{^}}}.(!{{o,o}}_((={{{o,{o,t{^}}},{o,t{^}}}}_[\x{t{
^}}{t{^}}.T{o}]{{o,t{^}}}){{o,{o,t{^}}}}_[\x{t{^}}{t{^}}.(!{{o,o}}_(p{{o,t{^}}}{{o,t
{^}}}_x{t{^}}{t{^}}){o}){o}]{{o,t{^}}}){o}){o}]{{o,{o,t{^}}}}]

## Definition of inequality
:= != [\x{@}{@}.[\y{@}{@}.(!{{o,o}}_((={{{o,@},@}}_x{@}{@}){{o,@}}_y{@}{@}){o}){o}]{
{o,@}}]
```

2.2.151 File definitions2.r0.txt

```
##
##  Further Definitions
##
##
##  Source: [Andrews 2002 (ISBN 1-4020-0763-9), pp. 231, 233]
##
##  Copyright (c) 2017 Owl of Minerva Press GmbH. All rights reserved.
##  Written by Ken Kubota (<mail@kenkubota.de>).
##
##  This file is part of the publication of the mathematical logic R0.
##  For more information, visit: <http://doi.org/10.4444/100.10>
##

<< definitions1.r0.txt

## Definition of the subset
:= SBSET [\t{^}{^}.[\x{{o,t{^}}}{{o,t{^}}}.[\y{{o,t{^}}}{{o,t{^}}}.((A{{{o,{o,\3{^}}
},^}}_t{^}{^}){{o,{o,t{^}}}}_[\z{t{^}}{t{^}}.((=>{{{o,o},o}}_(x{{o,t{^}}}{{o,t{^}}}_
z{t{^}}{t{^}}){o}){{o,o}}_(y{{o,t{^}}}{{o,t{^}}}_z{t{^}}{t{^}}){o}){o}]{{o,t{^}}}){o
}]{{o,{o,t{^}}}}]{{{o,{o,t{^}}},{o,t{^}}}}]

## Definition of the power set
:= PWSET [\t{^}{^}.[\y{{o,t{^}}}{{o,t{^}}}.[\x{{o,t{^}}}{{o,t{^}}}.(((SBSET{{{{o,{o,
\4{^}}},{o,\3{^}}},^}}_t{^}{^}){{{o,{o,t{^}}},{o,t{^}}}}_x{{o,t{^}}}{{o,t{^}}}){{o,{
o,t{^}}}}_y{{o,t{^}}}{{o,t{^}}}){o}]{{o,{o,t{^}}}}]{{{o,{o,t{^}}},{o,t{^}}}}]

## Definition of the uniqueness quantifier (with type abstraction)
:= E1 [\t{^}{^}.[\p{{o,t{^}}}{{o,t{^}}}.((E{{{o,{o,\3{^}}},^}}_t{^}{^}){{o,{o,t{^}}}
}_[\y{t{^}}{t{^}}.((={{{o,{o,t{^}}},{o,t{^}}}}_p{{o,t{^}}}{{o,t{^}}}){{o,{o,t{^}}}}_
(={{{o,t{^}},t{^}}}_y{t{^}}{t{^}}){o,t{^}}){o}]{{o,t{^}}}){o}]{{o,{o,t{^}}}}]
```

2.2.152 File definitions3.r0.txt

```
##
##  New Definitions
##
##
##  Source: [Kubota 2017 (doi: 10.4444/100.10)]
##
##  Copyright (c) 2017 Owl of Minerva Press GmbH. All rights reserved.
##  Written by Ken Kubota (<mail@kenkubota.de>).
##
##  This file is part of the publication of the mathematical logic R0.
##  For more information, visit: <http://doi.org/10.4444/100.10>
##

<< definitions2.r0.txt

## Definition of the universal set
:= V [\x{@}{@}.T{o}]

## Definition of the empty set
:= O [\x{@}{@}.F{o}]

## Definition of the polymorphic identity relation helper function
:= == [\t{^}{^}.[\x{t{^}}{t{^}}.[\y{t{^}}{t{^}}.((={{{o,t{^}},t{^}}}_x{t{^}}{t{^}}){
{o,t{^}}}_y{t{^}}{t{^}}){o}]{{o,t{^}}}]{{{o,t{^}},t{^}}}]

## Definition of the polymorphic non-identity relation helper function
:= !== [\t{^}{^}.[\x{t{^}}{t{^}}.[\y{t{^}}{t{^}}.(!{{o,o}}_((={{{o,t{^}},t{^}}}_x{t{
^}}{t{^}}){{o,t{^}}}_y{t{^}}{t{^}}){o}){o}]{{o,t{^}}}]{{{o,t{^}},t{^}}}]

## Definition of the polymorphic descriptor helper function
:= I [\t{^}{^}.[\x{{o,t{^}}}{{o,t{^}}}.(i{{t{^},{o,t{^}}}}_x{{o,t{^}}}{{o,t{^}}}){t{
^}}]{{t{^},{o,t{^}}}}]

## Definition of exclusive disjunction (logical exclusive "or", XOR)
:= XOR [\x{o}{o}.[\y{o}{o}.(!{{o,o}}_((={{{o,o},o}}_x{o}{o}){{o,o}}_y{o}{o}){o}){o}]
{{o,o}}]

## Definition of commutativity
:= COMMT [\t{^}{^}.[\f{{{t{^},t{^}},t{^}}}{{{t{^},t{^}},t{^}}}.((={{{o,t{^}},t{^}}}_
((f{{{t{^},t{^}},t{^}}}{{{t{^},t{^}},t{^}}}_x{t{^}}{t{^}}){{t{^},t{^}}}_y{t{^}}{t{^}
}){t{^}}){{o,t{^}}}_((f{{{t{^},t{^}},t{^}}}{{{t{^},t{^}},t{^}}}_y{t{^}}{t{^}}){{t{^}
,t{^}}}_x{t{^}}{t{^}}){t{^}}){o}]{{o,{{t{^},t{^}},t{^}}}}]

## Definition of associativity
:= ASSOC [\t{^}{^}.[\f{{{t{^},t{^}},t{^}}}{{{t{^},t{^}},t{^}}}.((={{{o,t{^}},t{^}}}_
((f{{{t{^},t{^}},t{^}}}{{{t{^},t{^}},t{^}}}_((f{{{t{^},t{^}},t{^}}}{{{t{^},t{^}},t{^
}}}_x{t{^}}{t{^}}){{t{^},t{^}}}_y{t{^}}{t{^}}){t{^}}){{t{^},t{^}}}_z{t{^}}{t{^}}){t{
```

```
^}})({o,t{^}}}_((f{{{t{^},t{^}},t{^}}}{{{t{^},t{^}},t{^}}}_x{t{^}}{t{^}})({t{^},t{^}
}}_((f{{{t{^},t{^}},t{^}}}{{{t{^},t{^}},t{^}}}_y{t{^}}{t{^}})({t{^},t{^}}}_z{t{^}}{t
{^}})({t{^}})({t{^}})({o}]{{o,{{t{^},t{^}},t{^}}}}]
```

2.2.153 File group.r0.txt

```
##
##  Groups
##
##
##  Source: [Kubota 2017 (doi: 10.4444/100.10)]
##
##  Copyright (c) 2017 Owl of Minerva Press GmbH. All rights reserved.
##  Written by Ken Kubota (<mail@kenkubota.de>).
##
##  This file is part of the publication of the mathematical logic R0.
##  For more information, visit: <http://doi.org/10.4444/100.10>
##

<< basics.r0.txt


## .1:  Associativity
:= GrpAsc ((A{{{o,{o,\3{^}}},^}}_g{^}{^}){{o,{o,g{^}}}}_[\a{g{^}}{g{^}}.((A{{{o,{o,\
3{^}}},^}}_g{^}{^}){{o,{o,g{^}}}}_[\b{g{^}}{g{^}}.((A{{{o,{o,\3{^}}},^}}_g{^}{^}){{o
,{o,g{^}}}}_[\c{g{^}}{g{^}}.((={{{o,g{^}},g{^}}}_((l{{{g{^},g{^}},g{^}}}{{{g{^},g{^}
},g{^}}}_((l{{{g{^},g{^}},g{^}}}{{{g{^},g{^}},g{^}}}_a{g{^}}{g{^}}){{g{^},g{^}}}_b{g
{^}}{g{^}}){g{^}}){{g{^},g{^}}}_c{g{^}}{g{^}}){g{^}}){{o,g{^}}}_((l{{{g{^},g{^}},g{^
}}}{{{g{^},g{^}},g{^}}}_a{g{^}}{g{^}}){{g{^},g{^}}}_((l{{{g{^},g{^}},g{^}}}{{{g{^},g
{^}},g{^}}}_b{g{^}}{g{^}}){{g{^},g{^}}}_c{g{^}}{g{^}}){g{^}}){g{^}}){o}]{{o,g{^}}}){
o}]{{o,g{^}}}){o}]{{o,g{^}}})

## .2:  Identity element
:= GrpIdy ((A{{{o,{o,\3{^}}},^}}_g{^}{^}){{o,{o,g{^}}}}_[\a{g{^}}{g{^}}.((&{{{o,o},o
}}_((={{{o,g{^}},g{^}}}_((l{{{g{^},g{^}},g{^}}}{{{g{^},g{^}},g{^}}}_a{g{^}}{g{^}}){{
g{^},g{^}}}_e{g{^}}{g{^}}){g{^}}){{o,g{^}}}_a{g{^}}{g{^}}){o}){{o,o}}_((={{{o,g{^}},
g{^}}}_((l{{{g{^},g{^}},g{^}}}{{{g{^},g{^}},g{^}}}_e{g{^}}{g{^}}){{g{^},g{^}}}_a{g{^
}}{g{^}}){g{^}}){{o,g{^}}}_a{g{^}}{g{^}}){o}){o}]{{o,g{^}}})

## .3:  Inverse element
:= GrpInv ((A{{{o,{o,\3{^}}},^}}_g{^}{^}){{o,{o,g{^}}}}_[\a{g{^}}{g{^}}.((E{{{o,{o,\
3{^}}},^}}_g{^}{^}){{o,{o,g{^}}}}_[\b{g{^}}{g{^}}.((&{{{o,o},o}}_((={{{o,g{^}},g{^}}
}_((l{{{g{^},g{^}},g{^}}}{{{g{^},g{^}},g{^}}}_a{g{^}}{g{^}}){{g{^},g{^}}}_b{g{^}}{g{
^}}){g{^}}){{o,g{^}}}_e{g{^}}{g{^}}){o}){{o,o}}_((={{{o,g{^}},g{^}}}_((l{{{g{^},g{^}
},g{^}}}{{{g{^},g{^}},g{^}}}_b{g{^}}{g{^}}){{g{^},g{^}}}_a{g{^}}{g{^}}){g{^}}){{o,g{
^}}}_e{g{^}}{g{^}}){o}){o}]{{o,g{^}}}){o}]{{o,g{^}}})


##
```

```
##  Definition of group (all three group properties combined)
##


:= Grp [\g{^}{^}.[\l{{{g{^},g{^}},g{^}}}{{{g{^},g{^}},g{^}}}.((&{{{o,o},o}}_GrpAsc{o
}){{o,o}}_((E{{{o,{o,\3{^}}},^}}_g{^}{^}){{o,{o,g{^}}}}_[\e{g{^}}{g{^}}.((&{{{o,o},o
}}_GrpIdy{o}){{o,o}}_GrpInv{o}){o}]{{o,g{^}}}){o}){o}]{{o,{{g{^},g{^}},g{^}}}}]


## Group property identity element only (with identity element abstracted)
:= GrpId0 [\g{^}{^}.[\l{{{g{^},g{^}},g{^}}}{{{g{^},g{^}},g{^}}}.[\e{g{^}}{g{^}}.GrpI
dy{o}]{{o,g{^}}}]{{{o,g{^}},{{g{^},g{^}},g{^}}}}]
```

2.2.154 File group_identity_element_unique.r0.txt

```
##
##  Uniqueness of the Group Identity Element
##
##
##  Source: [Kubota 2017 (doi: 10.4444/100.10)]
##
##  Copyright (c) 2017 Owl of Minerva Press GmbH. All rights reserved.
##  Written by Ken Kubota (<mail@kenkubota.de>).
##
##  This file is part of the publication of the mathematical logic R0.
##  For more information, visit: <http://doi.org/10.4444/100.10>
##

<< basics.r0.txt
<< K8005.r0.txt
<< group.r0.txt

## shorthands
:= $HYPTH ((&{{{o,o},o}}_((&{{{o,o},o}}_((Grp{{{o,{{\4{^},\4{^}},\3{^}}}},^}}_g{^}{^}
){{o,{{g{^},g{^}},g{^}}}}}_l{{{g{^},g{^}},g{^}}}{{{g{^},g{^}},g{^}}}){o}){{o,o}}_(((G
rpId0{{{{o,\3{^}},{{\4{^},\4{^}},\3{^}}}},^}}_g{^}{^}){{{o,g{^}},{{g{^},g{^}},g{^}}}}
_l{{{g{^},g{^}},g{^}}}{{{g{^},g{^}},g{^}}}){{o,g{^}}}_e{g{^}}{g{^}}){o}){o}){{o,o}}_
(((GrpId0{{{{o,\3{^}},{{\4{^},\4{^}},\3{^}}}},^}}_g{^}{^}){{{o,g{^}},{{g{^},g{^}},g{^
}}}}_l{{{g{^},g{^}},g{^}}}{{{g{^},g{^}},g{^}}}){{o,g{^}}}_f{g{^}}{g{^}}){o})
:= $TMPDED ((A{{{o,{o,\3{^}}},^}}_g{^}{^}){{o,{o,g{^}}}}_[\a{g{^}}{g{^}}.((&{{{o,o},
o}}_((={{{o,g{^}},g{^}}}_((l{{{g{^},g{^}},g{^}}}{{{g{^},g{^}},g{^}}}_a{g{^}}{g{^}}){
{g{^},g{^}}}_f{g{^}}{g{^}}){g{^}}){{o,g{^}}}_a{g{^}}{g{^}}){o}){{o,o}}_((={{{o,g{^}}
,g{^}}}_((l{{{g{^},g{^}},g{^}}}{{{g{^},g{^}},g{^}}}_f{g{^}}{g{^}}){{g{^},g{^}}}_a{g{
^}}{g{^}}){g{^}}){{o,g{^}}}_a{g{^}}{g{^}}){o}){o}]{{o,g{^}}})

## .1:  Let (g,l) be a group, and e and f identity elements of it

%K8005

## use Proof Template A5221 (Sub):  B  -->  B [x/A]
```

```
:= $B5221 %0
:= $T5221 o
:= $X5221 x{$T5221}
:= $A5221 $HYPTH
<< A5221.r0t.txt
:= $B5221; := $T5221; := $X5221; := $A5221

:= $FULLH %0

## .2:  Proof of H  =>  e * f = e

%$FULLH

## use Proof Template K8019H:  H => (A & B)  -->  H => A, H => B
:= $H8019H %0
<< K8019H.r0t.txt
:= $H8019H
%$B8019H
:= $A8019H; := $B8019H
%0

§\s /12
§\s /6
§\s /3

## use Proof Template A5215H (ALL I):  H => ALL x: B  -->  H => B [x/a]
:= $T5215H g{^}
:= $X5215H a{$T5215H}
:= $A5215H e{$T5215H}
:= $H5215H %0
<< A5215H.r0t.txt
:= $T5215H; := $X5215H; := $A5215H; := $H5215H
%0

## use Proof Template K8019H:  H => (A & B)  -->  H => A, H => B
:= $H8019H %0
<< K8019H.r0t.txt
:= $H8019H
%$A8019H
:= $A8019H; := $B8019H

:= $EIDTY %0

## .3:  Proof of H  =>  e * f = f

%$FULLH; := $FULLH

## use Proof Template K8019H:  H => (A & B)  -->  H => A, H => B
:= $H8019H %0
```

```
<< K8019H.r0t.txt
:= $H8019H
%$A8019H
:= $A8019H; := $B8019H

## use Proof Template K8019H:  H => (A & B)  -->  H => A, H => B
:= $H8019H %0
<< K8019H.r0t.txt
:= $H8019H
%$B8019H
:= $A8019H; := $B8019H

§\s /12
§\s /6
§\s /3

## use Proof Template A5215H (ALL I):  H => ALL x: B  -->  H => B [x/a]
:= $T5215H g{^}
:= $X5215H a{$T5215H}
:= $A5215H f{$T5215H}
:= $H5215H %0
<< A5215H.r0t.txt
:= $T5215H; := $X5215H; := $A5215H; := $H5215H
%0

## use Proof Template K8019H:  H => (A & B)  -->  H => A, H => B
:= $H8019H %0
<< K8019H.r0t.txt
:= $H8019H
%$B8019H
:= $A8019H; := $B8019H

:= $FIDTY %0

## .4:  Proof of H  =>  e = f

%$FIDTY; := $FIDTY
%$EIDTY; := $EIDTY
§s' %1 5 %0

## use Proof Template K8025 (Deduction Theorem):  (H & I) => A  -->  H => (I => A)
<< K8025.r0t.txt
%0

## use Proof Template K8025 (Deduction Theorem):  (H & I) => A  -->  H => (I => A)
<< K8025.r0t.txt
%0

:= GrpIdElUniq %0
```

```
## undefine local variables
:= $HYPTH; := $TMPDED


##
##  Print Result
##

%0
```

2.2.155 File natural_numbers.r0.txt

```
##
##  Peano's Postulates
##
##
##  Source: [Andrews 2002 (ISBN 1-4020-0763-9), pp. 258 f.]
##
##  Copyright (c) 2017 Owl of Minerva Press GmbH. All rights reserved.
##  Written by Ken Kubota (<mail@kenkubota.de>).
##
##  This file is part of the publication of the mathematical logic R0.
##  For more information, visit: <http://doi.org/10.4444/100.10>
##

<< basics.r0.txt


## variables used
## t: domain (type of the natural numbers)
## z: zero
## s: successor function
## n: set of natural numbers


## definition of the lambda abstraction as part of the universal quantifier on natur
al numbers,
## the universal quantifier with dot ([cf. Andrews 2002 (ISBN 1-4020-0763-9), p. 260
])
:= $DOT [\x{t{^}}{t{^}}.(=>{{{o,o},o}}_(n{{o,t{^}}}{{o,t{^}}}_x{t{^}}{t{^}}){o}){{o,
o}}]
:= DOT [\t{^}{^}.[\n{{o,t{^}}}{{o,t{^}}}.$DOT{{{o,o},t{^}}}]{{{{o,o},t{^}},{o,t{^}}}
}]
§\ ($DOT{{{o,o},t{^}}}_x{t{^}}{t{^}})
:= $DOTx %0/3
:= DOTx [\t{^}{^}.[\n{{o,t{^}}}{{o,t{^}}}.$DOTx{{o,o}}]{{{o,o},{o,t{^}}}}]
```

```
## "(P1) There is an entity called 0 which is a natural number."
:= $P1 (n{{o,t{^}}}{{o,t{^}}}_z{t{^}}{t{^}})
:= P1 [\t{^}{^}.[\z{t{^}}{t{^}}.[\s{{t{^},t{^}}}{{t{^},t{^}}}.[\n{{o,t{^}}}{{o,t{^}}
}.$P1{o}]{{o,{o,t{^}}}}]{{{o,{o,t{^}}},{t{^},t{^}}}}]{{{{o,{o,t{^}}},{t{^},t{^}}},t{
^}}}]

## "(P2) Every natural number n has a successor S[_]n which is also a natural number
."
:= $P2 ((A{{{o,{o,\3{^}}},^}}_t{^}{^}){{o,{o,t{^}}}}_[\x{t{^}}{t{^}}.((=>{{{o,o},o}}
_(n{{o,t{^}}}{{o,t{^}}}_x{t{^}}{t{^}}){o}){{o,o}}_(n{{o,t{^}}}{{o,t{^}}}_(s{{t{^},t{
^}}}{{t{^},t{^}}}_x{t{^}}{t{^}}){t{^}}){o}){o}]{{o,t{^}}})
:= P2 [\t{^}{^}.[\z{t{^}}{t{^}}.[\s{{t{^},t{^}}}{{t{^},t{^}}}.[\n{{o,t{^}}}{{o,t{^}}
}.$P2{o}]{{o,{o,t{^}}}}]{{{o,{o,t{^}}},{t{^},t{^}}}}]{{{{o,{o,t{^}}},{t{^},t{^}}},t{
^}}}]

## "(P3) 0 is not the successor of any natural number."
## (formula not verified yet, using a temporary definition)
:= $P3 T
:= P3 [\t{^}{^}.[\z{t{^}}{t{^}}.[\s{{t{^},t{^}}}{{t{^},t{^}}}.[\n{{o,t{^}}}{{o,t{^}}
}.$P3{o}]{{o,{o,t{^}}}}]{{{o,{o,t{^}}},{t{^},t{^}}}}]{{{{o,{o,t{^}}},{t{^},t{^}}},t{
^}}}]

## "(P4) If n and m are natural numbers with the same successors, then n and m are t
he same."
## (formula not verified yet, using a temporary definition)
:= $P4 T
:= P4 [\t{^}{^}.[\z{t{^}}{t{^}}.[\s{{t{^},t{^}}}{{t{^},t{^}}}.[\n{{o,t{^}}}{{o,t{^}}
}.$P4{o}]{{o,{o,t{^}}}}]{{{o,{o,t{^}}},{t{^},t{^}}}}]{{{{o,{o,t{^}}},{t{^},t{^}}},t{
^}}}]

## "(P5) Principle of Mathematical Induction"
:= $P5N ($DOTx{{o,o}}_((=>{{{o,o},o}}_(p{{o,t{^}}}{{o,t{^}}}_x{t{^}}{t{^}}){o}){{o,o
}}_(p{{o,t{^}}}{{o,t{^}}}_(s{{t{^},t{^}}}{{t{^},t{^}}}_x{t{^}}{t{^}}){t{^}}){o}){o})
:= $P5T ($DOTx{{o,o}}_(p{{o,t{^}}}{{o,t{^}}}_x{t{^}}{t{^}}){o})
:= $P5 ((A{{{o,{o,\3{^}}},^}}_{o,t{^}}{^}){{o,{o,{o,t{^}}}}}_[\p{{o,t{^}}}{{o,t{^}}}
.((=>{{{o,o},o}}_((&{{{o,o},o}}_(p{{o,t{^}}}{{o,t{^}}}_z{t{^}}{t{^}}){o}){{o,o}}_((A
{{{o,{o,\3{^}}},^}}_t{^}{^}){{o,{o,t{^}}}}_[\x{t{^}}{t{^}}.($DOTx{{o,o}}_((=>{{{o,o}
,o}}_(p{{o,t{^}}}{{o,t{^}}}_x{t{^}}{t{^}}){o}){{o,o}}_(p{{o,t{^}}}{{o,t{^}}}_(s{{t{^
},t{^}}}{{t{^},t{^}}}_x{t{^}}{t{^}}){t{^}}){o}){o}){o}]{{o,t{^}}}){o}){o}){{o,o}}_((
A{{{o,{o,\3{^}}},^}}_t{^}{^}){{o,{o,t{^}}}}_[\x{t{^}}{t{^}}.($DOTx{{o,o}}_(p{{o,t{^}
}}{{o,t{^}}}_x{t{^}}{t{^}}){o}){o}]{{o,t{^}}}){o}){o}]{{o,{o,t{^}}}})
:= P5 [\t{^}{^}.[\z{t{^}}{t{^}}.[\s{{t{^},t{^}}}{{t{^},t{^}}}.[\n{{o,t{^}}}{{o,t{^}}
}.$P5{o}]{{o,{o,t{^}}}}]{{{o,{o,t{^}}},{t{^},t{^}}}}]{{{{o,{o,t{^}}},{t{^},t{^}}},t{
^}}}]

## all of Peano's Postulates combined
:= $PEANO ((&{{{o,o},o}}_((&{{{o,o},o}}_((&{{{o,o},o}}_((&{{{o,o},o}}_$P1{o}){{o,o}}
_$P2{o}){o}){{o,o}}_$P3{o}){o}){{o,o}}_$P4{o}){o}){{o,o}}_$P5{o})
:= PEANO [\t{^}{^}.[\z{t{^}}{t{^}}.[\s{{t{^},t{^}}}{{t{^},t{^}}}.[\n{{o,t{^}}}{{o,t{
```

```
^}}}.$PEANO{o}]{{o,{o,t{^}}}}]{{{o,{o,t{^}}},{t{^},t{^}}}}]{{{{o,{o,t{^}}},{t{^},t{^
}}},t{^}}}]

## undefine local variables
:= $DOT; := $DOTx; := $P1; := $P2; := $P3; := $P4; := $P5; := $P5N; := $P5T; := $PEA
NO
```

2.2.156 File natural_numbers_andrews.r0.txt

```
##
##  Andrews' Definition of Natural Numbers
##
##
##  Source: [Andrews 2002 (ISBN 1-4020-0763-9), p. 260]
##
##  Copyright (c) 2017 Owl of Minerva Press GmbH. All rights reserved.
##  Written by Ken Kubota (<mail@kenkubota.de>).
##
##  This file is part of the publication of the mathematical logic R0.
##  For more information, visit: <http://doi.org/10.4444/100.10>
##

<< natural_numbers.r0.txt


## polymorphic sigma
:= SIGMA {o,{o,t{^}}}

## shorthand for polymorphic sigma
:= $S SIGMA


## zero
:= ATZERO (={{$S,{o,t{^}}}}_[\x{t{^}}{t{^}}.F{o}]{{o,t{^}}})
:= AZERO [\t{^}{^}.ATZERO{$S}]

## successor function
:= ATSUCC [\n{$S}{$S}.[\p{{o,t{^}}}{{o,t{^}}}.((E{{{o,{o,\3{^}}},^}}_t{^}{^}){$S}_[\
x{t{^}}{t{^}}.((&{{{o,o},o}}_(p{{o,t{^}}}{{o,t{^}}}_x{t{^}}{t{^}}){o}){{o,o}}_(n{$S}
{$S}_[\t{t{^}}{t{^}}.((&{{{o,o},o}}_(!{{o,o}}_((={{{o,t{^}},t{^}}}_t{t{^}}{t{^}}){{o
,t{^}}}_x{t{^}}{t{^}}){o}){o}){{o,o}}_(p{{o,t{^}}}{{o,t{^}}}_t{t{^}}{t{^}}){o}){o}]{
{o,t{^}}}){o}){o}]{{o,t{^}}}){o}]{$S}]
:= ASUCC [\t{^}{^}.ATSUCC{{$S,$S}}]

## set of natural numbers
:= $ANSETZ (p{{o,$S}}{{o,$S}}_ATZERO{$S})
:= $ANSETS ((A{{{o,{o,\3{^}}},^}}_$S{^}){{o,{o,$S}}}_[\x{$S}{$S}.((=>{{{o,o},o}}_(p{
{o,$S}}{{o,$S}}_x{$S}{$S}){o}){{o,o}}_(p{{o,$S}}{{o,$S}}_(ATSUCC{{$S,$S}}_x{$S}{$S})
{$S}){o}){o}]{{o,$S}})
```

```
:= ATNSET [\n{$S}{$S}.((A{{{o,{o,\3{^}}},^}}_{o,$S}{^}){{o,{o,{o,$S}}}}_[\p{{o,$S}}{
{o,$S}}.((=>{{{o,o},o}}_((&{{{o,o},o}}_(p{{o,$S}}{{o,$S}}_ATZERO{$S}){o}){{o,o}}_((A
{{{o,{o,\3{^}}},^}}_$S{^}){{o,{o,$S}}}_[\x{$S}{$S}.((=>{{{o,o},o}}_(p{{o,$S}}{{o,$S}
}_x{$S}{$S}){o}){{o,o}}_(p{{o,$S}}{{o,$S}}_(ATSUCC{{$S,$S}}_x{$S}{$S}){$S}){o}){o}]{
{o,$S}}){o}){o}){{o,o}}_(p{{o,$S}}{{o,$S}}_n{$S}{$S}){o}){o}]{{o,{o,$S}}}){o}]
:= ANSET [\t{^}{^}.ATNSET{{o,$S}}]

## set of finite sets
:= ATFINI [\p{{o,t{^}}}{{o,t{^}}}.((E{{{o,{o,\3{^}}},^}}_$S{^}){{o,{o,$S}}}_[\n{$S}{
$S}.((&{{{o,o},o}}_(ATNSET{{o,$S}}_n{$S}{$S}){o}){{o,o}}_(n{$S}{$S}_p{{o,t{^}}}{{o,t
{^}}}){o}){o}]{{o,$S}}){o}]
:= AFINI [\t{^}{^}.ATFINI{$S}]


## definition of the universal quantifier on (Andrews' definition of) natural number
s (with dot)
§= ((DOT{{{{{o,o},\3{^}},{o,\3{^}}},^}}_$S{^}){{{{o,o},$S},{o,$S}}}_(ANSET{{{o,{o,{o
,\4{^}}}},^}}_t{^}{^}){{o,$S}})
§\s /6
§\s /3
:= ADOT %0/3
§\ (ADOT{{{o,o},$S}}_x{$S}{$S})
:= ADOTx %0/3

## undefine local variables
:= $S; := $ANSETZ; := $ANSETS
```

2.2.157 File neumann.r0.txt

```
##
## Definition of natural numbers similar to the idea of John von Neumann
##
##
## Source: [Kubota 2017 (doi: 10.4444/100.10)]
##
## Copyright (c) 2017 Owl of Minerva Press GmbH. All rights reserved.
## Written by Ken Kubota (<mail@kenkubota.de>).
##
## This file is part of the publication of the mathematical logic R0.
## For more information, visit: <http://doi.org/10.4444/100.10>
##

<< basics.r0.txt
<< pair0.r0.txt


## zero (empty set)
:= NEUMNN0000 0
```

```
## successor function
:= NEUMNSUCCR [\x{@}{@}.((ODPRO{{{TYPRO,@},@}}_O{@}){{TYPRO,@}}_x{@}{@}){TYPRO}]

## predecessor function (= right element function)
:= NEUMNPREDR RELEO


##
##  Examples: Expand numbers zero, one, two and three
##

## .0
§= NEUMNNO000
:= NEUMNNO000EXPND %0

## .1
:= NEUMNNO001 (NEUMNSUCCR{{TYPRO,@}}_NEUMNNO000{@})
§= NEUMNNO001
§\s /3
:= NEUMNNO001EXPND %0

## .2
:= NEUMNNO002 (NEUMNSUCCR{{TYPRO,@}}_NEUMNNO001{@})
§= NEUMNNO002
§\s /3
§\s /7
:= NEUMNNO002EXPND %0

## .3
:= NEUMNNO003 (NEUMNSUCCR{{TYPRO,@}}_NEUMNNO002{@})
§= NEUMNNO003
§\s /3
§\s /7
§\s /15
:= NEUMNNO003EXPND %0


##
##  Expand 3 - 1 = 2 (via predecessor function)
##

## define 2 (expanded)
:= NM002 NEUMNNO002EXPND/3
## define 3 (expanded)
:= NM003 NEUMNNO003EXPND/3

## obtain predecessor of three
§= (NEUMNPREDR{{@,TYPRO}}_NM003{TYPRO})
```

```
## expand right element
§\s /3
§\s /12
§\s /6
§\s /3
§\s /6
§\s /3
```

2.2.158 File pair0.r0.txt

```
##
## Ordered Pairs With No Type Variable
##
##
## Source: [Andrews 2002 (ISBN 1-4020-0763-9), p. 208]
##
## Copyright (c) 2017 Owl of Minerva Press GmbH. All rights reserved.
## Written by Ken Kubota (<mail@kenkubota.de>).
##
## This file is part of the publication of the mathematical logic R0.
## For more information, visit: <http://doi.org/10.4444/100.10>
##


## definition of ordered pair (no type variable)
:= ODPR0 [\x{@}{@}.[\y{@}{@}.[\g{{{@,@},@}}{{{@,@},@}}.((g{{{@,@},@}}{{{@,@},@}}_x{@
}{@}){{@,@}}_y{@}{@}){@}]{{@,{{@,@},@}}}]{{{@,{{@,@},@}},@}}]

## type of ordered pair (no type variable)
:= TYPR0 {@,{{@,@},@}}

## example pair and (evaluated) standard pair (no type variable)
:= XLPR0 ((ODPR0{{{{@,{{@,@},@}},@},@}}_a{@}{@}){{{@,{{@,@},@}},@}}_b{@}{@})
§= XLPR0
§\s /6
§\s /3
:= SDPR0 %0/3
%0

## left element function (no type variable)
:= LELE0 [\p{TYPR0}{TYPR0}.(p{TYPR0}{TYPR0}_[\x{@}{@}.[\y{@}{@}.x{@}{@}]{{@,@}}]{{{@
,@},@}}){@}]

:= $L (LELE0{{@,TYPR0}}_XLPR0{TYPR0})
§= $L
§\s /3
§\s /12
§\s /6
§\s /3
```

```
§\s /6
§\s /3

## right element function (no type variable)
:= RELEO [\p{TYPRO}{TYPRO}.(p{TYPRO}{TYPRO}_[\x{@}{@}.[\y{@}{@}.y{@}{@}]{{@,@}}]{{{@
,@},@}}){@}]

:= $R (RELEO{{@,TYPRO}}_XLPRO{TYPRO})
§= $R
§\s /3
§\s /12
§\s /6
§\s /3
§\s /6
§\s /3

## undefine local variables
:= $L; := $R
```

2.2.159 File pair1.r0.txt

```
##
##  Ordered Pairs With One Type Variable
##
##
##  Source: [Andrews 2002 (ISBN 1-4020-0763-9), p. 208]
##
##  Copyright (c) 2017 Owl of Minerva Press GmbH. All rights reserved.
##  Written by Ken Kubota (<mail@kenkubota.de>).
##
##  This file is part of the publication of the mathematical logic R0.
##  For more information, visit: <http://doi.org/10.4444/100.10>
##


## definition of ordered pair (one type variable)
:= ODPR1 [\t{^}{^}.[\x{t{^}}{t{^}}.[\y{t{^}}{t{^}}.[\g{{{t{^},t{^}},t{^}}}{{{t{^},t{
^}},t{^}}}.((g{{{t{^},t{^}},t{^}}}{{{t{^},t{^}},t{^}}}_x{t{^}}{t{^}}){{t{^},t{^}}}_y
{t{^}}{t{^}}){t{^}}]{{t{^},{{t{^},t{^}},t{^}}}}]{{{t{^},{{t{^},t{^}},t{^}}},t{^}}}]{
{{{t{^},{{t{^},t{^}},t{^}}},t{^}},t{^}}}]

## type of ordered pair (one type variable)
:= TYPR1 [\t{^}{^}.{t{^},{{t{^},t{^}},t{^}}}{^}]

## example pair and (evaluated) standard pair (one type variable)
:= XLPR1 (((ODPR1{{{{{\4{^},{{\6{^},\6{^}},\5{^}}},\3{^}},\2{^}},^}}_u{^}{^}){{{{u{^
},{{u{^},u{^}},u{^}}},u{^}},u{^}}}_a{u{^}}{u{^}}){{{u{^},{{u{^},u{^}},u{^}}},u{^}}}_
b{u{^}}{u{^}})
§= XLPR1
```

```
§\s /12
§\s /6
§\s /3
:= SDPR1 %0/3
%0

## left element function (one type variable)
:= LELE1 [\t{^}{^}.[\p{{t{^},{{t{^},t{^}},t{^}}}}{{t{^},{{t{^},t{^}},t{^}}}}.(p{{t{^
},{{t{^},t{^}},t{^}}}}{{t{^},{{t{^},t{^}},t{^}}}}_[\x{t{^}}{t{^}}.[\y{t{^}}{t{^}}.x{
t{^}}{t{^}}]{{t{^},t{^}}}]{{{t{^},t{^}},t{^}}})(t{^}}]{{t{^},{t{^},{{t{^},t{^}},t{^}
}}}}]

:= $L ((LELE1{{{\2{^},{\3{^},{{\5{^},\5{^}},\4{^}}}},^}}_u{^}{^}){{u{^},{u{^},{{u{^}
,u{^}},u{^}}}}}_SDPR1{{u{^},{{u{^},u{^}},u{^}}}})
§= $L
§\s /6
§\s /3
§\s /3
§\s /6
§\s /3

## right element function (one type variable)
:= RELE1 [\t{^}{^}.[\p{{t{^},{{t{^},t{^}},t{^}}}}{{t{^},{{t{^},t{^}},t{^}}}}.(p{{t{^
},{{t{^},t{^}},t{^}}}}{{t{^},{{t{^},t{^}},t{^}}}}_[\x{t{^}}{t{^}}.[\y{t{^}}{t{^}}.y{
t{^}}{t{^}}]{{t{^},t{^}}}]{{{t{^},t{^}},t{^}}})(t{^}}]{{t{^},{t{^},{{t{^},t{^}},t{^}
}}}}]

:= $R ((RELE1{{{\2{^},{\3{^},{{\5{^},\5{^}},\4{^}}}},^}}_u{^}{^}){{u{^},{u{^},{{u{^}
,u{^}},u{^}}}}}_SDPR1{{u{^},{{u{^},u{^}},u{^}}}})
§= $R
§\s /6
§\s /3
§\s /3
§\s /6
§\s /3

## undefine local variables
:= $L; := $R
```

2.2.160 File pair3.r0.txt

```
##
##  Ordered Pairs With Three Type Variables
##
##
##  Source: [Andrews 2002 (ISBN 1-4020-0763-9), p. 208]
##
##  Copyright (c) 2017 Owl of Minerva Press GmbH. All rights reserved.
##  Written by Ken Kubota (<mail@kenkubota.de>).
```

```
##
##  This file is part of the publication of the mathematical logic R0.
##  For more information, visit: <http://doi.org/10.4444/100.10>
##



##
##  Comment
##
##  One might consider placing the two type variables of the pair elements first:
##      PROD := [\t.[\u.[\x:t.[\y:u.[\v.[\g:vut.(gxy)]]]]]],
##  hence
##      PROD a b
##  would represent the Cartesian product a x b.
##
##  Source: [cf. https://sourceforge.net/p/hol/mailman/message/35648326/ (Feb. 5, 20
17)]
##  [cf. https://sympa.inria.fr/sympa/arc/coq-club/2017-02/msg00024.html (Feb. 5, 20
17)]
##


## definition of ordered pair (three type variables)
:= ODPR3 [\t{^}{^}.[\x{t{^}}{t{^}}.[\u{^}{^}.[\y{u{^}}{u{^}}.[\v{^}{^}.[\g{{{v{^},u{
^}},t{^}}}{{{v{^},u{^}},t{^}}}.((g{{{v{^},u{^}},t{^}}}{{{v{^},u{^}},t{^}}}_x{t{^}}{t
{^}}){{v{^},u{^}}}_y{u{^}}{u{^}}){v{^}}]{{v{^},{{v{^},u{^}},t{^}}}}]{{{\2{^},{{\4{^}
,u{^}},t{^}}},^}}]{{{{\2{^},{{\4{^},u{^}},t{^}}},^},u{^}}}]{{{{{\2{^},{{\4{^},\6{^}}
,t{^}}},^},\2{^}},^}}]{{{{{{\2{^},{{\4{^},\6{^}},t{^}}},^},\2{^}},^},t{^}}}]

## type of ordered pair (three type variables)
:= TYPR3 [\t{^}{^}.[\u{^}{^}.[\v{^}{^}.{v{^},{{v{^},u{^}},t{^}}}{^}]{{^,^}}]{{{^,^},
^}}]

## example pair and (evaluated) standard pair (three type variables)
:= XLPR3 ((((ODPR3{{{{{{{\2{^},{{\4{^},\6{^}},\7{^}}},^},\2{^}},^},\2{^}},^}}_t{^}{^
}){{{{{{\2{^},{{\4{^},\6{^}},t{^}}},^},\2{^}},^},t{^}}}_a{t{^}}{t{^}}){{{{{\2{^},{{\
4{^},\6{^}},t{^}}},^},\2{^}},^}}_u{^}{^}){{{{\2{^},{{\4{^},u{^}},t{^}}},^},u{^}}}_b{
u{^}}{u{^}})
§= XLPR3
§\s /24
§\s /12
§\s /6
§\s /3
:= SDPR3 %0/3
%0

## left element function (three type variables)
:= LELE3 [\t{^}{^}.[\u{^}{^}.[\p{{{\2{^},{{\4{^},u{^}},t{^}}},^}}{{{\2{^},{{\4{^},u{
^}},t{^}}},^}}.((p{{{\2{^},{{\4{^},u{^}},t{^}}},^}}{{{\2{^},{{\4{^},u{^}},t{^}}},^}}
```

```
_t{^}{^}){{t{^},{{t{^},u{^}},t{^}}}}_[\x{t{^}}{t{^}}.[\y{u{^}}{u{^}}.x{t{^}}{t{^}}]{
{t{^},u{^}}}}]{{{t{^},u{^}},t{^}}}){t{^}}]{{t{^},{{\2{^},{{\4{^},u{^}},t{^}}},^}}}]{{
{t{^},{{\2{^},{{\4{^},\6{^}},t{^}}},^}},^}}]

:= $L (((LELE3{{{{\3{^},{{\2{^},{{\4{^},\6{^}},\6{^}}},^}},^},^}}_t{^}{^}){{{t{^},{{
\2{^},{{\4{^},\6{^}},t{^}}},^}},^}}_u{^}{^}){{t{^},{{\2{^},{{\4{^},u{^}},t{^}}},^}}}
_SDPR3{{{\2{^},{{\4{^},u{^}},t{^}}},^}})
§= $L
§\s /12
§\s /6
§\s /3
§\s /6
§\s /3
§\s /6
§\s /3

## right element function (three type variables)
:= RELE3 [\t{^}{^}.[\u{^}{^}.[\p{{{\2{^},{{\4{^},u{^}},t{^}}},^}}{{{\2{^},{{\4{^},u{
^}},t{^}}},^}}.((p{{{\2{^},{{\4{^},u{^}},t{^}}},^}}{{{\2{^},{{\4{^},u{^}},t{^}}},^}}
_u{^}{^}){{u{^},{{u{^},u{^}},t{^}}}}_[\x{t{^}}{t{^}}.[\y{u{^}}{u{^}}.y{u{^}}{u{^}}]{
{u{^},u{^}}}}]{{{u{^},u{^}},t{^}}}){u{^}}]{{u{^},{{\2{^},{{\4{^},u{^}},t{^}}},^}}}]{{
{\2{^},{{\2{^},{{\4{^},\6{^}},t{^}}},^}},^}}]

:= $R (((RELE3{{{{\2{^},{{\2{^},{{\4{^},\6{^}},\6{^}}},^}},^},^}}_t{^}{^}){{{\2{^},{
{\2{^},{{\4{^},\6{^}},t{^}}},^}},^}}_u{^}{^}){{u{^},{{\2{^},{{\4{^},u{^}},t{^}}},^}}
}_SDPR3{{{\2{^},{{\4{^},u{^}},t{^}}},^}})
§= $R
§\s /12
§\s /6
§\s /3
§\s /6
§\s /3
§\s /6
§\s /3

## undefine local variables
:= $L; := $R
```

2.2.161 File paradox_cantor.r0e.txt

```
##
## Cantor's paradox
##
##
## Source: [Kubota 2017 (doi: 10.4444/100.10)]
##
## Copyright (c) 2017 Owl of Minerva Press GmbH. All rights reserved.
## Written by Ken Kubota (<mail@kenkubota.de>).
##
```

```
##  This file is part of the publication of the mathematical logic R0.
##  For more information, visit: <http://doi.org/10.4444/100.10>
##

<< basics.r0.txt
<< A5200t.r0.txt


##
##  Demonstration of Positive Self-Reference: The universal set contains itself (not
 a paradox)
##

§= (V{{o,@}}_V{@})
§\ /3
§s %1 5 %0
%A5200t
§s %0 1 %1
%0

## demonstrate that V now has type V
§= {V} V
%0


##
##  Cantor's paradox: The power set of the universal set should be a subset of the u
niversal set
##

## obtain power set of universal set (resulting set has type 'o(ow)' -- is a set of
sets')
:= $PC ((PWSET{{{o,{o,\4{^}}},{o,\3{^}}},^}}_@{^}){{{o,{o,@}},{o,@}}}_V{{o,@}})

## power set of the universal set is a subset of ... (resulting function has type 'o
(o(ow))')
:= $SPC ((SBSET{{{o,{o,\4{^}}},{o,\3{^}}},^}}_{o,@}{^}){{{o,{o,{o,@}}},{o,{o,@}}}}_
$PC{{o,{o,@}}})

## ... the universal set (which has type 'ow')

## trying to apply the wff (will result in failure)

## interactive command for lambda application (with automatic type matching):
__ $SPC V


## undefine local variables
:= $PC; := $SPC
```

```
##
##  Q.E.D.
##

## It is not possible to express Cantor's paradox in the formulation R0 of higher-or
der logic.
```

2.2.162 File paradox_russell.r0e.txt

```
##
##  Russell's paradox
##
##
##  Source: [Kubota 2017 (doi: 10.4444/100.10)]
##
##  Copyright (c) 2017 Owl of Minerva Press GmbH. All rights reserved.
##  Written by Ken Kubota (<mail@kenkubota.de>).
##
##  This file is part of the publication of the mathematical logic R0.
##  For more information, visit: <http://doi.org/10.4444/100.10>
##

<< basics.r0.txt


##
##  The set of all sets that are not members of themselves
##

:= RUSSELL [\x{{o,@}}{{o,@}}.(!{{o,o}}_(x{{o,@}}{{o,@}}_x{{o,@}}{@}){o}){o}]

## trying to apply the wff onto itself (will result in failure)

## interactive command for lambda application (with automatic type matching):
__ RUSSELL RUSSELL


##
##  Q.E.D.
##

## It is not possible to express Russell's paradox in the formulation R0 of higher-o
rder logic.
```

2.2.163 File polymorphism.r0.txt

```
##
##  Polymorphism
##
##
##  Source: [Kubota 2017 (doi: 10.4444/100.10)]
##
##  Copyright (c) 2017 Owl of Minerva Press GmbH. All rights reserved.
##  Written by Ken Kubota (<mail@kenkubota.de>).
##
##  This file is part of the publication of the mathematical logic R0.
##  For more information, visit: <http://doi.org/10.4444/100.10>
##

<< basics.r0.txt


## testing the polymorphic identity relation (=) and
##         the polymorphic description operator (i) ...

## ... with type variable t
§= {t{^}} (i{{t{^},{o,t{^}}}}_p{{o,t{^}}}{{o,t{^}}})
%0

## ... with type variable a
§= {a{^}} (i{{a{^},{o,a{^}}}}_p{{o,a{^}}}{{o,a{^}}})
%0

## ... with type Boole
§= {o} (i{{o,{o,o}}}_p{{o,o}}{{o,o}})
%0
```

2.2.164 File scope_violation_in_lambda_conversion.r0e.txt

```
##
##  Scope Violation in Lambda Conversion
##
##
##  Source: [Kubota 2017 (doi: 10.4444/100.10)]
##
##  Copyright (c) 2017 Owl of Minerva Press GmbH. All rights reserved.
##  Written by Ken Kubota (<mail@kenkubota.de>).
##
##  This file is part of the publication of the mathematical logic R0.
##  For more information, visit: <http://doi.org/10.4444/100.10>
##
```

```
##
##  Condition "A is free for x in B"
##
##  [Andrews 2002 (ISBN 1-4020-0763-9), pp. 218 f. (5207) and p. 213 (definition of
term)]
##

§\ ([\x{@}{@}.[\y{@}{@}.((={{{o,@},@}}_x{@}{@}){{o,@}}_y{@}{@}){o}]{{o,@}}]{{{o,@},@
}}_y{@}{@})
```

2.2.165 File scope_violation_in_lambda_conversion_type.r0e.txt

```
##
##  Scope Violation in Lambda Conversion at Type Level
##
##
##  Source: [Kubota 2017 (doi: 10.4444/100.10)]
##
##  Copyright (c) 2017 Owl of Minerva Press GmbH. All rights reserved.
##  Written by Ken Kubota (<mail@kenkubota.de>).
##
##  This file is part of the publication of the mathematical logic R0.
##  For more information, visit: <http://doi.org/10.4444/100.10>
##



##
##  Condition "A is free for x in B"
##
##  [Andrews 2002 (ISBN 1-4020-0763-9), pp. 218 f. (5207) and p. 213 (definition of
term)]
##

§\ ([\t{^}{^}.[\u{^}{^}.x{t{^}}{t{^}}]{{t{^},^}}]{{{\2{^},^},^}}_u{^}{^})
```

2.2.166 File scope_violation_in_substitution.r0e.txt

```
##
##  Scope Violation in Substitution
##
##
##  Source: [Kubota 2017 (doi: 10.4444/100.10)]
##
##  Copyright (c) 2017 Owl of Minerva Press GmbH. All rights reserved.
##  Written by Ken Kubota (<mail@kenkubota.de>).
##
##  This file is part of the publication of the mathematical logic R0.
##  For more information, visit: <http://doi.org/10.4444/100.10>
```

```
##


##
##  Condition "the occurrence of A in C is not in a wf part [\x.E] of C,
##         where x is free in a member of H and free in [A = B]"
##  [Andrews 2002 (ISBN 1-4020-0763-9), p. 214 (Rule R')]
##

<< basics.r0.txt

## undefine V to see the formula in detail
:= V

## H => A = B
§! ((=>{{{o,o},o}}_(p{{o,@}}{{o,@}}_x{@}{@}){o}){{o,o}}_((={{{o,@},@}}_T{@}){{o,@}}_
(p{{o,@}}{{o,@}}_x{@}{@}){@}){o})

## H => C
§! ((=>{{{o,o},o}}_(p{{o,@}}{{o,@}}_x{@}{@}){o}){{o,o}}_((={{{o,@},@}}_[\x{@}{@}.T{o
}]{@}){{o,@}}_[\x{@}{@}.T{o}]{@}){o})

## now try to replace A (first T) in C
§s' %0 7 %1
```

2.2.167 File scope_violation_in_variable_renaming_conv.r0e.txt

```
##
##  Scope Violation in Variable Renaming (Lambda Conversion)
##
##
##  Source: [Kubota 2017 (doi: 10.4444/100.10)]
##
##  Copyright (c) 2017 Owl of Minerva Press GmbH. All rights reserved.
##  Written by Ken Kubota (<mail@kenkubota.de>).
##
##  This file is part of the publication of the mathematical logic R0.
##  For more information, visit: <http://doi.org/10.4444/100.10>
##



##
##  Condition "z is free for x in A"
##
##  [Andrews 2002 (ISBN 1-4020-0763-9), pp. 217 f. (5206) and p. 213 (definition of
term)]
##

§r [\x{t{^}}{t{^}}.[\z{t{^}}{t{^}}.((={{{o,t{^}},t{^}}}_x{t{^}}{t{^}}){{o,t{^}}}_z{t
```

```
{^}}{t{^}}){o}]{{o,t{^}}}] z{t{^}}
```

2.2.168 File scope_violation_in_variable_renaming_var.r0e.txt

```
##
##  Scope Violation in Variable Renaming (Free Variable)
##
##
##  Source: [Kubota 2017 (doi: 10.4444/100.10)]
##
##  Copyright (c) 2017 Owl of Minerva Press GmbH. All rights reserved.
##  Written by Ken Kubota (<mail@kenkubota.de>).
##
##  This file is part of the publication of the mathematical logic R0.
##  For more information, visit: <http://doi.org/10.4444/100.10>
##



##
##  Condition "z does not occur free in A"
##
##  [Andrews 2002 (ISBN 1-4020-0763-9), pp. 217 f. (5206)]
##

§r [\x{t{^}}{t{^}}.((={{{o,t{^}},t{^}}}_x{t{^}}{t{^}}){{o,t{^}}}_z{t{^}}{t{^}}){o}]
z{t{^}}
```

2.2.169 File vector.r0.txt

```
##
##  Vectors (Dependent Type Theory)
##
##
##  Source: [Kubota 2017 (doi: 10.4444/100.10)]
##
##  Copyright (c) 2017 Owl of Minerva Press GmbH. All rights reserved.
##  Written by Ken Kubota (<mail@kenkubota.de>).
##
##  This file is part of the publication of the mathematical logic R0.
##  For more information, visit: <http://doi.org/10.4444/100.10>
##

<< basics.r0.txt
<< pair3.r0.txt


##
##  Example: Define Three-Dimensional Vector as Nested Ordered Pair <a,<b,<c,0> > >
```

```
##

## level 1
:= TLVL1 {{\2{^},{{\4{^},@},s{^}}},^}
:= PLVL1 ((((ODPR3{{{{{{{\2{^},{{\4{^},\6{^}},\7{^}}},^},\2{^}},^},\2{^}},^}}_s{^}{^
}){{{{{{\2{^},{{\4{^},\6{^}},s{^}}},^},\2{^}},^},s{^}}}_c{s{^}}{s{^}}){{{{{\2{^},{{\
4{^},\6{^}},s{^}}},^},\2{^}},^}}_@{^}){{{{\2{^},{{\4{^},@},s{^}}},^},@}}_0{@})

## level 2
:= TLVL2 {{\2{^},{{\4{^},TLVL1},s{^}}},^}
:= PLVL2 ((((ODPR3{{{{{{{\2{^},{{\4{^},\6{^}},\7{^}}},^},\2{^}},^},\2{^}},^}}_s{^}{^
}){{{{{{\2{^},{{\4{^},\6{^}},s{^}}},^},\2{^}},^},s{^}}}_b{s{^}}{s{^}}){{{{{\2{^},{{\
4{^},\6{^}},s{^}}},^},\2{^}},^}}_TLVL1{^}){{{{\2{^},{{\4{^},TLVL1},s{^}}},^},TLVL1}}
_PLVL1{TLVL1})

## level 3
:= TLVL3 {{\2{^},{{\4{^},TLVL2},s{^}}},^}
:= PLVL3 ((((ODPR3{{{{{{{\2{^},{{\4{^},\6{^}},\7{^}}},^},\2{^}},^},\2{^}},^}}_s{^}{^
}){{{{{{\2{^},{{\4{^},\6{^}},s{^}}},^},\2{^}},^},s{^}}}_a{s{^}}{s{^}}){{{{{\2{^},{{\
4{^},\6{^}},s{^}}},^},\2{^}},^}}_TLVL2{^}){{{{\2{^},{{\4{^},TLVL2},s{^}}},^},TLVL2}}
_PLVL2{TLVL2})


##
##  Type Depending on Level/Dimension (Dependent Type Theory)
##

## type successor function
:= TZERO @
:= TSUCC [\t{^}{^}.[\x{^}{^}.{{\2{^},{{\4{^},x{^}},t{^}}},^}{^}]{{^,^}}]

## evaluate type successor function for type s
§\ (TSUCC{{{^,^},^}}_s{^}{^})
:= TSUCCTYPES %0/3

## evaluate types for all three levels
:= TSUCCN0000 TZERO
:= TSUCCN0001 (TSUCCTYPES{{^,^}}_TSUCCN0000{^})
:= TSUCCN0002 (TSUCCTYPES{{^,^}}_TSUCCN0001{^})
:= TSUCCN0003 (TSUCCTYPES{{^,^}}_TSUCCN0002{^})

## level 1
§= TSUCCN0001
§\s /3
:= TSUCCN0001EXPND %0

## level 2
§= TSUCCN0002
§\s /3
```

```
%TSUCCN0001EXPND
§s %1 53 %0
:= TSUCCN0002EXPND %0

## level 3
§= TSUCCN0003
§\s /3
%TSUCCN0002EXPND
§s %1 53 %0


##
##  Obtain Vector Elements
##

## first element (left element at top level)
§= (((LELE3{{{{\3{^},{{\2{^},{{\4{^},\6{^}},\6{^}}}},^}},^},^}}_s{^}{^}){{{s{^},{{\2{
^},{{\4{^},\6{^}},s{^}}}},^}},^}}_TLVL2{^}){{s{^},TLVL3}}_PLVL3{TLVL3})
§\s /12
§\s /6
§\s /3
§\s /96
§\s /48
§\s /24
§\s /12
§\s /6
§\s /3
§\s /6
§\s /3

## etc.


##
##  Finally, one may use the recursion operator R to implement vectors and vector
##  access via an index number, and thus obtain a fully dependent type theory,
##  in which the type depends on an object (the dimension or the index number).
##
##  For the formal definition of R and some of its applications,
##  see [Andrews 2002 (ISBN 1-4020-0763-9), pp. 281 f., 284].
##
```

2.2.170 File xor_associativity.r0.txt

```
##
##  Associativity of Exclusive Disjunction (Exclusive OR, XOR)
##
##
##  Source: [Kubota 2017 (doi: 10.4444/100.10)]
```

```
##
##  Copyright (c) 2017 Owl of Minerva Press GmbH. All rights reserved.
##  Written by Ken Kubota (<mail@kenkubota.de>).
##
##  This file is part of the publication of the mathematical logic R0.
##  For more information, visit: <http://doi.org/10.4444/100.10>
##

<< basics.r0.txt
<< xor_table.r0.txt


:= $L [\a{o}{o}.[\b{o}{o}.[\c{o}{o}.((={{{o,o},o}}_((XOR{{{o,o},o}}_((XOR{{{o,o},o}}
_a{o}{o}){{o,o}}_b{o}{o}){o}){{o,o}}_c{o}{o}){o}){{o,o}}_((XOR{{{o,o},o}}_a{o}{o}){{
o,o}}_((XOR{{{o,o},o}}_b{o}{o}){{o,o}}_c{o}{o}){o}){o}){o}]{{o,o}}]{{{o,o},o}}]

## .1:  subcase TT

:= $TT (($L{{{{o,o},o},o}}_T{o}){{{o,o},o}}_T{o})
§= $TT
§\s /6
§\s /3

## use Proof Template A5222 (Rule of Cases):  [\x.A]T, [\x.A]F  -->  A
:= $L5222 %0/3
:= $X5222 c{o}
:= $T5222 ($L5222{{o,o}}_T{o})
:= $F5222 ($L5222{{o,o}}_F{o})

## case T
§= {o} $T5222
§\s /3
%XorTableTTisF
§s %1 53 %0
%XorTableFTisT
§s %1 13 %0
%XorTableTTisF
§s %1 15 %0
%XorTableTFisT
§s %1 7 %0
%A5230a
§s %1 3 %0
## use Proof Template A5201b (Swap):  A = B  -->  B = A
<< A5201b.r0t.txt
%0
%T
§s %0 1 %1

## case F
```

```
§= $F5222
§\s /3
%XorTableTTisF
§s %1 53 %0
%XorTableFFisF
§s %1 13 %0
%XorTableTFisT
§s %1 15 %0
%XorTableTTisF
§s %1 7 %0
%A5230d
§s %1 3 %0
## use Proof Template A5201b (Swap):  A = B  -->  B = A
<< A5201b.r0t.txt
%0
%T
§s %0 1 %1

<< A5222.r0t.txt
:= $L5222; := $X5222; := $T5222; := $F5222

:= $TT
:= $TT %0

## .2:  subcase TF

:= $TF (($L{{{o,o},o},o}}_T{o}){{{o,o},o}}_F{o})
§= $TF
§\s /6
§\s /3

## use Proof Template A5222 (Rule of Cases):  [\x.A]T, [\x.A]F  -->  A
:= $L5222 %0/3
:= $X5222 c{o}
:= $T5222 ($L5222{{o,o}}_T{o})
:= $F5222 ($L5222{{o,o}}_F{o})

## case T
§= $T5222
§\s /3
%XorTableTFisT
§s %1 53 %0
%XorTableTTisF
§s %1 13 %0
%XorTableFTisT
§s %1 15 %0
%XorTableTTisF
§s %1 7 %0
%A5230d
```

```
§s %1 3 %0
## use Proof Template A5201b (Swap):  A = B  -->  B = A
<< A5201b.r0t.txt
%0
%T
§s %0 1 %1

## case F
§= $F5222
§\s /3
%XorTableTFisT
§s %1 53 %0
%XorTableTFisT
§s %1 13 %0
%XorTableFFisF
§s %1 15 %0
%XorTableTFisT
§s %1 7 %0
%A5230a
§s %1 3 %0
## use Proof Template A5201b (Swap):  A = B  -->  B = A
<< A5201b.r0t.txt
%0
%T
§s %0 1 %1

<< A5222.r0t.txt
:= $L5222; := $X5222; := $T5222; := $F5222

:= $TF
:= $TF %0

## .3:  subcase FT

:= $FT (($L{{{o,o},o},o}}_F{o}){{{o,o},o}}_T{o})
§= $FT
§\s /6
§\s /3

## use Proof Template A5222 (Rule of Cases):  [\x.A]T, [\x.A]F  -->  A
:= $L5222 %0/3
:= $X5222 c{o}
:= $T5222 ($L5222{{o,o}}_T{o})
:= $F5222 ($L5222{{o,o}}_F{o})

## case T
§= $T5222
§\s /3
%XorTableFTisT
```

```
§s %1 53 %0
%XorTableTTisF
§s %1 13 %0
%XorTableTTisF
§s %1 15 %0
%XorTableFFisF
§s %1 7 %0
%A5230d
§s %1 3 %0
## use Proof Template A5201b (Swap):  A = B  -->  B = A
<< A5201b.r0t.txt
%0
%T
§s %0 1 %1

## case F
§= $F5222
§\s /3
%XorTableFTisT
§s %1 53 %0
%XorTableTFisT
§s %1 13 %0
%XorTableTFisT
§s %1 15 %0
%XorTableFTisT
§s %1 7 %0
%A5230a
§s %1 3 %0
## use Proof Template A5201b (Swap):  A = B  -->  B = A
<< A5201b.r0t.txt
%0
%T
§s %0 1 %1

<< A5222.r0t.txt
:= $L5222; := $X5222; := $T5222; := $F5222

:= $FT
:= $FT %0

## .4:  subcase FF

:= $FF (($L{{{{o,o},o},o}}_F{o}){{{o,o},o}}_F{o})
§= $FF
§\s /6
§\s /3

## use Proof Template A5222 (Rule of Cases):  [\x.A]T, [\x.A]F  -->  A
:= $L5222 %0/3
```

```
:= $X5222 c{o}
:= $T5222 ($L5222{{o,o}}_T{o})
:= $F5222 ($L5222{{o,o}}_F{o})

## case T
§= $T5222
§\s /3
%XorTableFFisF
§s %1 53 %0
%XorTableFTisT
§s %1 13 %0
%XorTableFTisT
§s %1 15 %0
%XorTableFTisT
§s %1 7 %0
%A5230a
§s %1 3 %0
## use Proof Template A5201b (Swap):  A = B  -->  B = A
<< A5201b.r0t.txt
%0
%T
§s %0 1 %1

## case F
§= $F5222
§\s /3
%XorTableFFisF
§s %1 53 %0
%XorTableFFisF
§s %1 13 %0
%XorTableFFisF
§s %1 15 %0
%XorTableFFisF
§s %1 7 %0
%A5230d
§s %1 3 %0
## use Proof Template A5201b (Swap):  A = B  -->  B = A
<< A5201b.r0t.txt
%0
%T
§s %0 1 %1

<< A5222.r0t.txt
:= $L5222; := $X5222; := $T5222; := $F5222

:= $FF
:= $FF %0

## .5:  case T
```

```
:= $T [\b{o}{o}.((($L{{{{o,o},o},o}}_T{o}){{{o,o},o}}_b{o}{o}){{o,o}}_c{o}{o}){o}]
§= $T
§\s /28
§\s /14
§\s /7

## use Proof Template A5222 (Rule of Cases):  [\x.A]T, [\x.A]F  -->  A
:= $L5222 %0/3
:= $X5222 b{o}
:= $T5222 ($L5222{{o,o}}_T{o})
:= $F5222 ($L5222{{o,o}}_F{o})

## case T
§= $T5222
§\s /3
## use Proof Template A5201b (Swap):  A = B  -->  B = A
<< A5201b.r0t.txt
%0
%$TT; := $TT
§s %0 1 %1

## case F
§= $F5222
§\s /3
## use Proof Template A5201b (Swap):  A = B  -->  B = A
<< A5201b.r0t.txt
%0
%$TF; := $TF
§s %0 1 %1

<< A5222.r0t.txt
:= $L5222; := $X5222; := $T5222; := $F5222

:= $T
:= $T %0

## .6:  case F

:= $F [\b{o}{o}.((($L{{{{o,o},o},o}}_F{o}){{{o,o},o}}_b{o}{o}){{o,o}}_c{o}{o}){o}]
§= $F
§\s /28
§\s /14
§\s /7

## use Proof Template A5222 (Rule of Cases):  [\x.A]T, [\x.A]F  -->  A
:= $L5222 %0/3
:= $X5222 b{o}
:= $T5222 ($L5222{{o,o}}_T{o})
```

```
:= $F5222 ($L5222{{o,o}}_F{o})

## case T
§= $T5222
§\s /3
## use Proof Template A5201b (Swap):  A = B  -->  B = A
<< A5201b.r0t.txt
%0
%$FT; := $FT
§s %0 1 %1

## case F
§= $F5222
§\s /3
## use Proof Template A5201b (Swap):  A = B  -->  B = A
<< A5201b.r0t.txt
%0
%$FF; := $FF
§s %0 1 %1

<< A5222.r0t.txt
:= $L5222; := $X5222; := $T5222; := $F5222

:= $F
:= $F %0

## .7:  general case

:= $R [\a{o}{o}.((($L{{{{o,o},o},o}}_a{o}{o}){{{o,o},o}}_b{o}{o}){{o,o}}_c{o}{o}){o}
]
§= $R
§\s /28
§\s /14
§\s /7

## use Proof Template A5222 (Rule of Cases):  [\x.A]T, [\x.A]F  -->  A
:= $L5222 %0/3
:= $X5222 a{o}
:= $T5222 ($L5222{{o,o}}_T{o})
:= $F5222 ($L5222{{o,o}}_F{o})

## case T
§= $T5222
§\s /3
## use Proof Template A5201b (Swap):  A = B  -->  B = A
<< A5201b.r0t.txt
%0
%$T; := $T
§s %0 1 %1
```

```
## case F
§= $F5222
§\s /3
## use Proof Template A5201b (Swap):  A = B  -->  B = A
<< A5201b.r0t.txt
%0
%$F; := $F
§s %0 1 %1

<< A5222.r0t.txt
:= $L5222; := $X5222; := $T5222; := $F5222

:= $R
:= $L

## .8:  match general definition

## use Proof Template A5220 (Gen):  A  -->  ALL x: A
:= $T5220 o
:= $X5220 c{$T5220}
:= $A5220 %0
<< A5220.r0t.txt
:= $T5220; := $X5220; := $A5220

## use Proof Template A5220 (Gen):  A  -->  ALL x: A
:= $T5220 o
:= $X5220 b{$T5220}
:= $A5220 %0
<< A5220.r0t.txt
:= $T5220; := $X5220; := $A5220

## use Proof Template A5220 (Gen):  A  -->  ALL x: A
:= $T5220 o
:= $X5220 a{$T5220}
:= $A5220 %0
<< A5220.r0t.txt
:= $T5220; := $X5220; := $A5220

:= XorAssociativity %0
```

2.2.171 File xor_case_f.r0.txt

```
##
##  Proof:  (F X A) = A;  (A X F) = A;  (F X A) = (A X F)  ; X = XOR
##
##
##  Source: [Kubota 2017 (doi: 10.4444/100.10)]
##
```

```
##  Copyright (c) 2017 Owl of Minerva Press GmbH. All rights reserved.
##  Written by Ken Kubota (<mail@kenkubota.de>).
##
##  This file is part of the publication of the mathematical logic R0.
##  For more information, visit: <http://doi.org/10.4444/100.10>
##

<< basics.r0.txt
<< xor_table.r0.txt


## .a (case left):  (F X A) = A

## use Proof Template A5222 (Rule of Cases):  [\x.A]T, [\x.A]F  -->  A
:= $L5222 [\x{o}{o}.((={{{o,o},o}}_((XOR{{{o,o},o}}_F{o}){{o,o}}_x{o}{o}){o}){{o,o}}
_x{o}{o}){o}]
:= $X5222 x{o}
:= $T5222 ($L5222{{o,o}}_T{o})
:= $F5222 ($L5222{{o,o}}_F{o})

## subcase T:  (F X T) = T
§\ {o} $T5222
%XorTableFTisT
§s %1 13 %0
%A5230a
§s %1 3 %0

## use Proof Template A5219d (Rule T):  A = T  -->  A
:= $A5219d %0
<< A5219d.r0t.txt
:= $A5219d
%0

## subcase F:  (F X F) = F
§\ {o} $F5222
%XorTableFFisF
§s %1 13 %0
%A5230d
§s %1 3 %0

## use Proof Template A5219d (Rule T):  A = T  -->  A
:= $A5219d %0
<< A5219d.r0t.txt
:= $A5219d
%0

<< A5222.r0t.txt
:= $L5222; := $X5222; := $T5222; := $F5222
```

```
:= XorCaseFLeft %0

## .b (case right):  (A X F) = A

## use Proof Template A5222 (Rule of Cases):  [\x.A]T, [\x.A]F  -->  A
:= $L5222 [\x{o}{o}.((={{{o,o},o}}_((XOR{{{o,o},o}}_x{o}{o}){{o,o}}_F{o}){o}){{o,o}}
_x{o}{o}){o}]
:= $X5222 x{o}
:= $T5222 ($L5222{{o,o}}_T{o})
:= $F5222 ($L5222{{o,o}}_F{o})

## subcase T:  (T X F) = T
§\ {o} $T5222
%XorTableTFisT
§s %1 13 %0
%A5230a
§s %1 3 %0

## use Proof Template A5219d (Rule T):  A = T  -->  A
:= $A5219d %0
<< A5219d.r0t.txt
:= $A5219d
%0

## subcase F:  (F X F) = F
§\ {o} $F5222
%XorTableFFisF
§s %1 13 %0
%A5230d
§s %1 3 %0

## use Proof Template A5219d (Rule T):  A = T  -->  A
:= $A5219d %0
<< A5219d.r0t.txt
:= $A5219d
%0

<< A5222.r0t.txt
:= $L5222; := $X5222; := $T5222; := $F5222

:= XorCaseFRight %0

## .c:  (F X A) = (A X F)

%XorCaseFRight

## use Proof Template A5201b (Swap):  A = B  -->  B = A
<< A5201b.r0t.txt
%0
```

```
%XorCaseFLeft
§s %0 3 %1

:= XorCaseFLeftRight %0
```

2.2.172 File xor_case_t.r0.txt

```
##
##  Proof:  (T X A) = ~A;  (A X T) = ~A;  (T X A) = (A X T)  ; X = XOR
##
##
##  Source: [Kubota 2017 (doi: 10.4444/100.10)]
##
##  Copyright (c) 2017 Owl of Minerva Press GmbH. All rights reserved.
##  Written by Ken Kubota (<mail@kenkubota.de>).
##
##  This file is part of the publication of the mathematical logic R0.
##  For more information, visit: <http://doi.org/10.4444/100.10>
##

<< basics.r0.txt
<< xor_table.r0.txt
<< A5231.r0.txt


## .a (case left):  (T X A) = ~A

## use Proof Template A5222 (Rule of Cases):  [\x.A]T, [\x.A]F  -->  A
:= $L5222 [\x{o}{o}.((={{{o,o},o}}_((XOR{{{o,o},o}}_T{o}){{o,o}}_x{o}{o}){o}){{o,o}}
_(!{{o,o}}_x{o}{o}){o}){o}]
:= $X5222 x{o}
:= $T5222 ($L5222{{o,o}}_T{o})
:= $F5222 ($L5222{{o,o}}_F{o})

## subcase T:  (T X T) = ~T
§\ {o} $T5222
%A5231a
§s %1 7 %0
%XorTableTTisF
§s %1 13 %0
%A5230d
§s %1 3 %0

## use Proof Template A5219d (Rule T):  A = T  -->  A
:= $A5219d %0
<< A5219d.r0t.txt
:= $A5219d
%0
```

```
## subcase F:  (T X F) = ~F
§\ {o} $F5222
%A5231b
§s %1 7 %0
%XorTableTFisT
§s %1 13 %0
%A5230a
§s %1 3 %0

## use Proof Template A5219d (Rule T):  A = T  -->  A
:= $A5219d %0
<< A5219d.r0t.txt
:= $A5219d
%0

<< A5222.r0t.txt
:= $L5222; := $X5222; := $T5222; := $F5222

:= XorCaseTLeft %0

## .b (case right):  (A X T) = ~A

## use Proof Template A5222 (Rule of Cases):  [\x.A]T, [\x.A]F  -->  A
:= $L5222 [\x{o}{o}.((={{{o,o},o}}_((XOR{{{o,o},o}}_x{o}{o}){{o,o}}_T{o}){o}){{o,o}}
_(!{{o,o}}_x{o}{o}){o}){o}]
:= $X5222 x{o}
:= $T5222 ($L5222{{o,o}}_T{o})
:= $F5222 ($L5222{{o,o}}_F{o})

## subcase T:  (T X T) = ~T
§\ {o} $T5222
%A5231a
§s %1 7 %0
%XorTableTTisF
§s %1 13 %0
%A5230d
§s %1 3 %0

## use Proof Template A5219d (Rule T):  A = T  -->  A
:= $A5219d %0
<< A5219d.r0t.txt
:= $A5219d
%0

## subcase F:  (F X T) = ~F
§\ {o} $F5222
%A5231b
§s %1 7 %0
```

```
%XorTableFTisT
§s %1 13 %0
%A5230a
§s %1 3 %0

## use Proof Template A5219d (Rule T):  A = T  -->  A
:= $A5219d %0
<< A5219d.r0t.txt
:= $A5219d
%0

<< A5222.r0t.txt
:= $L5222; := $X5222; := $T5222; := $F5222

:= XorCaseTRight %0

## .c:  (T X A) = (A X T)

%XorCaseTRight

## use Proof Template A5201b (Swap):  A = B  -->  B = A
<< A5201b.r0t.txt
%0

%XorCaseTLeft
§s %0 3 %1

:= XorCaseTLeftRight %0
```

2.2.173 File xor_group.r0.txt

```
##
##  Group Property of Exclusive Disjunction (Exclusive OR, XOR)
##
##
##  Source: [Kubota 2017 (doi: 10.4444/100.10)]
##
##  Copyright (c) 2017 Owl of Minerva Press GmbH. All rights reserved.
##  Written by Ken Kubota (<mail@kenkubota.de>).
##
##  This file is part of the publication of the mathematical logic R0.
##  For more information, visit: <http://doi.org/10.4444/100.10>
##

<< A5229.r0.txt
<< group.r0.txt
<< xor_associativity.r0.txt
<< xor_identity_element.r0.txt
<< xor_inverse_element.r0.txt
```

```
## shorthands
:= $Xab ((XOR{{{o,o},o}}_a{o}{o}){{o,o}}_b{o}{o})
:= $Xbc ((XOR{{{o,o},o}}_b{o}{o}){{o,o}}_c{o}{o})
:= $GrpAsc ((A{{{o,{o,\3{^}}},^}}_o{^}){{o,{o,o}}}_[\a{o}{o}.((A{{{o,{o,\3{^}}},^}}_
o{^}){{o,{o,o}}}_[\b{o}{o}.((A{{{o,{o,\3{^}}},^}}_o{^}){{o,{o,o}}}_[\c{o}{o}.((={{{o
,o},o}}_((l{{{o,o},o}}{{{o,o},o}}_((l{{{o,o},o}}{{{o,o},o}}_a{o}{o}){{o,o}}_b{o}{o})
{o}){{o,o}}_c{o}{o}){o}){{o,o}}_((l{{{o,o},o}}{{{o,o},o}}_a{o}{o}){{o,o}}_((l{{{o,o}
,o}}{{{o,o},o}}_b{o}{o}){{o,o}}_c{o}{o}){o}){o}){o}]{{o,o}}){o}]{{o,o}}){o}]{{o,o}})
:= $GrpIdy ((A{{{o,{o,\3{^}}},^}}_o{^}){{o,{o,o}}}_[\a{o}{o}.((&{{{o,o},o}}_((={{{o,
o},o}}_((l{{{o,o},o}}{{{o,o},o}}_a{o}{o}){{o,o}}_e{o}{o}){o}){{o,o}}_a{o}{o}){o}){{o
,o}}_((={{{o,o},o}}_((l{{{o,o},o}}{{{o,o},o}}_e{o}{o}){{o,o}}_a{o}{o}){o}){{o,o}}_a{
o}{o}){o}){o}]{{o,o}})
:= $GrpInv ((A{{{o,{o,\3{^}}},^}}_o{^}){{o,{o,o}}}_[\a{o}{o}.((E{{{o,{o,\3{^}}},^}}_
o{^}){{o,{o,o}}}_[\b{o}{o}.((&{{{o,o},o}}_((={{{o,o},o}}_((l{{{o,o},o}}{{{o,o},o}}_a
{o}{o}){{o,o}}_b{o}{o}){o}){{o,o}}_e{o}{o}){o}){{o,o}}_((={{{o,o},o}}_((l{{{o,o},o}}
{{{o,o},o}}_b{o}{o}){{o,o}}_a{o}{o}){o}){{o,o}}_e{o}{o}){o}){o}]{{o,o}}){o}]{{o,o}})
:= $XAsc XorAssociativity
:= $XIdy ((A{{{o,{o,\3{^}}},^}}_o{^}){{o,{o,o}}}_[\a{o}{o}.((&{{{o,o},o}}_((={{{o,o}
,o}}_((XOR{{{o,o},o}}_a{o}{o}){{o,o}}_e{o}{o}){o}){{o,o}}_a{o}{o}){o}){{o,o}}_((={{{
o,o},o}}_((XOR{{{o,o},o}}_e{o}{o}){{o,o}}_a{o}{o}){o}){{o,o}}_a{o}{o}){o}){o}]{{o,o}
})
:= $XInv ((A{{{o,{o,\3{^}}},^}}_o{^}){{o,{o,o}}}_[\a{o}{o}.((E{{{o,{o,\3{^}}},^}}_o{
^}){{o,{o,o}}}_[\b{o}{o}.((&{{{o,o},o}}_((={{{o,o},o}}_((XOR{{{o,o},o}}_a{o}{o}){{o,
o}}_b{o}{o}){o}){{o,o}}_e{o}{o}){o}){{o,o}}_((={{{o,o},o}}_((XOR{{{o,o},o}}_b{o}{o})
{{o,o}}_a{o}{o}){o}){{o,o}}_e{o}{o}){o}){o}]{{o,o}}){o}]{{o,o}})
:= $XFIdy XorIdentityElement
:= $XFInv XorInverseElement

## .1

§= ((Grp{{{o,{{\4{^},\4{^}},\3{^}}},^}}_o{^}){{o,{{o,o},o}}}_XOR{{{o,o},o}})
§\s /6
§\s /3

:= $T1 %0

## .2

§= (/15{{o,o}}_F{o})
§\s /3

:= $T2 %0

## .3

%XorIdentityElement
## use Proof Template A5219b (Rule T):  A  -->  A = T
```

```
:= $A5219b %0
<< A5219b.r0t.txt
:= $A5219b

:= $E %0

%$T2; := $T2
%$E; := $E
§s %1 13 %0

:= $T3 %0

## .4

%XorInverseElement
## use Proof Template A5219b (Rule T):  A  -->  A = T
:= $A5219b %0
<< A5219b.r0t.txt
:= $A5219b

:= $E %0

%$T3; := $T3
%$E; := $E
§s %1 7 %0

## .5

%A5229a
§s %1 3 %0
## use Proof Template A5201b (Swap):  A = B  -->  B = A
<< A5201b.r0t.txt
%0
%T
§s %0 1 %1

## .6

## use Proof Template K8031 (EXI Gen):  ([\x.B]A)  -->  EXI x: B
:= $T8031 o
:= $B8031 %0/2
:= $A8031 %0/3
:= $P8031 ($B8031{{o,$T8031}}_$A8031{$T8031})
<< K8031.r0t.txt
:= $T8031; := $B8031; := $A8031

:= $T6 %0

## .7
```

```
%$T1
%$T6; := $T6

## use Proof Template A5219b (Rule T):  A  -->  A = T
:= $A5219b %0
<< A5219b.r0t.txt
:= $A5219b

:= $TMP %0

%$T1; := $T1
%$TMP; := $TMP
§s %1 7 %0

:= $TMP %0

%XorAssociativity

## use Proof Template A5219b (Rule T):  A  -->  A = T
:= $A5219b %0
<< A5219b.r0t.txt
:= $A5219b

%$TMP; := $TMP
%1
§s %1 13 %0

%A5229a
§s %1 3 %0
## use Proof Template A5201b (Swap):  A = B  -->  B = A
<< A5201b.r0t.txt
%0
%T
§s %0 1 %1

:= XorGroup %0

## demonstrate that XOR now has type Grp_o
§= {(Grp{{{o,{{\4{^},\4{^}},\3{^}}},^}}_o{^})} XOR
%0

## demonstrate that Grp_o now has type tau (type "type")
§= {^} (Grp{{{o,{{\4{^},\4{^}},\3{^}}},^}}_o{^})
%0

## undefine local variables
:= $Xab; := $Xbc; := $GrpAsc; := $GrpIdy; := $GrpInv; := $XAsc; := $XIdy; := $XInv;
:= $XFIdy; := $XFInv
```

2.2.174 File xor_group_identity_element_unique.r0.txt

```
##
##  Uniqueness of the Group Identity Element of the XOR Group
##
##
##  Source: [Kubota 2017 (doi: 10.4444/100.10)]
##
##  Copyright (c) 2017 Owl of Minerva Press GmbH. All rights reserved.
##  Written by Ken Kubota (<mail@kenkubota.de>).
##
##  This file is part of the publication of the mathematical logic R0.
##  For more information, visit: <http://doi.org/10.4444/100.10>
##

<< A5223.r0.txt
<< group_identity_element_unique.r0.txt
<< xor_group.r0.txt


## shorthands
:= $GIdOXe (((GrpIdO{{{{o,\3{^}},{{\4{^},\4{^}},\3{^}}},^}}_o{^}){{{o,o},{{o,o},o}}}
_XOR{{{o,o},o}}){{o,o}}_e{o}{o})
:= $GIdOXf (((GrpIdO{{{{o,\3{^}},{{\4{^},\4{^}},\3{^}}},^}}_o{^}){{{o,o},{{o,o},o}}}
_XOR{{{o,o},o}}){{o,o}}_f{o}{o})

## .1

%GrpIdElUniq

## use Proof Template A5221 (Sub):  B  -->  B [x/A]
:= $B5221 %0
:= $T5221 ^
:= $X5221 g{$T5221}
:= $A5221 o
<< A5221.r0t.txt
:= $B5221; := $T5221; := $X5221; := $A5221
%0

## use Proof Template A5221 (Sub):  B  -->  B [x/A]
:= $B5221 %0
:= $T5221 {{{o,o},o}}
:= $X5221 l{$T5221}
:= $A5221 XOR
<< A5221.r0t.txt
:= $B5221; := $T5221; := $X5221; := $A5221

:= $TMP %0
```

```
## .2

%XorGroup

## use Proof Template A5219b (Rule T):  A  -->  A = T
:= $A5219b %0
<< A5219b.r0t.txt
:= $A5219b
%0

%$TMP; := $TMP
%1
§s %1 5 %0

:= $TMP %0

## use Proof Template A5221 (Sub):  B  -->  B [x/A]
:= $B5221 %A5223
:= $T5221 o
:= $X5221 y{o}
:= $A5221 %0/3
<< A5221.r0t.txt
:= $B5221; := $T5221; := $X5221; := $A5221
%0

%$TMP; := $TMP
%1
§s %1 1 %0

## use Proof Template K8026 (Deduction Theorem Reversed):  H => (I => A)  -->  (H &
I) => A
<< K8026.r0t.txt
%0

:= XorGrpIdElUniq %0

## undefine local variables
:= $GIdOXe; := $GIdOXf
```

2.2.175 File xor_identity_element.r0.txt

```
##
##  Neutral Element of Exclusive Disjunction (Exclusive OR, XOR)
##
##
##  Source: [Kubota 2017 (doi: 10.4444/100.10)]
##
##  Copyright (c) 2017 Owl of Minerva Press GmbH. All rights reserved.
##  Written by Ken Kubota (<mail@kenkubota.de>).
```

```
##
##  This file is part of the publication of the mathematical logic R0.
##  For more information, visit: <http://doi.org/10.4444/100.10>
##

<< basics.r0.txt
<< xor_case_f.r0.txt


%XorCaseFRight
%XorCaseFLeft

## use Proof Template K8020:  A, B  -->  A & B
:= $A8020 %1
:= $B8020 %0
<< K8020.r0t.txt
:= $A8020; := $B8020
%0

## use Proof Template A5220 (Gen):  A  -->  ALL x: A
:= $T5220 o
:= $X5220 x{$T5220}
:= $A5220 %0
<< A5220.r0t.txt
:= $T5220; := $X5220; := $A5220
%0

§rs /3 a{o}

:= XorIdentityElement %0
```

2.2.176 File xor_inverse_element.r0.txt

```
##
##  Inverse Element of Exclusive Disjunction (Exclusive OR, XOR)
##
##
##  Source: [Kubota 2017 (doi: 10.4444/100.10)]
##
##  Copyright (c) 2017 Owl of Minerva Press GmbH. All rights reserved.
##  Written by Ken Kubota (<mail@kenkubota.de>).
##
##  This file is part of the publication of the mathematical logic R0.
##  For more information, visit: <http://doi.org/10.4444/100.10>
##

<< basics.r0.txt
<< A5229.r0.txt
<< xor_table.r0.txt
```

```
<< group.r0.txt

## shorthands
:= $T1 (([\g{^}{^}.[\l{{{g{^},g{^}},g{^}}}{{{g{^},g{^}},g{^}}}.[\e{g{^}}{g{^}}.GrpIn
v/15{{o,g{^}}}]{{{o,g{^}},g{^}}}]{{{{o,g{^}},g{^}},{{g{^},g{^}},g{^}}}}]{{{{{o,\4{^}
},\3{^}},{{\4{^},\4{^}},\3{^}}},^}}_o{^}){{{{o,o},o},{{o,o},o}}}_XOR{{{o,o},o}})
:= $T1a ([\l{{{o,o},o}}{{{o,o},o}}.[\e{o}{o}.[\b{o}{o}.((&{{{o,o},o}}_((={{{o,o},o}}
_((l{{{o,o},o}}{{{o,o},o}}_a{o}{o}){{o,o}}_b{o}{o}){o}){{o,o}}_e{o}{o}){o}){{o,o}}_(
(={{{o,o},o}}_((l{{{o,o},o}}{{{o,o},o}}_b{o}{o}){{o,o}}_a{o}{o}){o}){{o,o}}_e{o}{o})
{o}){o}]{{o,o}}]{{{o,o},o}}]{{{{o,o},o},{{o,o},o}}}_XOR{{{o,o},o}})
:= $T1b [\e{o}{o}.[\b{o}{o}.((&{{{o,o},o}}_((={{{o,o},o}}_((XOR{{{o,o},o}}_a{o}{o}){
{o,o}}_b{o}{o}){o}){{o,o}}_e{o}{o}){o}){{o,o}}_((={{{o,o},o}}_((XOR{{{o,o},o}}_b{o}{
o}){{o,o}}_a{o}{o}){o}){{o,o}}_e{o}{o}){o}){o}]{{o,o}}]

## .1

§= $T1
§\s /6
§\s /3

§= ((%0/3{{{o,o},o}}_F{o}){{o,o}}_a{o}{o})
§\s /6
§\s /3

:= $ATMP %0

## .2

## use Proof Template A5222 (Rule of Cases):  [\x.A]T, [\x.A]F  -->  A
:= $L5222 [\a{o}{o}./7{o}]
:= $X5222 x{o}
:= $T5222 ($L5222{{o,o}}_T{o})
:= $F5222 ($L5222{{o,o}}_F{o})

## case T
§\ {o} $T5222
%XorTableTTisF
§s %1 13 %0
%A5230d
§s %1 3 %0
## use Proof Template A5201b (Swap):  A = B  -->  B = A
<< A5201b.r0t.txt
%0
%T
§s %0 1 %1

## case F
§\ {o} $F5222
%XorTableFFisF
```

```
§s %1 13 %0
%A5230d
§s %1 3 %0
## use Proof Template A5201b (Swap):  A = B  -->  B = A
<< A5201b.r0t.txt
%0
%T
§s %0 1 %1

<< A5222.r0t.txt
:= $L5222; := $X5222; := $T5222; := $F5222
%0

## use Proof Template A5221 (Sub):  B  -->  B [x/A]
:= $B5221 %0
:= $T5221 o
:= $X5221 x{$T5221}
:= $A5221 a{$T5221}
<< A5221.r0t.txt
:= $B5221; := $T5221; := $X5221; := $A5221
%0

## use Proof Template A5219b (Rule T):  A  -->  A = T
:= $A5219b %0
<< A5219b.r0t.txt
:= $A5219b

:= $ETMP %0

%$ATMP; := $ATMP

%$ETMP
§s %1 13 %0

%$ETMP; := $ETMP
§s %1 7 %0

%A5229a
§s %1 3 %0
## use Proof Template A5201b (Swap):  A = B  -->  B = A
<< A5201b.r0t.txt
%0
%T
§s %0 1 %1

§\s /2

## .3
```

```
## use Proof Template K8031 (EXI Gen):  ([\x.B]A)  -->  EXI x: B
:= $T8031 o
:= $B8031 %0/2
:= $A8031 %0/3
:= $P8031 ($B8031{{o,$T8031}}_$A8031{$T8031})
<< K8031.r0t.txt
:= $T8031; := $B8031; := $A8031
%0

## .4

## use Proof Template A5220 (Gen):  A  -->  ALL x: A
:= $T5220 o
:= $X5220 a{$T5220}
:= $A5220 %0
<< A5220.r0t.txt
:= $T5220; := $X5220; := $A5220
%0

:= XorInverseElement %0

## undefine local variables
:= $T1; := $T1a; := $T1b
```

2.2.177 File xor_table.r0.txt

```
##
##  Proof:  (T X T) = F;  (T X F) = T;  (F X T) = T;  (F X F) = F  ; X = XOR
##
##
##  Source: [Kubota 2017 (doi: 10.4444/100.10)]
##
##  Copyright (c) 2017 Owl of Minerva Press GmbH. All rights reserved.
##  Written by Ken Kubota (<mail@kenkubota.de>).
##
##  This file is part of the publication of the mathematical logic R0.
##  For more information, visit: <http://doi.org/10.4444/100.10>
##

<< basics.r0.txt
<< A5230.r0.txt
<< A5231.r0.txt


##
##  Proof
##

## .a:  (T X T) = F
```

```
§= {o} ((XOR{{{o,o},o}}_T{o}){{o,o}}_T{o})
§\s /6
§\s /3

%A5230a
§s %1 7 %0

%A5231a
§s %1 3 %0

:= XorTableTTisF %0

## .b:  (T X F) = T

§= {o} ((XOR{{{o,o},o}}_T{o}){{o,o}}_F{o})
§\s /6
§\s /3

%A5230b
§s %1 7 %0

%A5231b
§s %1 3 %0

:= XorTableTFisT %0

## .c:  (F X T) = T

§= {o} ((XOR{{{o,o},o}}_F{o}){{o,o}}_T{o})
§\s /6
§\s /3

%A5230c
§s %1 7 %0

%A5231b
§s %1 3 %0

:= XorTableFTisT %0

## .d:  (F X F) = F

§= {o} ((XOR{{{o,o},o}}_F{o}){{o,o}}_F{o})
§\s /6
§\s /3

%A5230d
§s %1 7 %0
```

```
%A5231a
§s %1 3 %0

:= XorTableFFisF %0
```

Chapter 3

Page References for [Andrews, 2002]

definitions I ($\forall$, $\wedge$, etc.)	212
axioms for $\mathcal{Q}_0$	213
definitions II ($\subseteq$, $\mathcal{P}$, etc.)	231
definition of the uniqueness quantifier ($\exists_1$)	233
declaration of the description operator (ι)	211
definition of the definite-description operator ($\imath$)	234
Rule R	213
Rule R′	214
Rule RR	216
λ-conversion	207, 218 f. (5207)
substitution (S, $\mathsf{\d{S}}$)	81 f.
renaming of bound variables	217 f. (5206)
renaming/substitution of free variables	222 f. (5221_{n} = Sub)
universal generalization	222 (5220 = Gen)
universal instantiation	221 (5215 = $\forall$I)
self-reference	313
object language/meta-language	11
meta-language (Tarski)	336 f.

References

Andrews, Peter B. (2002). *An Introduction to Mathematical Logic and Type Theory: To Truth Through Proof*. Second edition. Dordrecht, Boston, London: Kluwer Academic Publishers. ISBN: 1-4020-0763-9. DOI: 10.1007/978-94-015-9934-4.

Kubota, Ken (2015). *On the Theory of Mathematical Forms (draft from May 18, 2015)*. Unpublished manuscript. SHA-512: a0dfe205eb1a2cb29efaa579d68fa2e5 45af74d8cd6c270cf4c95ed1ba6f7944 fdcffaef2e761c8215945a9dcd535a50 011d8303fd59f2c8a4e6f64125867dc4. DOI: 10.4444/100.10.

All previous pages shall be made available online shortly after the publication of this book, which is scheduled for March 15, 2017.

The online file will be accessible via the link http://doi.org/10.4444/100.10 and has the following SHA-512 checksum:

```
2ca7be176113ddd687ad8f7ef07b6152 770327ea7993423271b84e399fe8b507
67a071408594ec6a40159e14c85b97d2 168462157b22017d701e5c87141157d8
```

www.ingramcontent.com/pod-product-compliance
Lightning Source LLC
Chambersburg PA
CBHW082310210326
41599CB00030B/5750